Signals, Systems and Stochastic Processes

Signals, Systems and Stochastic Processes

Dr. D. Ganesh Rao
(Formerly Professor and Head, MSRIT, Bangalore)
Professor, Department of Electronics and Communication Engineering,
R R Institute of Technology, Bangalore - 560 090

Dr. K. N. Hari Bhat
(Formerly Professor and Head, NITK, Suratkal)
Professor, Department of Electronics and Communication Engineering
NITTE Meenakshi Institute of Technology, Bangalore - 560 064

Dr. K. Anitha Sheela
Professor, Department of Electronics and Communication Engineering
JNTU, Hyderabad - 500 085

CENGAGE

Andover • Melbourne • Mexico City • Stamford, CT • Toronto • Hong Kong • New Delhi • Seoul • Singapore • Tokyo

CENGAGE

Signals, Systems, and Stochastic Processes

D. Ganesh Rao
K.N. Hari Bhat
K. Anitha Sheela

ISBN-13: 978-93-866-5089-4
ISBN-10: 93-866-5089-4

Cengage Learning India Pvt. Ltd.
418, F.I.E., Patparganj
Delhi 110092

Cengage Learning is a leading provider of customized learning solutions with office locations around the globe, including Australia, Brazil, India, Mexico, Singapore, United Kingdom, and United States. Locate your local office at:

www.cengage.com/global

Cengage Learning products are represented in Canada by Nelson Education, Ltd.

For product information, visit **www.cengage.co.in**

Preface

This book is a text on Signals, Systems and Stochastic Processes, at a second-year-degree level. The purpose of writing this book was to provide the reader with a precise, practical up-to-date exposition of subject. Accordingly, this book contains a wealth of material that trains a student to face the challenges posed by growing trends in communication, control and signal processing areas.

The book starts with basic continuous and discrete-time signals and different operations performed on signals. In chapter 2 trigonometric Fourier series and exponential Fourier series are introduced. Mathematical development and Properties of Fourier transform are discussed in chapter 3 and the chapter ends with a treaties on Hilbert transform. Chapter 4 starts the discussion on various properties of systems and continues with the response of the system, concepts like transfer function and ends with with detailed study on filters. Different operations like convolution and correlation are discussed in chapter 5. Chapter 6 explains Laplace transform and Waveform synthesis in detail. Sampling of signals is dealt in chapter 7. Chapter 8 is devoted to study on z-transform in detail. Temporal characteristics and Spectral characteristics of Random Process are elaborated in the last two chapters that should guide the students for further studies related to communication engineering and related areas.

This book offers special techniques that other books in the same area don't. There are many reasons that make this book an excellent companion to be kept with you till your retirement from the field of Engineering:

- Theory and problems are well balanced. However, a slight bias is towards problems; the problem solving ability can enhance the understanding of the concepts in a much better way.
- The derivations of various concepts are explained in a simple manner without sacrificing the rigor of the concepts.
- The concepts are strengthened only by exercising the brain with a variety of problems. This issue is well addressed in the text by solving a number of problems of varied nature.
- Problems are well chosen to impart the analytical skills without forgetting the need to emphasize the engineering needs. Also, the systematic problem solving skills give an action plan that motivates the students to read this book without much ado.

- Problems are graded to make the process of assimilation gradual but stable. Problems of greater difficulty are addressed by techniques that are not intimidatory.
- This book is a confidence booster to students who have the desire to graduate to higher levels of Engineering. This desire must be supported by inspiration and perspiration. Our book does both. The end result is a better grade in the competitive examinations like GATE, IES etc. Preserve the book!!

We have tried to make this book suitable for practicing engineers as well. Many engineers from industry want to change their career by adapting to the fields of demand like digital communication and signal processing. But they are obsessed with a myth an engineer needs practice but not theoretical skills". But it is true that sound theoretical skills make the practice perfect. This aspect is addressed by making mathematics not too abominable, without cutting short the rigor of Maths. Considering the feedback from students and teachers, theory has been modified for more clarity, wherever necessary and additional examples have been added. And also their suggestions have been considered to improve the total outlook of the book.

The first author acknowledges the encouragement given by the Management of R R Institute of Technology, Bangalore during the preparation of this book. The second author profusely thanks the Management of NITTE Meenakshi Institute of Technology, Bangalore for their support during the preparation of the book. Authors express their gratitude to their family members for their co-operation while penning the manuscript. The authors also acknowledge the contribution of Mr. R. Subramanian, MSRIT in different stages of development of the book and thank M/s. Cengage Learnig India for bringing out the book in an excellent form.

Any design is not complete and liable for improvement. Therefore we request the readers to mail their valuable comments, suggestions and criticism to ***dganeshrao2011@gmail.com***.

D. Ganesh Rao

K. N. Hari bhat

K. Anitha Sheela

Contents

Chapter 1

Fundamentals

1.1 Signal

A Signal is a function of a set of independent variables, with time perhaps the most prevalent single variable. A signal itself carries some kind of information available for observation. In general, a signal is a function or sequence of values that represents information.

1.1.1 Examples of signals

Fig. 1.1 shows an example of a continuous signal. The diagram demonstrates the temperature curves for a house wall, not with respect to time, but in relation to the location. The curves shows the temperature profile inside a 14cm thick wall, where the air temperature at the right side suddenly rose by 10° K. One hour later the local temperature follows the curve represented by the broken line. At another time we would have a different temperature curve.

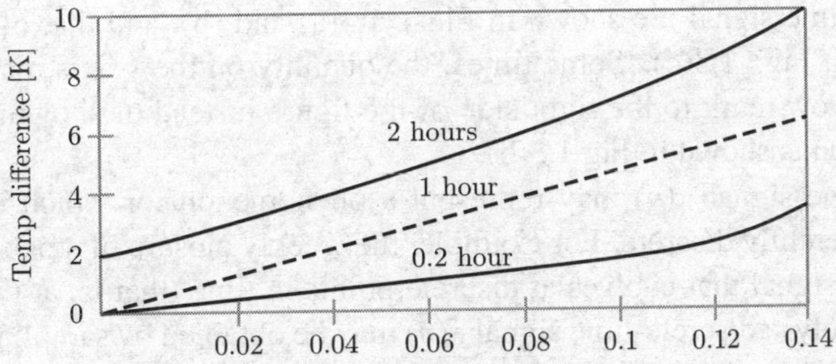

Fig. 1.1 *Temperature curve of a house wall.*

Fig. 1.2 shows another kind of variable quantity, the stock market index over a span of time. We make an assumption that though the index changes while the stock market is open, the diagram depicts only the weekly averages. Thus, the depicted value does not change continuously, but only once in a week. When the signal amplitude occurs only at certain fixed points in time, but not for times in between, we call the signal discrete or more explicitly as discrete-time signal.

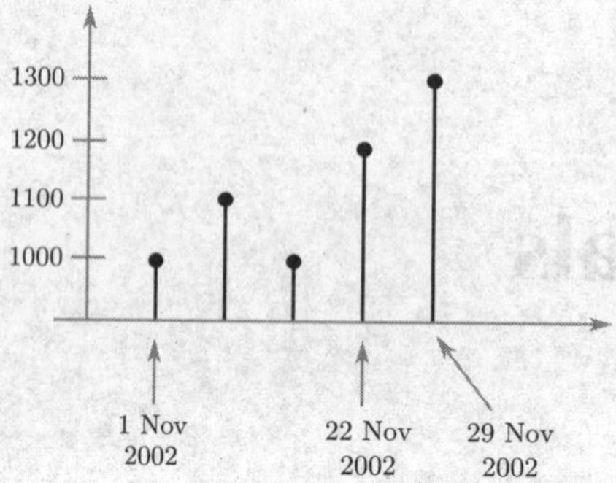

Fig. 1.2 *Weekly stock market index*

Additional examples of signals are: (i) a voltage as a function of time, (ii) a potential as a function of position in three-dimensional space, (iii) a force as a function of time and position, and (iv) an electric intensity as a function of x and y coordinates and time.

1.1.2 Classification of signals

A. Continuous-time and discreate-time signals

A signal $x(t)$ is a continuous-time signal, if t is a continuous variable. If t is a discrete variable, that is, $x(t)$ is defined at discrete-times, then $x(t)$ is a discrete-time signal. Since a discrete-time signal is defined at discrete-times, a discrete-time signal is often identified as a sequence of numbers, denoted by $x(n)$, where n is an integer. Graphical representations of a continuous-time signal are shown in Figs. 1.3(a) and (b), and that of a discrete-time signal is shown in Fig. 1.3(c). Some times, the quantity on the y-axis may be mentioned with a vertical arrow mark to the right side of the figure instead of showing on the y-axis. This representation is shown in Fig 1.3(b).

A discrete-time signal $x(n)$ may represent a phenomenon for which the independent variable, n, is inherently discrete. For example, the weekly closure of stock market average by its nature is a signal that evolves at discrete points in time (that is, at the close of each week). Alternatively, a discrete-time signal $x(n)$ may be obtained by sampling a continuous-time signal $x(t)$.

A discrete-time signal could be obtained from a continuous-time signal by replacing t by nT_s, where T_s is the reciprocal of the sampling rate or time interval between the adjacent samples. This procedure is referred to as *Sampling*.

A sequence $x(n)$ is complex, if it can be written in the form $x_R(n) + jx_I(n)$, where $x_R(n)$ and $x_I(n)$ are the real and imaginary parts of the sequence respectively, and $x_I(n)$ does not

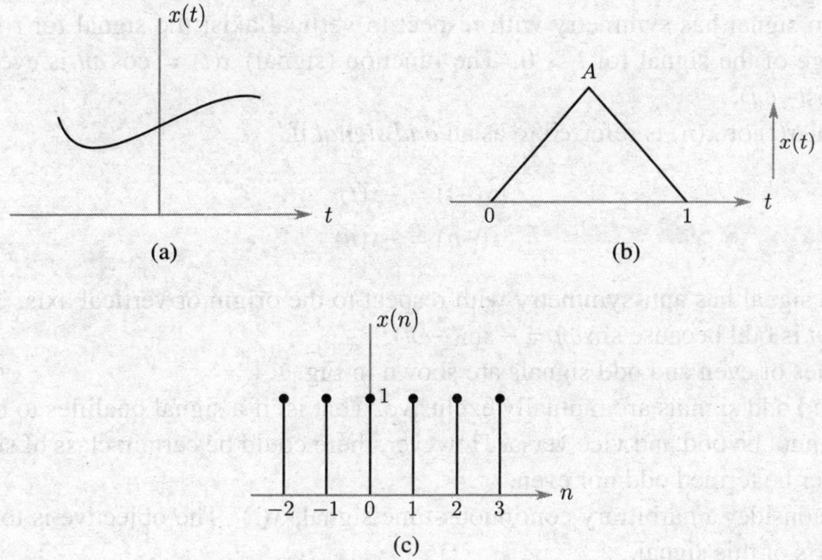

Fig. 1.3 : *Graphical representation of (a) & (b) Continuous-time and (c) Discrete-time signals*

equal to zero for all n. If $x_I(n)$ equals zero for all n, the sequence will be referred to as a *real sequence*.

B. Analog and digital signals

If a continuous-time signal $x(t)$ can take on any value in the continuous interval (a, b), where a may be $-\infty$ and b may be $+\infty$, then the continuous-time signal $x(t)$ is called as an *analog signal*. If a discrete-time signal $x(n)$ can take on only a finite number of distinct values, then we call this signal as a *digital signal*.

C. Deterministic and random signals

Signals are termed *deterministic*, if their behaviour is known and can be represented, eg., by a formula. The deflection voltage of an oscilloscope is a *deterministic signal*, for its behaviour is known and can be represented by a sawtooth wave. By contrast, we cannot define the amplitude values of a voice signal by means of formulae or graphical elements; in addition, their continued behaviour is not known. Such signals are termed *stochaistic* (random). Since, it is impossible to specify their behaviour in terms of functions, such signals are described by expected values such as mean, variance and many others.

D. Even and odd signals

A signal $x(t)$ or $x(n)$ is referred to as an *even signal* if

$$x(-t) = x(t)$$
$$x(-n) = x(n)$$

An even signal has symmetry with respect to vertical axis; the signal for $t < 0$ is the mirror image of the signal for $t > 0$. The function (signal) $x(t) = \cos \omega t$ is even because $\cos \omega t = \cos(-\omega t)$.

A signal $x(t)$ or $x(n)$ is referred to as an *odd signal* if

$$x(-t) = -x(t)$$
$$x(-n) = -x(n)$$

An odd signal has anti-symmetry with respect to the origin or vertical axis. The signal $x(t) = \sin \omega t$ is odd because $\sin \omega t = -\sin(-\omega t)$.

Examples of even and odd signals are shown in Fig. 1.4.

Even and odd signals are mutually exclusive. That is, if a signal qualifies to be an even signal, it cannot be odd and vice versa. However, there could be certain class of signals that could neither be termed odd nor even.

Let us consider an arbitrary continuous-time signal, $x(t)$. The objective is to find even and odd parts of this signal.

Fig. 1.4 *Examples of even signals (a and b) and odd signals (c and d).*

Let

$$x(t) = x_{ev}(t) + x_{od}(t) \tag{1.1}$$

where $x_{ev}(t)$ is the even part of $x(t)$ and $x_{od}(t)$ is the odd part of $x(t)$. Changing t to $-t$ in Eq. (1.1), we get,

$$x(-t) = x_{ev}(-t) + x_{od}(-t) \tag{1.2}$$

Since, $x_{ev}(t)$ is even, we can write,

$$x_{ev}(-t) = x_{ev}(t) \tag{1.3}$$

Also, since, $x_{od}(t)$ is *odd*, we can write

$$x_{od}(-t) = -x_{od}(t) \tag{1.4}$$

Substituting Eqs. (1.3) and (1.4) in Eq. (1.2), we get,

$$x(-t) = x_{ev}(t) - x_{od}(t) \tag{1.5}$$

Adding Eqs. (1.1) and (1.5), we get

$$2x_{ev}(t) = x(t) + x(-t)$$

Hence,

$$x_{ev}(t) = \frac{1}{2}[x(t) + x(-t)] \tag{1.6}$$

Subtracting Eq. (1.5) from Eq. (1.1) we get,

$$2x_{od}(t) = x(t) - x(-t)$$

Hence,

$$x_{od}(t) = \frac{1}{2}[x(t) - x(-t)] \tag{1.7}$$

Above Eqs. (1.6) and (1.7) in discrete-time become,

$$x_{ev}(n) = \frac{1}{2}[x(n) + x(-n)] \tag{1.8}$$

and,

$$x_{od}(n) = \frac{1}{2}[x(n) - x(-n)] \tag{1.9}$$

The average value A_x of a signal $x(t)$ is defined as

$$A_x = \lim_{T\to\infty} \frac{1}{2T} \int_{-T}^{T} x(t)dt$$

The average value of a signal is contained in its even function since the average value of an odd function is zero.

Example 1.1. Find the even and odd components of $x(t) = e^{jt}$.

Solution: Let $x_{ev}(t)$ and $x_{od}(t)$ be the even and odd components of $x(t)$.
Then,

$$x_{ev}(t) = \frac{1}{2}[x(t) + x(-t)]$$
$$= \frac{1}{2}[e^{jt} + e^{-jt}] = \cos t$$
$$x_{od}(t) = \frac{1}{2}[x(t) - x(-t)]$$
$$= \frac{1}{2}[e^{jt} - e^{-jt}] = j\sin t.$$

Example 1.2. Show that the product of two even signals or two odd signals is an even signal while the product of an even and odd signal is an odd signal.

Solution: Let $y(t) = y_1(t)y_2(t)$. If $y_1(t)$ and $y_2(t)$ are both even, then

$$y(-t) = y_1(-t)y_2(-t)$$
$$= y_1(t)y_2(t) = y(t)$$

and thus $y(t)$ is even. If $y_1(t)$ and $y_2(t)$ are both odd, then

$$y(-t) = y_1(-t)y_2(-t) = -y_1(t)[-y_2(t)]$$
$$= y_1(t)y_2(t) = y(t)$$

and thus $y(t)$ is even. If $y_1(t)$ is even and $y_2(t)$ is odd, then

$$y(-t) = y_1(-t)y_2(-t)$$
$$= y_1(t)[-y_2(t)]$$
$$= y_1(t)y_2(t) = -y(t)$$

and thus $y(t)$ is odd. Note that in the above analysis, variable t represents time in continuous-domain. The same results are valid in discrete-domain too.

E. Periodic and nonperiodic signals

A continuous-time signal $x(t)$ is said to be periodic with period T, if there is a positive nonzero value of T for which

$$x(t + T) = x(t) \quad \text{for all}\ ^{1}t \tag{1.10}$$

[1] All t means t ranging from $-\infty$ to ∞.

An example of such a signal is shown in Fig. 1.5(a). A signal that is not periodic is said to be aperiodic. From Eq. (1.10) and Fig. 1.5(a), it follows that

$$x(t + mT_0) = x(t)$$

for all t and any integer m. The fundamental period T_0 of $x(t)$ is the smallest positive value of T for which Eq. (1.10) holds. With T_0 in seconds, the fundamental frequency in Hertz (the number of periods per second) and the fundamental frequency in rad/second are given by

$$f_0 = \frac{1}{T_0}\ \text{Hz}$$

$$\omega_0 = 2\pi f_0 = \frac{2\pi}{T_0}\ \text{rad/second}$$

Periodic discrete-time signals are defined on a similar basis. A sequence or a discrete-time signal $x(n)$ is periodic with a period N, if there is a positive integer N for which

$$x(n + N) = x(n), \quad \text{all} \quad n \tag{1.11}$$

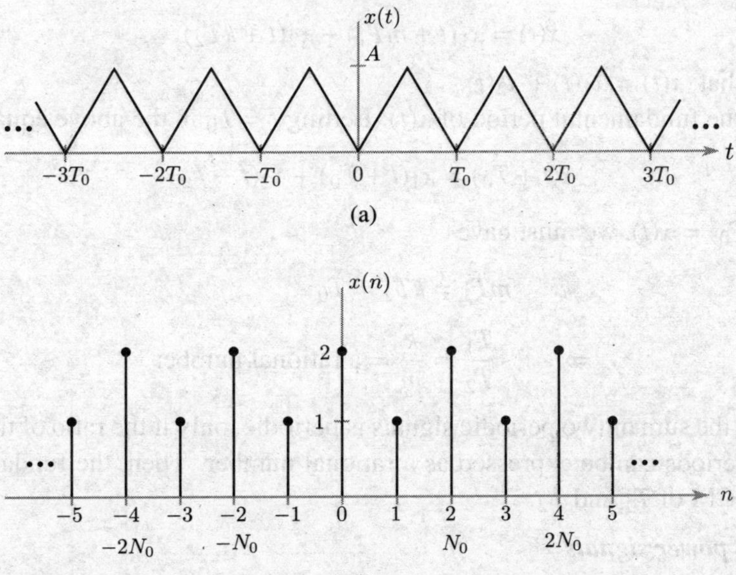

Fig. 1.5 *Examples of periodic signals.*

An example of such a sequence is given in Fig. 1.5(b). From Eq. (1.11) and Fig. 1.5(b) it follows that

$$x(n + mN_0) = x(n) \tag{1.12}$$

for all n and any integer m.

The fundamental period N_0 of $x(n)$ is the smallest integer N for which Eq. (1.11) holds. Any sequence which is not periodic is called a *nonperiodic (aperiodic)* sequence.

Example 1.3. What is the period of a constant signal $x(t)$ (dc signal)?

Solution: For a constant signal $x(t)$, the fundamental period is undefined since $x(t)$ is periodic for any choice of T and so there is no smallest positive value. In other words, a dc signal is nonperiodic.

Example 1.4. Let $x_1(t)$ and $x_2(t)$ be two periodic signals with fundamental periods T_1 and T_2 respectively. Under what conditions, the sum $x(t) = x_1(t) + x_2(t)$ is periodic and what is the fundamental period of $x(t)$, if it is periodic?

Solution: Since $x_1(t)$ and $x_2(t)$ are periodic with fundamental periods T_1 and T_2 respectively, we have

$$x_1(t) = x_1(t + mT_1), \quad m = \text{a positive integer}$$

$$x_2(t) = x_2(t + kT_2), \quad k = \text{a positive integer}$$

Thus,

$$x(t) = x_1(t + mT_1) + x_2(t + kT_2)$$

We know that, $x(t) = x_1(t) + x_2(t)$.
Let T_0 be the fundamental period of $x(t)$. Letting $t = T_0$ in the above equation, we get

$$x(t + T_0) = x_1(t + T_0) + x_2(t + T_0)$$

For $x(t + T_0) = x(t)$, we must have

$$mT_1 = kT_2 = T_0 \tag{1.13}$$

$$\Rightarrow \quad \frac{T_1}{T_2} = \frac{k}{m} = \text{a rational number} \tag{1.14}$$

Therefore, the sum of two periodic signals is periodic, only if the ratio of their respective fundamental periods can be expressed as a rational number. Then, the fundamental period of $x(t)$ is the LCM of T_1 and T_2.

F. Energy and power signals

Consider a resistance, R excited by a voltage, $v(t)$ producing a current, $i(t)$. Let us assume that, $R = 1$ ohm. Then, we have

$$i(t) = v(t) = x(t)$$

The instantaneous power expended in $R = 1$ ohm is given by

$$p(t) = x^2(t) \quad \text{Watts} \tag{1.15}$$

The power dissipated in 1 ohm resistor is called *normalized power*.

The total energy, E and the average power, P of the signal $x(t)$ are

$$E = \int_{-\infty}^{\infty} x^2(t)dt \tag{1.16}$$

$$P = \lim_{T\to\infty} \frac{1}{T} \int_{-\frac{T}{2}}^{\frac{T}{2}} x^2(t)dt \tag{1.17}$$

If $x(t)$ is complex, then Eqs. (1.16) and (1.17) are modified as

$$E = \int_{-\infty}^{\infty} |x(t)|^2 dt \tag{1.18}$$

$$P = \lim_{T\to\infty} \frac{1}{T} \int_{-\frac{T}{2}}^{\frac{T}{2}} |x(t)|^2 dt \tag{1.19}$$

If signal $x(t)$ is periodic, then Eq. (1.19) gets modified as

$$P = \frac{1}{T_0} \int_{0}^{T_0} |x(t)|^2 dt \tag{1.20}$$

Similarly, for a discrete-time signal $x(n)$, the total energy content, E and the average power, P are defined as

$$E = \sum_{n=-\infty}^{\infty} |x(n)|^2 \tag{1.21}$$

$$P = \lim_{N\to\infty} \frac{1}{2N+1} \sum_{n=-N}^{N} |x(n)|^2 \tag{1.22}$$

If $x(n)$ is periodic, then its average power is given as

$$P = \frac{1}{N_0} \sum_{n=0}^{N_0-1} |x(n)|^2 \tag{1.23}$$

Based on definitions in Eq. (1.18) to Eq. (1.23), the following classifications and inferences are made:

1. $x(t)$ or $x(n)$ is defined to be an energy signal, if and only if the the total energy content of the signal is a finite quantity i.e., $0 < E < \infty$. All energy signals will have zero average power i.e. $P = 0$. Energy signals generally include non-periodic signals that have a finite time duration (such as rectangular function - discussed later) and signals that approach zero asymptotically, that is, $x(t)$ approaches zero as t approaches infinity.

2. $x(t)$ or $x(n)$ is defined to be a power signal, if and only if the average power content of the signal is a finite quantity i.e., $0 < P < \infty$. All power signals will have $E = \infty$. The step function, the signum function, and periodic functions (all discussed later) are power signals.

3. Power and energy signals are mutually exclusive. However, there could be certain class of signals which do not satisfy both the classifications referred above. Such signals are neither energy nor power signals.

Example 1.5. Calculate the energy of the signal,

$$x(t) = \begin{cases} 1, & \frac{-\epsilon}{2} \leq t \leq +\frac{\epsilon}{2} \\ 0, & \text{otherwise} \end{cases}$$

Solution:

$$E \triangleq \int_{-\infty}^{\infty} |x(t)|^2 dt = \int_{-\frac{\epsilon}{2}}^{+\frac{\epsilon}{2}} 1^2 dt = \epsilon$$

The symbols $\triangleq$ and $\Rightarrow$ are used to indicate "by the definition" and "which implies" respectively.

The average power in $x(t)$ is

$$P = \lim_{T \to \infty} \frac{1}{T} \int_{-\frac{T}{2}}^{\frac{T}{2}} |x(t)|^2 dt = \lim_{T \to \infty} \frac{1}{T} \int_{-\frac{\epsilon}{2}}^{+\frac{\epsilon}{2}} 1^2 dt = \lim_{T \to \infty} \frac{1}{T}(\epsilon) = 0$$

This example shows that if a signal has finite duration and it is bounded, then the power of the signal is zero. Also, since $0 < E < \infty$, $x(t)$ is an energy signal.

Example 1.6. Calculate the average power of the signal,

$$x(t) = A\cos(\omega_0 t + \theta), \quad -\infty \leq t \leq \infty$$

Solution: The sinusoidal signal $x(t)$ is periodic with $T_0 = \frac{2\pi}{\omega_0}$. Then, the average power of $x(t)$ is

$$P = \frac{1}{T_0} \int_{0}^{T_0} |x(t)|^2 dt = \frac{\omega_0}{2\pi} \int_{0}^{\frac{2\pi}{\omega_0}} A^2 \cos^2(\omega_0 t + \theta) dt$$

$$= \frac{A^2\omega_0}{2\pi}\int_0^{\frac{2\pi}{\omega_0}} \frac{1}{2}[1+\cos(2\omega_0 t + 2\theta)]dt = \frac{A^2}{2}$$

Since $P = \frac{A^2}{2} < \infty$, $x(t)$ is a power signal.

Example 1.7. Find the average power of the signal, $x(n) = u(n)$, where $u(n)$ is as shown in Fig. Ex. 1.7.

Solution:

$$P = \lim_{N\to\infty} \frac{1}{2N+1}\sum_{n=-N}^{N} |x(n)|^2$$

$$= \lim_{N\to\infty} \frac{1}{2N+1}\sum_{n=0}^{N} 1^2$$

$$= \lim_{N\to\infty} \frac{1}{2N+1}(N+1) = \frac{1}{2} < \infty$$

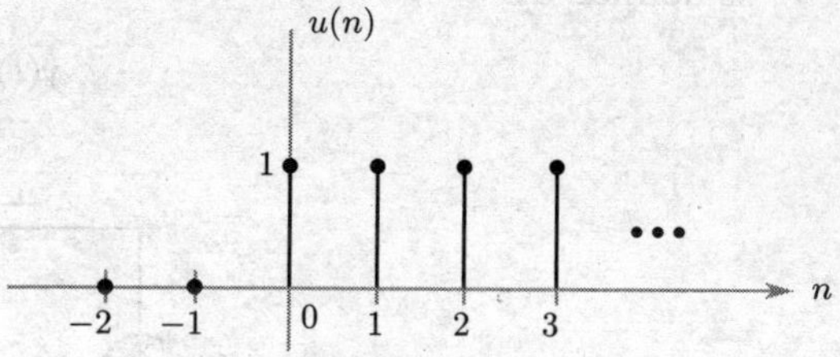

Fig. Ex. 1.7

Thus, $x(n)$ is a power signal.

1.2 Basic Operations on signals

A. Operations performed on dependent variables

(i) Amplitude scaling

Let $x(t)$ be a continuous-time signal. The signal $y(t)$ resulting from amplitude scaling is given by

$$y(t) = Cx(t)$$

where, C is the scaling factor.

A physical example of a device that performs amplitude scaling is an electronic amplifer.

(ii) Addition

Let $x_1(t)$ and $x_2(t)$ denote, a pair of continuous-time signals. The signal $y(t)$ obtained by addition of $x_1(t)$ and $x_2(t)$ is defined by

$$y(t) = x_1(t) + x_2(t)$$

A physical example of a device that performs addition is a frequency mixer, which combines low frequency and high frequency signals.

(iii) Multiplication

Let $x_1(t)$ and $x_2(t)$ denote, a pair of continuous-time signals. The signals $y(t)$ resulting from the multiplication of $x_1(t)$ by $x_2(t)$ is defined by $y(t) = x_1(t)x_2(t)$.

The multiplicative operation is often encountered in analog communication, where an audio frequency signal is multiplied by a high frequency sinusoid known as *carrier*. The resulting signal is known as *amplitude modulated wave*.

(iv) Differentiation

Let $x(t)$ denote, a continuous-time signal. The derivative of $x(t)$ with respect to time is defined by

$$y(t) = \frac{dx(t)}{dt}$$

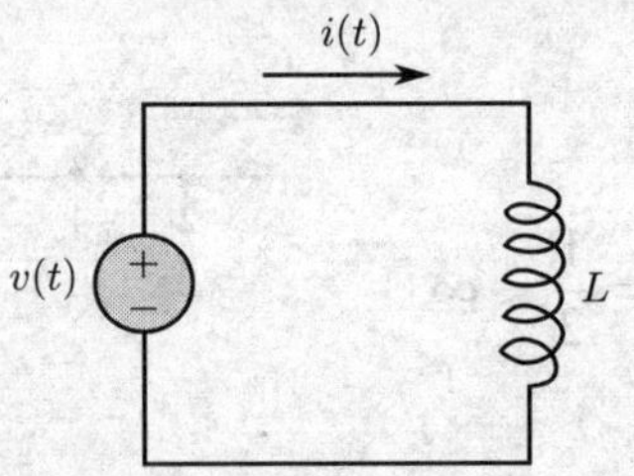

Fig. 1.6 *Inductor demonstrating the concept of differentiation.*

A physical device that performs differentiation is an inductor. The voltage across the inductor shown in Fig. 1.6 is given by,

$$v(t) = L\frac{di(t)}{dt}$$

(v) Integration

Let $x(t)$ denote a continuous-time signal. The integral of $x(t)$ with respect to t is defined by

$$y(t) = \int_{-\infty}^{t} x(\tau)d\tau$$

This type of operation is encountered in a capacitor. The voltage across the capacitor shown in Fig. 1.7 is given by

$$v(t) = \frac{1}{C}\int_{-\infty}^{t} i(\tau)d\tau$$

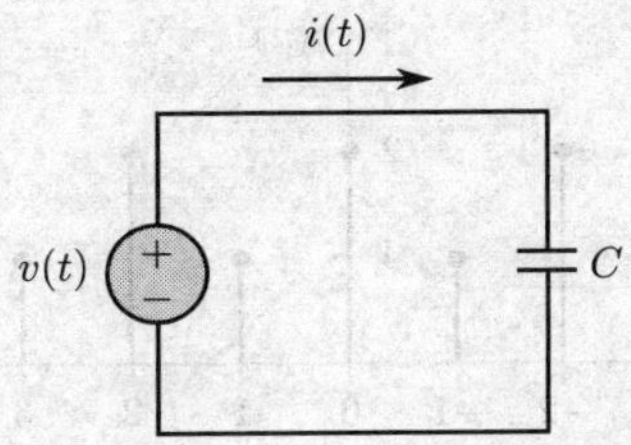

Fig. 1.7 *Capacitor demonstrating the concept of integration.*

B. Operations performed on the independent variable

(i) Time-scaling

Let $x(t)$ be a continuous-time signal. The signal $y(t)$ obtained by scaling independent variable, t, by a factor a is given by

$$y(t) = x(at)$$

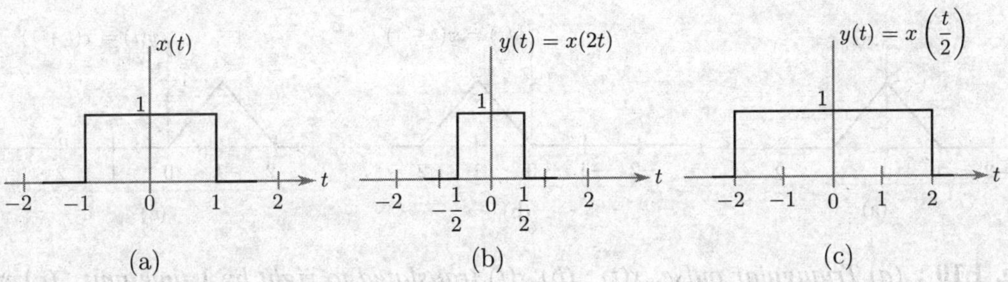

Fig. 1.8 : *(a) Rectangular pulse, x(t) (b) Compression by a factor of 2 (c) Expansion by a factor of 2.*

Fig. 1.8 demonstrates the concept of time-scaling. Note that, when $a > 1, y(t)$, is the compressed version of $x(t)$ and if $a < 1, y(t)$ is the expanded version of $x(t)$.

In the discrete-time case, we write $y(n) = x(mn), m > 0$ and m is an integer.

Fig. 1.9(a) shows an arbitrary discrete-time signal $x(n)$ and Fig. 1.9(b) shows the time-scaled version of $x(n)$.

Comparing Figs. 1.9(a) and 1.9(b), we find that in $y(n)$, samples of amplitude 1 are totally missing. Hence, time-scaling of a discrete-time signal leads to inherent loss of information.

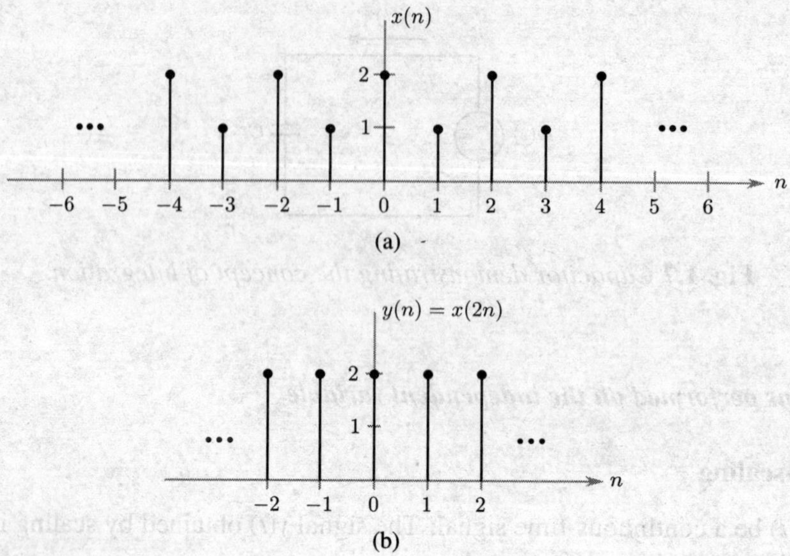

Fig. 1.9 *(a) Discrete-time signal x(n) (b) Time-scaled version of x(n)*

(ii) Time shifting

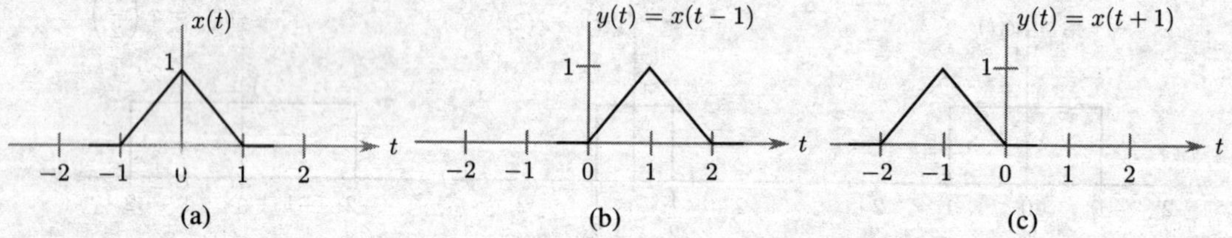

Fig. 1.10 : *(a) Triangular pulse, x(t) (b) x(t) translated to right by* 1 *time unit (c) x(t) translated to left by* 1 *time unit*

Let $x(t)$ be a continuous-time signal. Replacing t by $t + b$ results in a time shifted signal $y(t)$ defined below:

$$y(t) = x(t + b)$$

Figs. 1.10(b) and 1.10(c) denote the time shifted versions of $x(t)$ for $b < 0$ and $b > 0$, respectively. If $b < 0$, $x(t)$ is shifted to right by an amount, b secs and if $b > 0$, $x(t)$ is translated to left by b secs.

(iii) Reflection

Let $x(t)$ denote a continuous-time signal. The reflected version of $x(t)$ is obtained by replacing t with $-t$. The signal, $y(t) = x(-t)$ represents a reflected version of $x(t)$ about the vertical axis.

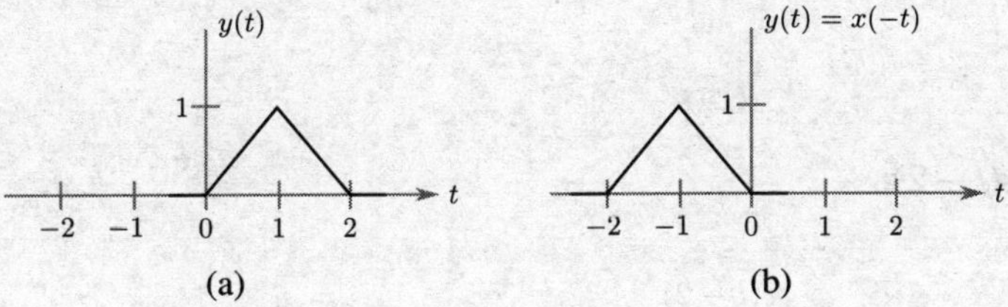

Fig. 1.11 *(a) Triangular pulse, x(t) (b) Reflection of x(t) about vertical axis*

The following important points may be noted.

1. For an even signal $x(t)$, the signal and its reflected version $x(-t)$ are identical.
2. For an odd signal $x(t)$, the signal and its reflected version $x(-t)$ are negative of each other. These properties are illustrated in Fig. 1.12. A similar concept holds good for discrete-time signals also.

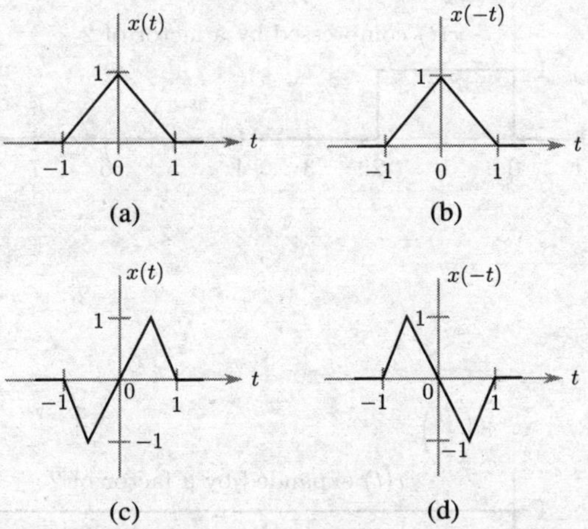

Fig. 1.12 : *(a) & (b) Even signal x(t) and its reflection x(−t); (c) & (d) odd signal x(t) and its reflection x(−t).*

Example 1.8. A continuous-time signal $x(t)$ is shown in Fig. Ex. 1.8. Sketch and label each of the following signals.

(a) $x(t-2)$ (b) $x(2t)$ (c) $x\left(\frac{t}{2}\right)$ (d) $x(-t)$

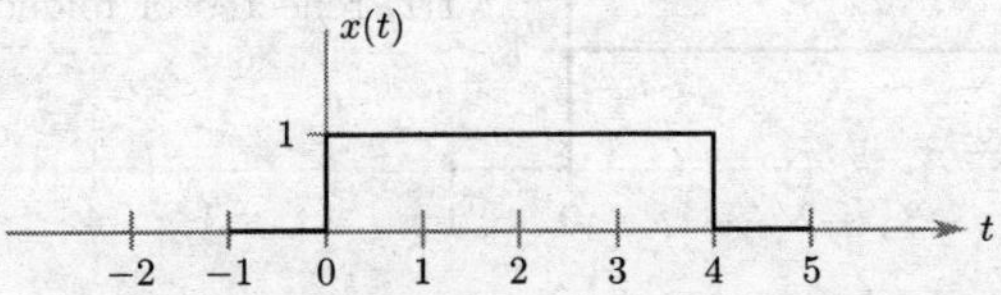

Fig. Ex. 1.8

Solution:

(a) Sketch of $x(t-2)$:

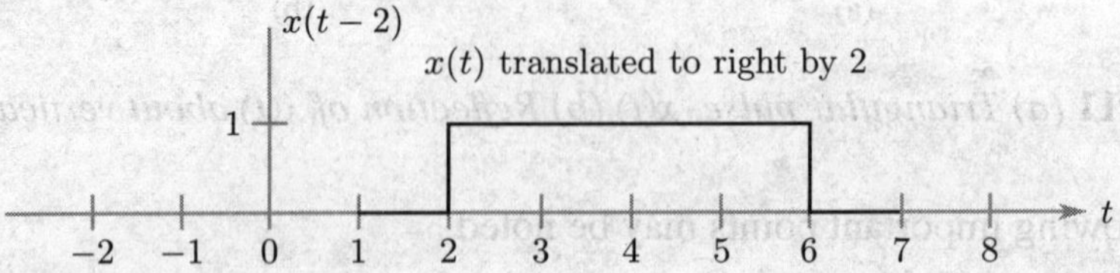

(b) Sketch of $x(2t)$:

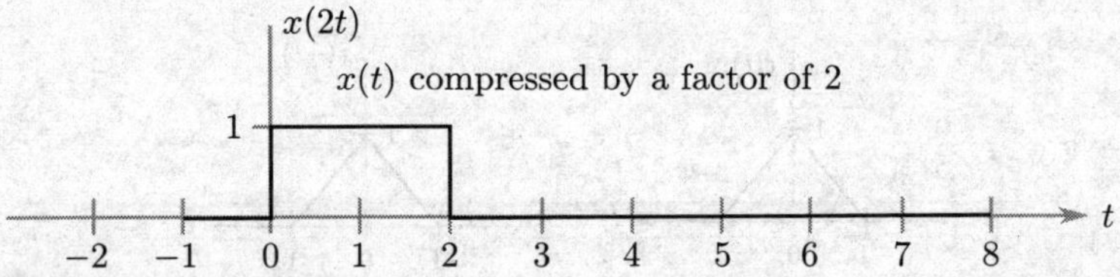

(c) Sketch of $x\left(\frac{t}{2}\right)$:

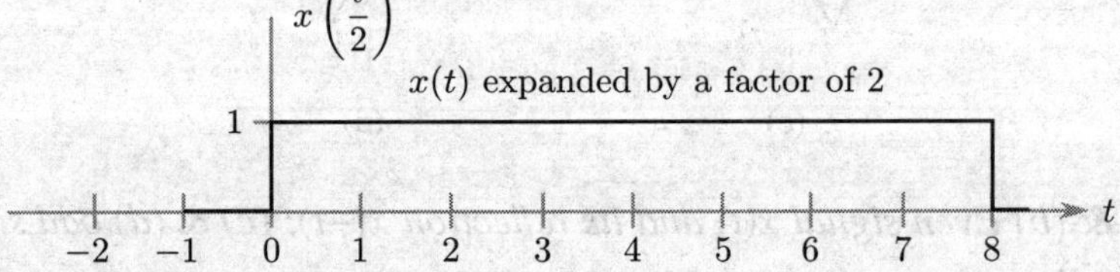

(d) Sketch of $x(-t)$:

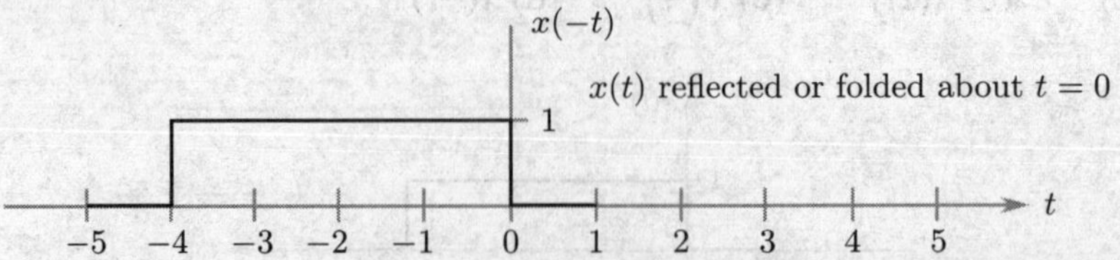

Example 1.9. A discrete-time signal $x(n)$ is shown in Fig. Ex. 1.9. Sketch and label each of the following signals.

(a) $x(n-2)$ (b) $x(2n)$ (c) $x(-n)$

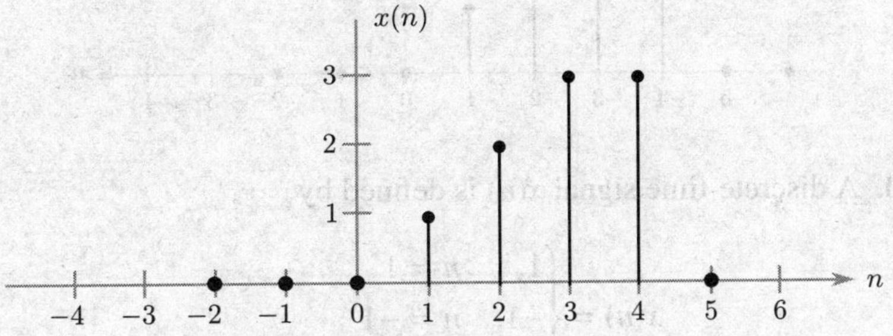

Fig. Ex. 1.9

Solution:

(a) Sketch of $x(n-2)$:

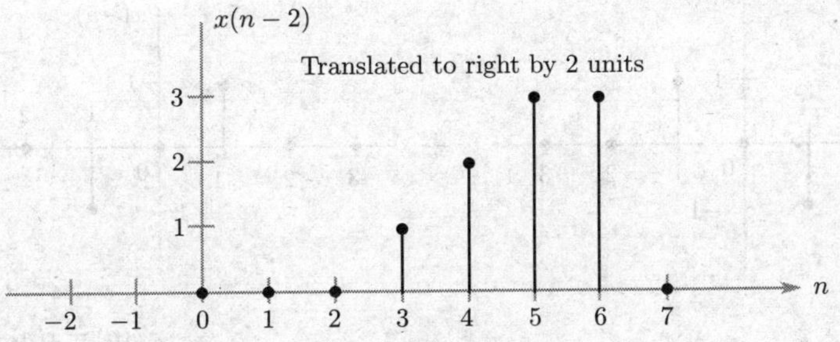

(b) Sketch of $x(2n)$:

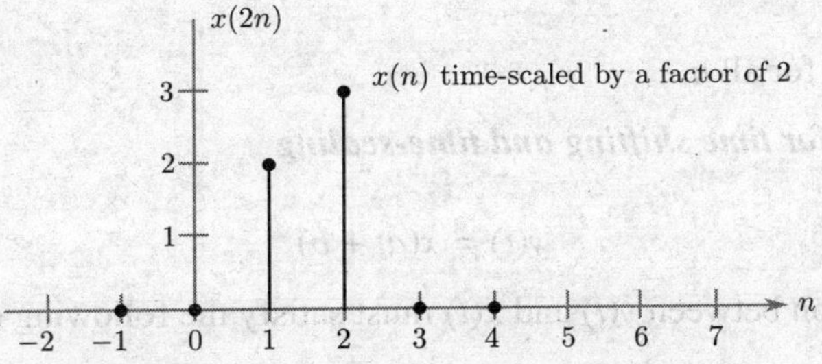

(c) Sketch of $x(-n)$:

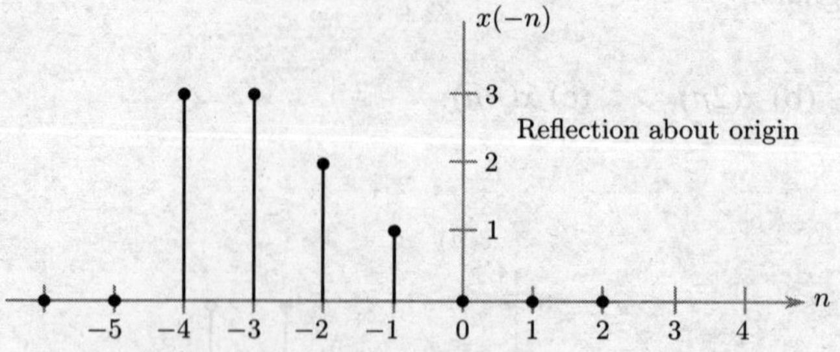

Example 1.10. A discrete-time signal $x(n)$ is defined by

$$x(n) = \begin{cases} 1, & n = 1 \\ -1, & n = -1 \\ 0, & n = 0 \ \& \ |n| > 1 \end{cases}$$

Find the composite signal $y(n)$ defined in terms of $x(n)$ by

$$y(n) = x(n) + x(-n)$$

Solution:

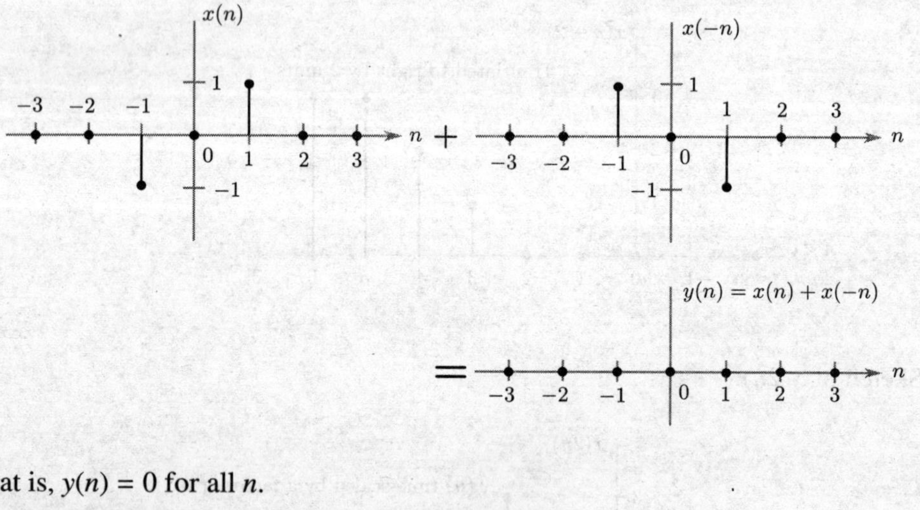

That is, $y(n) = 0$ for all n.

C. Precedence rule for time shifting and time-scaling

Let

$$y(t) = x(at + b) \tag{1.24}$$

The above relation between $y(t)$ and $x(t)$ must satisfy the following conditions.

$$y(0) = x(b) \tag{1.25}$$

$$\text{and} \qquad y\left(\frac{-b}{a}\right) = x(0) \tag{1.26}$$

The above relations (1.25) and (1.26) are useful checks in terms of corresponding values of $x(t)$.

As an example, let us take, $y(t) = x(2t + 3)$. First, let us perform time-shifting operation on $x(t)$, resulting in an intermediate signal $v(t)$ defined by

$$v(t) = x(t + 3)$$

Next, the time-scaling operation is performed on $v(t)$. This repalces t by $2t$, resulting in the desired output,

$$y(t) = v(2t) = x(2t + 3)$$

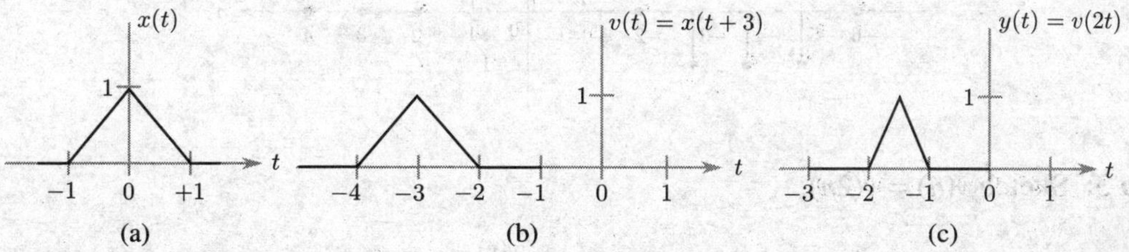

Fig. 1.13 : *(a) triangular waveform, $x(t)$ (b) time-shifted waveform, $v(t)$ (c) time-scaled waveform, $y(t)$*

Verification:

$$y(0) = x(3) = 0$$

$$y\left(-\frac{3}{2}\right) = x(0) = 1$$

Hence, the precedence rule is to do the time-shifting operation first and then the time-scaling operation.

Example 1.11. A discrete-time signal $x(n)$ is described by

$$x(n) = \begin{cases} 1, & n = 1, 2, 3 \\ -1, & n = -1, -2, -3 \\ 0, & n = 0, |n| > 3 \end{cases}$$

Find $y(n) = x(2n + 2)$.

Solution:

***Step* 1:** Sketch $x(n)$.

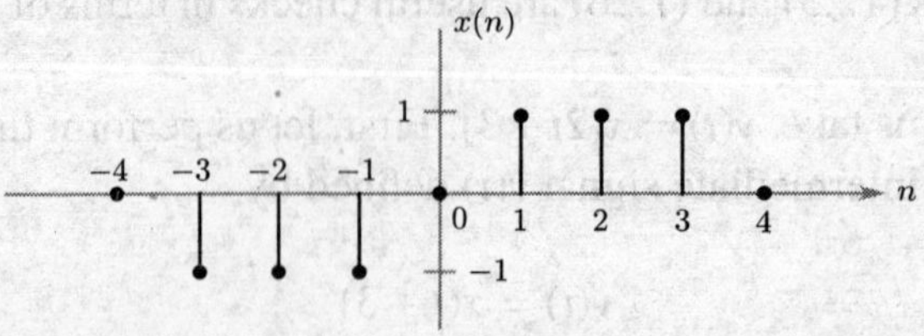

***Step* 2:** Sketch $v(n) = x(n+2)$.

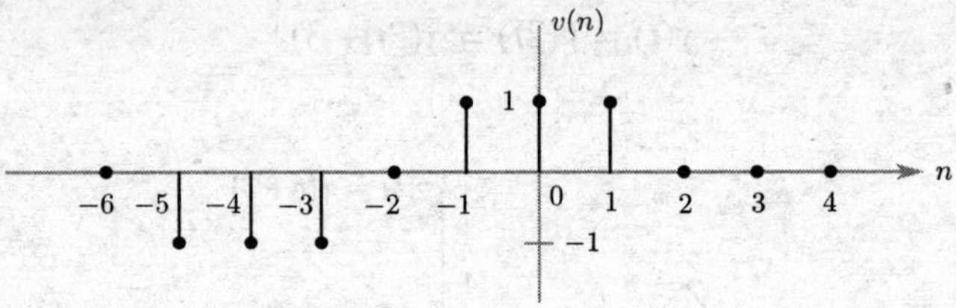

***Step* 3:** Sketch $y(n) = v(2n)$.

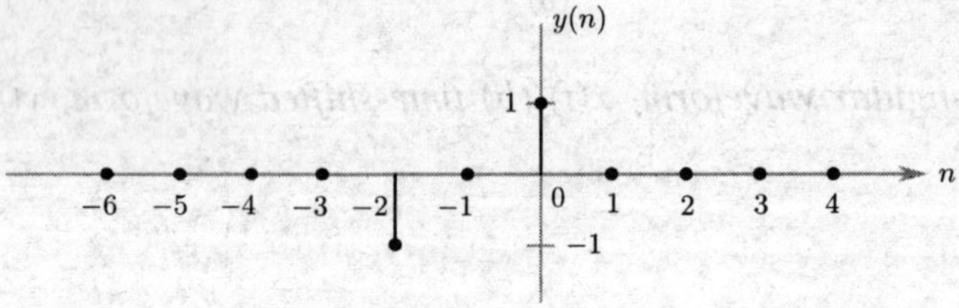

1.3 Basic Continuous-Time Signals

1.3.1 Unit step function

The unit step function, $u(t)$ is defined as follows:

$$u(t) = \begin{cases} 1, & t > 0 \\ 0, & t < 0 \end{cases} \tag{1.27}$$

The unit step function, $u(t)$ is sketched in Fig. 1.14(a). Note that $u(t)$ is discontinuous at $t = 0$. In order to avoid the confusion at $t = 0$ regarding the amplitude of $u(t)$, it is often convenient to define $u(t)$ as follows:

$$u(t) = \begin{cases} 1, & t \geq 0^+ \\ 0, & t \leq 0^- \end{cases}$$

Similarly, the shifted unit step function is defined as follows:

$$u(t+t_0) = \begin{cases} 1, & t+t_0 > 0 \quad \text{or} \quad t > -t_0 \\ 0, & t+t_0 < 0 \quad \text{or} \quad t < -t_0 \end{cases}$$

The shifted unit step function is shown in Fig. 1.14(b).

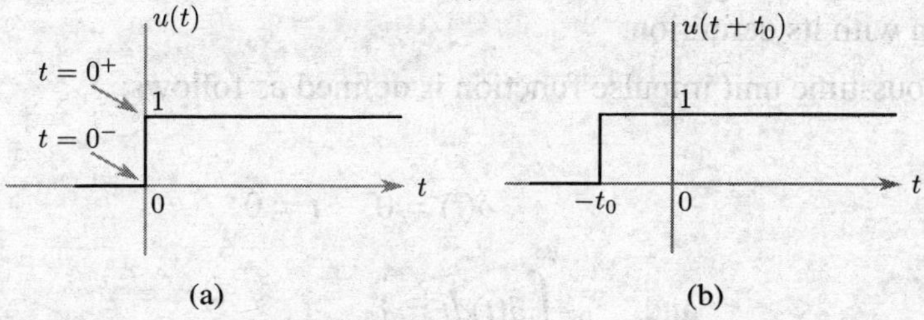

Fig. 1.14 *(a) Unit step function (b) Shifted unit step funciton*

A step function is used as a building block to generate several discontinuous waveforms. An example to this effect is shown in Fig. 1.15.

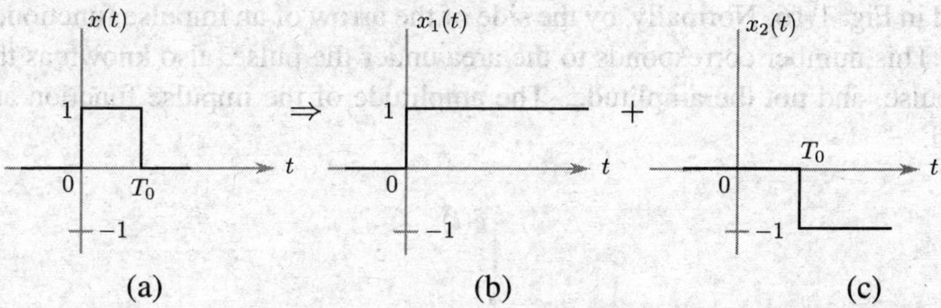

Fig. 1.15 : *(a) Discontinuous waveform (b) Unit step funciton (c) Shifted and inverted unit step function*

Hence,

$$\begin{aligned} x(t) &= x_1(t) + x_2(t) \\ &= u(t) - u(t-T_0) \end{aligned}$$

The unit step function has the property

$$u(t-t_0) = [u(t-t_0)]^2 = [u(t-t_0)]^k$$

with k being any positive integer. This property is based on the relations $(0)^k = 0$ and $(1)^k = 1, k = 1, 2, 3 \ldots$.

A second property is related to time-scaling:

$$u(at-t_0) = u\left(t - \frac{t_0}{a}\right); \; a \neq 0$$

1.3.2 Unit impulse function

Engineers have found very good use for $j = \sqrt{-1}$ inspite of the fact that it is not a real number and does not appear in nature. Electrical and electronic engineering analysis makes use of j extensively. In the same token, engineers have found tremendous use for the unit impulse function, even though this funciton cannot exist in nature. We introduce a unit impulse function with its definition.

The continuous-time unit impulse function is defined as follows:

$$\delta(t) = 0, \quad t \neq 0 \tag{1.28}$$

and

$$\int_{-\infty}^{\infty} \delta(t)dt = 1 \tag{1.29}$$

From Eq. (1.28), we find that, $\delta(t) = \delta(-t)$.

The inference is that, $\delta(t)$ is an even function of time, t. The unit impulse function $\delta(t)$ is sketched in Fig. 1.16. Normally, by the side of the arrow of an impulse function, we write a number. This number corresponds to the area under the pulse, also known as the weight of the impulse, and not the amplitude. The amplitude of the impulse function at $t = 0$ is unbounded.

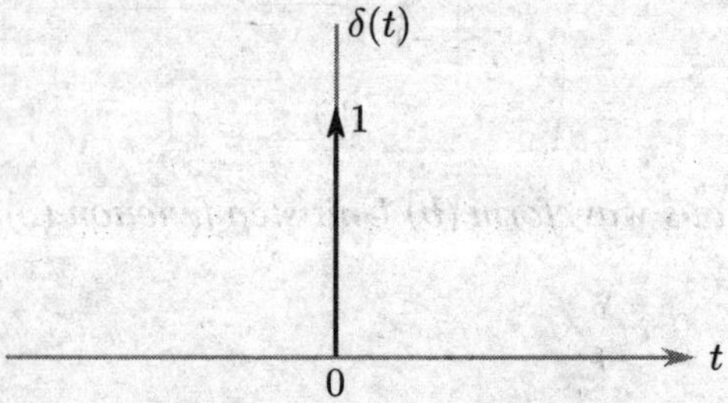

Fig. 1.16 *Unit impulse function.*

We can visualize an impulse as a tall, narrow rectangular pulse of unit area, as shown in Fig. 1.17. The width of the rectangular pulse is a very small value ($T \to 0$). Consequently, its height is a very large value $\left(\frac{1}{T}\right)$. The unit impulse therefore can be regarded as a rectangular pulse with a width that has become infinitesimally small and a height that has become infinitely large, maintaining the overall area at unity.

That is,

$$\delta(t) = \lim_{T \to 0} g_T(t)$$

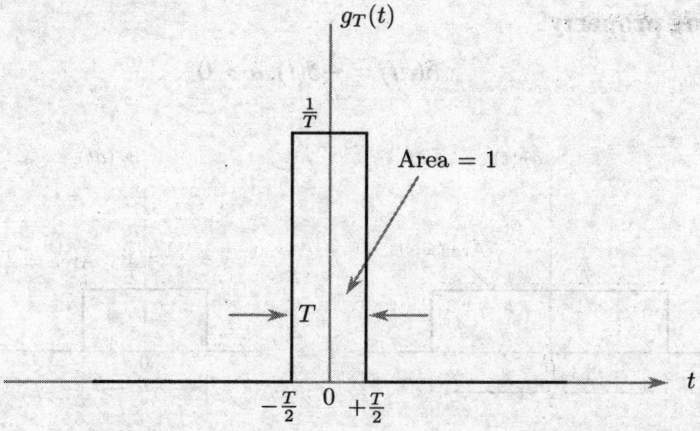

Fig. 1.17 *Approximation of unit impulse by a rectangular pulse.*

Following are some of the important properties of the unit impulse function:

(i) Product property

Let us consider, what happens when we multiply the unit impulse $\delta(t)$ by a function $\phi(t)$ that is known to be continuous at $t = 0$. Since the impulse exists only at $t = 0$, and the value $\phi(t)$ at $t = 0$ is $\phi(0)$, we obtain

$$\phi(t)\delta(t) = \phi(0)\delta(t)$$

Similarly,

$$\phi(t)\delta(t - t_0) = \phi(t_0)\delta(t - t_0)$$

(ii) Sifting property of the unit impulse function

$$\int_{t_1}^{t_2} \phi(t)\delta(t - t_0)dt = \begin{cases} \phi(t_0), & t_1 < t_0 < t_2 \\ 0, & \text{otherwise} \end{cases}$$

or

$$\int_{-\infty}^{\infty} \phi(t)\delta(t - t_0)dt = \phi(t_0)$$

That is, the function $\phi(t)$ gets evaluated at the point where the impulse is located.

(iii) Time-scaling property

$$\delta(at) = \frac{1}{a}\delta(t), a > 0$$

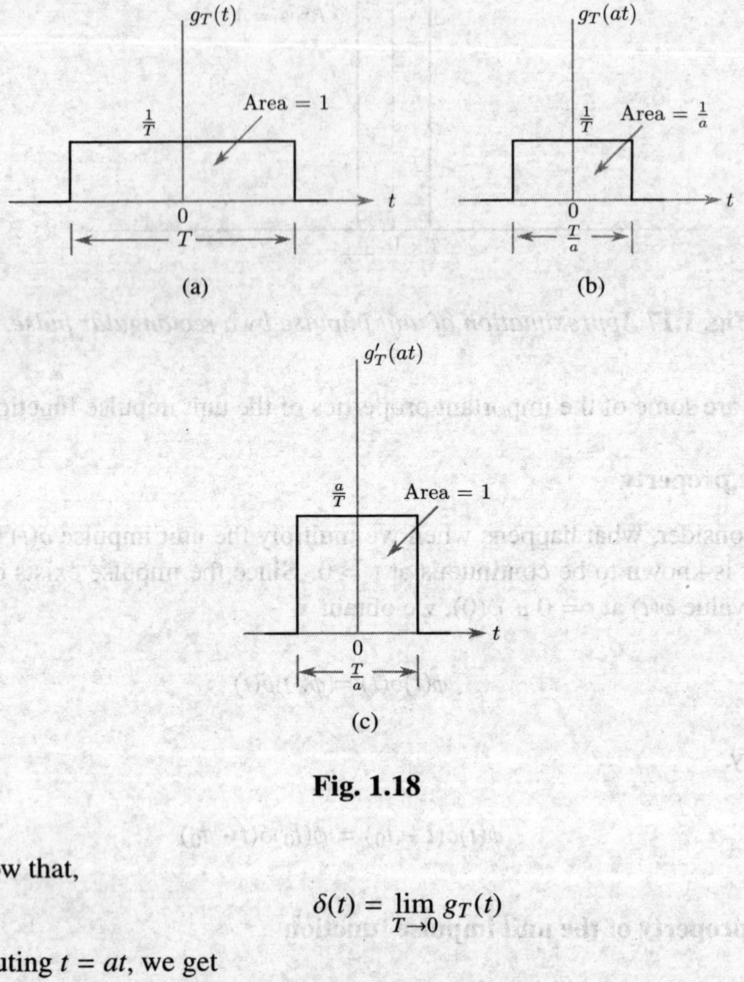

Fig. 1.18

We know that,

$$\delta(t) = \lim_{T\to 0} g_T(t)$$

Substituting $t = at$, we get

$$\delta(at) = \lim_{T\to 0} g_T(at) \tag{1.30}$$

From Figs. 1.18(b) and 1.18(c), we can write

$$g'_T(at) = a\, g_T(at) \tag{1.31}$$

Substituting Eq. (1.31) in Eq. (1.30), we get

$$\delta(at) = \lim_{T\to 0} \frac{1}{a} g'_T(at)$$

Since, the area under the function $g'_T(at)$ is unity, it follows that,

$$\lim_{T\to 0} g'_T(at) = \delta(t)$$

Therefore,

$$\delta(at) = \frac{1}{a}\delta(t)$$

Hence, the proof.

It may be noted that an impulse function, $\delta(t)$ is the derivative of the step function.

Additional Properties:

1. $\delta(t - t_0) = \frac{d}{dt}u(t - t_0)$

2. $u(t - t_0) = \int_{-\infty}^{t} \delta(\tau - t_0)d\tau = \begin{cases} 1, & t > t_0 \\ 0, & t < t_0 \end{cases}$

3. $\int_{-\infty}^{\infty} \delta(at - t_0)dt = \frac{1}{|a|}\int_{-\infty}^{\infty} \delta\left(t - \frac{t_0}{a}\right)dt$

4. $\delta(x^2 - a^2) = \frac{1}{2|a|}[\delta(x + a) + \delta(x - a)]$

1.3.3 Ramp function

The unit step function is the integral of unit impulse function with respect to time. By the same token, the integral of the step function $u(t)$ is a ramp function of unit slope. Mathematically, a ramp function is defined as follows:

$$r(t) = \begin{cases} t, & t \geq 0 \\ 0, & t < 0 \end{cases}$$

Equivalently, we may write

$$r(t) = t\,u(t)$$

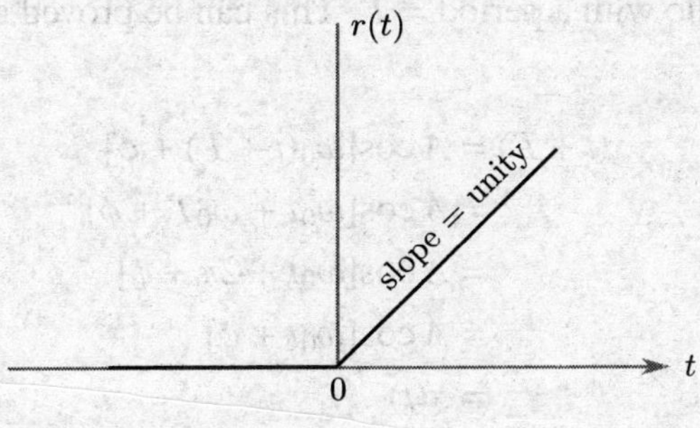

Fig. 1.19 *Unit ramp function.*

1.3.4 Exponential signals

A real exponential signal in its most general form is written as

$$x(t) = Be^{at}$$

where B and a are real parameters.

The parameter B is the amplitude of the signal $x(t)$ at $t = 0$.

If $a < 0$, $x(t)$ is said to be a decaying exponential. On the other hand, if $a > 0$, $x(t)$ is said to be a growing exponential. Figs. 1.20(a) and 1.20(b) illustrate the decaying and growing exponential functions respectively.

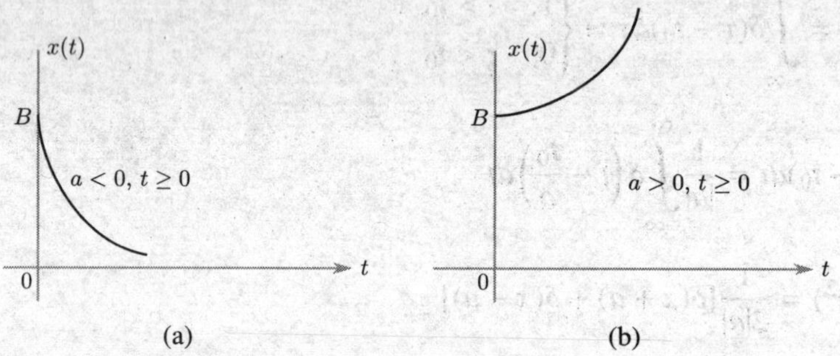

Fig. 1.20 *(a) Decaying exponential function (b) Growing exponential funciton*

1.3.5 Sinusoidal signals

General expression for a continuous-time sinusoid is

$$x(t) = A\cos[\omega_0 t + \phi]$$

where A is the amplitude, ω_0 is the angular frequency in radians/sec and ϕ is the phase angle in radians.

The fundamental period of a sinusoidal signal is defined by, $T_0 = \frac{2\pi}{\omega_0}$. A continuous-time sinusoid is always periodic with a period = T. This can be proved as follows:

Consider,

$$\begin{aligned}
x(t+T) &= A\cos[\omega_0(t+T)+\phi] \\
&= A\cos[\omega_0 t + \omega_0 T + \phi] \\
&= A\cos[\omega_0 t + 2\pi + \phi] \\
&= A\cos[\omega_0 t + \phi] \\
&= x(t)
\end{aligned}$$

which defines the condition for a periodic signal. Therefore, a continuous-time sinusoidal signal is always periodic.

Consider the sum of two sinusoids given by $y(t) = x_1(t) + x_2(t)$. The sum $y(t)$ will be periodic when the ratio of each pair of individual frequencies (or periods) is a rational number, if periodic, the fundamental frequency $f_0 = \text{GCD}(f_1, f_2)$. The common fundamental period is $T_0 = \frac{1}{f_0} = \text{LCM}(T_1, T_2)$.

1.3.6 Exponentially damped sinusoidal signals

The multiplication of a sinusoidal signal, $A\sin(\omega_0 t + \theta)$ by a real-valued decaying exponential signal, $e^{-\alpha t}$ results in a new signal referred to as exponentially damped sinusoidal signal as shown in Fig. 1.21.

That is,

$$x(t) = \begin{cases} Ae^{-\alpha t}\sin(\omega_0 t + \theta), \ \alpha > 0, & t \geq 0 \\ 0, & t < 0 \end{cases}$$

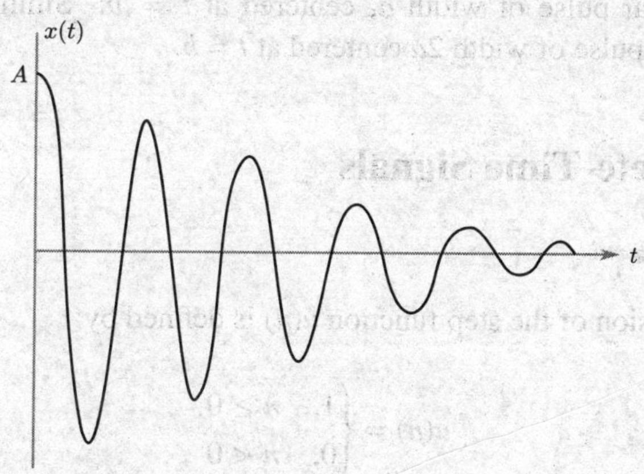

Fig. 1.21 *Exponentially damped sinusoidal signal shown for t ≥ 0*

1.3.7 Pulse signals

The rectangular pulse rect(t) and triangular pulse tri(t) are defined as follows:

$$\text{rect}(t) = \begin{cases} 1, & |t| < 0.5 \\ 0, & \text{elsewhere} \end{cases}$$

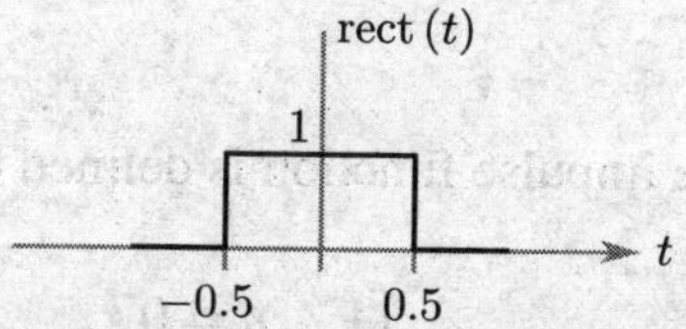

Height = 1, Area = 1, Width = 1.

$$\text{tri}\,(t) = \begin{cases} 1 - |t|, & |t| \leq 1 \\ 0, & \text{elsewhere} \end{cases}$$

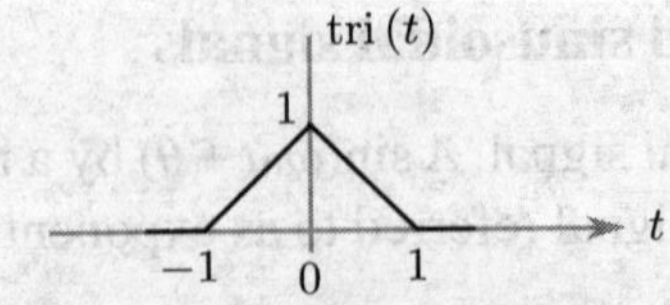

Height = 1, Area = 1, Width = 2.

Both the pulses exhibit even symmetry. It may be noted that the signal, $x(t) = \text{rect}\left(\frac{t-b}{a}\right)$ describes a rectangular pulse of width a, centered at $t = b$. Similarly, $y(t) = \text{tri}\left(\frac{t-b}{a}\right)$ describes a triangular pulse of width $2a$ centered at $t = b$.

1.4 Basic Discrete-Time Signals

1.4.1 Step function

The discrete-time version of the step function $u(n)$ is defined by

$$u(n) = \begin{cases} 1, & n \geq 0 \\ 0, & n < 0 \end{cases}$$

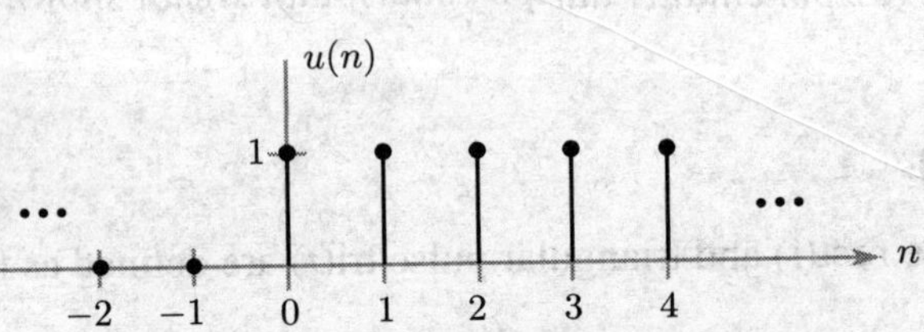

Fig. 1.22 *Discrete-time unit step function*

1.4.2 Impulse function

The discrete-time version of the impulse function is defined by

$$\delta(n) = \begin{cases} 1, & n = 0 \\ 0, & n \neq 0 \end{cases}$$

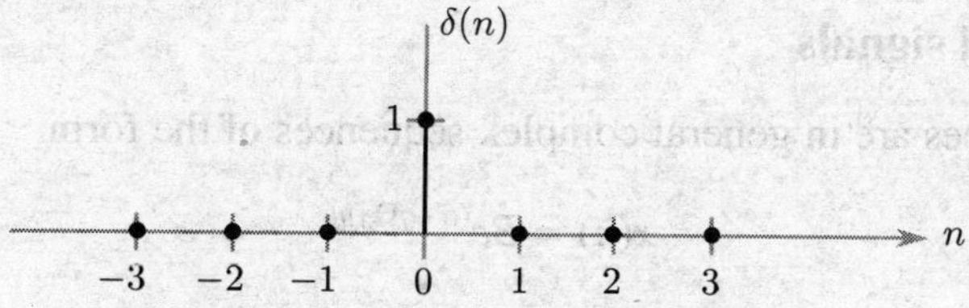

Fig. 1.23 *Discrete-time impulse function*

Following are some of the important properties of unit impulse function:

(i) Product property

$$x(n)\delta(n-k) = x(k)\delta(n-k)$$

(ii) Sifting property

$$\sum_{n=N_1}^{N_2} x(n)\delta(n-k) = \begin{cases} x(k), & N_1 \le k \le N_2 \\ 0, & \text{otherwise} \end{cases}$$

or

$$\sum_{n=-\infty}^{\infty} x(n)\delta(n-k) = x(k)$$

1.4.3 Ramp function

The discrete-time version of a ramp function is defined by

$$r(n) = \begin{cases} n, & n \ge 0 \\ 0, & n < 0 \end{cases}$$

or equivalently, $r(n) = nu(n)$.

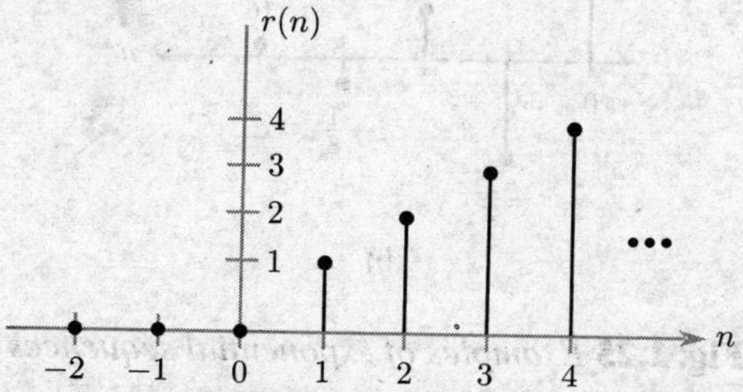

Fig. 1.24 *Discrete-time unit ramp function*

1.4.4 Exponential signals

Exponential sequences are in general complex sequences of the form

$$x(n) = Be^{(\sigma + j\Omega_0)n}$$

where B is the amplitude, σ is the damping constant, and Ω_0 is the angular frequency.

In discrete-time, it is common to write a real exponential signal ($\Omega_0 = 0$) as

$$x(n) = Be^{\sigma n} = Br^n, \quad \text{where} \quad r = e^{\sigma}$$

If $0 < |r| < 1$, $x(n)$ is called decaying exponential signal and if $|r| > 1$, then $x(n)$ is known as the growing exponential sequence.

Two examples for unilateral exponential sequences with real values are given by

(i) $B = 1$ and $r = \frac{1}{2}$ leading to $x_1(n) = \left(\frac{1}{2}\right)^n$, $n \geq 0$ and

(ii) $B = -1$ and $r = \frac{1}{2}$ leading to $x_2(n) = \left(-\frac{1}{2}\right)^n$, $n \geq 0$

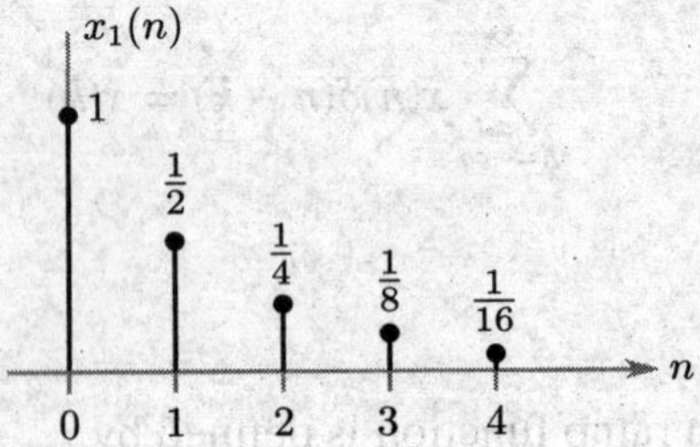

(a)

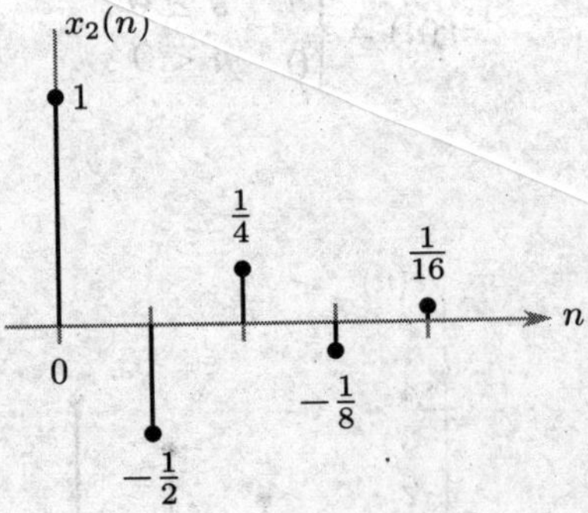

(b)

Fig. 1.25 *Examples of exponential sequences*

A complex exponential signal is defined by $x(n) = Be^{j\Omega_0 n}$.

In contrast to continuous exponential functions, different values of the angular frequency Ω_0 can lead to same exponential series. As $e^{j\Omega_0 n}$ is only evaluated for integers n, adding multiples of 2π to Ω_0 does not change the sequence:

$$e^{j\Omega_0 n} = e^{j(\Omega_0+2\pi)n}$$

The complex exponential sequence $x(n) = e^{j\Omega_0 n}$ is periodic only if $\frac{\Omega_0}{2\pi}$ is a rational number. This can be proved as follows:

$x(n)$ will be periodic if

$$e^{j\Omega_0(n+N)} = e^{j\Omega_0 n}$$

$$\Rightarrow \qquad e^{j\Omega_0 N} = 1$$

The above equation holds good only if

$$\Omega_0 N = 2\pi m; \quad \text{where } m = \text{positive integer}$$

$$\text{or} \qquad \frac{\Omega_0}{2\pi} = \frac{m}{N} = \text{rational number}$$

Hence, $x(n)$ is periodic only if $\frac{\Omega_0}{2\pi}$ is a rational number. Note this property is quite different from the property that the continuous-time signal $e^{j\omega_0 t}$ is periodic for any value of ω_0. Thus, if $\frac{\Omega_0}{2\pi}$ is a rational number, and N and m have no factors in common, then the period, N of $e^{j\Omega_0 n}$ given by

$$N = \frac{2\pi m}{\Omega_0}$$

1.4.5 Sinusoidal signals

The discrete-time version of a sinusoidal signal is written as

$$x(n) = A\cos[\Omega_0 n + \phi]$$

where Ω_0 is the angular frequency in radians.

The period of a discrete-time sinusoid is measured in samples. Let the period of $x(n)$ be N. Then,

$$x(n+N) = A\cos[\Omega_0 n + \Omega_0 N + \phi]$$

For $x(n)$ and $x(n+N)$ to be identical,

$$\Omega_0 N = 2\pi m \quad \text{radians}$$

$$\Rightarrow \qquad \Omega_0 = \frac{2\pi m}{N}; \quad m \text{ and } N \text{ are integers}$$

Unlike continuous-time sinusoidal signals, not all discrete-time sinusoidal signals with arbitrary values of Ω_0 are periodic.

Alternatively, a discrete-time sinusoid, $x(n)$ is periodic, if and only if $\frac{\Omega_0}{2\pi}$ is a rational number, that is, $\frac{\Omega_0}{2\pi} = \frac{p}{q}$ where p and q are integers. If $\frac{\Omega_0}{2\pi}$ is a rational number, then fundamental period, $N_0 = q$ samples. On the other hand, if $\frac{\Omega_0}{2\pi}$ is not a rational number, then $x(n)$ is not periodic.

If $x_1(n)$ and $x_2(n)$ are two sinusoidal periodic sequences with fundamental periods N_1 and N_2 respectively, then the sum $x(n) = x_1(n)+x_2(n)$ is periodic only if $\frac{N_1}{N_2}$ is a rational number. If $\frac{N_1}{N_2}$ is a rational number, then the fundamental period of $x(n)$ is $N_0 = \text{LCM}(N_1, N_2)$.

Relation between sinusoidal and complex exponential signals:

The Euler's identity allows us to write

$$e^{j\theta} = \cos\theta + j\sin\theta$$

This result indictes that we may express the discrete-time sinusoidal signal as the real part of complex exponential signal $Be^{j\Omega_0 n}$, that is

$$A\cos(\Omega_0 n + \phi) = \text{Re}\,\{Be^{j\Omega_0 n}\}$$

where B is defined in terms of A and ϕ: $B = Ae^{j\phi}$.

Fig 1.26 depicts the two dimensional representation of the complex exponential $e^{j\Omega_0 n}$ for $\Omega_0 = \frac{\pi}{4}$ and $n = 0, 1, 2 \ldots 7$. The projection of each value on the real axis is $\cos\Omega_0 n$, while the projection on the imaginary axis is $\sin\Omega_0 n$.

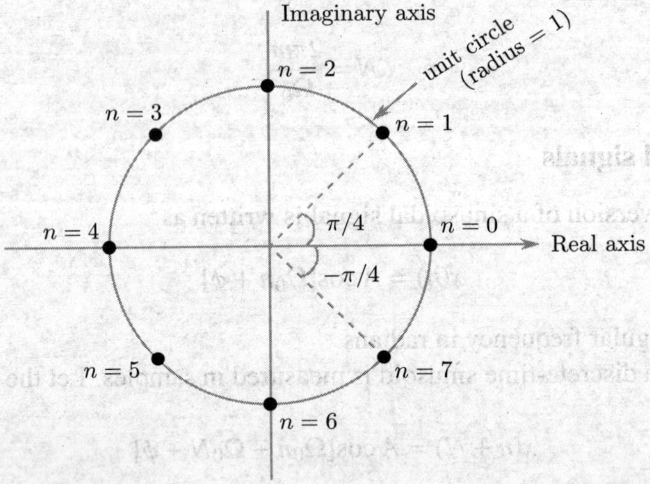

Fig. 1.26 *Representation of $e^{j\Omega_0 n}$ in the complex plane for $\Omega_0 = \pi/4$, showing eight points*

1.4.6 Exponentially damped sinusoidal signals

The discrete-time version of the exponentially damped sinusoidal signal is described by

$$x(n) = Br^n \sin[\Omega_0 n + \phi]$$

For the signal to decay exponentially with time, the parameter r must lie the range, $0 < |r| < 1$.

1.4.7 Pulse signals

The discrete version of a rectangular pulse rect $\left(\frac{n}{2N}\right)$ is defined by

$$\text{rect}\left(\frac{n}{2N}\right) = \begin{cases} 1, & |n| \le N \\ 0, & \text{elsewhere} \end{cases}$$

The signal rect$\left(\frac{n}{2N}\right)$ has $(2N + 1)$ unit samples over $-N \le n \le N$.

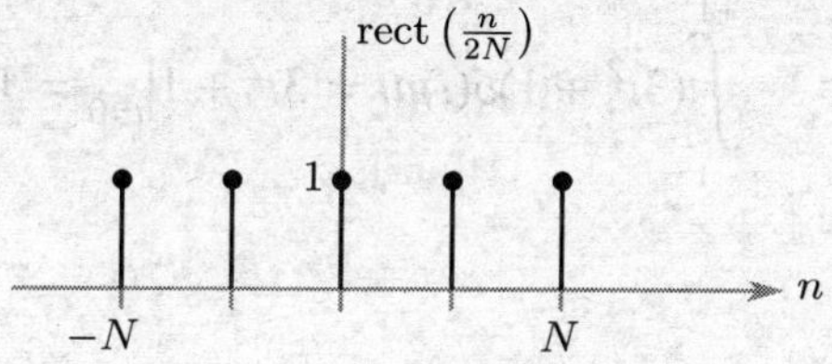

The discrete version of a triangular pulse tri $\left(\frac{n}{N}\right)$ is defined by

$$\text{tri}\left(\frac{n}{N}\right) = \begin{cases} 1 - \frac{|n|}{N}, & |n| \le N \\ 0, & \text{elsewhere} \end{cases}$$

The signal $x(n) = \text{tri}\left(\frac{n}{N}\right)$ has $(2N + 1)$ samples over $-N \le n \le N$, with the terminal samples $x(N)$ and $x(-N)$ being zero.

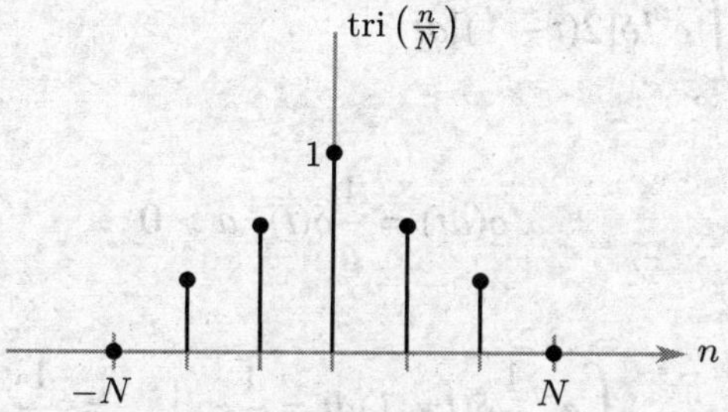

Example 1.12. Evaluate the following integrals:

(a) $\displaystyle\int_{-1}^{1} (3t^2 + 1)\delta(t)dt$ (b) $\displaystyle\int_{1}^{2} (3t^2 + 1)\delta(t)dt$

(c) $\displaystyle\int_{-\infty}^{\infty} (t^2 + \cos \pi t)\delta(t - 1)dt$ (d) $\displaystyle\int_{-\infty}^{\infty} e^{-t}\delta(2t - 2)dt$

Solution:

(a) Sifting property of an impulse function is

$$\int_{-\infty}^{\infty} x(t)\delta(t-t_0)dt = x(t_0)$$

or

$$\int_{a}^{b} x(t)\delta(t-t_0)dt = \begin{cases} x(t_0), & a < t_0 < b \\ 0, & \text{otherwise} \end{cases}$$

Hence,

$$\int_{-1}^{+1} (3t^2+1)\delta(t)dt = 3t^2+1\Big|_{t=0} = 1$$

(b) $\int_{1}^{2} (3t^2+1)\delta(t)dt = 0$

The integral vanishes due to the fact that the impulse $\delta(t)$ does not appear in the range of integration.

(c) $\int_{-\infty}^{\infty} [t^2+\cos\pi t]\delta(t-1)dt = t^2+\cos\pi t\Big|_{t=1} = 1+\cos\pi = 0$

(d) $\int_{-\infty}^{\infty} e^{-t}\delta(2t-2)dt = \int_{-\infty}^{\infty} e^{-t}\delta[2(t-1)]dt$

Since,

$$\delta(at) = \frac{1}{a}\delta(t),\ a > 0$$

we get,

$$\int_{-\infty}^{\infty} e^{-t}\frac{1}{2}\delta(t-1)dt = \frac{1}{2}e^{-t}\Big|_{t=1} = \frac{1}{2e}$$

Example 1.13. Determine whether each of the following signals is periodic. If a signal is periodic, determine its fundamental period.

(a) $x(t) = \cos\left(t+\frac{\pi}{4}\right)$ (b) $x(t) = \cos\left(\frac{\pi}{3}t\right)+\sin\left(\frac{\pi}{4}t\right)$

(c) $x(t) = \cos t + \sin\sqrt{2}t$ (d) $x(t) = \sin^2 t$

(e) $x(t) = e^{j\left(\frac{\pi}{2}t-1\right)}$

Solution:

(a) $x(t) = \cos\left(t + \frac{\pi}{4}\right)$

$\omega_0 = 1$

$x(t)$ is periodic with a fundamental period $T_0 = \frac{2\pi}{\omega_0} = 2\pi$ seconds.

(b) $x(t) = \cos\left(\frac{\pi}{3}t\right) + \sin\left(\frac{\pi}{4}t\right)$

$= x_1(t) + x_2(t)$

$$T_1 = \frac{2\pi}{\frac{\pi}{3}} = 6, \quad T_2 = \frac{2\pi}{\frac{\pi}{4}} = 8$$

Since, $\frac{T_1}{T_2} = \frac{6}{8} = \frac{3}{4}$ is a rational number, $x(t)$ is periodic with a fundamental period $T_0 = 4T_1 = 3T_2 = 24$ seconds.

(c) $x(t) = \cos t + \sin \sqrt{2}t$

$= x_1(t) + x_2(t)$

$x_1(t)$ is periodic with $T_1 = \frac{2\pi}{1} = 2\pi$ seconds and $x_2(t)$ is periodic with $T_2 = \frac{2\pi}{\sqrt{2}} = \sqrt{2}\pi$ seconds. Since, $\frac{T_1}{T_2} = \sqrt{2}$ is an irrational number, $x(t)$ is nonperiodic.

(d) $x(t) = \sin^2 t$

$$\Rightarrow \quad x(t) = \frac{1}{2} - \frac{1}{2}\cos 2t$$

The first term on the RHS is a *dc* term, while the second term on the RHS is periodic with a period, $T_2 = \frac{2\pi}{\omega_2} = \frac{2\pi}{2} = \pi$ seconds. Therefore, $x(t)$ is periodic with a fundamental period $T_0 = \pi$ seconds.

(e)

$$x(t) = e^{j(\frac{\pi}{2}t - 1)}$$

$$\Rightarrow \quad x(t) = e^{-j}e^{j\frac{\pi}{2}t}$$

$$= e^{-j}e^{j\omega_0 t}$$

where $\omega_0 = \frac{\pi}{2}$

Therefore, $x(t)$ is periodic with a fundamental period, $T_0 = \frac{2\pi}{\omega_0} = \frac{2\pi}{\frac{\pi}{2}} = 4$ seconds.

Example 1.14. Determine, whether each of the following signals is periodic. If a signal is periodic, find its fundamental period.

(a) $x(n) = 5\cos(0.2\pi n)$

(b) $x(n) = \sin(2n)$

Solution:

(a)

$$x(n) = 5\cos(0.2\pi n)$$
$$\Omega_0 = 0.2\pi$$

Also, $$\frac{\Omega_0}{2\pi} = \frac{0.2\pi}{2\pi} = \frac{1}{10} = \frac{p}{q} = \text{a rational number}$$

Therefore, fundamental period of $x(n)$ is $N_0 = q = 10$ samples.

(b)

$$x(n) = \sin(2n)$$
$$\Omega_0 = 2$$

Also, $$\frac{\Omega_0}{2\pi} = \frac{2}{2\pi} = \frac{1}{\pi} = \text{an irrational number}$$

Therefore, $x(n)$ is not periodic.

Example 1.15. Consider the continuous-time signal $x(t)$ defined by:

$$x(t) = \begin{cases} \frac{t}{T}, & \frac{-T}{2} \le t \le \frac{T}{2} \\ 1, & t \ge \frac{T}{2} \\ 0, & t < -\frac{T}{2} \end{cases}$$

The signal $x(t)$ defined above is applied to a differentiator. Show that the output of the differentiator approaches the unit impulse, $\delta(t)$ as $T \to 0$.

Solution:

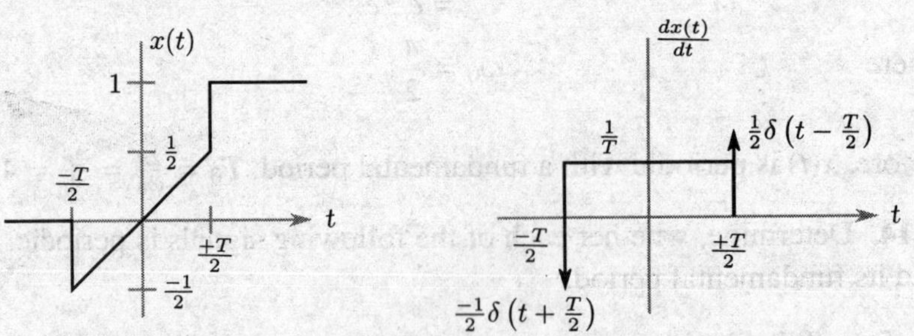

As $T \to 0$, the impulses at $t = \pm\frac{T}{2}$ will appear at the origin and will get cancelled out. Also, the rectangular pulse of unit area as $T \to 0$ becomes $\delta(t)$. Hence, the solution.

Example 1.16. Find and sketch the first derivatives of the following signals:

(a) $x(t) = u(t) - u(t-a), \quad a > 0$

(b) $y(t) = t[u(t) - u(t-a)], \quad a > 0$

Solution:

(a)

$$x(t) = u(t) - u(t-a)$$

$$\Rightarrow \qquad \frac{dx(t)}{dt} = \delta(t) - \delta(t-a)$$

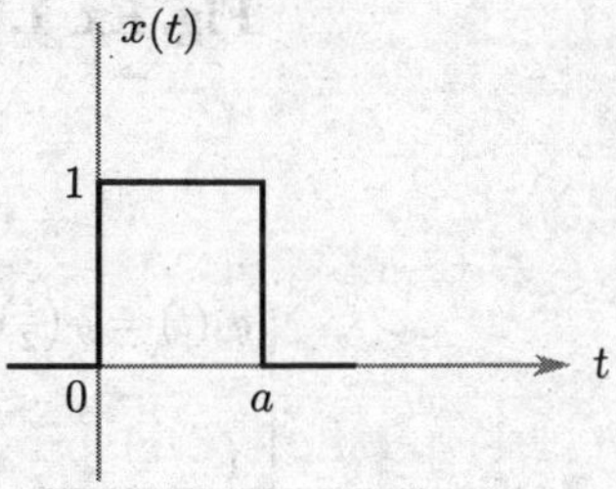

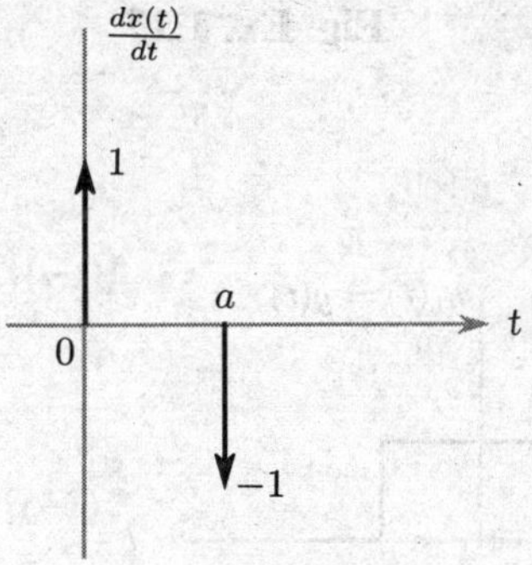

(b)

$$y(t) = t[u(t) - u(t-a)]$$

$$\begin{aligned}\Rightarrow \qquad y'(t) &= \frac{dy(t)}{dt} \\ &= t[\delta(t) - \delta(t-a)] + [u(t) - u(t-a)] \times 1 \\ &= t\delta(t) - t\delta(t-a) + u(t) - u(t-a) \\ &= t\big|_{t=0}\delta(t) - t\big|_{t=a}\delta(t-a) + u(t) - u(t-a) \\ &= 0 - a\delta(t-a) + [u(t) - u(t-a)]\end{aligned}$$

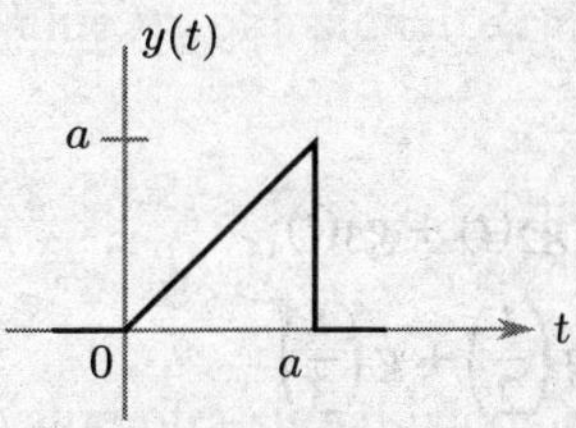

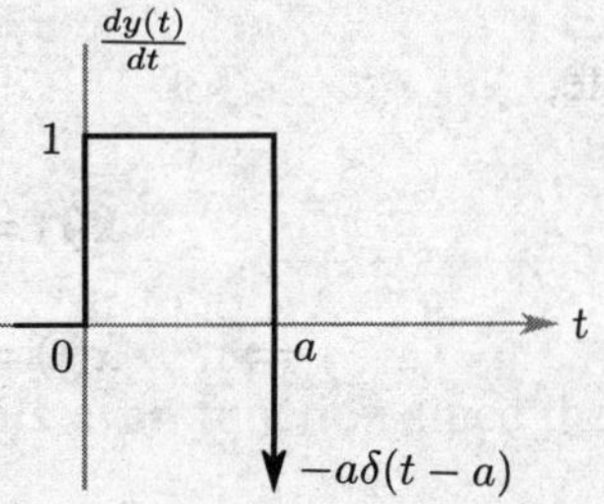

Example 1.17. Fig. Ex. 1.17 shows a pulse $x(t)$ that may be viewed as the superposition of three rectangular pulses. Using the rectangular pulse $g(t)$ shown in Fig. Ex. 1.17(a) as a building block, construct the waveform $x(t)$ and express $x(t)$ in terms of $g(t)$.

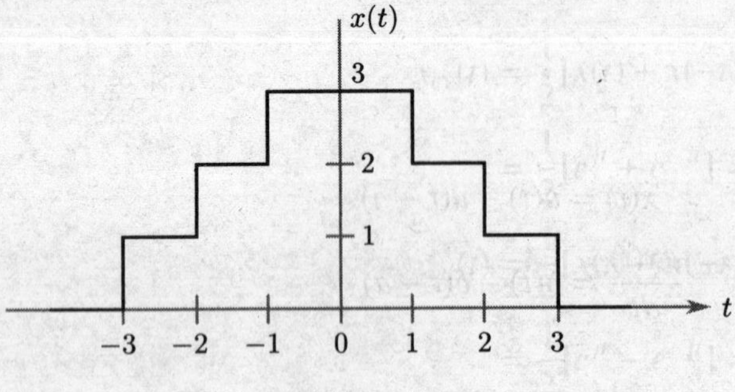

Fig. Ex. 1.17

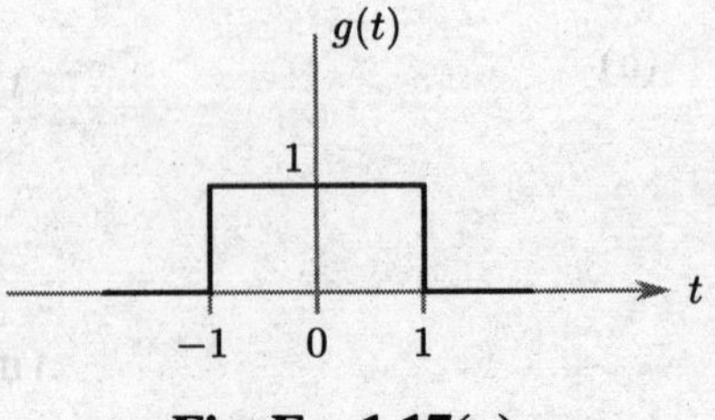

Fig. Ex. 1.17(a)

Solution:

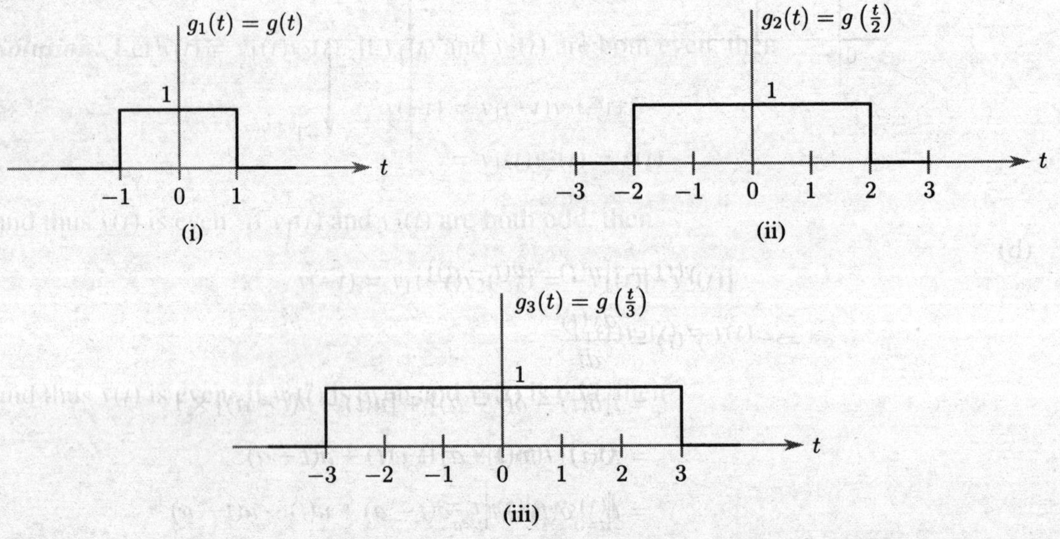

Fig. Ex. 1.17(b) *(i) Rectangular pulse g(t) reproduced verbatim (ii) Rectangular pulse g(t) expanded by a factor of 2 (iii) Rectangular pulse g(t) expanded by a factor of 3*

We can write,

$$x(t) = g_1(t) + g_2(t) + g_3(t)$$

$$\Rightarrow \quad x(t) = g(t) + g\left(\frac{t}{2}\right) + g\left(\frac{t}{3}\right)$$

Example 1.18. Find the even and odd parts of the following signals:

(a) $x(n) = u(n)$

(b) $y(n) = \alpha^n u(n)$

Solution:

(a) The even part of a signal $x(n)$ is given by

$$x_{ev}(n) = \frac{1}{2}[x(n) + x(-n)]$$

Therefore,

$$u_{ev}(n) = \frac{1}{2}[u(n) + u(-n)]$$

$$= \begin{cases} 1, & n = 0 \\ \frac{1}{2}, & n \neq 0 \end{cases}$$

which may also be written as

$$u_{ev}(n) = \frac{1}{2} + \frac{1}{2}\delta(n)$$

The odd part of a signal $x(n)$ is given by

$$x_{od}(n) = \frac{1}{2}[x(n) - x(-n)]$$

$$\Rightarrow \quad u_{od}(n) = \frac{1}{2}[u(n) - u(-n)]$$

$$= \begin{cases} \frac{1}{2}, & n > 0 \\ 0, & n = 0 \\ -\frac{1}{2}, & n < 0 \end{cases}$$

or $\quad u_{od}(n) = \frac{1}{2}\,\text{sgn}\,(n)$

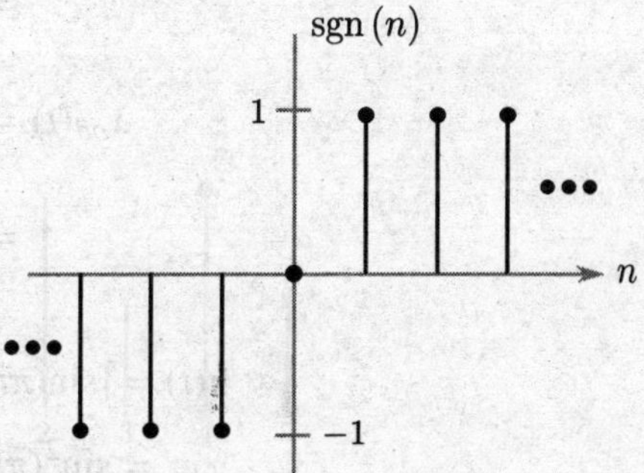

where sgn(n) is the signum function.

(b)

$$y(n) = \alpha^n u(n)$$

$$y_{ev}(n) = \frac{1}{2}[y(n) + y(-n)]$$

$$= \frac{1}{2}[\alpha^n u(n) + \alpha^{-n} u(-n)]$$

$$= \begin{cases} \frac{1}{2}\alpha^n, & n > 0 \\ 1, & n = 0 \\ \frac{1}{2}\alpha^{-n}, & n < 0 \end{cases}$$

$$\text{or} \quad y_{ev}(n) = \frac{1}{2}\alpha^{|n|} + \frac{1}{2}\delta(n)$$

$$y_{od}(n) = \frac{1}{2}[y(n) - y(-n)]$$

$$= \frac{1}{2}[\alpha^n u(n) - \alpha^{-n}u(-n)]$$

$$= \frac{1}{2}\alpha^{|n|}\,\text{sgn}\,(n)$$

Example 1.19. Find the even and odd parts of the following signals:

(a) $x(t) = e^{-t}u(t)$

(b) $x(t) = [\sin(\pi t) + \cos(\pi t)]^2$

Solution:

(a)

$$x_{ev}(t) = \frac{1}{2}[x(t) + x(-t)]$$

$$= \frac{1}{2}[e^{-t}u(t) + e^{t}u(-t)]$$

$$x_{od}(t) = \frac{1}{2}[x(t) - x(-t)]$$

$$= \frac{1}{2}[e^{-t}u(t) - e^{t}u(-t)]$$

(b)

$$x(t) = [\sin(\pi t) + \cos(\pi t)]^2$$

$$= \sin^2(\pi t) + \cos^2 \pi t + 2\sin \pi t \cos \pi t$$

$$= x_{ev}(t) + x_{od}(t)$$

Thus,

$$x_{ev}(t) = \sin^2(\pi t) + \cos^2(\pi t) = 1$$

$$x_{od}(t) = 2\sin \pi t \cos \pi t$$

$$= \sin(2\pi t)$$

Example 1.20. For the signal $x(t)$ shown in Fig. Ex. 1.20, sketch the following.

(a) $y_1(t) = x(-t)$ (b) $y_2(t) = x(t+3)$ (c) $y_3(t) = x(2t-2)$

(d) $y_4(t) = x[0.5(t-2)]$ (e) $y_5(t) = x(-0.5\,t - 1)$ (f) $x_{ev}(t)$

(g) $x_{od}(t)$

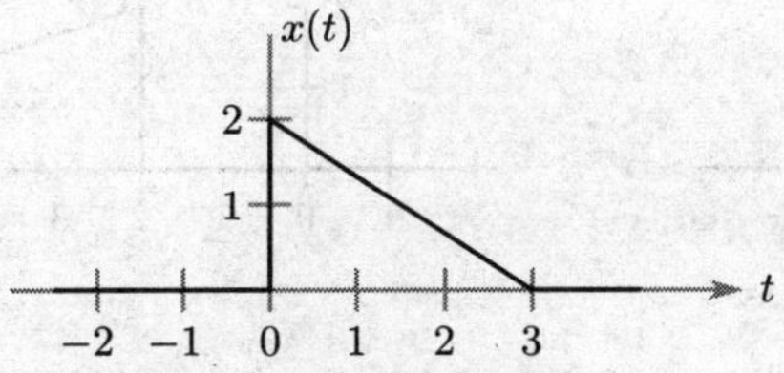

Fig. Ex. 1.20

Solution:

(a) $y_1(t) = x(-t)$.

Fold $x(t)$ about the vertical axis through the origin.

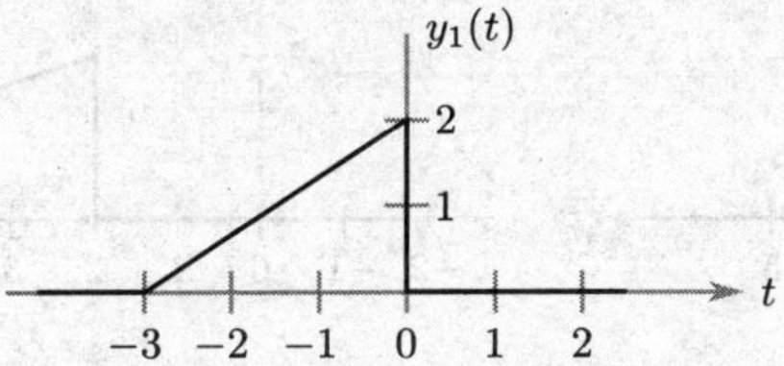

(b) $y_2(t) = x(t+3)$.

Shift $x(t)$ left by 3

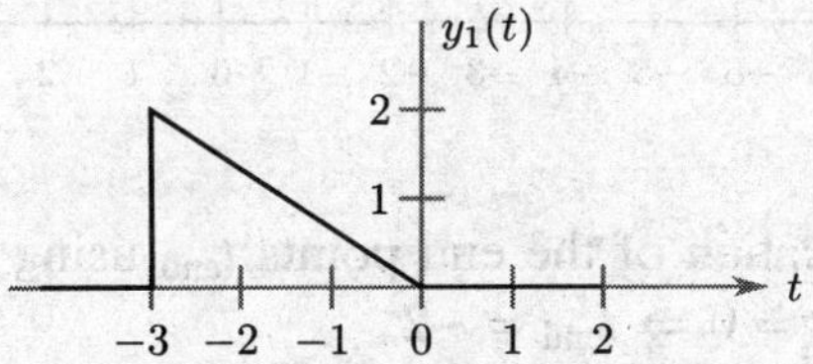

(c) $y_3(t) = x(2t-2)$.

First get $v(t) = x(t-2)$ and then $y_3(t) = v(2t)$ [compression by 2]

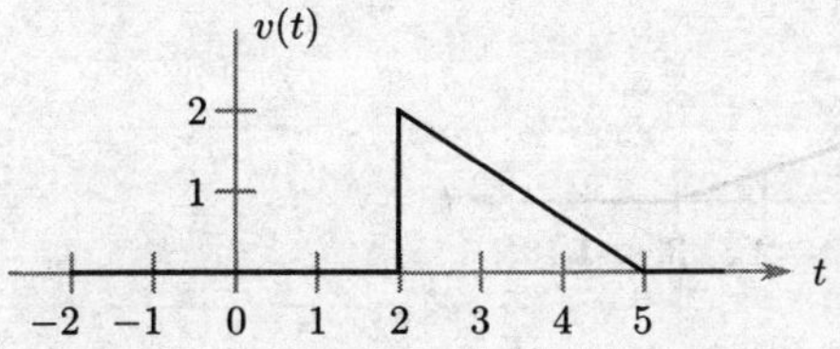

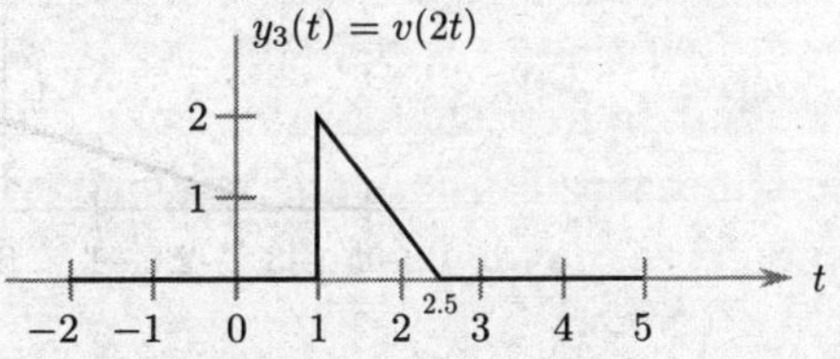

(d) $y_4(t) = x(0.5(t-2)) = x(0.5\,t-1)$

Translate $x(t)$ to right by 1 time unit to get $v(t)$ and then expand $v(t)$ by a factor of 2 to get $y_4(t)$.

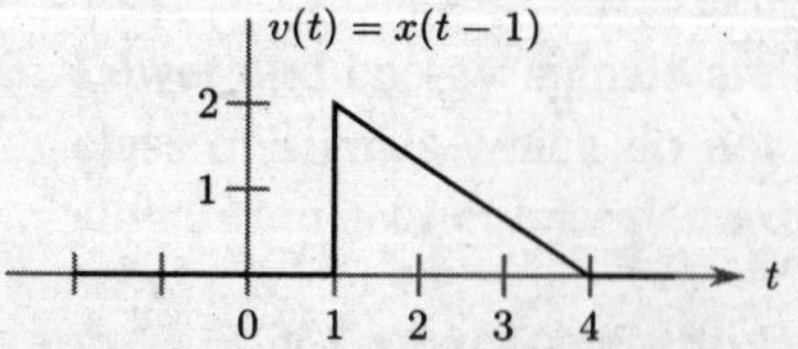

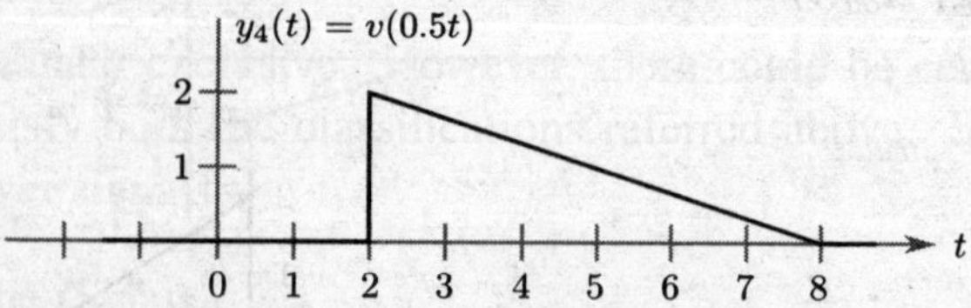

We can confirm the location of the end points t_{end} using $0.5\,t_{\text{end}} - 1 = t$. Hence, for $t = 3 \Rightarrow t_{\text{end}} = 8$ and $t = 0 \Rightarrow t_{\text{end}} = 2$.

(e) $y_5(t) = x(-0.5t-1)$

Shift $x(t)$ to right by 1 unit, then expand by 2 and fold.

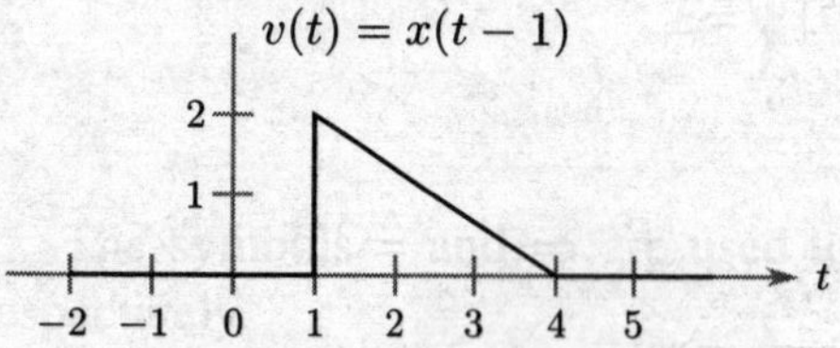

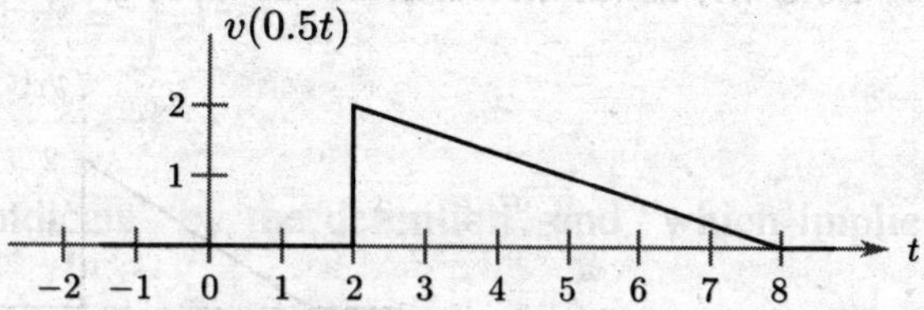

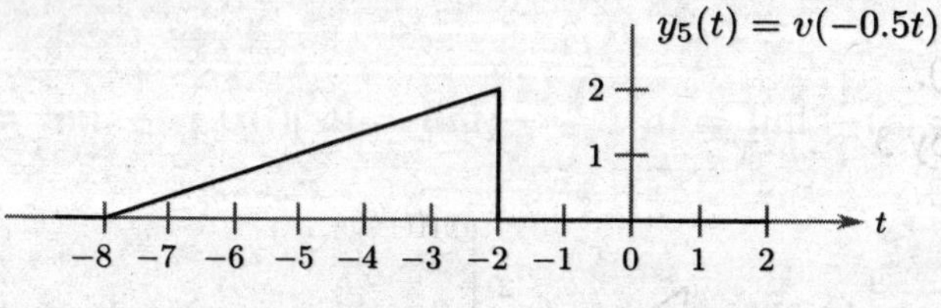

We can confirm the location of the end points t_{end} using $-0.5\,t_{\text{end}} - 1 = t$. Hence, $t = 3 \Rightarrow t_{\text{end}} = -8$ and $t = 0 \Rightarrow t_{\text{end}} = -2$.

(f) $x_{ev}(t) = 0.5x(t) + 0.5x(-t)$

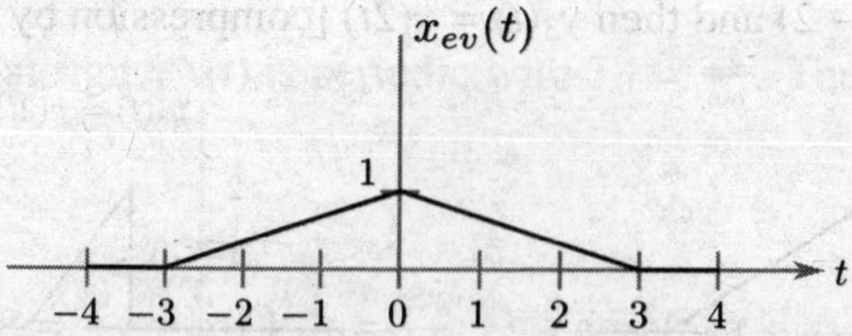

(g) $x_{od}(t) = 0.5x(t) - 0.5x(-t)$

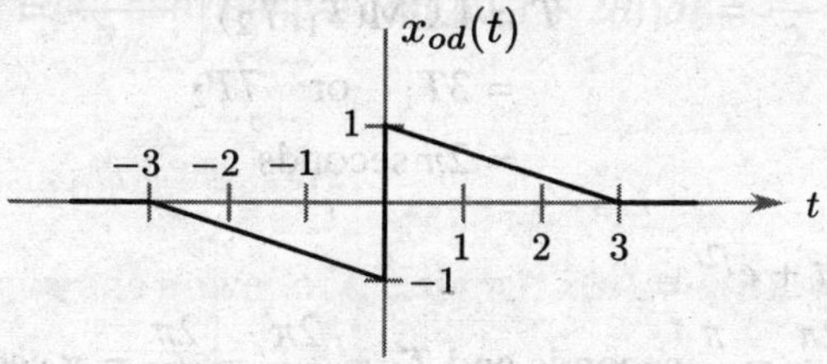

Example 1.21. If the signal $x(t) = e^{j2\pi f_0 t}$ is periodic with period $T = \frac{1}{f_0}$, verify the following:

(a) Is the signal $y(t) = x(2t) + 6x(0.5t)$ periodic? If so, what is its period?

(b) Is the signal $f(t) = 2e^{j16\pi t} + 4e^{-j7\pi t}$ periodic? If so, what is its period?

Solution:

(a) $y(t) = x(2t) + 6x(0.5t)$

The frequencies are $f_1 = 2f_0$ and $f_2 = 0.5f_0$. Hence, $y(t)$ is periodic with fundamental frequency = $\text{GCD}(f_1, f_2) = 0.5f_0$ and period = $2T$.

(b) $f(t) = 2e^{j16\pi t} + 4e^{-j7\pi t}$

Comparing each term on the right side of $f(t)$ with $e^{j2\pi ft}$, we get

$$f_1 = 8\text{Hz and } f_2 = 3.5\text{Hz}$$

Hence, fundamental frequency = $\text{GCD}(f_1, f_2) = 0.5\text{Hz}$ and period = 2 seconds.

Example 1.22. For each signal given below, check for periodicity. If periodic, find the fundamental period T_0, in each case.

(a) $x(t) = \cos 3t + \sin 7t$ (b) $x(t) = \cos 6t + e^{j2t}$ (c) $x(t) = \cos 2t + \sin \pi t$

Solution:

(a) Given: $x(t) = \cos 3t + \sin 7t$

Here, $\omega_1 = 3\text{rad/sec}$ and $\omega_2 = 7\text{rad/sec}$.

Thus, $T_1 = \frac{2\pi}{3}$ seconds and $T_2 = \frac{2\pi}{7}$ seconds

Consider the ratio, $\frac{T_1}{T_2}$:

$$\frac{T_1}{T_2} = \frac{7}{3} = \text{rational number}$$

Hence,

$$\begin{aligned} T &= \text{LCM}(T_1, T_2) \\ &= 3T_1 \quad \text{or} \quad 7T_2 \\ &= 2\pi \text{ seconds} \end{aligned}$$

(b) Given: $x(t) = \cos 6t + e^{j2t}$

Here, $T_1 = \dfrac{2\pi}{\omega_1} = \dfrac{2\pi}{6} = \dfrac{\pi}{3}$ seconds and $T_2 = \dfrac{2\pi}{\omega_2} = \dfrac{2\pi}{2} = \pi$ seconds

Consider the ratio $\dfrac{T_1}{T_2}$:

$$\frac{T_1}{T_2} = \frac{1}{3} = \text{rational number}$$

Hence,

$$\begin{aligned} T &= \text{LCM}(T_1, T_2) \\ &= 3T_1 \text{ or } T_2 \\ &= \pi \text{ seconds} \end{aligned}$$

(c) Given: $x(t) = \cos 2t + \sin \pi t$

Here, $T_1 = \dfrac{2\pi}{\omega_1} = \dfrac{2\pi}{2} = \pi$ seconds and $T_2 = \dfrac{2\pi}{\omega_2} = \dfrac{2\pi}{\pi} = 2$ seconds

Consider the ratio $\dfrac{T_1}{T_2}$:

$$\frac{T_1}{T_2} = \frac{\pi}{2} = \text{ Not a rational number}$$

Hence, $x(t)$ is not periodic.

Example 1.23. Evaluate the integral given below:

$$I = \int_{-\infty}^{\infty} \delta(at - b)\sin^2(t - 4)dt$$

Solution: Recall the product property of an impulse function:

$$f(t)\delta(t - t_0) = f(t_0)\delta(t - t_0)$$

Applying product property, we get

$$I = \int_{-\infty}^{\infty} \sin^2(t - 4)\big|_{t=\frac{b}{a}} \delta(at - b)dt = \sin^2\left(\frac{b}{a} - 4\right)\int_{-\infty}^{\infty} \delta(at - b)dt$$

Letting $at - b = x$, we get $dt = \frac{dx}{a}$.

Thus,

$$I = \sin^2\left(\frac{b}{a} - 4\right) \times \frac{1}{a}\int_{-\infty}^{\infty} \delta(x)dx = \frac{1}{a}\sin^2\left(\frac{b}{a} - 4\right)$$

Example 1.24. For the discrete-time signals given below, find the fundamental period, N_0 if the signal is periodic; otherwise, show that the signal is not periodic.

(a) $x(n) = e^{j5\pi n/7}$ (b) $x(n) = e^{j3n}$ (c) $x(n) = e^{j\frac{0.2n}{\pi}}$

Solution:

(a) $x(n+N) = e^{j\frac{5\pi}{7}(n+N)} = e^{j\frac{5\pi}{7}n}e^{j\frac{5\pi}{7}N}$

For $x(n+N)$ to be equal to $x(n)$, it is required that

$$e^{j\frac{5\pi}{7}N} = 1$$

$$\Rightarrow \frac{5\pi}{7}N = 2\pi k$$

$$\Rightarrow N = \frac{14}{5}k; \quad k \text{ and } N \text{ are integers}$$

Hence for $k = 5$ (smallest integer), we get $N_0 = 14$

(b) $x(n+N) = e^{j3(n+N)} = e^{j3n}.e^{j3N}$

For $x(n+N)$ to be identical to $x(n)$, it is required that

$$e^{j3N} = 1$$

$$\Rightarrow 3N = 2\pi k$$

$$\Rightarrow N = \frac{2\pi k}{3}; \quad k \text{ and } N \text{ are integers}$$

Since, no integer k can give an integer N, $x(n)$ is not periodic.

(c) $x(n+N) = e^{j\frac{0.2}{\pi}(n+N)}$

$= e^{j\frac{0.2n}{\pi}} \quad e^{j\frac{0.2}{\pi}N}$

For $x(n+N)$ to be identical to $x(n)$, it is required that

$$e^{j\frac{0.2N}{\pi}} = 1$$

$$\Rightarrow \frac{0.2N}{\pi} = 2\pi k$$

$$\Rightarrow N = \frac{2\pi^2}{0.2}k$$

Since no integer k can give an integer N, it implies that $x(n)$ is not periodic.

Example 1.25. Given:

$$x_1(n) = \cos(0.2\pi n)$$
$$x_2(n) = \cos(0.125\pi n)$$
$$\text{and} \quad x_3(n) = \cos(0.4\pi n)$$

Find the fundamental period of the sum, $x(n) = x_1(n) + x_2(n) + x_3(n)$.

Solution:

1. To find fundamental period of $x_1(n)$:

   Here,

   $$\Omega_0 = 0.2\pi$$

   Hence,

   $$\frac{\Omega_{0_1}}{2\pi} = \frac{1}{10} = \frac{p}{q}$$

   Since p and q are integers, $\frac{\Omega_{0_1}}{2\pi}$ is a rational number. Hence, $N_{0_1} = q = 10$.

2. To find fundamental period of $x_2(n)$:

   Here,

   $$\Omega_{0_2} = 0.125\pi$$

   $$\Rightarrow \quad \frac{\Omega_{0_2}}{2\pi} = \frac{1}{16} = \frac{p}{q}$$

   Since p and q are integers, it implies $\frac{\Omega_{0_2}}{2\pi}$ is a rational number. Hence, $N_{0_2} = q = 16$.

3. To find fundamental period of $x_3(n)$:

   Here,

   $$\Omega_{0_3} = 0.4\pi$$

   $$\Rightarrow \quad \frac{\Omega_{0_3}}{2\pi} = \frac{1}{5} = \frac{p}{q}$$

   Since p and q are integers, $\frac{\Omega_{0_3}}{2\pi}$ is a rational number. Hence, $N_{0_3} = q = 5$

4. To find the fundamental period of the sum, $x(n)$:

   $$N_0 = \text{LCM}\{N_{0_1}, N_{0_2}, N_{0_3}\}$$
   $$= \text{LCM}\{5, 10, 16\}$$
   $$= 80$$

Example 1.26. Given:

$$x_1(n) = e^{j0.2n\pi} \quad \text{and} \quad x_2(n)=e^{-j0.3n\pi}$$

What is the period of the signal, $x(n) = x_1(n) + x_2(n)$?

Solution:

1. To find fundamental period of $x_1(n)$:

$$x_1(n + N) = e^{j0.2\pi(n+N)} = e^{j0.2\pi n} \cdot e^{j0.2\pi N}$$

For $x_1(n + N)$ to be identical to $x_1(n)$, it is rquired that

$$e^{j0.2\pi N} = 1$$
$$\Rightarrow \quad 0.2\pi N = 2\pi k$$
$$\Rightarrow \quad N = 10k; \quad \text{N and } k \text{ are integers}$$

For $k = 1$, we get $N_{0_1} = 10$.

2. To find fundamental period of $x_2(n)$:

$$x_2(n + N) = e^{-j0.3\pi(n+N)} = e^{-j0.3\pi n} \cdot e^{-j0.3\pi N}$$

For $x_2(n + N)$ to be identical to $x_2(n)$, it is required that

$$e^{-j0.3\pi N} = 1$$
$$\Rightarrow \quad 0.3\pi N = 2\pi k$$
$$\Rightarrow \quad N = \frac{20}{3}k; \quad k \text{ and } N \text{ are integers.}$$

For $k = 3$, we get $N_{0_2} = 20$.

3. To find fundamental period of the sum signal, $x(n)$:

$$N_0 = \text{LCM}\{10, 20\}$$
$$= 20$$

Example 1.27. Find the fundamental period of the complex signal given by

$$x(n) = (\sqrt{j})^n + (-\sqrt{j})^n$$

Solution: Rewrite $x(n)$ as

$$x(n) = e^{j\pi n/4} + e^{-j\pi n/4}$$
$$= 2\cos\left(\frac{\pi n}{4}\right)$$

Here,

$$\Omega_0 = \frac{\pi}{4}$$
$$\Rightarrow \quad \frac{\Omega_0}{2\pi} = \frac{1}{8} = \frac{p}{q}$$

Since p and q are integers, $\frac{\Omega_0}{2\pi}$ is a rational number. Hence, $N_0 = q = 8$.

Example 1.28. Let $x(n) = 2^n[u(n+1) - u(n-4)]$. Sketch the following signals.

(a) $y_1(n) = x(n-3)$ (b) $y_2(n) = x(n+1)$

(c) $y_3(n) = x(-n+4)$ (d) $y_4(n) = x(-n-2)$

Solution:

$$x(n) = \left(\frac{1}{2}, \underset{\Uparrow}{1}, 2, 4, 8\right)$$

It may be noted that the index arrow ($\Uparrow$) represents the reference, $n = 0$.

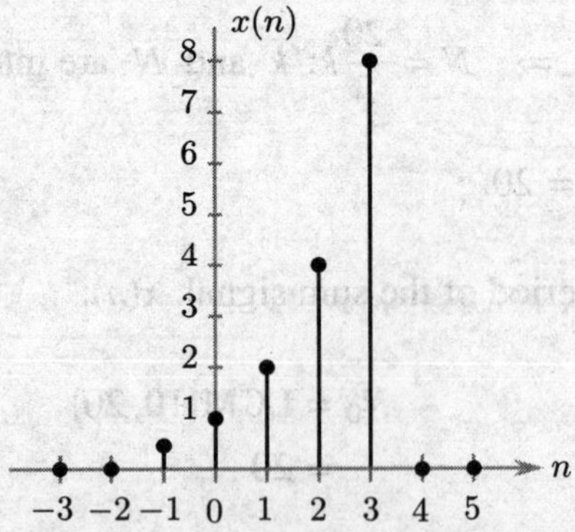

(a) $y_1(n) = x(n-3) = \left(\underset{\Uparrow}{0}, 0, \frac{1}{2}, 1, 2, 4, 8 \right)$ (Shift right by 3)

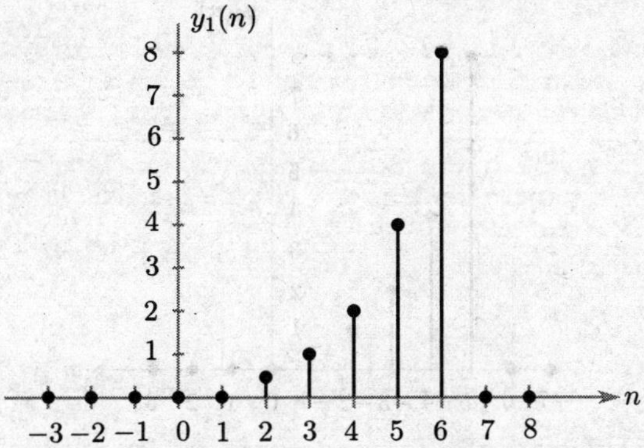

(b) $y_2(n) = x(n+1) = \left(\frac{1}{2}, 1, \underset{\Uparrow}{2}, 4, 8 \right)$ (Shift left by 1)

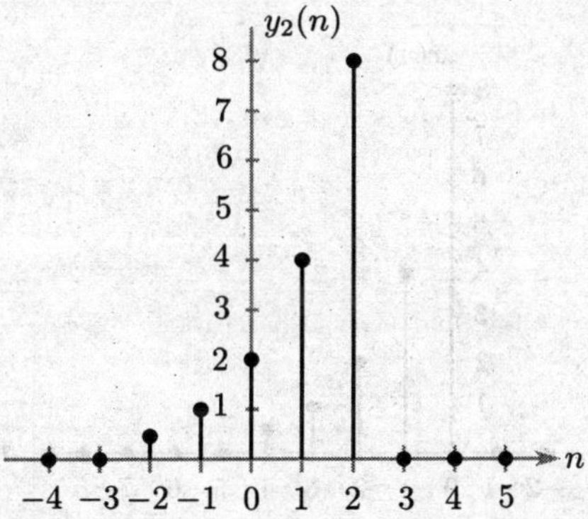

(c) $y_3(n) = x(-n+4) = \left(\underset{\Uparrow}{0}, 8, 4, 2, 1, \frac{1}{2} \right)$ (Shift left by 4, then fold)

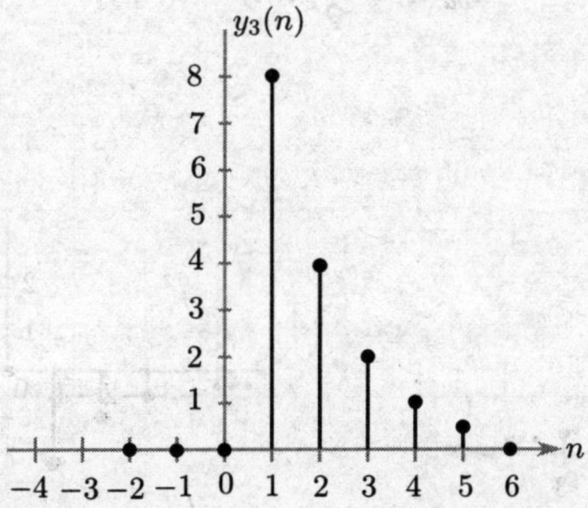

(d) $y_4(n) = x(-n-2) = \left(8, 4, 2, 1, \frac{1}{2}, \underset{\Uparrow}{0}\right)$ (Shift right by 2, then fold)

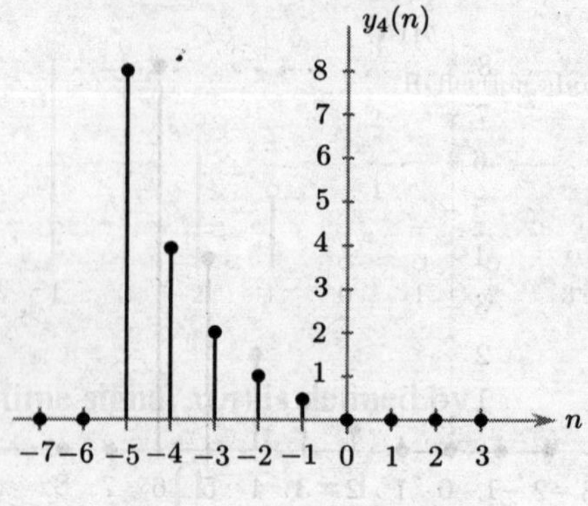

Example 1.29. Sketch $x(n)$ and then its odd and even parts.

$$x(n) = 8(0.5)^n u(n)$$

Solution:

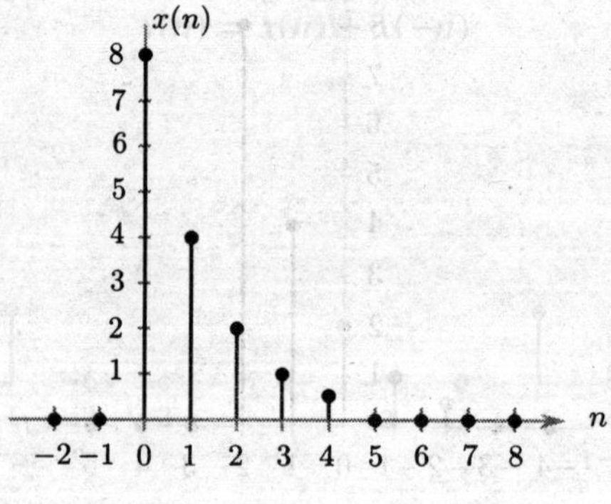

$$x_{ev}(n) = \frac{1}{2}[x(n) + x(-n)]$$

$$x_{od}(n) = \frac{1}{2}[x(n) - x(-n)]$$

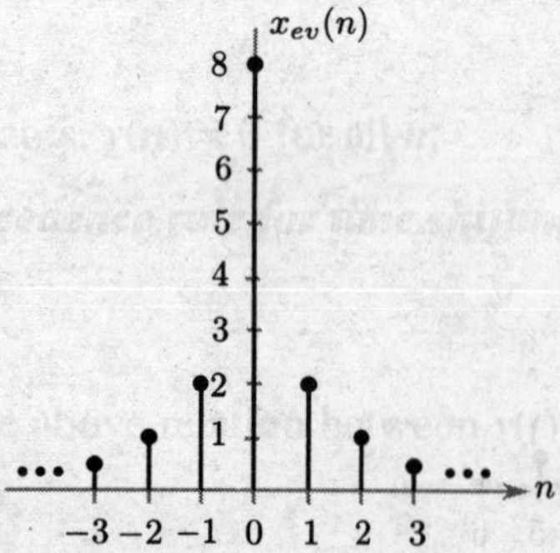

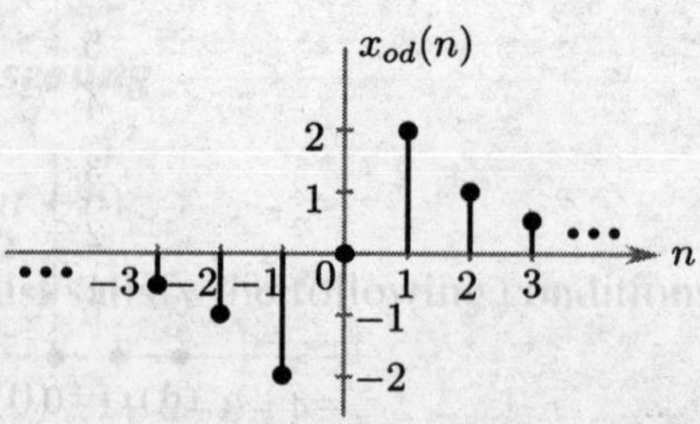

Example 1.30. Sketch $x(n)$ and then plot its even and odd parts.

$$x(n) = 1 + u(n)$$

Solution:

$$x_{ev}(n) = \frac{1}{2}[x(n) + x(-n)]$$

$$x_{od}(n) = \frac{1}{2}[x(n) - x(-n)]$$

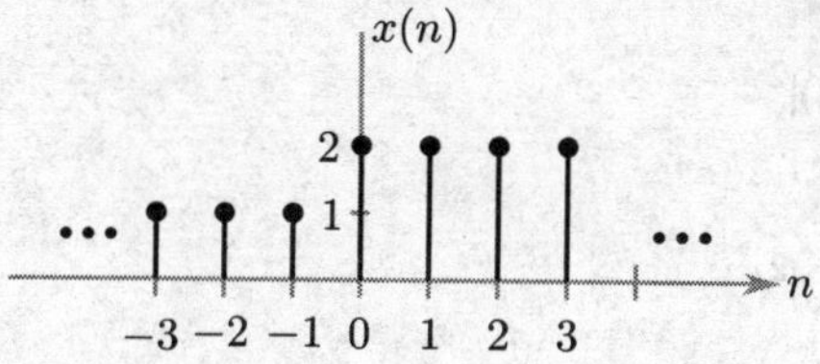

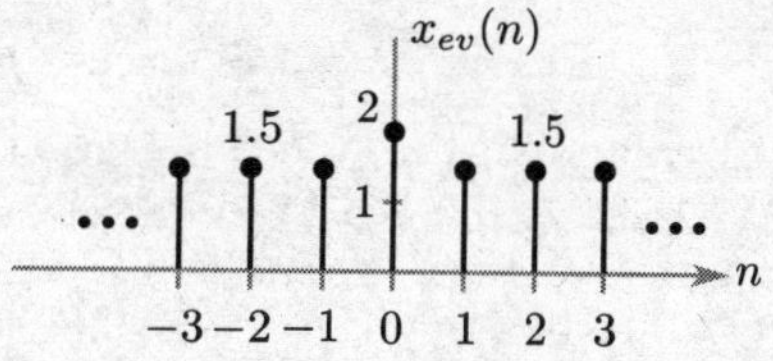

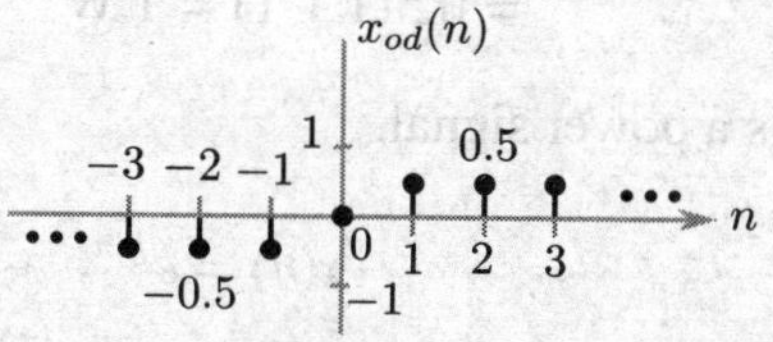

Example 1.31. Classify the following signals as energy signals, power signals or neither.

(a) $x(n) = 2^n u(-n)$ (b) $x(n) = \cos(n\pi)$

(c) $x(n) = e^{jn\pi}$ (d) $x(n) = (j)^n + (j)^{-n}$

Solution:

(a)

$$x(n) = 2^n u(-n)$$

$$E \triangleq \sum_{n=-\infty}^{\infty} |x(n)|^2$$

$$= \sum_{n=-\infty}^{0} 4^n$$

Putting $n = -m$, we get

$$E = \sum_{m=+\infty}^{0} 4^{-m} = \sum_{m=0}^{\infty} \left(\frac{1}{4}\right)^m = \frac{1}{1 - \frac{1}{4}} = \frac{4}{3}\text{J}$$

Since E is finite, $x(n)$ is an energy signal.

(b)

$$x(n) = \cos(n\pi)$$

$$\frac{\Omega_0}{2\pi} = \frac{\pi}{2\pi} = \frac{1}{2} = \frac{p}{q} = \text{ a rational number}$$

Hence,

$$N_0 = q = 2$$

Thus, $x(n) = (\underset{\Uparrow}{1}, -1)$ for one period.

So,

$$P = \frac{1}{4}\sum_{n=0}^{N-1} |x(n)|^2$$

$$= \frac{1}{2}\sum_{n=0}^{1} |x(n)|^2$$

$$= 0.5(1 + 1) = 1 \text{ W}$$

Since, P is finite, $x(n)$ is a power signal.

(c)

$$x(n) = e^{jn\pi}$$

$$\Rightarrow \quad x(n + N) = e^{j\pi n} e^{j\pi N}$$

For $x(n + N)$ to be identical to $x(n)$, It is required that

$$e^{j\pi N} = 1 \Rightarrow \pi N = 2\pi k \Rightarrow N = 2k$$

where N and k are integers.

Hence, $N_0 = 2$ (for $k = 1$).

$$P = \frac{1}{N}\sum_{n=0}^{N-1} |x(n)|^2$$

$$= \frac{1}{2}\sum_{n=0}^{1} 1 = \frac{1}{2}(1 + 1) = 1 \text{ W}$$

Since P is finite, $x(n)$ is a power signal.

(d)

$$x(n) = (j)^n + (j)^{-n} = e^{jn\frac{\pi}{2}} + e^{-jn\frac{\pi}{2}}$$

$$= 2\cos\left(\frac{n\pi}{2}\right)$$

$$\frac{\Omega_0}{2\pi} = \frac{1}{4} = \frac{p}{q} = \text{ a rational number}$$

Hence, $N_0 = q = 4$.

Hence,

$$x(n) = (\underset{\Uparrow}{2}, 0, -2, 0) \text{ for one period.}$$

So

$$P = \frac{1}{N}\sum_{n=0}^{N-1} |x(n)|^2$$
$$= \frac{1}{4}\sum_{n=0}^{3} |x(n)|^2 = \frac{1}{4}(4+4) = 2\text{ W}$$

Since P is finite, $x(n)$ is a power signal.

Example 1.32. The energy in a real-valued signal, $x(n)$, is defined as the sum of the squares of the sequence values:

$$E = \sum_{n=-\infty}^{\infty} x^2(n)$$

Suppose $x(n)$ has an even part $x_{ev}(n)$ equal to

$$x_{ev}(n) = \left(\frac{1}{2}\right)^{|n|}$$

and the energy in $x_{ev}(n)$ is $E_{ev} = 5\text{ J}$. Find the energy in the odd part, $x_{od}(n)$ of the signal $x(n)$.

Solution: Let

$$x(n) = x_{ev}(n) + x_{od}(n)$$
$$E = \sum_{n=-\infty}^{\infty} x^2(n)$$
$$= \sum_{n=-\infty}^{\infty} [x_{ev}(n) + x_{od}(n)]^2$$
$$= \sum_{n=-\infty}^{\infty} x_{ev}^2(n) + \sum_{n=-\infty}^{\infty} x_{od}^2(n) + \sum_{n=-\infty}^{\infty} 2x_{ev}(n)x_{od}(n)$$

Note that $x_{ev}(n)x_{od}(n)$ is the product of an even and an odd sequences, the product is odd. Hence,

$$\sum_{n=-\infty}^{\infty} x_{ev}(n)x_{od}(n) = 0$$

Accordingly, we get

$$E = \sum_{n=-\infty}^{\infty} x_{ev}^2(n) + \sum_{n=-\infty}^{\infty} x_{od}^2(n) = E_{ev} + E_{od}$$

The next step in the analysis is to find E_{ev}.

$$\begin{aligned}
E_{ev} &= \sum_{n=-\infty}^{\infty} x_{ev}^2(n) \\
&= \sum_{n=-\infty}^{\infty} \left(\frac{1}{2}\right)^{2|n|} \\
&= -1 + 2\sum_{n=0}^{\infty} \left(\frac{1}{2}\right)^{2n} = \frac{5}{3} \text{ J}
\end{aligned}$$

Hence,

$$\begin{aligned}
E_{od} &= E - E_{ev} \\
&= 5 - \frac{5}{3} = \frac{10}{3} \text{ J}
\end{aligned}$$

Example 1.33. Sketch each of the following signals.

(a) $x_1(n) = \text{rect}\left(\frac{n}{6}\right)$

(b) $x_2(n) = \text{rect}\left(\frac{n-2}{4}\right)$

(c) $x_3(n) = 6 \text{ tri}\left(\frac{n-4}{3}\right)$

Solution:

(a) $$x_1(n) = \text{rect}\left(\frac{n}{6}\right)$$

$x_1(n)$ is a 7 sample rectangular pulse from $n = -3$ to $n = +3$.

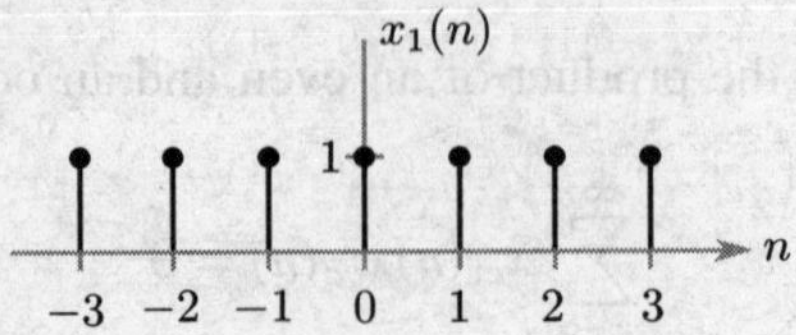

(b)

$$x_2(n) = \text{rect}\left(\frac{n-2}{4}\right)$$

$x_2(n)$ is a 5 sample rectangular pulse centered at $n = 2$.

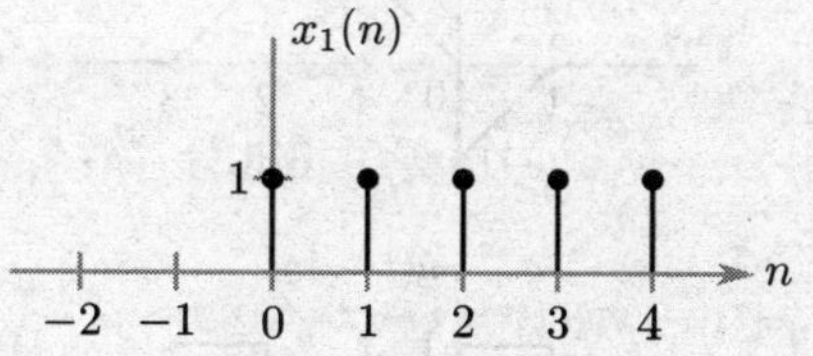

(c)

$$x_3(n) = 6\,\text{tri}\left(\frac{n-4}{3}\right)$$

$x_3(n)$ is a 7 sample triangular pulse centered at $n = 4$, with end values $x_3(1)$ and $x_3(7)$ being zero.

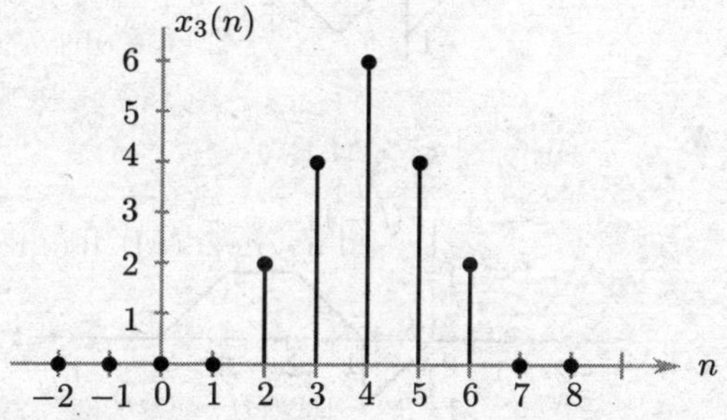

Example 1.34. Let $x(t)$ and $y(t)$ be given as in Fig. Ex 1.34 and Fig. 1.34(a). Carefully sketch the following signals.

(a) $x(t)y(t-1)$

(b) $x(t-1)y(-t)$

(c) $z(t) = x(t)[\delta(t-1) + \delta(t-2)] + y(t)\delta(t+2)$

(d) $x(t)y(-t-1)$

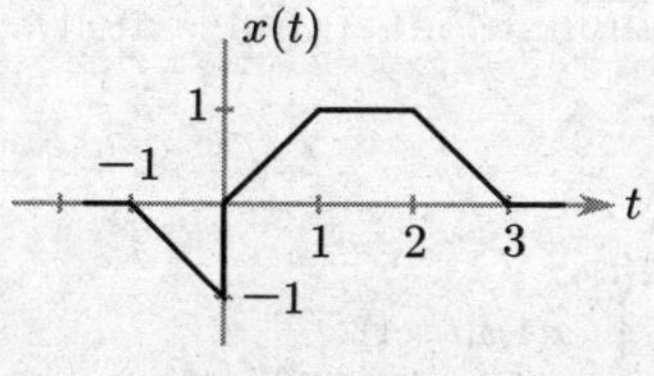

Fig. Ex. 1.34

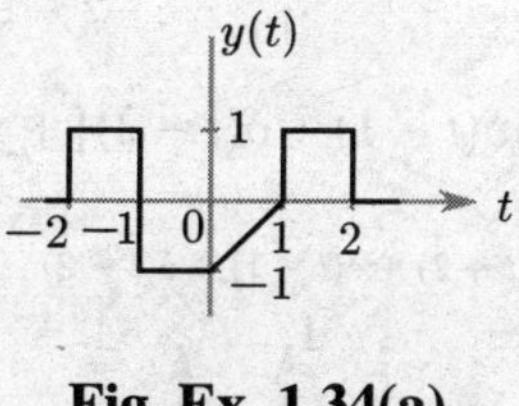

Fig. Ex. 1.34(a)

Solution:

(a) $x(t)y(t-1)$

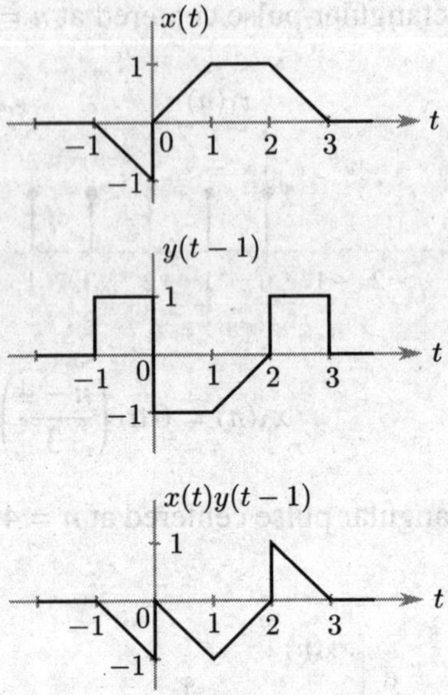

(b) $x(t-1)y(-t)$

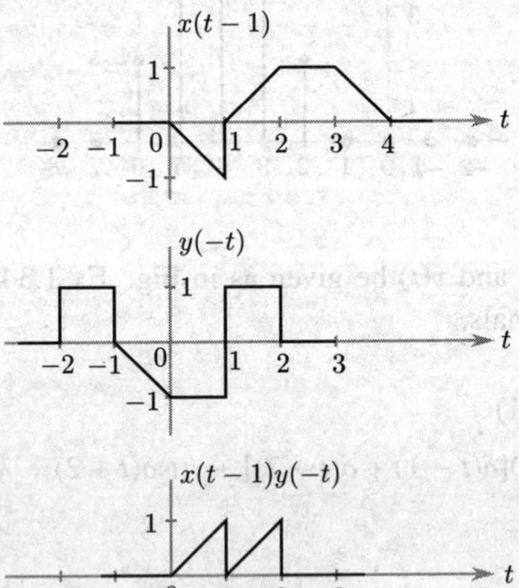

(c) $z(t) = x(t)[\delta(t-1) + \delta(t-2)] + y(t)\delta(t+2)$

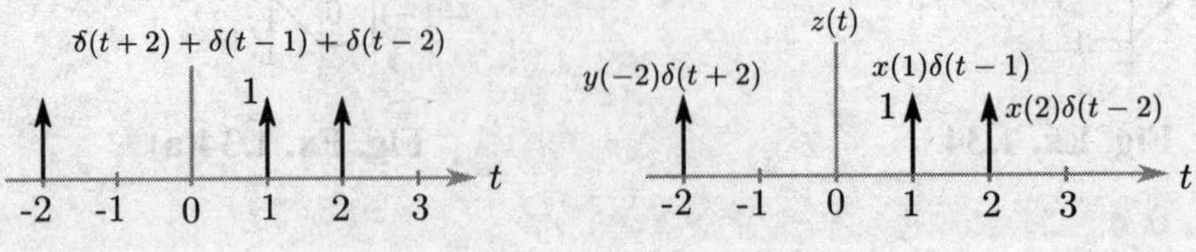

(d) $x(t)y(-t-1)$

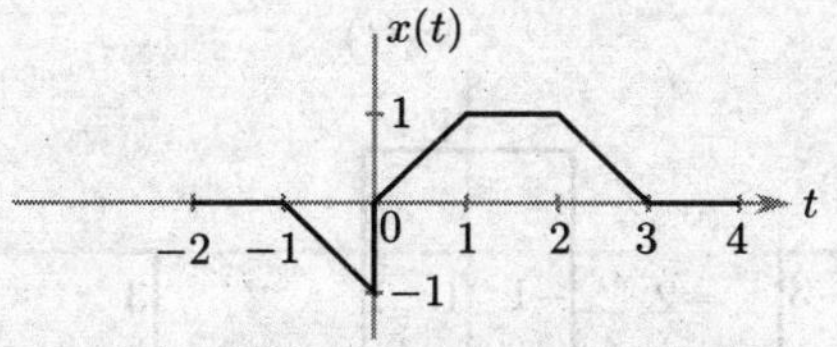

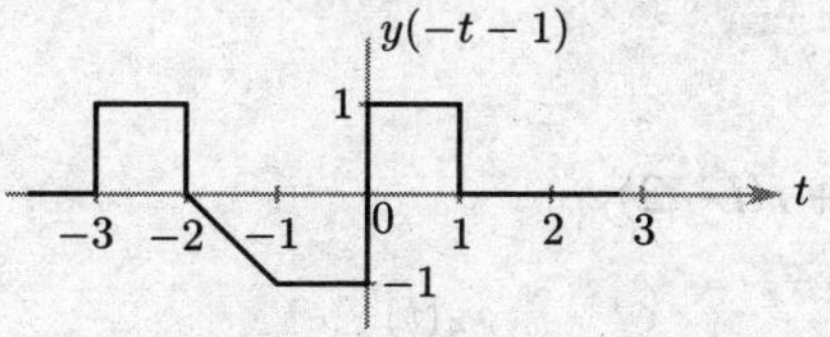

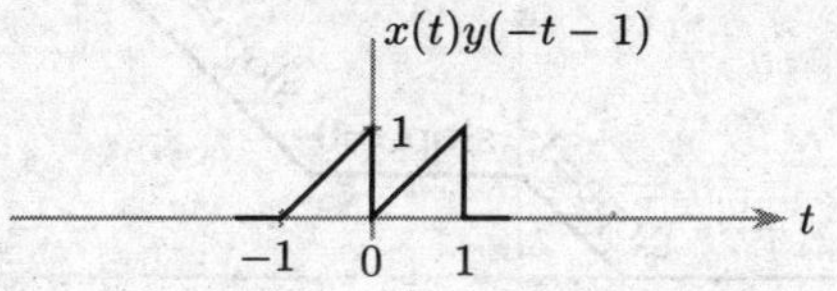

Example 1.35. Sketch the waveforms for the following signals:

(a) $x_1(t) = u(t) - u(t-3)$

(b) $x_2(t) = u(t+2) - 2u(t) + u(t-2)$

(c) $x_3(t) = -u(t+3) + 2u(t+1) - 2u(t-1) + u(t-3)$

(d) $x_4(t) = r(t+1) - r(t) + r(t-2)$

(e) $x_5(t) = r(t+2) - r(t+1) - r(t-1) + r(t-2)$

Solution:

(a) $x_1(t) = u(t) - u(t-3)$

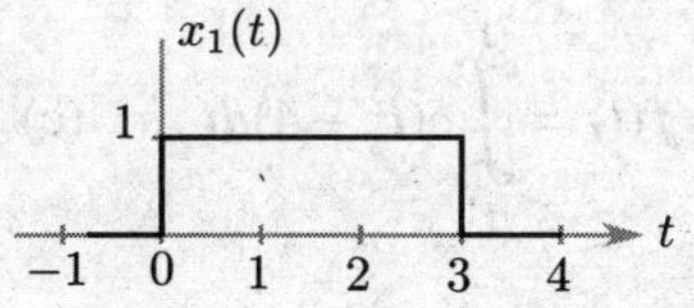

(b) $x_2(t) = u(t+2) - 2u(t) + u(t-2)$

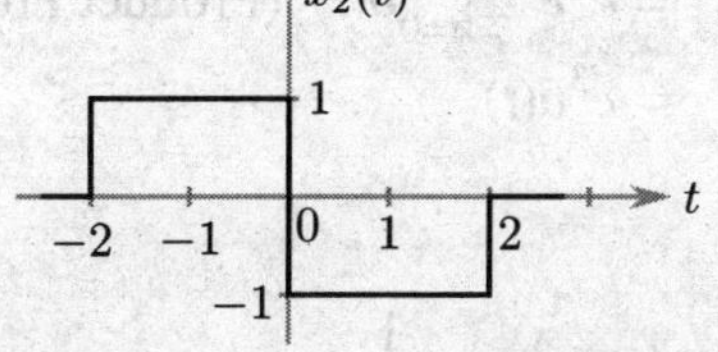

(c) $x_3(t) = -u(t+3) + 2u(t+1) - 2u(t-1) + u(t-3)$

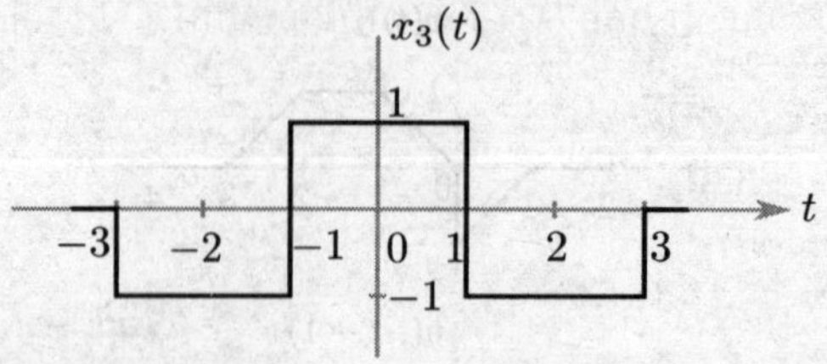

(d) $x_4(t) = r(t+1) - r(t) + r(t-2)$

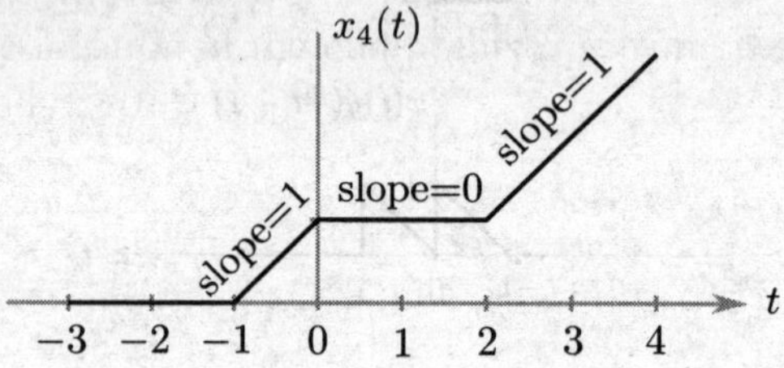

(e) $x_5(t) = r(t+2) - r(t+1) - r(t-1) + r(t-2)$

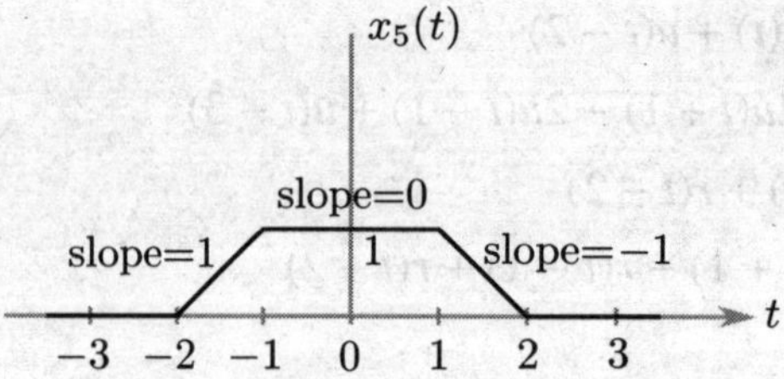

Example 1.36. Determine each of the following:

(a) $f(t) = e^{-3(t-1)}\delta(t)$ (b) $f(t) = \int_{-1}^{1} \delta(t^2 - 4)dt$ (c) $f(t) = \int_{-\infty}^{\infty} \delta(t^2 - 4)dt$

Solution:

(a)

$$\begin{aligned} f(t) &= e^{-3t}e^{3}\delta(t) \\ &= e^{3}e^{-3t}\Big|_{t=0}\delta(t) \text{ (Product Property)} \\ &= e^{3}\delta(t) \end{aligned}$$

(b) Since

$$\delta(x^2 - a^2) = \frac{1}{2|a|}[\delta(x+a) + \delta(x-a)],$$

we have $$f(t) = \int_{-1}^{1} \frac{1}{4}[\delta(t-2) + \delta(t+2)]dt$$

$$= \frac{1}{4}\int_{-1}^{1} \delta(t-2)dt + \frac{1}{4}\int_{-1}^{1} \delta(t+2)dt$$

$$= 0 + 0 = 0$$

(c)

$$f(t) = \int_{-\infty}^{\infty} \delta(t^2 - 4)dt = \frac{1}{4}\int_{-\infty}^{\infty} [\delta(t-2) + \delta(t+2)]dt$$

$$= \frac{1}{4}(1+1) = \frac{1}{2}$$

Example 1.37. Support the argument that the following signals describe the impulse

$$x(t) = A\delta(t) \quad \text{as } a \to 0$$

What is the constant A for each signal?

(a) $x(t) = \frac{1}{a}e^{\frac{-\pi t^2}{a^2}}$ and $\int_0^{\infty} x(t)dt = \frac{1}{2}$

(b) $x(t) = \dfrac{a}{a^2 + t^2}$ and $\int_0^{\infty} x(t)dt = \frac{\pi}{2}$

(c) $x(t) = \frac{1}{a}\,\text{sinc}\left(\frac{t}{a}\right)$ and $\int_0^{\infty} x(t)dt = \frac{1}{2}$

Solution:

(a) As $a \to 0$, the signal $x(t)$ becomes more compressed and tall. Also, since $x(t)$ is even symmetric, we find

$$\int_{-\infty}^{\infty} x(t)dt = 2\int_0^{\infty} x(t)dt = 2\left(\frac{1}{2}\right) = 1$$

Thus, the limiting form of $x(t)$ is $\delta(t)$.

(b) $x(t) = \frac{a}{a^2+t^2}$. As $a \to 0$, the signal $x(t)$ becomes more compressed and tall. Also, since $x(t)$ is even symmetric, we find

$$\int_{-\infty}^{\infty} x(t)dt = 2\int_{0}^{\infty} x(t)dt = \frac{2 \times \pi}{2} = \pi$$

Hence, the limiting form of $x(t)$ is $\pi\delta(t)$.

(c) $x(t) = \frac{1}{a}\,\text{sinc}\left(\frac{t}{a}\right)$. As $a \to 0$, $x(t)$ becomes very much compressed and tall. Also, since $x(t)$ is even symmetric,

$$\int_{-\infty}^{\infty} x(t)dt = 2\int_{0}^{\infty} x(t)dt = 2\left(\frac{1}{2}\right) = 1$$

The limiting form of $x(t)$ is $\delta(t)$.

Summary of Important points

- Signals can be classified as continuous-time and discrete-time signals.
- A discrete-time signal is defined only at discrete instants of time.
- A discrete-time signal is usually represented as a sequence of values of $x(n)$, for integer values of n.
- Continuous-time signals that satisfy the condition $x(t) = x(t+T)$ are periodic with fundamental period T_0.
- A discrete-time signal $x(n)$ is periodic with period N, if $x(n+N) = x(n)$ for some integer N.
- The fundamental radian frequency of the periodic signal is related to the fundamental period T_0, by the relationship $\omega_0 = \frac{2\pi}{T_0}$.
- The complex exponential $x(t) = e^{j\omega_0 t}$ is periodic with a fundamental period $T_0 = \frac{2\pi}{\omega_0}$ for all ω_0.
- Harmonically related continuous-time exponentials,

$$x_k(t) = e^{jk\omega_0 t}$$

are periodic with common period $T_0 = \frac{2\pi}{\omega_0}$.

- A complex exponential $x(n) = e^{j\Omega_0 n}$ is periodic, only if $\frac{\Omega_0}{2\pi}$ is a rational number.

- The discrete-time unit step and impulse functions are related as

$$u(n) = \sum_{k=-\infty}^{n} \delta(k)$$

$$\delta(n) = u(n) - u(n-1)$$

- Any discrete-time signal $x(n)$ can be expressed in terms of shifted impulse functions as

$$x(n) = \sum_{k=-\infty}^{\infty} x(k)\delta(n-k)$$

- Time-scaling of discrete-time signals may yield a signal that is completely different from the original signal.
- Time-scaling of continuous-time signals preserves the shape of the signal. However, it may compress or expand the original continuous-time signal $x(t)$.
- The total energy E of a continuous-time signal $x(t)$ is defined by

$$E = \lim_{T\to\infty} \int_{\frac{-T}{2}}^{\frac{T}{2}} |x(t)|^2 dt$$

The average power P of the signal $x(t)$ is defined by

$$P = \lim_{T\to\infty} \frac{1}{T} \int_{\frac{-T}{2}}^{\frac{T}{2}} |x(t)|^2 dt$$

- The total energy E of a discrete-time signal $x(n)$ is defined by

$$E = \lim_{N\to\infty} \sum_{n=-N}^{N} |x(n)|^2$$

The average power P of the signal $x(n)$ is defined by

$$P = \lim_{N\to\infty} \frac{1}{2N+1} \sum_{n=-N}^{N} |x(n)|^2$$

- The signal $x(t)$ or $x(n)$ is an energy signal, if $0 < E < \infty$.
- The signal $x(t)$ or $x(n)$ is a power signal, if $0 < P < \infty$.
- Energy and power signals are mutually exclusive. However, it is quite possible that it may be neither.
- The signal $x(t)$ is even symmetric if $x(t) = x(-t)$

- The signal $x(n)$ is even symmetric if $x(n) = x(-n)$
- The signal $x(t)$ is odd symmetric if $x(t) = -x(-t)$
- The signal $x(n)$ is odd symmetric if $x(n) = -x(-n)$
- The sifting property of the impulse function is

$$\int_{t_1}^{t_2} x(t)\delta(t-t_0)dt = \begin{cases} x(t_0), & t_1 < t_0 < t_2 \\ 0, & \text{otherwise} \end{cases}$$

$$\sum_{N_1}^{N_2} x(n)\delta(n-k) = \begin{cases} x(k), & N_1 \le k \le N_2 \\ 0, & \text{otherwise} \end{cases}$$

- The product property of impulse function is

$$x(t)\delta(t-t_0) = x(t_0)\delta(t-t_0)$$
$$x(n)\delta(n-N_0) = x(N_0)\delta(n-N_0)$$

Exercise Problems

EP. 1.1 Determine and sketch the even and odd parts of the signals depicted in Fig. EP.1.1 and Fig. EP.1.1(a). Label your sketches carefully.

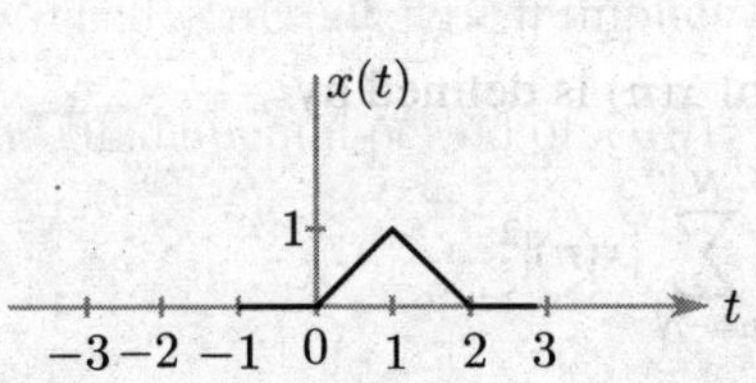

Fig. EP. 1.1

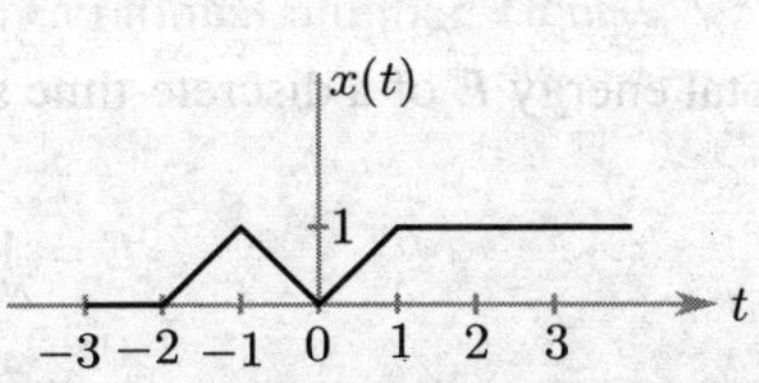

Fig. EP. 1.1(a)

Ans:

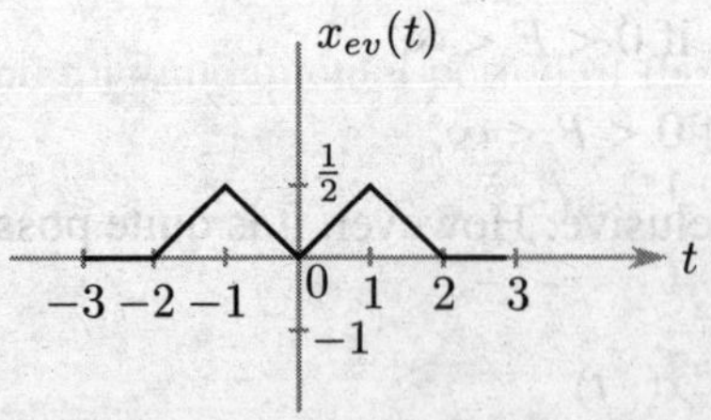

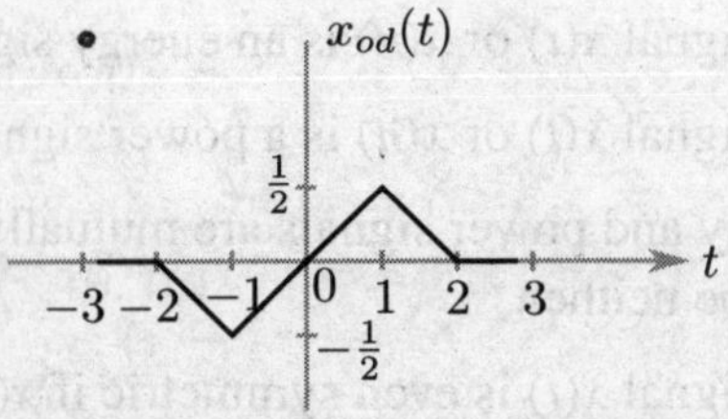

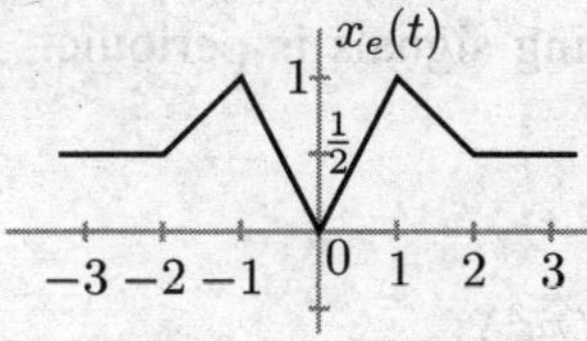

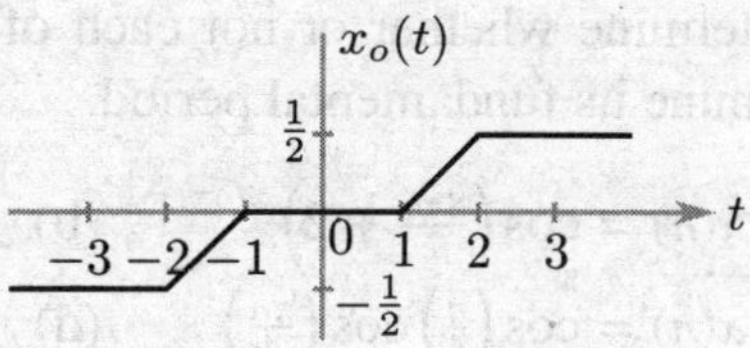

EP. 1.2 Determine and sketch the even and odd parts of the signal shown in Fig. EP.1.2.

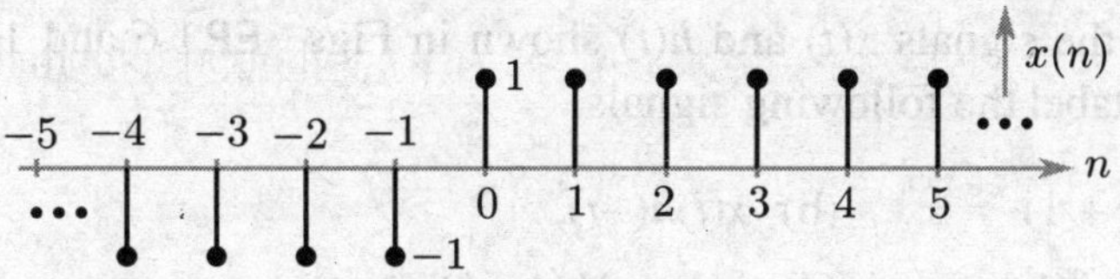

Fig. EP. 1.2

Ans:

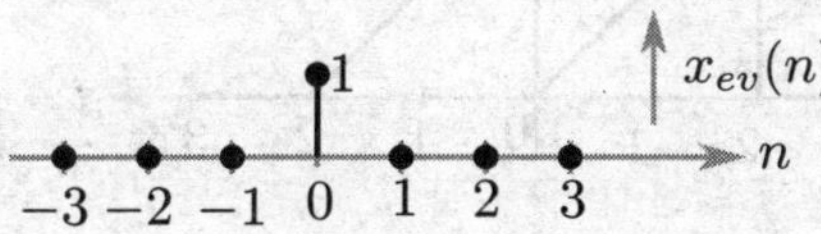

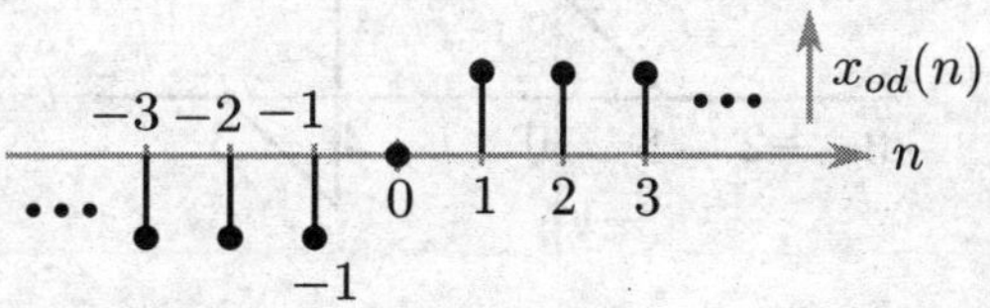

EP. 1.3 Let $x(t)$ be an arbitrary continuous-time signal having the even and odd parts defined by $x_{ev}(t)$ and $x_{od}(t)$ respectively. Then show that,

$$\int_{-\infty}^{\infty} x^2(t)dt = \int_{-\infty}^{+\infty} x_{ev}^2(t)dt + \int_{-\infty}^{+\infty} x_{od}^2(t)dt$$

EP. 1.4 Determine whether or not each of the following signals is periodic. If periodic, determine its fundamental period.

(a) $x(t) = 2\cos\left(3t + \frac{\pi}{4}\right)$ (b) $x(t) = e^{j(\pi t - 1)}$

(c) $x(t) = \left[\sin\left(t - \frac{\pi}{6}\right)\right]^2$ (d) $x(t) = \sum_{n=-\infty}^{\infty} e^{-(t-3n)^2}$

Ans: (a) Periodic, period = $\frac{2\pi}{3}$ (b) Periodic, period = 2

(c) Periodic, period = π (d) Periodic, period = 3

EP. 1.5 Determine whether or not each of the following signals is periodic. If periodic, determine its fundamental period.

(a) $x(n) = \cos\left(\frac{8\pi n}{7} + 2\right)$ (b) $x(n) = e^{j\left(\frac{n}{8} - \pi\right)}$

(c) $x(n) = \cos\left(\frac{n}{4}\right)\cos\left(\frac{\pi n}{4}\right)$ (d) $x(n) = \cos\left(\frac{\pi n^2}{8}\right)$

Ans: (a) Periodic, $N_0 = 7$ (b) Aperiodic
(c) Aperiodic (d) Periodic, $N_0 = 4$

EP. 1.6 Consider the signals $x(t)$ and $h(t)$ shown in Figs. EP.1.6 and 1.6(a) respectively. Sketch and label the following signals.

(a) $x(t)h(t+1)$ (b) $x(t)h(-t)$

(c) $x(t-1)h(1-t)$ (d) $x(1-t)h(t-1)$

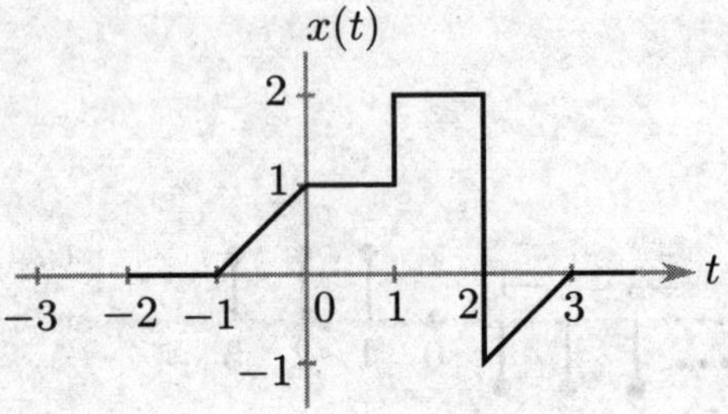

Fig. EP. 1.6

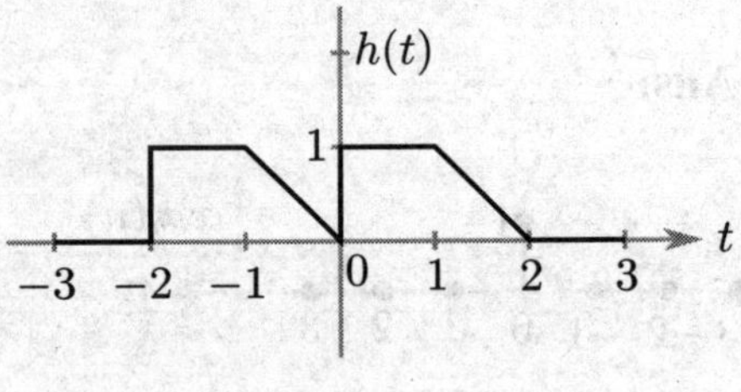

Fig. EP. 1.6(a)

Ans:

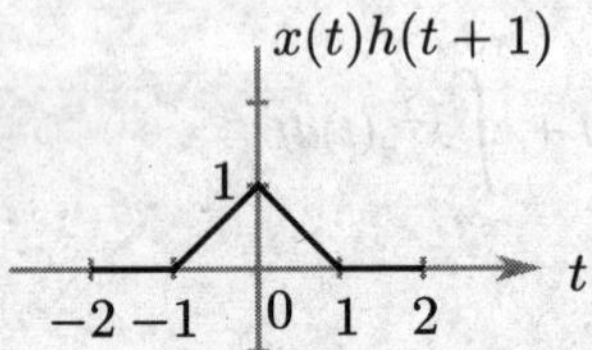

(a)

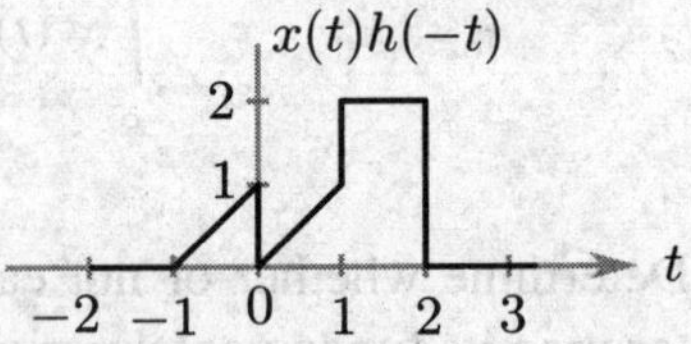

(b)

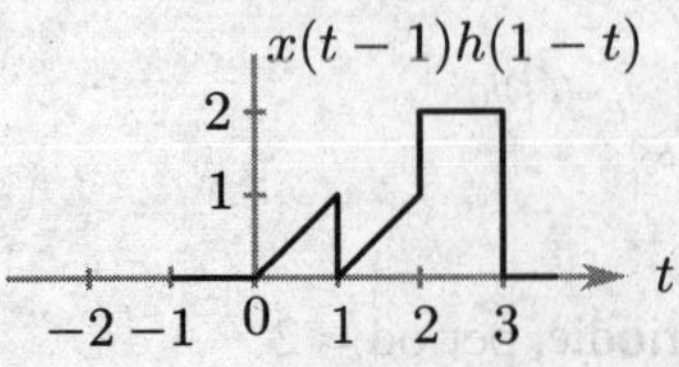

(c)

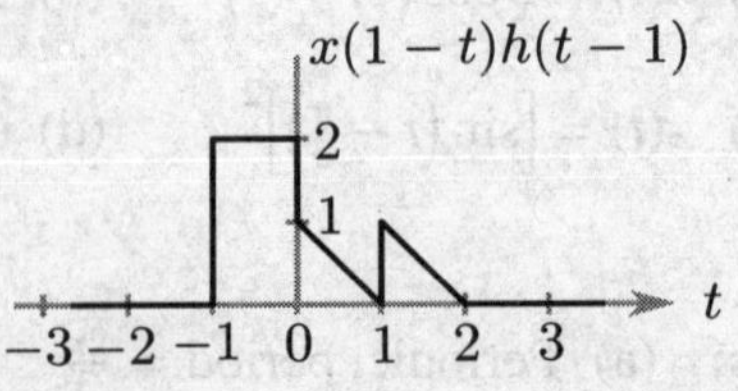

(d)

EP. 1.7 Let

$$x(t) = u(t+2) - 2u(t+1) + 2u(t) - u(t-2) - 2u(t-3) + 2u(t-4)$$

This signal $x(t)$ is applied to an integrator resulting in a response $y(t)$. Find $y(t)$.

Ans:

$$y(t) = r(t+2) - 2r(t+1) + 2r(t) - r(t-2) - 2r(t-3) + 2r(t-4).$$

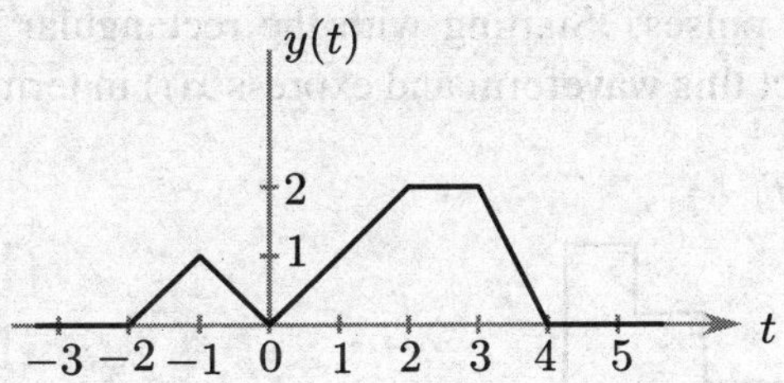

EP. 1.8 Evaluate the following integrals.

(a) $\int_{-2}^{1} (t + t^2)\delta(t-3)dt$ (b) $\int_{-2}^{4} (t + t^2)\delta(t-3)dt$

(c) $\int_{0}^{3} e^{t-2}\delta(2t-4)dt$ (d) $\int_{-\infty}^{t} \delta(\tau)d\tau$

Ans: (a) 0 (b) 12 (c) $\frac{1}{2}$ (d) $u(t)$

EP. 1.9 Find the even and odd parts of each of the following signals.

(a) $x(t) = \cos t + \sin t + \sin t \cos t$ (b) $x(t) = 1 + t + 3t^2 + 5t^3 + 9t^4$

(c) $x(t) = (1 + t^3)\cos^3(10t)$

Ans:

(a) $x_{ev}(t) = \cos t, \quad x_{od}(t) = \sin t + \sin t \cos t$

(b) $x_{ev}(t) = 1 + 3t^2 + 9t^4, \quad x_{od}(t) = t + 5t^3$

(c) $x_{ev}(t) = \cos^3 10t, \quad x_{od}(t) = t^3 \cos^3 10t$

EP. 1.10 A rectangular pulse $x(t)$ is defined by

$$x(t) = \begin{cases} +A, & 0 \le t \le T \\ 0, & \text{otherwise} \end{cases}$$

The pulse $x(t)$ is applied to an integrator defined by

$$y(t) = \int_0^t x(\tau)d\tau$$

Find the total energy of the output $y(t)$.

Ans: $E = \frac{A^2T^3}{3}$ Joules.

EP. 1.11 Fig. EP.1.11 shows a staircase-like signal $x(t)$ that may be viewed as superposition of four rectangular pulses. Starting with the rectangular pulse $g(t)$ shown in Fig. EP.1.11(a), construct this waveform and express $x(t)$ in terms of $g(t)$.

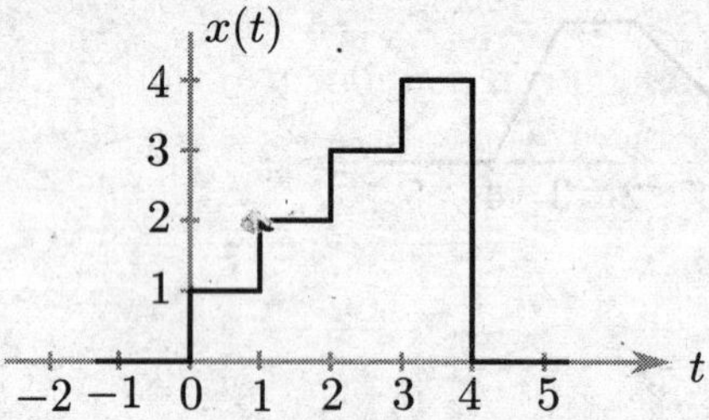

Fig. EP. 1.11

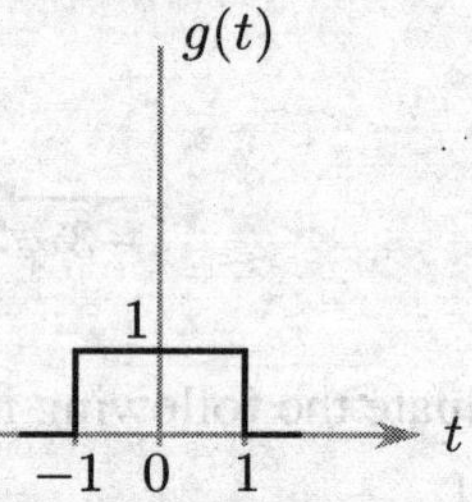

Fig. EP. 1.11(a)

Ans: $x(t) = g\left(\frac{t}{2} - 1\right) + g\left(\frac{2}{3}t - \frac{5}{3}\right) + g(t-3) + g(2t-7)$

EP. 1.12 Consider a sequence $x(n) = x_{ev}(n) + x_{od}(n)$, where $x_{ev}(n)$ is the even part of the sequence $x(n)$ and $x_{od}(n)$ is the odd part of the sequence $x(n)$. Show that the average power of the sequence $x(n)$ is

$$P = P_{ev} + P_{od}$$

where P_{ev} is the average power of $x_{ev}(n)$ and P_{od} is the average power of $x_{od}(n)$.

Chapter 2

Signal Analysis and Fourier Series Representation of Signals

Joseph Fourier (1768–1830) was a French mathematician and physicist who triggered the investigation of Fourier series and its application. J.L. Lagrange objected the discovery of Fourier and ultimately Fourier's investigations were supported by P.L. Dirichlet's rigorous mathematical proof in 1829.

It can be shown that a signal can be represented as a vector having orthonormal/orthogonal functions as the co-ordinate axes in an n-dimensional space. Representation of a signal as a vector plays a very important role in detecting a signal buried in noise.

According to Fourier that most continuous-time signals can be decomposed as an infinite sum of complex harmonics using Fourier series, these include finite–power signals and signals satisfying the Dirichlet conditions.

The objective of this chapter is to represent a given function of time as a sum of various exponential functions. This study is very essential for understanding the intricacies of communication systems. Fourier analysis helps us to understand the various practical aspects of a communication system with a good theoretical support.

2.1 Analogy between vectors and signals

For detecting a transmitted signal in the presence of noise, it is essential that a signal be represented by a vector. Representing a signal as a vector is better understood because a perfect analogy exists between the two.

A vector has both magnitude and direction, and is represented by a boldface letter. Let $\boldsymbol{A_1}$ and $\boldsymbol{A_2}$ be two coplanar vectors having an angular seperation, θ as depicted in Fig. 2.1. The component of vector, $\boldsymbol{A_1}$ along vector, $\boldsymbol{A_2}$ is $C_{12}\,\boldsymbol{A_2}$. Infact, $C_{12}\,\boldsymbol{A_2}$ is the projection of $\boldsymbol{A_1}$ along $\boldsymbol{A_2}$. Referring Fig. 2.1, we can write,

$$\mathbf{A_1} = C_{12}\mathbf{A_2} + \mathbf{A_e} \tag{2.1}$$

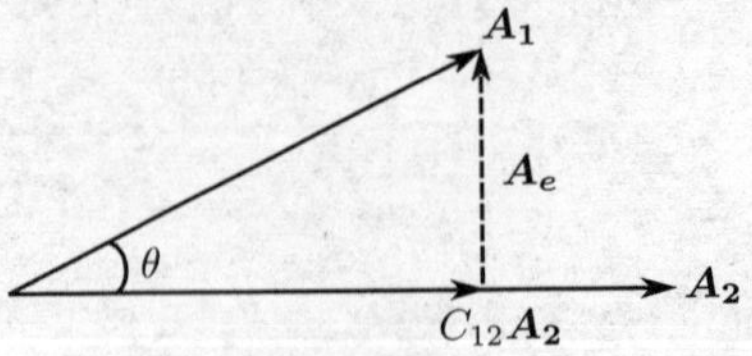

Fig. 2.1

In Fig. 2.1, $\boldsymbol{A_e}$ has the smallest magintude, because it is perpendicular to $\boldsymbol{A_2}$. However Eq. 2.1 is not the only way of expressing vector, $\boldsymbol{A_1}$ in terms of $\boldsymbol{A_2}$. In Fig. 2.2, we have represented vector $\boldsymbol{A_1}$ in terms of $\boldsymbol{A_2}$ by two different methods. Infact, vector $\boldsymbol{A_1}$ can be represented by vector, $\boldsymbol{A_2}$ by infinite alternate methods. Thus, referring Fig. 2.2, we can write

$$\mathbf{A_1} = C_1\mathbf{A_2} + \mathbf{A_{e_1}} \tag{2.2}$$

$$\mathbf{A_1} = C_2\mathbf{A_2} + \mathbf{A_{e_2}} \tag{2.3}$$

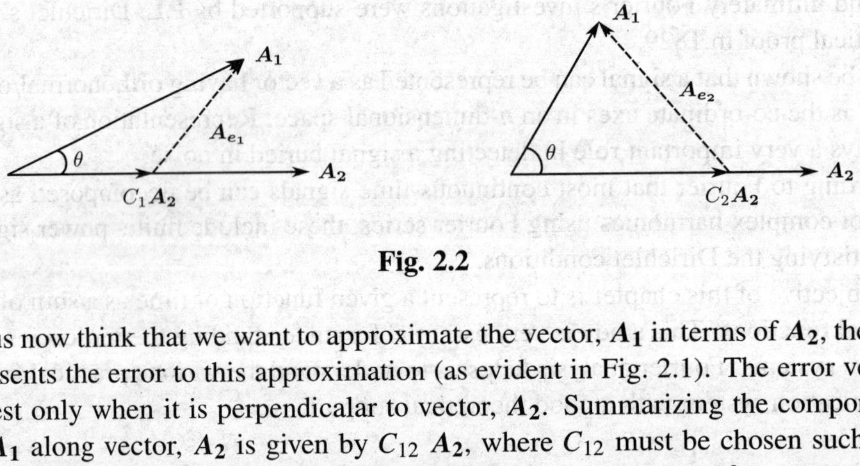

Fig. 2.2

Let us now think that we want to approximate the vector, $\boldsymbol{A_1}$ in terms of $\boldsymbol{A_2}$, then vector $\boldsymbol{A_e}$ repersents the error to this approximation (as evident in Fig. 2.1). The error vector, $\boldsymbol{A_e}$ is smallest only when it is perpendicalar to vector, $\boldsymbol{A_2}$. Summarizing the component of a vector, $\boldsymbol{A_1}$ along vector, $\boldsymbol{A_2}$ is given by C_{12} $\boldsymbol{A_2}$, where C_{12} must be chosen such that the error vector is minimum. It is clear that the larger the component of a vector along the other, the more closely they resemble each other in their directions. The factor, C_{12} is an indication of the similarity of the two vectors. If the angle θ is 90°, then $C_{12} = 0$ and the vector $\boldsymbol{A_1}$ has no component along the other vector, $\boldsymbol{A_2}$. When $\boldsymbol{A_1}$ and $\boldsymbol{A_2}$ are perpendicular to each other, they are known as orthogonal vectors. It may be noted that the orthogonal vectors are independent vectors. Thus, if the vectors are orthogonal, then the parameter C_{12} is zero.

The concept of orthogonality can be extended to signals. Let us consider two signals $x_1(t)$ and $x_2(t)$. Just like we did for vector, we want to approximate $x_1(t)$ in terms of $x_2(t)$ over a finite time interval (t_1, t_2). That is,

$$x_1(t) \simeq C_{12}x_2(t) \quad \text{for} \quad t_1 < t < t_2.$$

Let us define an error function, $x_e(t)$ as

$$x_e(t) = x_1(t) - C_{12}x_2(t) \tag{2.4}$$

The best approximation results when mean of the square of the error instead of the error itself is minimized. Let us designate the average of $x_e^2(t)$ by ε.

$$\varepsilon = \frac{1}{(t_2 - t_1)} \int_{t_1}^{t_2} x_e^2(t)\,dt$$

$$= \frac{1}{(t_2 - t_1)} \int_{t_1}^{t_2} \{x_1(t) - C_{12}x_2(t)\}^2\,dt \tag{2.5}$$

To find the value of C_{12} which will minimize ε, we set

$$\frac{d\varepsilon}{dC_{12}} = 0$$

That is,

$$\frac{d}{dC_{12}}\left\{\frac{1}{(t_2 - t_1)} \int_{t_1}^{t_2} \{x_1(t) - C_{12}x_2(t)\}^2\,dt\right\} = 0$$

Changing the order of integration and differentiation, we get

$$\frac{1}{(t_2 - t_1)}\left[\int_{t_1}^{t_2} \frac{d}{dC_{12}}\left\{x_1^2(t) - 2C_{12}x_1(t)x_2(t) + C_{12}^2x_2^2(t)\right\}\,dt\right] = 0$$

$$\Rightarrow \quad \frac{1}{(t_2 - t_1)}\left[\int_{t_1}^{t_2} \frac{d}{dC_{12}}x_1^2(t)\,dt - 2\int_{t_1}^{t_2} x_1(t)x_2(t)\,dt + 2C_{12}\int_{t_1}^{t_2} x_2^2(t)\,dt\right] = 0$$

$$\Rightarrow \quad \frac{1}{(t_2 - t_1)}\left[0 - 2\int_{t_1}^{t_2} x_1(t)x_2(t)\,dt + 2C_{12}\int_{t_1}^{t_2} x_2^2(t)\,dt\right] = 0$$

Since $t_2 - t_1 \neq 0$, we get

$$-2\int_{t_1}^{t_2} x_1(t)x_2(t)\,dt + 2C_{12}\int_{t_1}^{t_2} x_2^2(t)\,dt = 0$$

Hence,

$$C_{12} = \frac{\int_{t_1}^{t_2} x_1(t)x_2(t)\, dt}{\int_{t_1}^{t_2} x_2^2(t)\, dt} \tag{2.6}$$

By comparsion with vectors, if $C_{12} = 0$, then the signal $x_1(t)$ has no component of the signal $x_2(t)$ and we say that the two vectors are orthogonal over the interval (t_1, t_2). Summarizing, we define 'the condition for two real signals to be orthogonal as

$$\int_{t_1}^{t_2} x_1(t)x_2(t)\, dt = 0 \tag{2.7}$$

2.2 Orthogonal Signal Space

The space bounded by a set of 'n' orthogonal functions forming the coordinate axes is called orthogonal signal space. For example, let us consider a signal $x(t)$ given by $x(t) = 2x_1(t) + 4x_2(t)$, then $x(t)$ is viewed as a vector as depicted in Fig. 2.3.

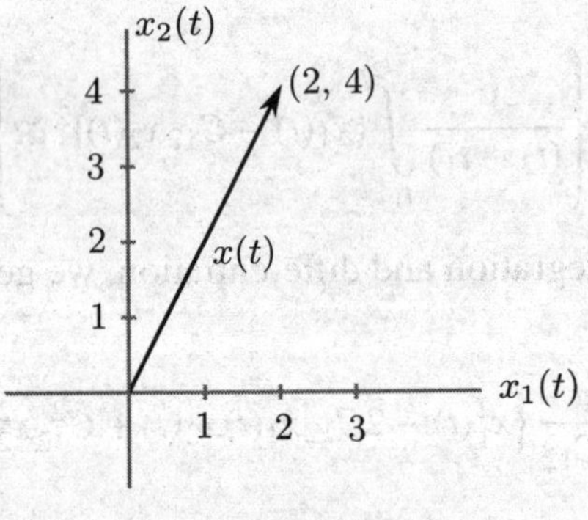

Fig. 2.3

In the example chosen, it may be noted that the signal $x(t)$ is expressed as a linear combination of the orthogonal functions $x_1(t)$ and $x_2(t)$. Let us next consider the technique of expressing an arbitrary signal $x(t)$ as linear combination of 'n' orthogonal functions. This is done in the section to follow.

Approximations of a function by a set of mutually orthogonal functions

Consider a set of 'n' orthogonal functions $x_1(t), x_2(t), \ldots x_n(t)$ over an intirval (t_1, t_2). Then, we have

$$\int_{t_1}^{t_2} x_j(t)\;\; x_k(t)\;\; dt = \begin{cases} 0, & j \neq k \\ E_j, & j = k \end{cases} \tag{2.8}$$

where E_j is the energies of the signals $x_j(t)$, $j = 1, 2, \ldots n$. If $E_j = 1$, then the set is normalized and is called an orthonormal set. An orthogonal set can always be normalized by dividing each orthogonal function by $\sqrt{E_j}$ for $j = 1, 2, \ldots n$.

Let an arbitrary function $x(t)$ be approximated over an interval (t_1, t_2) by a linear combination of $x_1(t), x_2(t), \ldots x_n(t)$. That is,

$$x(t) \simeq C_1 x_1(t) + C_2 x_2(t) + \cdots + C_n x_n(t)$$
$$\simeq \sum_{j=1}^{n} C_j x_j(t) \tag{2.9}$$

By definition,

$$x_e(t) = x(t) - \sum_{j=1}^{n} C_j x_j(t)$$

and

$$\varepsilon = \frac{1}{(t_2 - t_1)} \int_{t_1}^{t_2} \left[x(t) - \sum_{j=1}^{n} C_j x_j(t) \right]^2 dt$$

The best approximation results when ε is minimized with respect to C_j. That is,

$$\frac{\partial \varepsilon}{\partial C_j} = 0 \quad \text{for } j = 1, 2, \ldots n$$

$$\Rightarrow \quad \frac{\partial}{\partial C_j} \int_{t_1}^{t_2} \left[x(t) - \sum_{j=1}^{n} C_j x_j(t) \right]^2 dt = 0$$

$$\Rightarrow \quad \int_{t_1}^{t_2} \frac{\partial}{\partial C_j} \left\{ x^2(t) - 2 \sum_{j=1}^{n} C_j x_j(t) x(t) + \sum_{j=1}^{n} C_j^2 x_j^2(t) \right\} dt = 0$$

$$\Rightarrow \quad -2 \int_{t_1}^{t_2} \frac{\partial}{\partial C_j} \left[C_1 x_1(t) x(t) + \cdots + C_j x_j(t) x(t) + \cdots + C_n x_n(t) x(t) \right] dt$$

$$+ \int_{t_1}^{t_2} \frac{\partial}{\partial c_j} \left[C_1^2 x_1^2(t) + \cdots + C_j^2 x_j^2(t) + \cdots + C_n^2 x_n^2(t) \right] dt = 0$$

$$\Rightarrow \quad -2 \int_{t_1}^{t_2} x_j(t) x(t)\, dt + \int_{t_1}^{t_2} 2 C_j x_j^2(t)\, dt = 0$$

Hence,

$$C_j = \frac{\int_{t_1}^{t_2} x_j(t)x(t)\,dt}{\int_{t_1}^{t_2} x_j^2(t)\,dt}, \quad j = 1, 2, \cdots n \tag{2.10}$$

Summary: Given a set of 'n' mutually orthogonal functions $x_1(t), x_2(t), \ldots x_n(t)$ over the interval (t_1, t_2), it is possible to approximate an arbitrary function $x(t)$ over this interval by a linear combination of these functions. That is,

$$x(t) \simeq \sum_{j=1}^{n} C_j x_j(t)$$

For best approximation, we have to choose C_j as per the Eq. (2.10).

Mean square error (MSE)

Let us now proceed to find the value of MSE, ε when the optimum values of the coefficients $C_1, C_2, \ldots C_n$ are chosen as per the Eq. (2.10). By definition,

$$\varepsilon = \frac{1}{(t_2 - t_1)} \int_{t_1}^{t_2} \left[x(t) - \sum_{j=1}^{n} C_j x_j(t) \right]^2 dt$$

$$= \frac{1}{(t_2 - t_1)} \left[\int_{t_1}^{t_2} x^2(t)\,dt + \sum_{j=1}^{n} C_j^2 \int_{t_1}^{t_2} x_j^2(t)\,dt - 2 \sum_{j=1}^{n} C_j \int_{t_1}^{t_2} x(t)x_j(t)\,dt \right] \tag{2.11}$$

Recall Eq. (2.10):

$$C_j = \frac{\int_{t_1}^{t_2} x_j(t)x(t)\,dt}{\int_{t_1}^{t_2} x_j^2(t)\,dt}$$

Cross-multiplying, we get

$$\int_{t_1}^{t_2} x_j(t)x(t)\,dt = C_j \int_{t_1}^{t_2} x_j^2(t)\,dt \tag{2.12}$$

From Eq. (2.8), when $j = k$, we get

$$\int_{t_1}^{t_2} x_j^2(t)\,dt = E_j \tag{2.13}$$

Substituting Eq. (2.13) in Eq. (2.12) gives

$$\int_{t_1}^{t_2} x_j(t)x(t)\,dt = C_j E_j \tag{2.14}$$

Substituting Eq. (2.14) and Eq. (2.13) in Eq. (2.11), we get

$$\begin{aligned}
\varepsilon &= \frac{1}{(t_2 - t_1)}\left[\int_{t_1}^{t_2} x^2(t)\,dt + \sum_{j=1}^{n} C_j^2 E_j - 2\sum_{j=1}^{n} C_j^2 E_j\right] \\
&= \frac{1}{(t_2 - t_1)}\left[\int_{t_1}^{t_2} x^2(t)\,dt - \sum_{j=1}^{n} C_j^2 E_j\right] \\
&= \frac{1}{(t_2 - t_1)}\left[\int_{t_1}^{t_2} x^2(t)\,dt - \left\{C_1^2 E_1 + C_2^2 E_2 + \cdots + C_n^2 E_n\right\}\right]
\end{aligned} \tag{2.15}$$

Thus, by using Eq. (2.15) the mean-square error can be evaluated.

2.3 Representation of a function by a complete set of orthogonal functions

From Eq. (2.15), it can be seen that if the number of orthogonal functions increase, the error, ε will become smaller. Hence, hypothetically, if $n \to \infty$, it is possible that $\sum_{j=1}^{n} C_j^2 E_j^2$ may become equal to $\int_{t_1}^{t_2} x^2(t)\,dt$ and eventually, $\varepsilon \to 0$. Hence, as $n \to \infty$, we may write

$$\int_{t_1}^{t_2} x^2(t)\,dt = \sum_{j=1}^{\infty} C_j^2 E_j \tag{2.16}$$

As $n \to \infty$, Eq. (2.9) becomes

$$x(t) = \sum_{j=1}^{\infty} C_j x_j(t) \tag{2.17}$$

$$\Rightarrow \qquad x(t) = C_1 x_1(t) + C_2 x_2(t) + \cdots + C_j x_j(t) + \cdots \tag{2.18}$$

It may be noted that representation of $x(t)$ is now exact. This is because, if $x(t)$ is represented by Eq. (2.18), we have $\varepsilon = 0$.

A set of functions $x_1(t), x_2(t) \dots x_n(t)$ that are mutually orthogonal over the interval (t_1, t_2) is said to be a complete or a closed set, if there exists no function $x(t)$ for which it is true that

$$\int_{t_1}^{t_2} x(t)x_j(t)\,dt = 0 \quad \text{for} \quad j = 1, 2, \dots$$

In otherwords, it means that $x(t)$ must be a member of orthogonal set defined by $\{x_1(t),\ x_2(t), \dots x_n(t)\}$.

Summary: For a set of functions $x_1(t)x_2(t), \dots x_n(t)$ that are mutually orthogonal over the interval (t_1, t_2), we have

$$\int_{t_1}^{t_2} x_j(t)x_k(t)\,dt = \begin{cases} 0, & j \neq k \\ E_j, & j = k \end{cases}$$

If this orthogonal set is complete, then any function $x(t)$ can be written as

$$x(t) = C_1x_1(t) + C_2x_2(t) + \cdots + C_jx_j(t) + \cdots \tag{2.19(a)}$$

$$\text{where,} \qquad C_j = \frac{\int_{t_1}^{t_2} x(t)x_j(t)\,dt}{E_j}$$

$$= \frac{\int_{t_1}^{t_2} x(t)x_j(t)\,dt}{\int_{t_1}^{t_2} x_j^2(t)\,dt} \tag{2.19(b)}$$

Eq. (2.19(b)) reveals that $x(t)$ contains a component $x_j(t)$ and this component has magnitutde C_j. Representation of a signal by a set of infinite mutually orthogonal functions is known as generalized Fourier series representation of $x(t)$.

Anology with vectors:

If a vector is represented by its mutually orthogonal components, the square of the length of the vector equals the sum of the squares of the lengths of its orthogonal components. For example, if $\mathbf{A} = A_x\mathbf{a_x} + A_y\mathbf{a_y} + A_z\mathbf{a_z}$ then $|\mathbf{A}|^2 = A^2 = A_x^2 + A_y^2 + A_z^2$. It may be noted that $\mathbf{a_x}$, $\mathbf{a_y}$ and $\mathbf{a_z}$ represent unit vectors along x, y and z directions, respectively. A similar result holds good for signals also. This anology can be visualized by having a look at Eq. (2.19). That is, from Eq. (2.19(a)), we find that

$$|x(t)|^2 = C_1^2 + C_2^2 + \cdots + C_j^2 + \cdots$$

2.4 Orthogonality in Complex Functions

Let $x_1(t)$ and $x_2(t)$ be two Complex functions. Consider again the problem of approximating a signal $x_1(t)$ by a signal $x_2(t)$ over an interval (t_1, t_2):

$$x_1(t) = C_{12}x_2(t) \tag{2.20}$$

The energy of the signal $x_2(t)$ is

$$E_2 = \int_{t_1}^{t_2} |x_2(t)|^2\, dt = \int_{t_1}^{t_2} x_2(t)x_2^*(t)\, dt \tag{2.21}$$

where, $x_2^*(t)$ is the complex congugate of $x_2(t)$.

In the present context,

$$x_e(t) = x_1(t) - C_{12}x_2(t)$$

It may be noted that $x_e(t)$ and C_{12} are both complex. For best approximation, we select C_{12} so that ε, is minimized. The result is

$$C_{12} = \frac{\int_{t_1}^{t_2} x_1(t)x_2^*(t)\, dt}{\int_{t_1}^{t_2} x_2(t)x_2^*(t)\, dt} \tag{2.22}$$

For orthogonality of $x_1(t)$ and $x_2(t)$, C_{12} must be zero. Letting $C_{12} = 0$ in Eq. (2.22) gives

$$\int_{t_1}^{t_2} x_1(t)x_2^*(t)\, dt = 0 \tag{2.23}$$

Taking complex conjugates on both sides of Eq. (2.23) gives

$$\int_{t_1}^{t_2} x_1^*(t)x_2(t)\, dt = 0 \tag{2.24}$$

Thus, the conjugation can be on either $x_1(t)$ or $x_2(t)$.

Let us now generalize the result. For a set of complex functions $x_1(t), x_2(t), \ldots,$ that are mutually orthogonal over the enterval (t_1, t_2):

$$\int_{t_1}^{t_2} x_j(t)x_k^*(t)dt = \begin{cases} 0, & \text{if } j \neq k \\ E_j, & \text{if } j = k \end{cases} \tag{2.25}$$

If this set of orthogonal functions is complete, then any arbitrary function, $x(t)$ can be expressed as a linear combination of these orthogonal functions. That is

$$x(t) = C_1 x_1(t) + C_2 x_2(t) + \cdots + C_j x_j(t) + \cdots \tag{2.26}$$

where

$$C_j = \frac{1}{E_j} \int_{t_1}^{t_2} x(t) x_j^*(t)\, dt$$

Example 2.1. Consider the rectangular function $x(t)$ defined by

$$x(t) = \begin{cases} A, & 0 < t < \pi \\ -A, & \pi < t < 2\pi \end{cases}$$

Approximate $x(t)$ by a waveform $\sin t$ over the interval $(0, 2\pi)$. The criterion for approximation is that MSE is minimized.

Solution: The function $x(t)$ will be approximated over the interval $(0, 2\pi)$ as

$$x(t) = C_{12} \sin t$$

Recall:

$$C_j = \frac{\int_{t_1}^{t_2} x(t) x_j(t)\, dt}{\int_{t_1}^{t_2} x_j^2(t)\, dt}$$

In the present context, $x_j(t) = \sin t$ and $C_j = C_{12}$

Thus,

$$C_{12} = \frac{\int_0^{2\pi} x(t) \sin t\, dt}{\int_0^{2\pi} \sin^2(t)\, dt}$$

$$= \frac{1}{\pi}\left[\int_0^{2\pi} x(t) \sin t\, dt\right]$$

$$= \frac{1}{\pi}\left[\int_0^{\pi} A \sin t\, dt + \int_{\pi}^{2\pi} -A \sin t\, dt\right]$$

$$= \frac{4A}{\pi}$$

Hence, $x(t) \simeq \frac{4A}{\pi} \sin t$ represents the best approximation of $x(t)$ by a function $\sin t$ that will minimize MSE. The same has been shown in Fig. Ex. 2.1.

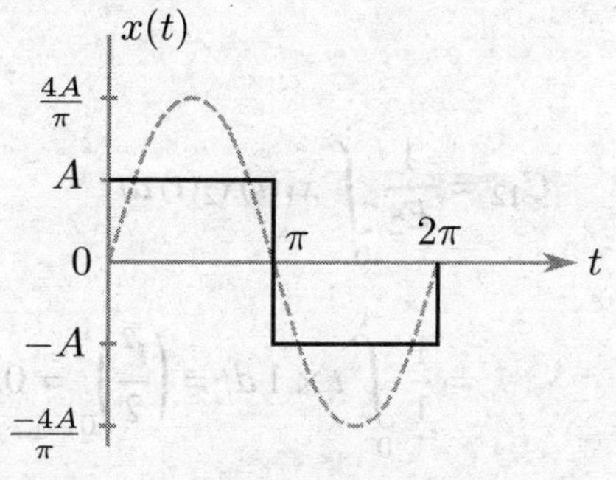

Fig. Ex. 2.1

Example 2.2. (a) Refer the signals $x_1(t)$ and $x_2(t)$ shown in Fig. Ex. 2.2. If $x_1(t) = C_{12}x_2(t)$, find the optimum value of C_{12} so that error signal energy is minimized.

(b) Find the error signal $x_e(t)$ and its energy E_e.

(c) Show that the error signal is orthogonal and that $E_1 = C_{12}^2 E_2 + E_e$.

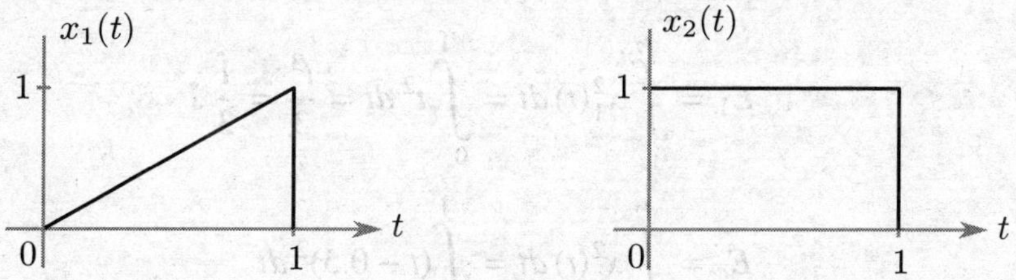

Fig. Ex. 2.2

Solution:

(a) It may be noted that minimization of energy of the error signal or minimization of MSE gives identical result (same value for C_{12}).

In the present problem, energy of the signal, $x_2(t)$ is

$$E_2 = \int_0^1 x_2^2(t)\,dt = \int_0^1 1^2\,dt = 1\text{ J}$$

The mathematical descriptions of $x_1(t)$ and $x_2(t)$ are as follows:

$$x_1(t) = \begin{cases} t, & 0 < t < 1 \\ 0, & \text{otherwise} \end{cases}$$

$$\text{and} \qquad x_2(t) = \begin{cases} 1, & 0 < t < 1 \\ 0, & \text{otherwise} \end{cases}$$

Hence,

$$C_{12} = \frac{1}{E_2} \int_{t_1}^{t_2} x_1(t) x_2(t)\, dt$$

$$= \frac{1}{1} \int_0^1 t \times 1\, dt = \left(\frac{t^2}{2}\right)_0^1 = 0.5$$

(b) Hence, $x_1(t) = 0.5x_2(t)$ and the error signal is

$$x_e(t) = x_1(t) - 0.5x_2(t)$$

$$= \begin{cases} t - 0.5 \times 1, & 0 < t < 1 \\ 0, & \text{otherwise} \end{cases}$$

Let us now proceed to find the energies of the signals $x_1(t)$ and $x_e(t)$.

$$E_1 = \int_{t_1}^{t_2} x_1^2(t)\, dt = \int_0^1 t^2\, dt = \frac{t^3}{3} = \frac{1}{3}\ \text{J}$$

$$E_e = \int_{t_1}^{t_2} x_e^2(t)\, dt = \int_0^1 (t - 0.5)^2\, dt$$

$$= \left[\frac{(t - 0.5)^3}{3}\right]_0^1 = \frac{1}{12}\ \text{J}$$

(c) The error signal $x_e(t)$ is orthogonal to $x_2(t)$ because

$$\int_{t_1}^{t_2} x_e(t) x_2(t) = \int_0^1 (t - 0.5) \times 1\, dt = 0$$

It may be noted that $E_1 = C_{12}^2 E_2 + E_e$.

Analogy between signals and vectors: The error vector $\boldsymbol{x_e}$ is orthogonal to the component $C_{12}\boldsymbol{x_2}$. Hence, the square of the length of vector $\boldsymbol{x_1}$ (energy of $x_1(t)$) is equal to sum of the square of the lengths of $C_{12}\boldsymbol{x_2}$ and $\boldsymbol{x_e}$ (sum of the energies of $C_{12}x_2(t)$ and $x_e(t)$).

Example 2.3. Refer the signals $x_1(t)$ and $x_2(t)$ depicted in Example 2.2. Determine the component of the form $x_1(t)$ contained in $x_2(t)$ That is, find optimum value of C_{21} in the approximation $x_2(t) = C_{21}x_1(t)$ so that MSE is minimized. Also find the energy of the error signal.

Solution: In the present context,

$$E_1 = \int_{t_1}^{t_2} x_1^2(t)\,dt = \int_0^1 t^2\,dt$$

$$= \left(\frac{t^3}{3}\right)_0^1 = \frac{1}{3}\text{ J}$$

Hence,

$$C_{21} = \frac{1}{E_1}\int_{t_1}^{t_2} x_2(t)x_1(t)\,dt$$

$$= \frac{1}{\frac{1}{3}}\int_0^1 1 \times t\,dt = \frac{3}{2}$$

Thus, $x_2(t) = C_{21}x_1(t) = 1.5x_1(t)$

The corresponding error signal is

$$x_e(t) = x_2(t) - 1.5x_1(t)$$

$$= \begin{cases} 1 - 1.5\,t, & 0 < t < 1 \\ 0, & \text{otherwise} \end{cases}$$

Let us now proceed to find the energy of the error signal, $x_e(t)$.

$$E_e = \int_{t_1}^{t_2} x_e^2(t)\,dt$$

$$= \int_0^1 (1 - 1.5t)^2\,dt$$

$$= 0.25\text{ J}$$

Example 2.4. (a) For the signals $x_1(t)$ and $x_2(t)$ shown in Fig. Ex. 2.4, find the component of the waveform $x_2(t)$ contained in $x_1(t)$. That is, find the optimum value of C_{12} in the approximation $x_1(t) \approx C_{12}x_2(t)$ so that MSE is minimized.

(b) Find the error signal $x_e(t)$ and its energy.

(c) Also show that $x_e(t)$ is orthogonal to $x_2(t)$.

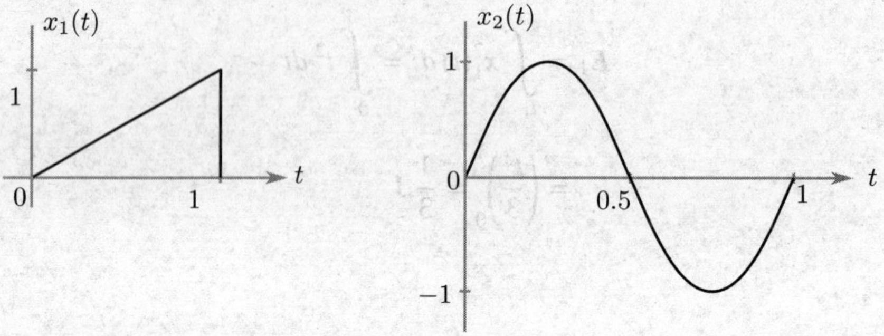

Fig. Ex. 2.4

Solution:

(a) The time-domain descriptions of $x_1(t)$ and $x_2(t)$ are given below:

$$x_1(t) = \begin{cases} t, & 0 < t < 1 \\ 0, & \text{otherwise} \end{cases}$$

and

$$x_2(t) = \begin{cases} A \sin \omega t, & 0 < t < 1 \\ 0, & \text{otherwise} \end{cases}$$

$$= \begin{cases} A \sin\left(\dfrac{2\pi t}{T}\right), & 0 < t < 1 \\ 0, & \text{otherwise} \end{cases}$$

$$= \begin{cases} 1 \cdot \sin\left(\dfrac{2\pi}{1}t\right), & 0 < t < 1 \\ 0, & \text{otherwise} \end{cases}$$

$$= \begin{cases} \sin 2\pi t, & 0 < t < 1 \\ 0, & \text{otherwise} \end{cases}$$

The energy of the signal $x_2(t)$ is

$$E_2 = \int_{t_1}^{t_2} x_2^2(t)\,dt = \int_0^1 \sin^2 2\pi t\,dt$$

$$= \frac{1}{2}\int_0^1 (1 + \cos 4\pi t)\, dt = \frac{1}{2}\left[t + \frac{\sin 4\pi t}{4\pi}\right]_0^1$$

$$= \frac{1}{2}\ \text{J}$$

Also,

$$C_{12} = \frac{1}{E_2}\int_{t_1}^{t_2} x_1(t)x_2(t)\, dt$$

$$= \frac{1}{\frac{1}{2}}\int_0^1 t \sin 2\pi t\, dt$$

$$= -\frac{1}{\pi}$$

(b) Therefore,

$$x_1(t) \simeq C_{12}x_2(t)$$

$$= -\frac{1}{\pi}\sin 2\pi t$$

Thus, the error signal is

$$x_e(t) = x_1(t) - C_{12}x_2(t)$$

$$= \begin{cases} t + \dfrac{1}{\pi}\sin 2\pi t, & 0 < t < 1 \\ 0, & \text{otherwise} \end{cases}$$

The energy of the error signal is

$$E_e = \int_{t_1}^{t_2} x_e^2(t)\, dt$$

$$= \int_0^1 \left[t + \frac{1}{\pi}\sin 2\pi t\right]^2 dt$$

$$= \int_0^1 \left[t^2 + \frac{2}{\pi}t\sin 2\pi t + \frac{1}{\pi^2}\sin^2 2\pi t\right] dt$$

$$= \int_0^1 t^2\,dt + \frac{2}{\pi}\int_0^1 t\sin 2\pi t\,dt + \frac{1}{\pi^2}\int_0^1 \left(\frac{1+\cos 4\pi t}{2}\right) dt$$

$$= \left(\frac{t^3}{3}\right)_0^1 + \frac{2}{\pi}\left[-\frac{0.5}{\pi}\right] + \frac{1}{2\pi^2}\left[t + \frac{\sin 4\pi t}{4\pi}\right]_0^1$$

$$= \frac{1}{3} - \frac{1}{\pi^2} + \frac{1}{2\pi^2}$$

$$= 0.28\ \text{J}$$

(c) The error signal, $x_e(t)$ is orthogonal to $x_2(t)$ because

$$\int_0^1 x_e(t)x_2(t)\,dt = \int_0^1 \left[t + \frac{1}{\pi}\sin 2\pi t\right]\sin 2\pi t\,dt = 0$$

Example 2.5. (a) It is found that two real signals $x_1(t)$ and $x_2(t)$ are mutually orthogonal to each other. Show that the sum: $x_1(t) + x_2(t)$ and the difference: $x_1(t) - x_2(t)$ have identical energies given by $E_1 + E_2$.

(b) Also show that $C_1x_1(t) + C_2x_2(t)$ and $C_1x_1(t) - C_2x_2(t)$ have identical energies given by $C_1^2E_1 + C_2^2E_2$, where C_1 and C_2 are any pair of arbitrary constants.

Solution:

(a) The energy of the sum signal is

$$E_s = \int_{t_1}^{t_2} [x_1(t) + x_2(t)]^2\,dt$$

$$= \int_{t_1}^{t_2} x_1^2(t)\,dt + \int_{t_1}^{t_2} x_2^2(t)\,dt + 2\int_{t_1}^{t_2} x_1(t)x_2(t)\,dt$$

The third intergral on the right-hand side of the above equation is zero because $x_1(t)$ and $x_2(t)$ are orthogonal to each other. Thus,

$$E_s = E_1 + E_2$$

Let us next proceed to find the energy of the difference signal

$$E_d = \int_{t_1}^{t_2} [x_1(t) - x_2(t)]^2\,dt$$

$$= \int_{t_1}^{t_2} x_1^2(t)\,dt + \int_{t_1}^{t_2} x_2^2(t)\,dt - 2\int_{t_1}^{t_2} x_1(t)x_2(t)\,dt$$

$$= E_1 + E_2 - 0$$

$$= E_1 + E_2$$

Hence, we find that, $E_s = E_d$.

(b) The energy of the sum signal is

$$E_s = \int_{t_1}^{t_2} [C_1x_1(t) + C_2x_2(t)]^2\,dt$$

$$= C_1^2 \int_{t_1}^{t_2} x_1^2(t)\,dt + C_2^2 \int_{t_1}^{t_2} x_2^2(t)\,dt + 2C_1C_2 \int_{t_1}^{t_2} x_1(t)x_2(t)\,dt$$

$$= C_1^2E_1 + C_2^2E_2 + 2C_1C_2 \times 0$$

$$= C_1^2E_1 + C_2^2E_2$$

The energy of the difference signal is

$$E_d = \int_{t_1}^{t_2} [C_1x_1(t) - C_2x_2(t)]^2\,dt$$

$$= C_1^2 \int_{t_1}^{t_2} x_1^2(t)\,dt + C_2^2 \int_{t_1}^{t_2} x_2^2(t)\,dt - 2C_1C_2 \int_{t_1}^{t_2} x_1(t)x_2(t)\,dt$$

$$= C_1^2E_1 + C_2^2E_2$$

Hence, we find that, $E_s = E_d$.

Example 2.6. If $x_1(t)$ and $x_2(t)$ are two complex mutually orthogonal signals, then show that $x_1(t) + x_2(t)$ and $x_1(t) - x_2(t)$ have identical energies given by $E_1 + E_2$, where E_1 and E_2 are the energies of the signals $x_1(t)$ and $x_2(t)$, respectively.

Solution: We know that, for any complex signal $x(t)$,

$$|x(t)|^2 = x(t)x^*(t)$$

Letting $x(t) = x_1(t) \pm x_2(t)$ gives

$$|x_1(t) \pm x_2(t)|^2 = [x_1(t) \pm x_2(t)][x_1(t) \pm x_2(t)]^*$$

$$\Rightarrow \quad |x_1(t) \pm x_2(t)|^2 = [x_1(t) \pm x_2(t)][x_1^*(t) \pm x_2^*(t)]$$
$$= x_1(t)x_1^*(t) + x_2(t)x_2^*(t) \pm x_1(t)x_2^*(t) \pm x_1^*(t)x_2(t)$$
$$= |x_1(t)|^2 + |x_2(t)|^2 \pm x_1(t)x_2^*(t) \pm x_2(t)x_1^*(t)$$

It may be noted that for a complex signal, $x(t)$, the total energy content of the signal is computed using the expression, $E_x = \int_{-\infty}^{\infty} |x(t)|^2\, dt$. Thus, the energy of the sum signal is

$$E_s = \int_{t_1}^{t_2} |x_1(t) + x_2(t)|^2\, dt$$

$$\Rightarrow \quad E_s = \int_{t_1}^{t_2} |x_1(t)|^2\, dt + \int_{t_1}^{t_2} |x_2(t)|^2\, dt + \int_{t_1}^{t_2} x_1(t)x_2^*(t)\, dt + \int_{t_1}^{t_2} x_1^*(t)x_2(t)\, dt$$

The third and fourth integrals on the right-hand side of the above equation are zero because $x_1(t)$ and $x_2(t)$ are orthogonal signals. Thus, we get,

$$E_s = \int_{t_1}^{t_2} |x_1(t)|^2\, dt + \int_{t_1}^{t_2} |x_2(t)|^2\, dt$$
$$= E_1 + E_2$$

Let us now proceed to find the energy of the difference signal.

$$E_d = \int_{t_1}^{t_2} |x_1(t) - x_2(t)|^2\, dt$$
$$= \int_{t_1}^{t_2} |x_1(t)|^2\, dt + \int_{t_1}^{t_2} |x_2(t)|^2\, dt - \int_{t_1}^{t_2} x_1(t)x_2^*(t)\, dt - \int_{t_1}^{t_2} x_1^*(t)x_2(t)\, dt$$
$$= E_1 + E_2 - 0 - 0$$
$$= E_1 + E_2$$

Hence, we find that, $E_s = E_d$.

Example 2.7. An arbitary signal $x_1(t)$ is approximated in terms of another signal $x_2(t)$ over a finite time interval (t_1, t_2). That is,

$$x_1(t) = Cx_2(t), \quad t_1 < t < t_2$$

The constant C is found by minimizing mean square error (MSE) or the energy of the error signal. Show that $x_2(t)$ and the error signal $x_e(t)$ are orthogonal over the interval (t_1, t_2).

Solution: The error signal is given by

$$x_e(t) = x_1(t) - Cx_2(t)$$

Also,

$$\int_{t_1}^{t_2} x_2(t)x_e(t)\,dt = \int_{t_1}^{t_2} x_2(t)[x_1(t) - Cx_2(t)]\,dt$$

$$= \int_{t_1}^{t_2} x_1(t)x_2(t)\,dt - C\int_{t_1}^{t_2} x_2^2(t)\,dt \tag{2.27}$$

Also, we know that,

$$C = \frac{\int_{t_1}^{t_2} x_1(t)x_2(t)\,dt}{\int_{t_1}^{t_2} x_2^2(t)\,dt} \tag{2.28}$$

Substituting Eq. (2.28) in Eq. (2.27) gives

$$\int_{t_1}^{t_2} x_2(t)[x_1(t) - Cx_2(t)]\,dt = \int_{t_1}^{t_2} x_1(t)x_2(t)\,dt - \int_{t_1}^{t_2} x_1(t)x_2(t)\,dt$$

$$= 0$$

Hence, $x(t)$ and $x_e(t)$ are orthogonal to each other.

Example 2.8. Refer the rectangular waveform defined by

$$x_1(t) = \begin{cases} 1, & 0 < t < \pi \\ -1, & \pi < t < 2\pi \end{cases}$$

Approximate $x_1(t)$ in terms of $\sin t$ and find the energy of the error signal.

Solution: Let $x_2(t) = \sin t$.

Then, the energy stored in $x_2(t)$ is

$$E_2 = \int_{t_1}^{t_2} x_2^2(t)\,dt$$

$$= \int_{0}^{2\pi} \sin^2 t\,dt$$

$$= \pi \text{ J}$$

Hence,

$$C = \frac{1}{E_2}\int_0^{2\pi} x_1(t)x_2(t)\,dt$$
$$= \frac{1}{\pi}\left[\int_0^{\pi} \sin t\,dt + \int_{\pi}^{2\pi} -\sin t\,dt\right]$$
$$= \frac{4}{\pi}$$

Thus, we get,

$$x_1(t) \simeq \frac{4}{\pi}\sin t$$

The error signal is now given as

$$x_e(t) = x_1(t) - \frac{4}{\pi}\sin t$$
$$= \begin{cases} 1 - \frac{4}{\pi}\sin t, & 0 < t < \pi \\ -1 - \frac{4}{\pi}\sin t, & \pi < t < 2\pi \end{cases}$$

Hence, the energy stored in the error signal is

$$E_e = \int_{t_1}^{t_2} x_e^2(t)\,dt$$
$$= \int_0^{\pi}\left[1 - \frac{4}{\pi}\sin t\right]^2 dt + \int_{\pi}^{2\pi}\left(-1 - \frac{4}{\pi}\sin t\right)^2 dt$$
$$= -\frac{8}{\pi}\left[\int_0^{\pi} \sin t\,dt + \int_{\pi}^{2\pi} \sin t\,dt\right]$$
$$= \frac{-8}{\pi}\int_0^{2\pi} \sin t\,dt$$
$$= 0\ \text{J}$$

Example 2.9. Let $\phi_1(t)$ and $\phi_2(t)$ be two orthonormal functions defined over an interval $(0, T)$. Determine the vector representation for the following three signals:

$$x_1(t) = 2\phi_1(t) - \phi_2(t)$$

$$x_2(t) = 3\phi_2(t)$$
$$x_3(t) = -\phi_2(t)$$

Also find the energy stored in these signals.

Solution: The given signals can be represented by vectors in the two-dimensional space bounded by the axes $\phi_1(t)$ and $\phi_2(t)$.

(1) $x_1(t) = 2\phi_1(t) - \phi_2(t) \Rightarrow \quad \boldsymbol{x_1} = (2, -1)$

(2) $x_2(t) = 3\phi_2(t) \quad \Rightarrow \quad \boldsymbol{x_2} = (0, 3)$

(3) $x_3(t) = -\phi_2(t) \quad \Rightarrow \quad \boldsymbol{x_3} = (0, -1)$

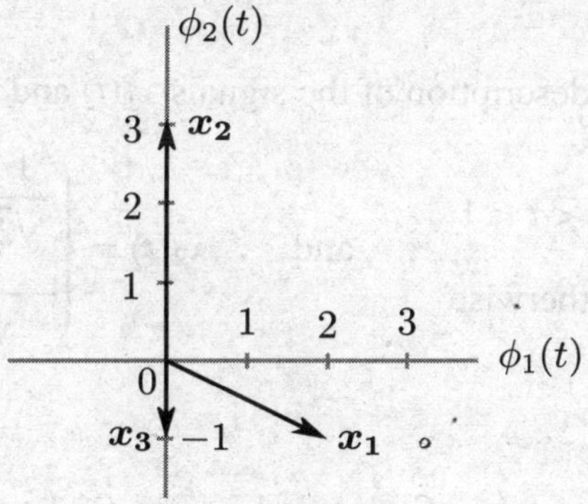

Fig. Ex. 2.8

The signal space diagram is as shown in Fig. Ex. 2.8.

The energy stored in a signal is the magnitude square of its length in the signal space diagram.

Then,

$$E_1 = \int_0^T x_1^2(t)\,dt = (2-0)^2 + (-1-0)^2 = 5 \text{ J}$$

$$E_2 = \int_0^T x_2^2(t)\,dt = (0-0)^2 + (3-0)^2 = 9 \text{ J}$$

and

$$E_3 = \int_0^T x_3^2(t)\,dt = (0-0)^2 + (0+1)^2 = 1 \text{ J}$$

Example 2.10. Refer the signals given in Fig. Ex. 2.10. Check for orthogonality. If they are found to be orthogonal, find the corresponding orthonormal functions.

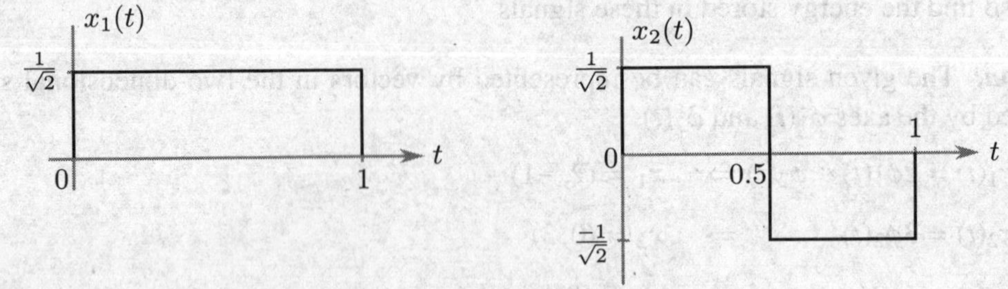

Fig. Ex. 2.10

Solution: The time-domain description of the signals $x_1(t)$ and $x_2(t)$ are given below:

$$x_1(t) = \begin{cases} \dfrac{1}{\sqrt{2}}, & 0 < t < 1 \\ 0, & \text{otherwise} \end{cases} \quad \text{and} \quad x_2(t) = \begin{cases} \dfrac{1}{\sqrt{2}}, & 0 < t < 0.5 \\ -\dfrac{1}{\sqrt{2}}, & 0.5 < t < 1 \end{cases}$$

Also,

$$\int_0^1 x_1(t)x_2(t)\,dt = \int_0^{0.5} \frac{1}{\sqrt{2}} \times \frac{1}{\sqrt{2}}\,dt + \int_{0.5}^{1} \frac{1}{\sqrt{2}} \times -\frac{1}{\sqrt{2}}\,dt$$

$$= \left[(t)_0^{0.5} - (t)_{0.5}^{1}\right]\frac{1}{2}$$

$$= [0.5 - 0.5]\frac{1}{2} = 0$$

Hence, $x_1(t)$ and $x_2(t)$ are orthogonal to each other.
Let us now proceed to find the energy stored in the signals $x_1(t)$ and $x_2(t)$.

$$E_1 = \int_0^1 x_1^2(t)\,dt = \int_0^1 \left(\frac{1}{\sqrt{2}}\right)^2 dt = \frac{1}{2}\text{ J}$$

and

$$E_2 = \int_0^1 x_2^2(t)\,dt = \int_0^{0.5} \left(\frac{1}{\sqrt{2}}\right)^2 dt + \int_{0.5}^{1} \left(-\frac{1}{\sqrt{2}}\right)^2 dt$$

$$= \frac{1}{2}[t]_0^{0.5} + \frac{1}{2}[t]_{0.5}^{1}$$

$$\Rightarrow \quad E_2 = \frac{1}{2} \times 0.5 + \frac{1}{2} \times 0.5$$
$$= \frac{1}{2} \text{ J}$$

The corresponding orthonormal signals can be obtained by dividing $x_1(t)$ by $\sqrt{E_1}$ and $x_2(t)$ by $\sqrt{E_2}$. Then,

$$y_1(t) = \frac{x_1(t)}{\sqrt{E_1}} \quad \text{and} \quad y_2(t) = \frac{x_2(t)}{\sqrt{E_2}}$$

will have unit energy. The orthonormal functions $y_1(t)$ and $y_2(t)$ are sketched in Fig. Ex. 2.10 (a).

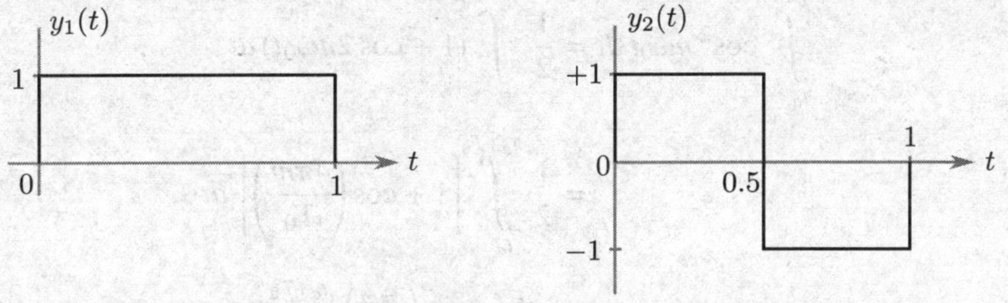

Fig. Ex. 2.10(a)

2.5 Trignometric Fourier Series (real form)

Fourier used the trignometric Fourier series in his paper in 1807. It was also used by both Euler and d'Alembert as early as 1747, some 60 years before him. The complex form (to be discussed later), which Fourier never used, was first discussed by Laplace is 1782, about 25 years before Fourier submitted his paper. Fourier's name has become attached to both forms, not because he discovered them, but he was the first person to use the trignometric form for solving partial differential equations.
It can be shown that the trignometric signal set

$$\{1, \cos\omega_0 t,\ \cos 2\omega_0 t, \ldots \cos n\omega_0 t,\ \sin\omega_0 t,\ \sin 2\omega_0 t, \ldots\ \sin n\omega_0 t\} \text{ is}$$

orthogonal over an interval $(0, T_0)$, where $T_0 = \dfrac{1}{f_0}$ is the period of the sinusoid of frequency f_0. The above trignometric set is a closed or complete set for a class of energy signals, that is, signals having finite energies. Thus, we can express a signal $x(t)$ by a trignometric Fourier series over an interval $(t_a, t_a + T_0)$ as

$$x(t) = a_0 + a_1 \cos\omega_0 t + a_2 \cos 2\omega_0 t + \cdots + b_1 \sin\omega_0 t + b_2 \sin 2\omega_0 t + \cdots \quad (2.29)$$

$$\Rightarrow \quad x(t) = a_0 + \sum_{n=1}^{\infty} a_n \cos n\omega_0 t + \sum_{n=1}^{\infty} b_n \sin n\omega_0 t; \quad t_a < t < t_a + T_0 \quad (2.30)$$

In Eq. (2.30), we have $\omega_0 = \frac{2\pi}{T_0}$. Let us use Eq. (2.19(b)) to find the Fourier coefficients a_0, a_n and b_n.

Thus,

$$a_n = \frac{\int_{t_a}^{t_a+T_0} x(t)\cos n\omega_0 t\,dt}{\int_{t_a}^{t_a+T_0} \cos^2 n\omega_0 t\,dt} \tag{2.31}$$

The next step in the analysis is to evaluate the integral in the denominator of Eq. (2.31).

$$\begin{aligned}\int_{t_a}^{t_a+T_0} \cos^2 n\omega_0 t\,dt &= \frac{1}{2}\int_{t_a}^{t_a+T_0} (1+\cos 2n\omega_0 t)\,dt \\ &= \frac{1}{2}\int_{t_a}^{t_a+T_0} \left[1+\cos\left(\frac{4\pi nt}{T_0}\right)\right] dt \\ &= \frac{1}{2}\left[t + \frac{\sin\left(\frac{4\pi nt}{T_0}\right)}{\frac{4\pi n}{T_0}}\right]_{t_a}^{t_a+T_0} \\ &= \frac{1}{2}T_0;\quad n \neq 0\end{aligned} \tag{2.32}$$

When $n = 0$, the integral in the denominator of Eq. (2.31) becomes

$$\int_{t_a}^{t_a+T_0} dt = T_0 \tag{2.33}$$

Thus, letting $n = 0$ in Eq. (2.31), we get

$$a_0 = \frac{\int_{t_a}^{t_a+T_0} x(t)\,dt}{T_0} = \frac{1}{T_0}\int_{t_a}^{t_a+T_0} x(t)\,dt \tag{2.34}$$

When $n \neq 0$, Eq. (2.31) reduces to

$$a_n = \frac{2}{T_0}\int_{t_a}^{t_a+T_a} x(t)\cos n\omega_0 t\,dt;\; n = 1, 2, 3, 4, \ldots \tag{2.35}$$

By following a similar proceduce, we find that

$$b_n = \frac{2}{T_0} \int_{t_a}^{t_a+T_0} x(t) \sin n\omega_0 t \, dt; \quad n = 1, 2, 3, 4, \cdots \tag{2.36}$$

Eq. (2.30) is known as *synthesis equation*. From the synthesis equation, we find that a_n and b_n multiply their respective cosine and sine terms. Since the magnitude of sine and cosine terms are less than or equal to unity, the rate of convergence of a Fourier series solely depends on how fast the coefficients converge (die out).

2.5.1 Effect of symmetry

If a periodic signal $x(t)$ is even, then $b_n = 0$ and its trignometric Fourier series contains only cosine terms:

$$x(t) = a_0 + \sum_{n=1}^{\infty} a_n \cos n\omega_0 t; \quad \omega_0 = \frac{2\pi}{T_0} \tag{2.37}$$

where

$$a_0 = \frac{2}{T_0} \int_0^{T_0/2} x(t) \, dt \tag{2.38}$$

$$a_n = \frac{4}{T_0} \int_0^{T_0/2} x(t) \cos n\omega_0 t \, dt \tag{2.39}$$

It $x(t)$ is odd, then $a_n = 0$ and its Fourier series contains only sine terms:

$$x(t) = \sum_{n=1}^{\infty} b_n \sin n\omega_0 t; \quad \omega_0 = \frac{2\pi}{T_0} \tag{2.40}$$

where

$$b_n = \frac{4}{T_0} \int_0^{T_0/2} x(t) \sin n\omega_0 t \, dt \tag{2.41}$$

The advantage of even and odd symmetry is that it is sufficient to know the waveform $x(t)$ over a half-period. This is the reason of having the limits of integration in Eq. (2.38), Eq. (2.39) and Eq. (2.41) between 0 and $T_0/2$.

Also, for a periodic signal $x(t)$, if $x(t - T_0/2) = -x(t)$, the signal is said to have half-wave symmetry. It may be noted that for a signal with half-wave symmetry, all cosine and sine terms with even 'n' vanish. For detailed analysis refer Appendix-A.

2.6 Harmonic form Fourier series (Compact form)

Another form of the Fourier series representation of a real periodic signal $x(t)$ with fundamental period T_0 is

$$x(t) = c_0 + \sum_{n=1}^{\infty} c_n \cos[n\omega_0 t + \theta_n] \tag{2.42}$$

Eq. (2.42) can be derived from Eq. (2.30) (Refer Example 2.35) and is known as harmonic form Fourier series of $x(t)$. The term c_0 is known as *dc* component and the term $c_n \cos[n\omega_0 t + \theta_n]$ is referred to as n^{th} harmonic component of $x(t)$. The first harmonic component ($n = 1$) is known as the fundamental component since it has the same fundamental period as $x(t)$. The coefficients c_n and the angles θ_n are called harmonic amplitudes and phase angles respectively. c_n and θ_n are related to a_n and b_n as

$$c_0 = a_0$$

$$c_n = \sqrt{a_n^2 + b_n^2}$$

$$\theta_n = \begin{cases} -\tan^{-1}\left(\frac{b_n}{a_n}\right); & a_n > 0 \\ -\pi + \tan^{-1}\left(\frac{b_n}{a_n}\right); & a_n < 0 \end{cases}$$

2.7 Convergence of Fourier series

A sufficient condition, perhaps the most famous regarding Fourier series, was proved in two memoirs published in 1830 and 1837 by a German mathematician, Dirichlet.

A periodic signal $x(t)$ has a Fourier series representation if it satisfies the following Dirichlet conditions:

1. $x(t)$ is absolutely integrable over any period, that is,

$$\int_{<T_0>} |x(t)|\, dt < \infty$$

   where $\int_{<T_0>}$ denotes integration over any one period and the range 0 to T_0 or $-\frac{T_0}{2}$ to $+\frac{T_0}{2}$ is commonly used for integration.

2. $x(t)$ has a finite number of maxima and minima within any finite interval of t.

3. $x(t)$ has a finite number of discontinuities within any finite interval of t, and each of these discontinuties are finite.

2.8 Amplitude and Phase spectra

A periodic time function $x(t)$ is defined by its Fourier coefficients and its period.

In other words, when we know a_0, a_n, b_n and T_0, we can construct $x(t)$, at least theoretically. When we know a_n and b_n, we also know the amplitude (c_n) and phase angle (θ_n) of each harmonic. We can present graphically the description of a periodic function in terms of the amplitude and phase angle of each term in the harmonic form Fourier Series. The plot of amplitude (c_n) of each term versus the frequency is called amplitude spectrum of $x(t)$, and the plot of phase angle (θ_n) versus the frequency is called phase spectrum of $x(t)$. Because the amplitude and phase angle data occur at discrete values of frequency (that is, ω_0, $2\omega_0 \ldots$), these plots are also referred to as line spectra. It may be noted that $n = 1$ corresponds to ω_0, $n = 2$ corresponds to $2\omega_0$ and so on.

Example 2.11. Find the trignometric Fourier series representation for the periodic sawtooth waveform shown in Fig. Ex. 2.11. Also sketch amplitude and phase spectra.

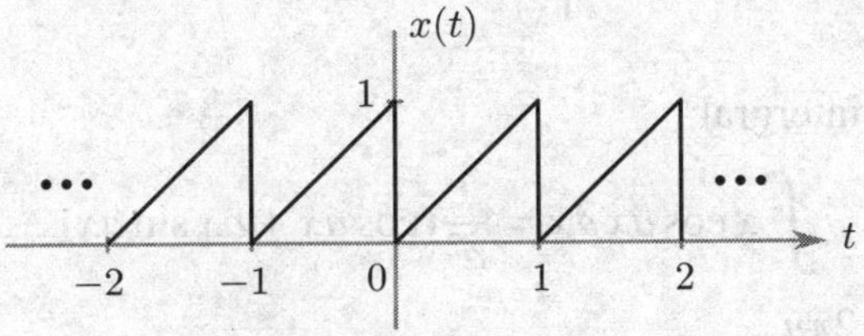

Fig. Ex. 2.11

Solution: From the given waveform, $x(t)$, we find that:

$$T_0 = 1 \text{ sec}$$

$$\text{Hence,} \qquad \omega_0 = \frac{2\pi}{T_0}$$

$$= 2\pi \text{ rad / sec}$$

The mathematical description of the waveform, $x(t)$ over one period is

$$x(t) = t, \quad 0 < t < 1$$

Let us now proceed to find the Fourier coefficients a_0, a_n and b_n:

(i)
$$a_0 = \frac{1}{T_0} \int_{t_0}^{t_0+T_0} x(t)\, dt$$

$$= \frac{1}{T_0} \int_{<T_0>} x(t)\, dt$$

$$\Rightarrow \qquad a_0 = \frac{1}{1}\int_0^1 t\,dt$$

$$= \left(\frac{t^2}{2}\right)_0^1 = \frac{1}{2}$$

(ii)
$$a_n = \frac{2}{T_0}\int_{t_0}^{t_0+T_0} x(t)\cos n\omega_0 t\,dt$$

$$= \frac{2}{T_0}\int_{<T_0>} x(t)\cos n\omega_0 t\,dt$$

$$= \frac{2}{1}\int_0^1 t\cos 2\pi nt\,dt$$

Recall the standard intergral:

$$\int x\cos ax\,dx = \frac{1}{a^2}(\cos ax + ax\sin ax)$$

Here, $x = t$ and $a = 2\pi n$.
Thus,

$$a_n = 2\left[\frac{1}{(2\pi n)^2}(\cos 2\pi nt + 2\pi nt\sin 2\pi nt)\right]_0^1$$

$$= \frac{2}{(2\pi n)^2}[\cos 2\pi n + 2\pi n\sin 2\pi n - 1]$$

Since $\cos 2\pi n = 1$ and $\sin 2\pi n = 0$, we get

$$a_n = \frac{2}{(2\pi n)^2}[1 + 0 - 1] = 0$$

(iii)
$$b_n = \frac{2}{T_0}\int_{t_a}^{t_a+T_0} x(t)\sin n\omega_0 t\,dt$$

$$= \frac{2}{T_0}\int_{<T_0>} x(t)\sin n\omega_0 t\,dt$$

$$= \frac{2}{1}\int_0^1 t\sin 2\pi nt\,dt$$

Recall the standard integral:

$$\int x \sin ax\, dx = \frac{1}{a^2}(\sin ax - ax \cos ax)$$

Here, $x = t$ and $a = 2\pi n$.
Hence,

$$\begin{aligned} b_n &= 2\left[\frac{1}{(2\pi n)^2}(\sin 2\pi nt - 2\pi nt \cos 2\pi nt)\right]_0^1 \\ &= \frac{2}{(2\pi n)^2}[\sin 2\pi n - 2\pi n \cos 2\pi n] \\ &= -\frac{1}{\pi n}; \quad n = 1, 2, 3, 4, \ldots . \end{aligned}$$

Hence, the trignometric Fouries series representation of $x(t)$ is

$$\begin{aligned} x(t) &= a_0 + \sum_{n=1}^{\infty} a_n \cos n\omega_0 t + \sum_{n=1}^{\infty} b_n \sin n\omega_0 t \\ &= \frac{1}{2} + 0 + \sum_{n=1}^{\infty}\left(-\frac{1}{\pi n}\right)\sin 2\pi nt \\ &= \frac{1}{2} - \frac{1}{\pi}\sum_{n=1}^{\infty}\frac{\sin 2\pi nt}{n} \end{aligned}$$

For plotting the amplitude and phase spectra, let us find c_n and θ_n.

$$\begin{aligned} c_0 &= a_0 = \frac{1}{2} \\ c_n &= \sqrt{a_n^2 + b_n^2} \\ &= \sqrt{0^2 + b_n^2} \\ &= \pm b_n = |b_n| = \frac{1}{\pi n}; n = 1, 2, 3 \ldots \end{aligned}$$

and $$\begin{aligned} \theta_n &= -\tan^{-1}\left(\frac{b_n}{a_n}\right) \\ &= -\tan^{-1}\left[\frac{-\frac{1}{\pi n}}{0}\right] = +\tan^{-1}\left[\frac{\frac{1}{\pi n}}{0}\right] \\ &= +\pi/2, \quad n = 1, 2, 3, \ldots . \end{aligned}$$

Since $c_0 = \frac{1}{2}$, it implies $\theta_0 = \angle 1/2 = 0°$. The sketch of amplitude and phase spectra are shown in Fig. Ex. 2.11 (a) and Fig. Ex. 2.11 (b), respectrively.

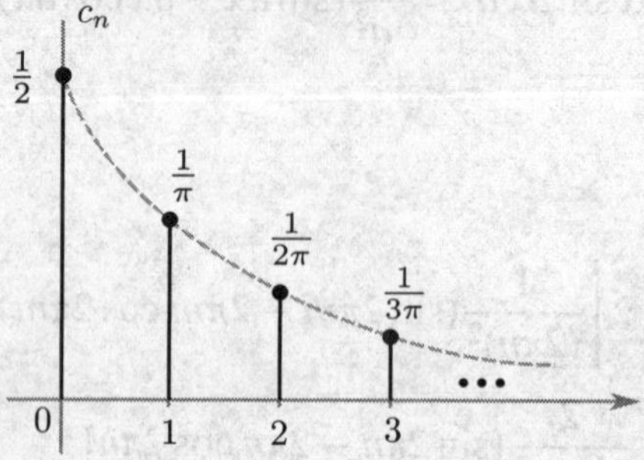

Fig. Ex. 2.11(a): *Amplitude spectrum*

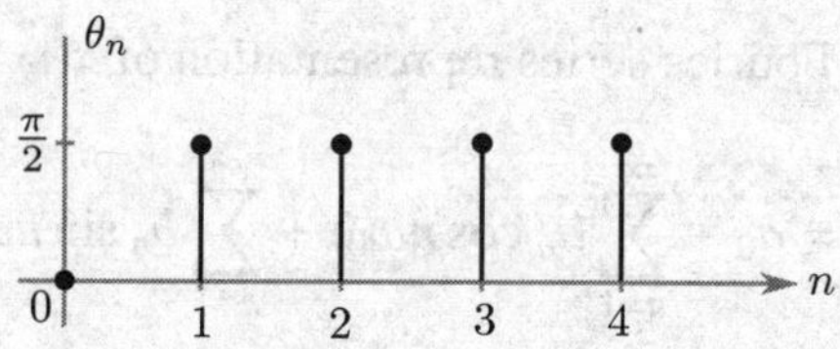

Fig. Ex. 2.11(b): *Phase spectrum*

Example 2.12. Find the trignometric Fourier series representation for the periodic signal $x(t)$ shown in Fig. Ex. 2.12. Also, sketch amplitude and phase spectra.

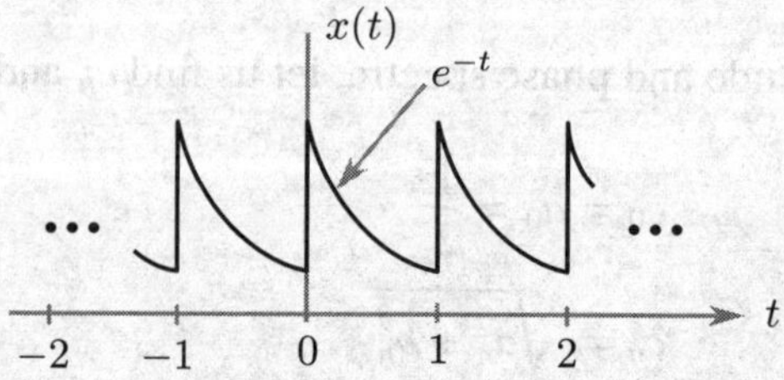

Fig. Ex. 2.12

Solution: From the given waveform, $x(t)$, we find that, $T_0 = 1$ sec.

Hence,

$$\omega_0 = \frac{2\pi}{T_0} = 2\pi \text{ rad / sec}$$

The mathematical description of $x(t)$ over one period is

$$x(t) = e^{-t}, \quad 0 < t < 1$$

The trignometric Fourier series representation for $x(t)$ is

$$x(t) = a_0 + \sum_{n=1}^{\infty} a_n \cos n\omega_0 t + \sum_{n=1}^{\infty} b_n \sin n\omega_0 t$$

Let us now proceed to find the values of a_0, a_n and b_n.

(i)

$$\begin{aligned} a_0 &= \frac{1}{T_0} \int_{<T_0>} x(t)\, dt \\ &= \frac{1}{1} \int_0^1 e^{-t}\, dt \\ &= \left[-e^{-t}\right]_0^1 \\ &= 0.632 \end{aligned}$$

(ii)

$$\begin{aligned} a_n &= \frac{2}{T_0} \int_{<T_0>} x(t) \cos n\omega_0 t\, dt \\ &= \frac{2}{1} \int_0^1 e^{-t} \cos 2\pi n t\, dt \end{aligned}$$

Recall the standard integral:

$$\int e^{ax} \cos bx\, dx = \frac{e^{ax}}{a^2 + b^2} [a \cos bx + b \sin bx]$$

Here, $x = t$, $\quad a = -1$ and $b = 2\pi n$.

Hence,

$$a_n = \left\{ \frac{2e^{-t}}{(-1)^2 + (2\pi n)^2} [(-1) \cos 2\pi n t + 2\pi n \sin 2\pi n t] \right\}_{t=0}^{1}$$

Noting that $\cos 2\pi n = 1$ and $\sin 2\pi n = 0$, we get

$$a_n = \frac{2 \times 0.632}{1 + (2\pi n)^2}; \quad n = 1, 2, 3, \ldots$$

(iii)
$$b_n = \frac{2}{T_0} \int_{<T_0>} x(t) \sin n\omega_0 t \, dt$$
$$= \frac{2}{1} \int_0^1 e^{-t} \sin 2\pi nt \, dt$$

Recall the standard integral:

$$\int e^{ax} \sin bx \, dx = \frac{e^{ax}}{a^2 + b^2} [a \sin bx - b \cos bx]$$

Here, $x = t$, $a = -1$ and $b = 2\pi n$.

Thus,

$$b_n = \left\{ \frac{2e^{-t}}{(-1)^2 + (2\pi n)^2} [(-1) \sin 2\pi nt - 2\pi n \cos 2\pi nt] \right\}_{t=0}^{1}$$
$$= \frac{4\pi n \times 0.632}{1 + (2\pi n)^2}; \quad n = 1, 2, 3, \ldots$$

Hence, the trignometric Fourier series representation for the signal $x(t)$ is

$$x(t) = 0.632 + \sum_{n=1}^{\infty} \left[\frac{2 \times 0.632}{1 + (2\pi n)^2} \cos 2\pi nt + \frac{4\pi n \times 0.632}{1 + (2\pi n)^2} \sin 2\pi nt \right]$$

For plotting the amplitude and phase spectra, we find the Fourier coefficients of the compact form of Fourier series. They are found as follows:

$$c_0 = a_0 = 0.632$$

For a real positive number, the associated angle is zero. Hence, the angle of c_0, that is, $\theta_0 = 0°$.

$$c_n = \sqrt{a_n^2 + b_n^2} = \sqrt{\left[\frac{2 \times 0.632}{1 + (2\pi n)^2} \right]^2 + \left[\frac{4\pi n(0.632)}{1 + (2\pi n)^2} \right]^2}$$
$$= \frac{2 \times 0.632}{1 + (2\pi n)^2} \sqrt{1 + (2\pi n)^2} = \frac{2 \times 0.632}{\sqrt{1 + (2\pi n)^2}}; \quad n = 1, 2, 3, \ldots$$

Since $a_n > 0$, we have

$$\theta_n = -\tan^{-1}\left(\frac{b_n}{a_n}\right) = -\tan^{-1}(2\pi n); \quad n = 1, 2, 3, \ldots$$

The sketch of amplitude and phase spectra are depicted in Fig. Ex. 2.12 (a) and Fig. Ex. 2.12 (b), respectively.

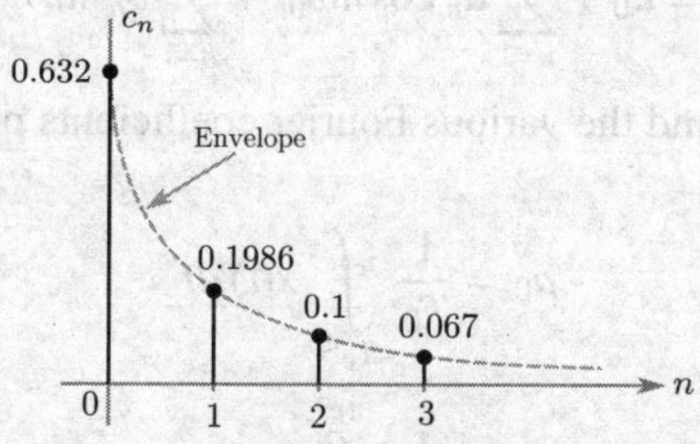

Fig. Ex. 2.12(a)

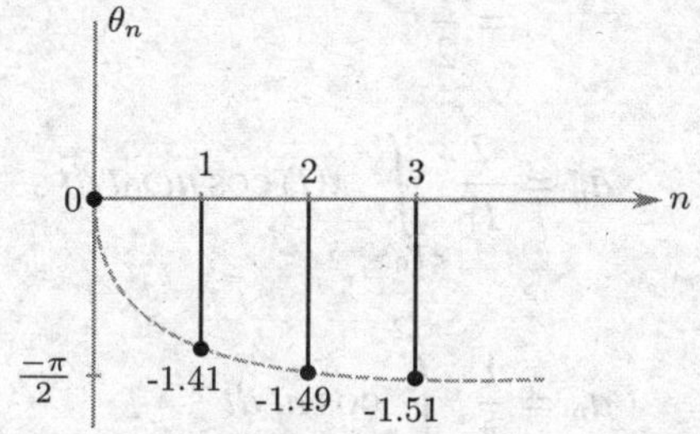

Fig. Ex. 2.12(b)

Example 2.13. Find the trignometric Fourier series in compact form for the periodic waveform shown in Fig. Ex. 2.13. Also sketch amplitude and phase spectra.

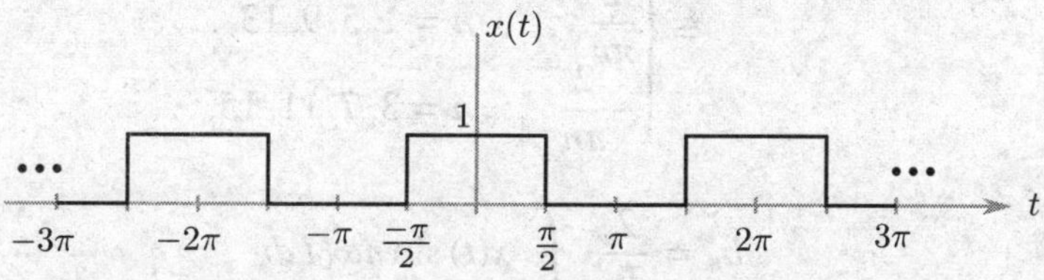

Fig. Ex. 2.13

Solution: From the given waveform, $x(t)$, we find that fundamental period, $T_0 = 2\pi$ secs and hence, $\omega_0 = \dfrac{2\pi}{T_0} = 1$ rad / sec.

The mathematical description of the given waveform over one period: $(-\pi, \pi)$ is

$$x(t) = \begin{cases} 0, & -\pi < t < -\pi/2 \\ 1, & -\pi/2 < t < \pi/2 \\ 0, & \pi/2 < t < \pi \end{cases}$$

The trignometric Fourier series is given by

$$x(t) = a_0 + \sum_{n=1}^{\infty} a_n \cos n\omega_0 t + \sum_{n=1}^{\infty} b_n \sin n\omega_0 t$$

Let us now proceed to find the various Fourier coefficients present in the above Fourier series.

(i)

$$a_0 = \frac{1}{T_0} \int_{<T_0>} x(t)\, dt$$

$$\Rightarrow \quad a_0 = \frac{1}{2\pi} \int_{-\pi/2}^{\pi/2} 1 \cdot dt$$

$$= \frac{1}{2}$$

(ii)

$$a_n = \frac{2}{T_0} \int_{<T_0>} x(t) \cos n\omega_0 t\, dt$$

$$\Rightarrow \quad a_n = \frac{1}{\pi} \int_{-\pi/2}^{\pi/2} \cos nt\, dt$$

$$= \frac{2}{n\pi} \sin\left(\frac{n\pi}{2}\right)$$

$$= \begin{cases} 0 & ;\ n = 2, 4, 6, 8, \ldots \\ \dfrac{2}{\pi n} & ;\ n = 1, 5, 9, 13, \ldots \\ -\dfrac{2}{\pi n} & ;\ n = 3, 7, 11, 15, \ldots \end{cases}$$

(iii)

$$b_n = \frac{2}{T_0} \int_{<T_0>} x(t) \sin n\omega_0 t\, dt$$

$$= \frac{1}{\pi} \int_{-\pi/2}^{\pi/2} \sin nt\, dt$$

$$= 0$$

Hence, the trignometric Fourier series is

$$x(t) = \frac{1}{2} + \frac{2}{\pi}\left[\cos t - \frac{1}{3}\cos 3t + \frac{1}{5}\cos 5t - \frac{1}{7}\cos 7t + \cdots\right]$$

The compact trignometric Fourier series is given by

$$x(t) = c_0 + \sum_{n=1}^{\infty} c_n \cos[n\omega_0 t + \theta_n]$$

where, $$c_0 = a_0$$

$$c_n = \sqrt{a_n^2 + b_n^2}$$

and $$\theta_n = \begin{cases} -\tan^{-1}(b_n/a_n) & ; a_n > 0 \\ -\pi + \tan^{-1}(b_n/a_n) & ; a_n < 0 \end{cases}$$

Hence, for the present problem, we have $c_0 = a_0 = \frac{1}{2}$ and $c_n = \sqrt{a_n^2} = \pm a_n = |a_n|$

$$\Rightarrow \qquad c_n = |a_n| = \begin{cases} 0 & ; n = 2, 4, 6, 8, \ldots \\ \dfrac{2}{\pi n} & ; n = 1, 3, 5, 7, 9, 11, \ldots \end{cases}$$

$$\theta_n = \begin{cases} -\tan^{-1}\left(\frac{b_n}{a_n}\right) & ; a_n > 0 \\ -\pi + \tan^{-1}\left(\frac{b_n}{a_n}\right) & ; a_n < 0 \end{cases}$$

$$= \begin{cases} 0 & ; n \neq 3, 7, 11, 15, \ldots \\ -\pi & ; n = 3, 7, 11, 15, \ldots \end{cases}$$

Since $b_n = 0$, the trignometric Fourier series and compact Fourier series are identical.

The amplitude and phase spectra are as shown in Fig. Ex. 2.13 (a) and Fig. Ex. 2.13(b), respectively.

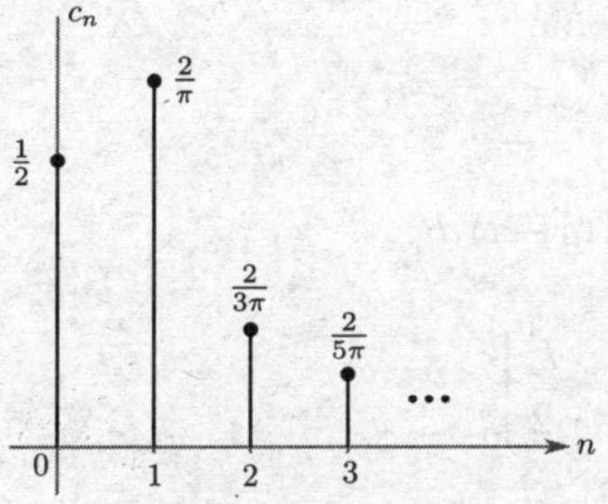

Fig. Ex. 2.13(a)

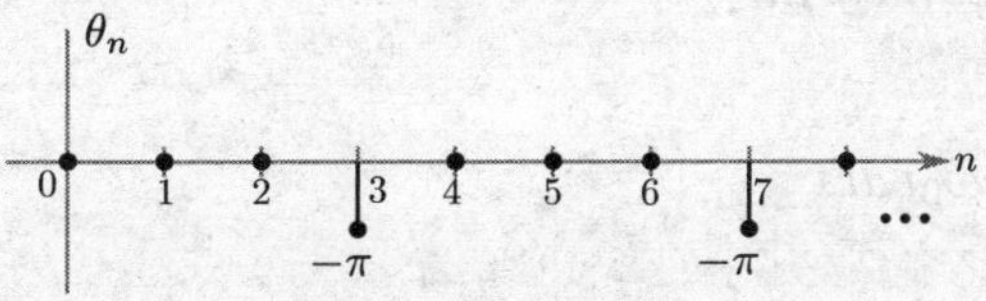

Fig. Ex. 2.13(b)

Example 2.14. Draw the amplitude and phase spectra for the time-domain signal $x(t)$ shown in Fig. Ex. 2.14.

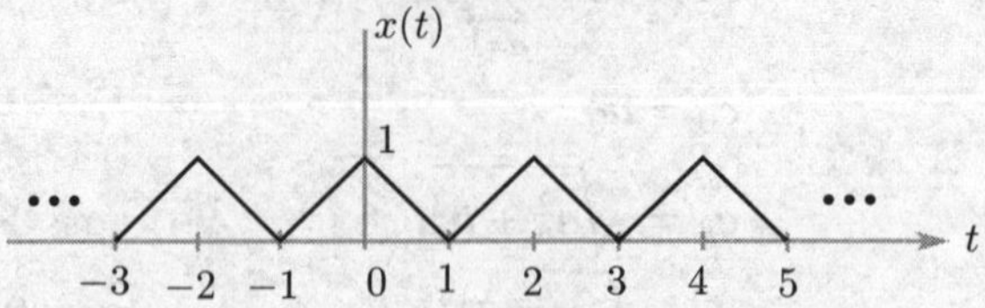

Fig. Ex. 2.14

Solution: We have fundamental period, $T_0 = 2$ sec.

Hence,

$$\omega_0 = \frac{2\pi}{T_0} = \pi \text{ rad / sec}$$

The mathematical discription of $x(t)$ over one period: $(-1, 1)$ is

$$x(t) = \begin{cases} 1 + t, & -1 < t < 0 \\ 1 - t, & 0 < t < 1 \end{cases}$$

The trignometric Fourier series is given by

$$x(t) = a_0 + \sum_{n=1}^{\infty} a_n \cos n\omega_0 t + \sum_{n=1}^{\infty} b_n \sin n\omega_0 t$$

The next step in the analysis is to find the Fourier coefficients a_0, a_n and b_n.

(i) $a_0 = \dfrac{1}{T_0} \displaystyle\int_{<T_0>} x(t)\, dt$

$$= \frac{1}{2}\int_{-1}^{0} (1 + t)\, dt + \frac{1}{2}\int_{0}^{1} (1 - t)\, dt$$

$$= \frac{1}{2}\left[t + \frac{t^2}{2}\right]_{-1}^{0} + \frac{1}{2}\left[t - \frac{t^2}{2}\right]_{0}^{1}$$

$$= \frac{1}{2}$$

(ii) $a_n = \dfrac{2}{T_0} \displaystyle\int_{<T_0>} x(t) \cos n\omega_0 t\, dt$

$$= \frac{2}{2}\int_{-1}^{1} x(t) \cos n\omega_0 t\, dt$$

$$= \int_{-1}^{0} (1 + t) \cos n\pi t\, dt + \int_{0}^{1} (1 - t) \cos n\pi t\, dt$$

Hence,

$$a_n = \int_{-1}^{0} \cos n\pi t\, dt + \int_{-1}^{0} t \cos n\pi t\, dt + \int_{0}^{1} \cos n\pi t\, dt - \int_{0}^{1} t \cos n\pi t\, dt$$

$$= \int_{-1}^{1} \cos n\pi t\, dt + \int_{-1}^{0} t \cos n\pi t\, dt - \int_{0}^{1} t \cos n\pi t\, dt$$

Recall the Standard Integral:

$$\int x \cos ax dx = \frac{1}{a^2}[\cos ax + ax \sin ax]$$

Therefore,

$$a_n = \left[\frac{\sin n\pi t}{n\pi}\right]_{-1}^{1} + \frac{1}{\pi^2 n^2}\Big[\cos n\pi t + n\pi t \sin n\pi t\Big]_{-1}^{0} + \frac{1}{\pi^2 n^2}\Big[\cos n\pi t + n\pi t \sin n\pi t\Big]_{0}^{1}$$

Noting $\cos n\pi = (-1)^n$ and $\sin n\pi = 0$, we get

$$a_n = \frac{2}{\pi^2 n^2}\,[1 - (-1)^n] = \begin{cases} \dfrac{4}{\pi^2 n^2}; & n = 1, 3, 5, 7, \ldots \\ 0 \quad ; & n = 2, 4, \ldots \end{cases}$$

(iii) Since $x(-t) = x(t)$, it implies that $x(t)$ is an even signal. For an even signal, we have $b_n = 0$. Hence, the trignometric Fourier series representation for $x(t)$ is

$$x(t) = \frac{1}{2} + \frac{4}{\pi^2}\left[\cos \pi t + \frac{\cos 3\pi t}{9} + \frac{\cos 5\pi t}{25} + \cdots\right]$$

For plotting amplitude and phase spectra, the compact form is more apt and is given by

$$x(t) = c_0 + \sum_{n=1}^{\infty} c_n \cos[n\omega_0 t + \theta_n]$$

where,

$$c_0 = a_0 = \frac{1}{2}$$

$$c_n = \sqrt{a_n^2 + b_n^2} = |a_n| = \begin{cases} \dfrac{4}{\pi^2 n^2}; & n = 1, 3, 5, 7, \ldots \\ 0 \quad ; & n = 2, 4, 6, 8, \ldots \end{cases}$$

and

$$\theta_n = \begin{cases} -\tan^{-1}\left(\frac{b_n}{a_n}\right) & ; \ a_n > 0 \\ -\pi + \tan^{-1}\left(\frac{b_n}{a_n}\right) & ; \ a_n < 0 \end{cases}$$

In the present problem, $a_n > 0$ and $b_n = 0$. Thus, $\theta_n = 0$. Hence, the phase spectrum is a null spectrum.

The sketch of amplitude spectrum is shown is Fig. Ex. 2.14 (a).

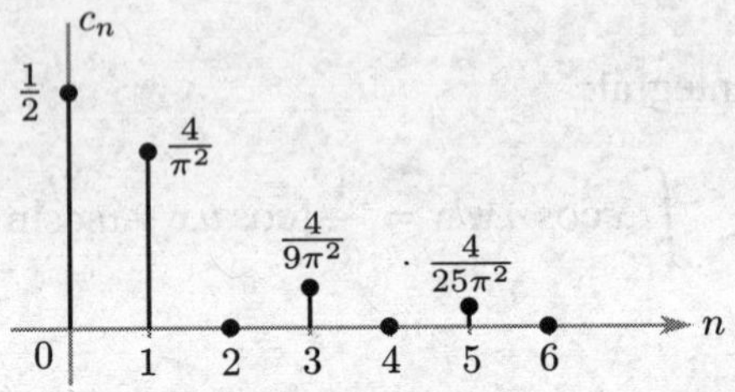

Fig. Ex. 2.14(a)

Example 2.15. Draw the amplitude and phase spectra of the fully rectified sine function shown in Fig. Ex. 2.15. Use compact trignometric Fourier series.

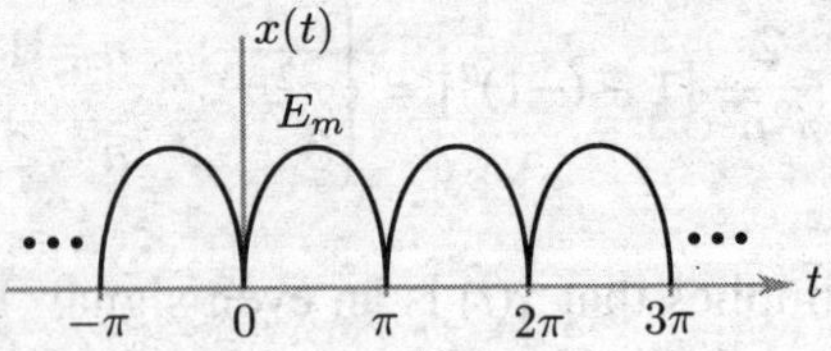

Fig. Ex. 2.15

Solution: We find, fundamental period, $T_0 = \pi$ sec.

Hence, $$\omega_0 = \frac{2\pi}{T_0} = 2 \text{ rad / sec}$$

The description of $x(t)$ over one period: $(0, \pi)$ is found as follows:

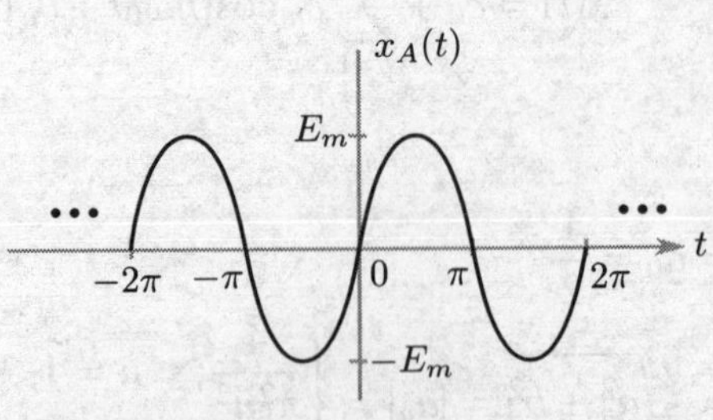

Fig. Ex. 2.15(a)

Consider a periodic signal $x_A(t)$ as shown in Fig. Ex. 2.15 (a). The fundamental period of $x_A(t)$ is $T = 2\pi$ sec. The time-domain discription of $x_A(t)$ is

$$x_A(t) = E_m \sin \omega t$$

$$= E_m \sin \frac{2\pi}{T} t$$

$$= E_m \sin t$$

The waveform $x(t)$ over the first cycle is

$$x(t) = x_A(t), \quad 0 < t < \pi$$

$$= E_m \sin t$$

The trignometric Fourier series representation of $x(t)$ is

$$x(t) = a_0 + \sum_{n=1}^{\infty} a_n \cos n\omega_0 t + \sum_{n=1}^{\infty} b_n \sin n\omega_0 t$$

To find a_0, a_n and b_n:-

(i) $a_0 = \frac{1}{T_0} \int_{<T_0>} x(t)\, dt$

$$= \frac{1}{\pi} \int_0^{\pi} E_m \sin t\, dt$$

$$= \frac{E_m}{\pi} \left[-\cos t \right]_0^{\pi}$$

$$= \frac{2E_m}{\pi}$$

(ii) $a_n = \frac{2}{T_0} \int_{<T_0>} x(t) \cos n\omega_0 t\, dt$

$$= \frac{2}{\pi} \int_0^{\pi} E_m \sin t \cos 2nt\, dt$$

Recall the trignometric identity:

$$\sin A \cos B = \frac{1}{2} \left[\sin(A + B) + \sin(A - B) \right]$$

Hence,

$$a_n = \frac{2E_m}{\pi}\int_0^{\pi}\frac{1}{2}\{\sin(t+2nt)+\sin(t-2nt)\}\,dt$$

$$= \frac{E_m}{\pi}\int_0^{\pi}[\sin(1+2n)t+\sin(1-2n)t]\,dt$$

$$\Rightarrow \quad a_n = \frac{-E_m}{\pi}\left[\frac{\cos(1+2n)t}{1+2n}+\frac{\cos(1-2n)t}{1-2n}\right]_0^{\pi}$$

$$= \frac{-E_m}{\pi}\left[\frac{\cos(1+2n)\pi}{1+2n}+\frac{\cos(1-2n)\pi}{1-2n}-\frac{1}{1+2n}-\frac{1}{1-2n}\right]$$

$$= \frac{-E_m}{\pi}\left[-\frac{1}{1+2n}-\frac{1}{1-2n}-\frac{1}{1+2n}-\frac{1}{1-2n}\right]$$

$$= \frac{E_m}{\pi}\left[\frac{2}{1-2n}+\frac{2}{1+2n}\right]$$

$$= \frac{4E_m}{\pi(1-4n^2)};\ n = 1,2,3\ldots$$

(iii) Since, $x(-t) = -x(t)$, it implies $x(t)$ is even. Hence, $b_n = 0$.

Hence, the trignometric Fourier series is

$$x(t) = \frac{2E_m}{\pi}+\sum_{n=1}^{\infty}\frac{4E_m}{\pi(1-4n^2)}\cos 2nt$$

$$= \frac{2E_m}{\pi}-\frac{4E_m}{\pi}\left[\frac{\cos 2t}{3}+\frac{\cos 4t}{15}+\frac{\cos 6t}{35}+\cdots\right]$$

For sketching amplitude and phase spectra, the compact form of Fourier series is more suitable. The compact form is

$$x(t) = c_0+\sum_{n=1}^{\infty}c_n\cos[n\omega_0 t+\theta_n]$$

where $\quad c_0 = a_0 = \dfrac{2E_m}{\pi}$

$$c_n = \sqrt{a_n^2+b_n^2} = \sqrt{a_n^2} = \pm a_n = |a_n|$$

$$\Rightarrow c_n = \left|\frac{4E_m}{\pi(1-4n^2)}\right|;\ n = 1,2,3\ldots$$

We have,

$$\theta_n = \begin{cases} -\tan^{-1}\left(\frac{b_n}{a_n}\right) & ; \; a_n > 0 \\ -\pi + \tan^{-1}\left(\frac{b_n}{a_n}\right) & ; \; a_n < 0 \end{cases}$$

Thus,

$$\theta_n = \begin{cases} 0 & ; \; n = 0 \\ -\pi & ; \; n \neq 0 \end{cases}$$

The amplitude and phase spectra are as shown in Fig. Ex. 2.15 (b) and Fig. Ex. 2.15(c), respectively.

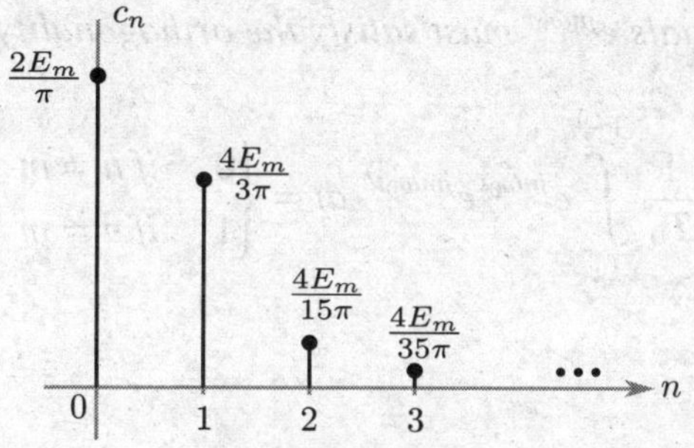

Fig. Ex. 2.15(b)

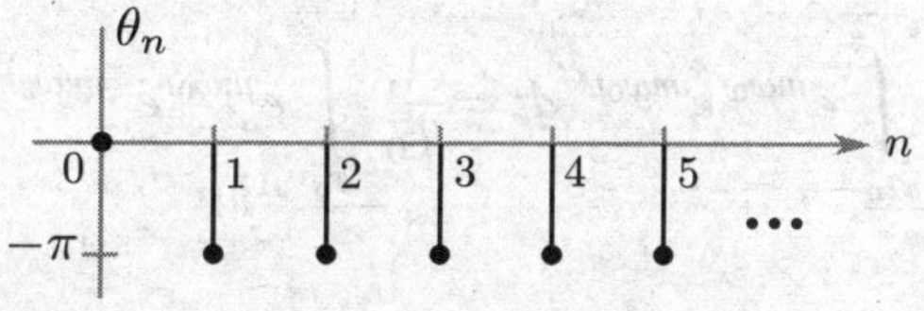

Fig. Ex. 2.15(c)

2.9 Exponential Fourier series Representation of Continuous-Time Periodic Signals

Let $x(t)$ be periodic with a fundamental period, T_0, defined analytically. Then, it can also be represented by the infinite series of complex exponentials, that is,

$$x(t) = \sum_{k=-\infty}^{\infty} X(k) e^{jk\omega_0 t}; \qquad \omega_0 = \frac{2\pi}{T_0} \tag{2.43}$$

where the coefficients $X(k)$ can be found from the analytical definition of $x(t)$ as follows:

$$X(k) = \frac{1}{T_0} \int_{<T_0>} x(t) e^{-jk\omega_0 t} \, dt \tag{2.44}$$

The symbol $\int\limits_{<T_0>}$ means that the integration is performed over one period of $x(t)$.

It should be noted that any physically realizable periodic waveform can be expressed as a sum of complex exponentials. Eq. (2.43) is called the *synthesis equation,* and Eq. (2.44) is called the *analysis equation*. The constants $X(k)$ are called *Complex Fourier Coefficients* and the term $X(0)$ is called the *average value* or *dc value* of the waveform.

A periodic waveform $x(t)$ and its Fourier coefficient $X(k)$ can be represented symbolically as shown below:

$$x(t) \overset{FS}{\longleftrightarrow} X(k)$$

Theorem 2.1. *Orthogonality of the complex exponentials:*

The complex exponentials $e^{jn\omega_0 t}$ *must satisfy the orthogonality condition*

$$\frac{1}{T_0}\int_{-\frac{T_0}{2}}^{+\frac{T_0}{2}} e^{jn\omega_0 t} e^{jm\omega_0 t^\star}\, dt = \begin{cases} 0 & \text{if } n \neq m \\ 1 & \text{if } n = m \end{cases} \tag{2.45}$$

where $T_0 = \dfrac{2\pi}{\omega_0}$.

Please note that $f^\star(t)$ *denotes the complex conjugate of* $f(t)$.

Proof:

$$\frac{1}{T_0}\int_{-\frac{T_0}{2}}^{+\frac{T_0}{2}} e^{jn\omega_0 t} e^{jm\omega_0 t^\star}\, dt = \frac{1}{T_0}\int_{-\frac{T_0}{2}}^{+\frac{T_0}{2}} e^{jn\omega_0 t} e^{-jm\omega_0 t}\, dt$$

$$= \frac{1}{T_0}\int_{-\frac{T_0}{2}}^{+\frac{T_0}{2}} e^{j(n-m)\omega_0 t}\, dt \tag{2.46}$$

when $n \neq m$, let $n - m = p$. Then Eq. (2.46) becomes

$$\frac{1}{T_0}\int_{-\frac{T_0}{2}}^{+\frac{T_0}{2}} e^{jn\omega_0 t} e^{jm\omega_0 t^\star}\, dt = \frac{1}{T_0}\int_{-\frac{T_0}{2}}^{+\frac{T_0}{2}} e^{jp\omega_0 t}\, dt$$

$$= \left.\frac{e^{jp\omega_0 t}}{T_0 jp\omega_0}\right|_{-\frac{T_0}{2}}^{+\frac{T_0}{2}}$$

$$= \frac{e^{jp\pi} - e^{-jp\pi}}{T_0 jp\omega_0} = 0$$

On the other hand when $n = m$, then Eq. (2.46) becomes

$$\frac{1}{T_0}\int_{-\frac{T_0}{2}}^{+\frac{T_0}{2}} e^{jn\omega_0 t} e^{jm\omega_0 t^\star}\, dt = \frac{1}{T_0}\int_{-\frac{T_0}{2}}^{+\frac{T_0}{2}} e^0\, dt = 1$$

Note that in the above derivation, we have used the symmetric interval $-\dfrac{T_0}{2}$ to $+\dfrac{T_0}{2}$ over which the integration is carried out. There is nothing special about the interval and the above theorem is true as long as the integration is done over one complete interval of length T_0. Hence, the theorem may be stated in a generalized manner as follows:

$$\frac{1}{T_0}\int_{<T_0>} e^{jn\omega_0 t} e^{jm\omega_0 t^\star}\, dt = \begin{cases} 0 & \text{if } n \neq m \\ 1 & \text{if } n = m \end{cases} \tag{2.47}$$

2.10 Derivation of Complex Coefficients of Exponential Fourier Series

Let us make use of the orthogonality property as follows. Let $x(t)$ be a periodic waveform with a fundamental period T_0, and assume that it can be expressed as an infinite sum of complex exponentials, that is,

$$x(t) = \sum_{k=-\infty}^{\infty} X(k) e^{jk\omega_0 t} \tag{2.48}$$

where the values of $X(k)$ are yet to be found. Multiplying both the sides of Eq. (2.48) by $e^{-jm\omega_0 t}$ and integrating, we get

$$\frac{1}{T_0}\int_{<T_0>} x(t)e^{-jm\omega_0 t}\, dt = \frac{1}{T_0}\int_{<T_0>} \sum_{k=-\infty}^{\infty} X(k) e^{jk\omega_0 t} e^{-jm\omega_0 t}\, dt \tag{2.49}$$

Interchanging the orders of summation and integration on the right-side of Eq. (2.49), we get

$$\frac{1}{T_0}\int_{<T_0>} x(t)e^{-jm\omega_0 t}\, dt = \sum_{k=-\infty}^{\infty} X(k) \left[\frac{1}{T_0}\int_{<T_0>} e^{jk\omega_0 t} e^{-jm\omega_0 t}\, dt\right] \tag{2.50}$$

Then by Theorem (2.1), the quantity in square brackets is equal to zero as long as $k \neq m$. However, when $k = m$, it is equal to 1, and hence Eq. (2.50) reduces down to a single term, $X(m)$. We have thus shown that

$$\frac{1}{T_0}\int_{<T_0>} x(t)e^{-jm\omega_0 t} dt = X(m)$$

Replacing m by k, we obtain

$$X(k) = \frac{1}{T_0} \int_{<T_0>} x(t)e^{-jk\omega_0 t}\, dt \tag{2.51}$$

which is the required formula for finding $X(k)$ in Eq. (2.48).

Example 2.16. Obtain the exponential **FS** representation for the waveform $x(t)$ shown in Fig. Ex. 2.16.

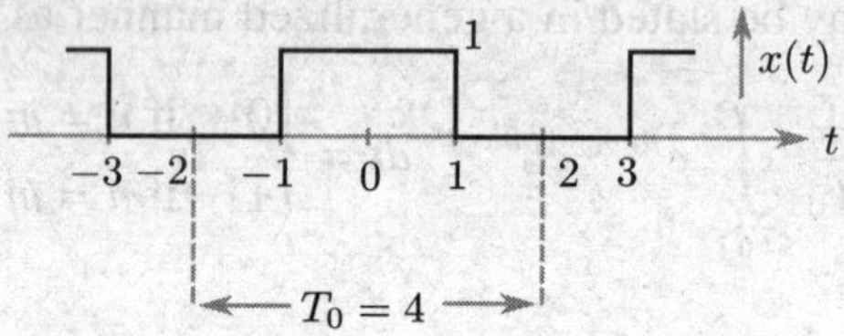

Fig. Ex. 2.16

Solution: The analytical definition of the waveform over one period is as follows:

$$x(t) = \begin{cases} 0 & \text{for } -2 < t < -1 \\ 1 & \text{for } -1 < t < 1 \\ 0 & \text{for } 1 < t < 2 \end{cases}$$

$$X(k) \triangleq \frac{1}{T_0} \int_{<T_0>} x(t)e^{-jk\omega_0 t}\, dt \quad \text{where } \omega_0 = \frac{2\pi}{T_0} = \frac{2\pi}{4} = \frac{\pi}{2}$$

$$= \frac{1}{4} \int_{-1}^{1} x(t)e^{-jk\omega_0 t}\, dt$$

$$= \frac{1}{4} \int_{-1}^{1} 1 \times e^{-jk\frac{\pi}{2}t}\, dt$$

$$= \frac{1}{4} \left[\frac{e^{-jk\pi\frac{t}{2}}}{-jk\frac{\pi}{2}} \right]_{-1}^{1}$$

$$= \frac{1}{4} \left[\frac{e^{jk\frac{\pi}{2}} - e^{-jk\frac{\pi}{2}}}{+j(k\frac{\pi}{2})} \right]$$

$$= \frac{1}{2} \frac{\sin(k\frac{\pi}{2})}{(k\frac{\pi}{2})}$$

Table 2.1. Fourier coefficients of Example 2.16

k	0	± 1	± 2	± 3	± 4	± 5
$X(k)$	$\frac{1}{2}$	$\frac{1}{\pi}$	0	$-\frac{1}{3\pi}$	0	$\frac{1}{5\pi}$

The exponential **FS** representation for the function $x(t)$ is given by

$$\begin{aligned} x(t) &= \sum_{k=-\infty}^{\infty} X(k)e^{jk\omega_0 t} \\ &= \sum_{k=-\infty}^{\infty} X(k)e^{jk\frac{\pi}{2}t} \\ &= \frac{1}{\pi}\left[\cdots + \frac{1}{5}e^{-j5\frac{\pi}{2}t} - \frac{1}{3}e^{-j3\frac{\pi}{2}t} + \frac{1}{1}e^{-j\pi\frac{t}{2}} \right. \\ &\quad \left. + \frac{\pi}{2} + \frac{1}{1}e^{j\frac{\pi}{2}t} - \frac{1}{3}e^{j3\frac{\pi}{2}t} + \frac{1}{5}e^{j5\frac{\pi}{2}t} + \cdots\right] \end{aligned}$$

From the above exponential **FS**, we can precisely see what our waveform is comprised of. The average or dc value of $x(t)$ is $X(0) = \dfrac{1}{2}$. The first few nonzero harmonics are

$$\begin{aligned} h(1) &= X(1)e^{j\omega_0 t} + X(-1)e^{-j\omega_0 t} \\ &= \frac{1}{\pi}e^{j\frac{\pi}{2}t} + \frac{1}{\pi}e^{-j\frac{\pi}{2}t} \\ &= \frac{2}{\pi}\cos\left(\frac{\pi}{2}t\right) \\ h(3) &= -\frac{1}{3\pi}e^{j3\frac{\pi}{2}t} - \frac{1}{3\pi}e^{-j3\frac{\pi}{2}t} \\ &= -\frac{2}{3\pi}\cos\left(3\frac{\pi}{2}t\right) \end{aligned}$$

and so on.

Theorem 2.2. *Let*

$$x(t) \overset{\mathbf{FS}}{\longleftrightarrow} X(k).$$

Then

$$X^*(k) = X(-k)$$

for all k, if and only if $x(t)$ is a real function of t.

Proof: Let $x(t)$ be real. Then,

$$\begin{aligned} X^*(k) &= \left[\frac{1}{T_0}\int_{<T_0>} x(t)e^{-jk\omega_0 t}\,dt\right]^* \\ &= \frac{1}{T_0}\int_{<T_0>} x^*(t)e^{jk\omega_0 t}\,dt \\ &= \frac{1}{T_0}\int_{<T_0>} x(t)e^{jk\omega_0 t}\,dt = X(-k) \end{aligned} \tag{2.52}$$

Hence, the theorem.

Corollary: *Let*

$$X(k) = A(k) + jB(k) \tag{2.53}$$

Then,

$$\begin{aligned} X(-k) &= X^*(k) \\ &= A(k) - jB(k) \end{aligned} \tag{2.54}$$

From Eqs. (2.53) and (2.54), we find that $|X(k)| = |X(-k)|$. Hence, the plot of $|X(k)|$ versus k, known as the amplitude spectrum exhibits even symmetry with respect to vertical axis. Also, we find from Eqs. (2.53) and (2.54) that $\measuredangle X(k) = -\measuredangle X(-k)$, implying that the plot of $\measuredangle X(k)$ versus k known as the phase spectrum exhibits odd symmetry with respect to vertical axis.

Theorem 2.3. *Let $x(t)$ be a real periodic function defined by $x(t) = x_{ev}(t) + x_{od}(t)$. Also, the Fourier coefficient of $x(t)$ denoted by $X(k)$ is complex and is defined by $X(k) = A(k) + jB(k)$.*

Then,

$$x_{ev}(t) \overset{\textbf{FS}}{\longleftrightarrow} A(k)$$

$$x_{od}(t) \overset{\textbf{FS}}{\longleftrightarrow} jB(k)$$

Proof:

$$x_{ev}(t) = \frac{1}{2}[x(t) + x(-t)]$$

Replacing $x(t)$ and $x(-t)$ by their respective synthesis equations, we get

$$x_{ev}(t) = \frac{1}{2}\sum_{k=-\infty}^{\infty} X(k)e^{jk\omega_0 t} + \frac{1}{2}\sum_{k=-\infty}^{\infty} X(k)e^{-jk\omega_0 t}$$

$$= \sum_{k=-\infty}^{\infty} X(k) \times \frac{1}{2}\left[e^{jk\omega_0 t} + e^{-jk\omega_0 t}\right]$$

$$= \sum_{k=-\infty}^{\infty} X(k)\cos(k\omega_0 t) \tag{2.55}$$

Substituting $X(k) = A(k) + jB(k)$ in Eq. (2.55), we get

$$x_{ev}(t) = \sum_{k=-\infty}^{\infty} A(k)\cos(k\omega_0 t) + j\sum_{k=-\infty}^{\infty} B(k)\cos(k\omega_0 t) \tag{2.56}$$

We know that,

$$X(k) = A(k) + jB(k) \tag{2.57}$$

$$\Rightarrow X^*(k) = A(k) - jB(k) \tag{2.58}$$

and

$$X(-k) = A(-k) + jB(-k) \tag{2.59}$$

Since, $X^*(k) = X(-k)$, we get

$$A(k) = A(-k)$$

and

$$B(k) = -B(-k)$$

which shows that $A(k)$ is even and $B(k)$ is odd.

Thus, $B(k)$ has been shown to be an odd function of k, whereas $\cos(k\omega_0 t)$ is an even function. Thus, their product is an odd function and hence the second term on the RHS of Eq. (2.56) must be zero. Hence, Eq. (2.56) reduces to

$$x_{ev}(t) = \sum_{k=-\infty}^{\infty} A(k)\cos(k\omega_0 t) + 0 \tag{2.60}$$

The zero on the RHS of Eq. (2.60) is replaced by an odd function,

$$\sum_{k=-\infty}^{\infty} jA(k)\sin(k\omega_0 t)$$

As a consequence of this, Eq. (2.60) becomes

$$x_{ev}(t) = \sum_{k=-\infty}^{\infty} A(k)[\cos(k\omega_0 t) + j\sin(k\omega_0 t)] \tag{2.61}$$

$$\Rightarrow \quad x_{ev}(t) = \sum_{k=-\infty}^{\infty} A(k)e^{jk\omega_0 t} \tag{2.62}$$

That is,

$$x_{ev}(t) \overset{\textbf{FS}}{\longleftrightarrow} A(k)$$

A similar argument shows that $x_{od}(t) \overset{\textbf{FS}}{\longleftrightarrow} jB(k)$, and hence the proof.

Corollary:

(a) *$X(k)$ is real and even, if $x(t)$ is real and even.*

(b) *$X(k)$ is purely imaginary and odd, if $x(t)$ is real and odd.*

Example 2.17. Find the Complex Fourier Coefficents for the function $x(t)$ shown in Fig. Ex. 2.17.

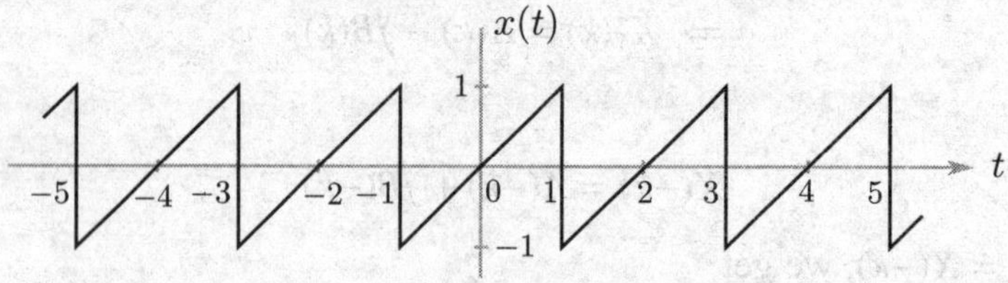

Fig. Ex. 2.17

Solution: We make the following predictions:

(i) The waveform $x(t)$ is real and hence, $X(-k) = X^*(k)$.

(ii) The waveform $x(t)$ is odd and hence $X(k)$ is purely imaginary; that is, $X(k) = jB(k)$ and $A(k) = 0$.

(iii) The imaginary part of $X(k)$, that is $B(k)$ is odd.

The analytic definition of $x(t)$ over one period is as follows:

$$x(t) = t \text{ for } -1 < t < 1 \quad \text{and} \quad x(t+2) = x(t)$$

Since, $T_0 = 2$,

$$\omega_0 = \frac{2\pi}{T_0} = \pi$$

Then,

$$X(k) = \frac{1}{T_0} \int_{<T_0>} x(t) e^{-jk\omega_0 t}\, dt$$

$$= \frac{1}{2} \int_{-1}^{1} t[\cos(k\pi t) - j\sin(k\pi t)]\, dt \tag{2.63}$$

Note that, $t\cos(k\pi t)$ is odd, $t\sin(k\pi t)$ is even and the limits of integration are symmetric. The implications of these, reduce Eq. (2.63) to

$$X(k) = \frac{2}{2}\int_{0}^{1} -jt\sin(k\pi t)\,dt$$

$$= -j\left[-t\frac{\cos(k\pi t)}{k\pi} + \frac{\sin(k\pi t)}{k^2\pi^2}\right]_0^1$$

$$= j\frac{\cos(k\pi)}{k\pi} = j\frac{(-1)^k}{k\pi}(k \neq 0) \tag{2.64}$$

when $k = 0$, Eq. (2.63) reduces to

$$X(k) = \frac{1}{2}\int_{-1}^{1} t\,dt = 0$$

Thus, $X(0) = 0$.

Therefore,

$$X(k) = \begin{cases} j\frac{(-1)^k}{k\pi}, & k \neq 0 \\ 0, & k = 0 \end{cases}$$

Table 2.2

k	$X(k)$	$\theta(k) = \angle X(k)$ **Radians**
− 2	$\frac{-j}{2\pi}$	$-\frac{\pi}{2}$
− 1	$\frac{j}{\pi}$	$\frac{\pi}{2}$
0	0	0
1	$\frac{-j}{\pi}$	$-\frac{\pi}{2}$
2	$\frac{j}{2\pi}$	$+\frac{\pi}{2}$

From the above Table 2.2, we infer the following:

(i) $X(-k) = X^*(k)$.

(ii) $X(k)$ is purely imaginary, that is, $X(k) = jB(k) = \frac{j(-1)^k}{k\pi}$.

(iii) $A(k) =$ Real part of $X(k) = 0$

$B(k) =$ Imaginary part of $X(k)$

$$= \frac{(-1)^k}{k\pi} \Rightarrow B(k) = -B(-k)$$

Thus, $B(k)$ is odd.

(iv) $\angle X(k) = \theta(k)$ exhibits odd symmetry with respect to vertical axis.

2.11 Convergence of Exponential Fourier Series

It is known that a periodic signal $x(t)$ has a Fourier series representation if it satisfies the following Dirichlet conditions:

Let $x(t)$ be a single-valued periodic function. Then,

1. Over any period $[t_0, t_0 + T]$, $x(t)$ must have the property that $\int_{t_0}^{t_0+T} |x(t)|dt < \infty$. In the language of mathematics, this means that $x(t)$ is absolutely integrable. The consequence of this property is that $X(k)$ is finite.

2. Over any period of the signal, there must be only a finite number of maxima and minima. In other words, functions like $\sin\left(\frac{1}{t}\right)$ are excluded.

3. Over any period, $x(t)$ must have only a finite number of discontinuities.

Any function of time that appears in physical systems will satisfy these conditions. Note that Dirichlet conditions are sufficient but not necessary conditions for a periodic signal to have Fourier series representation.

The Gibbs phenomenon: Immediately on each side of a discontinuity, the Fourier series shows a certain behaviour known as *Gibbs phenomenon.*

In practice, we always use the truncated version of the infinite Fourier series. As a result of this, if we plot the values of the function $x(t)$ using truncated version, we find an oscillatory behaviour immediately on each side of a discontinuity.

However, as we extend the truncation point of the Fourier series and take more and more terms into account, the oscillations move closer and closer to the discontinuity, and we can always find a truncation point for which they are insignificant, no matter how closely we position ourselves to it. (The overshoot is about 9% of the value of the waveform and becomes narrower and narrower as we increase the terms in the truncated Fourier series).

2.12 Amplitude and Phase Spectra of a Periodic Signal

The Complex Fourier Coefficient is given by

$$X(k) = A(k) + jB(k)$$

The magnitude of $X(k)$ is given by

$$|X(k)| = \sqrt{A^2(k) + B^2(k)}$$

and the phase of $X(k)$ is given by

$$\theta(k) = \angle X(k) = \tan^{-1}\left[\frac{B(k)}{A(k)}\right]$$

A plot of $|X(k)|$ versus k is called *amplitude spectrum* of the periodic signal $x(t)$ and the plot of $\theta(k)$ versus k is called the *phase spectrum* of $x(t)$. Since the index k assumes only integers, the amplitude and phase spectra are not continuous curves but appear only at discrete values of k. They are therefore referred to as discrete frequency spectra or line spectra.

For a real periodic signal, $x(t)$, $X(-k) = X^*(k)$, which implies $|X(-k)| = |X(k)|$ and $\theta(-k) = -\theta(k)$. Hence as discussed in theorem 2.2, the amplitude spectrum is an even function of k and the phase spectrum is an odd function of k for a real periodic signal.

2.13 Properties of Fourier Series

In this section, we will discuss some of the important properties of Fourier series.

2.13.1 Linearity

If

$$x(t) \overset{\textbf{FS}}{\longleftrightarrow} X(k)$$

and

$$y(t) \overset{\textbf{FS}}{\longleftrightarrow} Y(k)$$

then,

$$z(t) = ax(t) + by(t) \overset{\textbf{FS}}{\longleftrightarrow} Z(k) = aX(k) + bY(k)$$

In this case, both $x(t)$ and $y(t)$ are assumed to have same fundamental period, $T_0 = \dfrac{2\pi}{\omega_0}$.

Proof:

$$Z(k) \triangleq \frac{1}{T_0} \int_{<T_0>} z(t) e^{-jk\omega_0 t}\, dt$$

$$= \frac{1}{T_0} \int_{<T_0>} [ax(t) + by(t)] e^{-jk\omega_0 t}\, dt$$

$$= a \frac{1}{T_0} \int_{<T_0>} x(t) e^{-jk\omega_0 t}\, dt + b \frac{1}{T_0} \int_{<T_0>} y(t) e^{-jk\omega_0 t}\, dt$$

$$\Rightarrow \qquad Z(k) = aX(k) + bY(k)$$

Hence, the proof.

2.13.2 Translation or time shift

If

$$x(t) \overset{\text{FS}}{\longleftrightarrow} X(k)$$

then,

$$z(t) = x(t - t_0) \overset{\text{FS}}{\longleftrightarrow} Z(k) = e^{-jk\omega_0 t_0} X(k)$$

Proof:

$$Z(k) \triangleq \frac{1}{T_0} \int_{<T_0>} z(t) e^{-jk\omega_0 t}\, dt$$

$$= \frac{1}{T_0} \int_{<T_0>} x(t - t_0) e^{-jk\omega_0 t}\, dt$$

Putting $t - t_0 = \lambda$, we get,

$$Z(k) = \frac{1}{T_0} \int_{<T_0>} x(\lambda) e^{-jk\omega_0 (t_0 + \lambda)}\, d\lambda$$

$$= e^{-jk\omega_0 t_0} \frac{1}{T_0} \int_{<T_0>} x(\lambda) e^{-jk\omega_0 \lambda}\, d\lambda$$

$$= e^{-jk\omega_0 t_0} X(k)$$

Therefore, $\quad Z(k) = e^{-jk\omega_0 t_0} X(k)$

Hence, the proof.

2.13.3 Frequency shift

If

$$x(t) \overset{\textbf{FS}}{\longleftrightarrow} X(k)$$

then,

$$z(t) = e^{jk_0\omega_0 t}x(t) \overset{\textbf{FS}}{\longleftrightarrow} Z(k) = X(k - k_0)$$

Proof:

$$Z(k) \overset{\Delta}{=} \frac{1}{T_0}\int_{<T_0>} z(t)e^{-jk\omega_0 t}\,dt$$

$$= \frac{1}{T_0}\int_{<T_0>} e^{jk_0\omega_0 t}x(t)e^{-jk\omega_0 t}\,dt$$

$$= \frac{1}{T_0}\int_{<T_0>} x(t)e^{-j(k-k_0)\omega_0 t}\,dt$$

$$= X(k - k_0)$$

Hence, the proof.

2.13.4 Scaling

If

$$x(t) \overset{\textbf{FS}}{\longleftrightarrow} X(k)$$

then,

$$z(t) = x(at) \overset{\textbf{FS}}{\longleftrightarrow} Z(k) = X(k); \quad a > 0$$

Proof: It $x(t)$ is periodic, then $z(t) = x(at)$ is also periodic. If $x(t)$ has a fundamental period T_0, then $z(t) = x(at)$ has the fundamental period $\frac{T_0}{a}$. Consequently, if $x(t)$ has a fundamental frequency equal to ω_0, then the fundamental frequency of $z(t)$ is $a\omega_0$.

We have,

$$X(k) = \frac{1}{T_0}\int_{<T_0>} x(t)e^{-jk\omega_0 t}\,dt$$

Therefore,

$$Z(k) = \frac{1}{\left(\frac{T_0}{a}\right)} \int_{\left\langle \frac{T_0}{a} \right\rangle} z(t) e^{-jka\omega_0 t}\, dt$$

$$= \frac{a}{T_0} \int_{\left\langle \frac{T_0}{a} \right\rangle} x(at) e^{-jka\omega_0 t}\, dt$$

Put $at = \lambda$, then $dt = \frac{1}{a}\, d\lambda$.

Therefore,

$$Z(k) = \frac{a}{T_0} \int_{<T_0>} x(\lambda) e^{-jk\omega_0 \lambda} \frac{1}{a}\, d\lambda$$

$$= X(k)$$

Hence, the proof.

Therefore, the **FS** coefficients of $x(t)$ and $x(at)$ are identical. However, the harmonic spacing changes from ω_0 to $a\omega_0$.

2.13.5 Time-differentiation

If

$$x(t) \overset{\mathbf{FS}}{\longleftrightarrow} X(k)$$

then,

$$\frac{dx(t)}{dt} \overset{\mathbf{FS}}{\longleftrightarrow} jk\omega_0 X(k)$$

Proof: We know that,

$$x(t) = \sum_{k=-\infty}^{\infty} X(k) e^{jk\omega_0 t} \tag{2.65}$$

Differentiating with respect to t, we get,

$$\frac{dx(t)}{dt} = \sum_{k=-\infty}^{\infty} X(k)\, jk\omega_0 e^{jk\omega_0 t}$$

$$\Rightarrow \quad \frac{dx(t)}{dt} = \sum_{k=-\infty}^{\infty} [jk\omega_0 X(k)] e^{jk\omega_0 t} \tag{2.66}$$

Comparing Eq. (2.66) with Eq. (2.65), we get,

$$\frac{dx(t)}{dt} \overset{\textbf{FS}}{\longleftrightarrow} jk\omega_0 X(k)$$

Hence, the proof.

2.13.6 Time-domain convolution

If

$$x(t) \overset{\textbf{FS}}{\longleftrightarrow} X(k)$$

and

$$y(t) \overset{\textbf{FS}}{\longleftrightarrow} Y(k)$$

then,

$$z(t) = x(t) \circledast y(t) \overset{\textbf{FS}}{\longleftrightarrow} z(k) = T_0 X(k) Y(k)$$

where $T_0 = \dfrac{2\pi}{\omega_0}$.

Hence, $\circledast$ denotes periodic convolution.

Proof: Convolution is the process of overlapping two signals. The convolution of two time functions $f_1(t)$ and $f_2(t)$ is denoted as $f_1(t) * f_2(t)$ and by definition,

$$f_1(t) * f_2(t) = \int_{-\infty}^{\infty} f_1(\tau) f_2(t-\tau)\, d\tau$$

where τ is a dummy variable. When the limits of integration are over one period, then convolution is known as periodic convolution.

Students are advised to read this property after studying convolution, which is discussed in detail in chapter 5.

It is assumed that $x(t)$, $y(t)$ and $z(t)$ have the same fundamental period $T_0 = \dfrac{2\pi}{\omega_0}$.

$$Z(k) \triangleq \frac{1}{T_0} \int_{t=<T_0>} z(t) e^{-jk\omega_0 t}\, dt$$

$$= \frac{1}{T_0} \int_{t=<T_0>} [x(t) \circledast y(t)] e^{-jk\omega_0 t}\, dt \tag{2.67}$$

From the definition of periodic convolution, we have

$$x(t) \circledast y(t) = \int_{l=<T_0>} x(l)y(t-l)\,dl \tag{2.68}$$

Substituting Eq. (2.68) in Eq. (2.67), we get

$$Z(k) = \frac{1}{T_0} \int_{t=<T_0>} \left[\int_{l=<T_0>} x(l)y(t-l)\,dl \right] e^{-jk\omega_0 t}\,dt \tag{2.69}$$

Changing the order of integrations on the RHS of Eq. (2.69), we get

$$Z(k) = \frac{1}{T_0} \int_{l=<T_0>} x(l) \int_{t=<T_0>} y(t-l)e^{-jk\omega_0 t}\,dt\,dl$$

Put $t - l = m \Rightarrow dt = dm$.

Then,

$$Z(k) = \frac{1}{T_0} \left[\int_{l=<T_0>} x(l) \int_{m=<T_0>} y(m)e^{-jk\omega_0 (m+l)}\,dm\,dl \right]$$

$$= \frac{1}{T_0} \left[\int_{l=<T_0>} x(l)e^{-jk\omega_0 l}dl \int_{m=<T_0>} y(m)e^{-jk\omega_0 m}\,dm \right]$$

$$= \frac{1}{T_0}[T_0X(k)T_0Y(k)]$$

$$\Rightarrow \quad Z(k) = T_0X(k)Y(k)$$

Hence, the proof.

2.13.7 Modulation or multiplication theorem

If

$$x(t) \overset{\textbf{FS}}{\longleftrightarrow} X(k)$$

and

$$y(t) \overset{\textbf{FS}}{\longleftrightarrow} Y(k)$$

then,

$$z(t) = x(t)y(t) \overset{\textbf{FS}}{\longleftrightarrow} Z(k) = X(k) * Y(k)$$

Proof:

$$Z(k) \triangleq \frac{1}{T_0} \int_{<T_0>} z(t)e^{-jk\omega_0 t}\, dt$$

$$= \frac{1}{T_0} \int_{<T_0>} x(t)y(t)e^{-jk\omega_0 t}\, dt \tag{2.70}$$

We have the synthesis equation,

$$x(t) = \sum_{l=-\infty}^{\infty} X(l)e^{jl\omega_0 t} \tag{2.71}$$

Substituting Eq. (2.71) in Eq. (2.70), we get

$$Z(k) = \frac{1}{T_0} \int_{<T_0>} \left[\sum_{l=-\infty}^{\infty} X(l)e^{jl\omega_0 t} \right] y(t)e^{-jk\omega_0 t}\, dt \tag{2.72}$$

Changing the orders of summation and integration, we get

$$Z(k) = \frac{1}{T_0} \left[\sum_{l=-\infty}^{\infty} X(l) \int_{<T_0>} y(t)e^{-j(k-l)\omega_0 t}\, dt \right]$$

$$= \sum_{l=-\infty}^{\infty} X(l)Y(k-l)$$

$$\Rightarrow \qquad Z(k) = X(k) * Y(k)$$

Hence, the proof.

Notes and comments

1. The *sa* function is defined as follows:

$$sa(x) = \frac{\sin x}{x}$$

"*sa*" stands for "sine over argument" and is often pronounced as "sah". sa(x) is sketched in Fig. 3.4. and is seen to have a maximum value of unity at $x = 0$ and approaches zero in an oscillatory manner as x goes from $-\infty$ to ∞. Since, $\sin x = 0$ for $x = \pm n\pi$, sa(x) crosses the x-axis whenever x is an integral multiple of π.

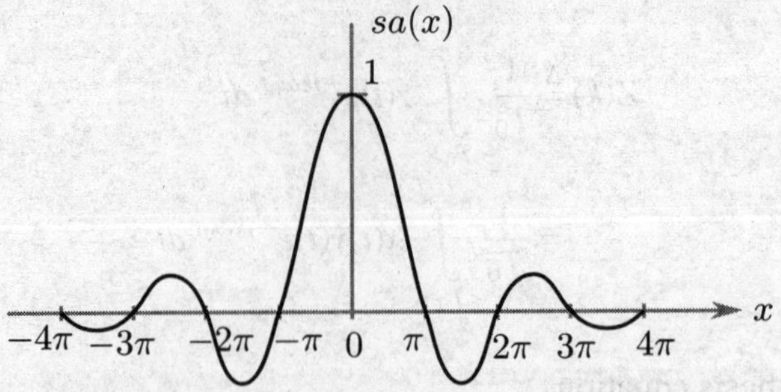

Fig. 2.4 *The sa function.*

2. The sinc function is defined as follows:

$$\text{sinc}(x) = \frac{\sin(\pi x)}{\pi x}$$

Thus, $\text{sinc}(x) = \text{sa}(\pi x)$ and $\text{sa}(x) = \text{sinc}\left(\frac{x}{\pi}\right)$.

Example 2.18. Find the Complex Fourier Coefficient for $x(t)$ given below:

$$x(t) = \cos\left(\frac{2\pi}{3}t\right) + 2\cos\left(\frac{5\pi}{3}t\right)$$

Solution: Given

$$x(t) = \cos\left(\frac{2\pi}{3}t\right) + 2\cos\left(\frac{5\pi}{3}t\right) \tag{2.73}$$

$$\Rightarrow \quad x(t) = \frac{1}{2}e^{j\frac{2\pi}{3}t} + \frac{1}{2}e^{-j\frac{2\pi}{3}t} + e^{j\frac{5\pi}{3}t} + e^{-j\frac{5\pi}{3}t} \tag{2.74}$$

From Eq. (2.73), we find that,

$$\omega_{0_1} = \frac{2\pi}{3} \quad \text{and} \quad \omega_{0_2} = \frac{5\pi}{3}$$

Therefore,

$$\omega_0 = GCD[\omega_{0_1}, \omega_{0_2}] = \frac{\pi}{3} \Rightarrow T_0 = 6$$

Substituting $\frac{\pi}{3}$ as ω_0 in Eq. (2.74), we get

$$x(t) = \frac{1}{2}e^{j2\omega_0 t} + \frac{1}{2}e^{-j2\omega_0 t} + e^{j5\omega_0 t} + e^{-j5\omega_0 t} \tag{2.75}$$

Comparing Eq. (2.75) with

$$x(t) = \sum_{k=-\infty}^{\infty} X(k)e^{jk\omega_0 t}$$

we get $X(-5) = 1, \quad X(-2) = \frac{1}{2}, \quad X(2) = \frac{1}{2}, \quad X(5) = 1.$

Example 2.19. Find the Complex Fourier Coefficient for $x(t)$ given below:

$$x(t) = \cos(2t) + \sin(\pi t)$$

Solution: $x(t)$ is not periodic. Hence, no Fourier series, no ω_0 and no T_0.

Example 2.20. Find the Complex Fourier Coefficient for $x(t)$ given below:

$$x(t) = \frac{1}{4} + \sum_{n=1}^{4}\left[\frac{1}{n^2}\cos\left(\frac{nt}{5}\right) + \frac{(-1)^n}{(2n+1)^2}\sin\left(\frac{nt}{5}\right)\right]$$

Solution:

$$\omega_0 = \frac{1}{5}, \quad T_0 = \frac{2\pi}{\omega_0} = 10\pi$$

$$x(t) = \frac{1}{4} + \sum_{n=1}^{4}\left[\frac{1}{2n^2}e^{j\frac{nt}{5}} + \frac{1}{2n^2}e^{-j\frac{nt}{5}}\right]$$

$$+ \sum_{n=1}^{4}\left[\frac{(-1)^n}{2j(2n+1)^2}e^{j\frac{nt}{5}} - \frac{(-1)^n}{2j(2n+1)^2}e^{-j\frac{nt}{5}}\right]$$

$$= \frac{1}{4} + \sum_{n=1}^{4}\left[\left(\frac{1}{2n^2} - \frac{j(-1)^n}{2(2n+1)^2}\right)e^{j\frac{nt}{5}}\right]$$

$$+ \sum_{n=1}^{4}\left[\left(\frac{1}{2n^2} + \frac{j(-1)^n}{2(2n+1)^2}e^{-j\frac{nt}{5}}\right)\right]$$

Replacing n by k and $\frac{1}{5}$ by ω_0, we get

$$x(t) = \frac{1}{4} + \sum_{k=1}^{4}\left[\left(\frac{1}{2k^2} - \frac{j(-1)^k}{2(2k+1)^2}\right)e^{jk\omega_0 t}\right]$$

$$+ \sum_{k=1}^{4}\left[\left(\frac{1}{2k^2} + \frac{j(-1)^k}{2(2k+1)^2}\right)e^{-jk\omega_0 t}\right]$$

Therefore,

$$X(k) = \frac{1}{2k^2} - \frac{j(-1)^k}{2(2k+1)^2}$$

$$X(-k) = \frac{1}{2k^2} + \frac{j(-1)^k}{2(2k+1)^2}$$

where $k = 1, 2, 3, 4$ and $X(0) = \dfrac{1}{4}$.

Example 2.21. Find the Complex Fourier Coefficient for the periodic waveform $x(t)$ shown in Fig. Ex. 2.21. Also, draw the amplitude and phase spectra.

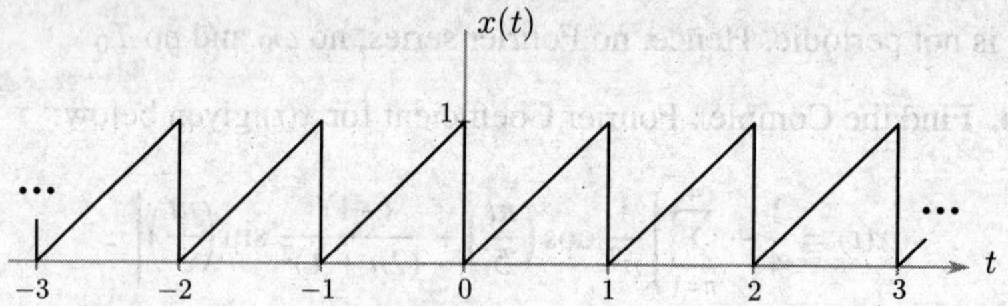

Fig. Ex. 2.21

Solution:

$$T_0 = 1 \text{ sec}, \quad \omega_0 = \frac{2\pi}{T_0} = \frac{2\pi}{1} = 2\pi$$

The analytical expression for $x(t)$ over one period is

$$x(t) = t, \quad 0 < t < 1$$

$$X(k) \triangleq \frac{1}{T_0} \int_{<T_0>} x(t) e^{-jk\omega_0 t}\, dt$$

$$= \frac{1}{1} \int_0^1 t e^{-j2\pi kt}\, dt \tag{2.76}$$

We know that,

$$\int t e^{at}\, dt = \frac{e^{at}}{a^2}[at - 1] \tag{2.77}$$

Therefore,

$$X(k) = \left[\frac{e^{-j2\pi kt}}{(-j2\pi k)^2}(-j2\pi kt - 1) \right]_{t=0}^{1}$$

$$= -j\frac{2\pi k}{j^2(2\pi k)^2} = \frac{j}{2\pi k}, \quad k \neq 0$$

Putting $k = 0$ in Eq. (2.76), we get

$$X(0) = \int_0^1 t\,dt = \frac{1}{2}$$

Summarizing,

$$X(k) = \begin{cases} \frac{j}{2\pi k} & ; \; k \neq 0 \\ \frac{1}{2} & ; \; k = 0 \end{cases}$$

$$\Rightarrow \qquad |X(k)| = \begin{cases} \frac{1}{|2\pi k|} & ; \; k \neq 0 \\ \frac{1}{2} & ; \; k = 0 \end{cases}$$

and

$$\measuredangle X(k) = \theta(k) = \begin{cases} \frac{\pi}{2} & ; \; k > 0 \\ -\frac{\pi}{2} & ; \; k < 0 \\ 0 & ; \; k = 0 \end{cases}$$

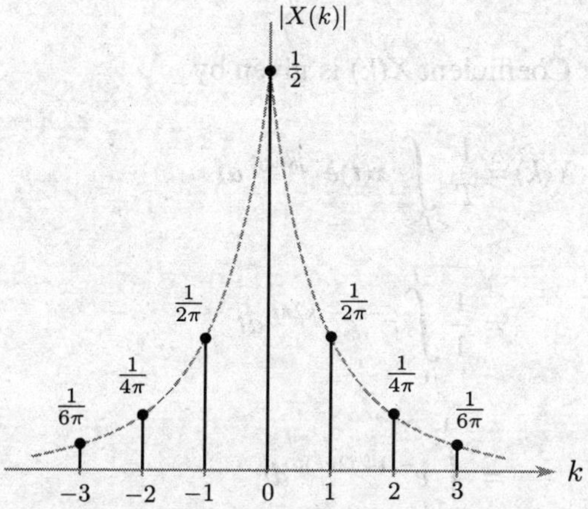

Fig. Ex. 2.21(a): *Amplitude spectrum.*

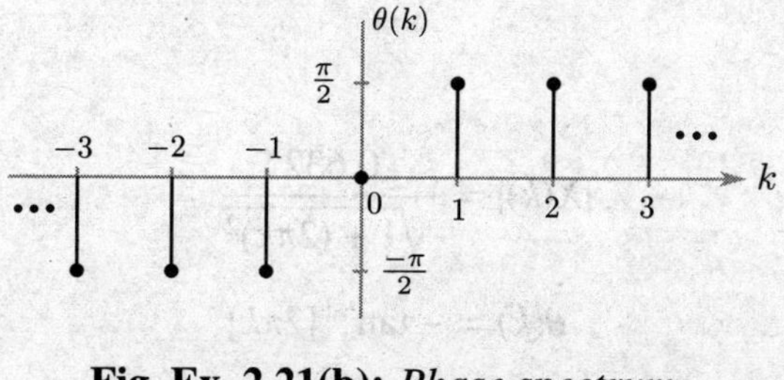

Fig. Ex. 2.21(b): *Phase spectrum.*

Example 2.22. Find the Complex Fourier Coefficient $X(k)$ for $x(t)$ given below. Also draw the amplitude and phase spectra.

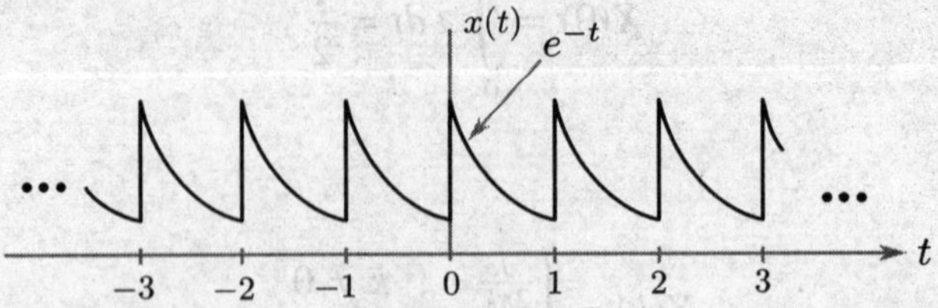

Fig. Ex. 2.22

Solution:

$$T_0 = 1 \text{ sec}, \quad \omega_0 = \frac{2\pi}{T_0} = 2\pi$$

The analytical expression for $x(t)$ over one period is

$$x(t) = e^{-t}, \quad 0 < t < 1$$

The Complex Fourier Coefficient $X(k)$ is given by

$$X(k) \triangleq \frac{1}{T} \int_{<T>} x(t)e^{-jk\omega_0 t}\, dt$$

$$= \frac{1}{1} \int_0^1 e^{-t} e^{-jk2\pi t}\, dt$$

$$= \int_0^1 e^{-(1+j2\pi k)t}\, dt$$

$$= \frac{1}{(1 + j2\pi k)}[1 - e^{-1}] = \frac{0.632}{1 + j2\pi k}$$

Therefore,

$$|X(k)| = \frac{0.632}{\sqrt{1 + (2\pi k)^2}}$$

$$\theta(k) = -\tan^{-1}[2\pi k]$$

Table 2.3

k	$\|X(k)\|$	$\theta(k)$ Rad
− 3	0.03	+ 1.51
− 2	0.05	+ 1.49
− 1	0.099	+ 1.14
0	0.632	0
1	0.099	− 1.14
2	0.05	− 1.49
3	0.03	− 1.51

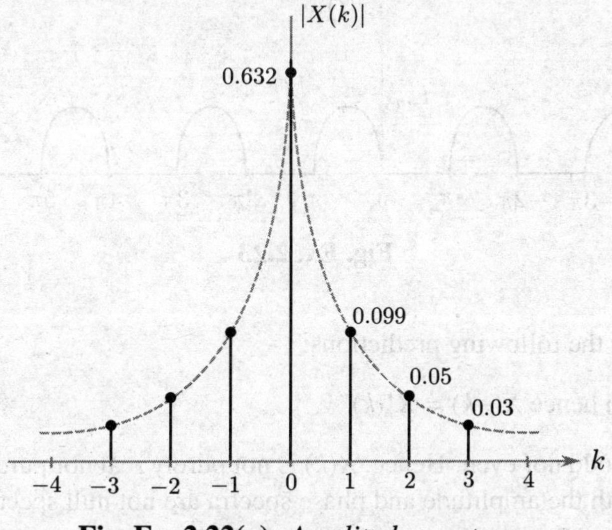

Fig. Ex. 2.22(a): *Amplitude spectrum.*

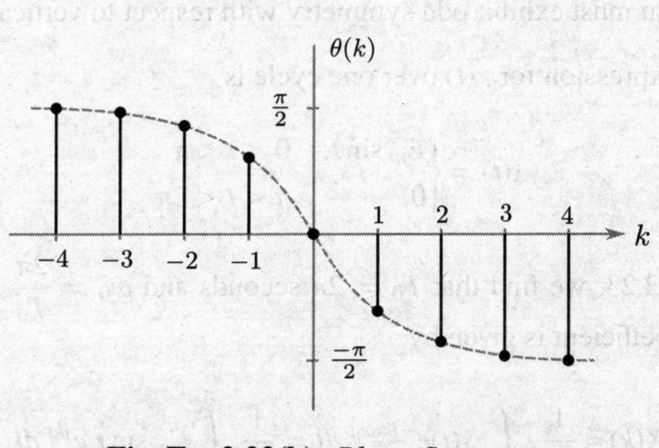

Fig. Ex. 2.22(b): *Phase Spectrum.*

Points to be remembered:

(i) $x(t)$ is neither odd nor even. Hence, $X(k)$ is complex, that is, it has a real part and an imaginary part.

(ii) $x(t)$ is real. Hence, $X(-k) = X^*(k)$.

(iii) The amplitude spectrum exhibits even symmetry with respect to vertical axis and the phase spectrum exhibits odd symmetry with respect to vertical axis. This is true for any real $x(t)$.

Example 2.23. Find the Complex Fourier Coefficient $X(k)$ for the periodic waveform shown in Fig. Ex. 2.23.

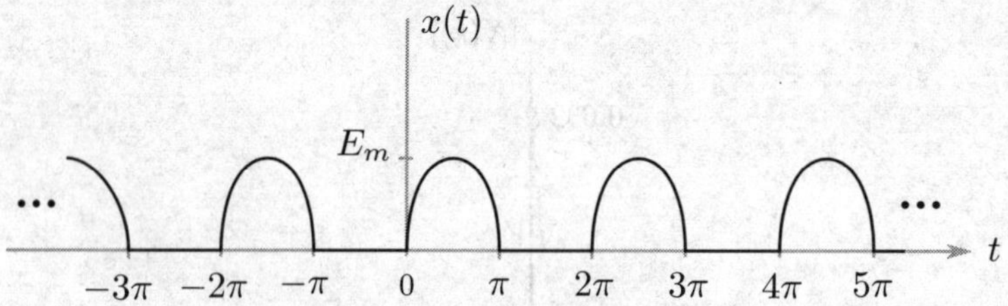

Fig. Ex. 2.23

Solution: Check for the following predictions:

(i) $x(t)$ is real and hence $X(-k) = X^*(k)$.

(ii) $x(t)$ is neither odd nor even. Hence, $X(k)$ is not purely real nor purely imaginary. This means that both the amplitude and phase spectra are not null spectra.

(iii) The amplitude spectrum must exhibit even symmetry with respect to vertical axis and phase spectrum must exhibit odd symmetry with respect to vertical axis.

The analytical expression for $x(t)$ over one cycle is

$$x(t) = \begin{cases} E_m \sin t, & 0 < t < \pi \\ 0, & \pi < t < 2\pi \end{cases}$$

From Fig. Ex. 2.23, we find that $T_0 = 2\pi$ seconds and $\omega_0 = \frac{2\pi}{T_0} = 1$ rad / sec. The Complex Fourier Coefficient is given by

$$X(k) \triangleq \frac{1}{T_0} \int_{<T_0>} x(t) e^{-jk\omega_0 t}\, dt = \frac{1}{2\pi} \int_0^{\pi} E_m \sin t\, e^{jkt}\, dt \tag{2.78}$$

We know that,

$$\int e^{ax}\sin(bx+c)\,dx = \frac{e^{ax}}{a^2+b^2}[a\sin(bx+c) - b\cos(bx+c)] \tag{2.79}$$

In the present context,

$$a = -jk, \quad b = 1, \quad c = 0, \quad x = t$$

Therefore,

$$X(k) = \frac{E_m}{2\pi}\left[\frac{e^{-jkt}}{(-jk)^2+1}(-jk\sin t - \cos t)\right]_{t=0}^{\pi} \tag{2.80}$$

$$= \frac{E_m}{2\pi(1-k^2)}[(-1)^k + 1], \quad k \neq \pm 1$$

Please note that,

$$e^{\pm j\pi k} = (-1)^k$$

and

$$e^{\pm j2\pi k} = 1$$

To find $X(\pm 1)$, proceed as follows:
Using Euler's formula for $\sin t$ and letting $k = 1$, in Eq. (2.78), we get

$$X(1) = \frac{E_m}{2\pi}\int_0^{\pi}\left[\frac{e^{jt} - e^{-jt}}{2j}\right]e^{-jt}\,dt$$

$$= \frac{E_m}{4\pi j}\int_0^{\pi}(1 - e^{-j2t})\,dt$$

$$= \frac{E_m}{4\pi j} \times \pi = \frac{E_m}{4j}$$

Similarly, by putting $k = -1$ in Eq. (2.78) and proceeding in a similar manner as above, we get

$$X(-1) = \frac{-E_m}{4j}$$

Please note that $X(-1) = X^*(1)$.

Summarizing the results, we have

$$X(k) = \begin{cases} 0, & k = \pm 3, \quad \pm 5, \quad \pm 7, \quad \pm 9, \ldots \\ \dfrac{E_m}{\pi(1-k^2)}, & k = 0, \quad \pm 2, \quad \pm 4, \quad \pm 6, \ldots \\ \dfrac{E_m}{4j}, & k = 1 \\ \dfrac{-E_m}{4j}, & k = -1 \end{cases}$$

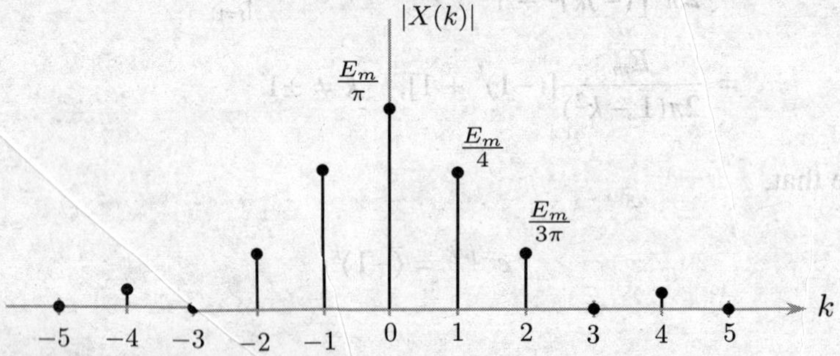

Fig. Ex. 2.23(a): *Amplitude spectrum.*

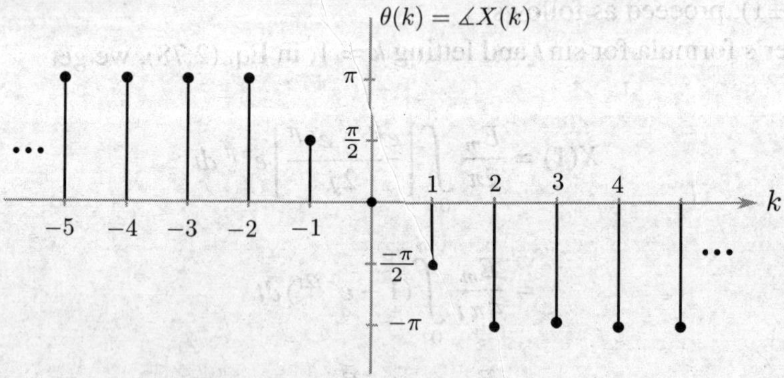

Fig. Ex. 2.23(b): *Phase spectrum.*

Following tips are useful for drawing the phase spectrum.
Let A be some arbitrary constant.

(i) $$-A = A\angle -\pi \quad \text{for} \quad k > 0$$

and $$-A = A\angle +\pi \quad \text{for} \quad k < 0$$

(ii) $$jA = A\angle + \frac{\pi}{2}$$

and $$-jA = A\angle - \frac{\pi}{2}$$

Example 2.24. What is the time-domain representation for the periodic function $x(t)$ that has the following Fourier Coefficient? State if $x(t)$ is real or not and give its average value. Assume $T_0 = 2$ secs.

$$X(k) = \begin{cases} \frac{1}{2}, & k = 0 \\ \frac{\cos(k\pi) - 1}{k^2\pi^2}, & k \neq 0 \end{cases}$$

Solution:

$$x(t) = \sum_{k=-\infty}^{\infty} X(k)e^{jk\omega_0 t}$$

where,

$$\omega_0 = \frac{2\pi}{T_0} = \frac{2\pi}{2} = \pi$$

Therefore

$$x(t) = \sum_{k=-\infty}^{\infty} X(k)e^{jk\pi t}$$

Substituting the given vlaue of $X(k)$, we get

$$x(t) = \frac{1}{2} + \sum_{\substack{k=-\infty \\ k\neq 0}}^{\infty} \left[\frac{\cos(k\pi) - 1}{k^2\pi^2}\right] e^{jk\pi t}$$

Average value $= \frac{1}{2}$.

To test if $x(t)$ is real or not, check whether $X^*(k) = X(-k)$?

Given,

$$X(k) = \begin{cases} \frac{1}{2}, & k = 0 \\ \frac{\cos k\pi - 1}{k^2\pi^2}, & k \neq 0 \quad T = 2 \end{cases}$$

$X^*(k)$ is same as $X(k)$

$$X(-k) = \begin{cases} \frac{1}{2}, & k = 0 \\ \frac{\cos k\pi - 1}{k^2\pi^2}, & k \neq 0 \end{cases}$$

which is also same as $X(k)$. Thus, $X^*(k) = X(-k) \Rightarrow x(t)$ is real.

Example 2.25. Find the Complex Fourier Coefficient $X(k)$ for the periodic waveform shown in Fig. Ex. 2.25.

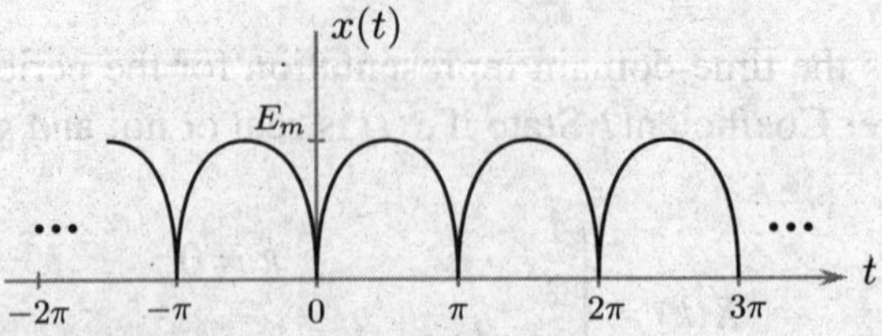

Fig. Ex. 2.25

Solution: We predict the following:

(i) $x(t)$ is real. Hence, $X(-k) = X^*(k)$.

(ii) $x(t)$ real and is an even function of t. Hence, $X(k)$ is purely real and is an even function of k. The implication of this is that the phase spectrum will be either a null spectrum or a spectrum with samples having amplitude $-\pi$ for $k > 0$ and amplitude $+\pi$ for $k < 0$.

(iii) The magnitude spectrum will exhibit an even symmetry with respect to vertical axis and phase spectrum will exhibit an odd symmetry with respect to vertical axis.

From Fig. Ex. 2.25, we find that $T_0 = \pi$ and hence,

$$\omega_0 = \frac{2\pi}{T_0} = 2$$

The analytical expression for $x(t)$ over one period is as follows:

$$x(t) = E_m \sin t, \quad 0 < t < \pi$$

The Complex Fourier Coefficient is given by

$$X(k) = \frac{1}{T_0} \int_{<T_0>} x(t) e^{-jk\omega_0 t}\, dt = \frac{1}{\pi} \int_0^{\pi} E_m \sin t e^{-j2kt}\, dt \tag{2.81}$$

Using Eq. (2.79), we get

$$X(k) = \frac{E_m}{\pi} \left[\frac{e^{-2jkt}}{(-j2k)^2 + 1} (-j2k \sin t - \cos t) \right]_0^{\pi}$$

$$= \frac{2E_m}{\pi(1 - 4k^2)}$$

For all value of $k (k \neq 0)$, $X(k)$ is negative and real.

Hence,

$$\measuredangle X(k) = \theta(k) = \begin{cases} -\pi & \text{for} \quad k > 0 \\ 0 & \text{for} \quad k = 0 \\ +\pi & \text{for} \quad k < 0 \end{cases}$$

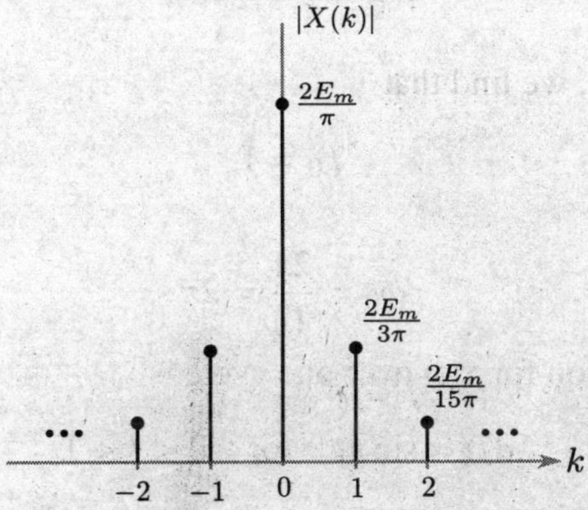

Fig. Ex. 2.25(a): *Amplitude spectrum.*

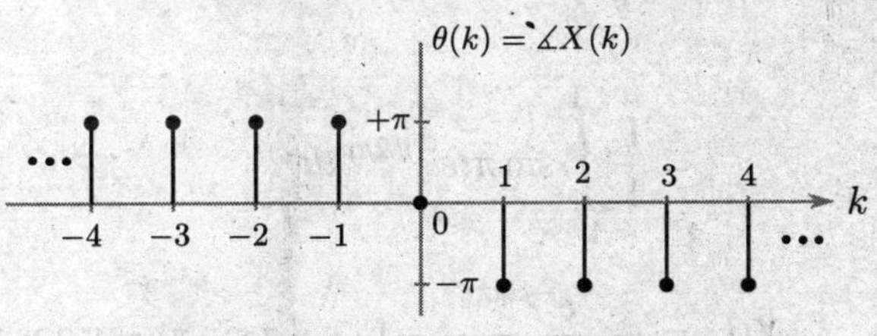

Fig. Ex. 2.25(b): *Phase spectrum.*

Example 2.26. Find the Complex Fourier Coefficient $X(k)$ for $x(t) = |\sin \pi t|$.

Solution:

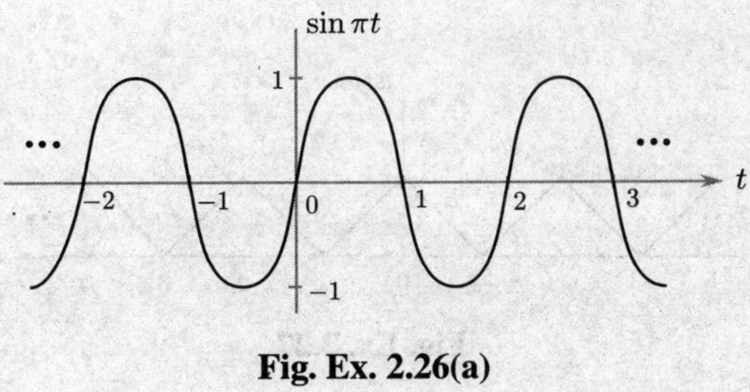

Fig. Ex. 2.26(a)

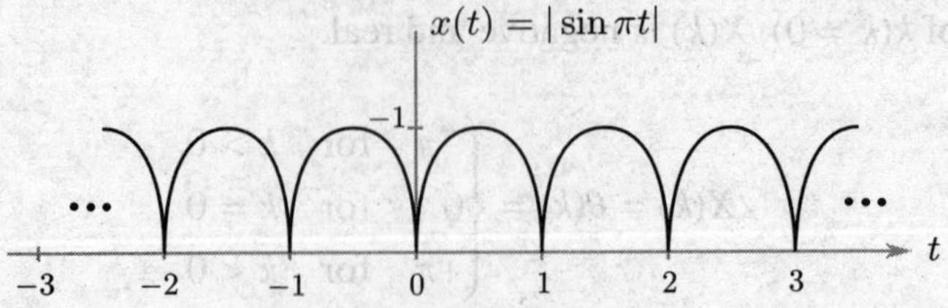

Fig. Ex. 2.26(b)

From Fig. Ex. 2.26 (b), we find that

$$T_0 = 1$$

and

$$\omega_0 = \frac{2\pi}{T_0} = 2\pi$$

The analytical expression for $x(t)$ over one cycle is

$$x(t) = \sin \pi t \quad \text{for} \quad 0 < t < 1$$

The Complex Fourier Coefficient $X(k)$ is given by

$$X(k) \triangleq \frac{1}{T_0} \int\limits_{<T_0>} x(t) e^{-jk\omega_0 t} \, dt$$

$$= \frac{1}{1} \int_1^1 \sin \pi t e^{-jk2\pi t} \, dt$$

$$\Rightarrow \qquad X(k) = \frac{e^{-j2k\pi t}}{(-jk2\pi)^2 + \pi^2} [-2jk\pi \sin \pi t - \pi \cos \pi t]_0^1$$

$$= \frac{2}{\pi(1 - 4k^2)}$$

Example 2.27. Find the Complex Fourier Coefficient $X(k)$ for the waveform $x(t)$ shown in Fig. Ex. 2.27.

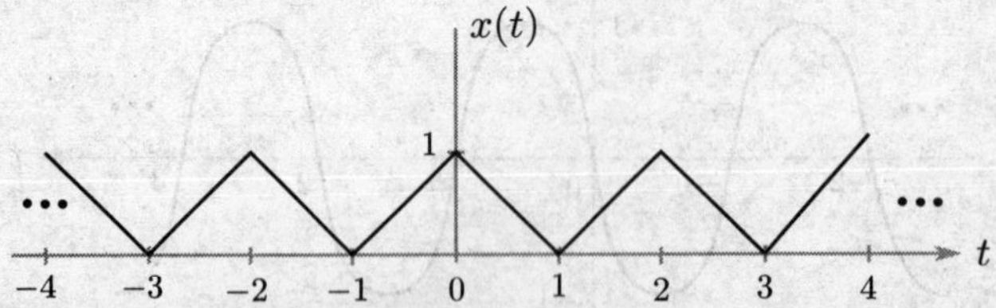

Fig. Ex. 2.27

Solution: We predict the following:

(i) $x(t)$ is real. Hence, $X(-k) = X^*(k)$.

(ii) $x(t)$ is an even function of t. Hence, $X(k)$ is purely real and is an even function of k.

(iii) The magnitude spectrum exhibits even symmetry and phase spectrum exhibits odd symmetry with respect to vertical axis.

(iv) Since, $X(k)$ is purely real, the phase spectrum could be a null spectrum or a spectrum with samples having a magnitude $-\pi$ for $k > 0$ and $+\pi$ for $k < 0$.

From Fig. Ex. 2.27, we find that,

$$T_0 = 2 \text{ and } \omega_0 = \frac{2\pi}{T_0} = \pi$$

The analytical expression for the waveform $x(t)$ over one period is given by

$$x(t) = \begin{cases} 1 + t, & -1 < t < 0 \\ 1 - t, & 0 < t < 1 \end{cases}$$

The Complex Fourier Coefficient is given by

$$X(k) = \frac{1}{T_0} \int_{<T_0>} x(t) e^{-jk\omega_0 t}\, dt$$

Hence,

$$X(k) = \frac{1}{2}\left[\int_{-1}^{0} (1+t) e^{-jk\pi t}\, dt + \int_{0}^{1} (1-t) e^{-jk\pi t}\, dt\right] \tag{2.82}$$

Performing integration by parts, we get

$$X(k) = \frac{1}{\pi^2 k^2}[1 - (-1)^k], \quad k \neq 0$$

To find $X(0)$, put $k = 0$ in Eq. (2.82).

Then,

$$X(0) = \frac{1}{2}\left[\int_{-1}^{0} (1+t)\, dt + \int_{0}^{1} (1-t)\, dt\right]$$

$$= \frac{1}{2}$$

Therefore,

$$X(k) = \begin{cases} \frac{1}{2}, & k = 0 \\ 0, & k = \pm 2, \pm 4, \pm 6, \pm 8 \cdots \\ \frac{2}{\pi^2 k^2}, & k = \pm 1, \pm 3, \pm 5, \pm 7 \cdots \end{cases}$$

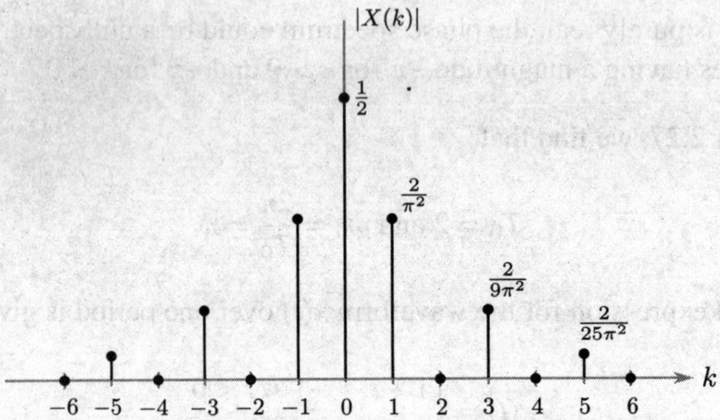

Fig. Ex. 2.27(a): *Amplitude spectrum.*

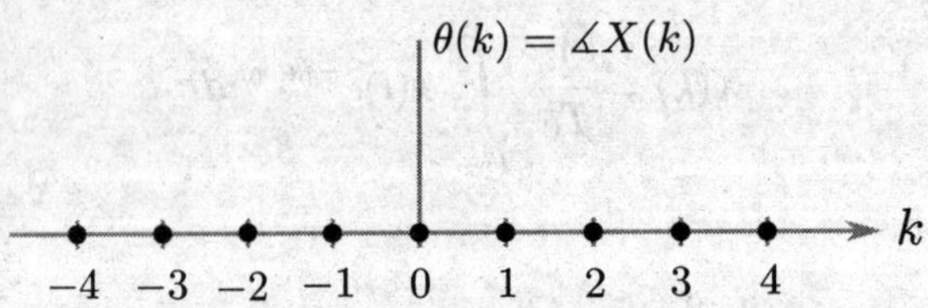

Fig. Ex. 2.27(b): *Phase spectrum.*

Since, k is positive and real for all values of k from $-\infty$ to ∞, $\theta(k) = \measuredangle X(k) = 0$. This means that the phase spectrum is a null spectrum and the same is illustrated in Fig. Ex. 2.27(b).

Example 2.28. Find the exponential **FS** representation for a periodic impulse train (Dirac comb) shown in Fig. Ex. 2.28.

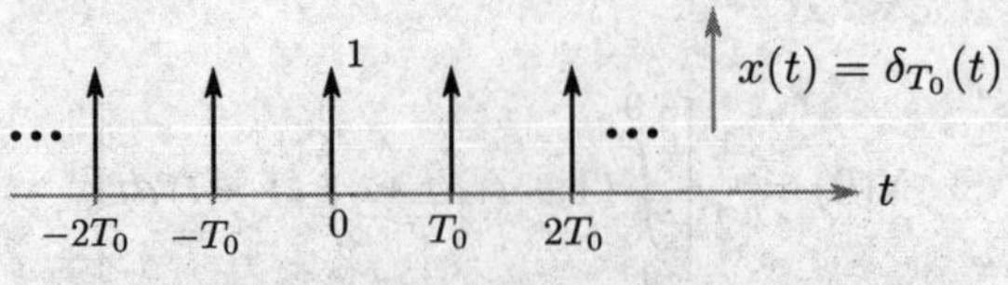

Fig. Ex. 2.28: *Dirac Comb.*

Solution: The exponential **FS** representation for $x(t)$ is given by

$$x(t) = \delta_{T_0}(t) = \sum_{k=-\infty}^{\infty} X(k)e^{-jk\omega_0 t} \tag{2.83}$$

where $\omega_0 = \dfrac{2\pi}{T_0}$.

The Complex Fourier Coefficient, $X(k)$ is computed as follows:

$$X(k) \triangleq \frac{1}{T} \int_{<T>} x(t)e^{-jk\omega_0 t}\, dt$$

$$\Rightarrow \quad X(k) = \frac{1}{T_0} \int_0^{T_0} \delta(t)e^{-jk\omega_0 t}\, dt$$

$$= \frac{1}{T} e^{-jk\omega_0 t}\Big|_{t=0} = \frac{1}{T_0}$$

Therefore,

$$x(t) = \frac{1}{T_0} \sum_{k=-\infty}^{\infty} e^{-jk\frac{2\pi}{T_0}t}$$

Example 2.29. Find the Complex Fourier Coefficient $X(k)$ for $x(t)$ defined by

$$x(t) = \sum_{m=-\infty}^{\infty} \left[\delta\left(t - \frac{1}{2}m\right) + \delta\left(t - \frac{3}{2}m\right)\right]$$

Solution:

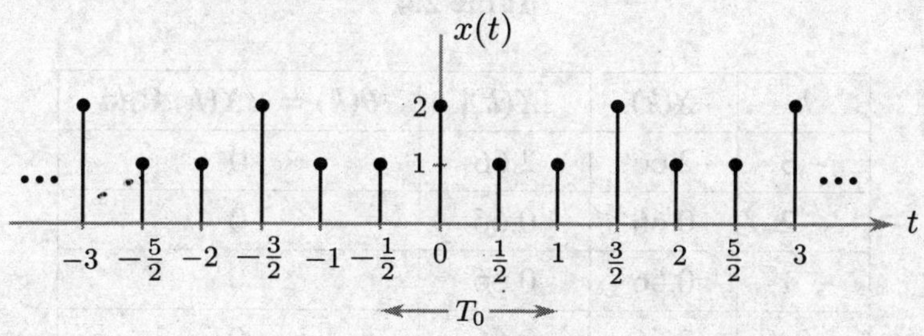

Fig. Ex. 2.29(a)

$$T_0 = \frac{3}{2}, \quad \omega_0 = \frac{2\pi}{T_0} = \frac{4\pi}{3}$$

The analytical expression for $x(t)$ over one period is

$$x(t) = \begin{cases} 1, & t = -\frac{1}{2} \\ 2, & t = 0 \\ 1, & t = \frac{1}{2} \end{cases}$$

$$\text{or} \quad x(t) = \delta\left(t + \frac{1}{2}\right) + 2\delta(t) + \delta\left(t - \frac{1}{2}\right)$$

The Complex Fourier Coefficient $X(k)$ is given by

$$X(k) = \frac{1}{T_0} \int_{<T_0>} x(t) e^{-jk\omega_0 t}\, dt$$

$$= \frac{1}{\frac{3}{2}} \left[\int_{-\frac{1}{2}}^{1} \left[\delta\left(t + \frac{1}{2}\right) + 2\delta(t) - \frac{1}{2} + \delta\left(t - \frac{1}{2}\right) \right] e^{-jk\omega_0 t}\, dt \right]$$

$$\Rightarrow \quad X(k) = \frac{2}{3} \left[e^{-jk\omega_0 t}\Big|_{t=-\frac{1}{2}} + 2e^{-jk\omega_0 t}\Big|_{t=0} + e^{-jk\omega_0 t}\Big|_{t=\frac{1}{2}} \right]$$

$$= \frac{2}{3} \left[2\cos\left(\frac{k\omega_0}{2}\right) + 2 \right]$$

$$= \frac{4}{3} \left[1 + \cos\left(\frac{2\pi k}{3}\right) \right]$$

Table 2.4

k	$X(k)$	$\lvert X(k) \rvert$	$\theta(k) = \angle X(k)$ **Rad**
– 3	2.66	2.66	0
– 2	0.66	0.66	0
– 1	0.66	0.66	0
0	2.66	2.66	0
1	0.66	0.66	0
2	0.66	0.66	0
3	2.66	2.66	0

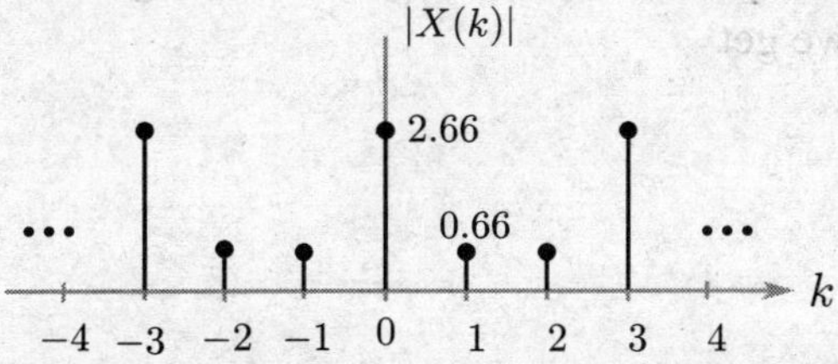

Fig. Ex. 2.29(b): *Amplitude spectrum.*

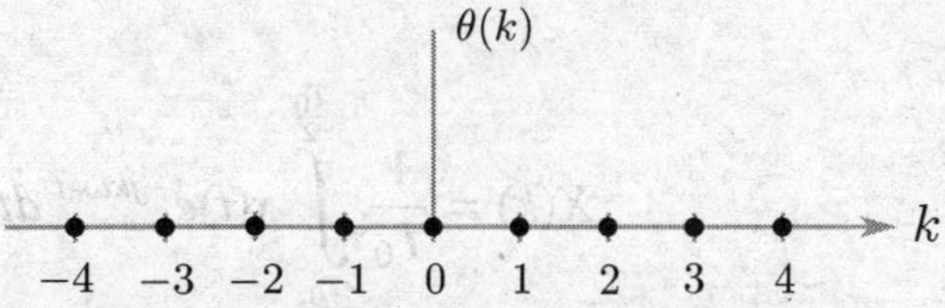

Fig. Ex. 2.29(c): *Phase spectrum.*

Example 2.30. Provide proofs showing which of the following statements are true and which are false.

(a) $x(t)$ being even implies $X(k) = \dfrac{2}{T_0} \displaystyle\int_0^{\frac{T_0}{2}} x(t)e^{-jk\omega_0 t}\, dt$

(b) $x(t)$ being even implies $X(k) = \dfrac{2}{T_0} \displaystyle\int_0^{\frac{T_0}{2}} x(t)\cos(k\omega_0 t)\, dt$

(c) $x(t)$ being odd implies $X(k) = \dfrac{-2j}{T_0} \displaystyle\int_0^{\frac{T_0}{2}} x(t)\sin(k\omega_0 t)\, dt$

Solution:

$$X(k) \triangleq \frac{1}{T_0} \int_{\frac{-T_0}{2}}^{\frac{T_0}{2}} x(t)e^{-jk\omega_0 t}\, dt$$

$$= \frac{1}{T_0} \int_{\frac{-T_0}{2}}^{\frac{T_0}{2}} [x_{ev}(t) + x_{od}(t)][\cos(k\omega_0 t) - j\sin(k\omega_0 t)]\, dt$$

$$\Rightarrow \qquad X(k) = \underbrace{\frac{1}{T_0} \int_{\frac{-T_0}{2}}^{\frac{T_0}{2}} x_{ev}(t)\cos(k\omega_0 t)\, dt}_{I} + \underbrace{\frac{1}{T_0} \int_{\frac{-T_0}{2}}^{\frac{T_0}{2}} x_{od}(t)\cos(k\omega_0 t)\, dt}_{II}$$

$$\underbrace{-\frac{j}{T_0} \int_{\frac{-T_0}{2}}^{\frac{T_0}{2}} x_{ev}(t)\sin(k\omega_0 t)\, dt}_{III} \underbrace{- \frac{j}{T_0} \int_{\frac{-T_0}{2}}^{\frac{T_0}{2}} x_{od}(t)\sin(k\omega_0 t)\, dt}_{IV}$$

In the above equation, parts II and III are equal to zero because their integrands (the function inside the integral) are odd. Hence, we get

$$X(k) = \frac{1}{T_0} \int_{\frac{-T_0}{2}}^{\frac{T_0}{2}} x(t) e^{-jk\omega_0 t}\, dt$$

$$= \underbrace{\frac{1}{T_0} \int_{\frac{-T_0}{2}}^{\frac{T_0}{2}} x_{ev}(t) \cos(k\omega_0 t)\, dt}_{I} - \underbrace{\frac{j}{T_0} \int_{\frac{-T_0}{2}}^{\frac{T_0}{2}} x_{od}(t) \sin(k\omega_0 t)\, dt}_{IV}$$

If $x(t)$ is even then $x_{od}(t)$ is zero and hence IV reduces to zero. Then, we get

$$X(k) = \underbrace{\frac{1}{T_0} \int_{\frac{-T_0}{2}}^{\frac{T_0}{2}} x_{ev}(t) \cos(k\omega_0 t)\, dt}_{I}$$

$$= \frac{2}{T_0} \int_{0}^{\frac{T_0}{2}} x_{ev}(t) \cos(k\omega_0 t)\, dt \tag{2.84}$$

Eq. (2.84) means that (a) is false and (b) is true. Similarly, if $x(t)$ is odd, then the part I is zero and so

$$X(k) = \underbrace{\frac{-j}{T_0} \int_{\frac{-T_0}{2}}^{\frac{T_0}{2}} x_{od}(t) \sin(k\omega_0 t)\, dt}_{IV}$$

$$= \frac{-2j}{T_0} \int_{0}^{\frac{T_0}{2}} x_{od}(t) \sin(k\omega_0 t)\, dt \tag{2.85}$$

This means that (c) is true.

Example 2.31. Determine the time signal $x(t)$ corresponding to the magnitude and phase spectra shown in Figs. Ex. 2.14 (a) and Ex. 2.15 (b) with $\omega_0 = \pi$.

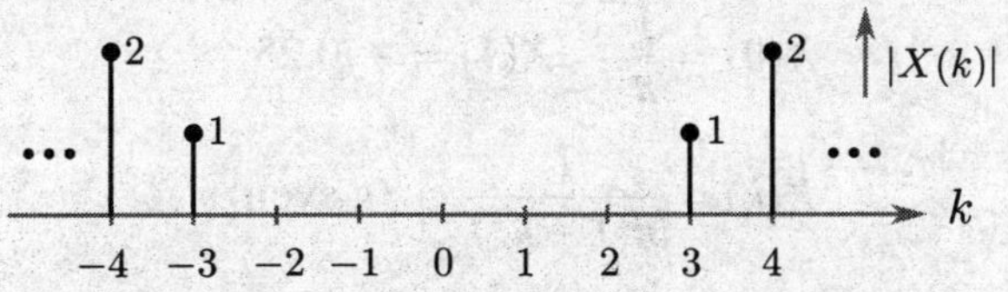

Fig. Ex. 2.31(a): *Magnitude spectrum.*

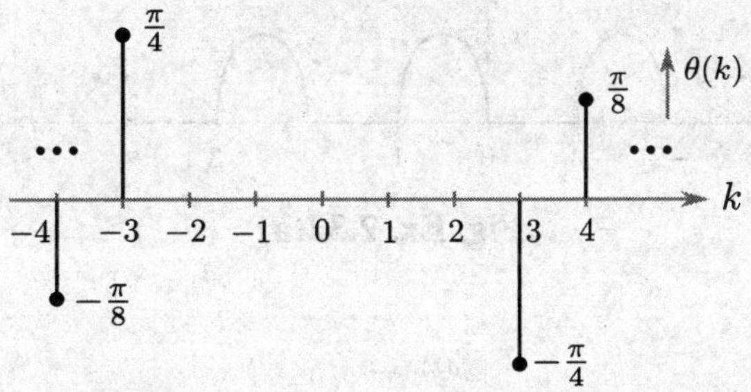

Fig. Ex. 2.31(b): *Phase spectrum.*

Solution: We know that,

$$X(k) = |X(k)|e^{j\theta(k)}$$

Hence, we can write from Figs. Ex. 2.31 (a) and Ex. 2.31 (b), the following Complex Fourier Coefficients.

$$X(3) = 1 \times e^{-\frac{j\pi}{4}}, \quad X(-3) = 1 \times e^{\frac{j\pi}{4}}$$

$$X(4) = 2 \times e^{\frac{j\pi}{8}}, \quad X(-4) = 2 \times e^{-\frac{j\pi}{8}}$$

Recall the synthesis equation:

$$x(t) = \sum_{k=-\infty}^{\infty} X(k)e^{jk\omega_0 t}$$

Hence,

$$\begin{aligned} x(t) &= e^{-\frac{j\pi}{4}}e^{j3\pi t} + e^{\frac{j\pi}{4}}e^{-j3\pi t} + 2e^{\frac{j\pi}{8}}e^{j4\pi t} + 2e^{-\frac{j\pi}{8}}e^{-j4\pi t} \\ &= \left[e^{j(3\pi t - \frac{\pi}{4})} + e^{-j(3\pi t - \frac{\pi}{4})}\right] + 2\left[e^{j(4\pi t + \frac{\pi}{8})} + e^{-j(4\pi t + \frac{\pi}{8})}\right] \end{aligned}$$

Therefore,

$$x(t) = 2\cos\left(3\pi t - \frac{\pi}{4}\right) + 4\cos\left(4\pi t + \frac{\pi}{8}\right)$$

Example 2.32. The Fourier series Coefficients of the signal $x(t)$ shown in Fig. Ex. 2.32 (a) are

$$X(0) = \frac{1}{\pi}, \quad X(1) = -j0.25$$

$$X(k) = \frac{1}{\pi(1-k^2)}, \quad (k \text{ even})$$

Use the properties to find the Fourier Coefficients of $y(t)$, $f(t)$ and $g(t)$.

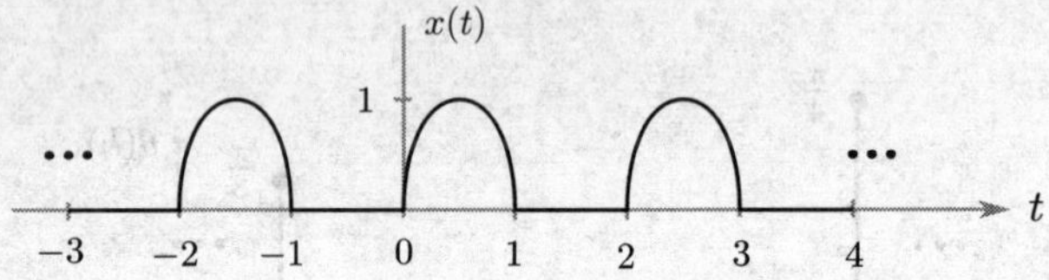

Fig. Ex. 2.32(a)

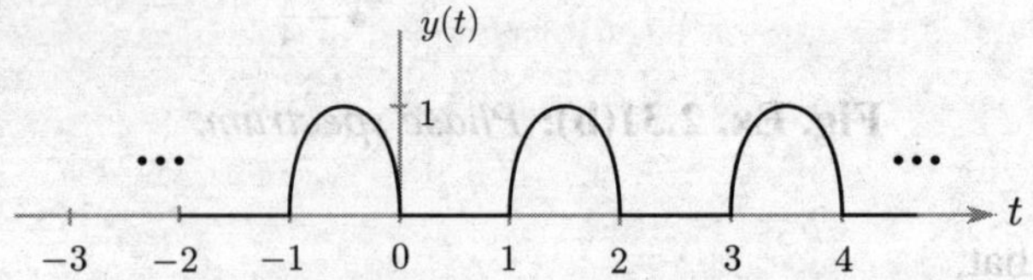

Fig. Ex. 2.32(b)

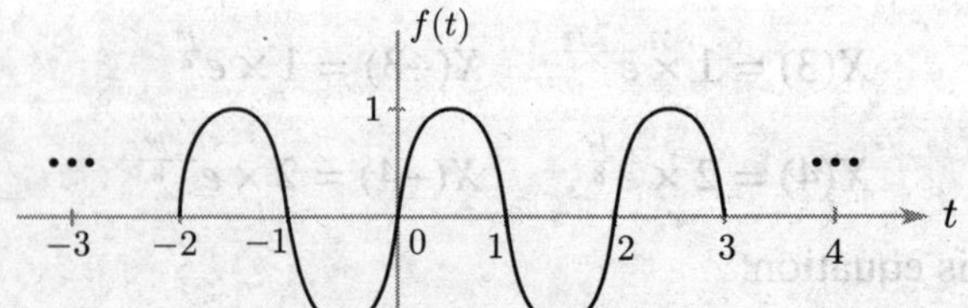

Fig. Ex. 2.32(c)

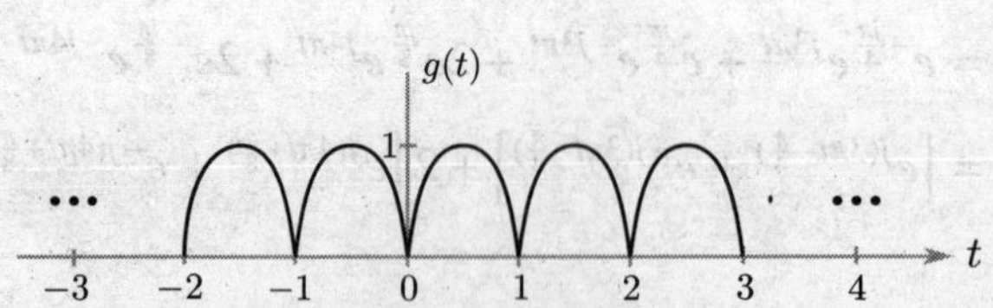

Fig. Ex. 2.32(d)

Solution:

(a) From Figs. Ex. 2.32 (a) and Ex. 2.32 (b), we find that,

$$\begin{aligned} y(t) &= x(t-1) \\ &= x(t-0.5T_0) \end{aligned}$$

where, $T_0 = 2$ seconds.

Applying time-shift property, we get

$$\begin{aligned} Y(k) &= X(k)e^{-jk\omega_0(0.5T_0)}, \quad (\text{Here } \ t_0 = 0.5T_0) \\ &= X(k)e^{-jk\frac{2\pi}{T_0}(0.5T_0)} \\ &= X(k)e^{-j\pi k} \\ &= (-1)^k X(k) \end{aligned}$$

Hence,

$$\begin{aligned} Y(0) &= X(0) = \frac{1}{\pi} \\ Y(1) &= -X(1) = j0.25 \end{aligned}$$

For even values of k, we find that $Y(k) = X(k)$.

Hence,

$$Y(k) = \frac{1}{\pi(1-k^2)}, \quad (k \text{ even})$$

(b) The sinusoid $f(t)$ shown in Fig. Ex. 2.32 (c) may be described as

$$f(t) = x(t) - y(t)$$

Applying linearity property, we get

$$F(k) = X(k) - Y(k)$$

Hence,

$$\begin{aligned} F(0) &= X(0) - Y(0) - 0 \\ F(1) &= X(1) - Y(1) = -j0.5 \\ F(k) &= X(k) - Y(k) = 0, \quad (k \neq 1 \text{ and } k \text{ even}) \end{aligned}$$

(c) The full-wave rectified sine wave $g(t)$ shown in Fig. Ex. 2.32(d) can be written as

$$g(t) = x(t) + y(t)$$

Again, applying linearity property, we get

$$G(k) = X(k) + Y(k)$$

$$\Rightarrow \qquad G(0) = X(0) + Y(0) = \frac{2}{\pi}$$

$$G(1) = X(1) + Y(1) = 0$$

$$G(k) = \frac{2}{\pi(1-k^2)}, \qquad (\text{k even})$$

Example 2.33. Consider the rectangular pulse train shown in Fig. Ex. 2.33. Using the derivative property, find $X(k)$.

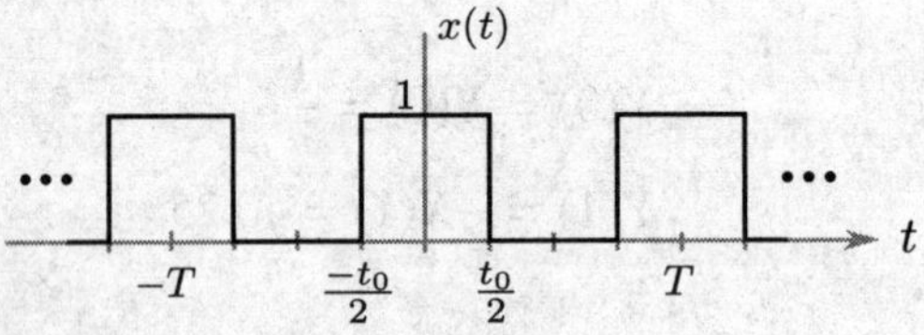

Fig. Ex. 2.33

Solution: Let us differentiate $x(t)$ so that $x'(t)$ has only impulses.

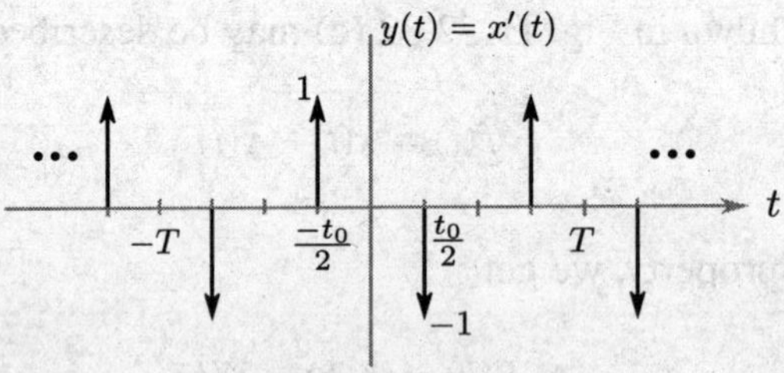

Fig. Ex. 2.33(a)

The analytical expression for $y(t)$ over the interval $\frac{-T}{2} < t < \frac{T}{2}$ is

$$y(t) = \delta\left(t + \frac{t_0}{2}\right) - \delta\left(t - \frac{t_0}{2}\right)$$

Hence,

$$Y(k) \stackrel{\Delta}{=} \frac{1}{T} \int_{-\frac{T}{2}}^{\frac{T}{2}} y(t) e^{-jk\omega_0 t}\, dt, \text{ where } \omega_0 = \frac{2\pi}{T}$$

$$= \frac{1}{T} \int_{\frac{-T}{2}}^{\frac{T}{2}} \left[\delta\left(t + \frac{t_0}{2}\right) - \delta\left(t - \frac{t_0}{2}\right)\right] e^{-jk\omega_0 t}\, dt$$

Applying sifting property, we find that

$$Y(k) = \frac{1}{T}\left[e^{-jk\omega_0 t}\Big|_{t=\frac{-t_0}{2}} - e^{-jk\omega_0 t}\Big|_{t=\frac{t_0}{2}} \right]$$

$$\Rightarrow \qquad Y(k) = \frac{1}{T}\left[e^{jk\omega_0 \frac{t_0}{2}} - e^{-jk\omega_0 \frac{t_0}{2}} \right]$$

$$= \frac{2j}{T} \sin\left(k\omega_0 \frac{t_0}{2}\right)$$

We know that,

$$y(t) = x'(t)$$

Applying the derivative property, we get,

$$Y(k) = jk\omega_0 X(k)$$

$$\Rightarrow \qquad X(k) = \frac{1}{jk\omega_0} Y(k)$$

$$= \frac{1}{\not{j}k\omega_0} \frac{2\not{j}}{T} \sin\left(k\omega_0 \frac{t_0}{2}\right)$$

$$= \frac{2}{k\omega_0 T} \sin\left(k\omega_0 \frac{t_0}{2}\right)$$

$$= \frac{2}{k(2\pi f_0 T)} \sin\left(k 2\pi f_0 \frac{t_0}{2}\right)$$

$$= \frac{t_0}{T} \frac{\sin(\pi k f_0 t_0)}{\pi k f_0 t_0}$$

$$= \frac{t_0}{T} \operatorname{sinc}(k f_0 t_0)\,; \; k \neq 0$$

Also,

$$X(0) = \frac{1}{T} \int_{\frac{-T}{2}}^{\frac{T}{2}} x(t)\, dt = \frac{1}{T} \int_{\frac{-t_0}{2}}^{\frac{t_0}{2}} 1 \cdot dt = \frac{t_0}{T}$$

Example 2.34. Derive the trignometric Fourier series from the complex exponential Fourier series.

Solution: The exponential Fourier series is given by

$$x(t) = \sum_{k=-\infty}^{\infty} X(k)e^{jk\omega_0 t}$$

$$= X(0) + \sum_{k=1}^{\infty}\left[X(k)e^{jk\omega_0 t} + X(-k)e^{-jk\omega_0 t}\right]$$

and using Euler's formula $e^{\pm jk\omega_0 t} = \cos k\omega_0 t \pm j\sin k\omega_0 t$,

we have

$$x(t) = X(0) + \sum_{k=1}^{\infty}\{[X(k) + X(-k)]\cos k\omega_0 t + j[X(k) - X(-k)]\sin k\omega_0 t\}$$

Letting

$$X(0) = a_0$$

$$X(k) + X(-k) = a_k$$

$$\text{and} \qquad j[X(k) - X(-k)] = b_k$$

we get

$$x(t) = a_0 + \sum_{k=1}^{\infty} a_k \cos k\omega_0 t + b_k \sin k\omega_0 t$$

Since k is a dummy variable, it can be replaced by n.

Thus,

$$x(t) = a_0 + \sum_{n=1}^{\infty}(a_n \cos n\omega_0 t + b_n \sin n\omega_0 t)$$

Example 2.35. Derive the expressions for c_n and θ_n in terms of a_n and b_n.

Solution: Recall the trignometric form of Fourier series:

$$x(t) = a_0 + \sum_{n=1}^{\infty} a_n \cos n\omega_0 t + b_n \sin n\omega_0 t$$

$$= a_0 + \sum_{n=1}^{\infty} a_n \cos n\omega_0 t + b_n \cos[n\omega_0 t - 90°]$$

Converting the above time-domain equation into phasor-domain equation, we get

$$\mathbf{X} = a_0 + \sum_{n=1}^{\infty} a_n\angle 0^\circ + b_n\angle -90^\circ$$

$$= a_0 + \sum_{n=1}^{\infty}(a_n - jb_n)$$

$$= a_0 + \sum_{n=1}^{\infty} c_n e^{j\theta_n}$$

where

$$c_n = \sqrt{a_n^2 + b_n^2} \quad \text{and} \quad \theta_n = \begin{cases} -\tan^{-1}\left(\frac{b_n}{a_n}\right), & a_n > 0 \\ -\pi + \tan^{-1}\left(\frac{b_n}{a_n}\right), & a_n < 0 \end{cases}$$

The time-domain expression is obtained by multiplying the phasor-domain equation by $e^{j\omega_0 t}$ and then taking the real part of this product.

Thus,

$$x(t) = \text{Real}\left\{\mathbf{X}e^{j\omega_0 t}\right\}$$

$$= \text{Real}\left\{a_0 e^{j\omega_0 t} + \sum_{n=1}^{\infty} c_n e^{j\omega_0 t}\cdot e^{j\theta_n}\right\}$$

$$= a_0\cos\omega_0 t + \sum_{n=1}^{\infty} c_n\cos(\omega_0 t + \theta_n)$$

$$= \sum_{n=0}^{\infty} c_0\cos(\omega_0 t + \theta_n); \quad \text{where } c_0 = a_0$$

Example 2.36. The trignometric Fourier series for the square wave shown in Fig. Ex. 2.36(a) is

$$x(t) = \frac{A}{2} + \frac{2A}{\pi}\left[\cos\omega_0 t - \frac{1}{3}\cos 3\omega_0 t + \frac{1}{5}\cos 5\omega_0 t - \cdots\right]$$

The square wave $f(t)$ shown in Fig. Ex. 2.36(b) is antisymmetrical with respect to the origin. Find the trignometric Fourier series for $f(t)$.

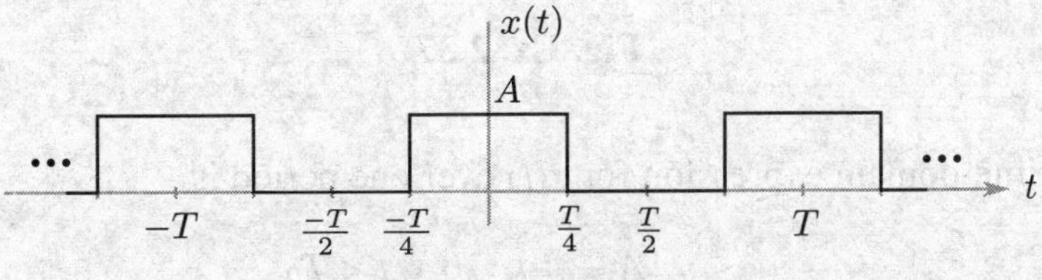

Fig. Ex. 2.36(a)

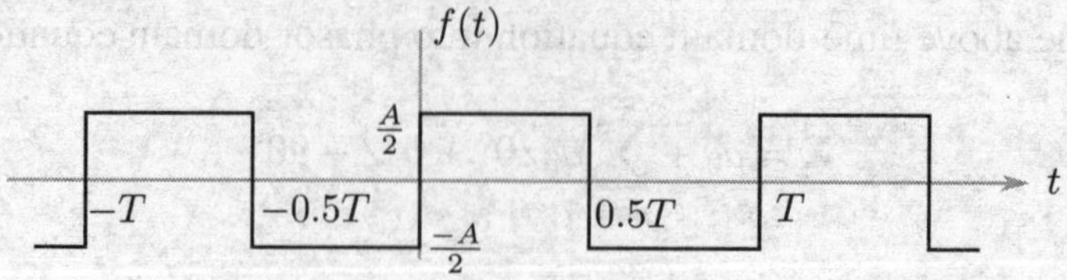

Fig. Ex. 2.36(b)

Solution: The waveform $f(t)$ is a translation of the waveform $x(t)$ of Fig. Ex. 2.36(a) plus $\left(-\frac{A}{2}\right)$. That is,

$$f(t) = x\left(t - \frac{T}{4}\right) - \frac{A}{2}$$

Substituting the Fourier series representation for $x(t)$ in the above equation, we get

$$f(t) = \frac{2A}{\pi}\left[\cos\left(\omega_0\left(t - \frac{T}{4}\right)\right) - \frac{1}{3}\cos\left(3\omega_0\left(t - \frac{T}{4}\right)\right) + \frac{1}{5}\cos\left(5\omega_0\left(t - \frac{T}{4}\right)\right) - \cdots\right]$$

Since $\omega_0 = \frac{2\pi}{T_0} = \frac{2\pi}{T}$, we get

$$\begin{aligned} f(t) &= \frac{2A}{\pi}\left[\cos\left(\omega_0 t - \frac{\pi}{2}\right) - \frac{1}{3}\cos\left(3\omega_0 t - \frac{3\pi}{2}\right) + \frac{1}{5}\cos\left(5\omega_0 t - \frac{5\pi}{2}\right) - \cdots\right] \\ &= \frac{2A}{\pi}\left[\sin\omega_0 t + \frac{1}{3}\sin 3\omega_0 t + \frac{1}{5}\sin 5\omega_0 t + \cdots\right] \end{aligned}$$

Comment: Since $f(t)$ is an odd function, i.e., if $f(t) = -f(-t)$, Fourier series has only sine terms.

Example 2.37. Using differentiation technique, find $X(k)$ for the triangular waveform shown in Fig. Ex. 2.37.

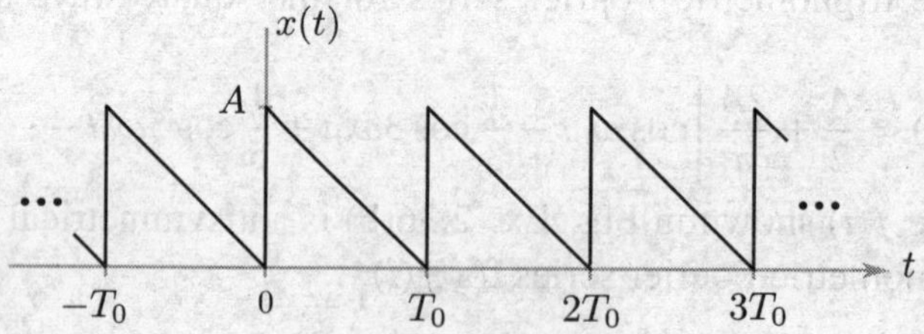

Fig. Ex. 2.37

Solution: The time-domain expression for $x(t)$ over one period is

$$x(t) = \begin{cases} A - \frac{A}{T_0}t, & 0 < t < T_0 \\ 0, & \text{otherwise} \end{cases}$$

Note that $x(t) = x(t + T)$, where $T = T_0, 2T_0, \ldots$
The above expression for $x(t)$ can also be written as

$$x(t) = \left[A - \frac{A}{T_0}t\right]\{u(t) - u(t - T_0)\}$$

Differentiating $x(t)$ with respect to t, we get

$$y(t) = \frac{dx(t)}{dt} = \left[A - \frac{A}{T_0}t\right]\{\delta(t) - \delta(t - T_0)\} + \{u(t) - u(t - T_0)\}\left[-\frac{A}{T_0}\right]$$

$$\Rightarrow \qquad y(t) = \left[A - \frac{A}{T_0}t\right]\delta(t) - \left[A - \frac{A}{T_0}t\right]\delta(t - T_0) - \frac{A}{T_0}[u(t) - u(t - T_0)]$$

Recall the product property:

$$f(t)\delta(t - t_0) = f(t_0)\delta(t - t_0)$$

Hence,

$$y(t) = \left[A - \frac{A}{T_0}t\right]_{t=0}\delta(t) - \left[A - \frac{A}{T_0}t\right]_{t=T_0}\delta(t - T_0) - \frac{A}{T_0}[u(t) - u(t - T_0)]$$

$$\Rightarrow \qquad y(t) = A\delta(t) - [A - A]\delta(t - T_0) - \frac{A}{T_0}[u(t) - u(t - T_0)]$$

$$\Rightarrow \qquad y(t) = A\delta(t) - \frac{A}{T_0}[u(t) - u(t - T_0)]$$

To find $Y(k)$:

$$Y(k) = \frac{1}{T_0}\int_0^{T_0} y(t)e^{-jk\omega_0 t}dt; \qquad \omega_0 = \frac{2\pi}{T_0}$$

$$\Rightarrow \qquad Y(k) = \frac{1}{T_0}\int_0^{T_0} A\delta(t)e^{-jk\frac{2\pi}{T_0}t}dt + \frac{1}{T_0}\int_0^{T_0} -\frac{A}{T_0}e^{-jk\frac{2\pi}{T_0}t}dt$$

$$\Rightarrow \qquad Y(k) = \frac{A}{T_0}e^{-j\frac{2\pi k}{T_0}t}\Big|_{t=0} - \frac{A}{T_0^2}\left[\frac{e^{-j\frac{2\pi k}{T_0}t}}{-j\frac{2\pi k}{T_0}}\right]_0^{T_0}$$

$$= \frac{A}{T_0} + \frac{A}{T_0}\left[\frac{e^{-j2\pi k} - 1}{j2\pi k}\right]$$

Since, $e^{-j2\pi k} = 1$, we get

$$Y(k) = \frac{A}{T_0}$$

To find $X(0)$:

$$X(0) = \frac{1}{T_0}\int_0^{T_0} x(t)dt$$

$$= \frac{1}{T_0}\int_0^{T_0} \left(A - \frac{A_0}{T}t\right)dt$$

$$= \frac{AT_0}{2}$$

The differentiation property impiles that $Y(k) = jk\omega_0 X(k)$. Hence, we may find $X(k)$ from $Y(k)$ as $X(k) = \frac{1}{jk\omega_0}Y(k)$, except for $k = 0$. The quantity $X(0)$ is the average value of $x(t)$ and is found above.

Therefore,

$$X(k) = \begin{cases} \frac{AT_0}{2}, & k = 0 \\ \frac{A/T_0}{jk\omega_0}, & k \neq 0 \end{cases}$$

$$= \begin{cases} \frac{AT_0}{2}, & k = 0 \\ \frac{A}{jk\frac{2\pi}{T_0}\times T_0} = \frac{A}{j2\pi k}, & k \neq 0 \end{cases}$$

Summary of Important points

- Two time functions $x(t)$ and $y(t)$ are said to be orthogonal over the interval (t_1, t_2) if

$$\int_{t_1}^{t_2} x(t)y^*(t)\, dt = 0$$

If, in addition,

$$\int_{t_1}^{t_2} |x(t)|^2\, dt = 1 = \int_{t_1}^{t_2} |y(t)|^2\, dt$$

the functions are said to be normalized and hence are called orthonormal. A set of functions $\{\phi_k(t)\}$ is called an orthogonal (orthonormal) set if each pair of functions in this set is orthogonal (orthonormal).

- The brilliant mathematician Jean–Baptiste–Joseph Fourier proposed that a periodic waveform could be represented by a series comprising of cosine and sine terms with appropriate coefficients. The integer multiple frequencies of the fundamental are called harmonic frequencies (or harmonics).

- The trignometric form of Fourier series is

$$x(t) = a_0 + \sum_{n=1}^{\infty} a_n \cos n\omega_0 t + \sum_{n=1}^{\infty} b_n \cos n\omega_0 t$$

The coefficients of the trignometric Fourier series can be found from

$$a_0 = \frac{1}{T} \int_{t_0}^{t_0+T_0} x(t)\, dt$$

$$a_n = \frac{2}{T} \int_{t_0}^{t_0+T_0} x(t) \cos n\omega_0 t\, dt, \quad n > 0$$

$$b_n = \frac{2}{T} \int_{t_0}^{t_0+T_0} x(t) \sin n\omega_0 t\, dt, \quad n > 0$$

- An alternate form of trignometric form of Fourier series is

$$x(t) = c_0 + \sum_{n=1}^{\infty} c_n \cos(n\omega_0 t + \theta_n)$$

where

$$c_0 = a_0$$

$$c_n = \sqrt{a_n^2 + b_n^2}$$

$$\theta_n = \begin{cases} -\tan^{-1}\left(\frac{b_n}{a_n}\right) & ; \; a_n > 0 \\ -\pi + \tan^{-1}\left(\frac{b_n}{a_n}\right) & ; \; a_n < 0 \end{cases}$$

- The exponential form of Fourier series is

$$x(t) = \sum_{k=-\infty}^{\infty} X(k) e^{jk\omega_0 t}$$

where $X(k)$ are the complex coefficients defined by

$$X(k) = \frac{1}{T_0} \int_{<T_0>} x(t) e^{-jk\omega_0 t}\, dt$$

- The line spectra consisting of the amplitude and phase of $X(k)$ when plotted against frequency are useful for portraying the frequencies that represent a waveform.

Exercise Problems

EP. 2.1 Refer the pair of signals shown in Fig. EP. 2.1. Find if $x(t)$ and $y(t)$ are orthogonal over the interval $(0, 4)$.

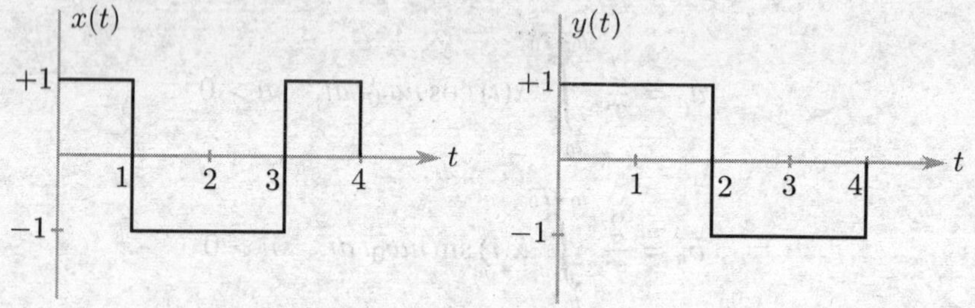

Fig. EP. 2.1

Ans: Orthogonal

EP. 2.2 Refer the signals $x(t)$ and $y(t)$ shown in EP. 2.2. Find if $x(t)$ and $y(t)$ are orthogonal over the interval $(0, 4)$.

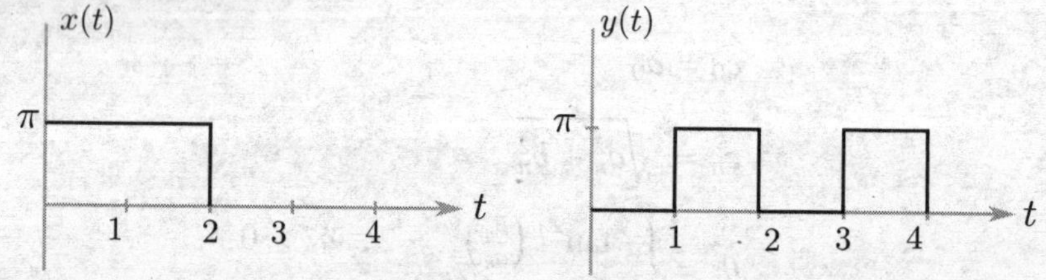

Fig. EP. 2.2

Ans: not orthogonal.

EP. 2.3 Given:

$$x(t) = \sin m\omega_0 t$$

$$y(t) = \sin n\omega_0 t$$

Are the functions $x(t)$ and $y(t)$ orthogonal over the interval $(0, T)$, where $t = \dfrac{2\pi}{\omega_0}$?

Ans: Orthogonal

EP. 2.4 Determine the trignometric Fourier series for the waveform $x(t)$ shown in Fig. EP. 2.4.

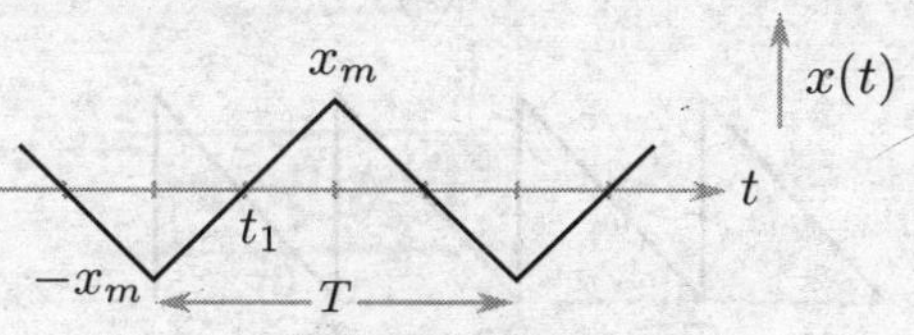

Fig. EP. 2.4

Ans: $x(t) = 3.24 \sum_{n=1}^{\infty} \frac{1}{n^2} \sin\left(\frac{n\pi}{2}\right) \sin n\omega_0 t$ for *odd n*.

EP. 2.5 Determine the trignometric Fourier series for the waveform $x(t)$ shown in Fig. EP. 2.5.

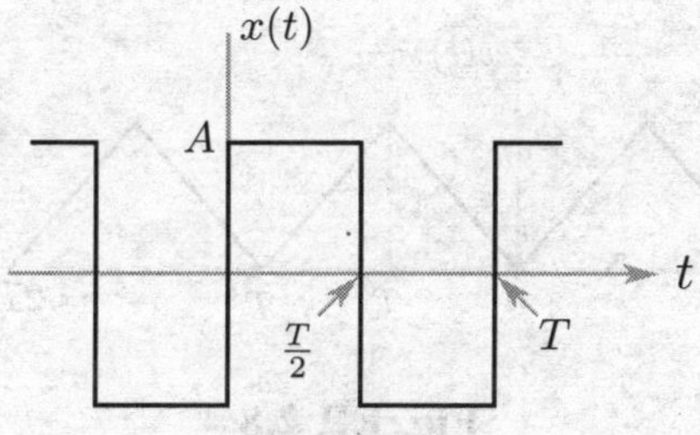

Fig. EP. 2.5

Ans: $x(t) = \frac{4A}{\pi} \sum_{n=1}^{\infty} \frac{1}{2n-1} \sin[(2n-1)\omega_0 t]$

EP. 2.6 Determine the trignometric Fourier series for the waveform shown in Fig. EP. 2.6.

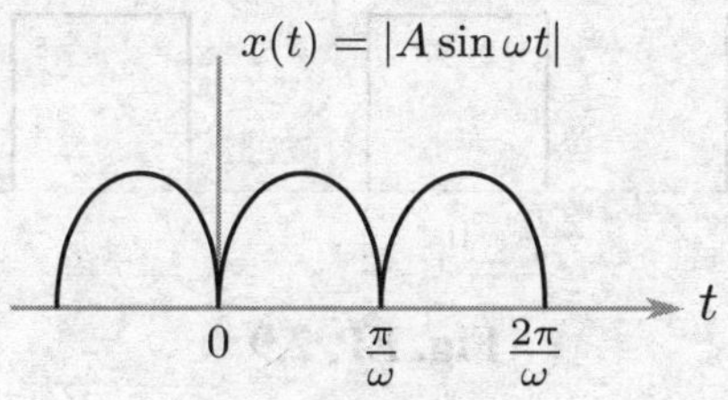

Fig. EP. 2.6

Ans: $x(t) = \frac{2A}{\pi} - \frac{4A}{\pi} \sum_{n=1}^{\infty} \frac{1}{4n^2-1} \cos n\omega_0 t$

EP. 2.7 Determine the trignometric Fourier series for the signal $x(t)$ shown in Fig. EP. 2.7.

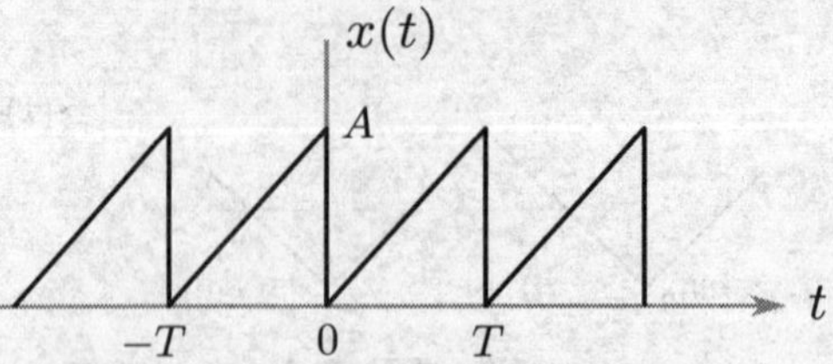

Fig. EP. 2.7

Ans: $$x(t) = \frac{A}{2} - \frac{A}{\pi}\sum_{n=1}^{\infty}\frac{\sin n\omega_0 t}{n}$$

EP. 2.8 Determine the trignometric Fourier series for the waveform shown in Fig. EP. 2.8.

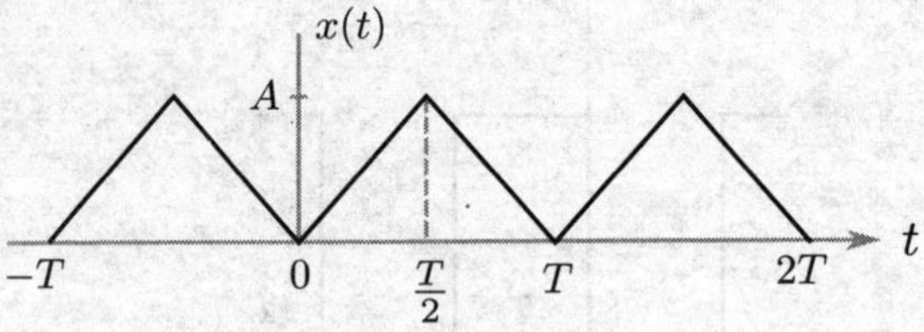

Fig. EP. 2.8

Ans: $$x(t) = \frac{A}{2} - \frac{4A}{\pi^2}\sum_{n=1}^{\infty}\frac{1}{(2n-1)^2}\cos[(2n-1)\omega_0 t]$$

EP. 2.9 Determine the complex Fourier coefficients for the square wave shown in Fig. EP. 2.9.

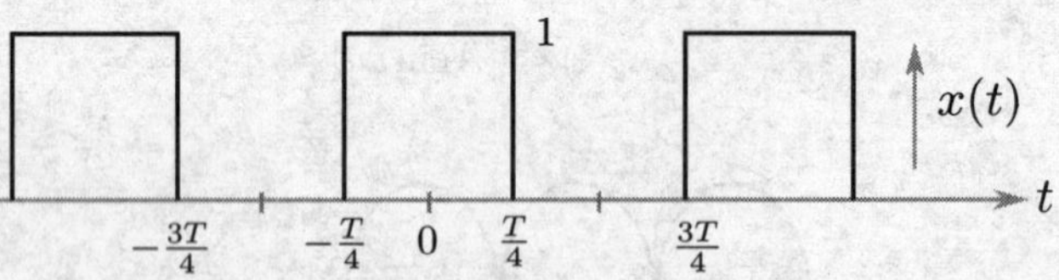

Fig. EP. 2.9

Ans: $$X(k) = \begin{cases} (-1)^{(k-1)/2}\dfrac{1}{k\pi} & ; k \text{ odd} \\ \dfrac{1}{2} & ; k = 0 \end{cases}$$

EP. 2.10 Determine the exponential Fourier coefficients for the waveform shown in EP. 2.10.

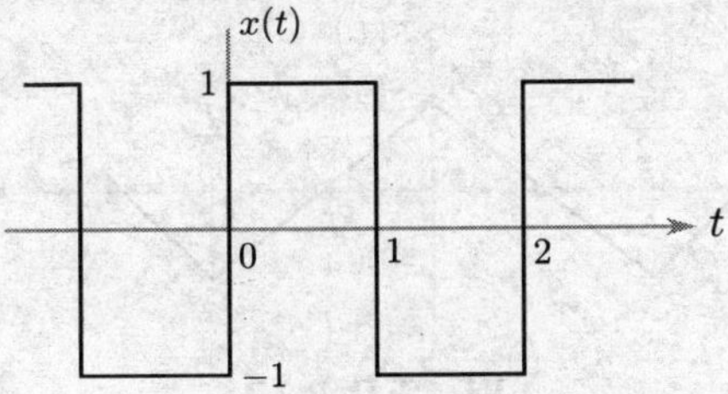

Fig. EP. 2.10

Ans: $X(k) = 0$ for even k, $X(k) = \dfrac{2}{jk\pi}$ for odd k

EP. 2.11 Find the exponential Fourier coefficients for the waveform $x(t)$ shown in Fig. EP. 2.11.

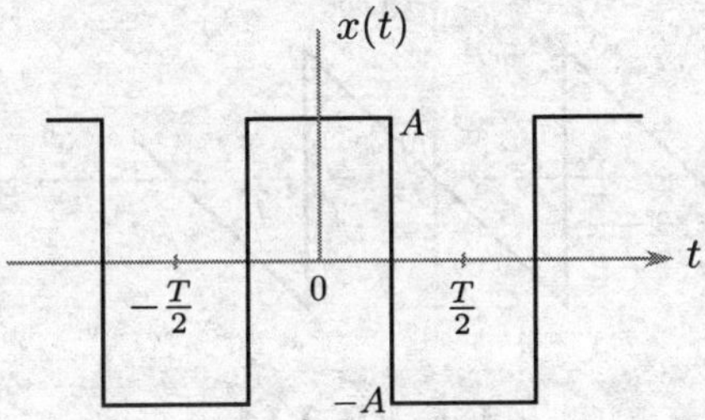

Fig. EP. 2.11

Ans: $$X(k) = \begin{cases} A\dfrac{\sin\left(\dfrac{k\pi}{2}\right)}{k\pi/2}, & k \text{ odd} \\ 0, & k = 0 \text{ and } k \text{ even} \end{cases}$$

EP. 2.12 Find the complex Fourier coefficients for the function $x(t)$ shown in Fig. EP. 2.12.

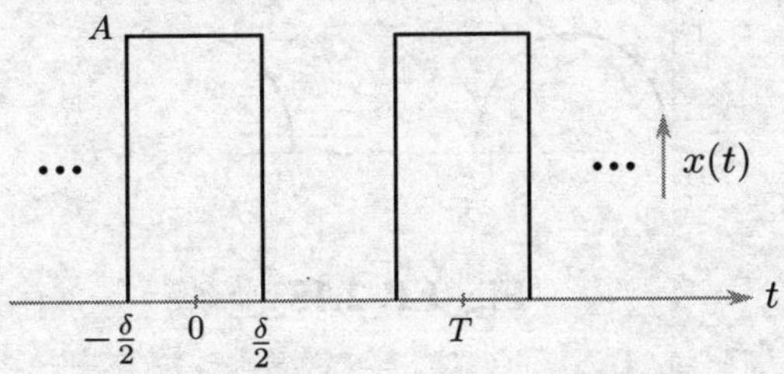

Fig. EP. 2.12

Ans: $X(k) = A\dfrac{\delta}{T}\dfrac{\sin(k\pi\delta/T)}{(k\pi\delta/T)}$

EP. 2.13 Find the complex Fourier coefficient for the function $x(t)$ shown in Fig. EP. 2.13.

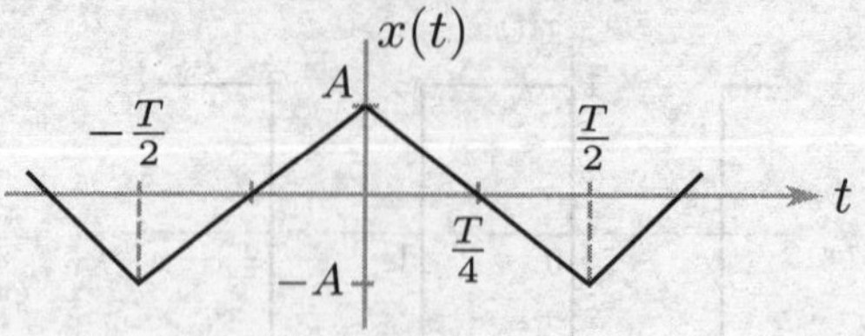

Fig. EP. 2.13

Ans: $$X(k) = \begin{cases} A\dfrac{\sin^2(k\pi/2)}{(k\pi/2)^2}, & n \neq 0 \\ 0, & n = 0 \end{cases}$$

EP. 2.14 Find the complex Fourier coefficient for the sawtooth wave shown in Fig. EP. 2.14

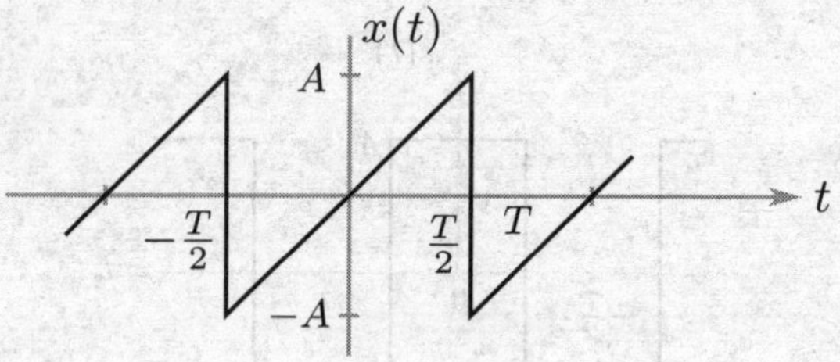

Fig. EP. 2.14

Ans: $$X(k) = \begin{cases} \dfrac{Aj(-1)^k}{k\pi}; & k \neq 0 \\ 0; & k = 0 \end{cases}$$

EP. 2.15 Find the complex Fourier coefficient for Half-wave rectified sinusoid shown in Fig. EP. 2.15.

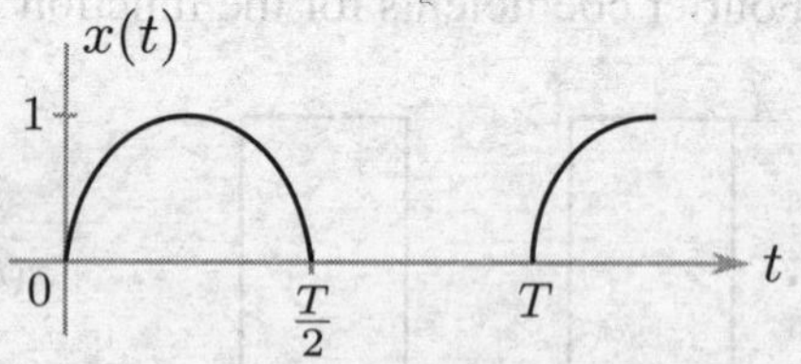

Fig. EP. 2.15

Ans: $$X(k) = \begin{cases} \dfrac{1}{\pi(1-k^2)}, & k \text{ even} \\ -j/4, & k = \pm 1 \\ 0, & \text{otherwise} \end{cases}$$

EP. 2.16 Find the complex Fourier coefficient for the Full-wave rectified sinusoid.

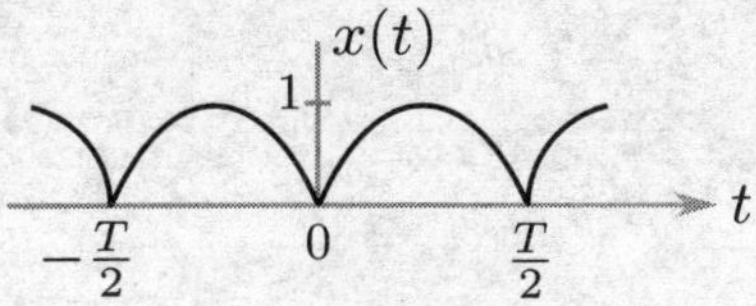

Fig. EP. 2.16

Ans: $X(k) = \begin{cases} \dfrac{2}{\pi(1-k^2)}, & k \text{ even} \\ 0, & \text{otherwise} \end{cases}$

EP. 2.17 Determine the complex Fourier coefficients for each of the following signals.

(a) $x(t) = \cos \omega_0 t$

(b) $x(t) = \cos 4t + \sin 6t$

Ans:

(a) $X(1)_* = \frac{1}{2}, \quad X(-1) = \frac{1}{2}, \quad X(k) = 0, \quad |k| \neq 1.$

(b) $X(-3) = X^*(3) = -1/2j$

$X(-2) = X(2) = 1/2$ and all other $X(k) = 0$.

EP. 2.18 Consider the Dirac-comb shown in EP. 2.18 and defined by

$$s_\delta(t) = \sum_{n=-\infty}^{\infty} \delta(t - nT_0)$$

(a) Determine the complex exponential Fourier coefficient $X(k)$.

(b) Determine the trignometric Fourier series of $s_\delta(t)$.

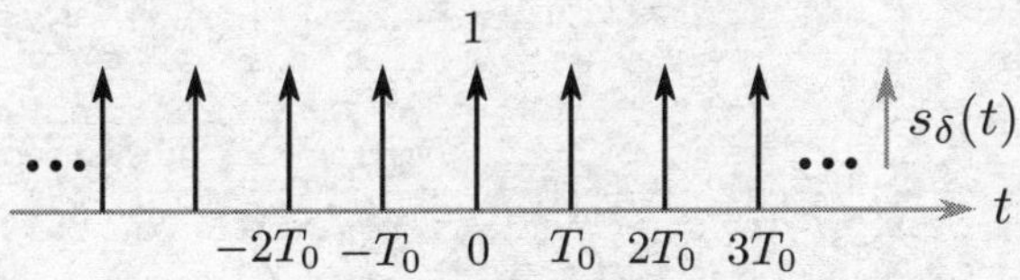

Fig. EP. 2.18

Ans:

(a) $X(k) = 1/T_0$

(b) $s_\delta(t) = \dfrac{1}{T_0} + \dfrac{2}{T_0} \sum_{n=1}^{\infty} \cos n\omega_0 t, \; \omega_0 = \dfrac{2\pi}{T_0}$

Chapter 3

Fourier Representation for Nonperiodic Signals: Fourier transform

In chapter 2, we have considered waveforms that have been periodic, for each of which there has always been a Fourier series representation. However, in practice, it is not always the case, the waveforms that we wish to analyse are not periodic. Most of the times, they are single pulses that occur just once and do not repeat themselves. Fourier analysis covers single pulses as well, and for that reason, we now commence the study of Fourier transform.

The popularity of the Fourier transform comes from the possibility of assigning a spectrum with the help of the delta impulse to many important practical functions for which there is no Laplace transform.

3.1 Mathematical development of Fourier transform

A periodic signal $x(t)$ with a fundamental period T_0 and exponential Fourier series coefficients $X(k)$ are related by

$$x(t) = \sum_{k=-\infty}^{\infty} X(k)e^{jk\omega_0 t} \tag{3.1}$$

where

$$X(k) = \frac{1}{T_0} \int_{<T_0>} x(t)e^{-jk\omega_o t}dt \tag{3.2}$$

If the period T_0 is stretched without limit, the periodic signal no longer remains periodic but becomes a single pulse $x(t)$, corresponding to one period of the periodic pulse, $x(t)$. The change over from periodic to aperiodic signal also represents a transistion from power signal to energy signal.

As $T_0 \to \infty$, $\omega_0 = \frac{2\pi}{T_0} \to 0$ and the line spectrum becomes a continuous spectrum. Also, if we replace ω_0 by an infinitesimally small quantity $d\omega \to 0$, the discrete frequency $k\omega_0$ may be replaced by a continuous frequency ω. The factor $\frac{1}{T_0}$ in Eq. (3.2) means that $X(k) \to 0$ as $T_0 \to \infty$. However, if we eliminate dependence of $X(k)$ on $\frac{1}{T_0}$ in the integral and work with $T_0X(k)$, we get

$$T_0X(k) = \int_{-T_0/2}^{T_0/2} x(t)e^{-jk\omega_0 t}dt \tag{3.3}$$

The integral on the right-side of Eq. (3.3) often exists as $T_0 \to \infty$ (even though, $T_0X(k)$ is an indeterminate form), and we obtain meaningful results. Further, because $k\omega_0 \to \omega$, the integral describes a function of ω. As a result, we define

$$\lim_{T_0\to\infty} T_0X(k) = X(\omega)$$

and obtain

$$X(\omega) = \lim_{T_0\to\infty} T_0X(k) = \int_{-\infty}^{\infty} x(t)e^{-j\omega t}dt \tag{3.4}$$

In Eq. (3.4), $X(\omega)$ is defined as the Fourier transform of the signal $x(t)$. The Fourier transform of $x(t)$ may also be written in the f-form by replacing ω by $2\pi f$ as

$$X(f) = \int_{-\infty}^{\infty} x(t)e^{-j2\pi ft}dt \tag{3.5}$$

The Fourier transform provides the frequency-domain representation of the aperiodic signal $x(t)$. Multiplying and dividing the right hand side of Eq. (3.1) by $T_0 = \frac{2\pi}{\omega_0}$, we get,

$$x(t) = \sum_{k=-\infty}^{\infty} T_0X(k)e^{jk\omega_0 t}\frac{1}{T_0} = \sum_{k=-\infty}^{\infty} T_0X(k)e^{jk\omega_0 t}\frac{\omega_0}{2\pi} \tag{3.6}$$

As the limit $T_0 \to \infty$ in Eq. (3.6), we get

- $\omega_0 \to d\omega$ (The fundamental frequency becomes infinitesimally small)
- $k\omega_0 \to \omega$ (Harmonic frequencies get so close together they become a continnum)
- The summation tends to an integral
- $x(t)$ which was periodic tends to become aperiodic.

All of these elements put together gives us an expression for the aperiodic signal in terms of its Fourier transform:

$$x(t) = \frac{1}{2\pi}\int_{-\infty}^{\infty} X(\omega)e^{j\omega t}d\omega \tag{3.7}$$

This is nothing but inverse Fourier transform, which allows us to obtain $x(t)$ from the spectrum $X(\omega)$. Also, inverse Fourier transform of $X(\omega)$ can be found as

$$x(t) = \frac{1}{2\pi}\int_{-\infty}^{\infty} X(f)e^{j2\pi ft}(2\pi df) = \int_{-\infty}^{\infty} X(f)e^{j2\pi ft}df \tag{3.8}$$

In Eq. (3.8), $X(f)$ means $X(2\pi f)$ and not by ω replaced by f in $X(\omega)$.

The signal $x(t)$ and its Fourier transform $X(\omega)$ or $X(f)$ form a unique transform pair and their relationship is symbolically denoted as follows:

$$x(t)\overset{\textbf{FT}}{\longleftrightarrow}X(\omega) \text{ or } x(t)\overset{\textbf{FT}}{\longleftrightarrow}X(f)$$

Sometimes, Fourier trasnsform of a signal is represented by $X(j\omega)$ in order to make explicit its relationship with Laplace transform.

Example 3.1. The cofficient $X(k)$ of the rectangular pulse train shown in Fig. 3.1(a) is

$$X(k) = \frac{1}{T_0}At_0 \sin c\left[\frac{k\omega_0 t_0}{2\pi}\right]$$

Find $X(\omega)$.

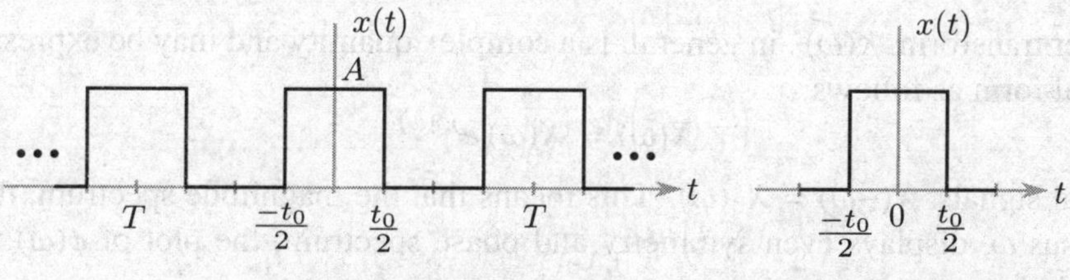

Fig. Ex. 3.1(a) **Fig. Ex. 3.1(b)**

Solution: The Fourier transform of the single pulse $x(t)$ that represents one period of $x(t)$ shown in Fig. Ex. 3.1(b) is then

$$X(\omega) = T_0 X(k)\big|_{k\omega_0=\omega}$$

$$\Rightarrow \quad X(\omega) = t_0 A \operatorname{sinc}\left(\frac{\omega t_0}{2\pi}\right)$$

3.2 Convergence/Existance of the Fourier transform

There are two important types of signals for which Fourier transform converges:

1. Signals of finite total energy, that is, for signals for which $E_x = \int_{-\infty}^{\infty} |x(t)|^2 dt < \infty$
2. Signals that obey Dirichlet conditions:
   (a) $x(t)$ is absolutely integrable, that is, $\int_{-\infty}^{\infty} |x(t)| dt < \infty$.
   (b) $x(t)$ has a finite number of maxima and minima over any finite interval of time.
   (c) $x(t)$ has a finite number of discontinuites over any finite interval of time.

The absolute integrable condition is sufficient but not necessary, and there are in fact time functions that although not integrable, have a Fourier transform. Examples of such time functions and their Fourier transforms are

$$x(t) = \frac{\sin t}{t} \overset{\textbf{FT}}{\longleftrightarrow} X(\omega) = \begin{cases} \pi, & |\omega| < 1 \\ 0, & |\omega| > 1 \end{cases}$$

and

$$x(t) = u(t) \overset{\textbf{FT}}{\longleftrightarrow} X(\omega) = \pi\delta(\omega) + \frac{1}{j\omega}$$

Later, we will extend the Fourier transform to include periodic signals (signals having finite average power) by allowing the use of impulse functions in the frequency-domain.

3.3 Magnitude and phase spectra

The Fourier transform, $X(\omega)$, in general, is a complex quantity and may be expressed in the exponential form as follows:

$$X(\omega) = |X(\omega)|e^{j\phi(\omega)}$$

For real signals, $X(-\omega) = X^*(\omega)$. This means that the magnitude spectrum, the plot of $|X(\omega)|$ versus ω, displays even symmetry and phase spectrum, the plot of $\phi(\omega)$ versus ω, displays odd symmetry.
The important facts about spectra:

1. Nonperiodic signals have continuous spectra.
2. Effect of symmetry on Fourier transform (**FT**) of real $x(t)$ are:
   (a) Even symmetry in $x(t)$: $X(\omega)$ is real and even symmetric.
   (b) Odd symmetry in $x(t)$: $X(\omega)$ is purely imaginary and odd symmetric.
   (c) No symmetry in $x(t)$: The real part of $X(\omega)$ is even symmetric and imaginary part of $X(\omega)$ is odd symmetric.

Example 3.2. Find the Fourier transform of the signal $x(t) = \delta(t + 0.5) - \delta(t - 0.5)$. Also plot the magnitude and phase spectra.

Solution: We predict the following:

(i) Since $x(t)$ is real, $X(-\omega) = X^*(\omega)$.

(ii) $x(t)$ exhibits odd symmetry. Hence, $X(\omega)$ is purely imaginary.

(iii) The magnitude spectrum displays even symmetry and phase spectrum exhibits odd symmetry.

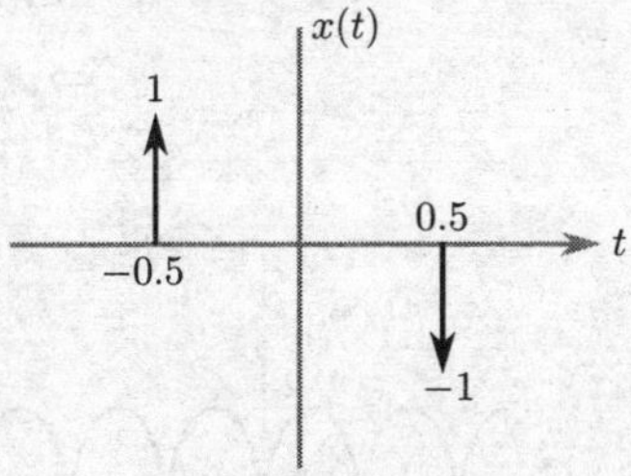

Fig. Ex. 3.2(a)

$$X(\omega) \triangleq \int_{-\infty}^{\infty} x(t)e^{-j\omega t}dt$$

$$\Rightarrow \quad X(\omega) = \int_{-\infty}^{\infty} [\delta(t + 0.5) - \delta(t - 0.5)]e^{-j\omega t}dt$$

$$= e^{-j\omega t}\Big|_{t=-0.5} - e^{-j\omega t}\Big|_{t=0.5}$$

$$= e^{j0.5\omega} - e^{-j0.5\omega} = 2j\sin(0.5\omega)$$

Hence,

$$|X(\omega)| = |2\sin 0.5\omega|$$

and

$$\theta(\omega) = \angle X(\omega) = \angle 2 + \angle j + \angle \sin(0.5\omega)$$

For positive values of $\sin(0.5\omega)$, the phase contributed by it is 0 radians and for negative values of $\sin(0.5\omega)$, the phase contributed by it is $-\pi$ radians. Hence,

$$\theta(\omega) = \begin{cases} \frac{\pi}{2} + 0 & \text{for} \quad \sin(0.5\omega) > 0 \\ \frac{\pi}{2} - \pi = -\frac{\pi}{2} & \text{for} \quad \sin(0.5\omega) < 0 \end{cases}$$

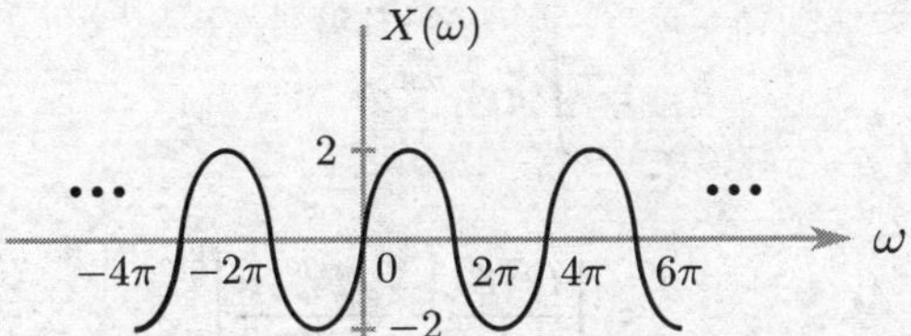

Fig. Ex. 3.2(b)

Please note that the zero-crossings of $X(\omega)$ are obtained by equating $\sin(0.5\omega)$ to zero. That is,

$$\sin(0.5\omega) = 0$$
$$\Rightarrow \quad \omega = \pm 2\pi m; \; m = 0, \pm 1, \pm 2, \ldots$$

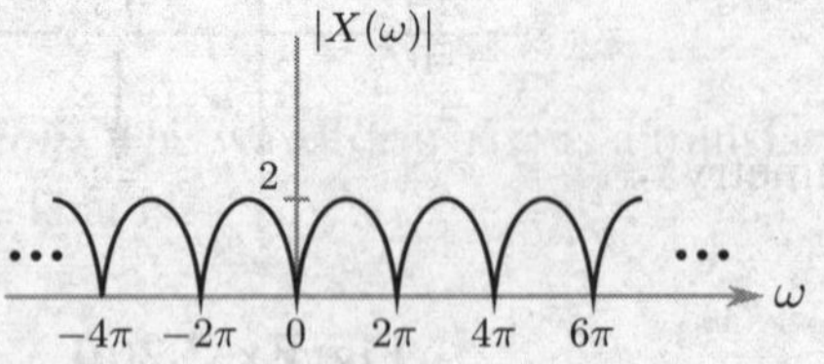

Fig. Ex. 3.2(c) *Amplitude spectrum.*

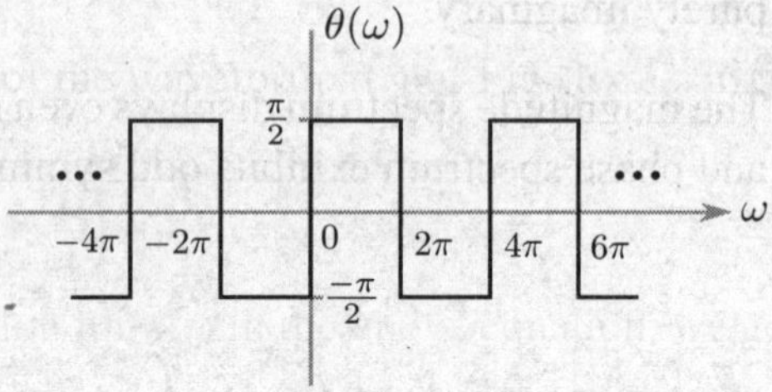

Fig. Ex. 3.2(d) *Phase spectrum.*

Example 3.3. Find the real and imaginary parts of the **FT** of the pulse $x(t)$ shown in Fig. Ex. 3.3.

Solution: We make the following predictions:

1. Since $x(t)$ is real, $X(-\omega) = X^*(\omega)$.
2. $x(t)$ is neither even nor odd. The implication is $X(\omega)$ contains both real and imaginary parts.

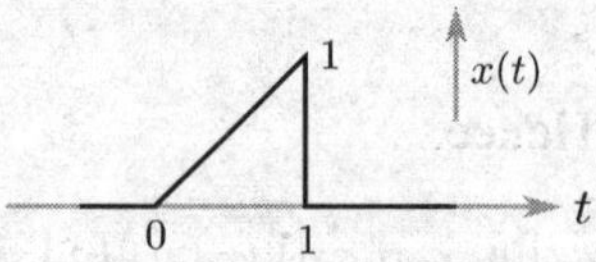

Fig. Ex. 3.3

The analytical definition of the pulse is

$$x(t) = \begin{cases} t, & \text{for } \; 0 < t < 1 \\ 0, & \text{otherwise} \end{cases}$$

$$X(\omega) \triangleq \int_{-\infty}^{\infty} x(t) e^{-j\omega t} dt$$

$$= \int_{0}^{1} t e^{-j\omega t} dt$$

$$= \left[\frac{t e^{j\omega t}}{-j\omega} - \frac{e^{-j\omega t}}{(-j\omega)^2} \right]_0^1$$

$$= \frac{e^{-j\omega}}{-j\omega} - \frac{(e^{-j\omega} - 1)}{(-j\omega)^2}$$

$$= \frac{1 - e^{-j\omega} - j\omega e^{-j\omega}}{(j\omega)^2}$$

Resolving $X(\omega)$ into real and imaginary parts, we get

$$A(\omega) = \frac{\omega \sin \omega + \cos \omega - 1}{\omega^2}$$

and

$$B(\omega) = \frac{\omega \cos \omega - \sin \omega}{\omega^2}$$

From the above expressions, we find that $A(\omega)$ is even and $B(\omega)$ is odd.

Also,

$$|X(\omega)| = \sqrt{A^2(\omega) + B^2(\omega)}$$

Note that, $B(\omega)$ is odd, but $B^2(\omega)$ is even. Since, both $A^2(\omega)$ and $B^2(\omega)$ are even, $|X(\omega)|$ is also even.

Phase of $X(\omega)$:

$$\theta(\omega) = \tan^{-1}\left[\frac{B(\omega)}{A(\omega)}\right]$$

Example 3.4. Find the Fourier transform of the following functions (known as Basic Fourier transform pairs):

(a) The unit impulse.

(b) The decaying exponential.

(c) The rect function.

Solution:

(a) The **FT** of $x(t) = \delta(t)$ is found by applying the sifting property of an impulse function.

$$X(\omega) \triangleq \int_{-\infty}^{\infty} x(t)e^{-j\omega t}dt$$

$$\Rightarrow \quad X(\omega) = \int_{-\infty}^{\infty} \delta(t)e^{-j\omega t}dt$$

$$= e^{-j\omega t}\Big|_{t=0} = 1$$

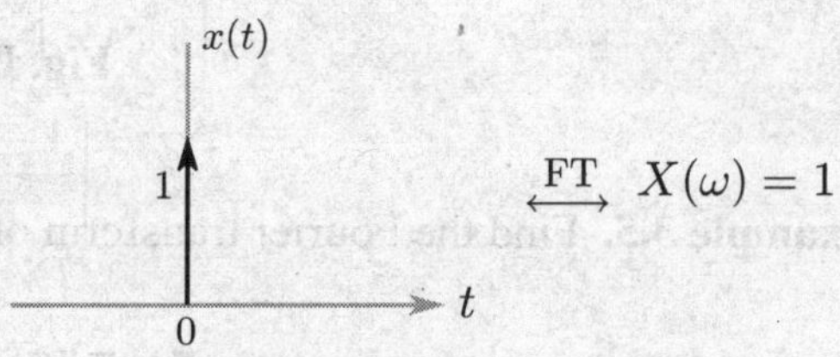

Fig. Ex. 3.4(a)

(b) Here, $x(t) = e^{-\alpha t}u(t)$.

$$X(\omega) = \int_{-\infty}^{\infty} e^{-\alpha t}u(t)e^{-j\omega t}dt$$

$$= \int_{0}^{\infty} e^{-(\alpha+j\omega)t}dt = \frac{1}{\alpha + j\omega}$$

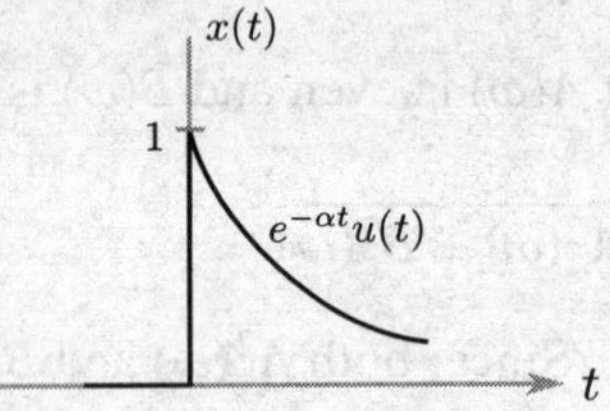

$$\xleftrightarrow{\text{FT}} \quad X(\omega) = \frac{1}{\alpha+j\omega}$$

Fig. Ex. 3.4(b)

(c) The signal $x(t) = \text{rect}(t)$ is unity over the interval −0.5 to +0.5 and zero elsewhere.

$$X(\omega) = \int_{-0.5}^{0.5} 1 \times e^{-j\omega t}dt = \left.\frac{-e^{j\omega t}}{-j\omega}\right|_{t=-0.5}^{t=0.5}$$

$$= \frac{\sin(0.5\omega)}{0.5\omega}$$

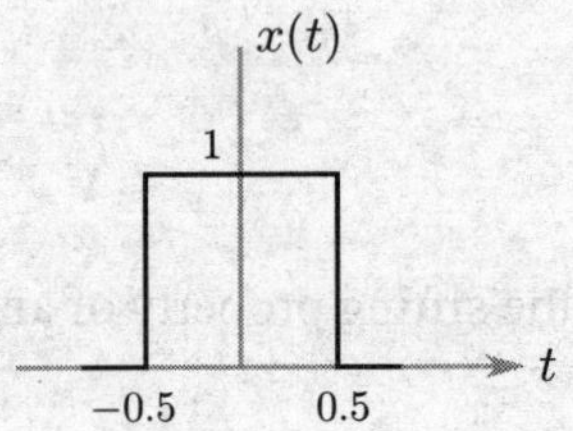

$$\xleftrightarrow{\text{FT}} \quad X(\omega) = \frac{(\sin 0.5\omega)}{0.5\omega}$$

Fig. Ex. 3.4(c)

Example 3.5. Find the Fourier transform of

$$x(t) = te^{-at}u(t); \quad a > 0$$

Solution: We know that

$$X(\omega) = \int_{-\infty}^{\infty} x(t)e^{-j\omega t}dt$$

$$\Rightarrow \quad X(\omega) = \int_{0}^{\infty} te^{-at}e^{-j\omega t}dt$$

$$= \int_{0}^{\infty} te^{-(a+j\omega)t}dt$$

Recall the standard integral:

$$\int xe^{ax}dx = \frac{e^{ax}}{a^2}[ax-1]$$

Then,

$$X(\omega) = \frac{e^{-(a+j\omega)t}}{(a+j\omega)^2}[-(j\omega+a)t-1]\Big|_0^{\infty}$$

Interchanging the limits of integration, we get

$$X(\omega) = \frac{(j\omega+a)t+1}{e^{(a+j\omega)t}(j\omega+a)^2}\Big|_{\infty}^{0}$$

Substituting the upper limit of integration, we get

$$X(\omega)\big|_{t=0} = \frac{1}{(a+j\omega)^2}$$

To find $X(\omega)$ at the lower limit, $t=\infty$, we apply L'Hôpital's rule, that is,

$$\frac{(j\omega+a)t+1}{e^{(j\omega+a)t}(j\omega+a)^2}\Big|_{t=\infty} = \lim_{t\to\infty}\frac{\frac{d}{dt}[(j\omega+a)t+1]}{\frac{d}{dt}[e^{(j\omega+a)t}(j\omega+a)^2]}$$

$$= \lim_{t\to\infty}\frac{(j\omega+a)}{(j\omega+a)e^{(j\omega+a)t}(j\omega+a)^2}$$

$$= 0$$

Hence,

$$X(\omega) = \frac{1}{(j\omega+a)^2}$$

That is,

$$x(t) = te^{-at}u(t) \overset{\textbf{FT}}{\longleftrightarrow} X(\omega) = \frac{1}{(j\omega+a)^2}$$

Example 3.6. Find the Fourier transform of $x(t) = u(t)$, the unit step fuction.

Solution: Since, $\int_{-\infty}^{\infty} |x(t)|dt = \int_{-\infty}^{\infty} |u(t)|dt = \infty$, we consider $x(t)$ as the limiting form of the decaying exponential signal, $e^{-at}u(t)$, as $a \to 0$.

$$\text{Recall: } e^{-at}u(t) \overset{\textbf{FT}}{\longleftrightarrow} \frac{1}{a + j\omega}$$

Hence, $x(t) = u(t)$ has the Fourier transform (**FT**) given by $X(\omega) = \lim_{a\to 0} \frac{1}{j\omega+a}$. Rationalizing, we get

$$\begin{aligned} X(\omega) &= \lim_{a\to 0} \frac{1}{j\omega + a} \times \frac{a - j\omega}{(a + j\omega)} \\ &= \lim_{a\to 0} \frac{a}{a^2 + \omega^2} - \lim_{a\to 0} \frac{j\omega}{a^2 + \omega^2} \\ &= \lim_{a\to 0} \frac{a}{a^2 + \omega^2} + \frac{1}{j\omega} \end{aligned}$$

Note that $\int_{-\infty}^{\infty} \frac{a}{a^2 + \omega^2} d\omega = \pi$ and as $a \to 0$, we find that $\frac{a}{a^2 + \omega^2}$ becomes compressed and tall. Furthermore, $\frac{a^2}{a^2 + \omega^2}$ is an even function of ω. For better understanding of this concept, refer Example 1.37.

Thus,

$$\lim_{a\to 0} \frac{a}{a^2 + \omega^2} = \pi\delta(\omega)$$

Therefore,

$$x(t) = u(t) \overset{\textbf{FT}}{\longleftrightarrow} X(\omega) = \frac{1}{j\omega} + \pi\delta(\omega)$$

3.4 Properties of the Fourier Transform

Properties present the frequency-domain effect of time-domain characteristics and operations on signals and vice-versa. Also, they are used to find new transform pairs more easily.

We denote the relationship between a signal and its Fourier transform as $x(t) \overset{\textbf{FT}}{\longleftrightarrow} X(\omega)$. We derive some of the useful properties using Eq. (3.4) and Eq. (3.7). Note that most of these properties are presented in table 3.1 in both ω-form and f-form. The conversion between the two form is sometimes not very straightforward and may include (or omit) factors of 2π. The f-form is required for studying the implications of sampling.

3.4.1 Linearity

$$\text{If } x(t) \overset{\textbf{FT}}{\longleftrightarrow} X(\omega)$$

$$\text{and } y(t) \overset{\textbf{FT}}{\longleftrightarrow} Y(\omega)$$

$$\text{then, } z(t) = ax(t) + by(t) \overset{\textbf{FT}}{\longleftrightarrow} Z(\omega) = aX(\omega) + bY(\omega)$$

Proof: By Definition,

$$Z(\omega) = \int_{-\infty}^{\infty} z(t)e^{-j\omega t}dt$$

$$= \int_{-\infty}^{\infty} [ax(t) + by(t)]e^{-j\omega t}dt$$

$$= a\int_{-\infty}^{\infty} x(t)e^{-j\omega t}dt + b\int_{-\infty}^{\infty} y(t)e^{-j\omega t}dt$$

$$= aX(\omega) + bY(\omega)$$

Thus, **FT** of a linear combination of a set of signals is the same linear combination of their individual **FT**.

3.4.2 Time shift

$$\text{If } \quad x(t) \overset{\textbf{FT}}{\longleftrightarrow} X(\omega), \quad \text{then}$$

$$y(t) = x(t - t_0) \overset{\textbf{FT}}{\longleftrightarrow} Y(\omega) = e^{-j\omega t_0}X(\omega)$$

That is, a time shift results in a phase shift in its Fourier transform.

Proof: By definition

$$Y(\omega) = \int_{-\infty}^{\infty} y(t)e^{-j\omega t}dt$$

$$= \int_{-\infty}^{\infty} x(t - t_0)e^{-j\omega t}dt$$

Put $t - t_0 = \lambda$, then $dt = \lambda$.

This gives

$$Y(\omega) = \int_{-\infty}^{\infty} x(\lambda)e^{-j\omega(\lambda+t_0)}d\lambda$$

$$= e^{-j\omega t_0} \int_{-\infty}^{\infty} x(\lambda)e^{-j\omega\lambda}d\lambda$$

$$= e^{-j\omega t_0}X(\omega)$$

This completes the proof.

3.4.3 Frequency shift

$$\text{If} \quad x(t) \overset{\textbf{FT}}{\longleftrightarrow} X(\omega),$$

$$\text{then} \quad y(t) = e^{j\beta t}x(t) \overset{\textbf{FT}}{\longleftrightarrow} Y(\omega) = X(\omega - \beta)$$

That is, multiplication of a signal in time-domain by an exponential, $e^{j\beta t}$ results in shifting of spectrum by β.

Proof: By definition,

$$Y(\omega) = \int_{-\infty}^{\infty} y(t)e^{-j\omega t}dt$$

$$\Rightarrow \quad Y(\omega) = \int_{-\infty}^{\infty} e^{j\beta t}x(t)e^{-j\omega t}dt$$

$$= \int_{-\infty}^{\infty} x(t)e^{-j(\omega-\beta)t}dt$$

Comparsion of this result with $X(\omega) = \int_{-\infty}^{\infty} x(t)e^{-j\omega t}dt$ gives

$$Y(\omega) = X(\omega - \beta)$$

It may be noted that time shift and frequency shift properties are duals of each other.

3.4.4 Scaling

$$\text{If } x(t) \overset{\textbf{FT}}{\longleftrightarrow} X(\omega), \quad \text{then}$$

$$y(t) = x(at) \overset{\textbf{FT}}{\longleftrightarrow} Y(\omega) = \frac{1}{|a|} X\left(\frac{\omega}{a}\right)$$

That is, scaling the time variable with 'a' either expands or contracts the Fourier transform.

Proof: By definition,

$$Y(\omega) = \int_{-\infty}^{\infty} y(t) e^{-j\omega t} dt$$

$$\Rightarrow \quad Y(\omega) = \int_{-\infty}^{\infty} x(at) e^{-j\omega t} dt$$

If the subutitution $\lambda = at$ is made with $a > 0$, then $d\lambda = a\,dt$ and the above equation can be written as

$$Y(\omega) = \frac{1}{a} \int_{-\infty}^{\infty} x(\lambda) e^{-j\left(\frac{\omega}{a}\right)\lambda} d\lambda$$

Comparsion of this result with $X(\omega) = \int_{-\infty}^{\infty} x(t) e^{-j\omega t} dt$ gives

$$Y(\omega) = \frac{1}{a} X\left(\frac{\omega}{a}\right)$$

If $a < 0$, it can be shown that

$$Y(\omega) = -\frac{1}{a} X\left(\frac{\omega}{a}\right)$$

Hence,

$$Y(\omega) = \frac{1}{|a|} X\left(\frac{\omega}{a}\right)$$

The absolute value sign on the scaling factor a allows the scaling property to be applicable when 'a' has either postive or negative values. The scaling property is its own dual because scaling by 'a' in one domain results in inverse scaling, $\frac{1}{a}$ and amplitude scaling by $\frac{1}{|a|}$ in the other. The scaling factor, $\frac{1}{|a|}$ ensures that the scaled signal in time-domain and the scaled spectrum in frequency-domain contain the same energy.

For example,

$$x(0.5t) \overset{\textbf{FT}}{\longleftrightarrow} 2X(2\omega)$$

Thus, expansion of $x(t)$ by a factor of 2 results in the compression of $X(\omega)$ by a factor of 2.

3.4.5 Time differentiation

$$\text{If } x(t) \overset{\mathbf{FT}}{\longleftrightarrow} X(\omega), \quad \text{then}$$

$$\frac{dx(t)}{dt} \overset{\mathbf{FT}}{\longleftrightarrow} j\omega X(\omega)$$

That is, differentiating a signal results in a multiplication of the Fourier transform by $j\omega$.

Proof: From the definition of inverse Fourier transform, we have

$$x(t) = \frac{1}{2\pi} \int_{-\infty}^{\infty} X(\omega) e^{j\omega t} d\omega$$

Differentiating both sides with respect to t, we get

$$\frac{dx(t)}{dt} = \frac{1}{2\pi} \int_{-\infty}^{\infty} [j\omega X(\omega)] e^{j\omega t} d\omega$$

Comparing the above result with the defining equation for inverse Fourier transform, we get

$$\frac{dx(t)}{dt} \overset{\mathbf{FT}}{\longleftrightarrow} j\omega X(\omega)$$

In general,

$$\frac{d^n x(t)}{dt^n} \overset{\mathbf{FT}}{\longleftrightarrow} (j\omega)^n X(\omega)$$

3.4.6 Integration or Accumulation

$$\text{If } x(t) \overset{\mathbf{FT}}{\longleftrightarrow} X(\omega), \quad \text{then}$$

$$y(t) = \int_{-\infty}^{t} x(\tau) d\tau \overset{\mathbf{FT}}{\longleftrightarrow} Y(\omega) = \frac{1}{j\omega} X(\omega) + \pi X(0) \delta(\omega)$$

Proof: This proof is better understood after going through convolution integral. However, we breifly define convolution. Convolution is the process of overlapping two signals. The convolution of two fuctions $x_1(t)$ and $x_2(t)$ is denoted by $x_1(t) * x_2(t)$ and by definition,

$$x_1(t) * x_2(t) = \int_{-\infty}^{\infty} x_1(\tau) x_2(t - \tau) d\tau$$

where τ is a dummy variable.

Letting $x_1(t) = x(t)$ and $x_2(t) = u(t)$, we get

$$x(t) * u(t) = \int_{-\infty}^{\infty} x(\tau)u(t-\tau)d\tau$$

Since,

$$u(t-\tau) = \begin{cases} 1, & t-\tau > 0 \text{ or } \tau < t \\ 0, & t-\tau < 0 \text{ or } \tau > t \end{cases}$$

we get

$$x(t) * u(t) = \int_{-\infty}^{t} x(\tau)d\tau$$

Also Fourier transform converts convolution of two signals into product of two signals. (refer chapter 5). That is,

$$x_1(t) * x_2(t) \overset{\textbf{FT}}{\longleftrightarrow} X_1(\omega)X_2(\omega)$$

Hence,

$$y(t) = \int_{-\infty}^{t} x(\tau)d\tau \overset{\textbf{FT}}{\longleftrightarrow} Y(\omega) = X(\omega)U(\omega)$$

Thus,

$$\begin{aligned} Y(\omega) &= X(\omega)U(\omega) \\ &= X(\omega)\left[\frac{1}{j\omega} + \pi\delta(\omega)\right] \\ &= \frac{1}{j\omega}X(\omega) + \pi X(\omega)\delta(\omega) \end{aligned}$$

Recall the product property of an impulse fucntion:

$$f(t)\delta(t-t_0) = f(t_0)\delta(t-t_0)$$

Hence,

$$Y(\omega) = \frac{1}{j\omega}X(\omega) + \pi X(\omega)\Big|_{\omega=0}\delta(\omega)$$

$$\Rightarrow \quad Y(\omega) = \frac{1}{j\omega}X(\omega) + \pi X(0)\delta(\omega)$$

3.4.7 Time Reversal

$$\text{If } x(t) \overset{\textbf{FT}}{\longleftrightarrow} X(\omega), \quad \text{then}$$

$$x(-t) \overset{\textbf{FT}}{\longleftrightarrow} X(-\omega)$$

That is, the time reversal of a signal results in its spectrum also reflected about the vertical axis at the origin.

Proof: By definition,

$$X(\omega) = \int_{t=-\infty}^{\infty} x(t)e^{-j\omega t}dt$$

Replacing t by $-t$ and ω by $-\omega$ in the above equation, we get

$$X(-\omega) = \int_{+\infty}^{-\infty} x(-t)e^{-j\omega t}(-dt)$$

$$= \int_{-\infty}^{\infty} x(-t)e^{-j\omega t}dt$$

Comparing the above result with the defining equation for **FT**, we get

$$x(-t) \overset{\textbf{FT}}{\longleftrightarrow} X(-\omega)$$

3.4.8 Conjugation

If $x(t) \overset{\textbf{FT}}{\longleftrightarrow} X(\omega)$, then $x^*(\pm t) \overset{\textbf{FT}}{\longleftrightarrow} X^*(\mp j\omega)$. This result is obtained if we replace t by $-t$ or ω by $-\omega$, in addition to conjugating both sides of the defining equation for **FT**.

3.4.9 Frequency differentiation

$$\text{If } x(t) \overset{\textbf{FT}}{\longleftrightarrow} X(\omega), \quad \text{then}$$

$$(-jt)x(t) \overset{\textbf{FT}}{\longleftrightarrow} \frac{dX(\omega)}{d\omega}$$

$$\text{or} \quad tx(t) \overset{\textbf{FT}}{\longleftrightarrow} j\frac{dX(\omega)}{d\omega}$$

Proof: By definition,

$$X(\omega) = \int_{-\infty}^{\infty} x(t)e^{-j\omega t}dt$$

Differentiating both sides with respect to ω, we get

$$\frac{dX(\omega)}{d\omega} = \int_{-\infty}^{\infty} [-jtx(t)]e^{-j\omega t}dt$$

Comparing the above result with the defining equation for **FT**, we get

$$-jtx(t) \overset{\textbf{FT}}{\longleftrightarrow} \frac{dX(\omega)}{d\omega} \quad \text{or} \quad tx(t) \overset{\textbf{FT}}{\longleftrightarrow} j\frac{dX(\omega)}{d\omega}$$

This property is valid only if the resulting function is Fourier transformable.
In general,

$$t^n x(t) \overset{\textbf{FT}}{\longleftrightarrow} (j)^n \frac{d^n X(\omega)}{d\omega^n}$$

3.4.10 Duality or the similarity theorem

$$\text{If} \quad x(t) \overset{\textbf{FT}}{\longleftrightarrow} X(\omega) \quad \text{then} \quad X(t) \overset{\textbf{FT}}{\longleftrightarrow} 2\pi x(-\omega)$$

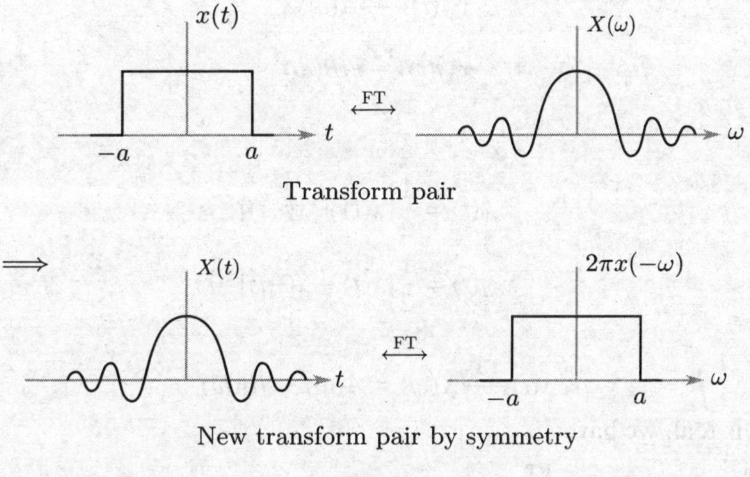

Fig. 3.1

Proof: From the definition of inverse Fourier transform, we have

$$x(t) = \frac{1}{2\pi} \int_{\omega=-\infty}^{\infty} X(\omega)e^{j\omega t}d\omega$$

Interchanging t and ω, we get

$$x(\omega) = \frac{1}{2\pi} \int_{t=-\infty}^{\infty} X(t)e^{jt\omega}dt$$

Replacing ω by $-\omega$, we get

$$x(-\omega) = \frac{1}{2\pi} \int_{t=-\infty}^{\infty} X(t) e^{-j\omega t} dt$$

$$\Rightarrow \quad 2\pi x(-\omega) = \int_{-\infty}^{\infty} X(t) e^{-j\omega t} dt$$

Hence,

$$X(t) \overset{\textbf{FT}}{\longleftrightarrow} 2\pi x(-\omega)$$

3.4.11 Symmetry

Consider a real signal $x(t)$ and let $X(\omega) = F\{x(t)\} = A(\omega) + jB(\omega)$ and $x(t) = x_{ev}(t) + x_{od}(t)$, where $x_{ev}(t)$ and $x_{od}(t)$ are the even and odd components of $x(t)$ respectively.

Then,

$$x_{ev}(t) \overset{\textbf{FT}}{\longleftrightarrow} A(\omega)$$

$$x_{od}(t) \overset{\textbf{FT}}{\longleftrightarrow} jB(\omega)$$

Proof:

$$x_{ev}(t) = \frac{1}{2}[x(t) + x(-t)]$$

$$x_{od}(t) = \frac{1}{2}[x(t) - x(-t)]$$

Given,

$$x(t) \overset{\textbf{FT}}{\longleftrightarrow} X(\omega) = A(\omega) + jB(\omega)$$

Since $x(t)$ is real, we have

$$x(-t) \overset{\textbf{FT}}{\longleftrightarrow} X(-\omega) = X^*(\omega) = A(\omega) - jB(\omega)$$

Thus, we find that

$$x_{ev}(t) \overset{\textbf{FT}}{\longleftrightarrow} \frac{1}{2}[X(\omega) + X^*(\omega)]$$

$$= A(\omega)$$

$$x_{od}(t) \overset{\textbf{FT}}{\longleftrightarrow} \frac{1}{2}[X(\omega) - X^*(\omega)]$$

$$= jB(\omega)$$

Thus, we have shown that the Fourier transform of a real even signal is a real function of ω, and that of a real odd signal is an imaginary function of ω.

3.1: Properties of Fourier Transform

Property	$x(t)$	$X(\omega)$	$X(f)$
Linearity	$a_1x_1(t) + a_2x_2(t)$	$a_1X_1(\omega) + a_2X_2(\omega)$	$a_1X_1(f) + a_2X_2(f)$
Time shift	$x(t - t_0)$	$e^{-j\omega t_0}X(\omega)$	$e^{-j2\pi f t_0}X(f)$
Frequency shift	$e^{j\beta t}x(t)$	$X(\omega - \beta)$	$X\left(f - \frac{\beta}{2\pi}\right)$
Scaling	$x(at)$	$\frac{1}{\lvert a\rvert}X\left(\frac{\omega}{a}\right)$	$\frac{1}{\lvert a\rvert}X\left(\frac{f}{a}\right)$
Time differentiation	$\frac{dx(t)}{dt}$	$j\omega X(\omega)$	$j2\pi f X(f)$
Integration	$\int_{-\infty}^{t} x(\tau)d\tau$	$\frac{1}{j\omega}X(\omega) + \pi X(0)\delta(\omega)$	$\frac{1}{j2\pi f}X(f) + 0.5X(0)\delta(f)$
Time reversal	$x(-t)$	$X(-\omega)$	$X(-f)$
Conjugation	$x^*(\pm t)$	$X^*(\mp\omega)$	$X^*(\mp f)$
Frequency differentiation	$-jtx(t)$	$\frac{dX(\omega)}{d\omega}$	$\frac{1}{2\pi}\frac{dX(f)}{df}$
Duality	$X(t)$	$2\pi x(-\omega)$	$x(-f)$

Example 3.7. Compute the Fourier transform $X(f)$ of the following signals and explain which transform will show conjugate symmetry.

(a) $x(t) = e^{-t-1}u(t)$, (b) $x(t) = e^{-t-j\pi t}u(t)$.

Solution:

(a)

$$X(\omega) \triangleq \int_{-\infty}^{\infty} x(t)e^{-j\omega t}dt$$

$$\Rightarrow \quad X(f) = \int_{-\infty}^{\infty} x(t)e^{-j2\pi f t}df$$

$$= \int_{-\infty}^{\infty} e^{-t-1}u(t)e^{-j2\pi f t}df$$

$$= e^{-1}\int_{0}^{\infty} e^{-(1+j2\pi f)t}dt$$

$$= \frac{e^{-1}}{1 + j2\pi f}$$

$X(-f) = X^*(f)$, because $x(t)$ is real.

(b) Let

$$y(t) = e^{-t}u(t)$$

Then,

$$Y(f) = \int_{-\infty}^{\infty} y(t)e^{-j2\pi ft}dt$$
$$= \int_{0}^{\infty} e^{-t}e^{-j2\pi ft}dt$$
$$= \frac{1}{1 + j2\pi f}$$

Now,

$$x(t) = e^{-j\pi t}y(t)$$
$$= e^{-j2\pi(\frac{1}{2})t}y(t)$$

Applying frequency shift property: $e^{j2\pi\beta t}x(t) \overset{\textbf{FT}}{\longleftrightarrow} X(f - \beta)$, we get

$$X(f) = Y(f + 0.5)$$
$$= \frac{1}{1 + j2\pi(f + 0.5)}$$

Since $x(t)$ is complex, $X(-f) \neq X^*(f)$, that is, no conjugate symmetry.

Example 3.8. Sketch the signal $x(t)$ and find its Fourier transform.

$$x(t) = r(t) - r(t - 1) - u(t - 1)$$

Solution:

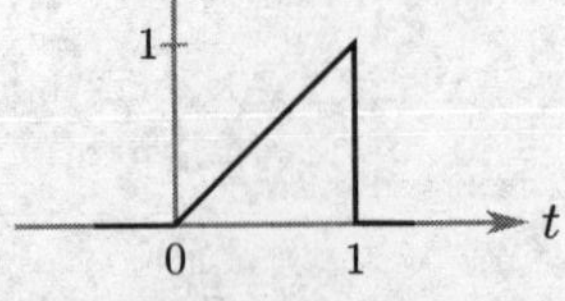

Fig. Ex. 3.8

$$X(\omega) \triangleq \int_{-\infty}^{\infty} x(t)e^{-j\omega t}dt$$

$$= \int_{0}^{1} te^{-j\omega t}dt$$

$$= \left.\frac{(1 + j\omega t)e^{-j\omega t}}{\omega^2}\right|_0^1$$

$$= \frac{(1 + j\omega)e^{-j\omega} - 1}{\omega^2}$$

Example 3.9. Find the Fourier transform of a rectangular pulse described below:

$$x(t) = \begin{cases} 1, & |t| < a \\ 0, & |t| > a \end{cases}$$

Solution:

$$X(\omega) \triangleq \int_{-\infty}^{\infty} x(t)e^{-j\omega t}dt$$

$$= \int_{-a}^{a} 1 \times e^{-j\omega t}dt \qquad (3.9)$$

$$= \left[\frac{e^{-j\omega t}}{-j\omega}\right]_{t=-a}^{a} = \frac{2\sin(a\omega)}{\omega}$$

$$= 2a\left[\frac{\sin(a\omega)}{(a\omega)}\right]$$

$$= 2a \text{ sa}(\omega), \quad \omega \neq 0$$

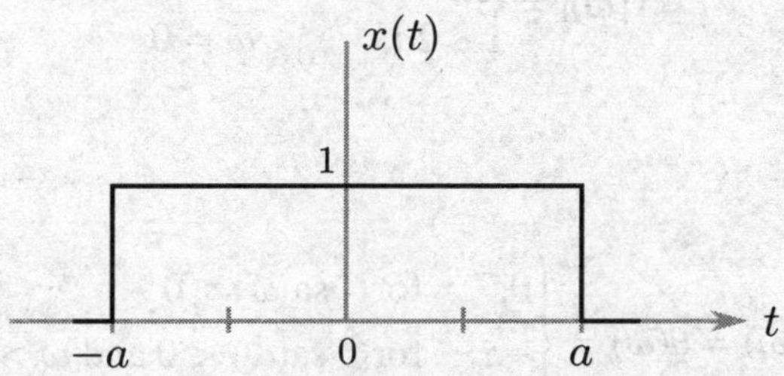

Fig. Ex. 3.9(a)

To find the value of $X(\omega)$ at $\omega = 0$, put $\omega = 0$ in Eq. (3.9).

$$X(0) = \int_{-a}^{a} 1 \times e^0 dt = 2a$$

To find the zero-crossings of $X(\omega)$, equate $X(\omega)$ to zero, that is,

$$\begin{aligned} &\sin(a\omega) = 0 \\ \Rightarrow \quad & a\omega = \pm n\pi; \; n = 1, 2, 3, 4 \ldots \\ \Rightarrow \quad & \omega = \pm \frac{n\pi}{a} \end{aligned}$$

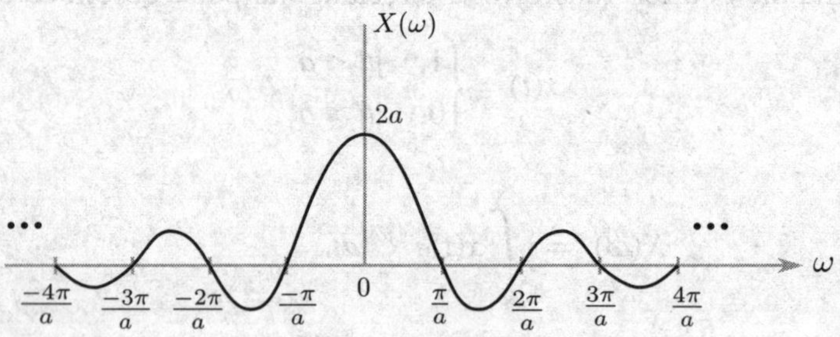

Fig. Ex. 3.9(b)

$$X(\omega) = \begin{cases} 2a \text{ sa}(\omega), & \omega \neq 0 \\ 2a, & \omega = 0 \end{cases}$$

Magnitude Spectrum:

$$|X(\omega)| = \begin{cases} |2a \text{ sa}(\omega)|, & \omega \neq 0 \\ 2a, & \omega = 0 \end{cases}$$

Phase Spectrum:

$$\measuredangle X(\omega) = \theta(\omega) = \begin{cases} 0, & \text{for } \text{sa}(\omega) \bullet 0 \\ -\pi, & \text{for } \text{sa}(\omega) < 0 \text{ and } \omega > 0 \\ +\pi, & \text{for } \text{sa}(\omega) < 0 \text{ and } \omega < 0 \end{cases}$$

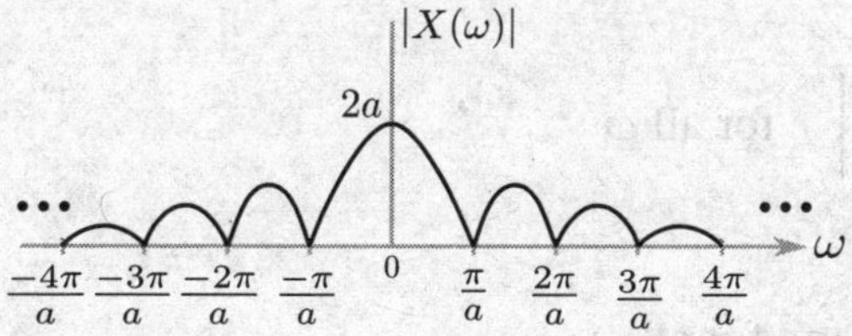

Fig. Ex. 3.9(c) *Magnitude spectrum.*

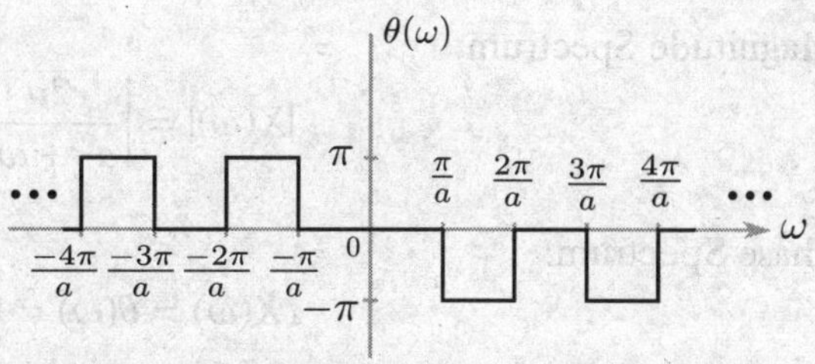

Fig. Ex. 3.9(d) *Phase spectrum.*

Example 3.10. Find the Fourier transform of the signal

$$x(t) = e^{-a|t|}, \quad a > 0$$

Also, sketch the magnitude and phase spectra.

Solution:

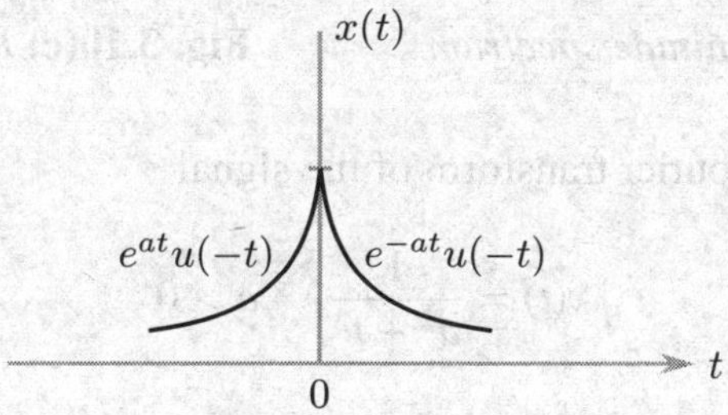

Fig. Ex. 3.10(a)

We predict the following:

(i) Since, $x(t)$ is a real function, we have, $X(-\omega) = X^*(\omega)$. Hence, $|X(\omega)|$ versus ω exhibits even symmetry and $\theta(\omega)$ versus ω exhibits odd symmetry.

(ii) Since, $x(t)$ is an even function, $X(\omega)$ will be purely real and an even function of ω.

(iii) Since, $X(\omega)$ is purely real, it implies $\angle X(\omega) = \theta(\omega)$ will be either 0 or $\pm\pi$.

$$X(\omega) \triangleq \int_{-\infty}^{\infty} x(t)e^{-j\omega t}dt$$

$$\Rightarrow \quad X(\omega) = \int_{-\infty}^{0} e^{at}e^{-j\omega t}dt + \int_{0}^{\infty} e^{-at}e^{-j\omega t}dt$$

$$= \frac{1}{a - j\omega} + \frac{1}{a + j\omega}$$

$$= \frac{2a}{a^2 + \omega^2}$$

Magnitude Spectrum:

$$|X(\omega)| = \left|\frac{2a}{a^2 + \omega^2}\right| \quad \text{for all } \omega$$

Phase Spectrum:

$$\angle X(\omega) = \theta(\omega) = 0 \text{ rad for all } \omega$$

⇒ phase spectrum is a null spectrum.

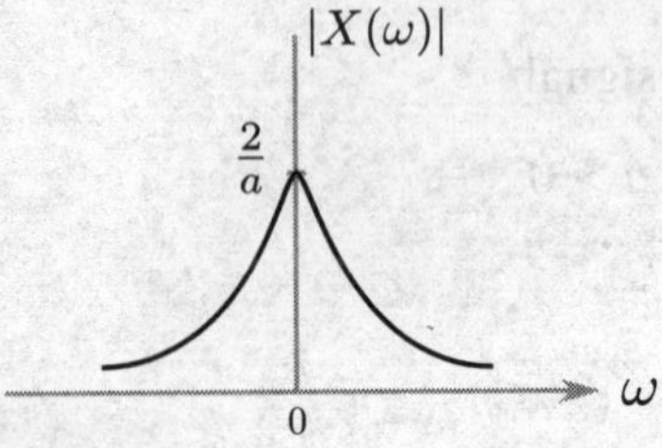

Fig. Ex. 3.10(b) *Magnitude spectrum.*

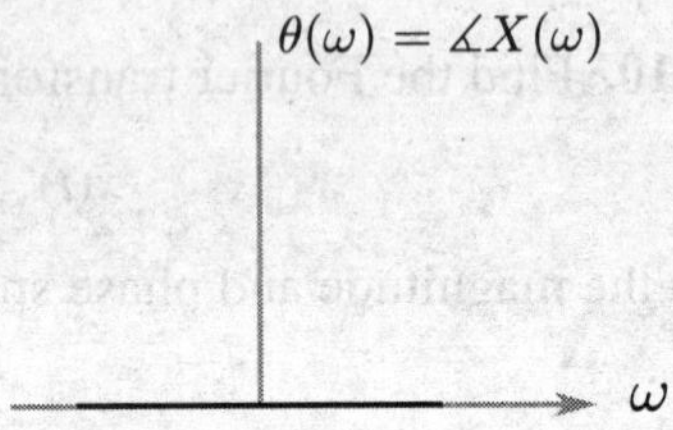

Fig. 3.10(c) *Phase spectrum.*

Example 3.11. Find the Fourier transform of the signal

$$y(t) = \frac{1}{a^2 + t^2}, \quad a > 0$$

Solution: From the immediate previous problem, we have

$$x(t) = e^{-a|t|} \overset{\textbf{FT}}{\longleftrightarrow} X(\omega) = \frac{2a}{a^2 + \omega^2}$$

Applying duality property,

$$X(t) \overset{\textbf{FT}}{\longleftrightarrow} 2\pi x(-\omega)$$

we get,

$$\frac{2a}{a^2 + t^2} \overset{\textbf{FT}}{\longleftrightarrow} 2\pi e^{-a|-\omega|}$$

$$\Rightarrow \quad \frac{1}{a^2 + t^2} \overset{\textbf{FT}}{\longleftrightarrow} \frac{2\pi}{2a} e^{-a|\omega|}$$

Hence,

$$\frac{1}{a^2 + t^2} \overset{\textbf{FT}}{\longleftrightarrow} \frac{\pi}{a} e^{-a|\omega|}$$

Example 3.12. Find the Fourier transform of $x(t) = 1$.

Solution: Since,

$$\int_{-\infty}^{\infty} |x(t)|dt = \int_{-\infty}^{\infty} dt \rightarrow \infty$$

the Dirichlet condition is not satisfied. We will now show that, Fourier transform of $x(t)$ still exists and has to be found using some of the properties of Fourier transaforms, already discussed.

$$\delta(t) \overset{\mathbf{FT}}{\longleftrightarrow} \delta(\omega) = \int_{-\infty}^{\infty} \delta(t)e^{-j\omega t}dt = e^{-j\omega t}\Big|_{t=0} = 1$$

The duality property is, if

$$g(t) \overset{\mathbf{FT}}{\longleftrightarrow} G(\omega)$$

then,

$$G(t) \overset{\mathbf{FT}}{\longleftrightarrow} 2\pi g(-\omega)$$

$$\Rightarrow \quad 1 \overset{\mathbf{FT}}{\longleftrightarrow} 2\pi\delta(-\omega)$$

Since, $\delta(\omega)$ is an even function of ω,

$$\delta(-\omega) = \delta(\omega)$$

Therefore,

$$1 \overset{\mathbf{FT}}{\longleftrightarrow} 2\pi\delta(\omega)$$

Corollary:

$$A \overset{\mathbf{FT}}{\longleftrightarrow} 2\pi A\delta(\omega)$$

Example 3.13. Find the Fourier transform of $x(t) = \cos\omega_0 t$ (Eternal sinuoid). Also, plot the magnitude and phase spectra.

Solution:

$$\begin{aligned} x(t) &= \cos\omega_0 t \\ &= \frac{1}{2}e^{j\omega_0 t} + \frac{1}{2}e^{-j\omega_0 t} \end{aligned}$$

We know that,

$$A \overset{\mathbf{FT}}{\longleftrightarrow} 2\pi A\delta(\omega)$$

Hence,

$$\frac{1}{2} \overset{\mathbf{FT}}{\longleftrightarrow} \pi\delta(\omega)$$

Applying frequency shift property, we get,

$$\frac{1}{2}e^{j\omega_0 t} \overset{\textbf{FT}}{\longleftrightarrow} \pi\delta(\omega - \omega_0)$$

and

$$\frac{1}{2}e^{-j\omega_0 t} \overset{\textbf{FT}}{\longleftrightarrow} \pi\delta(\omega + \omega_0)$$

Finally, applying linearity property, we get

$$x(t) = \frac{1}{2}e^{j\omega_0 t} + \frac{1}{2}e^{-j\omega_0 t} \overset{\textbf{FT}}{\longleftrightarrow} X(\omega) = \pi[\delta(\omega - \omega_0) + \delta(\omega + \omega_0)]$$

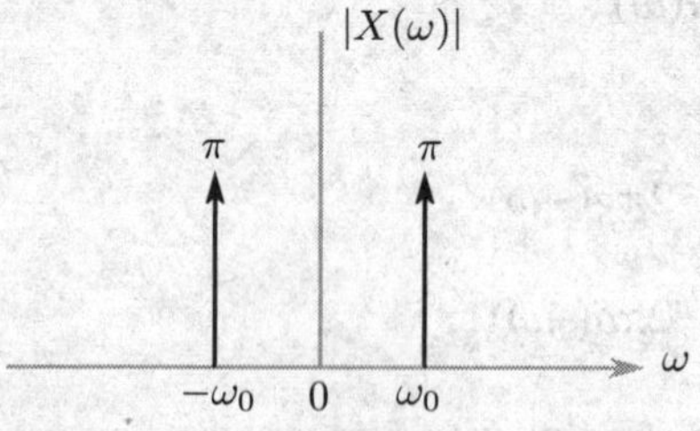

Fig. Ex. 3.13(a) *Amplitude spectrum.*

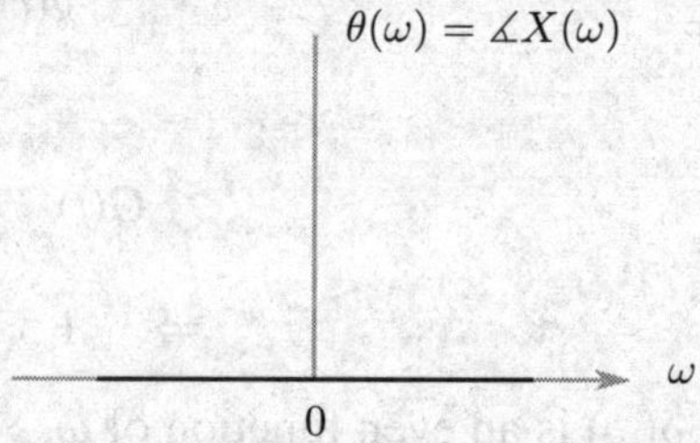

Fig. Ex. 3.13(b) *Phase spectrum.*

Example 3.14. Using the properties of Fourier transforms, find the Fourier transform of the following signals.

(a) $x(t) = \sin(\pi t)e^{-2t}u(t)$,

(b) $x(t) = e^{-3|t-2|}$ and

(c) $x(t) = \frac{d}{dt}[te^{-2t}\sin t\, u(t)]$.

Solution:

(a)

$$x(t) = \frac{e^{-2t}e^{j\pi t}}{2j}u(t) - \frac{e^{-2t}e^{-j\pi t}}{2j}u(t)$$

We know that,

$$e^{-at}u(t) \overset{\textbf{FT}}{\longleftrightarrow} \frac{1}{a + j\omega}$$

Hence,

$$e^{-2t}u(t) \overset{\textbf{FT}}{\longleftrightarrow} \frac{1}{2 + j\omega}$$

Also,

$$e^{j\beta t}s(t)\overset{\text{FT}}{\longleftrightarrow}S(\omega-\beta)$$

Hence,

$$\frac{1}{2j}e^{-2t}e^{j\pi t}u(t)\overset{\text{FT}}{\longleftrightarrow}\frac{1}{2j}\left[\frac{1}{2+j(\omega-\pi)}\right]$$

and

$$\frac{1}{2j}e^{-2t}e^{-j\pi t}u(t)\overset{\text{FT}}{\longleftrightarrow}\frac{1}{2j}\left[\frac{1}{2+j(\omega+\pi)}\right]$$

Finally, applying Linearity property we get

$$X(\omega)=\frac{1}{2j}\left[\frac{1}{2+j(\omega-\pi)}-\frac{1}{2+j(\omega+\pi)}\right]$$

(b) Let $y(t) = e^{-a|t|}$

$$\begin{aligned}Y(\omega) &= \int_{-\infty}^{0} e^{at}e^{-j\omega t}dt + \int_{0}^{\infty} e^{-at}e^{-j\omega t}dt\\ &= \frac{1}{a-j\omega}+\frac{1}{a+j\omega}\\ &= \frac{2a}{a^2+\omega^2}\end{aligned}$$

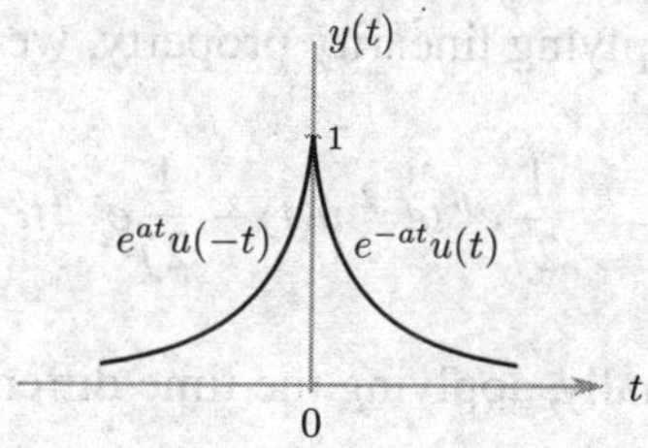

Then,

$$s(t)=e^{-3|t|}\overset{\text{FT}}{\longleftrightarrow}S(\omega)=\frac{6}{9+\omega^2}$$

Applying the time shift property, we get

$$x(t)=s(t-2)\overset{\text{FT}}{\longleftrightarrow}X(\omega)=e^{-j2\omega}S(\omega)$$

Hence,

$$X(\omega)=\frac{6e^{-j2\omega}}{9+\omega^2}$$

(c)

$$\begin{aligned}x(t) &= \frac{d}{dt}[te^{-2t}\sin(t)u(t)]\\ &= \frac{d}{dt}\left[te^{-2t}u(t)\left(\frac{e^{jt}-e^{-jt}}{2j}\right)\right]\end{aligned}$$

Since,

$$te^{-at}u(t)\overset{\textbf{FT}}{\longleftrightarrow}\frac{1}{(a+j\omega)^2}$$

we get,

$$te^{-2t}u(t)\overset{\textbf{FT}}{\longleftrightarrow}\frac{1}{(2+j\omega)^2}$$

Also,

$$e^{j\beta t}s(t)\overset{\textbf{FT}}{\longleftrightarrow}S(\omega-\beta)$$

Hence,

$$e^{jt}te^{-2t}u(t)\overset{\textbf{FT}}{\longleftrightarrow}\frac{1}{[2+j(\omega-1)]^2}$$

and

$$e^{-jt}te^{-2t}u(t)\overset{\textbf{FT}}{\longleftrightarrow}\frac{1}{[2+j(\omega+1)]^2}$$

Applying linearity property, we get

$$\frac{1}{2j}e^{jt}te^{-2t}u(t)-\frac{1}{2j}e^{-jt}te^{-2t}u(t)\overset{\textbf{FT}}{\longleftrightarrow}\frac{1}{2j}\left[\frac{1}{2+j(\omega-1)^2}-\frac{1}{2+j(\omega+1)^2}\right]$$

Finally, applying the time differentiation property,

$$\frac{d}{dt}f(t)\overset{\textbf{FT}}{\longleftrightarrow}j\omega F(\omega)$$

we get

$$\begin{aligned}X(\omega)&=j\omega\frac{1}{2j}\left[\frac{1}{[2+j(\omega-1)]^2}-\frac{1}{[2+j(\omega+1)]^2}\right]\\&=\frac{\omega}{2}\left[\frac{1}{[2+j(\omega-1)]^2}-\frac{1}{[2+j(\omega+1)]^2}\right]\end{aligned}$$

Example 3.15. Find the Fourier transform of the signal $x(t)$ defined below:

$$x(t)=\int_{-\infty}^{t}e^{-a\tau}u(\tau)d\tau$$

Solution: Property to be used:

$$\int_{-\infty}^{t} s(\tau)d\tau \overset{\text{FT}}{\longleftrightarrow} \frac{S(\omega)}{j\omega} + \pi S(0)\delta(\omega)$$

Using the Fourier-transform pair

$$s(t) = e^{-at}u(t) \overset{\text{FT}}{\longleftrightarrow} S(\omega) = \frac{1}{a + j\omega}$$

and, applying the property given above, we get

$$X(\omega) = \frac{1}{j\omega(a + j\omega)} + \pi\frac{1}{a}\delta(\omega)$$

Example 3.16. For the signal $x(t)$ shown in the figure below, evaluate the following quantities without explicity computing $X(\omega)$:

(a) $\int_{-\infty}^{\infty} X(\omega)d\omega$

(b) $\int_{-\infty}^{\infty} X(\omega)e^{j2\omega}d\omega$

(c) $\arg[X(\omega)]$

(d) $X(0)$

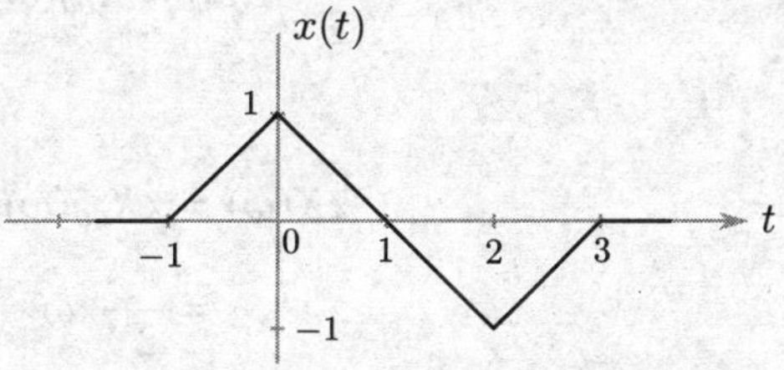

Solution:

(a) Recall: $x(t) = \frac{1}{2\pi}\int_{-\infty}^{\infty} X(\omega)e^{j\omega t}d\omega$.

Letting $t = 0$, we get

$$\int_{-\infty}^{\infty} X(\omega)d\omega = 2\pi x(0)$$

$$= 2\pi \times 1$$
$$= 2\pi$$

(b) We know that,

$$\frac{1}{2\pi}\int_{-\infty}^{\infty} X(\omega)e^{j2\omega}e^{j\omega t}d\omega = x(t + 2)$$

Putting $t = 0$, we get,

$$\frac{1}{2\pi}\int_{-\infty}^{\infty} X(\omega)e^{j2\omega}d\omega = x(2)$$

$$\Rightarrow \int_{-\infty}^{\infty} X(\omega)e^{j2\omega}d\omega = 2\pi x(2)$$

$$= 2\pi \times -1$$

$$= -2\pi$$

(c) $x(t)$ is nothing but an odd and real signal $x_{od}(t)$ shifted to right by 1 sec.

That is,

$$x(t) = x_{od}(t-1)$$

Hence,

$$X(\omega) = X_{od}(\omega)e^{-j\omega}$$

and,

$$\measuredangle X(\omega) = \measuredangle X_{od}(\omega) + \measuredangle e^{-j\omega}$$

$$= \frac{\pi}{2} - \omega$$

[Since, $X_{od}(\omega)$ is purely imaginary, its argument is $\frac{\pi}{2}$ rads.]

(d)

$$X(0) = \int_{-\infty}^{\infty} x(t)e^{-j\omega t}dt\Big|_{\omega=0}$$

$$= \int_{-\infty}^{\infty} x(t)dt = \int_{-1}^{3} x(t)dt = 0$$

Example 3.17. Given the transform pair:

$$x(t) = \begin{cases} 1, & |t| < 1 \\ 0, & (t) > 1 \end{cases} \xleftrightarrow{\textbf{FT}} X(\omega) = \frac{2\sin\omega}{\omega}$$

Using properties of Fourier transform, find the Fourier transforms of the functions shown in Fig. Ex. 3.17.

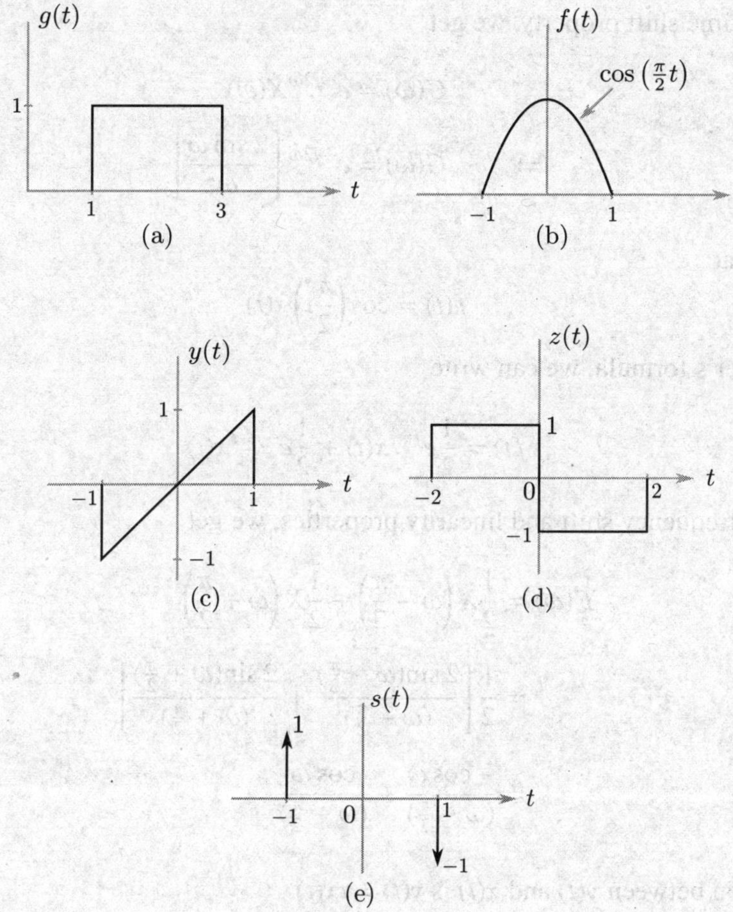

Fig. Ex. 3.17

Solution:

(a) The sketch of $x(t)$ is as shown in Fig. Ex. 3.17(f).

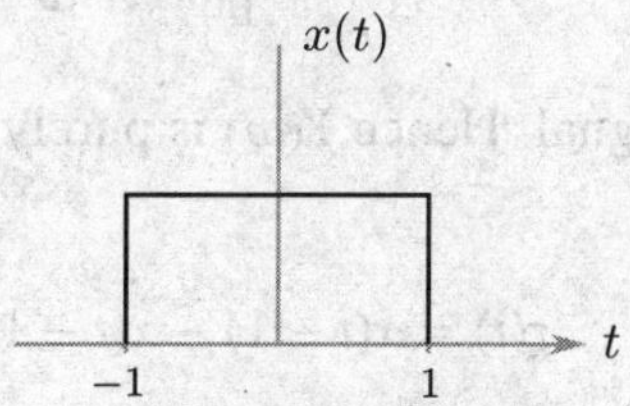

Fig. Ex. 3.17(f)

It may be noted that

$$g(t) = x(t - 2)$$

Applying time shift property, we get

$$G(\omega) = e^{-j2\omega}X(\omega)$$

$$\Rightarrow \qquad G(\omega) = e^{-j2\omega}\left[\frac{2\sin\omega}{\omega}\right]$$

(b) We find that

$$f(t) = \cos\left(\frac{\pi}{2}t\right)x(t)$$

Using Euler's formula, we can write

$$ft) = \frac{1}{2}e^{j\frac{\pi}{2}t}x(t) + \frac{1}{2}e^{-j\frac{\pi}{2}t}x(t)$$

Applying frequency shift and linearity properties, we get

$$\begin{aligned} F(\omega) &= \frac{1}{2}X\left(\omega - \frac{\pi}{2}\right) + \frac{1}{2}X\left(\omega + \frac{\pi}{2}\right) \\ &= \frac{1}{2}\left[\frac{2\sin(\omega - \frac{\pi}{2})}{(\omega - \frac{\pi}{2})} + \frac{2\sin(\omega + \frac{\pi}{2})}{(\omega + \frac{\pi}{2})}\right] \\ &= \frac{-\cos\omega}{(\omega - \frac{\pi}{2})} + \frac{\cos\omega}{(\omega + \frac{\pi}{2})} \end{aligned}$$

(c) The relation between $y(t)$ and $x(t)$ is $y(t) = tx(t)$.

Applying frequency differentiation property, we get

$$Y(\omega) = j\frac{d}{d\omega}X(\omega)$$

$$\Rightarrow \qquad Y(\omega) = j2\left[\frac{\omega\cos\omega - \sin\omega}{\omega^2}\right]$$

Comment: $y(t)$ is an odd signal. Hence, $Y(\omega)$ is purely imaginary.

(d) We find that

$$z(t) = x(t+1) - x(t-1)$$

Applying time shift and linearity properties, we get

$$\begin{aligned} Z(\omega) &= e^{j\omega}X(\omega) - e^{-j\omega}X(\omega) \\ &= 2jX(\omega)\left[\frac{e^{j\omega} - e^{-j\omega}}{2j}\right] \end{aligned}$$

$$= 2j \sin\omega \left[\frac{2\sin\omega}{\omega}\right]$$

$$= j4\frac{\sin^2\omega}{\omega}$$

Comment: Since $z(t)$ is an odd signal, $Z(\omega)$ is purely imaginary.

(e) The relation between $x(t)$ and $s(t)$ is

$$s(t) = \frac{dx(t)}{dt}$$

Applying time differentiation property, we get

$$S(\omega) = j\omega X(\omega)$$

$$\Rightarrow \qquad S(\omega) = j\omega\left[\frac{2\sin\omega}{\omega}\right]$$

Example 3.18. Find the **FT** of the pulse $x(t)$ shown in Fig. Ex. 3.18.

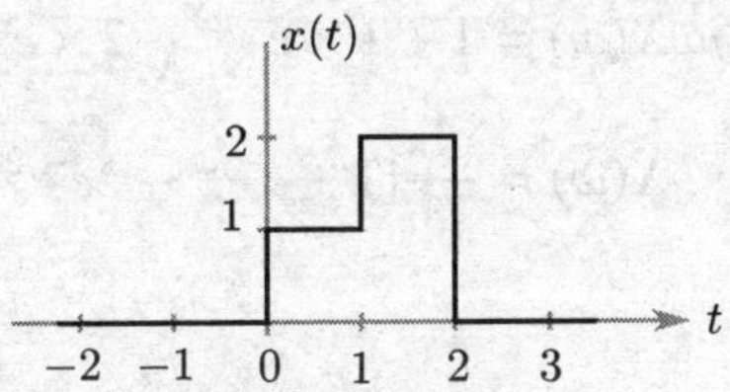

Fig. Ex. 3.18

Solution: The analytical expression for $x(t)$ is as follows:

$$x(t) = \begin{cases} 1, & 0 < t < 1 \\ 2, & 1 < t < 2 \end{cases}$$

$$X(\omega) \triangleq \int_{-\infty}^{\infty} x(t)e^{j\omega t}dt$$

$$= \int_0^1 1 \times e^{-j\omega t}dt + \int_1^2 2 \times e^{-j\omega t}dt$$

$$= \frac{1}{j\omega}[1 + e^{-j\omega} - 2e^{-j2\omega}]$$

An alternate method, known as the method of successive differentiation is often used to find the **FT** of a pulse.

In this method, we keep differentiating $x(t)$, till we get only impulses. In the present problem, $\frac{dx(t)}{dt}$ contains only impulses as shown in Fig. Ex. 3.18(a).

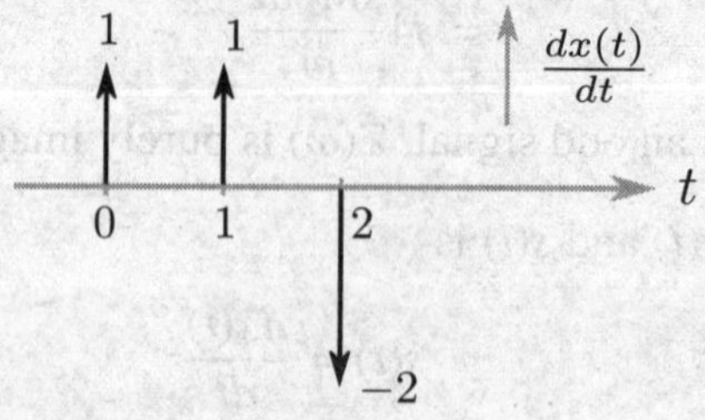

Fig. Ex. 3.18(a)

$$\frac{dx(t)}{dt} = \delta(t) + \delta(t-1) - 2\delta(t-2)$$

Taking **FT** on both sides, we get

$$j\omega X(\omega) = 1 + 1 \times e^{-j\omega} - 2 \times e^{-j2\omega}$$

$$\Rightarrow \qquad X(\omega) = \frac{1}{j\omega}[1 + e^{-j\omega} - 2e^{-j2\omega}]$$

Example 3.19. Find the **FT** of the triangular pulse $x(t)$, shown n Fig. Ex. 3.19.

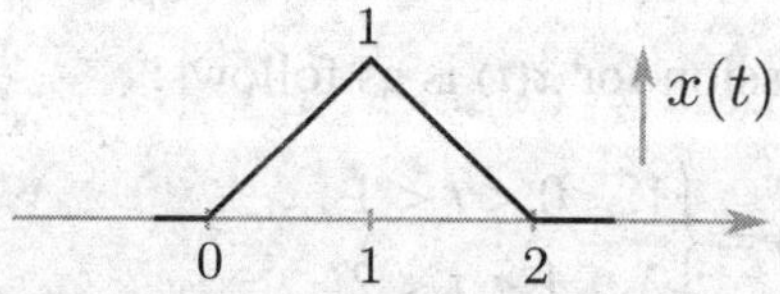

Fig. Ex. 3.19

Solution: This problem is solved using the technique of successive differentiation.

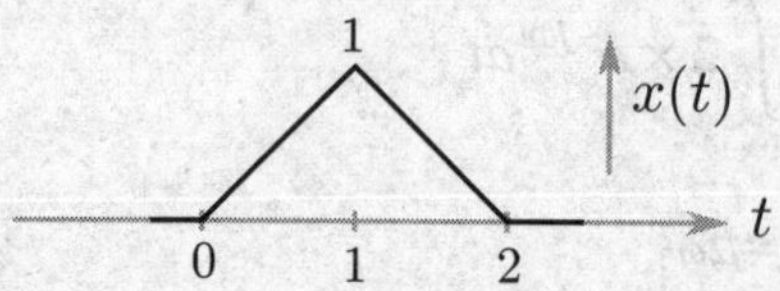

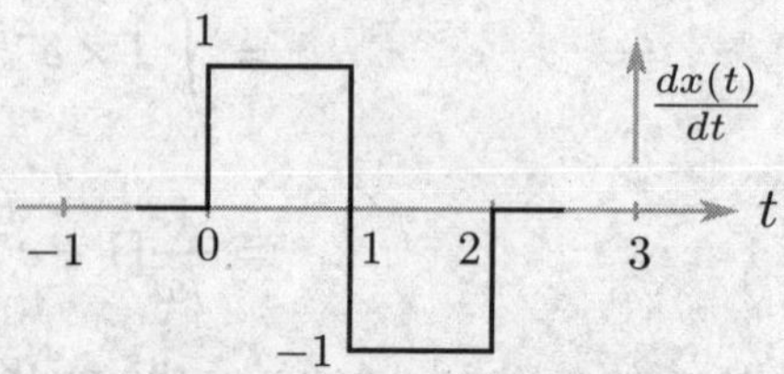

Fig. Ex. 3.19(a) **Fig. Ex. 3.19(b)**

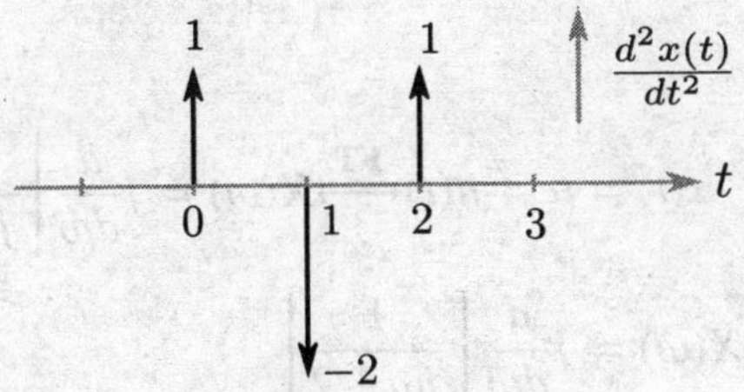

Fig. Ex. 3.19(c)

From Fig. Ex. 3.19(c), we can write

$$\frac{d^2x(t)}{dt^2} = \delta(t) - 2\delta(t-1) + \delta(t-2)$$

Taking **FT** on both the sides, we get

$$j^2\omega^2 X(\omega) = 1 - 2 \times e^{-j\omega} + e^{-j2\omega}$$

$$\Rightarrow \qquad X(\omega) = \frac{1}{j^2\omega^2}[1 - 2e^{-j\omega} + e^{-j2\omega}]$$

Example 3.20. Find the **FT** of the function

$$x(t) = te^{-2t}u(t)$$

Solution: We know that,

$$e^{-at}u(t) \xleftrightarrow{\textbf{FT}} \frac{1}{j\omega + a}$$

Hence,

$$e^{-2t}u(t) \xleftrightarrow{\textbf{FT}} \frac{1}{j\omega + 2}$$

Also, if

$$f(t) \xleftrightarrow{\textbf{FT}} F(\omega)$$

then

$$tf(t) \xleftrightarrow{\textbf{FT}} j\frac{d}{d\omega}F(\omega)$$

Here,

$$f(t) = e^{-2t}u(t)$$

Hence,

$$x(t) = te^{-2t}u(t) \overset{\textbf{FT}}{\longleftrightarrow} X(\omega) = j\frac{d}{d\omega}\left[\frac{1}{j\omega + 2}\right]$$

$$\Rightarrow \qquad X(\omega) = j\frac{d}{d\omega}\left[\frac{1}{j\omega + 2}\right]$$

$$= \frac{1}{(j\omega + 2)^2}$$

Magnitude Spectrum:

$$|X(\omega)| = \frac{1}{\left(\sqrt{\omega^2 + 4}\right)^2} = \frac{1}{\omega^2 + 4}$$

Phase Spectrum:

$$\theta(\omega) = \measuredangle X(\omega) = -2\tan^{-1}\left(\frac{\omega}{2}\right)$$

Note:

(i) $|(a + jb)^n| = \left[\sqrt{a^2 + b^2}\right]^n$

(ii) $\measuredangle(a + jb)^n = n\tan^{-1}\left(\frac{b}{a}\right)$

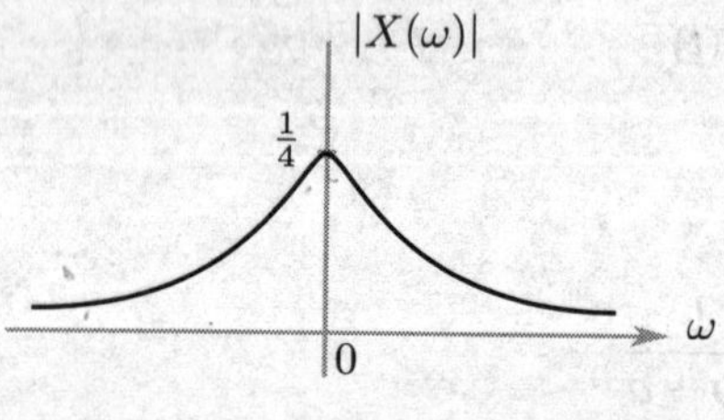

Fig. Ex. 3.20(a) *Amplitude spectrum.*

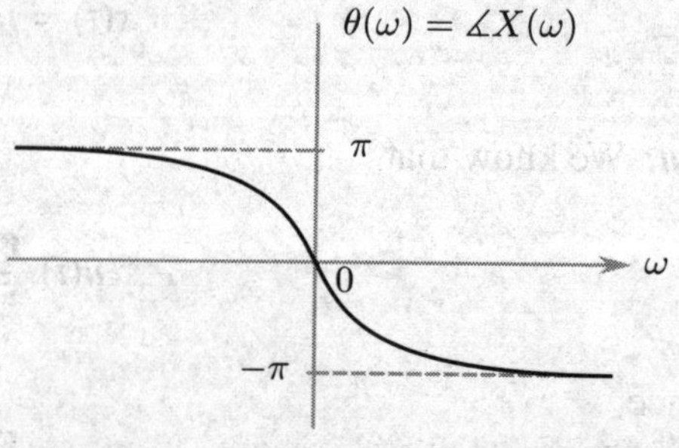

Fig. Ex. 3.20(b) *Phase spectrum.*

Please note that the Amplitude spectrum exhibits even symmetry and phase spectrum exhibits odd symmetry.

Example 3.21. Find the **FT** of the following signals:

(a) $x(t) = u(-t)$,

(b) $x(t) = e^{at}u(-t)$

Solution: Let $f(t)$ be real and

$$f(t) \overset{\textbf{FT}}{\longleftrightarrow} F(\omega)$$

then,

$$f(-t) \overset{\textbf{FT}}{\longleftrightarrow} F(-\omega)$$

(a) $x(t) = u(-t)$

In the present context $f(t) = u(t)$ and $x(t) = f(-t)$

We know that,

$$f(t) = u(t) \overset{\textbf{FT}}{\longleftrightarrow} F(\omega) = \pi\delta(\omega) + \frac{1}{j\omega}$$

Hence,

$$x(t) = f(-t) = u(-t) \overset{\textbf{FT}}{\longleftrightarrow} X(\omega) = F(-\omega) = \pi\delta(-\omega) + \frac{1}{j(-\omega)}$$

$$\Rightarrow \qquad X(\omega) = \pi\delta(\omega) - \frac{1}{j\omega}$$

Please note that, since $\delta(\omega)$ is an even function, we have taken $\delta(-\omega) = \delta(\omega)$.

(b) $x(t) = e^{at}u(-t)$

Again, let

$$f(t) = e^{-at}u(t)$$

$$\Rightarrow \qquad F(\omega) = \frac{1}{j\omega + a}$$

Hence,

$$x(t) = f(-t) = e^{at}u(-t) \overset{\textbf{FT}}{\longleftrightarrow} X(\omega) = F(-\omega) = \frac{1}{-j\omega + a}$$

$$\Rightarrow \qquad X(\omega) = \frac{1}{-j\omega + a}$$

Example 3.22. Find the **FT** of the function shown in Fig. Ex. 3.22 using recognizable transform pairs and properties.

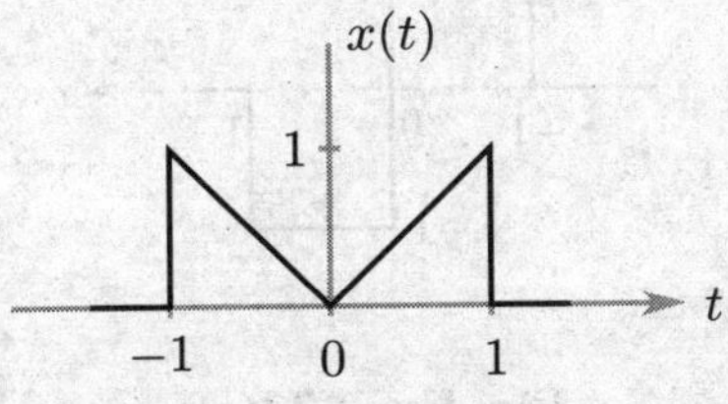

Fig. Ex. 3.22

Solution: The signal $x(t)$ is a linear combination of rectangular and triangular functions.

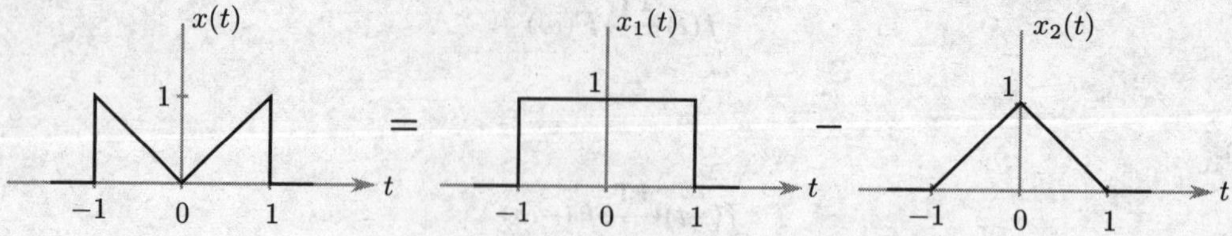

Fig. Ex. 3.22(a)

$$x(t) = x_1(t) - x_2(t)$$

$$\Rightarrow \quad X(\omega) = X_1(\omega) - X_2(\omega) \tag{3.10}$$

To find $X_1(\omega)$:

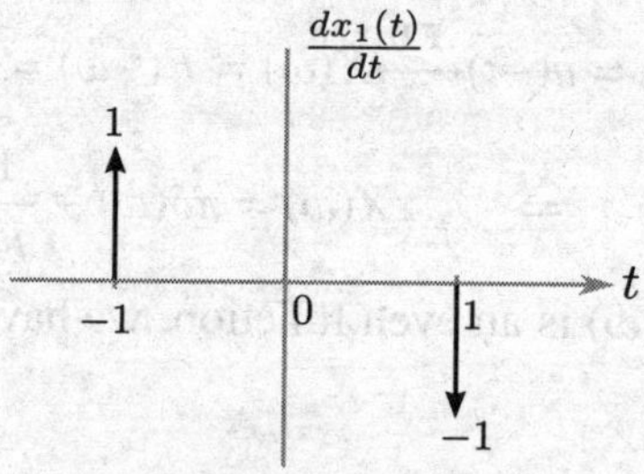

Fig. Ex. 3.22(b)

$$\frac{dx_1(t)}{dt} = \delta(t+1) - \delta(t-1)$$

$$\Rightarrow \ j\omega X_1(\omega) = e^{j\omega} - e^{-j\omega}$$

$$\Rightarrow \ X_1(\omega) = \frac{2}{\omega}\left[\frac{e^{j\omega} - e^{-j\omega}}{2j}\right] = \frac{2\sin\omega}{\omega} \tag{3.11}$$

To find $X_2(\omega)$:

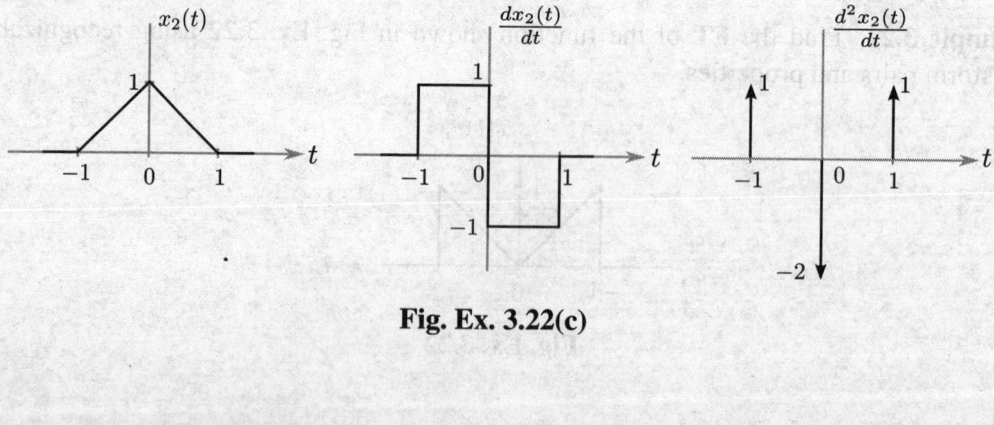

Fig. Ex. 3.22(c)

$$\frac{d^2x_2(t)}{dt^2} = \delta(t+1) - 2\delta(t) + \delta(t-1)$$

$$\Rightarrow j^2\omega^2 X_2(\omega) = e^{j\omega} - 2 + e^{-j\omega}$$

Therefore,

$$X_2(\omega) = \frac{e^{j\omega} + e^{-j\omega} - 2}{j^2\omega^2}$$

$$= \frac{2(\cos\omega - 1)}{j^2\omega^2} = \frac{2(1-\cos\omega)}{\omega^2} \tag{3.12}$$

Substituting Eqs. (3.11) and (3.12) in Eq. (3.10), we get

$$X(\omega) = \left(\frac{2\sin\omega}{\omega}\right) - \left(\frac{2 - 2\cos\omega}{\omega^2}\right)$$

Example 3.23. If $x(t) \overset{\textbf{FT}}{\longleftrightarrow} X(f)$, then show that

$$y(t) = e^{j\beta t}x(t) \overset{\textbf{FT}}{\longleftrightarrow} y(f) = X\left(f - \frac{\beta}{2\pi}\right)$$

Solution: By definition,

$$Y(f) = \int_{-\infty}^{\infty} y(t)e^{-j2\pi ft}dt$$

$$\Rightarrow \quad Y(f) = \int_{-\infty}^{\infty} e^{j\beta t}x(t)e^{-j2\pi ft}$$

$$= \int_{-\infty}^{\infty} x(t)e^{-j2\pi t\left(f - \frac{\beta}{2\pi}\right)}dt$$

Comparing the above result with the defining equation for **FT**, we get

$$Y(f) = X\left(f - \frac{\beta}{2\pi}\right)$$

Example 3.24. Find the Fourier transform of $x(t) = \frac{A}{T}t[u(t+T) - u(t-T)]$.

Solution: The function $x(t)$ can be viewed as the product of t and $x_1(t)$. That is, $x(t) = tx_1(t)$. The functions $x_1(t)$ and $x(t)$ are as shown in Fig. Ex. 3.24(a).

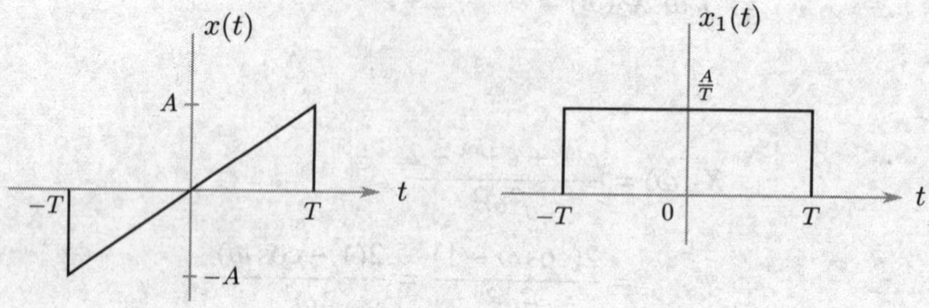

Fig. Ex. 3.24(a)

Let us now proceed to find **FT** of $x_1(t)$.

$$X_1(\omega) \triangleq \int_{-\infty}^{\infty} x_1(t)e^{-j\omega t}dt$$

$$\Rightarrow \quad X_1(\omega) = \int_{-T}^{T} \frac{A}{T}e^{-j\omega t}dt$$

$$= 2A\frac{\sin(\omega T)}{\omega T}$$

Recall the frequency differentiation property:

$$t^n f(t) \overset{\mathbf{FT}}{\longleftrightarrow} j^n \frac{d^n}{d\omega^n}F(\omega)$$

Then,

$$x(t) = tx_1(t) \overset{\mathbf{FT}}{\longleftrightarrow} X(\omega) = j\frac{d}{d\omega}X_1(\omega)$$

Hence,

$$X(\omega) = j\frac{d}{d\omega}\left[\frac{2A}{\omega T}\sin(\omega T)\right]$$

$$= j2A\frac{d}{d\omega}\left[\frac{\sin(\omega T)}{\omega T}\right]$$

$$= j\frac{2A}{\omega^2 T}[\omega^2 \cos\omega T - \sin\omega T]$$

Example 3.25. Find **FT** of signum function, $x(t) = \text{sgn}(t)$. Also sketch the phase spectrum.

Solution: A signum function is defined as follows:

$$\text{sgn}(t) = \begin{cases} 1, & t > 0 \\ 0, & t = 0 \\ -1, & t < 0 \end{cases}$$

A sketch of signum function is as shown in Fig. Ex. 3.25(a).

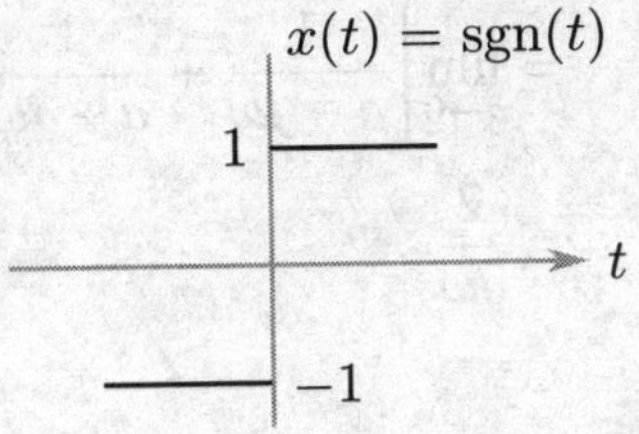

Fig. Ex. 3.25(a)

To derive the Fourier transform of signum function, it is convient to express it as an exponential function, $x_1(t)$ that approaches signum function as $a \to 0$, that is,

$$x(t) = \lim_{a \to 0} x_1(t)$$

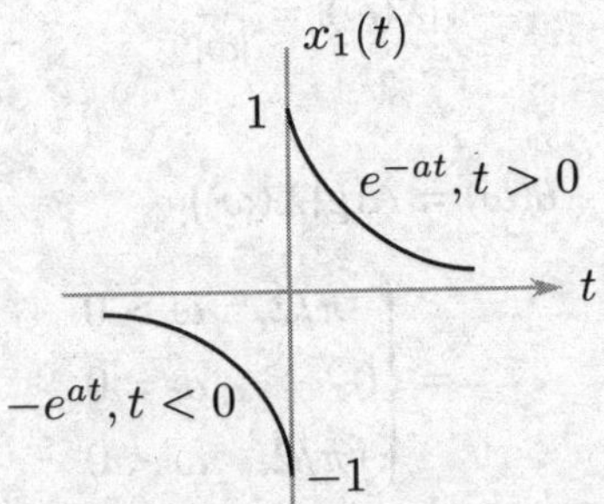

Fig. Ex. 3.25(b)

Let us now proceed to find **FT** of $x_1(t)$.

$$X_1(\omega) = \int_{-\infty}^{\infty} x_1(t) e^{-j\omega t} dt$$

$$= \int_{-\infty}^{0} -e^{at} e^{-j\omega t} dt + \int_{0}^{\infty} e^{-at} e^{-j\omega t} dt$$

$$= -\int_{-\infty}^{0} e^{(a-j\omega)t}dt + \int_{0}^{\infty} e^{-(a+j\omega)t}dt$$

$$= \frac{1}{a-j\omega} + \frac{1}{a+j\omega}$$

Hence,

$$X(\omega) = \lim_{a\to 0} X_1(\omega)$$

$$= \lim_{a\to 0}\left[\frac{1}{a-j\omega} + \frac{1}{a+j\omega}\right]$$

$$= \frac{2}{j\omega}$$

Aliter:

The signum function can be expressed in terms of unit step function as

$$\text{sgn}(t) = 2u(t) - 1$$

Hence,

$$\text{sgn}(t) \overset{\textbf{FT}}{\longleftrightarrow} 2\left[\pi\delta(\omega) + \frac{1}{j\omega}\right] - 2\pi\delta(\omega) = \frac{2}{j\omega}$$

Magnitude Spectrum:

$$|X(\omega)| = \frac{2}{|\omega|}$$

Phase Spectrum:

$$\theta(\omega) = \text{Arg}\{X(\omega)\}$$

$$= \begin{cases} -\pi/2, & \omega > 0 \\ 0, & \omega = 0 \\ +\pi/2, & \omega < 0 \end{cases}$$

The phase spectrum is as shown in Fig. 3.25(c).

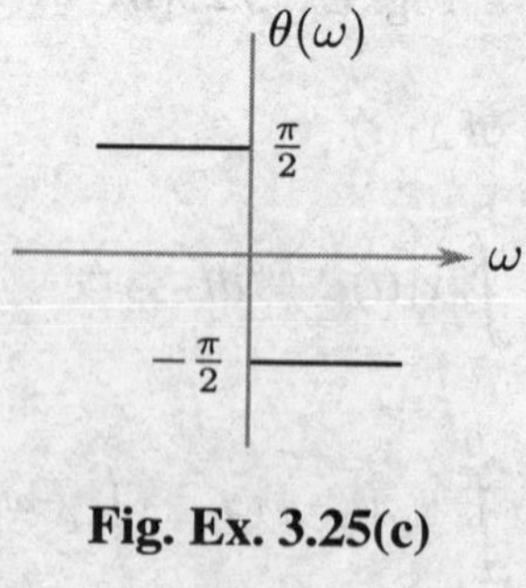

Fig. Ex. 3.25(c)

Example 3.26. Using properties, find **FT** of the following signals:

(a) $x(t) = \sin 2\pi t e^{-t} u(t)$

(b) $x(t) = e^{-4|t-2|}$

Solution:

(a) Using Euler's formula, rewrite $x(t)$ as $x(t) = \frac{1}{2j} e^{j2\pi t} e^{-t} u(t) - \frac{1}{2j} e^{-j2\pi t} e^{-t} u(t)$

$$\text{Recall}: \; e^{-at} u(t) \stackrel{\mathbf{FT}}{\longleftrightarrow} \frac{1}{a + j\omega}$$

Letting $a = 1$, we get

$$e^{-t} u(t) \stackrel{\mathbf{FT}}{\longleftrightarrow} \frac{1}{1 + j\omega}$$

Applying frequency shift property, we get

$$x_1(t) = e^{j2\pi t} e^{-t} u(t) \stackrel{\mathbf{FT}}{\longleftrightarrow} X_1(\omega) = \frac{1}{1 + j(\omega - 2\pi)}$$

$$\text{and} \quad x_2(t) = e^{-j2\pi t} e^{-t} u(t) \stackrel{\mathbf{FT}}{\longleftrightarrow} X_2(\omega) = \frac{1}{1 + j(\omega + 2\pi)}$$

Applying linearity property, we get

$$x(t) = \frac{1}{2j} x_1(t) - \frac{1}{2j} x_2(t) \stackrel{\mathbf{FT}}{\longleftrightarrow} X(\omega) = \frac{1}{2j}[X_1(\omega) - X_2(\omega)]$$

Hence,

$$X(\omega) = \frac{1}{2j}\left[\frac{1}{1 + j(\omega - 2\pi)} - \frac{1}{1 + j(\omega + 2\pi)}\right]$$

(b) Given, $x(t) = e^{-4|t-2|}$

$$\text{Recall}: \; e^{-a|t|} \stackrel{\mathbf{FT}}{\longleftrightarrow} \frac{2a}{a^2 + \omega^2}$$

Letting $a = 4$, we get

$$e^{-4|t|} \stackrel{\mathbf{FT}}{\longleftrightarrow} \frac{8}{16 + \omega^2}$$

Applying time shift property, we get

$$x(t) = e^{-4|t-2|} \stackrel{\mathbf{FT}}{\longleftrightarrow} X(\omega) = e^{-2\omega}\left[\frac{8}{16 + \omega^2}\right]$$

Example 3.27. Using properties, find **FT** of the signal given below:

$$x(t) = \frac{d}{dt}[te^{-3t} \sin t\, u(t)]$$

Solution: Let

$$x(t) = \frac{dx_1(t)}{dt}$$

where

$$x_1(t) = te^{-3t} \sin t\, u(t)$$

$$= \frac{1}{2j} te^{jt} e^{-3t} u(t) - \frac{1}{2j} te^{-jt} e^{-3t} u(t)$$

Recall:

$$e^{-at} u(t) \stackrel{\mathbf{FT}}{\longleftrightarrow} \frac{1}{a + j\omega}$$

Letting $a = 3$, we get

$$e^{-3t} u(t) \stackrel{\mathbf{FT}}{\longleftrightarrow} \frac{1}{3 + j\omega}$$

Applying frequency shift property, we get

$$x_2(t) = e^{jt} e^{-3t} u(t) \stackrel{\mathbf{FT}}{\longleftrightarrow} X_2(\omega) = \frac{1}{j(\omega - 1) + 3}$$

$$\text{and} \qquad x_3(t) = e^{-jt} e^{-3t} u(t) \stackrel{\mathbf{FT}}{\longleftrightarrow} X_3(\omega) = \frac{1}{j(\omega + 1) + 3}$$

Applying linearity property, we get

$$x_4(t) = \frac{1}{2j} x_2(t) - \frac{1}{2j} x_3(t) \stackrel{\mathbf{FT}}{\longleftrightarrow} X_4(\omega) = \frac{1}{2j}[X_2(\omega) - X_3(\omega)]$$

Hence,

$$X_4(\omega) = \frac{1}{2j}\left[\frac{1}{j(\omega - 1) + 3} - \frac{1}{j(\omega + 1) + 3}\right]$$

Applying frequency differentiation property, we get

$$x_1(t) = tx_4(t) \stackrel{\mathbf{FT}}{\longleftrightarrow} X_1(\omega) = j\frac{d}{d\omega} X_4(\omega)$$

Hence,

$$X_1(\omega) = j \times \frac{1}{2j}\left[\frac{-j}{[j(\omega - 1) + 3]^2} + \frac{j}{[j(\omega + 1) + 3]^2}\right]$$

$$= \frac{1}{2j}\left[\frac{1}{[j(\omega-1)+3]^2} - \frac{1}{[j(\omega+1)+3]^2}\right]$$

Applying time differentiation property, we get

$$x(t) = \frac{dx_1(t)}{dt} \overset{\mathbf{FT}}{\longleftrightarrow} X(\omega) = j\omega X_1(\omega)$$

Hence,

$$X(\omega) = \frac{\omega}{2}\left[\frac{1}{\{j(\omega-1)+3\}^2} - \frac{1}{\{j(\omega+1)+3\}^2}\right]$$

Example 3.28. Using properties, find the Fourier transform of the signal given below:

$$x(t) = e^{-2t+1}u\left(\frac{t-4}{2}\right)$$

Solution: Rewrite $x(t)$ as

$$x(t) = e^{-7}e^{-4(\frac{t}{2}-2)}u\left(\frac{t}{2}-2\right)$$

Recall:

$$e^{-at}u(t) \overset{\mathbf{FT}}{\longleftrightarrow} \frac{1}{j\omega + a}$$

Lettting $a = 4$, we get

$$e^{-4t}u(t) \overset{\mathbf{FT}}{\longleftrightarrow} \frac{1}{j\omega + 4}$$

Using time shift property, we get

$$x_1(t) = e^{-4(t-2)}u(t-2) \overset{\mathbf{FT}}{\longleftrightarrow} X_1(\omega) = e^{-j2\omega}\left[\frac{1}{j\omega+4}\right]$$

Applying scaling property, we get

$$x_2(t) = x_1\left(\frac{t}{2}\right) \overset{\mathbf{FT}}{\longleftrightarrow} X_2(\omega) = 2X_1(2\omega)$$

$$= 2e^{-j4\omega}\frac{1}{j2\omega+4}$$

$$= \frac{e^{-j4\omega}}{2+j\omega}$$

Applying linearity property, we get

$$x(t) = e^{-7}x_2(t) \overset{\mathbf{FT}}{\longleftrightarrow} X(\omega) = e^{-7}X_2(\omega)$$

Hence,

$$X(\omega) = e^{-7}\left[\frac{e^{-j4\omega}}{2 + j\omega}\right]$$

Example 3.29. Given the following transform pair:

$$x(t) = e^{-|t|} \overset{\textbf{FT}}{\longleftrightarrow} X(\omega) = \frac{2}{\omega^2 + 1}$$

Find the Fourier transform of the following signals:

(a) $\dfrac{d}{dt}e^{-|t|}$

(b) $\dfrac{1}{2\pi(t^2 + 1)}$

Solution:

(a) Let $x_1(t) = \frac{d}{dt}e^{-|t|} = \frac{d}{dt}x(t)$.

Applying time differentiation property, we get

$$X_1(\omega) = j\omega X(\omega)$$
$$= j\omega\left[\frac{2}{\omega^2 + 1}\right]$$

(b) Let $x_2(t) = \dfrac{1}{2\pi(t^2 + 1)}$

Recall: $e^{-|t|} \overset{\textbf{FT}}{\longleftrightarrow} \frac{2}{\omega^2+1}$.

Applying duality property, we get

$$\frac{2}{t^2 + 1} \overset{\textbf{FT}}{\longleftrightarrow} 2\pi e^{-|-\omega|}$$

$$\Rightarrow \quad \frac{1}{t^2 + 1} \overset{\textbf{FT}}{\longleftrightarrow} \frac{2\pi}{2}e^{-|\omega|}$$

Hence,

$$x_2(t) = \frac{1}{2\pi(t^2 + 1)} \overset{\textbf{FT}}{\longleftrightarrow} X_2(\omega) = \frac{\pi}{2\pi}e^{-|\omega|} = \frac{1}{2}e^{-|\omega|}$$

Example 3.30. Consider the transform pair given below:

$$x(t) \overset{\mathbf{FT}}{\longleftrightarrow} X(\omega) = \frac{j\omega}{-\omega^2 + 7j\omega + 6}$$

Find the Fourier transform of the following signals.

(a) $x(4t - 8)$

(b) $\int_{-\infty}^{t} x(\tau)d\tau$

(c) $e^{-j100t}x(t)$

Solution:

(a) Applying time shift property, we get

$$x_1(t) = x(t-8) \overset{\mathbf{FT}}{\longleftrightarrow} X_1(\omega) = e^{-j\omega 8}X(\omega) = e^{-j8\omega}\left[\frac{j\omega}{-\omega^2 + j7\omega + 6}\right]$$

Applying scaling property, we get

$$x_2(t) = x_1(4t) \overset{\mathbf{FT}}{\longleftrightarrow} X(\omega) = \frac{1}{4}X_1\left(\frac{\omega}{4}\right)$$

Hence,

$$X(\omega) = \frac{1}{4}e^{-j8\left(\frac{\omega}{4}\right)}\left[\frac{j\left(\frac{\omega}{4}\right)}{-\left(\frac{\omega}{4}\right)^2 + j7\left(\frac{\omega}{4}\right) + 6}\right]$$

$$= \frac{1}{4}e^{-j2\omega}\left[\frac{j4\omega}{-\omega^2 + j28\omega + 96}\right]$$

(b) Recall the integration property:

$$x_1(t) = \int_{-\infty}^{t} x(\tau)d\tau \overset{\mathbf{FT}}{\longleftrightarrow} X_1(\omega) = \frac{1}{j\omega}X(\omega) + \pi X(0)\delta(\omega)$$

Hence,

$$X_1(\omega) = \frac{1}{j\omega}\left[\frac{j\omega}{-\omega^2 + 7j\omega + 6}\right] + \pi \times 0 \times \delta(\omega)$$

$$= \frac{1}{-\omega^2 + j7\omega + 6}$$

Proof of this property is given in article 3.4.6.

(c) Applying frequency shift property, we get

$$x_1(t) = e^{-j100t}x(t) \overset{\text{FT}}{\longleftrightarrow} X_1(\omega) = X(\omega + 100)$$

Hence,

$$X_1(\omega) = \frac{j(\omega + 100)}{-(\omega + 100)^2 + j7(\omega + 100) + 6}$$

Example 3.31. Find Fourier transform of the signal given below:

$$x(t) = \frac{1}{4\pi(a - jt)^2}; a > 0$$

Solution: Recall:

$$x_1(t) = te^{-at}u(t) \overset{\text{FT}}{\longleftrightarrow} X_1(\omega) = \frac{1}{(a + j\omega)^2}$$

Recall the scaling property:

$$f(at) \overset{\text{FT}}{\longleftrightarrow} \frac{1}{|a|}F\left(\frac{\omega}{a}\right)$$

Letting $a = -1$, we get

$$f(-t) \overset{\text{FT}}{\longleftrightarrow} F(-\omega)$$

Hence,

$$x_2(t) = x_1(-t) = -te^{at}u(-t) \overset{\text{FT}}{\longleftrightarrow} X_2(\omega) = X_1(-\omega) = \frac{1}{(a - j\omega)^2}$$

Applying duality property, we get

$$X_2(t) = \frac{1}{(a - jt)^2} \overset{\text{FT}}{\longleftrightarrow} 2\pi x_2(-\omega) = 2\pi[\omega e^{-a\omega}U(\omega)]$$

Hence,

$$\frac{1}{4\pi(a - jt)^2} \overset{\text{FT}}{\longleftrightarrow} \frac{2\pi}{4\pi}[\omega e^{-a\omega}U(\omega)] = \frac{1}{2}[\omega e^{-a\omega}U(\omega)]$$

3.5 The Inverse Fourier Transform

In general, one would have to use the integral of Eq. (3.7) to find the inverse Fourier transform of $X(\omega)$. In most of the cases, the Fourier transform will be in the form of a rational function of $j\omega$ (a ratio of two polynomials in $j\omega$). In this case, it is quite simpler to perform a partial fraction expansion of the Fourier transform and then to identify each term in this expansion by using a Table of Fourier transforms and their corresponding time-domain counter parts. We summarize the most common Fourier transform pairs in Table 3.2.

Table 3.2: Useful Fourier transform pairs

$x(t)$	$X(\omega)$
$\delta(t)$	1
$\delta(t-t_0)$	$e^{-j\omega t_0}$
A	$2\pi A\delta(\omega)$
$e^{-j\omega_0 t}$	$2\pi\delta(\omega-\omega_0)$
$\text{sgn}(t)$	$\frac{2}{j\omega}$
$u(t)$	$\frac{1}{j\omega}+\pi\delta(\omega)$
$\cos\omega_0 t$	$\pi[\delta(\omega-\omega_0)+\delta(\omega+\omega_0)]$
$\sin\omega_0 t$	$j\pi[\delta(\omega-\omega_0)-\delta(\omega+\omega_0)]$
$e^{-at}u(t),\ a>0$	$\frac{1}{j\omega+a},\ a>0$
$te^{-at}u(t),\ a>0$	$\frac{1}{(j\omega+a)^2},\ a>0$
$e^{-at}\cos\omega_0 t\, u(t),\ a>0$	$\frac{j\omega+a}{(j\omega+a)^2+\omega_0^2},\ a>0$
$e^{-at}\sin\omega_0 t\, u(t),\ a>0$	$\frac{\omega_0}{(j\omega+a)^2+\omega_0^2},\ a>0$
$\begin{cases}1, & \lvert t\rvert<a\\0, & \lvert t\rvert>a\end{cases}$	$\frac{2\sin(a\omega)}{\omega}$

Example 3.32. Find the inverse Fourier transform of $X(\omega)=\dfrac{j\omega+12}{(j\omega)^2+5j\omega+6}$

Solution: Let $j\omega=s$. Then,

$$\begin{aligned}X(\omega)&=\frac{s+12}{s^2+5s+6}\\&=\frac{s+12}{(s+3)(s+2)}\\&=\frac{K_1}{s+2}+\frac{K_2}{s+3}\end{aligned}$$

By performing partial fraction expansion, we get

$$K_1 = 10$$
$$K_2 = -9$$

Therefore,

$$X(\omega) = \frac{10}{s+2} - \frac{9}{s+3}$$
$$= \frac{10}{j\omega+2} - \frac{9}{j\omega+3}$$

We know that,

$$e^{-at}u(t) \overset{\textbf{FT}}{\longleftrightarrow} \frac{1}{j\omega+a}$$

Hence,

$$x(t) = 10e^{-2t}u(t) - 9e^{-3t}u(t)$$

Example 3.33. Find the inverse **FT** of $X(\omega) = \dfrac{j\omega}{(j\omega+2)^2}$

Solution: Let $j\omega = s$ Then,

$$X(\omega) = \frac{s}{(s+2)^2}$$
$$= \frac{K_1}{(s+2)} + \frac{K}{(s+2)^2}$$

where,

$$K_1 = \frac{d}{ds}[s]_{s=-2} = 1$$
$$K_2 = s\Big|_{s=-2} = -2$$

Therefore,

$$X(\omega) = \frac{1}{(s+2)} - \frac{2}{(s+2)^2}$$
$$= \frac{1}{(j\omega+2)} - \frac{2}{(j\omega+2)^2} \qquad (3.13)$$

We know that,

$$e^{-at}u(t) \overset{\textbf{FT}}{\longleftrightarrow} \frac{1}{j\omega+a}$$

and

$$te^{-at}u(t) \overset{\mathbf{FT}}{\longleftrightarrow} \frac{1}{(j\omega + a)^2}$$

Therefore, the inverse **FT** of Eq. (3.13) yields

$$x(t) = e^{-2t}u(t) - 2te^{-2t}u(t)$$

Example 3.34. Find the inverse **FT** of the spectra shown in Figs. Ex. 3.34(a) and Ex. 3.34(b).

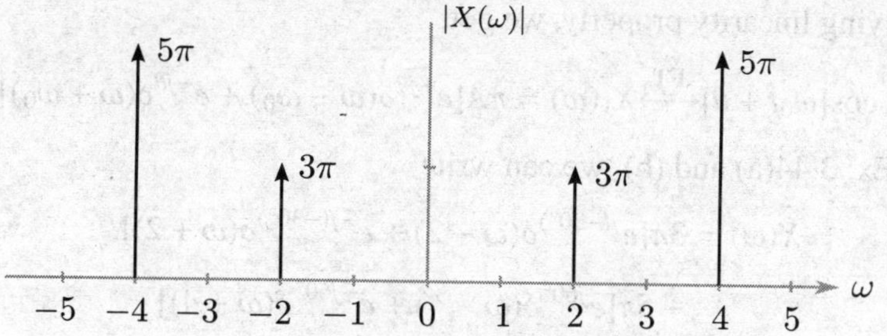

Fig. Ex. 3.34(a)

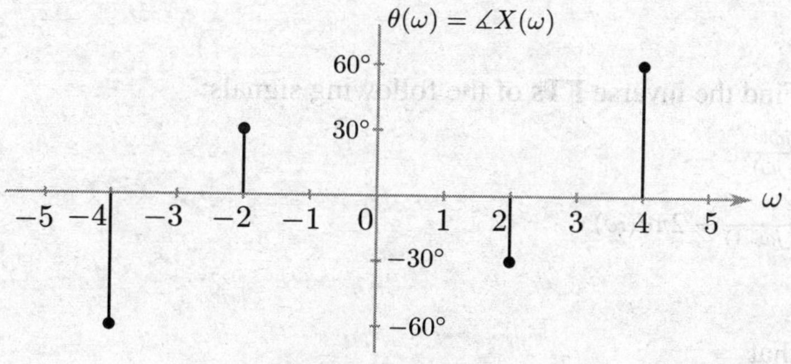

Fig. Ex. 3.34(b)

Solution: Let,

$$\begin{aligned} x_1(t) &= A\cos(\omega_0 t + \theta) \\ &= \frac{A}{2}e^{j(\omega_0 t + \theta)} + \frac{A}{2}e^{-j(\omega_0 t + \theta)} \\ &= \frac{A}{2}e^{j\theta}e^{j\omega_0 t} + \frac{A}{2}e^{-j\theta}e^{-j\omega_0 t} \end{aligned}$$

We know that,

$$K \overset{\textbf{FT}}{\longleftrightarrow} 2\pi K\delta(\omega)$$

$$\Rightarrow \quad \frac{A}{2}e^{j\theta} \overset{\textbf{FT}}{\longleftrightarrow} 2\pi\left(\frac{A}{2}e^{j\theta}\right)\delta(\omega) = \pi A e^{j\theta}\delta(\omega)$$

Applying frequency shift property, we get

$$\frac{A}{2}e^{j\theta}e^{j\omega_0 t} \overset{\textbf{FT}}{\longleftrightarrow} \pi A e^{j\theta}\delta(\omega - \omega_0)$$

$$\text{and} \quad \frac{A}{2}e^{-j\theta}e^{-j\omega_0 t} \overset{\textbf{FT}}{\longleftrightarrow} \pi A e^{-j\theta}\delta(\omega + \omega_0)$$

Finally, applying linearity property, we get,

$$x_1(t) = A\cos[\omega_0 t + \theta] \overset{\textbf{FT}}{\longleftrightarrow} X_1(\omega) = \pi A[e^{+j\theta}\delta(\omega - \omega_0) + e^{-j\theta}\delta(\omega + \omega_0)] \tag{3.14}$$

From Figs. Ex. 3.44(a) and (b), we can write

$$X(\omega) = 3\pi[e^{j(-30^\circ)}\delta(\omega - 2) + e^{-j(-30^\circ)}\delta(\omega + 2)]$$
$$+ 5\pi[e^{j60^\circ}\delta(\omega - 4) + e^{-j60^\circ}\delta(\omega + 4)]$$

Taking inverse **FT** (making use of Eq. (3.14)), we get

$$x(t) = 3\cos[2t - 30^\circ] + 5\cos[4t + 60^\circ]$$

Example 3.35. Find the inverse **FTs** of the following signals:

(a) $X(\omega) = \frac{j\omega}{(2+j\omega)^2}$

(b) $X(\omega) = \frac{1}{j\omega(j\omega+1)} + 2\pi\delta(\omega)$

Solution:

(a) We know that

$$\frac{1}{(2 + j\omega)^2} \overset{\textbf{FT}}{\longleftrightarrow} te^{-2t}u(t)$$

Also,

$$j\omega S(\omega) \overset{\textbf{FT}}{\longleftrightarrow} \frac{d}{dt}s(t)$$

Hence,

$$x(t) = \frac{d}{dt}[te^{-2t}u(t)]$$
$$= [-2te^{-2t} + e^{-2t}]u(t)$$

(b)

$$X(\omega) = \frac{1}{j\omega(j\omega+1)} + 2\pi\delta(\omega)$$

$$= \left[\frac{1}{j\omega(j\omega+1)} + \pi\delta(\omega)\right] + \pi\delta(\omega)$$

We need the following transform pairs:

$$\frac{1}{j\omega+1} \xleftrightarrow{\textbf{FT}} e^{-t}u(t)$$

and

$$\pi\delta(\omega) \xleftrightarrow{\textbf{FT}} \frac{1}{2}$$

Using the integration property:

$$\int_{-\infty}^{t} s(\tau)d\tau \xleftrightarrow{\textbf{FT}} \frac{1}{j\omega}S(\omega) + \pi S(0)\delta(\omega),$$

with, $s(t) = e^{-t}u(t)$ and $S(\omega) = \frac{1}{j\omega+1}$,

we get, $$\int_{-\infty}^{t} e^{-\tau}u(\tau)d\tau \xleftrightarrow{\textbf{FT}} \frac{1}{j\omega}\frac{1}{j\omega+1} + \pi \times 1 \times \delta(\omega)$$

This means that,

$$x(t) = \int_{-\infty}^{t} e^{-\tau}u(\tau)d\tau + \frac{1}{2}$$

$$= [1 - e^{-t}]u(t) + \frac{1}{2}$$

3.6 Fourier Transform of a Periodic Signal

Consider a continuous-time signal $x(t)$ with a fundamental period equal to T_0. Then, the signal $x(t)$ can be represented using exponential fourier series as

$$x(t) = \sum_{k=-\infty}^{\infty} X(k)e^{jk\omega_0 t}; \quad \omega_0 = \frac{2\pi}{T_0} \tag{3.15}$$

We know that,

$$A \xleftrightarrow{\textbf{FT}} 2\pi A\delta(\omega)$$

and

$$Ae^{j\omega_0 t} \xleftrightarrow{\textbf{FT}} 2\pi A\delta(\omega - \omega_0)$$

Therefore,

$$X(\omega) = \sum_{k=-\infty}^{\infty} 2\pi X(k)\delta(\omega)\Big|_{\omega\to\omega-k\omega_0}$$

$$\Rightarrow \quad X(\omega) = \sum_{k=-\infty}^{\infty} 2\pi X(k)\delta(\omega - k\omega_0)$$

Therefore,

$$X(\omega) = 2\pi \sum_{k=-\infty}^{\infty} X(k)\delta(\omega - k\omega_0) \tag{3.16}$$

Example 3.36. Show that the **FT** of a Dirac comb is also a Dirac comb.

Solution: A periodic impulse train is known as Dirac comb.

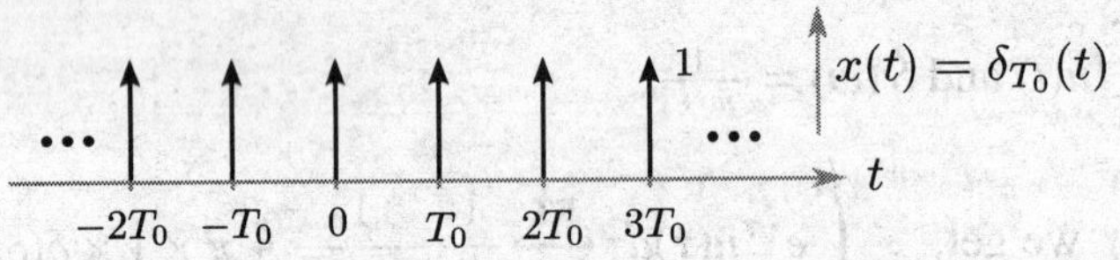

Fig. Ex. 3.36 *Dirac comb in time-domain.*

The Complex Fourier Coeffcient is given by

$$X(k) \triangleq \frac{1}{T_0} \int_{<T_0>} x(t)e^{-jk\omega_0 t}dt$$

$$\Rightarrow \quad X(k) = \frac{1}{T_0}\int_0^{T_0} \delta(t)e^{-jk\omega_0 t}dt$$

$$= \frac{1}{T_0}e^{-jk\omega_0 t}\Big|_{t=0}$$

$$= \frac{1}{T_0}$$

The **FT** of $x(t)$ is given by

$$X(\omega) = 2\pi \sum_{k=-\infty}^{\infty} X(k)\delta(\omega - k\omega_0) \tag{3.17}$$

$$= \frac{2\pi}{T_0}\sum_{k=-\infty}^{\infty} \delta(\omega - k\omega_0)$$

$$\Rightarrow \qquad X(\omega) = \omega_0 \sum_{k=-\infty}^{\infty} \delta(\omega - k\omega_0)$$

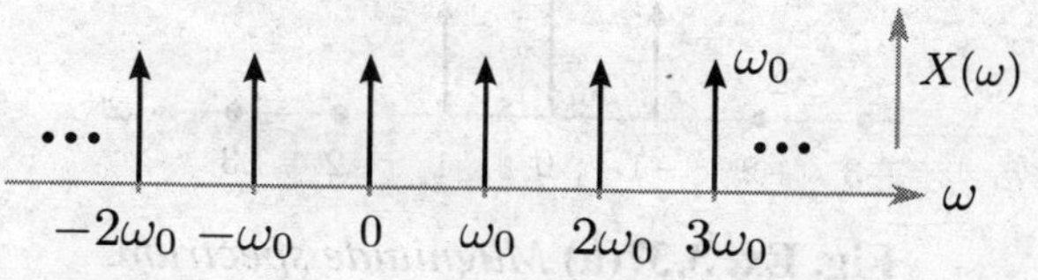

Fig. Ex. 3.36(a) *Dirac comb in frequency-domain.*

Example 3.37. Find the **FT**, $X(\omega)$ of the periodic signal $x(t)$ and sketch magnitude spectrum $|X(\omega)|$ and the phase spectrum $\angle X(\omega)$.

$$x(t) = 3 + 2\cos(10\pi t)$$

Solution:

$$x(t) = 3 + 2\left[\frac{1}{2}e^{j10\pi t} + \frac{1}{2}e^{-j10\pi t}\right]$$

$$= 3 + e^{j\omega_0 t} + e^{-j\omega_0 t} \tag{3.18}$$

where $\omega_0 = 10\pi$ rad/sec.

Comparing Eq. (3.18) with

$$x(t) = \sum_{k=-\infty}^{\infty} X(k)e^{jk\omega_0 t}$$

we get,

$$X(0) = 3, \quad X(1) = 1,$$

$$X(-1) = 1, \text{ and } X(k) = 0 \quad \text{for all other } k$$

Recall: The **FT** of a periodic signal $x(t)$ is given by

$$X(\omega) = 2\pi \sum_{k=-\infty}^{\infty} X(k)\delta(\omega - k\omega_0)$$

$$= 2\pi \sum_{k=-1}^{1} X(k)\delta(\omega - k\omega_0)$$

$$\Rightarrow \quad X(\omega) = 2\pi[X(-1)\delta(\omega + \omega_0) + X(0)\delta(\omega) + X(1)\delta(\omega - \omega_0)]$$

$$= 2\pi\delta(\omega + \omega_0) + 6\pi\delta(\omega) + 2\pi\delta(\omega - \omega_0)$$

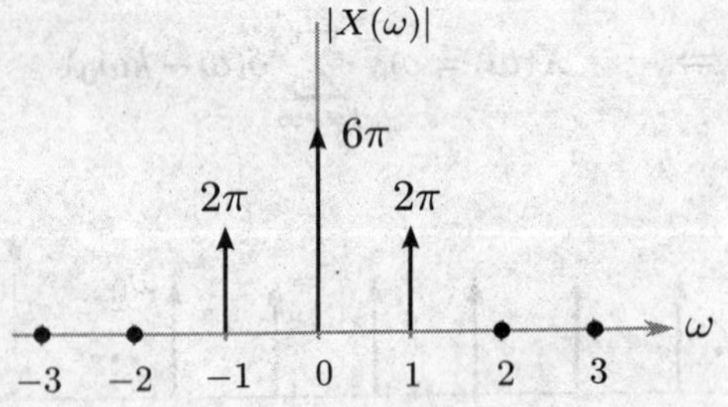

Fig. Ex. 3.37(a) *Magnitude spectrum.*

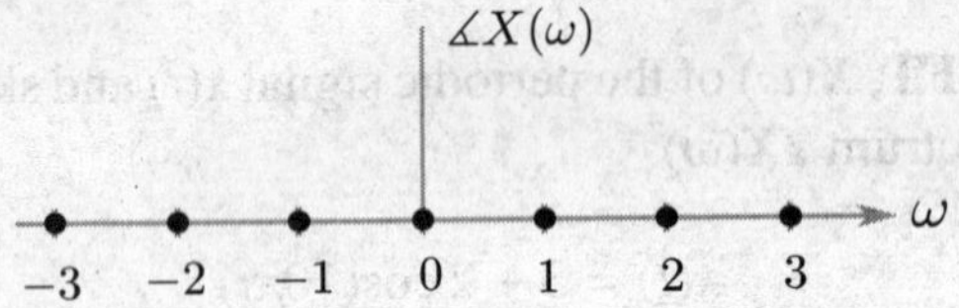

Fig. Ex. 3.37(b) *Phase spectrum (Null spectrum).*

3.7 Hilbert transform

Consider a filter (shown in Fig 3.2) that phase shifts all the frequency components of the input $x(t)$ by $\frac{-\pi}{2}$ radians; that is, its transfer function is

$$H(f) = -j\,\mathrm{sgn}(f)$$

where,

$$\mathrm{sgn}(f) = \begin{cases} 1, & f > 0 \\ 0, & f = 0 \\ -1 & f < 0 \end{cases}$$

Fig. 3.2 *Hilbert transform filter.*

It may be noted that $|H(f)| = 1$

$$\text{and} \quad \angle H(f) = \begin{cases} -\pi/2, & f > 0 \\ 0, & f = 0 \\ +\pi/2, & f < 0 \end{cases}$$

Then, $|H(f)|$ is an even function of f. If $X(f)$ is the Fourier transform of input signal (f-form) to the filter, then the output of the filter in frequency-domain is

$$\hat{X}(f) = X(f)H(f) = -j\,\mathrm{sgn}(f)X(f)$$

and the corresponding time-domain expression is

$$\hat{x}(t) = \mathscr{F}^{-1}\{H(f)X(f)\} \tag{3.19}$$

where $\mathscr{F}^{-1}\{\cdot\}$ represents an operator denoting inverse Fourier transform.

Pre-requiste:

1. Convolution is the process of overlapping two signals. The convolution of two time functions $g_1(t)$ and $g_2(t)$ is denoted by $g_1(t) * g_2(t)$ and by definition,

$$g_1(t) * g_2(t) = \int_{\tau=-\infty}^{\infty} g_1(\tau)g_2(t-\tau)d\tau \tag{3.20}$$

2. If $G_1(f)$ is the Fourier transform of $g_1(t)$, and $G_2(f)$ is the Fourier transform of $g_2(t)$, then

$$g_1(t) * g_2(t) \overset{\textbf{FT}}{\longleftrightarrow} G_1(f)G_2(f) \tag{3.21}$$

$$\text{or} \quad \mathscr{F}^{-1}\{G_1(f)G_2(f)\} = g_1(t) * g_2(t)$$

Thus, Eq. (3.19) can be written as

$$\hat{x}(t) = h(t) * x(t)$$

$$= \int_{-\infty}^{\infty} h(\tau)x(t-\tau)d\tau \tag{3.22}$$

where $h(t) = \mathscr{F}^{-1}\{-j\,\mathrm{sgn}(f)\}$ is the impulse response of the filter. We first find $\mathscr{F}^{-1}\{\mathrm{sgn}(f)\}$.

Consider the inverse Fourier transform of the function

$$M(f) = \begin{cases} e^{-\alpha f}, & f > 0 \\ -e^{\alpha f}, & f < 0 \end{cases}$$

It may be noted that $\lim_{\alpha \to 0} M(f) = \mathrm{sgn}(f)$

Thus, our strategy will be to find inverse Fourier transform of $M(f)$ and then take the limit of the result as $\alpha \to 0$. Inverse Fourier transform of $M(f)$ is

$$m(t) = \int_{-\infty}^{\infty} M(f)e^{j2\pi ft}df \qquad \text{(Ref Eq. (3.8))}$$

$$\Rightarrow \quad m(t) = \int_{-\infty}^{0} -e^{\alpha f} e^{j2\pi f t} df + \int_{0}^{\infty} e^{-\alpha f} e^{j2\pi f t} df$$

$$= j\frac{4\pi t}{\alpha^2 + (2\pi t)^2}$$

Taking the limit $\alpha \to 0$, we get the following transform pair:

$$\frac{j}{\pi t} \overset{\mathbf{FT}}{\longleftrightarrow} \operatorname{sgn}(f)$$

Hence,

$$h(t) = \frac{1}{\pi t} \overset{\mathbf{FT}}{\longleftrightarrow} H(f) = -j\operatorname{sgn}(f) \tag{3.23}$$

Using Eq. (3.23) in Eq. (3.22), we get

$$\hat{x}(t) = \int_{\tau=-\infty}^{\infty} \frac{x(\tau)}{\pi(t-\tau)} d\tau = x(t) * \frac{1}{\pi t} \tag{3.24}$$

Letting $t - \tau = \lambda$, we get

$$\hat{x}(t) = \int_{\lambda=+\infty}^{-\infty} \frac{x(t-\lambda)}{\pi\lambda}(-d\lambda)$$

$$= \int_{\lambda=-\infty}^{\infty} \frac{x(t-\lambda)}{\pi\lambda} d\lambda \tag{3.25}$$

Since λ is a dummy variable, it can be replaced by τ. Accordingly Eq. (3.25) becomes

$$\hat{x}(t) = \int_{\tau=-\infty}^{\infty} \frac{x(t-\tau)}{\pi\tau} d\tau \tag{3.26}$$

$$\Rightarrow \quad \hat{x}(t) = \frac{1}{\pi t} * x(t) \tag{3.27}$$

The function $\hat{x}(t)$ is defined as the Hilbert transform of $x(t)$.

Because Hilbert transform corresponds to a phase shift of $\frac{-\pi}{2}$ rad, it is of interest to note that the Hilbert transform of $\hat{x}(t)$ is $\hat{\hat{x}} = -x(t)$. This is because $\hat{\hat{x}}(t)$ undergoes a phase shift of $-\pi$ rad with respect to $x(t)$.

Some useful Hilbert transform pairs are listed in Table 3.3.

Table 3.3: Useful Hilbert transform pairs

Signal	Hilbert transform, $\hat{x}(t)$
$\cos 2\pi f t$	$\sin 2\pi f t$
$\sin 2\pi f t$	$-\cos 2\pi f t$
rect(t)	$-\frac{1}{\pi}\log_e\left\lvert\frac{t-0.5}{t+0.5}\right\rvert$
sinc(t)	$\frac{1-\cos\pi t}{\pi t}$
$\delta(t)$	$\frac{1}{\pi t}$
$\frac{1}{1+t^2}$	$\frac{t}{1+t^2}$
$\frac{1}{\pi t}$	$-\delta(t)$

Unlike Fourier transform, Hilbert transform converts a time-domain signal into another time-domian signal.

3.7.1 Useful properties of Hilbert transform

The three most useful properties of Hilbert transform are listed below:

1. The energy (or power) in a signal $x(t)$ and its Hilbert transform $\hat{x}(t)$ are equal.
2. A signal $x(t)$ and its Hilbert transform $\hat{x}(t)$ are orthogonal to each other.
3. If $c(t)$ and $x(t)$ are signals with nonoverlapping spectra, where $x(t)$ is low-pass and $c(t)$ is high-pass, then
$$\widehat{x(t)c(t)} = x(t)\hat{c}(t)$$

Example 3.38. Find the Hilbert transform of $x(t) = \cos 2\pi f_c t$

Solution:

$$\hat{x}(t) = \cos(2\pi f_c t - \pi/2)$$
$$= \sin 2\pi f_c t$$

That is, the Hilbert transform of $x(t)$ is found directly by adding $\frac{-\pi}{2}$ to the argument of the cosine.

Example 3.39. Find the Hilbert transforms of the signals given below:

(a) $x(t) = \cos 2\pi f t + \sin 2\pi f t$

(b) $x(t) = e^{-j2\pi f t}$

(c) $x(t) = \delta(t)$

(d) $x(t) = \cos 2\pi f_1 t \sin 2\pi f_2 t,\ f_2 > f_1$

Solution:

(a)

$$\begin{aligned}\hat{x}(t) &= \cos\left(2\pi ft - \frac{\pi}{2}\right) + \sin\left(2\pi ft - \frac{\pi}{2}\right)\\ &= \cos\left(\frac{\pi}{2} - 2\pi ft\right) - \sin\left(\frac{\pi}{2} - 2\pi ft\right)\\ &= \sin 2\pi ft - \cos 2\pi ft\end{aligned}$$

(b) Rewrite $x(t)$ as

$$x(t) = \cos 2\pi ft - j\sin 2\pi ft$$

Then,

$$\begin{aligned}\hat{x}(t) &= \cos\left(2\pi ft - \frac{\pi}{2}\right) - j\sin\left(2\pi ft - \frac{\pi}{2}\right)\\ &= \cos\left(\frac{\pi}{2} - 2\pi ft\right) + j\sin\left(\frac{\pi}{2} - 2\pi ft\right)\\ &= \sin 2\pi ft + j\cos 2\pi ft\\ &= j[\cos 2\pi ft - j\sin 2\pi ft]\\ &= je^{-j2\pi ft}\end{aligned}$$

(c)

$$\begin{aligned}\hat{x}(t) &= x(t) * \frac{1}{\pi t}\\ &= \delta(t) * \frac{1}{\pi t}\end{aligned}$$

For convolution operation, $\delta(t)$ is the identity operator (see chapter 5)

Hence,

$$\hat{x}(t) = \delta(t)$$

(d) Recall: $\widehat{x(t)c(t)} = x(t)\hat{c}(t)$

Given:

$$x(t) = \underbrace{\cos 2\pi f_1 t}_{\text{low-pass signal}}\ \underbrace{\sin 2\pi f_2 t}_{\text{high-pass signal}}$$

Hence,

$$\begin{aligned}\hat{x}(t) &= \cos 2\pi f_1 t\, \widehat{\sin 2\pi f_2 t}\\ &= \cos 2\pi f_1 t \sin\left(2\pi f_2 t - \frac{\pi}{2}\right)\\ &= -\cos 2\pi f_1 t \cos 2\pi f_2 t\end{aligned}$$

Example 3.40. Find Hilbert transform of a rectangular pulse shown in Fig. Ex. 3.40.

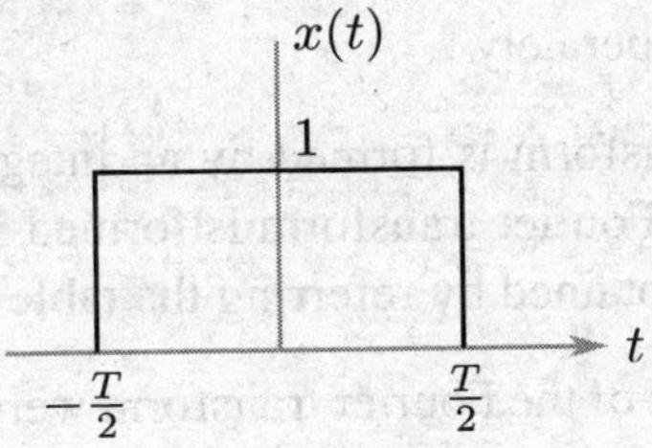

Fig. Ex. 3.40

Solution: Recall,

$$\hat{x}(t) = \int_{\tau=-\infty}^{\infty} \frac{x(\tau)}{\pi(t-\tau)} d\tau \quad \text{(Refer Eq. (3.24))}$$

Hence,

$$\hat{x}(t) = -\frac{1}{\pi} \int_{\tau=-T/2}^{T/2} \frac{1}{\tau - t} d\tau$$

$$= -\frac{1}{\pi} \log_e(\tau - t)\Big|_{\tau=-\frac{T}{2}}^{\frac{T}{2}}$$

$$= \frac{1}{\pi}\left[\log_e\left(\frac{T}{2} - t\right) - \log_e\left(-\frac{T}{2} - t\right)\right]$$

$$= \frac{1}{\pi} \log_e\left[\frac{t + \frac{T}{2}}{t - \frac{T}{2}}\right]$$

Summary of Important points

- In this chapter, we have extended Fourier series to aperiodic continuous-time signals, which resulted in the Fourier transform.
- We have defined the Fourier transform by Eq. (3.4) and Eq. (3.5) and inverse Fourier transform by Eq. (3.7) and Eq. (3.8).
- The sufficient conditions for the existance of Fourier transform are called Dirichlet conditions. In general, the Fourier transform of $x(t)$ exists if $x(t)$ is absolutely integrable. These conditions are sufficient but not necessary. We have seen that many useful signals used in practice violate these conditions, but still have Fourier transforms.

- The Fourier transform of a signal is called frequency spectrum. Frequency spectrum in most cases is a complex-valued continuous function of frequency. Its magnitude and phase are plotted seperately.
- The inverse Fourier transform is formed by an integral formula. However, a partial fraction expansion of a Fourier transform is formed in majority of the cases, and the time-domain signal is obtained by referring the table of transforms.
- Several useful properties of the Fourier transform were introduced and derived and are listed in Table 3.1. Fourier transforms of most commonly used signals were derived and are listed in Table3.2.
- The Fourier transform of a periodic signal can be defined by a train of impulses located at harmonic frequencies whose strengths are a constant times the Fourier series coefficients of the signal.
- Hilbert transform is the only known transform that transforms a time-domain signal into another time-domain signal.

Exercise Problems

EP. 3.1 Compute the Fourier transform of the doubled-sided exponential defined by

$$x(t) = e^{-a|t+3|}$$

using the defining equation of **FT**.

Ans: $X(\omega) = \dfrac{2a}{\omega^2 + a^2}\, e^{j3\omega}$

EP. 3.2 Find Fourier transform of the pulse shown in Fig. EP. 3.2.

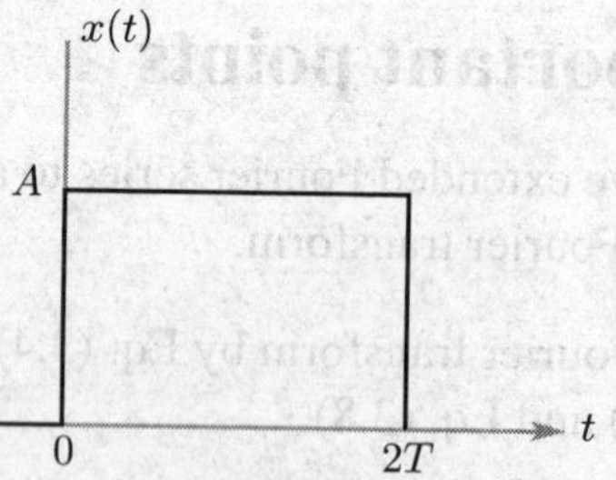

Fig. EP. 3.2

Ans: $X(\omega) = 2ATe^{-j\omega T}\left[\dfrac{\sin \omega T}{\omega T}\right]$

EP. 3.3 Find the Fourier transform of the waveform shown in Fig. EP. 3.3.

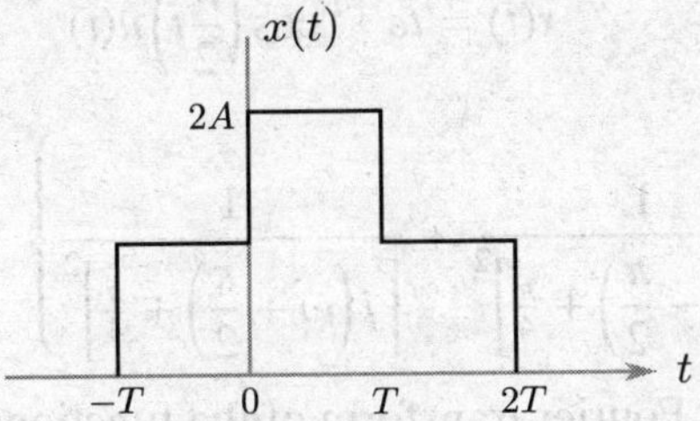

Fig. EP. 3.3

Ans: $X(\omega) = 4ATe^{-j\omega T/2}\cos\left(\frac{\omega T}{2}\right)\frac{\sin\omega T}{\omega T}$

EP. 3.4 Find the **FT** of

$$x(t) = \frac{A}{T_0}t\{u(t+T_0) - u(t-T_0)\}$$

Use defining equation only.

Ans: $X(\omega) = \frac{j2A}{\omega^2 T_0}[\omega T_0 \cos\omega T_0 - \sin\omega T_0]$

EP. 3.5 Derive the **FT** of the triangular pulse shown in Fig. EP. 3.5.

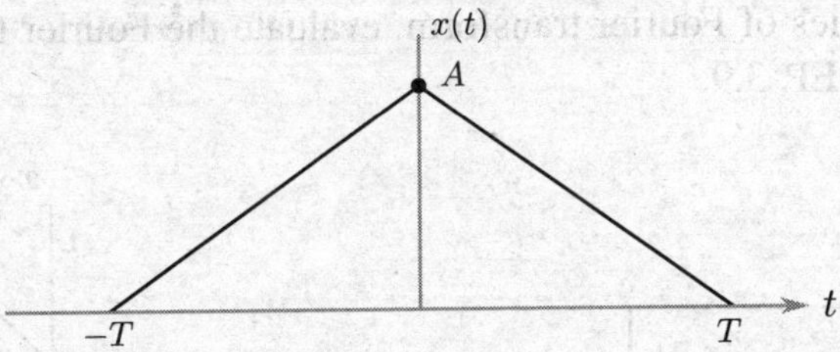

Fig. EP. 3.5

Ans: $X(\omega) = \frac{2A}{\omega^2 T/2}\sin^2(\omega T/2).$

EP. 3.6 Find **FT** of the functions given below:

(a) $x(t) = \cos\frac{\pi}{4}t$ (b) $x(t) = \cos\frac{\pi}{4}t\,u(t)$

Ans:

(a) $X(\omega) = \pi\delta(\omega - \pi/4) + \pi\delta(\omega + \pi/4)$

(b) $X(\omega) = \frac{1}{2}\left\{\pi\delta(\omega - \pi/4) + \frac{1}{j(\omega-\pi/4)} + \pi\delta(\omega + \frac{\pi}{4}) + \frac{1}{j(\omega+\pi/4)}\right\}$

EP. 3.7 Find Fourier transform of the function given below:

$$x(t) = te^{\frac{-\pi}{4}t}\cos\left(\frac{\pi}{2}t\right)u(t)$$

Ans: $X(\omega) = \frac{1}{2}\left\{\frac{1}{\left[j\left(\omega - \frac{\pi}{2}\right) + \frac{\pi}{4}\right]^2} + \frac{1}{\left[j\left(\omega + \frac{\pi}{2}\right) + \frac{\pi}{4}\right]^2}\right\}$

EP. 3.8 Using properties, find Fourier transform of the function given below:

$$x(t) = \int_{-\infty}^{t} \frac{\sin \pi\tau}{\pi\tau} d\tau$$

Ans:

$$X(\omega) = \begin{cases} \pi, & \omega = 0 \\ 1/j\omega, & -\pi < \omega < \pi \\ 0, & \text{otherwise} \end{cases}$$

EP. 3.9 Given:

$$x(t) = \begin{cases} 1, & -1 < t < 1 \\ 0, & \text{elsewhere} \end{cases} \overset{\textbf{FT}}{\longleftrightarrow} X(\omega) = \frac{2\sin\omega}{\omega}$$

Using the properties of Fourier transform, evaluate the Fourier transform of the signals given in Fig. EP. 3.9.

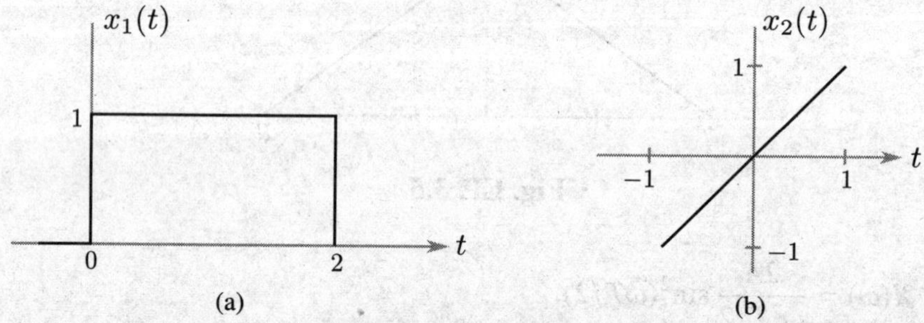

Fig. EP. 3.9

Ans:

$$X_1(\omega) = e^{-j\omega}\left[\frac{2\sin\omega}{\omega}\right]$$

$$X_2(\omega) = 2j\left[\frac{\omega\cos\omega - \sin\omega}{\omega^2}\right]$$

EP. 3.10 Find inverse **FT** of the transformed function given below:

$$X(j\omega) = \frac{1}{j\omega(j\omega + 1)} + 2\pi\delta(\omega)$$

Ans: $x(t) = [1 - e^{-t}]u(t) + \frac{1}{2}$

EP. 3.11 Using properties of Fourier transform, find inverse Fourier transform of $X(\omega)$ given below:

$$X(\omega) = \frac{j\omega}{(2 + j\omega)^2}$$

Ans: $x(t) = [-2te^{-2t} + e^{-2t}]u(t)$

EP. 3.12 Find Fourier transform of the signal defined by

$$x(t) = \frac{\sin at}{\pi t}$$

Ans:

$$X(\omega) = \begin{cases} 1, & |\omega| < a \\ 0, & |\omega| > a \end{cases}$$

EP. 3.13 A signal $x(t)$ has the Fourier transform

$$X(\omega) = \frac{j\omega}{-\omega^2 + 7j\omega + 6}$$

Evaluate the Fourier transform of the following functions:

(a) $y_1(t) = x(4t)$

(b) $y_2(t) = \frac{dx(t)}{dt}$

(c) $y_3(t) = x(-t)$

Ans:

(a) $Y_1(\omega) = \frac{j\omega}{-\omega^2 + 28j\omega + 96}$

(b) $Y_2(\omega) = \frac{-\omega^2}{-\omega^2 + j7\omega + 6}$

(c) $Y_3(\omega) = \frac{j\omega}{\omega^2 + 7j\omega - 6}$

EP. 3.14 Find Fourier transform of the signals given below:

(a) $x(t) = a\,\text{sinc}\left(\frac{at}{2}\right)$

(b) $y(t) = \text{sinc}^2(t)$

Ans:

(a) $X(\omega) = \begin{cases} 2\pi, & \frac{-a}{2} < \omega < \frac{a}{2} \\ 0, & \text{elsewhere} \end{cases}$

(b) $Y(\omega) = \begin{cases} \pi + \frac{\pi}{2}\omega, & -2 < \omega < 0 \\ \pi - \frac{\pi}{2}\omega, & 0 < \omega < 2 \end{cases}$

EP. 3.15 Find inverse fourier transform of $X(\omega)$ given below:

$$X(\omega) = \text{Im}\left\{e^{-j3\omega}\frac{1}{j\omega + 2}\right\}$$

Ans:

$$x(t) = \frac{e^{-2(t-3)}u(t-3) - e^{-2(-t-3)}u(-t-3)}{2}$$

EP. 3.16 Find inverse Fourier transform of

$$X(\omega) = \frac{1}{(j\omega + 1)(j\omega + 2)}$$

Ans: $x(t) = e^{-t}u(t) - e^{-2t}u(t)$

Chapter 4

Signal transmission through LTI systems

In this chapter, we shall study various properties of the system. In particular, we shall study two important properties, namely linearity and time-invariance. The response of a system to unit impulse excitation is called impulse response of the system. Using this impulse response, we shall investigate various properties of the system such as causality and stability by specifying the input-output relationships for linear time-invariant systems by differential equations along with their solutions will also be discussed.

The concept of filter was conceived early in human history. A paper filter was used to remove dirt and unwanted substances from water and wine. In a similar manner, an elelctric filter can be used to eliminate unwanted frequency components, such as electrical noise from an electrical signal. Physical realizability of such filters is also discussed.

4.1 System

In mathematical terms, a system may be viewed as an interconnection of operations that transforms an input signal to an output signal with properties different from those of the input signal. Let the overall operator $\mathbf{T}\{\cdot\}$ denote the action of a system. Then, $\mathbf{T}\{\cdot\}$ operates on the input of the system to yield an output signal.

That is,

$$y(t) = \mathbf{T}\{x(t)\}$$

The above equation is represented by a block diagram as depicted in Fig 4.1.

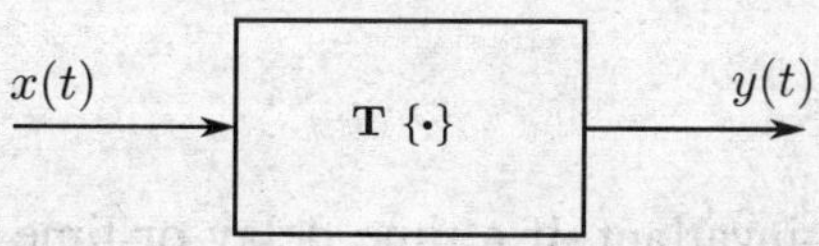

Fig. 4.1: *System viewed as an operator*

In order to understand the concept of system as an operator, let us consider a continuous-time system charaterized by an input-output relation defined by

$$y(t) = a \int_0^t x(\tau)\, d\tau$$

Then, operator $\mathbf{T}\{\cdot\}$ does the following activities on the input signal:

(i) It replaces t by τ in $x(t)$.

(ii) Then performs integration on $x(\tau)$ with respect to τ between the limits $(0, t)$.

(iii) Finally multiplies the integrated result by a scalar 'a' to give the output $y(t)$.

4.2 Properties of systems

4.2.1 Linearity

A continuous-time system described by $\mathbf{T}\{\cdot\}$ is linear, if the response to a weighted sum of inputs $x_1(t)$ and $x_2(t)$ is a weighted sum of the responses of the two inputs seperately, for all the weights and all the acceptable signals $x_1(t)$ and $x_2(t)$.

Therefore, a system described by $\mathbf{T}\{\cdot\}$ is linear, if the following relation is observed:

$$\mathbf{T}\{a_1x_1(t) + a_2x_2(t)\} = a_1\,\mathbf{T}\{x_1(t)\} + a_2\,\mathbf{T}\{x_2(t)\}$$

Superposition implies both homogenity and additivity.

$$\text{Homogenity: } \mathbf{T}\{ax(t)\} = a\,\mathbf{T}\{x(t)\}$$
$$\text{Additivity: } \mathbf{T}\{x_1(t) + x_2(t)\} = \mathbf{T}\{x_1(t)\} + \mathbf{T}\{x_2(t)\}$$

In precise, a system is linear if it obeys the principle of superposition.

4.2.2 Causality

A system is said to be causal, if the present value of the output signal depends only on the present and past values or a combination of present and past values of the input signal.

A continuous-time system is defined to be causal if the output at $t = t_0$ depends only on the input for $t \le t_0$. For example, an input-output relation of the form, $y(t) = x(t) + x(t-1)$ belongs to a causal system.

4.2.3 Time-invariance

A system is said to be time-invariant, if a time delay or time advance of the input signal leads to an identical time shifts in the output signal.

A continuous-time system specified by $\mathbf{T}\{\cdot\}$ is time-invariant, if for all $x(t)$ and t_0, we have

$$\mathbf{T}\{x(t-t_0)\} = y(t-t_0)$$

where

$$y(t) = \mathbf{T}\{x(t)\}$$

4.2.4 Stability

A continuous-time system is bounded input-bounded output stable (BIBO stable), if every bounded input produces a bounded output.

Mathematically, if $|x(t)| \leq M_x < \infty$, then for BIBO stability, the output $y(t)$ must obey the condition, $|y(t)| \leq M_y < \infty$.

As an example, let us investigate the BIBO stability of a continuous-time system whose input-output relation is given by

$$y(t) = x(t) + x(t-1)$$

Let $|x(t)| \leq M_x < \infty$ (Bounded input)
Then,

$$|y(t)| = |x(t) + x(t-1)|$$

Applying triangular inequality, we can write

$$|y(t)| \leq |x(t)| + |x(t-1)|$$

By translating a sequence, the bound on the translated sequence does not change.
Thus,

$$|y(t)| \leq M_x + M_x$$

$$\Rightarrow \quad |y(t)| \leq 2M_x < \infty \quad \text{(Bounded output)}$$

Hence, the system considered in the above example is BIBO stable.

Example 4.1. Consider the system shown in Fig. Ex. 4.1. Determine whether it is

(a) Causal (b) Linear (c) Time-invariant, and (d) Stable

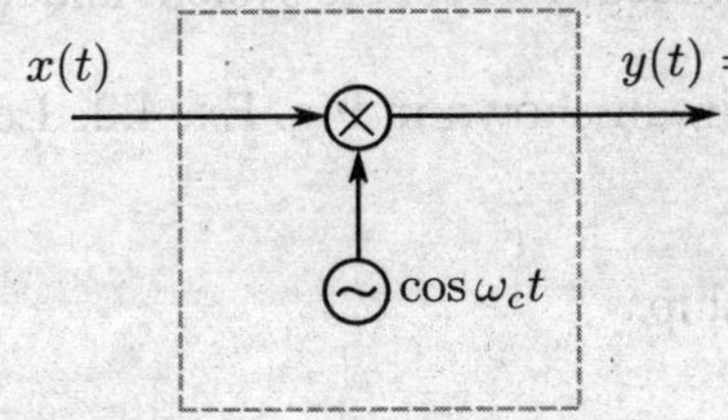

Fig. Ex. 4.1

Solution: From Fig. Ex. 4.1, we find that $y(t) = \mathbf{T}\{x(t)\} = x(t)\cos\omega_c t$.

(a) $y(t_0) = x(t_0)\cos\omega_c t_0$

The above equation means output at $t = t_0$ depends on the input at $t = t_0$. Hence, system is causal.

(b) Given $\mathbf{T}\{x(t)\} = x(t)\cos\omega_c t$.

The role of operator, $\mathbf{T}\{\cdot\}$ is to multiply the input signal, $x(t)$ by $\cos\omega_c t$.

Therefore,

$$\begin{aligned}\mathbf{T}\{a_1x_1(t) + a_2x_2(t)\} &= [a_1x_1(t) + a_2x_2(t)]\cos\omega_c t\\ &= a_1x_1(t)\cos\omega_c t + a_2x_2(t)\cos\omega_c t\\ &= a_1\,\mathbf{T}\{x_1(t)\} + a_2\,\mathbf{T}\{x_2(t)\}\end{aligned}$$

Thus, the system is linear.

(c) Given, $y(t) = \mathbf{T}\{x(t)\} = x(t)\cos\omega_c t$.

Hence,

$$\begin{aligned}\mathbf{T}\{x(t-t_0)\} &= x(t-t_0)\cos\omega_c t\\ &\neq y(t-t_0)\\ &= x(t-t_0)\cos[\omega_c(t-t_0)]\end{aligned}$$

Thus, the system is not time-invariant.

(d) let $|x(t)| \leq M_x < \infty$ (Bounded input)

$$\begin{aligned}\text{Given,}\quad & y(t) = x(t)\cos\omega_c t\\ \Rightarrow\quad & |y(t)| = |x(t)||\cos\omega_c t|\\ \Rightarrow\quad & |y(t)| \leq M_x|\cos\omega_c t|\end{aligned}$$

Since, $|\cos\omega_c t| \leq 1$, $y(t)$ is bounded. Hence, the system under investigation is stable.

Example 4.2. Refer the electrical circuit shown in Fig. Ex. 4.2. Let $x(t) = i(t)$ and $y(t) = v_C(t)$.

(a) Find the input-output relationship.

(b) Determine whether the system is

(i) Causal (ii) Linear (iii) Time-invariant (iv) Stable

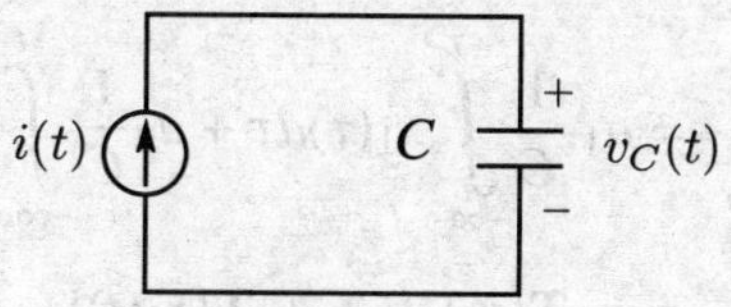

Fig. Ex. 4.2

Solution:

(a) $i(t) = \dfrac{C dv_C(t)}{dt}$

$$\text{or} \quad v_C(t) = \frac{1}{C}\int_{-\infty}^{t} i(\tau)\, d\tau$$

$$\Rightarrow \quad y(t) = \frac{1}{C}\int_{-\infty}^{t} x(\tau)\, d\tau$$

Hence,

$$y(t) = \mathbf{T}\{x(t)\} = \frac{1}{C}\int_{-\infty}^{t} x(\tau)\, d\tau$$

The above relation is put in the form of a block diagram as shown in Fig. Ex. 4.2(a).

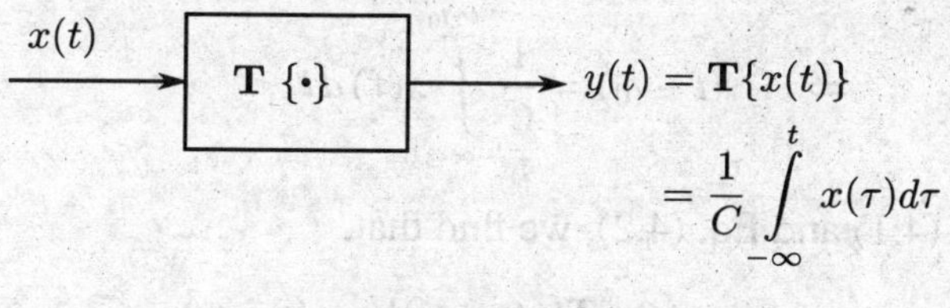

Fig. Ex. 4.2(a)

(b) (i) Since, the output depends only on the present and past values of the input, the system given in the problem is causal.

(ii) We know that $\mathbf{T}\{x(t)\} = \int_{-\infty}^{t} x(\tau)\, d\tau$. Hence, the role of operator, $\mathbf{T}\{\cdot\}$ is to replace t by τ in $x(t)$ and then integrate $x(\tau)$ with respect to τ between the limits $-\infty$ to t to produce the output.

Thus, $\mathbf{T}\{a_1x_1(t) + a_2x_2(t)\}$

$$= \frac{1}{C}\int_{-\infty}^{t} [a_1x_1(\tau) + a_2x_2(\tau)]d\tau$$

$$= a_1.\frac{1}{C}\int_{-\infty}^{t} x_1(\tau)d\tau + a_2\frac{1}{C}\int_{-\infty}^{t} x_2(\tau)d\tau$$

$$= a_1\, \mathbf{T}\{x_1(t)\} + a_2\, \mathbf{T}\{x_2(t)\}$$

Therefore, the system is linear.

(iii)

$$\mathbf{T}\{x(t-t_0)\} = \frac{1}{C}\int_{-\infty}^{t} x(\tau - t_0)\, d\tau$$

Using change of variables, $\lambda = \tau - t_0$, we get

$$\mathbf{T}\{x(t-t_0)\} = \frac{1}{C}\int_{-\infty}^{t-t_0} x(\lambda)\, d\lambda$$

Since, λ is a dummy variable, it can be replaced by τ. Accordingly, the above equation becomes

$$\mathbf{T}\{x(t-t_0)\} = \frac{1}{C}\int_{-\infty}^{t-t_0} x(\tau)\, d\tau \qquad (4.1)$$

Also, $y(t) = \frac{1}{C}\int_{-\infty}^{t} x(\tau)\, d\tau$

$$\Rightarrow \quad y(t-t_0) = \frac{1}{C}\int_{-\infty}^{t-t_0} x(\tau)\, d\tau \qquad (4.2)$$

From Eq. (4.1) and Eq. (4.2), we find that,

$$\mathbf{T}\{x(t-t_0)\} = y(t-t_0)$$

Thus, an identical shift in the input produces an identical shift in the output. This means that the system is time-invariant.

(iv) Let $x(t) = K_1 u(t)$

Fig. Ex. 4.2(b)

Then,

$$y(t) = \frac{1}{C}\int_{-\infty}^{t} K_1 u(\tau)d\tau$$

$$= \frac{K_1}{C}\int_{0}^{t} d\tau$$

$$= \frac{K_1}{C}t, t > 0$$

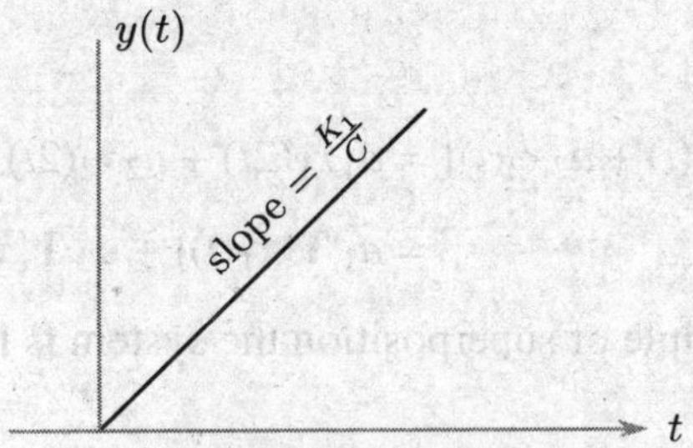

Fig. Ex. 4.2 (c)

Since, the output $y(t)$ is growing linearly without bound as shown in Fig. Ex. 4.2(c), the system is not stable.

Example 4.3. For the system described by the following input-output relationship, find which of the systems are linear and which are nonlinear. Also find time-invariant and time-variant systems.

(a) $\frac{dy}{dt} + 3ty(t) = t^2x(t)$

(b) $\left(\frac{dy}{dt}\right)^2 + 2y(t) = x(t)$

(c) $\frac{dy}{dt} + 3y(t) + 4 = x(t)$

Solution: If a system is described by a differential equation, then

(i) it is nonlinear if any term is a constant or a nonlinear function of $x(t)$ or $y(t)$ and

(ii) it is time varying if the coefficients of any term in $x(t)$ or $y(t)$ is an explicit function of t.

(a) (i) System is linear (ii) time-variant

(b) (i) System is nonlinear (ii) time-invariant

(c) (i) System is nonlinear (ii) time-invariant

Example 4.4. A system is described by the input-output relation given below:

$$y(t) = x(2t)$$

Check for linearity and time-invariance.

Solution: The system is represented as an operator $\mathbf{T}\{\cdot\}$. This aspect is shown in Fig. Ex. 4.4.

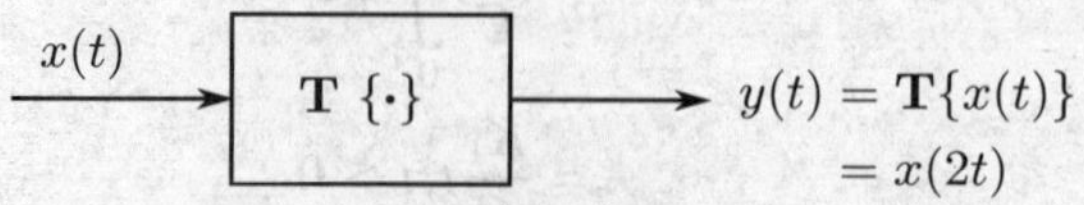

Fig. Ex. 4.4

Linearity:

$$\begin{aligned}\mathbf{T}\{a_1x_1(t) + a_2x_2(t)\} &= a_1x_1(2t) + a_2x_2(2t)\\ &= a_1\,\mathbf{T}\{x_1(t)\} + a_2\,\mathbf{T}\{x_2(t)\}\end{aligned}$$

Since system obeys the principle of superposition the system is linear.
Time-invariance:

$$\mathbf{T}\{x(t - t_0)\} = x(2t - t_0)$$

$$\text{But,}\qquad y(t - t_0) = x[2(t - t_0)]$$

Since, $\mathbf{T}\{x(t - t_0)\} \neq y(t - t_0)$, the system is time-variant.

Example 4.5. A system is described by the input-output relation given below:

$$y(t) = x(t)x'(t)$$

Check for linearity and time-invariance.

Solution: The system given in the problem is described by the block diagram shown in Fig. Ex. 4.5.

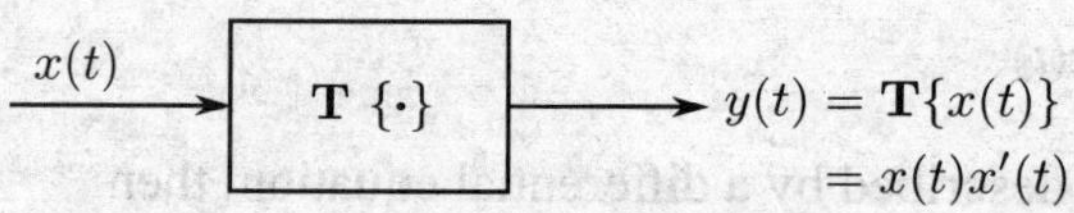

Fig. Ex. 4.5

Linearity:

$$\begin{aligned}\mathbf{T}\{a_1x_1(t) + a_2x_2(t)\} &= [a_1x_1(t) + a_2x_2(t)][a_1x_1'(t) + a_2x_2'(t)]\\ &\neq a_1\,\mathbf{T}\{x_1(t)\} + a_2\,\mathbf{T}\{x_2(t)\}\\ &= a_1x_1(t)x_1'(t) + a_2x_2(t)x_2'(t)\end{aligned}$$

Hence, system is nonlinear.

Time-invariance:

$$\mathbf{T}\{x(t-t_0)\} = x(t-t_0)x'(t-t_0)$$

and

$$y(t-t_0) = x(t-t_0)x'(t-t_0)$$

Since, $\mathbf{T}\{x(t-t_0)\} = y(t-t_0)$, the system given in the problem is time-invariant.

Example 4.6. A double-sideband suppressed carrier system is described by the following input-output relationship:

$$y(t) = x(t)\cos 2\pi f_c t$$

Check for causality, linearity and time-invariance.

Solution: The given system can be represented by an operator $\mathbf{T}\{\cdot\}$ as depicted in Fig. Ex. 4.6.

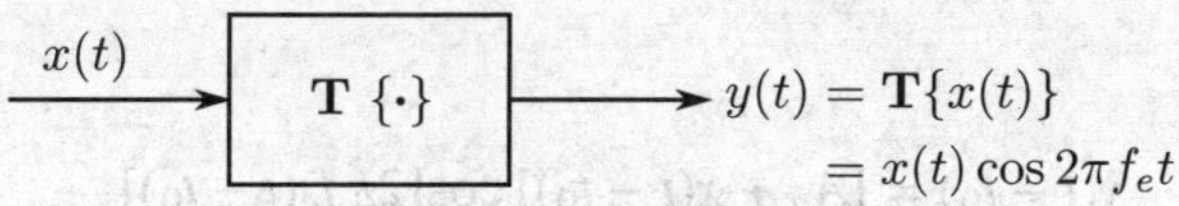

Fig. Ex. 4.6

Causality:

$$y(t_0) = x(t_0)\cos(2\pi f_c t_0)$$

Since output at $t = t_0$ depends on the input at $t = t_0$, the system is causal.
Linearity:

$$\begin{aligned}\mathbf{T}[a_1x_1(t) + a_2x_2(t)] &= [a_1x_1(t) + a_2x_2(t)]\cos 2\pi f_c t\\ &= a_1x_1(t)\cos 2\pi f_c t + a_2x_2(t)\cos 2\pi f_c t\\ &= a_1\,\mathbf{T}\{x_1(t)\} + a_2\,\mathbf{T}\{x_2(t)\}\end{aligned}$$

Hence, the system is linear.
Time-invariance:

$$\mathbf{T}\{x(t-t_0)\} = x(t-t_0)\cos 2\pi f_c t$$

and $\quad y(t-t_0) = x(t-t_0)\cos[2\pi f_c(t-t_0)]$

since, $\quad \mathbf{T}\{x(t-t_0)\} \neq y(t-t_0)$

The system given in the problem is time-variant.

Example 4.7. A standard amplitude modulation system is described by the following mathematical input-output equation:

$$y(t) = \mathbf{T}\{x(t)\} = [A_c + x(t)]\cos 2\pi f_c t$$

Check for causality, linearity and time-invariance.

Solution: Causality:

$$y(t_0) = [A_c + x(t_0)]\cos 2\pi f_c t_0$$

The above equation means that the output at $t = t_0$ depends on the input at $t = t_0$. That is, the present output is produced by present input only. Hence, system is causal.
Linearity:

$$\begin{aligned}\mathbf{T}\{a_1x_1(t) + a_2x_2(t)\} &= [A_c + a_1x_1(t) + a_2x_2(t)]\cos 2\pi f_c t \\ &\neq a_1\,\mathbf{T}\{x_1(t)\} + a_2\,\mathbf{T}\{x_2(t)\} \\ &= a_1[A_c + x_1(t)]\cos 2\pi f_c t + a_2[A_c + x_2(t)]\cos 2\pi f_c t\end{aligned}$$

Since, system does not obey the principle of superposition, the system is nonlinear.
Time-invariance:

$$\mathbf{T}\{x(t-t_0)\} = [A_c + x(t-t_0)]\cos 2\pi f_c t$$

and

$$y(t-t_0) = [A_c + x(t-t_0)]\cos[2\pi f_c(t-t_0)]$$

Since, $\mathbf{T}\{x(t-t_0\} \neq y(t-t_0)$, the system is time-variant.

4.3 Response of a linear system

A linear time-invariant (LTI) system is always described by a linear constant coefficient differential equation. Consider an LTI system described by an nth order constant coefficient differential equation with a single input, $x(t)$:

$$a_0\frac{d^n y(t)}{dt^n} + a_1\frac{d^{n-1}y(t)}{dt^{n-1}} + \cdots + a_n y(t) = x(t) \tag{4.3}$$

Let $s^n \equiv \frac{d^n}{dt^n}$

Then Eq. (4.3) gets transformed as $[a_0s^n + a_1s^{n-1} + \cdots + a_n]y(t) = x(t)$

4.3.1 The Zero-input, Zero-state and total responses

The structure of the Zero-input response depends only on the system details and is independent of the nature of the input. As the name suggests, it is the response of the system with input reduced to zero. In otherwords, it is the response of the system produced by initial conditions only.

The roots of an equation called characteristic equation defined in Eq. (4.4) determines the time-domain expression for the zero-input response.
Characteristic equation:

$$a_0s^n + a_1s^{n-1} + \cdots + a_n = 0 \tag{4.4}$$

Zero-input response:

$$y_{ZI}(t) = K_1 e^{s_1 t} + K_2 e^{s_2 t} + \cdots + K_n e^{s_n t} \tag{4.5}$$

where $s_1, s_2, \ldots s_n$ are the roots of the characteristic equation. Here, $K_1, K_2, \ldots K_n$ are constants and are determined by making use of the specified inital conditions.

The structure of zero-input response has to be modified for multiple roots of the characteristic equation. The zero-input response is also known as natural response of the system. Table 4.1 summarizes the form of the zero-input response for different root conditions of the characteristic equation defined by Eq. (4.4).

Table 4.1: Structure of the two-input response for LTI systems

Sl.no	Roots of the characteristic equation	Form of Zero-input response
1	Real and distinct: r	Ke^{rt}
2	Complex conjugate: $a \pm j\omega$	$e^{at}[K_1 \cos \omega t + K_2 \sin \omega t]$
3	Real, repeated: r^{p+1}	$e^{rt}[K_0 + K_1 t + \ldots + K_p t^p]$

The particular response has the same structure as the input and satisfies the given differential equation. As an example, if $x(t) = u(t)$, then the particular response is $y_p(t) = C$. The constants in the paticular solution are determined independently and uniquely by substituting the particular response in the given differential equation. We do not make use of zero-input response or initial conditions to find constants present in particular response. Table 4.2 gives the structure of the particular response for different input signals.

Table 4.2: Structure of particular response for LTI systems

Sl.no	Input	Structure of particular response
1	C_0	C_1(another constant)
2	$e^{\alpha t}$(See note below)	$Ce^{\alpha t}$
3	$\cos(\omega t + \beta)$	$C_1 \cos \omega t + C_2 \sin \omega t$
4	$e^{\alpha t} \cos(\omega t + \beta)$ (see note below)	$e^{\alpha t}[C_1 \cos \omega t + C_2 \sin \omega t]$
5	t	$C_0 + C_1 t$
6	t^p	$C_0 + C_1 t + C_2 t^2 + \cdots + C_p t^p$
7	$te^{\alpha t}$ (see note below)	$e^{\alpha t}[C_0 + C_1 t]$

Note: If input $x(t) = e^{\alpha t}$ and α is also a root of the characteristic equation repeated m times, then the structure of the particular response of Table 4.2 must be multiplied by t^m.

The Zero-state response is the response of the system produced by input only. It has the structure given below.

$$y_{ZS}(t) = y_{ZI}(t) + y_P(t)$$

The Zero-state response of the system is also known as forced response of the system. The constants present in the Zero-state response are evaluated by assuming zero inital conditions.

The system under investigation is an LTI system. Hence, obeys the principle of superposition. Thus, the total response of the system is the sum of zero-input response and zero-state response. That is,

$$y(t) = y_{ZI}(t) + y_{ZS}(t)$$

It may be noted that zero-state response and zero-input response also obey the principle of superposition.

Example 4.8. Evaluate Zero-state, zero-input and total response of the system given below:

$$y'(t) + 5y(t) = u(t), \quad y(0) = 2$$

Solution: The characteristic equation is

$$s + 5 = 0$$

$$\Rightarrow \quad s = -5$$

The zero-input response is given by

$$y_{ZI}(t) = K_1 e^{-5t}$$

The constant K_1 is evaluated by making use of prescribed inital condition.

$$y(0) = 2 \Rightarrow 2 = K_1 e^{-5t}\Big|_{t=0}$$

$$\Rightarrow K_1 = 2$$

Hence, the zero-input response of the system is $y_{ZI}(t) = 2e^{-5t}$.

The particular solution has the same structure as the input, $x(t)$. In the present example, $x(t) = 1, t > 0$, hence we take $y_P(t) = C$. The constant C is evaluated by substituting the particular solution in the given differential equation:

$$y_P'(t) + 5y_P(t) = 1$$

$$\Rightarrow \quad 0 + 5C = 1$$

$$\Rightarrow \quad C = \frac{1}{5}$$

Thus, the particular solution is

$$y_P(t) = \frac{1}{5}$$

The zero-state response of the system is given by

$$y_{ZS}(t) = y_{ZI}(t) + y_P(t)$$

$$= K_1 e^{-5t} + \frac{1}{5}$$

The constant K_1 is now evaluated by assuming zero initial condition.

$$y_{ZS}(0) = 0 \Rightarrow 0 = K_1 + \frac{1}{5} \quad \Rightarrow K_1 = -\frac{1}{5}$$

Hence, $y_{ZS}(t) = -\frac{1}{5}e^{-5t} + \frac{1}{5}$.

The total response of the system is

$$\begin{aligned} y(t) &= y_{ZI}(t) + y_{ZS}(t) \\ &= 2e^{-5t} - 0.2e^{-5t} + 0.2 \\ &= 1.8e^{-5t} + 0.2, \quad t \geq 0 \\ &= [1.8e^{-5t} + 0.2]u(t) \end{aligned}$$

Note that, since $y(t)$ is represented in terms of $u(t)$, it means that $t \geq 0$ is same as $t \geq 0^+$. This is because, $u(t) = 1$ for $t \geq 0^+$ and $u(t) = 0$ for $t \leq 0^-$.

Example 4.9. Evaluate the total response of an LTI system described by the differential equation given below:

$$y''(t) + 5y'(t) + 6y(t) = 2e^{-t}u(t)$$

$$y(0) = 0, \quad y'(0) = 1$$

Solution: The characteristic equation of the system is

$$s^2 + 5s + 6 = 0$$

The roots are $s = -3$ and $s = -2$. Hence, the zero-input response of the system is $y_{ZI}(t) = K_1^{-3t} + K_2e^{-2t}$

The constants K_1 and K_2 are evaluated by making use of the inital conditions given in the problem.

(i) $y(0) = 0 \Rightarrow K_1 + K_2 = 0$ (ii) $y'(0) = 1 \Rightarrow -3K_1 - 2K_2 = 1$

Solving, we get

$$K_1 = -1 \quad \text{and} \quad K_2 = 1$$

Thus,

$$y_{ZI}(t) = -e^{-3t} + e^{-2t}, \quad t \geq 0$$

Since input, $x(t) = 2e^{-t}u(t)$ and the roots of the characteristic equation are -2 and -3, we take the particular solution as $y_P(t) = Ce^{-t}$. The constant C is evaluated by substituting the particular solution in the given differential equation:

$$\begin{aligned} & y_p''(t) + 5y_p'(t) + 6y_p(t) = 2e^{-t} \\ \Rightarrow \quad & [C - 5C + 6C]e^{-t} = 2e^{-t} \\ \Rightarrow \quad & C = 1 \end{aligned}$$

Thus, the particular solution is

$$y_P(t) = e^{-t}$$

The Zero-state response of the system is

$$\begin{aligned} y_{ZS}(t) &= y_{ZI}(t) + y_P(t) \\ &= K_1 e^{-3t} + K_2 e^{-2t} + e^{-t} \end{aligned}$$

The constants K_1 and K_2 are found by assuming zero initial conditions.

(i) $y_{ZS}(0) = 0 \Rightarrow K_1 + K_2 + 1 = 0$

(ii) $y'_{ZS}(0) = 0 \Rightarrow -3K_1 - 2K_2 - 1 = 0$

Solving, we get, $K_1 = 1$ and $K_2 = -2$.

Hence, the zero-state response of the system is

$$y_{ZS}(t) = e^{-3t} - 2e^{-2t} + e^{-t}, \quad t \geq 0$$

Thus, the total response of the system is

$$\begin{aligned} y(t) &= y_{ZI}(t) + y_{ZS}(t) \\ &= e^{-t} - e^{-2t}, \quad t \geq 0 \\ &= [e^{-t} - e^{-2t}]u(t) \end{aligned}$$

Example 4.10. Evaluate the total response of an LTI system described by the differential equation given below:

$$y''(t) + 4y'(t) + 3y(t) = 36tu(t)$$

$y'(0) = 1$ and $y(0) = 0$.

Solution: The system characteristic equation is

$$s^2 + 4s + 3 = 0$$

The roots are $s = -1$ and $s = -3$.

Zero-input response:

$$y_{ZI}(t) = K_1 e^{-t} + K_2 e^{-3t}.$$

The constants K_1 and K_2 are evaluated by making use of the prescribed inital conditions.

(i) $y(0) = 0 \Rightarrow K_1 + K_2 = 0$

(ii) $y'(0) = 1 \Rightarrow -K_1 - 3K_2 = 1$

Solving, we get, $K_1 = 0.5$ and $K_2 = -0.5$.

Thus,

$$y_{ZI}(t) = 0.5e^{-t} - 0.5e^{-3t}, \quad t \geq 0$$

Particular response: Since, $x(t) = 36tu(t)$, we take the particular response of the system as $y_P(t) = C_0 + C_1 t$. The constants C_0 and C_1 are found by substituting the particular solution in the given differential equation:

$$y_P''(t) + 4y_P'(t) + 3y_P(t) = 36t$$
$$\Rightarrow \quad 0 + 4C_2 + 3C_1 + 3C_2 t = 36t$$

Comparing the coefficients of powers of t, we get

$$3C_1 + 4C_2 = 0$$

and

$$3C_2 = 36$$

Solving, we get $C_1 = -16$ and $C_2 = 12$. Hence, $y_P(t) = -16 + 12t$

Zero-state response: The zero-state response of the system is

$$y_{ZS}(t) = y_{ZI}(t) + y_P(t)$$
$$= K_1 e^{-t} + K_2 e^{-3t} - 16 + 12t$$

The constants K_1 and K_2 are evaluated by assuming zero initial conditions.

(i) $y_{ZS}(0) = 0 \Rightarrow K_1 + K_2 - 16 = 0$

(ii) $y_{ZS}'(0) = 0 \Rightarrow -K_1 - 3K_2 + 12 = 0$

Solving, we get, $K_1 = 18$ and $K_2 = -2$. Therefore, $y_{ZS}(t) = -16 + 12t + 18e^{-t} - 2e^{-3t}$

Total response: Total response of the system is

$$y(t) = y_{ZI}(t) + y_{ZS}(t)$$
$$= -16 + 12t + 18.5e^{-t} - 2.5e^{-3t}, \quad t \geq 0$$
$$= \left[-16 + 12t + 18.5e^{-t} - 2.5e^{-3t}\right] u(t)$$

Example 4.11. Find the total response of an LTI system described by the differential equation given below:

$$y''(t) + 3y'(t) + 2y(t) = 2x'(t) + x(t)$$

with $x(t) = 4u(t)$

$$y(0) = 2 \quad \text{and} \quad y'(0) = 1$$

Solution: The system characteristic equation is

$$s^2 + 3s + 2 = 0$$

The roots are $s = -1$ and $s = -2$

Zero-input response:

$$y_{ZI}(t) = K_1 e^{-t} + K_2 e^{-2t}$$

The constants K_1 and K_2 are evaluated by making use of the inital conditions given in the problem.

(i) $y(0) = 2 \Rightarrow K_1 + K_2 = 2$

(ii) $y'(0) = 1 \Rightarrow -K_1 - 2K_2 = 1$

Solving, we get, $K_1 = 5$ and $K_2 = -3$ Thus, $y_{ZI}(t) = 5e^{-t} - 3e^{-2t}, \quad t \geq 0$

Particular response:

Since $x(t) = 4u(t)$, let us choose the particular response as $y_P(t) = C$. The constant C is evaluated by substituting the particular solution in the differential equation given below:

$$y_l''(\ t) + 3y_p'(t) + 2y_p(t) = x(t)$$

$\Rightarrow 0 + 0 + 2K = 4 \Rightarrow k = 2.$

Thus, $y_P(t) = 2.$

Zero-state response: Let $y_1(t)$ be the Zero-state response of the system when the right-hand side of the differential equation is $x(t)$.

Thus,

$$y_1(t) = y_{ZI}(t) + y_P(t)$$
$$= K_1 e^{-t} + K_2 e^{-2t} + 2$$

The constants K_1 and K_2 are found by assuming zero initial conditions.

(i) $y_1(0) = 0 \Rightarrow K_1 + K_2 + 2 = 0$

(ii) $y_1'(0) = 0 \Rightarrow -K_1 - 2K_2 = 0$

Solving, we get, $K_1 = -4$ and $K_2 = 2$.

Thus,

$$y_1(t) = -4e^{-t} + 2e^{-2t} + 2, \quad t \geq 0.$$

The Zero-state response $y_{ZS}(t)$ of the system when the right-hand side of the differential equation is $2x'(t) + x(t)$ is found by applying the principle of superposition. That is,

$$y_{ZS}(t) = 2y_1'(t) + y_1(t)$$
$$= 2 + 4e^{-t} - 6e^{-2t}, \quad t \geq 0$$

Total response: Total response of the system is

$$y(t) = y_{ZI}(t) + y_{ZS}(t)$$
$$= 2 + 9e^{-t} - 9e^{-2t}, \quad t \geq 0$$
$$= [2 + 9e^{-t} - 9e^{-2t}]u(t)$$

Verification:

Putting back the solution in the differential equation given in the problem, we get

$$y''(t) + 3y'(t) + 2y(t) = 4 = 2x'(t) + x(t)$$

Example 4.12. Evaluate the total response of the system described by

$$y''(t) + 4y'(t) + 3y(t) = 3x''(t)$$

where, $x(t) = 4e^{-2t}u(t)$, $y(0) = 0$ and $y'(0) = 0$

Solution: The characteristic equation is

$$s^2 + 4s + 3 = 0$$

The roots are $s = -1$ and $s = -3$.
Zero-input response:

$$y_{ZI}(t) = K_1e^{-t} + K_2e^{-3t}$$

The constants K_1 and K_2 are zero because both the initial conditions given in the problem are zero. In other words, $y_{ZI}(t) = 0$

Particular response:

Since $x(t) = 4e^{-2t}u(t)$ and the roots of the characteristic equation are -1 and -3, we take the particular response as $y_P(t) = Ce^{-2t}$ The particular response has to satisfy the differential equation given below:

$$y_P''(t) + 4y_P'(t) + 3y_P(t) = x(t)$$
$$\Rightarrow 4Ce^{-2t} - 8Ce^{-2t} + 3Ce^{-2t} = 4e^{-2t}$$

Thus, $C = -4$
Hence, $y_P(t) = -4e^{-2t}$

Zero-state response:

Let $y_1(t)$ be the zero-state response of the system when the right-hand side of the given differential equation is $x(t)$. Then, we have,

$$y_1(t) = y_{ZI}(t) + y_P(t)$$
$$= K_1e^{-t} + K_2e^{-3t} - 4e^{-2t}$$

The constants K_1 and K_2 are found by assuming zero initial conditions.

(i) $y_1(0) = 0 \Rightarrow K_1 + K_2 - 4 = 0$

(ii) $y_1'(0) = 0 \Rightarrow -K_1 - 3K_2 + 8 = 0$

Solving, we get, $K_1 = 2$ and $K_2 = -2$

Hence,

$$y_1(t) = 2e^{-t} - 2e^{-3t} - 4e^{-2t}, \quad t \geq 0$$

Let $y_{ZS}(t)$ be the zero-state response of the system when the right-hand side of the differential equation is $3x''(t)$. Using the principle of superpostion, we get

$$\begin{aligned} y_{ZS}(t) &= 2y_1''(t) \\ &= 6e^{-t} - 48e^{-2t} + 54e^{-3t}, \quad t \geq 0 \end{aligned}$$

Total response:

Total response of the system is

$$\begin{aligned} y(t) &= y_{ZI}(t) + y_{ZS}(t) \\ &= 6e^{-t} - 48e^{-2t} + 54e^{-3t}, \quad t \geq 0 \end{aligned}$$

or

$$y(t) = [6e^{-t} - 48e^{-2t} + 54e^{-3t}]u(t)$$

Example 4.13. (a) Evaluate the total response of an LTI system described by the differential equation given below:

$$y''(t) + 4y'(t) + 4y(t) = 2e^{-2t}u(t)$$

$$y(0) = 0 \quad \text{and} \quad y'(0) = 1$$

(b) Find the total response if $x(t) = 2e^{-2t}u(t)$ but $y(0) = 0$ and $y'(0) = 2$

(c) Find the total response if $y(0) = 0$ and $y'(0) = 1$ but $x(t) = 4e^{-2t}u(t)$

Solution: (a) The system characteristic equation is $s^2 + 4s + 4 = 0$

Roots: $-2, -2$ (repeated roots)

Zero-input response:

Since roots of the characteristic equation are repeated, the zero-input response of the system is

$$y_{ZI}(t) = [K_1 + K_2 t]e^{-2t}, \quad t \geq 0$$

The constants K_1 and K_2 are evaluated by making use of the prescribed inital conditions.

(i) $y(0) = 0 \Rightarrow K_1 = 0$

(ii) $y'(0) = 1 \Rightarrow -2K_1 + K_2 = 1 \Rightarrow K_2 = 1$

Hence, the zero-input response of the system is

$$y_{ZI}(t) = te^{-2t}, \quad t \geq 0$$

Particular response:

Since $x(t) = 2e^{-2t}u(t)$ and the roots of the characteristic equation are $-2, -2$, we take $y_P(t) = t^2Ce^{2t}$. The particluar response has to satisfy the differential equation:

$$y_p''(t) + 4y_p'(t) + 4y_p(t) = 2e^{-2t}$$

$$\Rightarrow \quad [4Ct^2e^{-2t} - 8Cte^{-2t} + 2Ce^{-2t}]$$
$$+ 4[-2Ct^2e^{-2t} + 2Cte^{-2t}] + 4[t^2Ce^{-2t}] = 2e^{-2t}$$

$$\Rightarrow \quad [4Ct^2 - 8Ct + 2C - 8Ct^2 + 8Ct + 4Ct^2]e^{-2t} = 2e^{-2t}$$

$$\Rightarrow \quad 2Ce^{-2t} = 2e^{-2t}$$

$$\Rightarrow \quad C = 1$$

Hence, $y_P(t) = t^2e^{-2t}, \quad t \geq 0$

Zero-state response:

$$y_{ZS}(t) = y_{ZI}(t) + y_P(t)$$
$$= [K_1 + K_2t]e^{-2t} + t^2e^{-2t}$$

The contants K_1 and K_2 are evaluated with zero inital conditions.

(i) $y_{ZS}(0) = 0 \Rightarrow K_1 + 0 = 0 \Rightarrow K_1 = 0$

(ii) $y_{ZS}'(0) = 0 \Rightarrow -2K_1 + K_2 = 0 \Rightarrow K_2 = 0$

Hence, $y_{ZS}(t) = t^2e^{-2t}$

Total response:

$$y(t) = y_{ZI}(t) + y_{ZS}(t)$$
$$= te^{-2t} + t^2e^{-2t}, \quad t \geq 0$$

(b) Input is same as in part(a) but the initial conditions are doubled. Since initial conditions are doubled the principle of superposition indicates that $y_{ZI}(t)$ is doubled.

Hence,

$$y(t) = 2y_{ZI}(t) + y_{ZS}(t)$$
$$= 2te^{-2t} + t^2e^{-2t}, \quad t \geq 0$$

(c) Compared to part(a) input is doubled but initial conditions remain unchanged. Also, zero-state response obeys the principle of superposition. Hence the total response of the system is

$$y(t) = y_{ZI}(t) + 2y_{ZS}(t)$$
$$= te^{-2t} + 2t^2 e^{2t}, \qquad t \geq 0$$

4.4 Impulse response

The impulse response of a system is its response to a unit impulse applied at time $t = 0$ with all initial conditions of the system being zero. That is,

$$h(t) = \mathbf{T}\{\delta(t)\} \quad \text{(zero initial conditions)}$$

The above equation is represented by a block diagram as shown in Fig. 4.2.

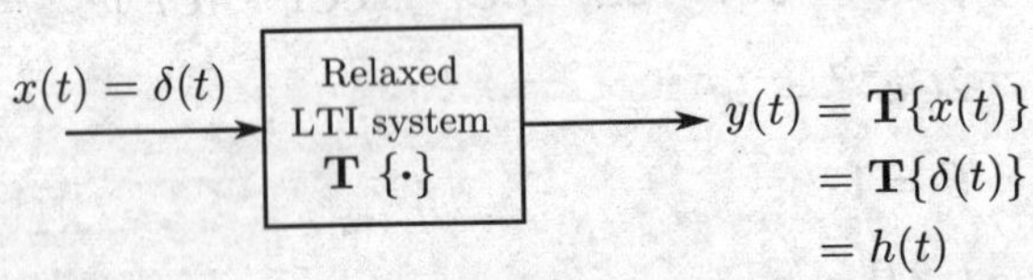

Fig. 4.2

If $s(t)$ is the response of a system (known as step response) to a unit step input $u(t)$, the impulse $\delta(t) = \frac{du(t)}{dt}$ produces a response $h(t) = \frac{ds(t)}{dt}$. That is, we first evaluate step response, $s(t)$ and then let $h(t) = \frac{ds(t)}{dt}$.

Alternatively, to find the impulse response directly, we assume that an impulse produces a sudden change in the initial conditions and then disappears. In other words, the impulse response $h(t)$ is found by solving a homogeneous differential equation but with nonzero inital conditions. That is,

(i) solving a first order nonhomogeneous differential equation:

$$y'(t) + ay(t) = \delta(t)$$

with $y(0) = 0$ is equivalent to solving a homogeneous differential equation:

$$h'(t) + a(h)(t) = 0$$

with $h(0) = 1$

(ii) and solving a second order nonhomogeneous differential equation:

$$y''(t) + ay'(t) + by(t) = \delta(t)$$

with $y(0) = 0$ and $y'(0) = 0$ is equivalent to solving a homogeneous differential equation:

$$h''(t) + ah'(t) + bh(t) = 0$$

with $h(0) = 0$ and $h'(0) = 1$.

These results can be generalized for an nth order single-input relaxed LTI system given by

$$y^{(n)}(t) + a_1 y^{(n-1)}(t) + \cdots + a_n y(t) = x(t)$$

The impulse response $h(t)$ is found as the solution to the homogeneous equation

$$h^{(n)}(t) + a_1 h^{(n-1)}(t) + \cdots + a_n h(t) = 0$$

with $h^{(n-1)}(0) = 1$ and other initial conditions are zero.

Example 4.14. Find the impulse response of an LTI system described by the differential equation given below:

$$y''(t) + 5y'(t) + 4y(t) = x(t)$$

Solution: The given nonhomogeneous differential equation with zero initial conditions is equivalent to the following homogeneous differential equation:

$$h''(t) + 5h'(t) + 4h(t) = 0$$

with inital conditions $h(0) = 0$ and $h'(0) = 1$.

Zero-input response:

The charcteristic equation is $s^2 + 5s + 4 = 0$

The roots are $s = -1$ and -4.

Hence,

$$h_{ZI}(t) = K_1 e^{-t} + K_2 e^{-4t}, \quad t \geq 0.$$

(i) $h(0) = 0 \Rightarrow K_1 + K_2 = 0$

(ii) $h'(0) = 1 \Rightarrow -K_1 - 4K_2 = 1$

Solving $K_1 = \frac{1}{3}$ and $K_2 = -\frac{1}{3}$.

Thus,

$$h_{ZI}(t) = \frac{1}{3}e^{-t} - \frac{1}{3}e^{-4t}, t \geq 0.$$

Zero-state response is zero. Thus,

$$\begin{aligned} h(t) &= h_{ZI}(t) + h_{ZS}(t) \\ &= \frac{1}{3}e^{-t} - \frac{1}{3}e^{-4t}, \quad t \geq 0 \end{aligned}$$

$$\text{or} \qquad h(t) = \left[\frac{1}{3}e^{-t} - \frac{1}{3}e^{-4t}\right] u(t)$$

Example 4.15. Find the impulse response of an LTI system described by the differential equation given below:

$$y'(t) + 2y(t) = 2x(t)$$

Solution: Let us start with single-input system defined by the differential equation:

$$y'(t) + 2y(t) = x(t)$$

The above nonhomogeneous differential equation with zero initial condition is equivalent to the following homogeneous differential equation:

$$h_0'(t) + 2h_0(t) = 0; \qquad h_0(0) = 1$$

Zero-input response:

The system characteristic equation is $s + 2 = 0 \Rightarrow s = -2$.

Hence,

$$[h_0]_{ZI}(t) = K_1 e^{-2t}$$

The constant K_1 is found as follows:

$$h_0(0) = 1 \Rightarrow K_1 = 1$$

Hence,

$$[h_0]_{ZI}(t) = e^{-2t}, \quad t \geq 0$$

Zero-state response:

For a homogeneous differential equation, the zero-state response is zero.

That is, $[h_0]_{ZS}(t) = 0$.

Total response:

$$h_0(t) = [h_0]_{ZI}(t) + [h_0]_{ZS}(t)$$
$$= e^{-2t}, \qquad t \geq 0$$

Finally, applying principle of superposition, we get

$$h(t) = 2h_0(t)$$
$$= 2e^{-2t}, \quad t \geq 0$$
$$= 2e^{-2t}u(t)$$

4.5 Transfer function

The transfer function is a frequency-domain representation of an LTI system with zero initial conditions.

The transfer function, $H(\omega)$ is defined as the Fourier transform of the system impulse response $h(t)$.

It is also defined as the ratio of Fourier transform of the output to Fourier transform of the input. That is ,

$$H(\omega) = \frac{Y(\omega)}{X(\omega)}$$

where $Y(\omega)$ and $X(\omega)$ are the Fourier transform of output $y(t)$ and input $x(t)$, respectively.

The transfer function $H(\omega)$ is used for finding zero-state response of an LTI system with zero initial conditions. Sometimes, it is also known as the frequency response of the system since the plots of $|H(\omega)|$ versus (ω) and $\angle H(\omega)$ versus ω represent magnitude and phase specta in the frequency-domain. The concept of frequency response is applicable only for stable systems whose zero-state response decays with time. For LTI systems, $H(\omega)$ is the ratio of polynomials in $j\omega$ and for this reason sometimes it is denoted as $H(j\omega)$. Also, for LTI systems, $H(\omega)$ exhibits conjugate symmetry, that is, $H(-\omega) = H^*(\omega)$. The conjugate symmetry implies that the plot of $|H(\omega)|$ versus ω exhibits even symmetry with respect to $\omega = 0$, while the plot of $\angle H(\omega)$ versus ω exhibits odd symmetry with respect to $\omega = 0$.

Example 4.16. Evaluate the frequency response for the LTI system, represented by impulse responses given below:

(a) $h(t) = -\delta(t+1) + \delta(t) - \delta(t-1)$

(b) $h(t) = e^{at}u(-t)$

Solution:

(a) Fourier transform of $h(t)$ is known as frequency response of the system.

That is,

$$H(\omega) = \int_{-\infty}^{\infty} h(t)e^{j\omega t}\, dt$$

$$\Rightarrow \quad H(\omega) = \int_{-\infty}^{\infty} [-\delta(t+1) + \delta(t) - \delta(t-1)]e^{-j\omega t}\, dt$$

Applying sifting property, we get

$$\begin{aligned} H(\omega) &= -e^{-j\omega t}\Big|_{t=-1} + e^{-j\omega t}\Big|_{t=0} - e^{-j\omega t}\Big|_{t=1} \\ &= -e^{j\omega} + 1 - e^{-j\omega} \\ &= 1 - 2\cos\omega \end{aligned}$$

(b) $H(\omega) = \int_{-\infty}^{\infty} h(t)e^{-j\omega t}\,dt$

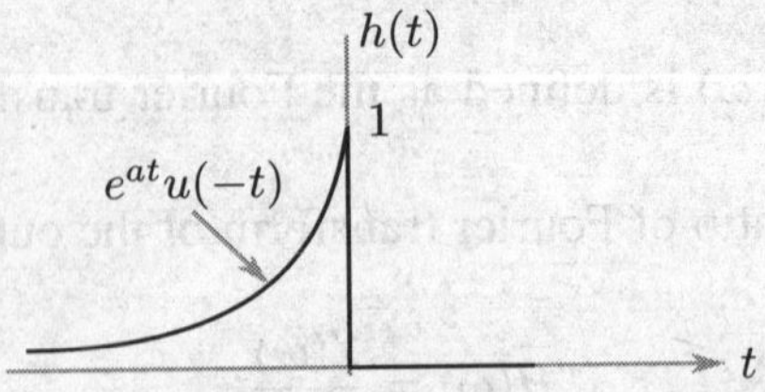

Fig. Ex. 4.16

Referring Fig. Ex. 4.16, we get

$$H(\omega) = \int_{-\infty}^{0} e^{at}e^{-j\omega t}\,dt$$

$$= \frac{1}{a - j\omega}$$

Example 4.17. The transfer function of an LTI system is $H(\omega) = \frac{16}{4+j\omega}$. Find the response, $y(t)$ for the input, $x(t) = u(t)$.

Solution: Given,

$$x(t) = u(t)$$

$$\Rightarrow \quad X(\omega) = \pi\delta(\omega) + \frac{1}{j\omega}$$

We know that,

$$H(\omega) = \frac{Y(\omega)}{X(\omega)}$$

$$\Rightarrow \quad Y(\omega) = H(\omega)X(\omega)$$

$$= \frac{16}{4 + j\omega}\left[\pi\delta(\omega) + \frac{1}{j\omega}\right]$$

$$= \pi\delta(\omega)\frac{16}{4 + j\omega} + \frac{16}{j\omega(4 + j\omega)}$$

Applying product property of an impulse function, we get

$$Y(\omega) = \pi \left. \frac{16}{4 + j\omega} \right|_{\omega=0} \delta(\omega) + \frac{16}{j\omega(4 + j\omega)}$$

$$= 4\pi\delta(\omega) + \left[\frac{4}{j\omega} - \frac{4}{4 + j\omega}\right]$$

$$= 4\left[\pi\delta(\omega) + \frac{1}{j\omega}\right] - 4\left[\frac{1}{4 + j\omega}\right]$$

Taking inverse **FT**, we get

$$y(t) = 4u(t) - 4e^{-4t}u(t)$$

Example 4.18. Consider the system described by

$$y'(t) + 2y(t) = x(t) + x'(t)$$

Find the impulse response, $h(t)$ of the system.

Solution: Taking Fourier transform on both sides of the given differential equation, we get

$$j\omega Y(\omega) + 2Y(\omega) = X(\omega) + j\omega X(\omega)$$

$$\Rightarrow \quad [j\omega + 2]Y(\omega) = [1 + j\omega]X(\omega)$$

Hence, the frequency response of the system is

$$H(\omega) = \frac{Y(\omega)}{X(\omega)} = \frac{1 + j\omega}{2 + j\omega}$$

$$= 1 - \frac{1}{2 + j\omega}$$

Taking the inverse Fourier transform of $H(\omega)$, the impulse response $h(t)$ is

$$h(t) = \delta(t) - e^{-2t}u(t)$$

Example 4.19. A linear time-invariant system has the response,

$$y(t) = [2e^{-t} - 2e^{-4t}]u(t)$$

when excited by an input

$$x(t) = [e^{-t} + e^{-3t}]u(t)$$

Find the following:

(a) the transfer function of the system

(b) system impulse response and

(c) the differential equation relating the input and output.

Solution:

(a) Recall: $e^{-at}u(t) \overset{FT}{\longleftrightarrow} \frac{1}{a+j\omega}$

Hence,

$$x(t) = [e^{-t} + e^{-3t}]u(t) \overset{FT}{\longleftrightarrow} X(\omega) = \frac{1}{1+j\omega} + \frac{1}{3+j\omega}$$

$$\Rightarrow X(\omega) = \frac{4+j2\omega}{(1+j\omega)(3+j\omega)}$$

and $$y(t) = [2e^{-t} - 2e^{-4t}]u(t) \overset{FT}{\longleftrightarrow} Y(\omega) = \frac{2}{1+j\omega} - \frac{2}{4+j\omega}$$

$$\Rightarrow Y(\omega) = \frac{8+j2\omega - 2 - 2j\omega}{(1+j\omega)(3+j\omega)} = \frac{6}{(1+j\omega)(4+j\omega)}$$

Hence,

$$H(\omega) = \frac{Y(\omega)}{X(\omega)}$$

$$= \frac{6}{(1+j\omega)(4+j\omega)} \times \frac{(1+j\omega)(3+j\omega)}{2(2+j\omega)}$$

$$= \frac{3(3+j\omega)}{(4+j\omega)(2+j\omega)}$$

(b) Inverse Fourier transform of $H(\omega)$ is the impulse response of the system and is found as follows:

Letting $j\omega = v$ we get

$$H(\omega) = \frac{3(v+3)}{(v+4)(v+2)}$$

Using the theory of partial fractions, we can write

$$H(\omega) = \frac{K_1}{v+4} + \frac{K_2}{v+2}$$

where, $$K_1 = \left.\frac{3(v+3)}{(v+2)}\right|_{v=-4} = \frac{3}{2}$$

and, $$K_2 = \left.\frac{3(v+3)}{(v+4)}\right|_{v=-2} = \frac{3}{2}$$

Hence, $$H(\omega) = \frac{3}{2}\left[\frac{1}{4+j\omega}\right] + \frac{3}{2}\left[\frac{1}{2+j\omega}\right]$$

Taking inverse Fourier transform on both sides, we get

$$h(t) = \frac{3}{2}\left[e^{-4t} + e^{-2t}\right]u(t)$$

(c) We know that,

$$\frac{Y(\omega)}{X(\omega)} = \frac{9+3j\omega}{8+6j\omega+(j\omega)^2}$$

$$\Rightarrow \quad (j\omega)^2 Y(\omega) + 6j\omega\, Y(\omega) + 8Y(\omega) = 9X(\omega) + 3j\omega X(\omega)$$

Taking inverse Fourier transform on both sides, we get

$$\frac{d^2y(t)}{dt^2} + 6\frac{dy(t)}{dt} + 8y(t) = 9x(t) + 3\frac{dx(t)}{dt}$$

Example 4.20. The input-output realtion of an LTI system is given by

$$\frac{dy(t)}{dt} + 2y(t) = x(t)$$

Determine $y(t)$, if $x(t)$ has the following the Fourier transforms:

(a) $X(\omega) = \frac{1+j\omega}{2+j\omega}$

(b) $X(\omega) = \frac{2+j\omega}{1+j\omega}$

(c) $X(\omega) = \frac{1}{(2+j\omega)(1+j\omega)}$

Solution:

$$\text{Given} \quad \frac{dy(t)}{dt} + 2y(t) = x(t)$$

Taking Fourier transform on both sides of the above differential equation, we get

$$j\omega Y(\omega) + 2Y(\omega) = X(\omega)$$

$$\Rightarrow \quad Y(\omega) = \frac{X(\omega)}{j\omega + 2}$$

(a)

$$\text{Given:} \quad X(\omega) = \frac{1+j\omega}{2+j\omega}$$

$$\text{Hence,} \quad Y(\omega) = \frac{(1+j\omega)}{(2+j\omega)^2}$$

Let $j\omega = v$

$$\text{Then,} \quad Y(\omega) = \frac{1+v}{(2+v)^2}$$

$$= \frac{K_1}{2+v} + \frac{K_2}{(2+v)^2}$$

We find that,

$$K_1 = \frac{d}{dv}[1+v]\Big|_{v=-2} = 1$$

and

$$K_2 = 1 + v\Big|_{v=-2} = -1$$

Hence,

$$Y(\omega) = \frac{1}{2+j\omega} - \frac{1}{(2+j\omega)^2}$$

Recall:

(i) $e^{-at}u(t) \xleftrightarrow{\textbf{FT}} \frac{1}{(a+j\omega)}$

(ii) $te^{-at}u(t) \xleftrightarrow{\textbf{FT}} \frac{1}{(a+j\omega)^2}$

Thus,

$$y(t) = e^{-2t}u(t) - te^{-2t}u(t)$$

(b) Given: $X(\omega) = \dfrac{2+j\omega}{1+j\omega}$.

Hence,

$$Y(\omega) = \frac{1}{1+j\omega}$$
$$\Rightarrow y(t) = e^{-t}u(t)$$

(c) Given: $X(\omega) = \dfrac{1}{(1+j\omega)(2+j\omega)}$

Therefore,

$$Y(\omega) = \frac{X(\omega)}{2+j\omega}$$
$$= \frac{1}{(2+j\omega)^2(j\omega+1)}$$

Letting $j\omega = v$, we get

$$Y(\omega) = \frac{1}{(v+2)^2(v+1)}$$
$$= \frac{K_1}{v+1} + \frac{K_2}{v+2} + \frac{K_3}{(v+2)^2}$$

where

$$K_1 = \left.\frac{1}{(v+2)^2}\right|_{v=-1} = 1$$

$$K_2 = \left.\frac{d}{dv}\left[\frac{1}{(v+1)}\right]\right|_{v=-2} = -\left.\frac{1}{(v+1)^2}\right|_{v=-2} = -1$$

$$K_3 = \left.\frac{1}{v+1}\right|_{v=-2} = -1$$

Hence,

$$Y(\omega) = \frac{1}{v+1} - \frac{1}{v+2} - \frac{1}{(v+2)^2}$$

$$= \frac{1}{1+j\omega} - \frac{1}{2+j\omega} - \frac{1}{(2+j\omega)^2}$$

Taking inverse Fourier transform on both sides, we get

$$y(t) = e^{-t}u(t) - e^{-2t}u(t) - te^{-2t}u(t)$$

Example 4.21. Find the frequency response and the impulse response of the LTI system described by the differential equation below:

$$\frac{d^2y(t)}{dt^2} + 5\frac{dy(t)}{dt} + 6y(t) = -\frac{dx(t)}{dt}$$

Solution: Given:

$$\frac{d^2y(t)}{dt^2} + 5\frac{dy(t)}{dt} + 6y(t) = \frac{-dx(t)}{dt}$$

Taking Fourier transform on the both sides, we get

$$(j\omega)^2Y(\omega) + 5j\omega Y(\omega) + 6Y(\omega) = -j\omega X(\omega)$$

Hence, the frequency response of the system is

$$H(\omega) = \frac{Y(\omega)}{X(\omega)}$$

$$= \frac{-j\omega}{(j\omega)^2 + 5j\omega + 6}$$

$$= \frac{-j\omega}{(j\omega+2)(j\omega+3)}$$

Letting $j\omega = v$, we get

$$H(\omega) = \frac{-v}{(v+2)(v+3)}$$
$$= \frac{K_1}{v+2} + \frac{K_2}{v+3}$$

where,

$$K_1 = \left.\frac{-v}{v+3}\right|_{v=-2} = 2$$

and

$$K_2 = \left.\frac{-v}{v+2}\right|_{v=-3} = -3$$

Hence,

$$H(\omega) = \frac{2}{2+j\omega} - \frac{3}{3+j\omega}$$

Recall: $e^{-at}u(t) \overset{\text{FT}}{\longleftrightarrow} \dfrac{1}{a+j\omega}$

Hence,

$$h(t) = 2e^{-2t}u(t) - 3e^{-3t}u(t).$$

Example 4.22. Find the differential equation description for an LTI system described by the impulse response given below:

$$h(t) = 2e^{-2t}u(t) - 2te^{-2t}u(t)$$

Solution: Recall:

$$e^{-at}u(t) \overset{\text{FT}}{\longleftrightarrow} \frac{1}{a+j\omega}$$

Hence,

$$H(\omega) = \frac{2}{2+j\omega} - \frac{2}{(2+j\omega)^2}$$
$$= \frac{2j\omega + 2}{(2+j\omega)^2}$$

Since, $H(\omega) = \dfrac{Y(\omega)}{X(\omega)}$, we get

$$\frac{Y(\omega)}{X(\omega)} = \frac{2j\omega + 2}{4 + 4j\omega + (j\omega)^2}$$

$$\Rightarrow \quad (j\omega)^2 Y(\omega) + 4j\omega Y(\omega) + 4Y(\omega) = 2j\omega X(\omega) + 2X(\omega)$$

Taking inverse Fourier transform on both sides, we get

$$\frac{d^2y(t)}{dt^2} + 4\frac{dy(t)}{dt} + 4y(t) = 2\frac{dx(t)}{dt} + 2x(t)$$

Example 4.23. Find the frequency response for the LTI systems described by the following impulse responses:

(a) $h(t) = e^{-|t|}$

(b) $h(t) = -\delta(t+2) + \delta(t) - \delta(t-2)$

Solution: Recall:

$$H(\omega) = \int_{-\infty}^{\infty} h(t)e^{-j\omega t}\,dt$$

(a) Given: $h(t) = e^{-|t|}$

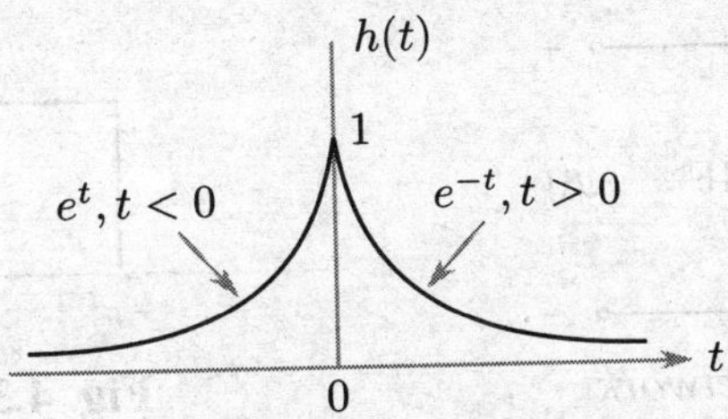

Fig. Ex. 4.23(a)

The given impulse response is a double-sided exponential signal and is as shown in Fig. Ex. 4.23(a).

Hence,

$$H(\omega) = \int_{-\infty}^{0} e^{t}e^{-j\omega t}\,dt + \int_{0}^{\infty} e^{-t}e^{-j\omega t}\,dt$$

$$= \frac{1}{1-j\omega} + \frac{1}{1+j\omega}$$

$$= \frac{2}{1+\omega^2}$$

(b) Given: $h(t) = -\delta(t+2) + \delta(t) - \delta(t-2)$.

Hence,

$$H(\omega) = \int_{-\infty}^{\infty} [-\delta(t+2) + \delta(t) - \delta(t-2)]e^{-j\omega t}\,dt$$

Applying sifting property, we get.

$$\begin{aligned} H(\omega) &= -e^{-j\omega t}\Big|_{t=-2} + e^{-j\omega t}\Big|_{t=0} - e^{-j\omega t}\Big|_{t=2} \\ &= -e^{j2\omega} + 1 - e^{-j2\omega} \\ &= 1 - 2\cos 2\omega \end{aligned}$$

4.6 The filter characteristics of linear systems

The electrical filter was independently invented in 1915 by George Campbell in the united states and K. W. Wagner in Germany. In order to understand the filtering process, let us consider an LTI system described by an *RC* network depicted in Fig. 4.3(a).

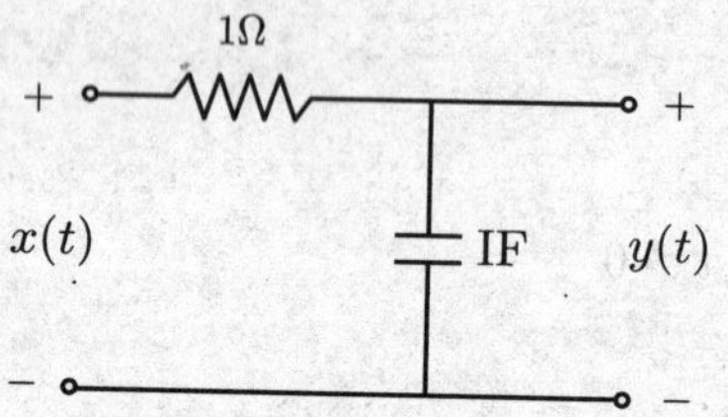

Fig. 4.3(a): *RC network*

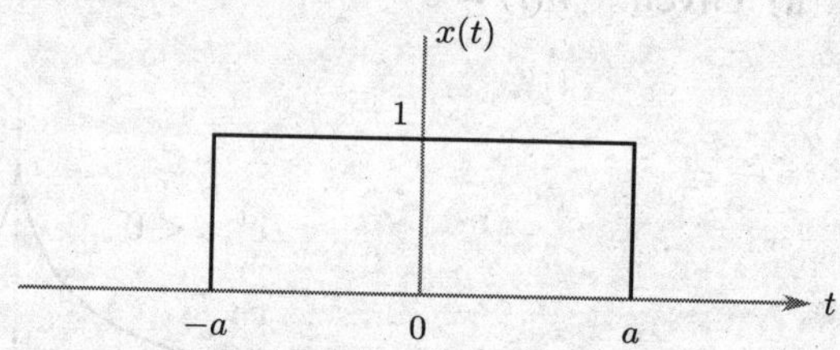

Fig. 4.3(b): *Input signal*

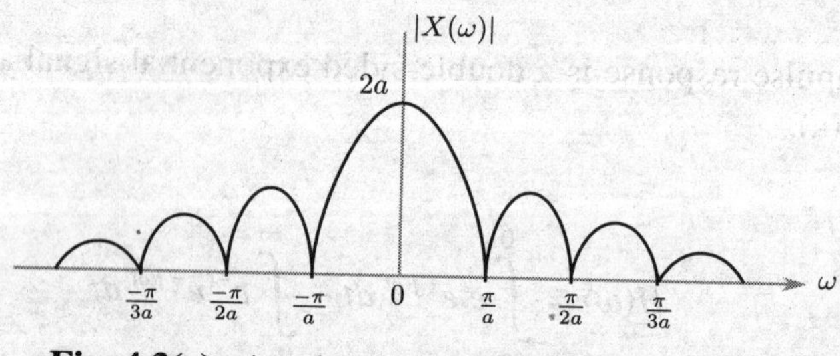

Fig. 4.3(c): *Amplitude spectrum of input signal*

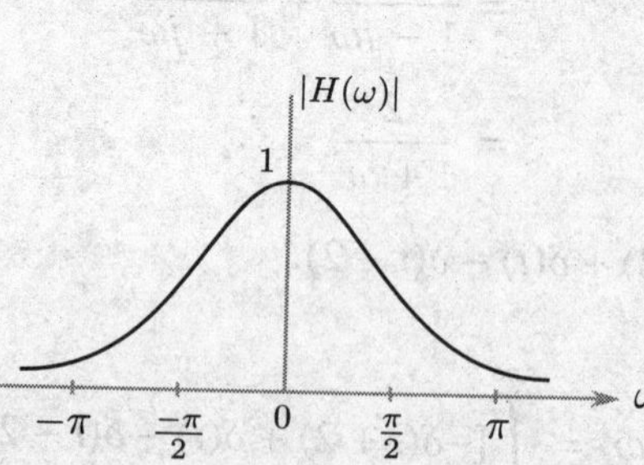

Fig. 4.3(d): *Magnitude response of RC network*

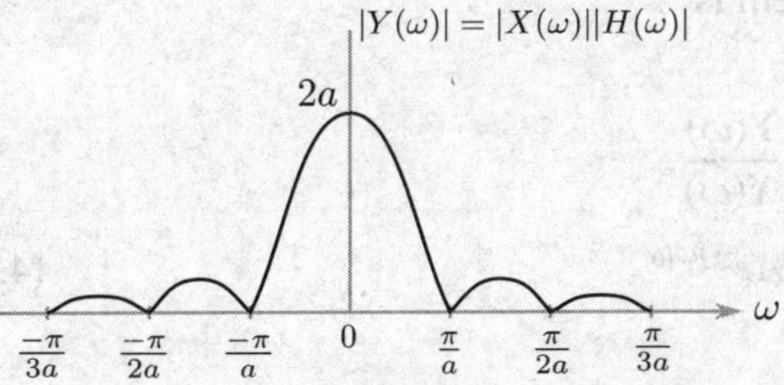

Fig. 4.3(e): *Amplitude spectrum of the output signal, y(t)*

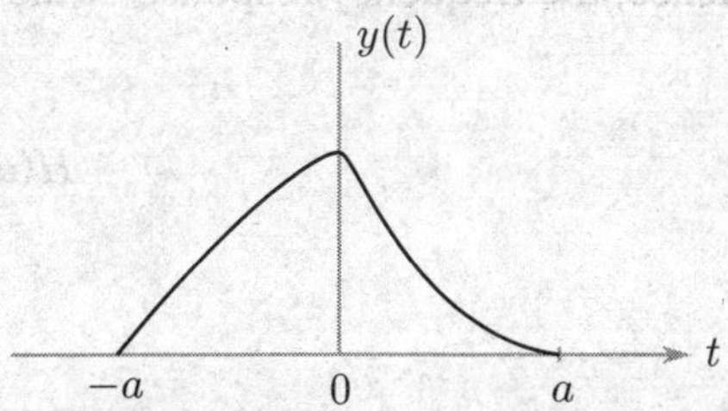

Fig. 4.3(f): *Response in time-domain.*

The transfer function of the *RC* network can be found as $H(\omega) = \frac{Y(\omega)}{X(\omega)} = \frac{1}{j\omega+1}$. The magnitude response of this network is $|H(\omega)| = \frac{1}{\sqrt{\omega^2+1}}$. The magnitude response is as shown in Fig. 4.3(d). From Fig. 4.3(d), it is found that the *RC* network shown in Fig. 4.3(a) attenuates the high frequency components present in the input signal and allows low frequency components present in $x(t)$ to appear at the output with a relatively small attenuation. Hence, the *RC* network behaves like a low-pass filter. The high frequency components present in the amplitude spectrum of $x(t)$ [shown in Fig. 4.3(c)] suffer severe attenuation as compared to low frequency components. The amplitude spectrum of the output signal, $y(t)$ is the product of $|X(\omega)|$ and $|H(\omega)|$, that is $Y(\omega)| = |X(\omega)||H(\omega)|$. In Fig. 4.3(e), the sketch of $|X(\omega)||H(\omega)|$ is shown. By comparing Fig. 4.3(c) and Fig. 4.3(e), we find that the high frequency components present in $x(t)$ are severly attenuated. Consequently, inverse Fourier transform of $Y(\omega)$, namely, $y(t)$ shown in Fig. 4.3(f) has a shape different from that of $x(t)$. Thus, the response, $y(t)$ is a distorted version of the input signal. This distortion is because the *RC* network does not offer same attenuation to all the frequency components present in the input signal, $x(t)$.

4.7 Distortionless transmission through a system

In the previous section, we dicussed the filtering process without giving regards to the phase response of the *RC* network. The phase response of a filter plays a very important role in preserving the shape of the input signal as it passes through the filter. For distortionless transmission through an LTI system, it is required that the output signal must have same shape as the input signal although its amplitude may be different and it may be delayed in time. Hence, if $x(t)$ is the input signal, the required response is

$$y(t) = Ax(t - t_0) \tag{4.6}$$

where t_0 is the time delay and $A(> 0)$ is an arbitrary constant known as the gain of the system. Taking Fourier transform on both sides of Eq. (4.6) gives

$$Y(\omega) = Ae^{-j\omega t_0}X(\omega) \tag{4.7}$$

Hence, the frequency response of the system is

$$H(\omega) = \frac{Y(\omega)}{X(\omega)}$$
$$= Ae^{-j\omega t_0} \tag{4.8}$$

In general, the frequency response of an LTI system is complex and can be written in the exponential form as

$$H(\omega) = |H(\omega)|e^{j\theta(\omega)} \tag{4.9}$$

Comparing Eq. (4.8) and Eq. (4.9), we find that

$$|H(\omega)| = A \tag{4.10}$$
$$\theta(\omega) = -\omega t_0 \tag{4.11}$$

Therefore, for distortionless transmission, the magnitude of $H(\omega)$ must be constant over the entire frequency range, and the phase of $H(\omega)$ must be linear with the frequency. These aspects are illustrated in Fig. 4.4.

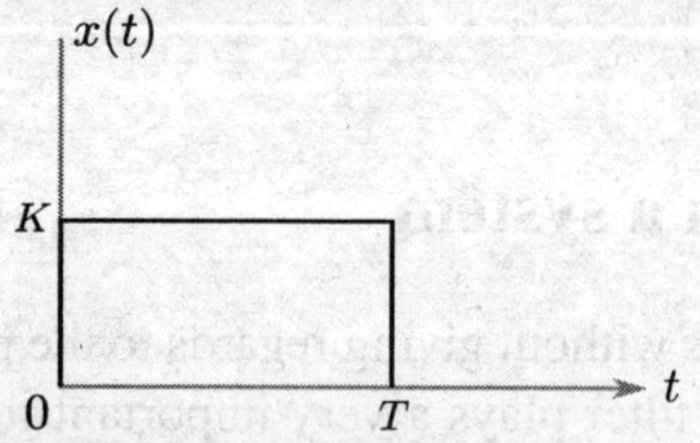

Fig. 4.4(a): *Input Signal*

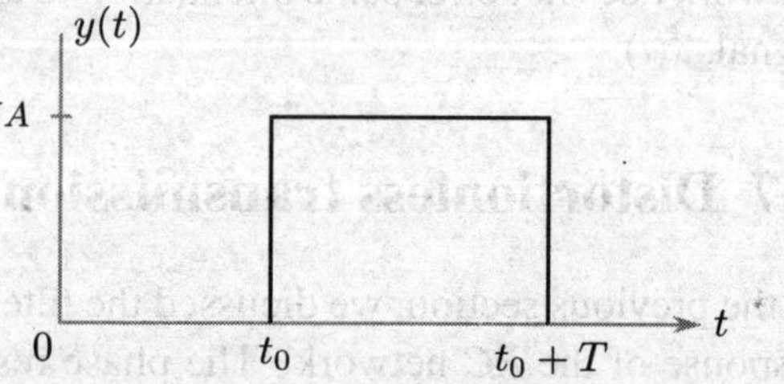

Fig. 4.4(b) *Output Signal translated by amount t_0*

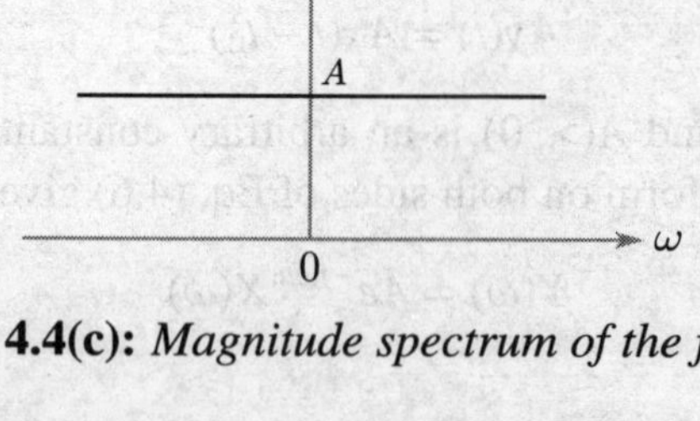

Fig. 4.4(c): *Magnitude spectrum of the filter*

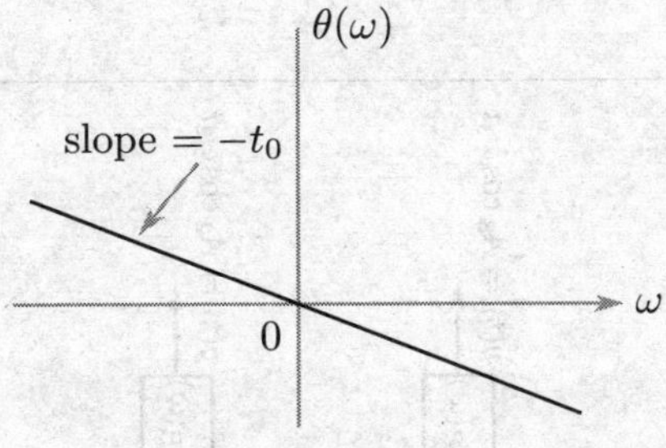

Fig. 4.4(d): *Phase spectrum of the filter*

4.8 Ideal filters

Let us assume that both the input and output of this filter are voltages. This ideal filter bifurcates its input voltage in two parts. One part is transmitted, unchanged to the output. The other part is completely eliminated. In other words, output of an ideal filter is an exact replica of a part of its input.

To understand how an electric filter works, consider an input voltage

$$x(t) = A_1 \cos \omega_1 t + A_2 \cos \omega_2 t + A_3 \cos \omega_3 t$$

The above input consists of a sum of three sinusoids, each at a different frequency. The filter bifurcates the input voltage into two parts, using frequency as the criterion for seperation. There are different ways of seperating the input $x(t)$ into two parts, and accordingly, several types of ideal filters. Table 4.1 illustrates the four types of ideal filters that exist.

Let us now consider the ideal low-pass filter shown in Table 4.1. The transfer function of the ideal low-pass filter is

$$H(\omega) = \begin{cases} 1, & |\omega| < \omega_C \\ 0, & \text{Otherwise} \end{cases} \tag{4.12}$$

The frequency, ω_c is called cutoff frequency. The frequency band, $\omega < \omega_c$ is called pass-band and the frequency band, $\omega > \omega_c$ is called stop-band. The components of the input that have frequencies in the pass-band are subjected to unity gain and zero phase shift. The components of the input that have frequencies in the stop-band are multiplied by a gain equal to zero and thus are killed. An ideal filter bifurcates its input signal into two parts: the terms that have frequencies in the pass-band and those terms that have frequencies in the stop-band. The output of the ideal filter contains only those terms with the frequencies in the pass-band.

Concept of negative frequency: There is nothing called negative frequency in practice. Then why do we draw spectra for negative frequency? The answer is, if the magnitude spectrum exhibits even symmetry with respect to $\omega = 0$ and phase spectrum exhibits odd symmetry with respect to $\omega = 0$, it implies that inverse **FT** of $H(\omega)$ is a real signal, $h(t)$.

The inverse **FT** of $H(\omega)$, namely $h(t)$ is non-zero for $t < 0$. This means ideal filters are noncausal systems and are not physically realizable.

Filter type	Ideal frequency response	Mathematical description	Filter input and output
1. Low-pass	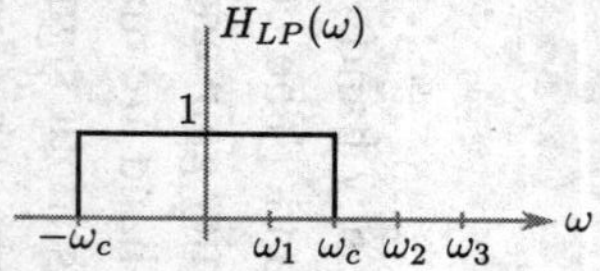	$H_{LP}(\omega) = \begin{cases} 1, & \lvert\omega\rvert < \omega_C \\ 0, & \text{otherwise} \end{cases}$	$x(t) = A_1\cos\omega_1 t + A_2\cos\omega_2 t + A_3\cos\omega_3 t \rightarrow H_{LP}(\omega) \rightarrow y(t) = A_1\cos\omega_1 t$
2. High-pass	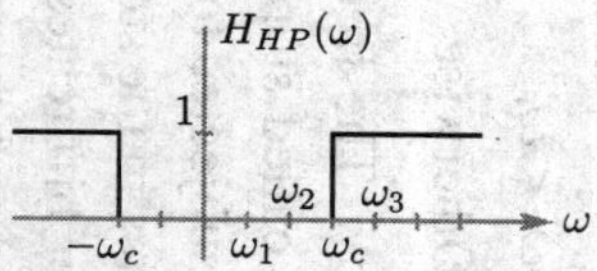	$H_{HP}(\omega) = \begin{cases} 1, & \lvert\omega\rvert > \omega_C \\ 0, & \text{otherwise} \end{cases}$	$x(t) = A_1\cos\omega_1 t + A_2\cos\omega_2 t + A_3\cos\omega_3 t \rightarrow H_{HP}(\omega) \rightarrow y(t) = A_3\cos\omega_3 t$
3. Band-pass	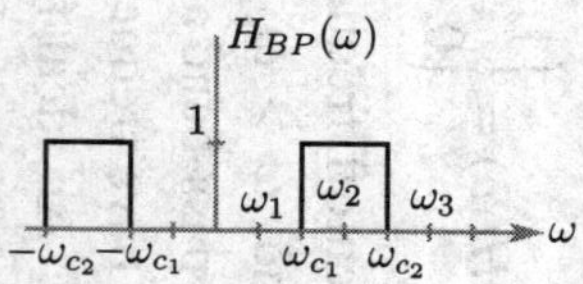	$H_{BP}(\omega) = \begin{cases} 1, & \omega_{C_1} < \lvert\omega\rvert < \omega_{C_2} \\ 0, & \text{otherwise} \end{cases}$	$x(t) = A_1\cos\omega_1 t + A_2\cos\omega_2 t + A_3\cos\omega_3 t \rightarrow H_{BP}(\omega) \rightarrow y(t) = A_2\cos\omega_2 t$
4. Band-stop/notch	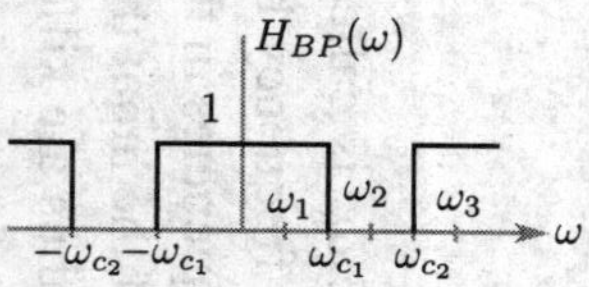	$H_{BS}(\omega) = \begin{cases} 0, & \omega_{C_1} < \lvert\omega\rvert < \omega_{C_2} \\ 1, & \text{otherwise} \end{cases}$	$x(t) = A_1\cos\omega_1 t + A_2\cos\omega_2 t + A_3\cos\omega_3 t \rightarrow H_{BS}(\omega) \rightarrow y(t) = A_1\cos\omega_1 t + A_3\cos\omega_3 t$

Table 4.1: Ideal filters

4.9 Practical filters

Practical filters are circuits that approximate ideal filters. Practical Fiters segregate their inputs into two parts: the terms in the pass-band and the terms in the stop-band. The terms in the pass-band are subjected to a gain that is approximately one and also experience some phase shift. These terms are passed to the output, but they are changed a little. The terms in the stop-band experience a small gain that isn't quite zero. Since these terms are not eliminated entirely, some small residue of these terms shows up in the filter output.

As an example of a practical frequency selective filter, we consider the *RC* circuit shown in Fig. 4.5.

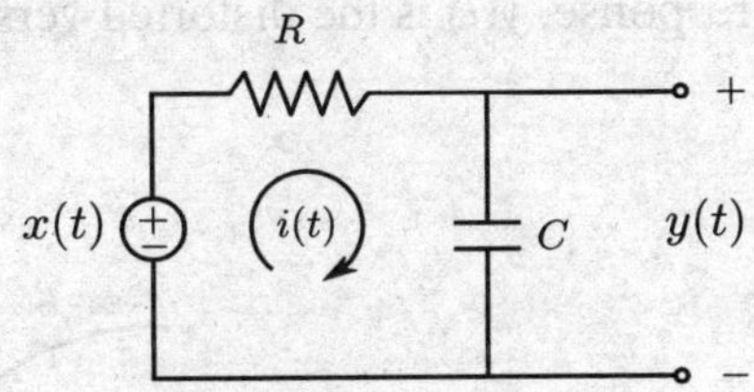

Fig. 4.5: *RC filter*

The current flowing through the circuit is

$$i(t) = C\frac{dy(t)}{dt}$$

Applying KVL to the input loop, we get

$$x(t) = Ri(t) + y(t)$$

$$\Rightarrow \quad x(t) = RC\frac{dy(t)}{dt} + y(t)$$

Taking Fouier transform on both sides of the above equation, we get

$$X(\omega) = RC\,j\omega Y(\omega) + Y(\omega)$$

Hence

$$\begin{aligned} H(\omega) &= \frac{Y(\omega)}{X(\omega)} \\ &= \frac{1}{RC\,j\omega + 1} \\ &= \frac{1}{1 + j\frac{\omega}{\omega_0}} \end{aligned}$$

where $\omega_0 = \frac{1}{RC}$. Thus, the amplitude response, $|H(\omega)|$ and phase response, $\theta(\omega)$ are given by

$$|H(\omega)| = \frac{1}{\sqrt{1 + \left(\frac{\omega}{\omega_0}\right)^2}}$$

and

$$\theta(\omega) = -\tan^{-1}\left(\frac{\omega}{\omega_0}\right)$$

The amplitude and phase responses are plotted in Fig. 4.5(a) and Fig. 4.5(b) respectively.

The amplitude response of the *RC* filter indicates that it is a low-pass filter. Also, phase response is not linear. Hence, response, $y(t)$ is the distorted version of the input, $x(t)$.

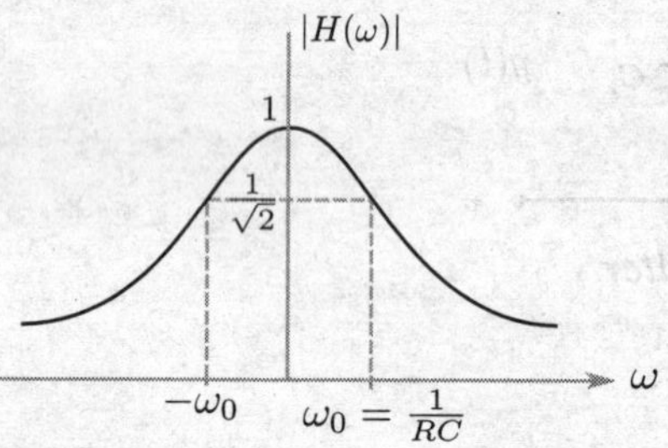

Fig. 4.5(a): *Amplitude response*

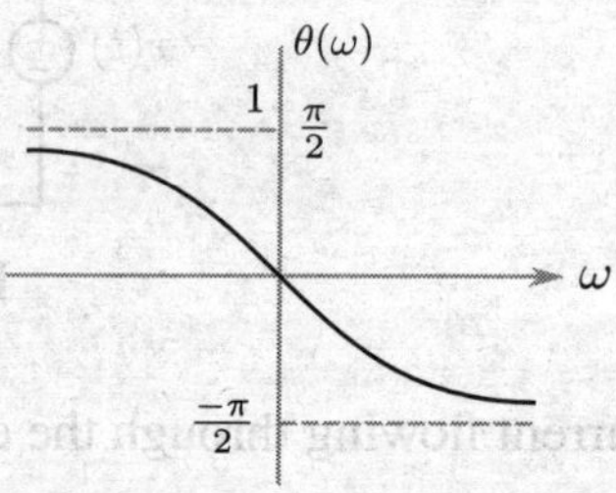

Fig. 4.5(b): *Phase response*

4.10 System bandwidth

For causal practical filters, a common definition of filter (or system) bandwidth is the 3-dB bandwidth, B. In the case of a low-pass *RC* filter, B is defined as the positive frequency at which the amplitude response, $|H(\omega)|$ drops to $\frac{|H(0)|}{\sqrt{2}}$, as illustrated in Fig. 4.6. Note that, $|H(0)|$ is the peak value of $H(\omega)$. The 3-dB bandwidth is also known as the half-power bandwidth because a voltage or current attenuation of 3 dB is equivalent to a power attenuation by a factor of 2.

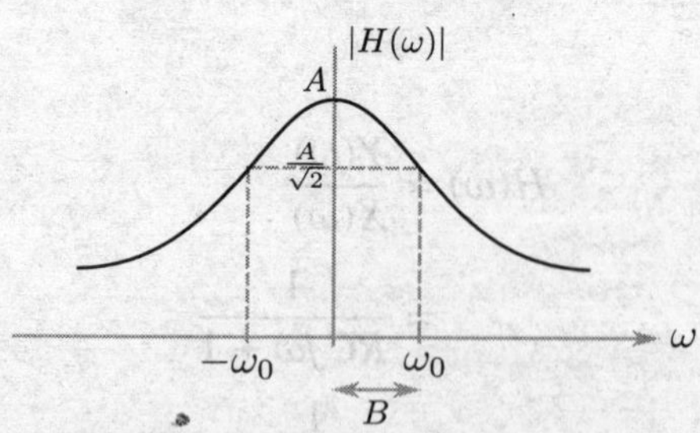

Fig. 4.6: *Amplitude response showing 3-dB bandwidth for a low-pass filter.*

In the case of a band-pass filter, B is defined as the difference between the frequencies at which $|H(\omega)|$ drops to a value equal to $\frac{1}{\sqrt{2}}$ times the peak value, $H|(\omega)|_{\omega=\omega_m}$, as illustrated in Fig. 4.7.

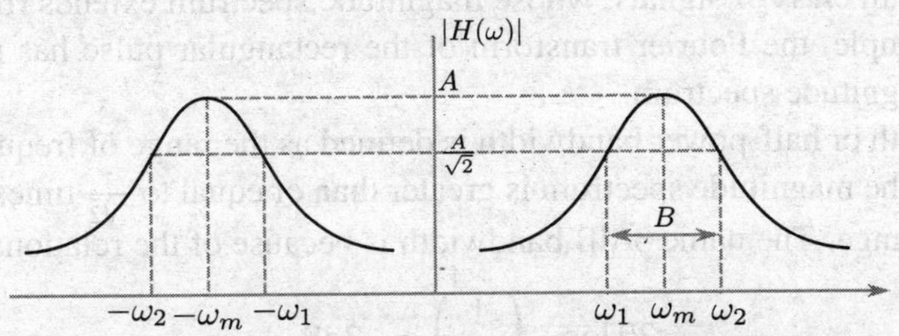

Fig. 4.7: *Magnitude response showing* 3-dB *bandwidth for a band-pass filter.*

4.11 Signal bandwidth

The Fouier transform provides a means of finding the bandwidths of signals. However, there is no single definition for bandwith and this problem is called bandwidth dilemma. We shall give three commonly used definitions for bandwidth.

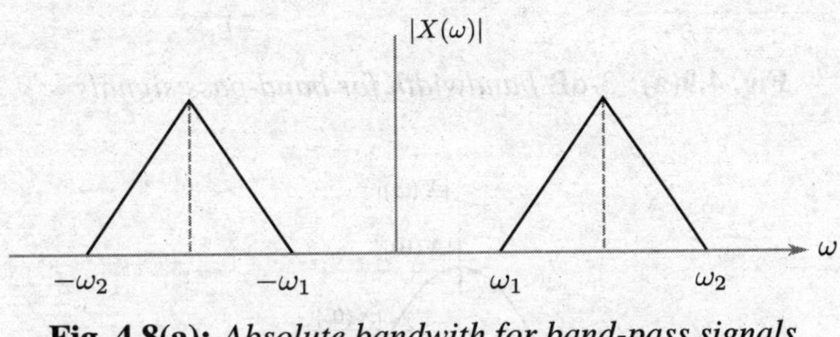

Fig. 4.8(a): *Absolute bandwith for band-pass signals*

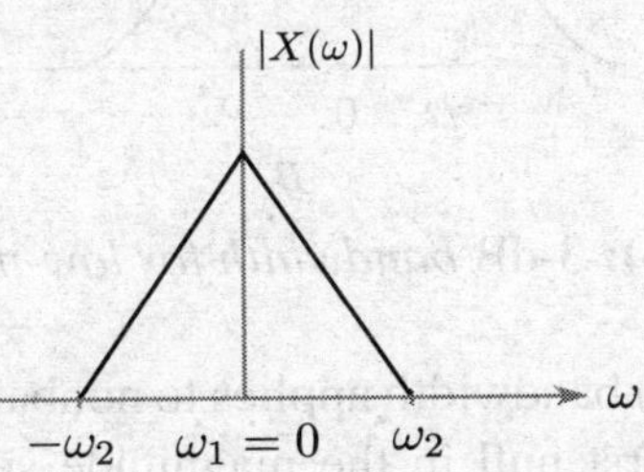

Fig. 4.8(b): *Absolute bandwith for low-pass signals*

Absolute bandwidth is $B = \omega_2 - \omega_1$, as depicted in Fig. 4.8. The magnitude spectrum is nonzero only within the band of frequencies, $\omega_1 \leq \omega \leq \omega_2$. In otherwords, absolute bandwidth is the width of the magnitude spectrum for positive frequencies. It may be noted that both ω_1 and ω_2 are greater than zero for band-pass signals, while for low-pass or base-band signal shown in Fig. 4.8(b), $\omega_1 = 0$. The definition of absolute bandwidth may not be applied to certain class of signals, whose magnitude spectrum extends from $\omega = -\infty$ to $\omega = \infty$. For example, the Fourier transform of the rectangular pulse has nonzero values over the entire magnitude spectrum.

3-dB bandwidth or half-power bandwidth is defined as the range of frequences at which the magnitude of the magnitude spectrum is greater than or equal to $\frac{1}{\sqrt{2}}$ times the maximum value within the range. The name 3-dB bandwidth is because of the relationship,

$$20 \log_{10}\left(\frac{1}{\sqrt{2}}\right) = -3\text{dB}$$

The 3-dB bandwidth is illustrated by Fig. 4.9.

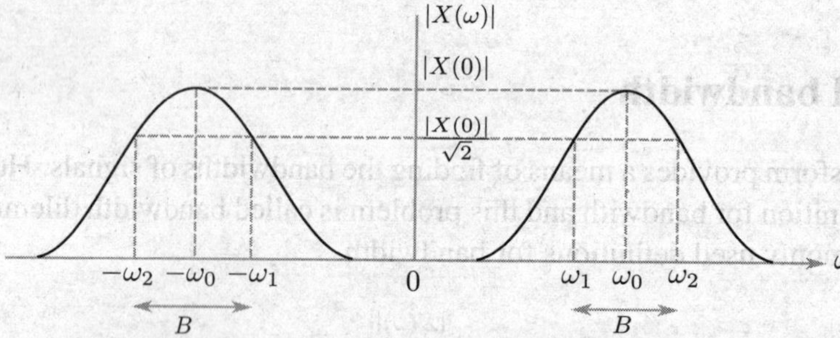

Fig. 4.9(a): 3-dB *bandwidth for band-pass signals*

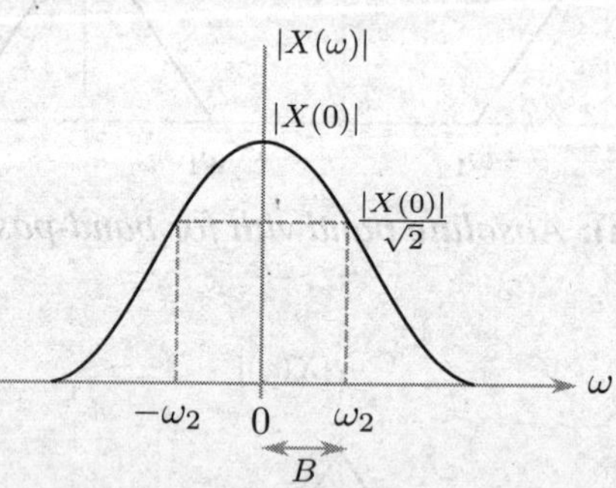

Fig. 4.9(b): 3-dB *bandwidth for low-pass signals*

Null-to null (zero-crossing) bandwidth applies to nonband-limited signals and is defined as the distance between the first null in the magnitude spectrum above ω_m and the first null in the magnitude spectrum below ω_m, where ω_m is the radian frequency at which the magnitude spectrum is maximum (refer Fig. 4.10(a)). For low-pass signals, the spectrum is maximum at $\omega = 0$, and the bandwidth is the distance between the first null and the origin (refer Fig. 4.10(b)), this bandwidth is sometimes called the first null bandwidth.

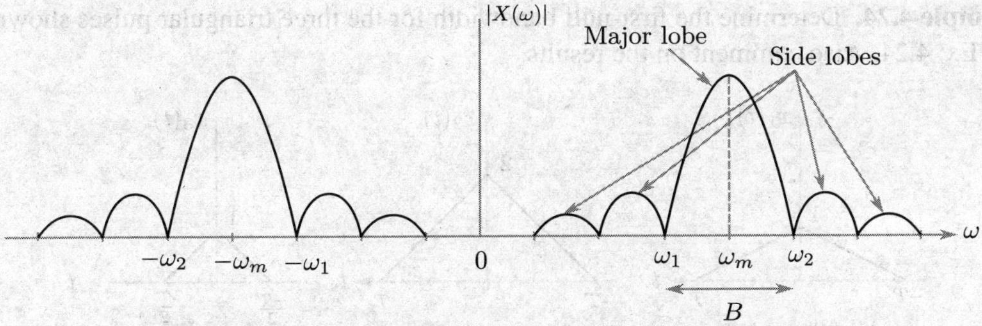

Fig. 4.10(a): *Null-to-null bandwidth for band-pass signals*

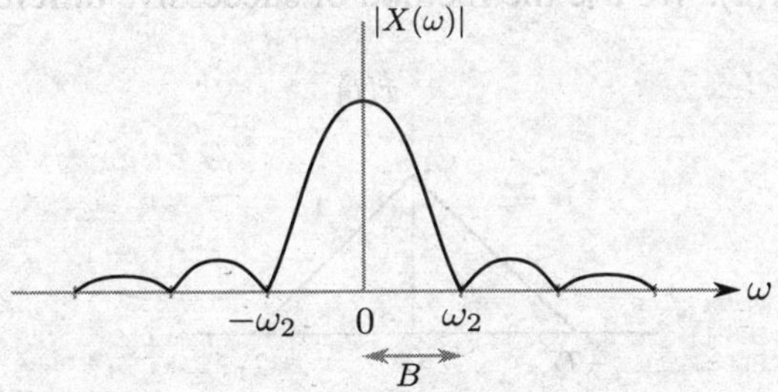

Fig. 4.10(b): *First-null bandwidth for low-pass signals*

$z\%$ bandwidth is defined such that

$$\int_{-B}^{B} |X(\omega)|^2 d\omega = \frac{z}{100} \int_{-\infty}^{\infty} |X(\omega)|^2 d\omega$$

For example, $z = 99$ defines the frequency band in which 99% of the total energy resides. The $z\%$ bandwidth is implicitly defined.

The RMS or Gabor bandwidth of a signal is defined as

$$B_{\text{RMS}}^2 = \frac{\int_{-\infty}^{\infty} \omega^2 |X(\omega)|^2 d\omega}{\int_{-\infty}^{\infty} |X(\omega)|^2 d\omega}$$

Example 4.24. Determine the first-null bandwidth for the three triangular pulses shown in Fig. Ex. 4.24. Also comment on the results.

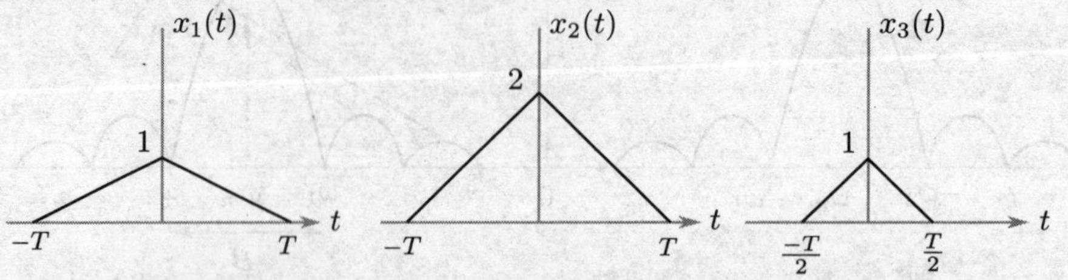

Fig. Ex. 4.24

Solution: The first step in the analysis is to find Fourier transform of a triangular pulse depicted in Fig. Ex. 4.24(a). We use the method of successive differentiation.

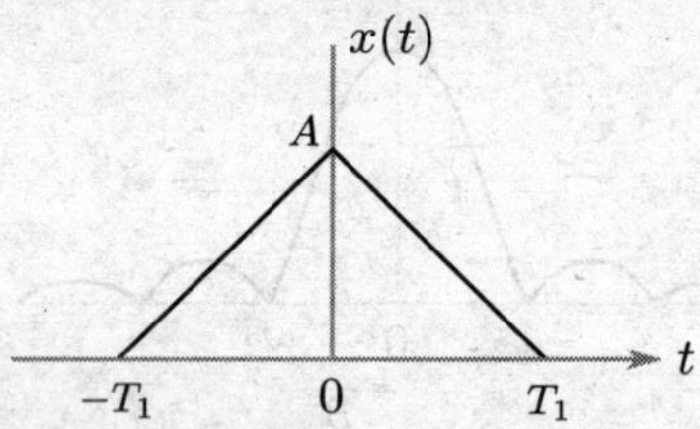

Fig. Ex. 4.24(a)

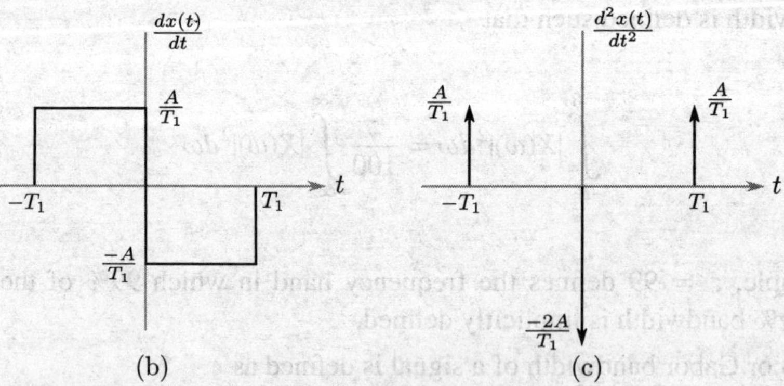

Fig. Ex. 4.24(b and c)

Referring Fig. Ex. 4.24(c), we can write

$$\frac{d^2x(t)}{dt^2} = \frac{A}{T_1}[\delta(t + T_1) - 2\delta(t) + \delta(t - T_1)]$$

Taking Fourier transform on both sides, we get

$$
\begin{aligned}
(j\omega)^2 X(\omega) &= \frac{A}{T_1}[e^{j\omega T_1} - 2 + e^{-j\omega T_1}] \\
&= \frac{A}{T_1}[2\cos\omega T_1 - 2] \\
&= \frac{2A}{T_1}[\cos\omega T_1 - 1] \\
&= \frac{-2A}{T_1}\sin^2\left(\frac{\omega T_1}{2}\right) \\
\Rightarrow \quad X(\omega) &= \frac{2A}{T_1\omega^2}\sin^2\left(\frac{\omega T_1}{2}\right) \\
&= \frac{4AT_1}{2(T_1^2\omega^2)}\sin^2\left(\frac{\omega T_1}{2}\right) \\
&= \frac{AT_1}{2}\quad\frac{sin^2\left(\frac{\omega T_1}{2}\right)}{\left(\frac{\omega T_1}{2}\right)^2} \\
&= \frac{AT_1}{2}\mathrm{sa}^2[\omega T_1/2]; \quad \omega \neq 0
\end{aligned}
$$

where $\mathrm{sa}[x] = \frac{sin x}{x}$.

Also note that $\lim_{x\to 0} \sin x/x = 1$.

Hence,

$$
|X(\omega)| = \begin{cases} \frac{AT_1}{2}\mathrm{sa}^2\left[\frac{\omega T_1}{2}\right]; & \omega \neq 0 \\ AT_1/2; & \omega = 0 \end{cases}
$$

To find zero-crossing of $|X(\omega)|$:

$$
\text{Let} \qquad |X(\omega)| = 0
$$

$$
\text{Then,} \qquad \sin\left(\frac{\omega T_1}{2}\right) = 0
$$

$$
\Rightarrow \qquad \frac{\omega T_1}{2} = \pm n\pi, \qquad n = 1, 2, 3\ldots
$$

$$
\Rightarrow \omega = \pm\frac{2n\pi}{T_1}
$$

Hence, the magnitude spectrum for $x_1(t)$, $x_2(t)$ and $x_3(t)$ are as shown in Fig. Ex. 4.24(d), (e) and (f), respectively.

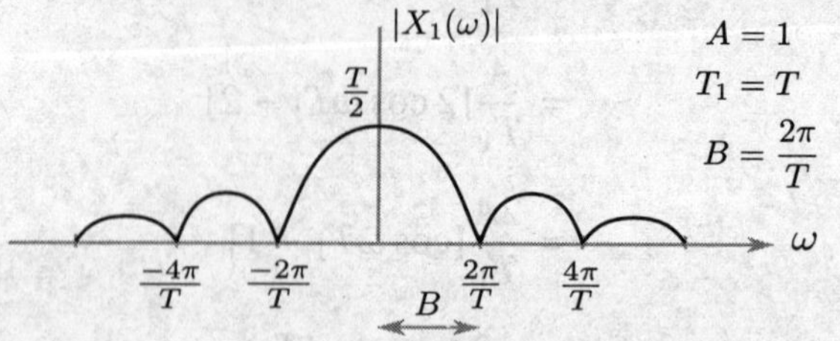

Fig. Ex. 4.24(d)

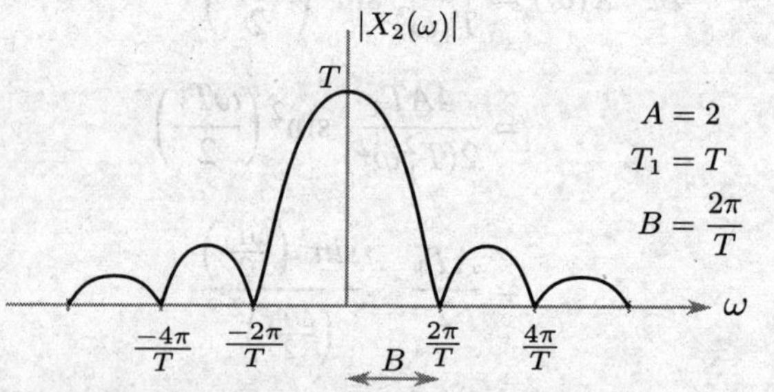

Fig. Ex. 4.24(e)

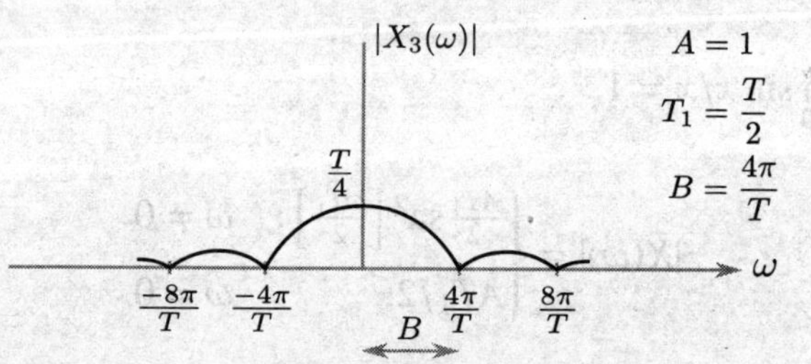

Fig. Ex. 4.24(f)

Comment: Shorter time duration results in wider bandwidth.

Example 4.25. For the circuit shown in Fig. Ex. 4.25, let $R = 1\Omega$ and $C = 1F$.

(a) Find the transfer fnction, $H(f)$.

(b) Find the magnitude and phase of $H(f)$ and identify the filter type.

(c) Find the resoponse, if the input $x(t) = \delta(t)$.

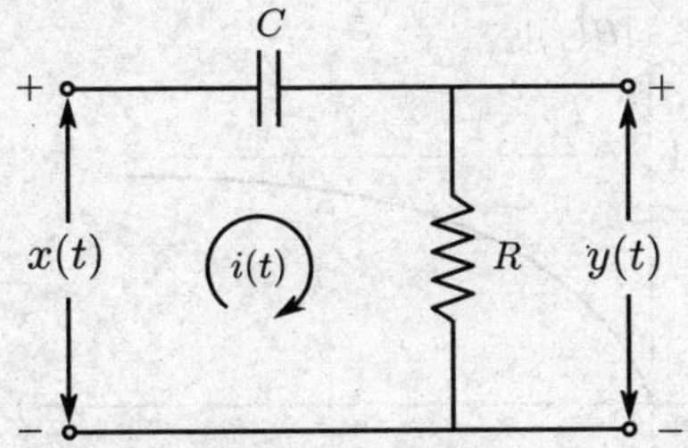

Fig. Ex. 4.25

Solution:

(a) The current in the circuit is given by

$$i(t) = \frac{1}{R} y(t)$$

Applying KVL to the input loop, we get

$$x(t) = \frac{1}{C} \int_0^t i(\tau)\, d\tau + y(t)$$

$$\Rightarrow \quad \frac{dx(t)}{dt} = \frac{1}{C} i(t) + \frac{dy(t)}{dt}$$

$$\Rightarrow \quad \frac{dx(t)}{dt} = \frac{1}{RC} y(t) + \frac{dy(t)}{dt}$$

Taking Fourier transform on both sides, we get,

$$j\omega X(\omega) = \frac{1}{RC} Y(\omega) + j\omega Y(\omega)$$

$$\Rightarrow \quad H(\omega) = \frac{Y(\omega)}{X(\omega)} = \frac{j\omega}{j\omega + \frac{1}{RC}}$$

Hence,

$$H(f) = \frac{j2\pi f}{j2\pi f + 1}$$

(b) Magnitude response:

$$|H(f)| = \frac{2\pi f}{\sqrt{(2\pi f)^2 + 1}}$$

Phase response:

$$\theta(f) = \pi/2 - \tan^{-1}(2\pi f), \quad f > 0$$

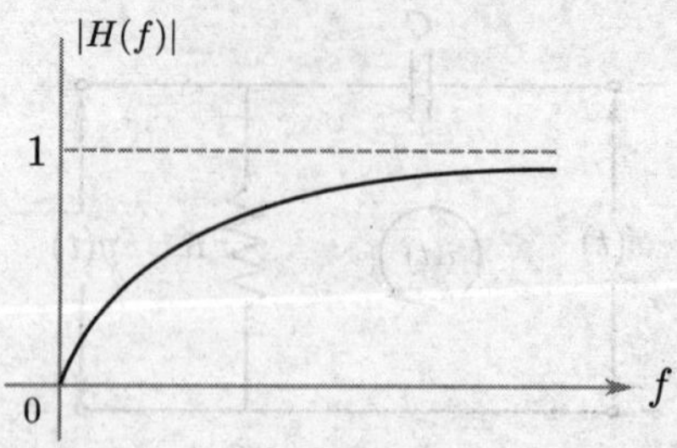

Fig. Ex. 4.25(a)

From the magnitude response sketched in Fig. Ex. 4.25(a), we find that the given *RC* circuit behaves like a high-pass filter.

(c) Given: $x(t) = \delta(t)$

$$X(f) = \int_{-\infty}^{\infty} x(t)e^{-j2\pi ft}\,dt$$

$$= \int_{-\infty}^{\infty} \delta(t)e^{-j2\pi ft}\,dt$$

$$= e^{-j2\pi ft}\Big|_{t=0} = 1$$

We know that,

$$H(f) = \frac{Y(f)}{X(f)}$$

Hence,

$$Y(f) = X(f)H(f)$$

$$= H(f)$$

$$= \frac{j2\pi f}{j2\pi f + 1}$$

$$= 1 - \frac{1}{1 + j2\pi f}$$

Recall:

1. $\delta(t) \overset{\text{FT}}{\longleftrightarrow} \delta(f) = 1$
2. $e^{-at}u(t) \overset{\text{FT}}{\longleftrightarrow} \dfrac{1}{a + j\omega} = \dfrac{1}{a + j2\pi f}$

Therefore, $y(t) = \delta(t) - e^{-t}u(t)$

4.12 Relationship between rise-time and bandwidth

The speed of response is measured by the rise-time, which is often defined as the time taken by the step response to rise from 10% to 90% of its final value. A larger rise-time implies a sluggish system because system takes a longer time to respond fully to an input. Such a system cannot respond effectively to rapid variations in the input. On the other hand, a smaller rise-time implies that a system is capable of responding to rapid variations in the input. Hence, there is a direct connection between rise-time and its filtering properties (bandwidth).

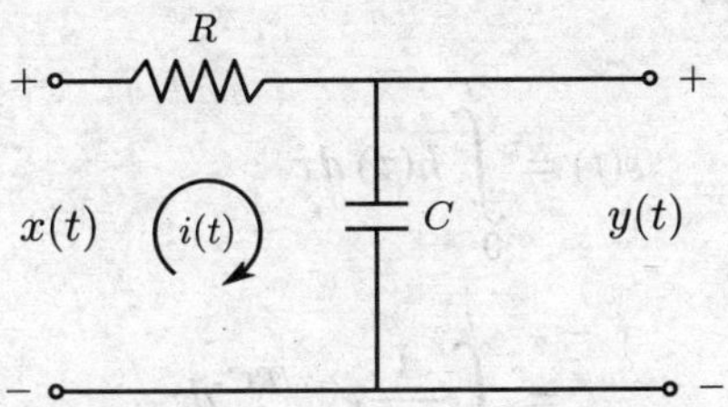

Fig. 4.11: *RC low-pass filter*

In order to know the relationship between rise-time and bandwidth, let us consider a low-pass *RC* filter as depicted in Fig. 4.11. Referring article 4.9, the frequency response of an *RC* low-pass filter is

$$H(\omega) = \frac{1}{1 + j\omega RC}$$

The magnitude response is $|H(\omega)| = \frac{1}{\sqrt{1+(\omega RC)^2}}$

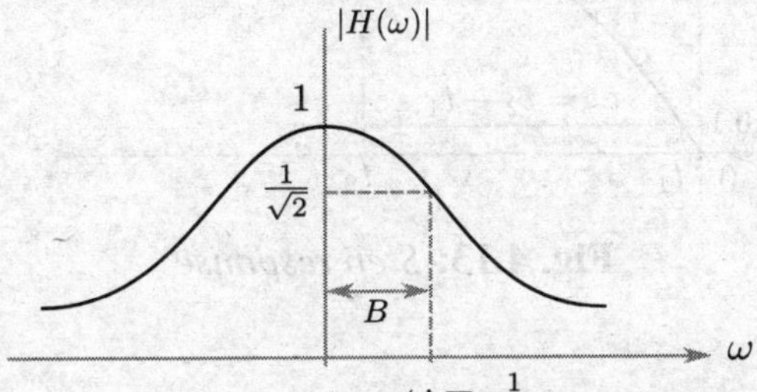

Fig. 4.12: *Magnitude response*

From the magnitude response shown in Fig. 4.12, we find that, 3-dB bandwidth of the low-pass filter is $B = \frac{1}{RC}$.

Rewrite the frequency response of the low-pass *RC* filter as

$$H(\omega) = \frac{1}{RC}\left[\frac{1}{\frac{1}{RC} + j\omega}\right]$$

Recall the Fourier-transform pair:

$$e^{-at}u(t) \xleftrightarrow{\mathbf{FT}} \frac{1}{a+j\omega}$$

Hence, the impulse response of the *RC* low-pass filter is

$$h(t) = \frac{1}{RC}e^{-t/RC}; \quad t \geq 0$$

The step response is nothing but the integral of impulse response. Thus, the step response of *RC* low-pass filter is

$$s(t) = \int_0^t h(\tau)\,d\tau$$

$$= \int_0^t \frac{1}{RC}e^{-\tau/RC}\,d\tau$$

$$= 1 - e^{-t/RC}; \quad t \geq 0$$

Using the above equation, the step response of the *RC* low-pass filter is sketched and is shown in Fig. 4.13.

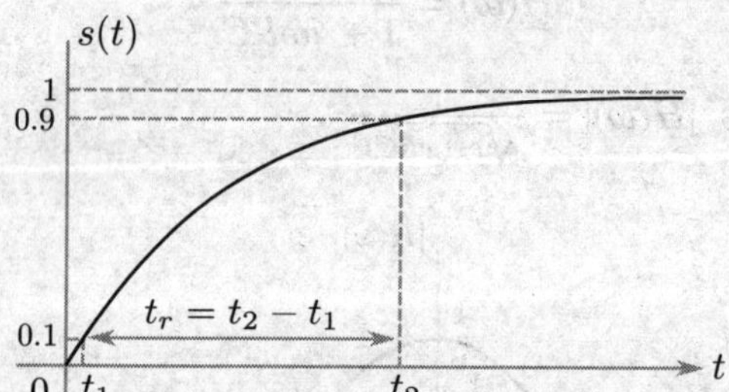

Fig. 4.13: *Step response*

Recall:

The rise-time, t_r is defined as the time taken by the step response to rise from 10% of its final value to 90% of its final value.

From Fig. 4.13, we have,

$$s(t_1) = 0.1$$

$$\Rightarrow \quad 1 - e^{-t_1/RC} = 0.1$$

$$\Rightarrow \quad e^{-t_1/RC} = 0.9$$

and

$$s(t_2) = 0.9$$

$$\Rightarrow \quad e^{-t_2/RC} = 0.1$$

Thus, we get $e^{-(t_2-t_1)/RC} = 9$.

Therefore,

$$t_r = t_2 - t_1$$

$$= RC \log_e 9$$

$$\Rightarrow \quad t_r = 2.197RC$$

$$= \frac{2.197}{B} \tag{4.13}$$

Thus, a system's bandwidth is inversely proportional to its rise-time. Although, Eq. (4.13) was derived for a first order *RC* low-pass filter, its implications are valid for higher order low-pass time-invariant systems in general.

4.13 Causality and Physical realizability

In the absence of any noise or disturbance; an ideal filter passes the desired signal. In other words, if the input signal is $x(t)$, then the desired output signal is

$$y(t) = Ax(t - t_d)$$

where A is any nonzero constant and t_d is the time delay. A delay, t_d in time-domain results in a phase shift, $e^{-j\omega t_d}$ in frequency-domain. Thus, the frequency response of the filter must be

$$H(\omega) = \frac{Y(\omega)}{X(\omega)} = \begin{cases} Ae^{-j\omega t_d}, & \text{if } \omega \text{ is in the pass-band} \\ 0, & \text{otherwise} \end{cases}$$

where the pass-band is the range of frequencies of the input signal that we want to pass.

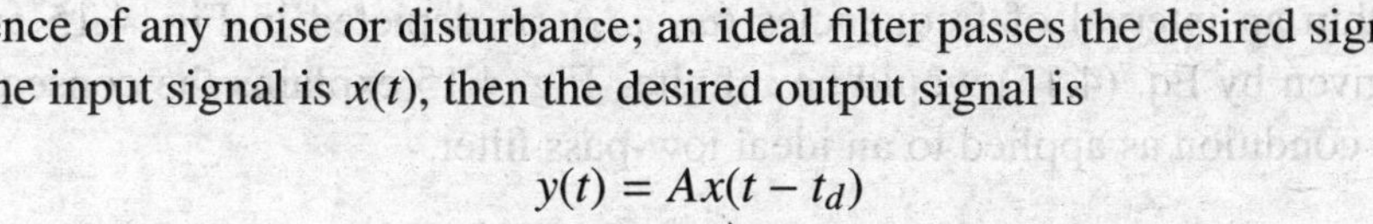

Fig. 4.14: *Frequency response of an ideal low-pass filter*

This means that the frequency response has to be zero outside the pass-band to attenuate the disturbances. The amplitude response and phase response of an ideal low-pass filter are shown in Fig. 4.14. The magnitude must be either zero or a constant, and the phase must be linear.

Theortically fine, but can we build an ideal filter practically? In otherwords, can we have an ideal filter which is causal? The answer is no; if we want the filter to be causal and implementable in real time, it cannot be ideal. This answer becomes more evident with the explaination offered below:

For a system to be causal, it is required that the system output at $t = t_0$ must be produced by an input $x(t)$ for $t \leq t_0$. This condition may also be expressed by using the impulse response of the system. For a physically realizable system, $h(t)$ must be causal; that is,

$$h(t) = 0 \quad \text{for} \quad t < 0 \tag{4.14}$$

In the frequency-domain, this condition is equivalent to Paley-Weiner criterion, which implies that necessary and sufficient condition for the amplitude response $|H(\omega)|$ to be realizable is

$$\int_{-\infty}^{\infty} \frac{\log_e |H(\omega)|}{1 + \omega^2}\, d\omega < \infty \tag{4.15}$$

The immediate consequence of this result is that an ideal filter cannot be causal, since it is zero within an interval of frequencies (ω_1, ω_2), as depicted in Fig. 4.15 and hence, the integral given by Eq. (4.15) would be infinite. Fig. 4.15 explains the consequences of paley-weiner condition as applied to an ideal low-pass filter.

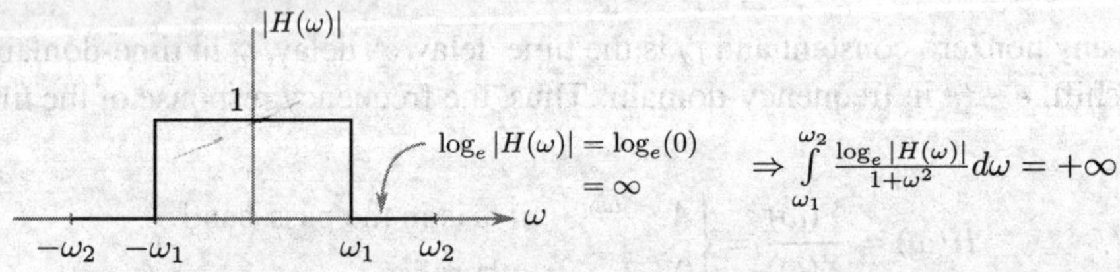

Fig. 4.15: *Ideal filter and Paley-Weiner condition*

Note that if $|H(\omega| = 0$ at a single frequency or a set of discrete frequencies, the integral in Eq. (4.15) may still be finite. Hence, for a physically realizable system, the magnitude response, $|H(\omega)|$ may be zero at some discrete frequencies, but it cannot be zero over any finite band of frequencies.

Summary of Important points

- A system which is both linear and time-invariant is called an LTI system.
- A continuous-time system is causal if its impulse response is zero for $t < 0$.
- A continuous-time system is stable if its impulse response is absolutely integrable.
- The transfer function of an LTI system is the ratio of Fourier transform of its output to Fourier transform of its input.
- Zero-state response of the system is the response produced when initial conditions are zero.
- Zero-input response of the system is the response produced when input is reduced to zero.
- Total response is the sum of Zero-input response and zero-state response.
- There is no single definition for signal bandwidth, which is known as bandwidth dilemma.
- Bandwidth and rise-time are inversely proportional to each other.
- Paley-Weiner criterion is a mathematical statement in frequency-domain for physical realization of a filter.

Exercise Problems

EP. 4.1 A continuous-time system is characterized by the following input-output relation:

$$y(t) = \frac{dx(t)}{dt}$$

Determine which of the properties listed below hold good:

a) Time-invariance
b) Linearity
c) Causality
d) Stability

Ans: System is time-invariant, linear and causal, but not stable.

EP. 4.2 Determine whether the system described by

$$y(t) = [\sin(5t)]x(t) \quad \text{is}$$

a) Causal

b) Time-invariant

c) Stable

d) Linear

Ans: Causal, time-variant, stable, and linear.

EP. 4.3 Determine whether the system described by $y(t) = \int_{-\infty}^{2t} x(\tau)\, d\tau$ is

a) Linear

b) Causal

c) Time-invariant

d) Stable

Ans: Linear, Noncausal, time-variant, unstable.

EP. 4.4 Determine whether the system described by $y(t) = x(t/4)$ is

a) Linear

b) Causal

c) Stable

d) Time-invariant

Ans: Linear, causal, stable, time-variant

EP. 4.5 Find the total response of an LTI system described by the differential equation given below:

$$y'(t) + 2y(t) = 2e^{-2t}u(t), \quad y(0) = 6$$

Ans: $y(t) = 2te^{-2t} + 6e^{-2t}$

EP. 4.6 Find the total response of an LTI sytem described by the differential equation given below:

$$y'(t) + 2y(t) = 2e^{-2t}\cos t, \quad t \geq 0, y(0) = 8$$

Ans: $y(t) = 2e^{-2t}\cos 2t + 6e^{-2t}$

EP. 4.7 Find the total response of an LTI sytem described by the differential equation given below:

$$y''(t) + 4y'(t) + 4y(t) = 8\cos 2t, \quad t \geq 0$$

$$y(0) = 0, \quad y'(0) = 1$$

Ans: $y(t) = \sin t - te^{-2t}, \quad t \geq 0$

EP. 4.8 The output $y(t)$ of a causal LTI system is related to the input $x(t)$ by the differnetial equation

$$\frac{d^2y(t)}{dt^2} + 4\frac{dy(t)}{dt} + 6y(t) = 2u(t)$$

Find the frequency response of the system.

Ans: $H(\omega) = \frac{2}{j^2\omega^2+4j\omega+6}$

EP. 4.9 An LTI system is excited by an input

$$x(t) = [e^{-t} + e^{-3t}]u(t)$$

the corresponding output is

$$y(t) = [2e^{-t} - 2e^{-4t}]u(t)$$

(a) Find the transfer function of the system

(b) Find the system impulse response

Ans:

$$H(\omega) = \frac{3(3 + j\omega)}{(4 + j\omega)(2 + j\omega)}$$

$$h(t) = \frac{3}{2}[e^{-4t} + e^{-2t}]u(t)$$

EP. 4.10 An LTI system is described by the following differential equation:

$$\frac{dy(t)}{dt} + ay(t) = \frac{-dx(t)}{dt} + ax(t)$$

(a) Find the transfer function of the system.

(b) Find the system impulse response.

Ans:

$$H(\omega) = \frac{a - j\omega}{a + j\omega}$$

$$h(t) = -\delta(t) + 2ae^{-at}u(t)$$

EP. 4.11 An LTI system is described by the differential equation given below:

$$\frac{dy(t)}{dt} + ay(t) = -\frac{dx(t)}{dt} + ax(t)$$

(a) Find the frequency response of the system.

(b) Let the input to the system be

$$x(t) = e^{-bt}u(t), \quad b > 0$$

What is the response when $b \neq a$? what is the response when $b = a$?

Ans:

$$y(t) = \left[\frac{2a}{b-a}e^{-at} + \frac{a+b}{a-b}e^{-bt}\right]u(t), \quad b \neq a$$

$$y(t) = [2ate^{-at} - e^{-at}]u(t), \quad b = a$$

EP. 4.12 For the electrical circuit shown in Fig. EP. 4.12, find the following:

a) Transfer function, $\frac{v_2(\omega)}{v_1(\omega)}$.

b) Impulse response, $h(t)$.

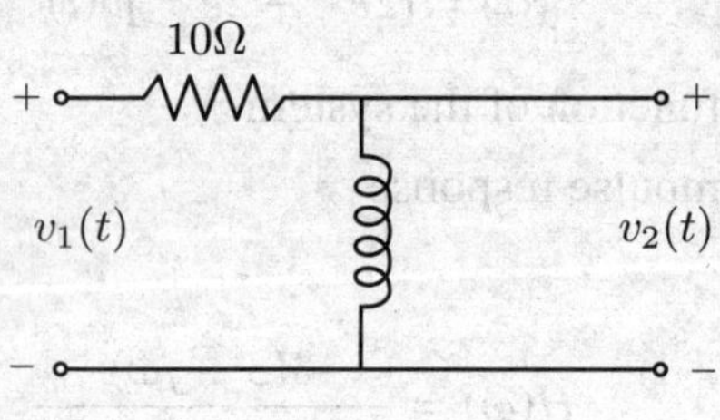

Ans:

$$H(\omega) = \frac{j\omega}{50 + j\omega}$$

$$h(t) = \delta(t) - 50e^{-50t}u(t)$$

EP. 4.13 Find the frequency response and impulse response of the system having output $y(t)$ and input $x(t)$ given below:

$$x(t) = e^{-2t}u(t)$$

$$y(t) = 2te^{-2t}u(t)$$

Ans:

$$H(\omega) = \frac{2}{2 + j\omega}; \quad h(t) = 2e^{-2t}u(t)$$

EP. 4.14 Find the frequency response and impulse response of an LTI system described by

$$\frac{d^2y(t)}{dt^2} + 5\frac{dy(t)}{dt} + by(t) = -\frac{dx(t)}{dt}$$

Ans:

$$H(\omega) = \frac{-j\omega}{(j\omega + 2)(j\omega + 3)}$$

$$h(t) = [2e^{-2t} - 3e^{-3t}]u(t)$$

EP. 4.15 Find the differential equation description for an LTI system having the impulse response given by

$$h(t) = 2e^{-2t}u(t) - 2te^{-2t}u(t)$$

Ans:

$$\frac{d^2y(t)}{dt^2} + 4\frac{dy(t)}{dt} + 4y(t) = 2\frac{dx(t)}{dt} + 2x(t)$$

Chapter 5

Convolution and Correlation of Signals

In chapter 4, we have studied the various properties of systems. Here in this chapter, we shall make use of two important properties, namely linearity and time-invariance. By introducing the concept of convolution operation, we shall find the system response to any arbitrary excitation, provided we know the impulse response of the system. Also, we shall investigate some of the important properties of systems, such as stability and causality by using the impulse response of the system. Also, we study signal similarity by using correlation and various other concepts required to understand the intricacies of communication systems.

5.1 Convolution of continuous-time signals

First, let us derive a relationship that expresses a general signal $x(t)$ as a function of an impulse function.

Let us recall the product property of an impulse function:

$$x(t)\,\delta(t-t_0) = x(t_0)\,\delta(t-t_0) \tag{5.1}$$

Letting $t_0 = \tau$ in Eq. (5.1) gives

$$x(t)\,\delta(t-\tau) = x(\tau)\,\delta(t-\tau)$$

The next step in the analysis is to consider the following integral:

$$\int_{-\infty}^{\infty} x(\tau)\,\delta(t-\tau)\,d\tau$$

Applying sifting property of impulse functions, the above integral gets evaluated as

$$\int_{-\infty}^{\infty} x(\tau)\,\delta(t-\tau)\,d\tau = x(\tau)\big|_{\tau=t} = x(t) \tag{5.2}$$

Eq. (5.2) is the desired result, in which a general signal $x(t)$ is expressed as an integral of shifted impulse function. We use this expression for $x(t)$ in the derivation of convolution integral.

The concept of convolution integral also demands the most important attributes of the systems, namely linearity and time-invariance. Let us consider a continuous-time system defined by an operator, $\mathbf{T}\{\cdot\}$ as shown in Fig. 5.1.

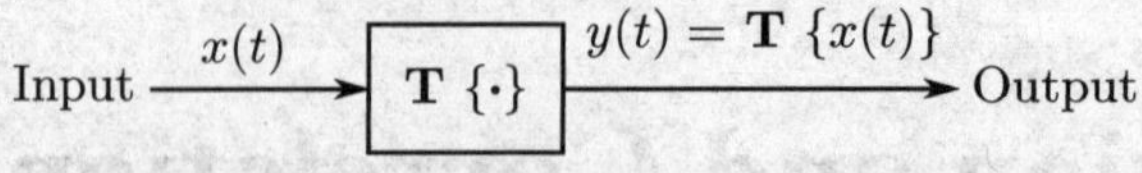

Fig. 5.1 *Continuous-time LTI system*

The response of the system shown in Fig. 5.1 is

$$y(t) = \mathbf{T}\{x(t)\} \tag{5.3}$$

Substituting Eq. (5.2) in Eq. (5.3) gives

$$y(t) = \mathbf{T}\left\{ \int_{-\infty}^{\infty} x(\tau)\delta(t-\tau)d\tau \right\} \tag{5.4}$$

Recall the homogenity property for linear systems:

$$\mathbf{T}\{a_1 x_1(t)\} = a_1\, \mathbf{T}\{x_1(t)\}$$

Applying the homogenity property to Eq. (5.4), we get

$$y(t) = \int_{-\infty}^{\infty} x(\tau)\, \mathbf{T}\{\delta(t-\tau)\}\, d\tau \tag{5.5}$$

The impulse response $h(t)$ of a continuous-time system is defined to be the response of the system when the input is $\delta(t)$, that is,

$$h(t) = \mathbf{T}\{\delta(t)\} \tag{5.6}$$

Recall the time-invariance property:

$$\mathbf{T}\{x(t-t_0)\} = y(t-t_0)$$

Applying time-invariance property to Eq. (5.6), we get

$$\mathbf{T}\{\delta(t-\tau)\} = h(t-\tau) \tag{5.7}$$

Substituting Eq. (5.7) in Eq. (5.5), we get

$$y(t) = \int_{-\infty}^{\infty} x(\tau)h(t-\tau)d\tau \tag{5.8}$$

Eq. (5.8) defines the convolution of two continuous-time signals $x(t)$ and $h(t)$ and is denoted symbolically by

$$\begin{aligned} y(t) &= x(t) * h(t) \\ &= \int_{-\infty}^{\infty} x(\tau)h(t-\tau)d\tau \end{aligned} \tag{5.9}$$

Eq. (5.9) is commonly called convolution integral. Thus, we have the fundamental result that the output of any continuous-time LTI system is the convolution of the input $x(t)$ with the impulse response $h(t)$ of the system.

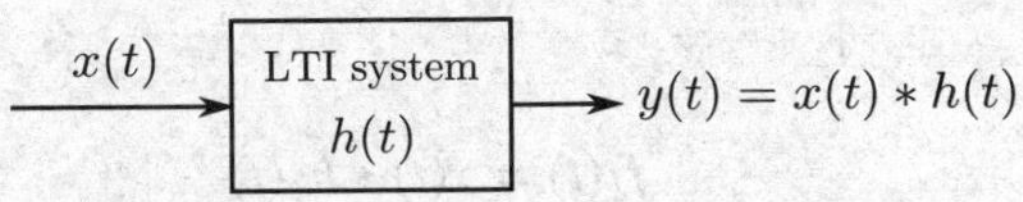

Fig. 5.2 *Continuous-time LTI system*

Fig. 5.2 illustrates the concept of Convolution integral.

5.2 Properties of convolution

The convolution integral of Eq. (5.9) has three important properties:

1. **Commutative property:**

   The convolution integral is symmetrical with respect to $x(t)$ and $h(t)$. This allows us to write

   $$x(t) * h(t) = h(t) * x(t)$$

   This property can be proved as follows:

   $$x(t) * h(t) = \int_{-\infty}^{\infty} x(\tau)h(t-\tau)d\tau$$

   Substitute, $t - \tau = \lambda \Rightarrow -d\tau = d\lambda$

Therefore,

$$x(t) * h(t) = -\int_{\lambda=\infty}^{-\infty} x(t-\lambda)h(\lambda)d\lambda$$

$$= \int_{-\infty}^{\infty} h(\lambda)x(t-\lambda)d\lambda$$

$$= h(t) * x(t)$$

2. **Associative property:**

   This property means the convolution of three or more functions is independent of the order in which the convolution is performed. That is,

$$[x(t) * h_1(t)] * h_2(t) = x(t) * [h_1(t) * h_2(t)]$$

   This property can be proved as follows:

   Let

$$f_1(t) = x(t) * h_1(t)$$

   and

$$f_2(t) = h_1(t) * h_2(t)$$

   Thus,

$$f_1(t) = \int_{\tau=-\infty}^{\infty} x(\tau)h_1(t-\tau)d\tau$$

$$\Rightarrow \quad f_1(\sigma) = \int_{\tau=-\infty}^{\infty} x(\tau)h(\sigma-\tau)d\tau \qquad (5.10)$$

   and

$$f_2(t) = \int_{\lambda=-\infty}^{\infty} h_1(\lambda)h_2(t-\lambda)d\lambda$$

$$\Rightarrow \quad f_2(t-\tau) = \int_{\lambda=-\infty}^{\infty} h_1(\lambda)h_2(t-\tau-\lambda)d\lambda \qquad (5.11)$$

Now we have,

$$[x(t) * h_1(t)] * h_2(t) = f_1(t) * h_2(t)$$

$$= \int_{\sigma=-\infty}^{\infty} f_1(\sigma) h_2(t-\sigma) d\sigma \qquad (5.12)$$

Substituting Eq. (5.10) in Eq. (5.12) gives

$$[x(t) * h_1(t)] * h_2(t) = \int_{\sigma=-\infty}^{\infty} \left[\int_{\lambda=-\infty}^{\infty} x(\tau) h_1(\sigma-\tau) d\tau \right] h_2(t-\sigma) d\sigma$$

Substituting $\sigma - \tau = \lambda$ and interchanging the order of integrations, we get

$$[x(t) * h_1(t)] * h_2(t) = \int_{\tau=-\infty}^{\infty} x(\tau) \left[\int_{\lambda=-\infty}^{\infty} h_1(\lambda) h_2(t-\tau-\lambda) d\lambda \right] d\tau \qquad (5.13)$$

Substituting Eq. (5.11) in Eq. (5.13), we get

$$[x(t) * h_1(t)] * h_2(t) = \int_{\tau=-\infty}^{\infty} x(\tau) f_2(t-\tau) d\tau$$

$$= x(t) * f_2(t) = x(t) * [h_1(t) * h_2(t)]$$

3. **Distributive property:**

The convolution integral satisfies the following relationship:

$$x(t) * [h_1(t) + h_2(t)] = x(t) * h_1(t) + x(t) * h_2(t)$$

This property can be proved as follows:

By using the definition of convolution integral, we can write

$$x(t) * [h_1(t) + h_2(t)] = \int_{\tau=-\infty}^{\infty} x(\tau)[h_1(t-\tau) + h_2(t-\tau)] d\tau$$

$$= \int_{\tau=-\infty}^{\infty} x(\tau) h_1(t-\tau) d\tau + \int_{\tau=-\infty}^{\infty} x(\tau) h_2(t-\tau) d\tau$$

$$= x(t) * h_1(t) + x(t) * h_2(t)$$

5.3 Graphical interpretation of convolution

Recall the convolution integral:

$$y(t) = x(t) * h(t) = \int_{-\infty}^{\infty} x(\tau)h(t-\tau)d\tau \tag{5.14}$$

Applying Commutative property of convolution, we get

$$y(t) = h(t) * x(t) = \int_{-\infty}^{\infty} h(\tau)x(t-\tau)d\tau \tag{5.15}$$

Eq. (5.15) at times may be easier to evaluate than Eq. (5.14).

In the convolution integral represented by Eq. (5.14), the time t decides the relative location of $h(t-\tau)$ with respect to $x(\tau)$. The convolution integral will give a nonzero result only when there is an overlap between $x(\tau)$ and $h(t-\tau)$. The computation of $y(t)$ for all t requires the evaluation of convolution integral for all t. We must evaluate the overlapping area of the product as t varies. Varying t means sliding the folded function $h(-\tau)$ past $x(\tau)$ by the chosen values of t, as shown in Fig. 5.3.

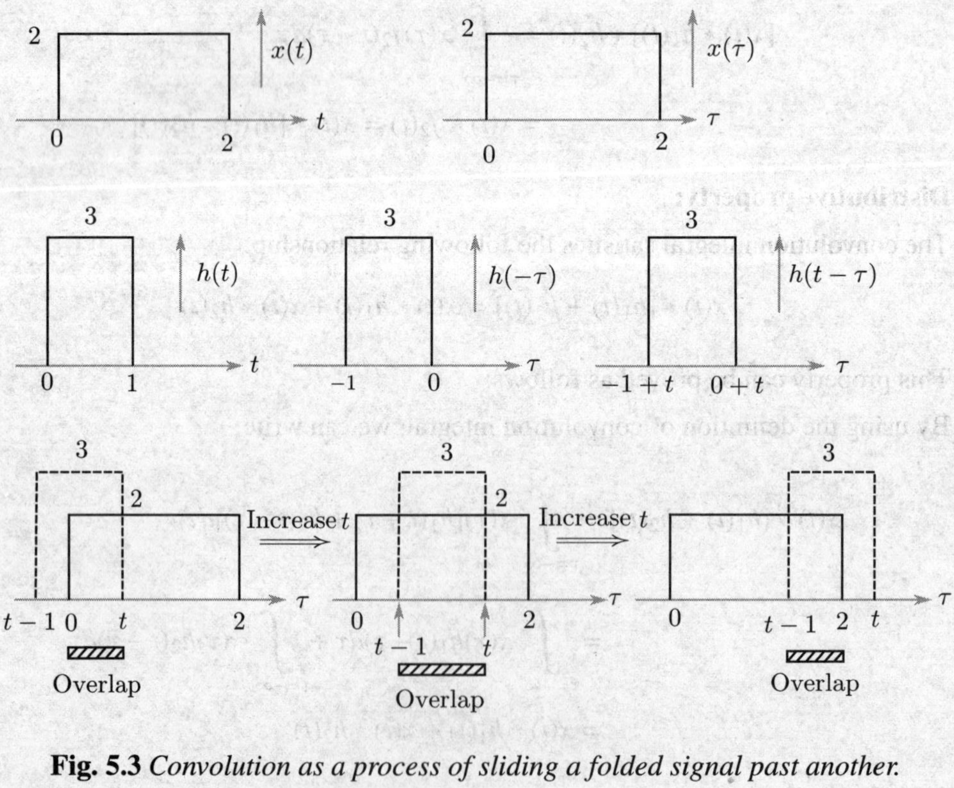

Fig. 5.3 *Convolution as a process of sliding a folded signal past another.*

Let us now summarize the various steps involved in the evaluation of a convolution integral:

1. The impulse response is folded or time reversed to obtain $h(-\tau)$ and then shifted by t, to give $h(t-\tau)$.
2. The signal $x(\tau)$ and $h(t-\tau)$ are multiplied together for all values of τ with t fixed at some value.
3. The product $x(\tau)h(t-\tau)$ is integrated over all τ to produce a single output value $y(t)$.
4. Steps 1 and 3 are repeated as t varies over $-\infty$ to ∞ to produce the entire output $y(t)$.

Example 5.1. Show that:

(a) $x(t) * \delta(t) = x(t)$

(b) $x(t) * \delta(t - t_0) = x(t_0)$

(c) $x(t) * u(t) = \int_{-\infty}^{t} x(\tau)d\tau$

Solution:

(a) $x(t) * \delta(t) = \int_{-\infty}^{\infty} x(\tau)\delta(t-\tau)d\tau$

Applying sifting property, we get

$$x(t) * \delta(t) = x(\tau)|_{\tau=t} = x(t)$$

(b) $x(t) * \delta(t - t_0) = \int_{-\infty}^{\infty} x(\tau)\delta(t-\tau-t_0)d\tau$

Applying sifting property, we get

$$x(t) * \delta(t-t_0) = x(\tau)|_{\tau=t-t_0} = x(t-t_0)$$

(c) Since, $u(t-\tau) = \begin{cases} 1, & t-\tau > 0 \text{ or } \tau < t \\ 0, & t-\tau < 0 \text{ or } \tau > t \end{cases}$, we can write

$$x(t) * u(t) = \int_{-\infty}^{\infty} x(\tau)u(t-\tau)d\tau$$

$$= \int_{-\infty}^{t} x(\tau)d\tau$$

Example 5.2. Convolute the two continuous-time signals given below:

$$x_1(t) = e^{-2t}u(t)$$

$$x_2(t) = u(t+2)$$

Solution: Given:

(1) $x_1(t) = e^{-2t}u(t) \Rightarrow x_1(\tau) = e^{-2\tau}u(\tau)$

(2) $x_2(t) = u(t+2) \Rightarrow x_2(\tau) = u(\tau+2) = \begin{cases} 1, & \tau+2 > 0 \text{ or } \tau > -2 \\ 0, & \tau+2 < 0 \text{ or } \tau < -2 \end{cases}$

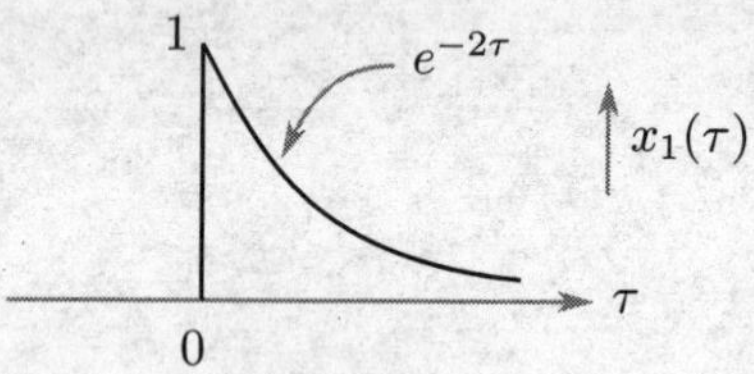

Fig. Ex. 5.2(a): *Signal,* $x_1(\tau)$

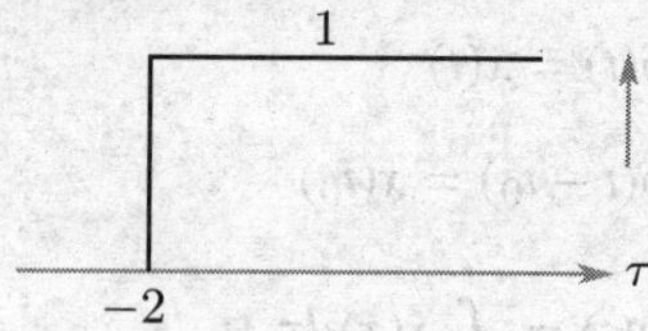

Fig. Ex. 5.2(b): *Signal,* $x_2(\tau)$

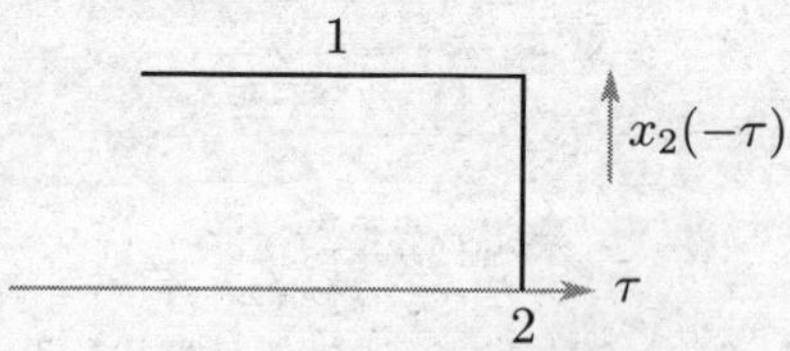

Fig. Ex. 5.2(c): *Folded signal,* $x_2(-\tau)$

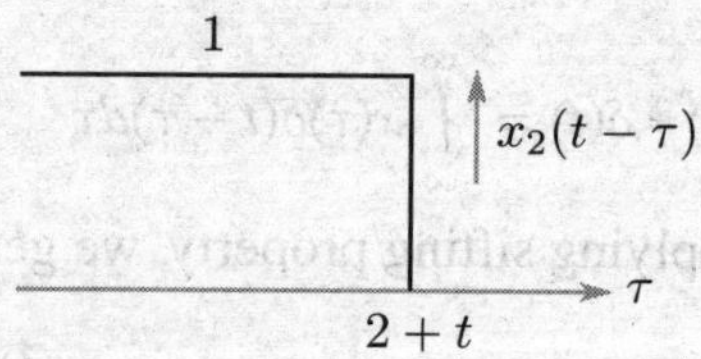

Fig. Ex. 5.2(d): *Folded and shifted signal,* $x_2(t-\tau)$

The sketch of $x_2(t-\tau)$ can also be formed as follows:
We know that

$$x_2(t) = \begin{cases} 1, & t > -2 \\ 0, & t < -2 \end{cases}$$

$$\Rightarrow \quad x_2(t-\tau) = \begin{cases} 1, & t-\tau > -2 \text{ or } \tau < t+2 \\ 0, & t-\tau < -2 \text{ or } \tau > t+2 \end{cases}$$

Let

$$y(t) = x_1(t) * x_2(t)$$

$$= \int_{-\infty}^{\infty} x_1(\tau)x_2(t-\tau)d\tau$$

Let us now proceed to find $y(t)$ for different values of t.

Case (i): For $t + 2 < 0$ or $t < -2$, the functions $x_1(\tau)$ and $x_2(t - \tau)$ do not overlap. Hence, $y(t) = y_1(t) = 0$.

Case (ii):

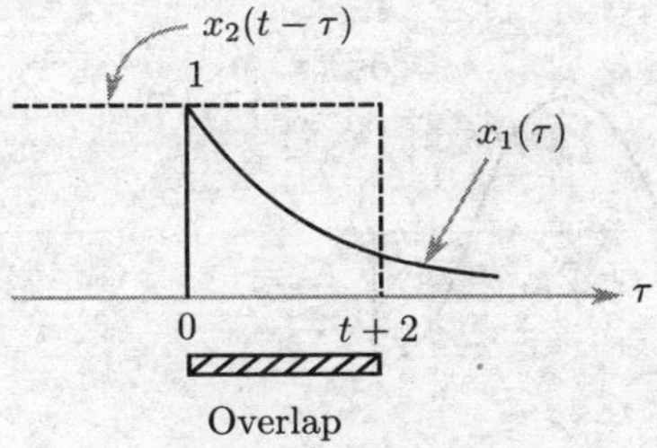

Fig. Ex. 5.2(e)

For $t + 2 > 0$ or $t > -2$, the functions $x_1(\tau)$ and $x_2(t - \tau)$ overlap as shown in Fig. Ex. 5.2(e).

Therefore,

$$y(t) = y_2(t) = \int_0^{t+2} e^{-2\tau} \times 1 d\tau$$

$$= \frac{1}{2} - \frac{1}{2}e^{-2(t+2)}$$

Summarizing the results, we get

$$y(t) = \begin{cases} 0, & t < -2 \\ \frac{1}{2} - \frac{1}{2}e^{-2(t+2)}, & t > -2 \end{cases}$$

Cross-Check:

At the end points $y_1(t)$ and $y_2(t)$ must be identical. This means $y_1(-2)$ and $y_2(-2)$ must be identical. In this case, we have $y_1(-2) = y_2(-2) = 0$.

Example 5.3. Convolute the two continuous-time signals given below:

$$x_1(t) = \cos \pi t[u(t + 1) - u(t - 3)]$$

$$x_2(t) = u(t)$$

Solution: It may be noted that

$$u(t + 1) - u(t - 3) = \begin{cases} 1, & -1 < t < 3 \\ 0, & \text{otherwise} \end{cases}$$

Hence,

$$x_1(t) = \begin{cases} \cos \pi t, & -1 < t < 3 \\ 0, & \text{otherwise} \end{cases}$$

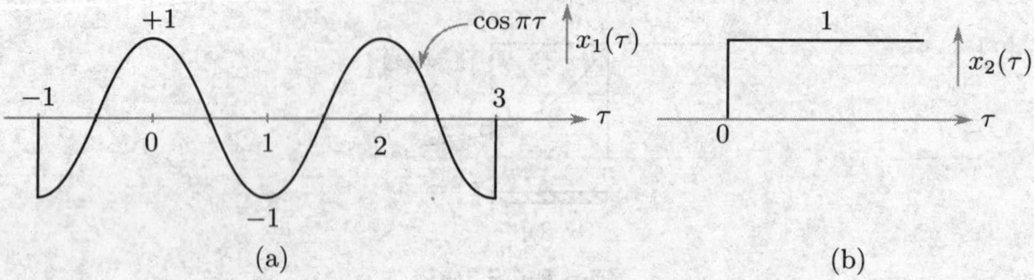

Fig. Ex. 5.3(a): *Signal,* $x_1(\tau)$ **(b):** *Signal,* $x_2(\tau)$

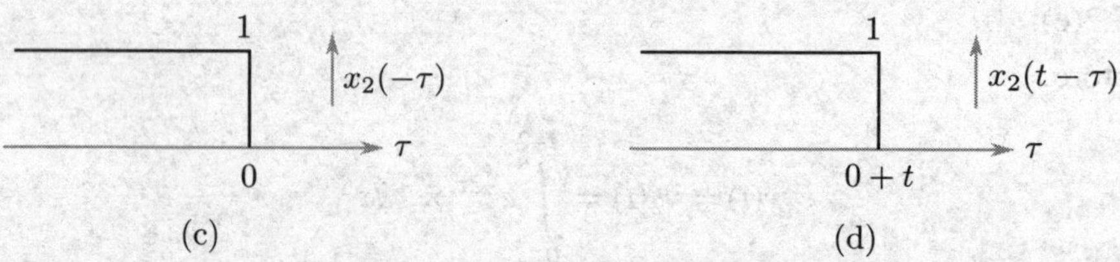

Fig. Ex. 5.3(c): *Folded signal,* $x_2(-\tau)$ **(d):** *Folded and shifted signal,* $x_2(t-\tau)$

The sketch of $x_2(t-\tau)$ can also be found as follows:

$$x_2(t) = u(t)$$

$$\Rightarrow \quad x_2(t-\tau) = u(t-\tau)$$

$$= \begin{cases} 1, & t-\tau > 0 \text{ or } \tau < t \\ 0, & t-\tau < 0 \text{ or } \tau > t \end{cases}$$

Let

$$y(t) = x_1(t) * x_2(t)$$

$$= \int_{-\infty}^{\infty} x_1(\tau) x_2(t-\tau) d\tau$$

Our next step in the analysis is to find $y(t)$ for different values of t.

Case (i): For $t < -1$, the functions $x_1(\tau)$ and $x_2(t-\tau)$ do not overlap and hence, $y(t) = y_1(t) = 0$.

Case (ii): For $t > -1$ and $t < 3$ or $-1 < t < 3$, the signals $x_1(\tau)$ and $x_2(t-\tau)$ overlap as illustrated in Fig. Ex. 5.3(e).

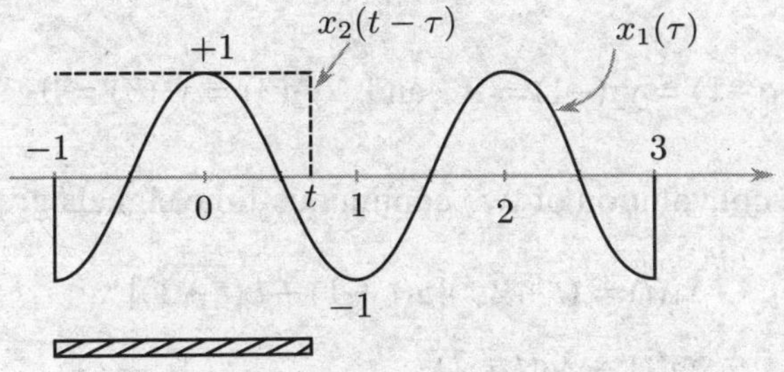

Fig. Ex. 5.3(e)

$$y(t) = y_2(t) = \int_{-1}^{t} \cos \pi\tau \times 1 d\tau$$

$$= \frac{\sin \pi t}{\pi}$$

Case (iii): For $t > 3$, the signals $x_1(\tau)$ and $x_2(t-\tau)$ overlap as shown in Fig. Ex. 5.3(f).

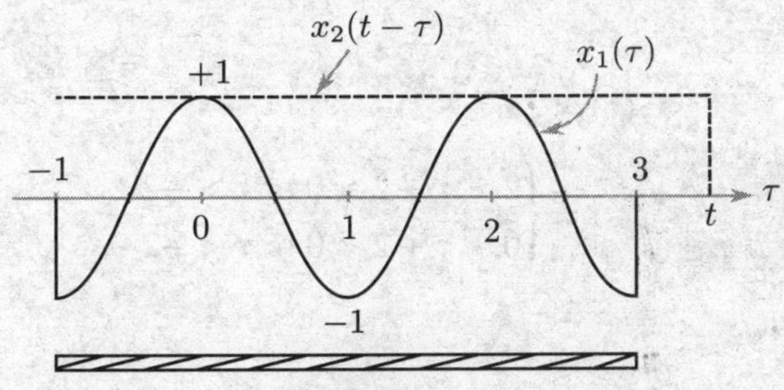

Fig. Ex. 5.3(f)

$$y(t) = y_3(t) = \int_{-1}^{3} \cos \pi\tau \times 1 d\tau$$

$$= 0$$

Summarizing, we get,

$$y(t) = \begin{cases} y_1(t) = 0, & t < -1 \\ y_2(t) = \dfrac{\sin \pi t}{\pi}, & -1 < t < 3 \\ y_3(t) = 0, & t > 3 \end{cases}$$

Cross-Check:

$$y_1(-1) = y_2(-1) = 0 \quad \text{and} \quad y_2(3) = y_3(3) = 0.$$

Example 5.4. Find the convolution of two continuous-time signals given below:

$$x_1(t) = (t + 2t^2)[u(t+1) - u(t-1)]$$

$$x_2(t) = 2u(t+2)$$

Solution: We note that,

$$u(t+1) - u(t-1) = \begin{cases} 1, & -1 < t < 1 \\ 0, & \text{otherwise} \end{cases}$$

Hence,

$$x_1(t) = \begin{cases} t + 2t^2, & -1 < t < 1 \\ 0, & \text{otherwise} \end{cases}$$

Also,

$$x_2(\tau) = 2u(\tau + 2)$$

$$= \begin{cases} 2, & \tau + 2 > 0 \text{ or } \tau > -2 \\ 0, & \tau + 2 < 0 \text{ or } \tau < -2 \end{cases}$$

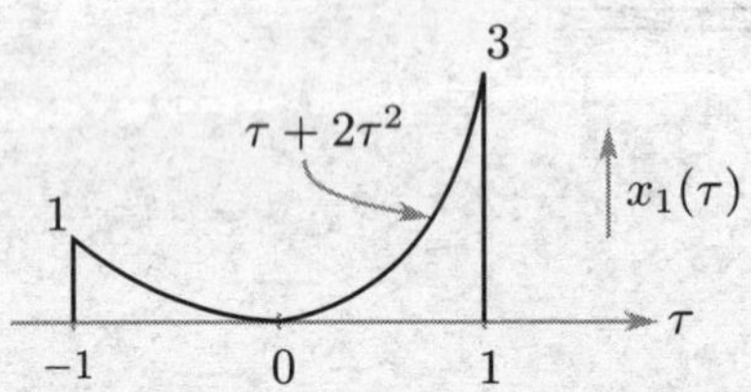

Fig. Ex. 5.4(a): *Signal,* $x_1(\tau)$

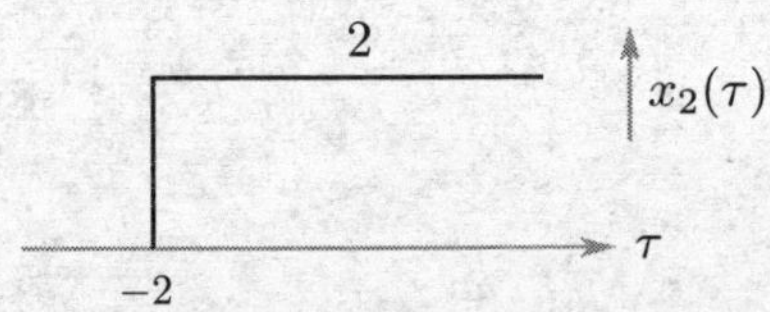

Fig. Ex. 5.4(b): *Signal,* $x_2(\tau)$

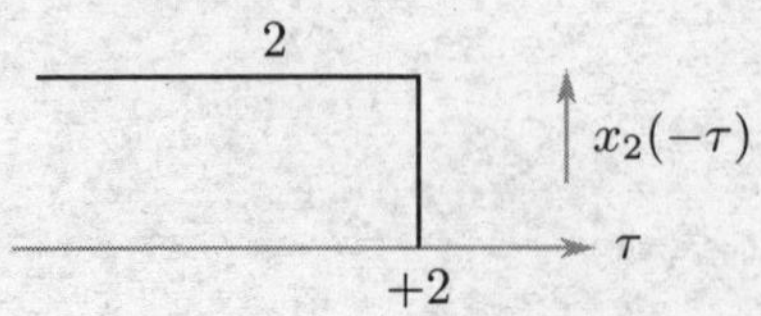

Fig. Ex. 5.4(c): *Folded signal,* $x_2(-\tau)$

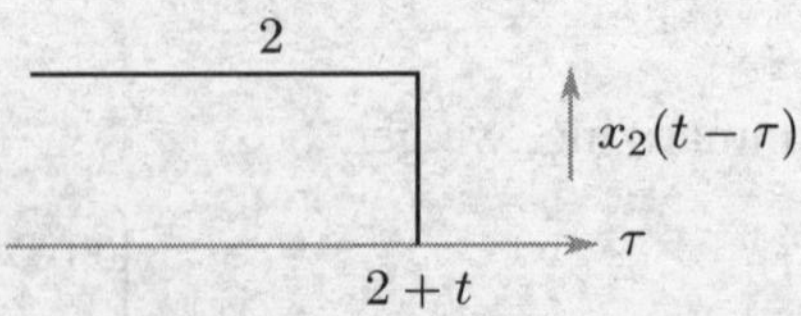

Fig. Ex. 5.4(d): *Folded and shifted signal* $x_2(t-\tau)$

Let

$$y(t) = x_1(t) * x_2(t) = \int_{-\infty}^{\infty} x_1(\tau)x_2(t-\tau)d\tau$$

Let us now find $y(t)$ for all t, that is, for $-\infty < t < \infty$.

Case (i): For $t+2 < -1$ or $t < -3$, the signals $x_1(\tau)$ and $x_2(t-\tau)$ do not overlap. Hence, $y(t) = 0$.

Case (ii): For $t+2 > -1$ and $t+2 < 1$ or $-3 < t < -1$, the functions $x_1(\tau)$ and $x_2(t-\tau)$ overlap as shown in Fig. Ex. 5.4(e).

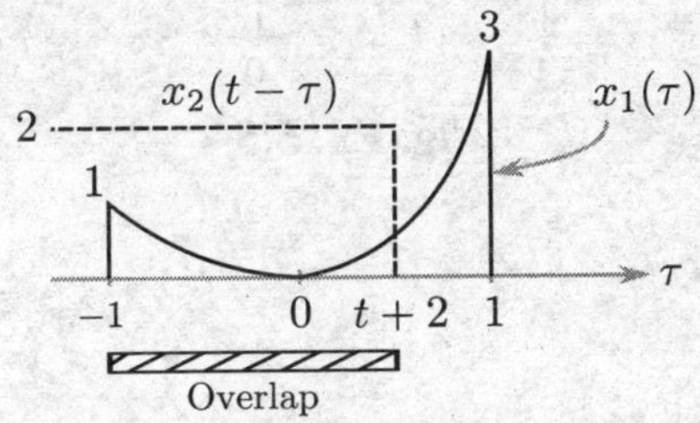

Fig. Ex. 5.4(e)

$$y(t) = \int_{-1}^{t+2} (\tau + 2\tau^2) \times 2\, d\tau = 2\left[\frac{\tau^2}{2} + 2\frac{\tau^3}{3}\right]_{-1}^{t+2}$$

$$= (t+2)^2 + \frac{4}{3}(t+2)^3 + \frac{1}{3}$$

Case (iii): For $t+2 > 1$ or $t > -1$, the signals $x_1(\tau)$ and $x_2(t-\tau)$ overlap as shown in Fig. Ex. 5.4(f).

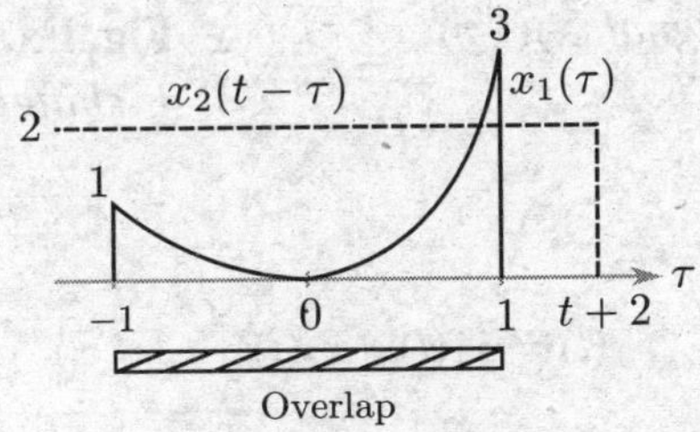

Fig. Ex. 5.4(f)

$$y(t) = \int_{-1}^{1} (\tau + 2\tau^2) \times 2d\tau = 2\left[\frac{\tau^2}{2} + 2\frac{\tau^3}{3}\right]_{-1}^{1} = \frac{8}{3}$$

Summarising the results, we get

$$y(t) = \begin{cases} 0, & t < -3 \\ (t+2)^2 + \frac{4}{3}(t+2)^3 + \frac{1}{3}, & -3 < t < -1 \\ 8/3, & t > -1 \end{cases}$$

Example 5.5. Convolute the two continuous-time signals shown in Fig. Ex. 5.5.

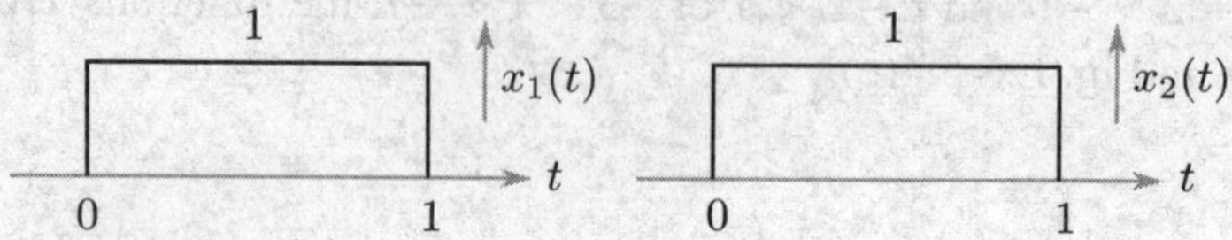

Fig. Ex. 5.5

Solution:

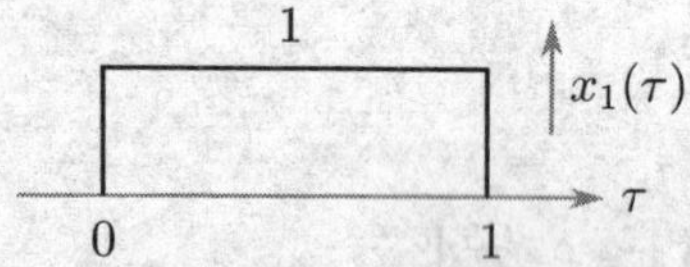

Fig. Ex. 5.5(a): *Signal,* $x_1(\tau)$

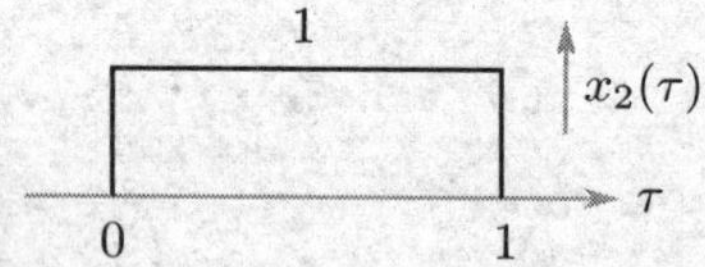

Fig. Ex. 5.5(b): *Signal,* $x_2(\tau)$

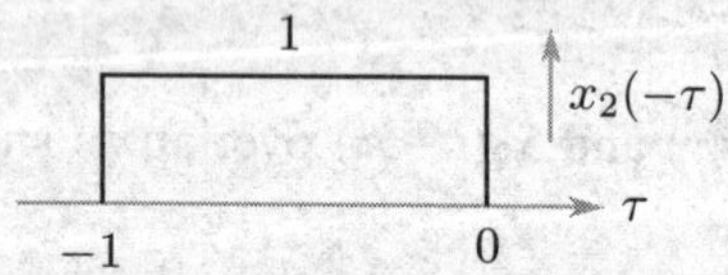

Fig. Ex. 5.5(c): *Folded signal,* $x_2(-\tau)$

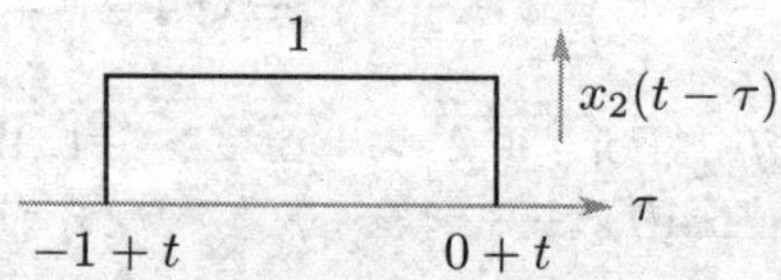

Fig. Ex. 5.5(d): *Folded and shifted signal,* $x_2(t-\tau)$

Let

$$y(t) = x_1(t) * x_2(t)$$

$$= \int_{-\infty}^{\infty} x_1(\tau) x_2(t-\tau) d\tau$$

The values of $y(t)$ as t is varied from $-\infty$ to ∞ is found as follows:

Case (i): For $t < 0$, the signals $x_1(\tau)$ and $x_2(t-\tau)$ do not overlap. Hence, the product, $x_1(\tau)x_2(t-\tau) = 0$. Thus, $y(t) = 0$.

Case (ii): For $t-1<0$ and $t>0$ or $0<t<1$, the functions $x_1(\tau)$ and $x_2(t-\tau)$ overlap as shown in Fig. Ex. 5.5(e).

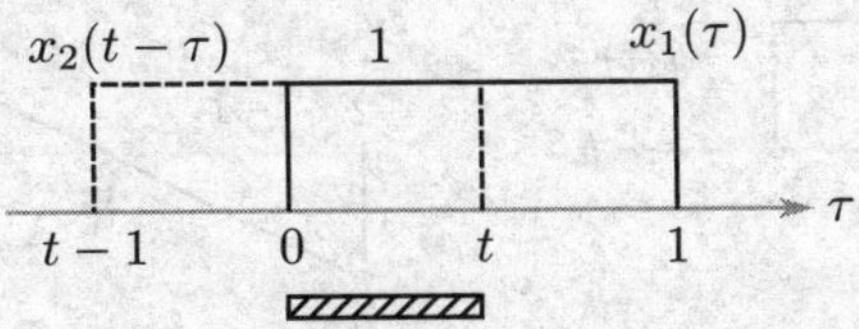

Fig. Ex. 5.5(e)

$$y(t)=\int_0^t 1\times 1 d\tau = t$$

Case (iii): For $t-1<1$ and $t>1$ or $1<t<2$, the functions $x_1(\tau)$ and $x_2(t-\tau)$ overlap as shown in Fig. Ex. 5.3(f).

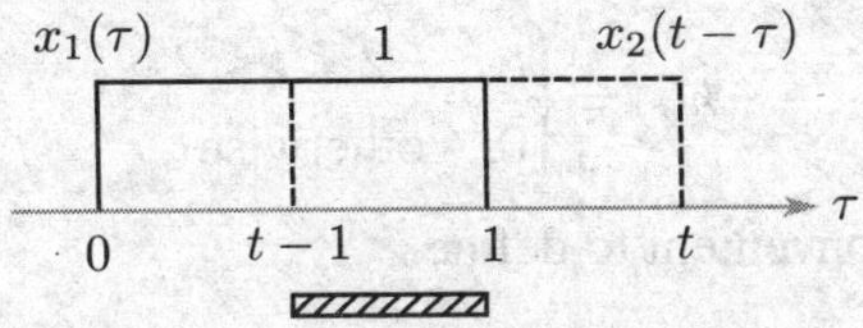

Fig. Ex. 5.5(f)

$$y(t)=\int_{t-1}^{1} 1\times 1 d\tau = 2-t$$

Case (iv): For $t-1>1$ or $t>2$, the functions $x_1(\tau)$ and $x_2(t-\tau)$ do not overlap.

Hence, $y(t)=0$.

Summarizing the results, we have

$$y(t)=\begin{cases}0, & t<0\\ t, & 0<t<1\\ 2-t, & 1<t<2\\ 0, & t>2\end{cases}$$

The output $y(t)$ is sketched in Fig. Ex. 5.5(g).

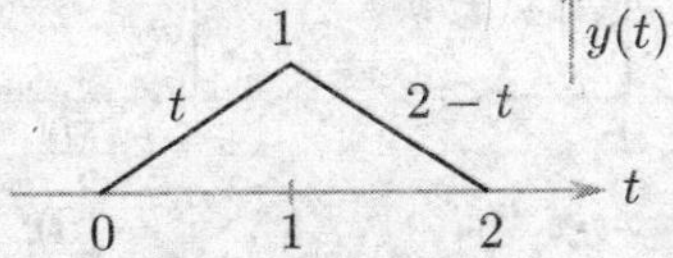

Fig. Ex. 5.5(g)

Example 5.6. Find the convolution of $x(t)$ and $h(t)$ sketched in Fig. Ex. 5.6.

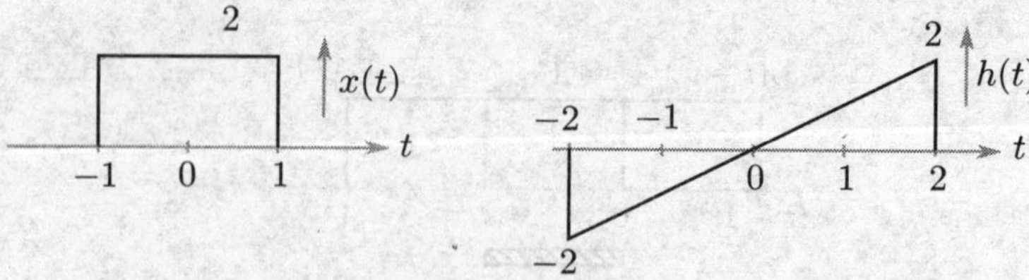

Fig. Ex. 5.6

Solution: To find expression for $h(t)$:

The equation of a straight line passing through the origin is $y = mx$.

Here,

$$y = h(t), \quad x = t \quad \text{and} \quad m = \tan\theta = \frac{2}{2} = 1$$

Thus,

$$h(t) = \begin{cases} t, & -2 < t < 2 \\ 0, & \text{otherwise} \end{cases}$$

For this problem, it is convenient to define

$$y(t) = h(t) * x(t)$$
$$= \int_{-\infty}^{\infty} h(\tau)x(t-\tau)d\tau$$

Hence, we require the sketches of $h(\tau)$ and $x(t-\tau)$.

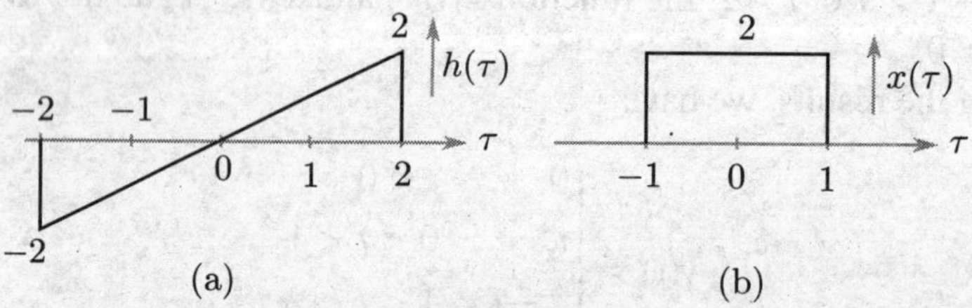

Fig. Ex. 5.6(a): *Sketch of* $h(\tau)$ **(b):** *Signal,* $x(\tau)$

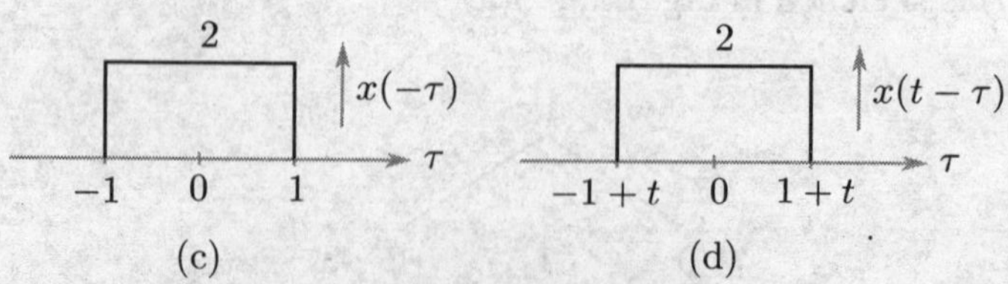

Fig. Ex. 5.6(c): *Folded signal,* $x(-\tau)$ **(d):** *Folded and shifted signal,* $x(t-\tau)$

Case (i): For $t-1 < -2$ and $t+1 > -2$ or $-3 < t < -1$, referring Fig. 5.6(e), the convolution integral becomes

$$y(t) = \int_{-2}^{t+1} 2\tau d\tau = t^2 + 2t - 3$$

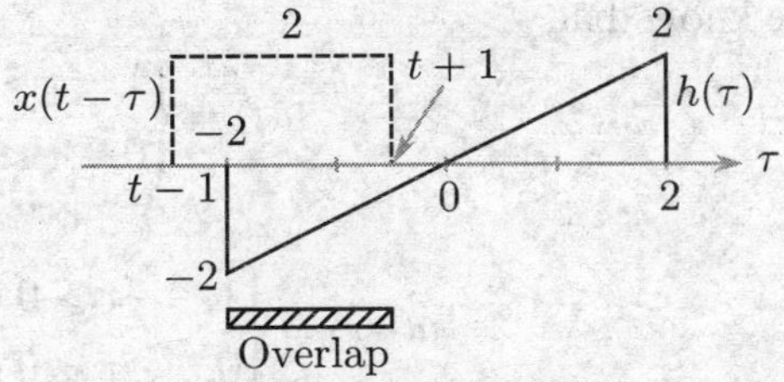

Fig. Ex. 5.6(e)

Case (ii): For $t - 1 < 0$ and $t + 1 > 0$ or $-1 < t < 1$, we get

$$y(t) = \int_{t-1}^{t+1} 2\tau d\tau = 4t$$

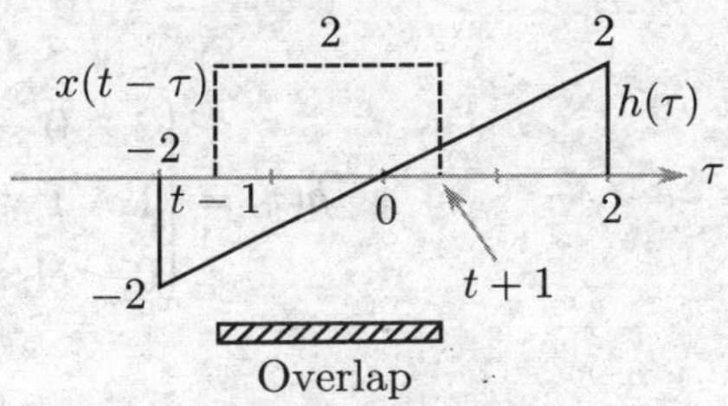

Fig. Ex. 5.6(f)

Case (iii): For $1 < t < 3$, we get

$$y(t) = \int_{t-1}^{2} 2\tau d\tau = -t^2 + 2t + 3$$

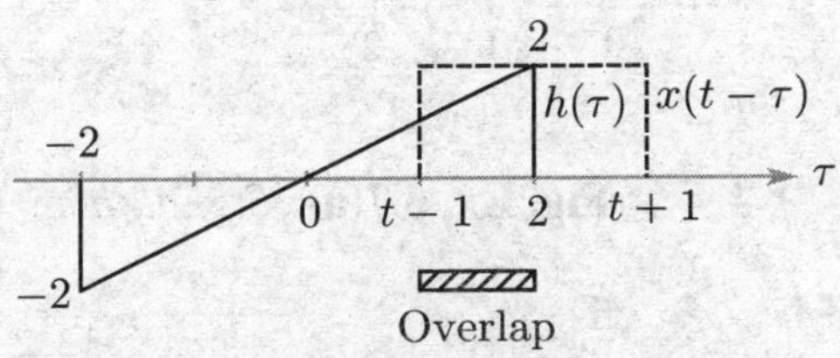

Fig. Ex. 5.6(g)

Case (iv): For $t > 3$ and for $t < -3$, $y(t) = 0$.

The sketch of $y(t) = x(t) * h(t)$ is as shown is Fig. 5.6(h).

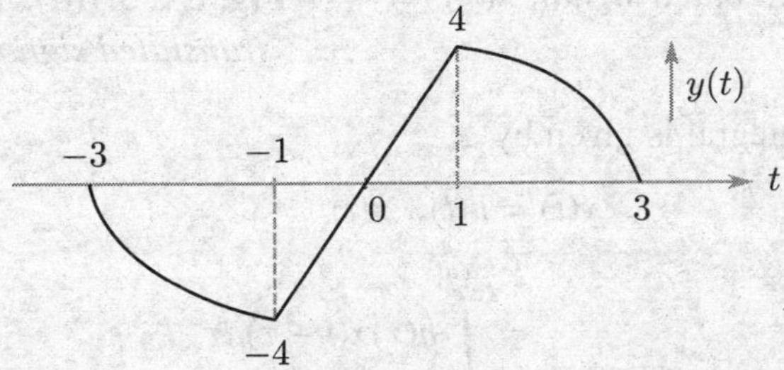

Fig. Ex. 5.6(h)

Example 5.7. A linear time-invariant system has an impulse response given by $h(t) = 2u(t) - u(t-1) - u(t-2)$. Find the system response for $x(t) = e^t u(-t)$

Solution: We know that,

$$u(t) = \begin{cases} 1, & t > 0 \\ 0, & t < 0 \end{cases}$$

Hence,

$$u(-\tau) = \begin{cases} 1, & -\tau > 0 \text{ or } \tau < 0 \\ 0, & -\tau < 0 \text{ or } \tau > 0. \end{cases}$$

Hence,

$$x(\tau) = \begin{cases} e^{\tau}, & \tau < 0 \\ 0, & \tau > 0 \end{cases}$$

Also,

$$h(\tau) = \begin{cases} 2, & 0 < \tau < 1 \\ 1, & 1 < \tau < 2 \\ 0, & \text{elsewhere} \end{cases}$$

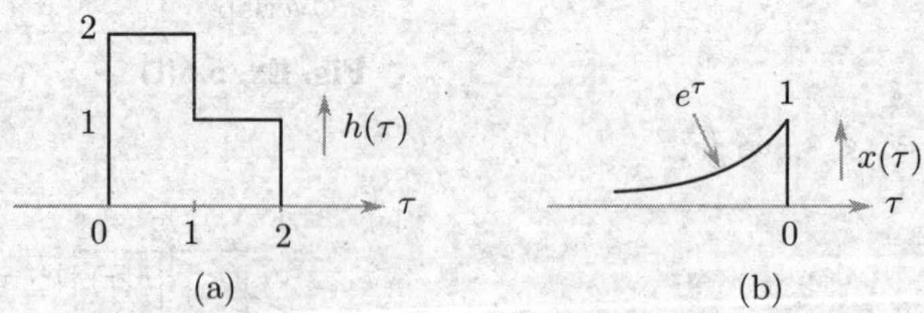

Fig. Ex. 5.7(a): *Signal, $h(\tau)$* **(b):** *Signal, $x(\tau)$*

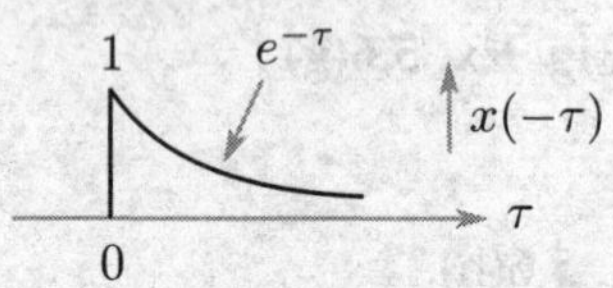

Fig. Ex. 5.7(c): *Folded signal, $x(-\tau)$*

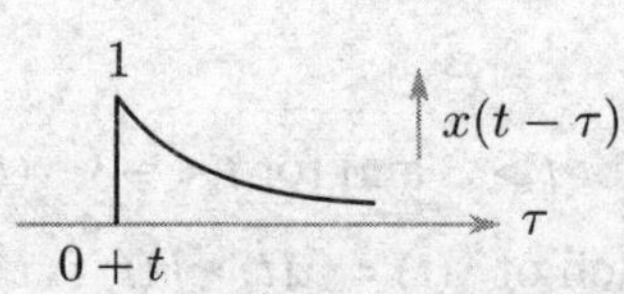

Fig. Ex. 5.7(d): *Folded and translated signal, $x(t-\tau)$*

The convolution integral is given by

$$y(t) = h(t) * x(t)$$

$$= \int_{-\infty}^{\infty} h(\tau)x(t-\tau)d\tau$$

Let us now evaluate the above convolution integral for different values of t.

(i) For $t > 2$, the signals $h(\tau)$ and $x(t-\tau)$ do not everlap. Hence, $y(t) = 0$.

(ii) For $1 < t < 2$, the signals $h(\tau)$ and $x(t-\tau)$ overlap as shown in Fig. Ex. 5.7(e).

We know that,

$$x(t) = \begin{cases} e^t, & t < 0 \\ 0, & t > 0 \end{cases}$$

Hence,

$$x(t-\tau) = \begin{cases} e^{t-\tau}, & t-\tau < 0 \text{ or } \tau > t \\ 0, & \tau < t \end{cases}$$

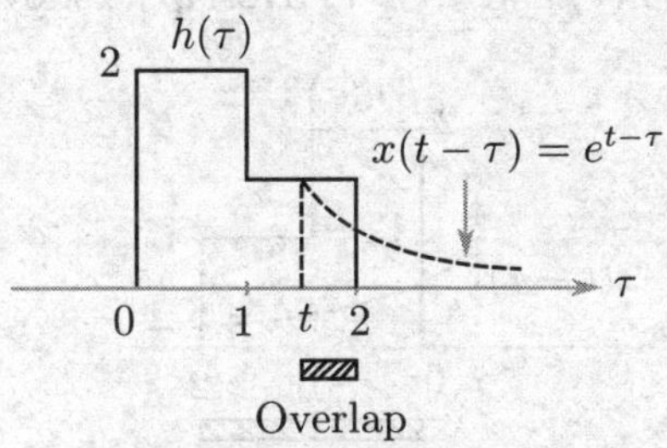

Fig. Ex. 5.7(e)

We have,

$$y(t) = \int_t^2 1 \times e^{t-\tau} d\tau$$

$$= e^t \left[\frac{e^{-\tau}}{-1}\right]_t^2$$

$$= 1 - e^{t-2}$$

(iii) For $0 < t < 1$, the signals $h(\tau)$ and $x(t-\tau)$ overlap as shown in Fig. Ex. 5.7(f).

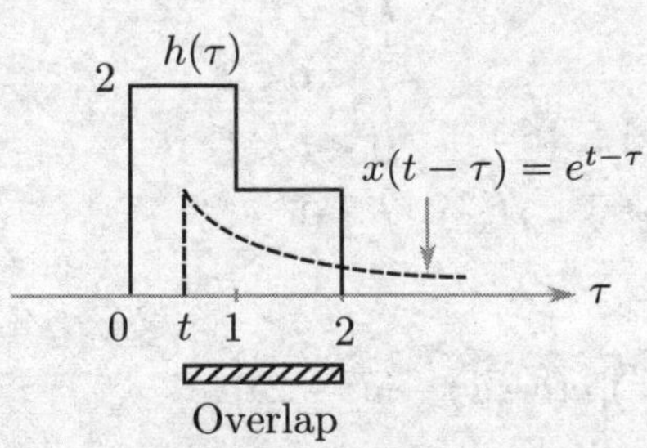

Fig. Ex. 5.7(f)

For this case,

$$y(t) = \int_t^2 h(\tau)x(t-\tau)d\tau$$

$$= \int_t^1 2 \times e^{t-\tau} d\tau + \int_1^2 1 \times e^{t-\tau} d\tau$$

$$= 2e^t \left[\frac{e^{-\tau}}{-1}\right]_t^1 + e^t \left[\frac{e^{-\tau}}{-1}\right]_1^2 = 2 - e^{t-1} - e^{t-2}$$

(iv) For $t < 0$, the functions $h(\tau)$ and $x(t-\tau)$ overlap as shown in Fig. Ex. 5.7(g).

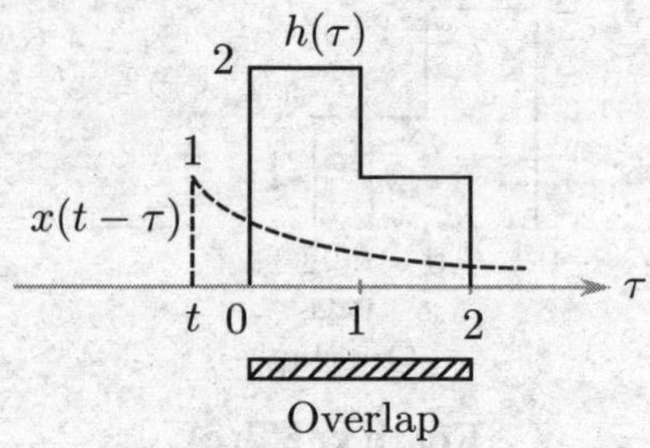

Fig. Ex. 5.7(g)

$$y(t) = \int_0^1 2 \times e^{t-\tau} d\tau + \int_1^2 1 \times e^{t-\tau} d\tau$$

$$= 2\left[e^t - e^{t-1}\right] + e^t\left[e^{-1} - e^{-2}\right]$$

$$= 2e^t - e^{t-1} - e^{t-2}$$

Summarizing,

$$y(t) = \begin{cases} 0, & t > 2 \\ 1 - e^{t-2}, & 1 < t < 2 \\ 2 - e^{t-1} - e^{t-2}, & 0 < t < 1 \\ 2e^t - e^{t-1} - e^{t-2}, & t < 0 \end{cases}$$

or

$$y(t) = (1 - e^{t-2})[u(t-1) - u(t-2)]$$

$$+ (2 - e^{-t-1} - e^{t-2})[u(t) - u(t-1)] + (2e^t - e^{t-1} - e^{t-2})u(-t)$$

Example 5.8. A linear time-invariant system has an impulse response given by

$$h(t) = \begin{cases} 0, & t < -4 \\ e^{-|t|}, & -4 \le t \le 4 \\ 0, & t > 4 \end{cases}$$

Find the response of the system to an input $x(t) = 2u(t+2) - 2u(t-2)$.

Solution: We have,

$$x(t) = 2u(t+2) - 2u(t-2)$$
$$= \begin{cases} 2, & -2 < t < 2 \\ 0, & \text{elsewhere} \end{cases}$$

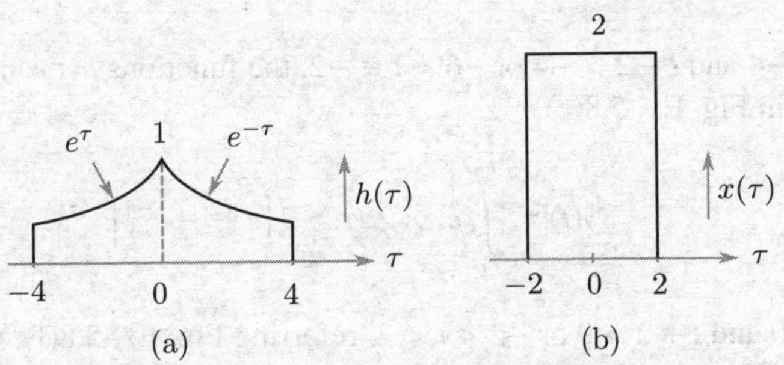

Fig. Ex. 5.8(a): *Signal,* $h(\tau)$ **(b):** *Signal,* $x(\tau)$

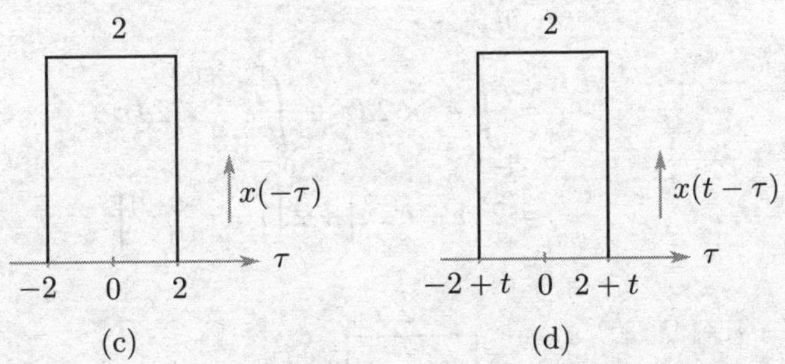

Fig. Ex. 5.8(c): *Folded signal,* $x(-\tau)$ **(d):** *Folded and shifted signal,* $x(t-\tau)$

Let

$$y(t) = h(t) * x(t)$$
$$= \int_{-\infty}^{\infty} h(\tau)x(t-\tau)d\tau$$

Let us now proceed to evaluate the above convolution integral, $y(t)$ for different values of t.

(i) For $t+2 < -4$ or $t < -6$, the signals $h(\tau)$ and $x(t-\tau)$ do not overlap. Hence, $y(t) = 0$.

(ii)

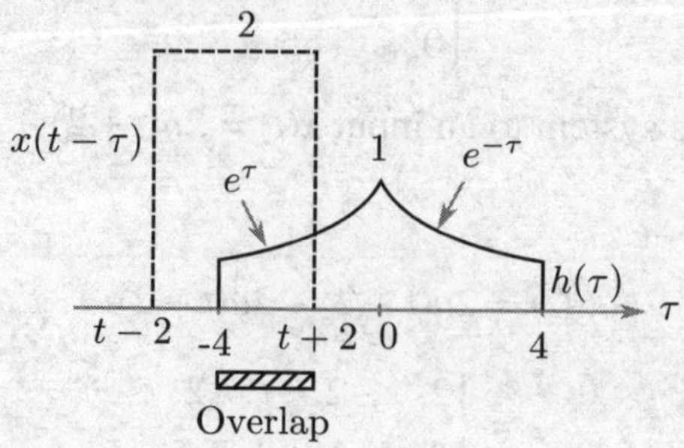

Fig. Ex. 5.8(e)

For $t-2 < -4$ and $t+2 > -4$ or $-6 < t < -2$, the functions $h(\tau)$ and $x(t-\tau)$ overlap as depicted in Fig. Ex. 5.8(e).

$$y(t) = \int_{-4}^{t+2} e^{\tau} \times 2d\tau = 2\left[e^{t+2} - e^{-4}\right]$$

(iii) For $t-2 < 0$ and $t+2 > 0$ or $-2 < t < 2$, referring Fig. Ex. 5.8(f), we have

$$y(t) = \int_{t-2}^{t+2} h(\tau)x(\tau)d\tau$$

$$= \int_{t-2}^{0} e^{\tau} \times 2d\tau + \int_{0}^{t+2} e^{-\tau} \times 2d\tau$$

$$= 2\left[1 - e^{t-2}\right] + 2\left[1 - e^{-(t+2)}\right]$$

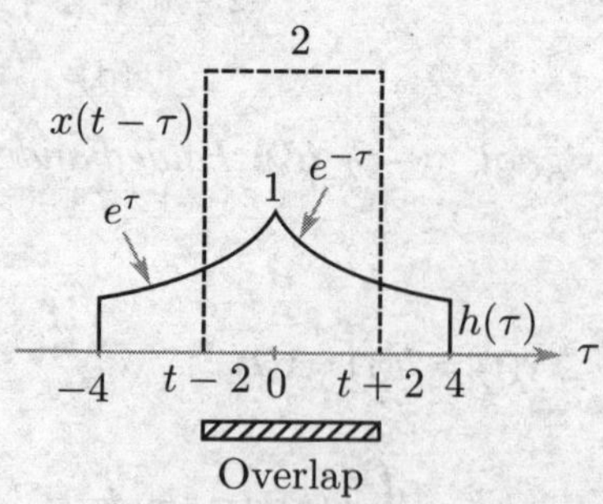

Fig. Ex. 5.8(f)

(iv)

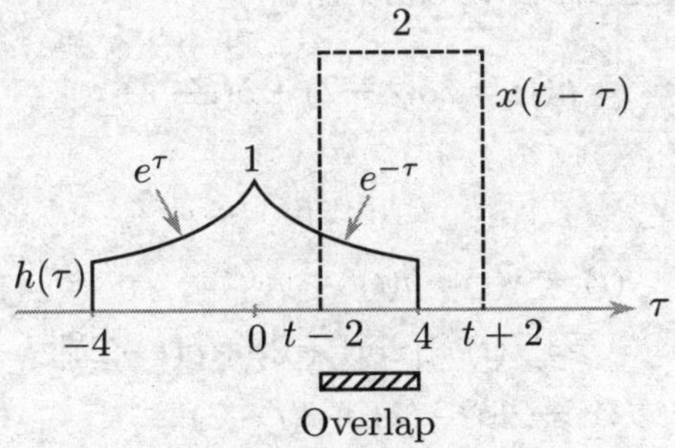

Fig. Ex. 5.8(g)

For $t - 2 < 4$ and $t + 2 > 4$ or $2 < t < 6$, the functions $h(\tau)$ amd $x(t - \tau)$ overlap as shown in Fig. Ex. 5.8(g).

$$y(t) = \int_{t-2}^{4} h(\tau)x(t-\tau)d\tau$$

$$= \int_{t-2}^{4} e^{-\tau} \times 2d\tau = 2\left[e^{-(t-2)} - e^{-4}\right]$$

(v) For $t - 2 > 4$ or $t > 6$, the functions $x(t - \tau)$ and $h(\tau)$ do not overlap. Hence, $y(t) = 0$.

Summarizing the results, we get

$$y(t) = \begin{cases} 0, & t < -6 \\ 2\left[e^{t+2} - e^{-4}\right], & -6 < t < -2 \\ 2[1 - e^{t-2}] + 2\left[1 - e^{-(t+2)}\right], & -2 < t < 2 \\ 2\left[e^{-(t-2)} - e^{-4}\right], & 2 < t < 6 \\ 0, & t > 6 \end{cases}$$

Example 5.9. Sketch the convolution, $y(t) = x(t) * h(t)$. The sketches of $x(t)$ and $h(t)$ are as shown in Fig. Ex. 5.9.

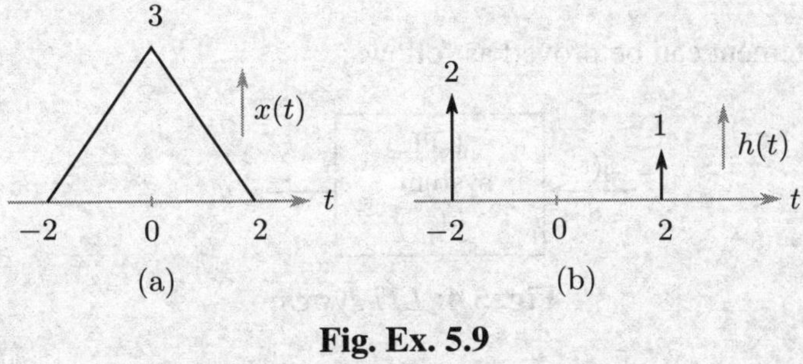

Fig. Ex. 5.9

Solution: Referring Fig. Ex. 5.9(b), we can write

$$h(t) = 2\delta(t+2) + \delta(t-2)$$

Let

$$\begin{aligned} y(t) &= x(t) * h(t) \\ &= x(t) * [2\delta(t+2) + \delta(t-2)] \\ &= 2x(t+2) + x(t-2) \end{aligned}$$

The sketch of $y(t)$ is as shown in Fig. Ex. 5.9(c).

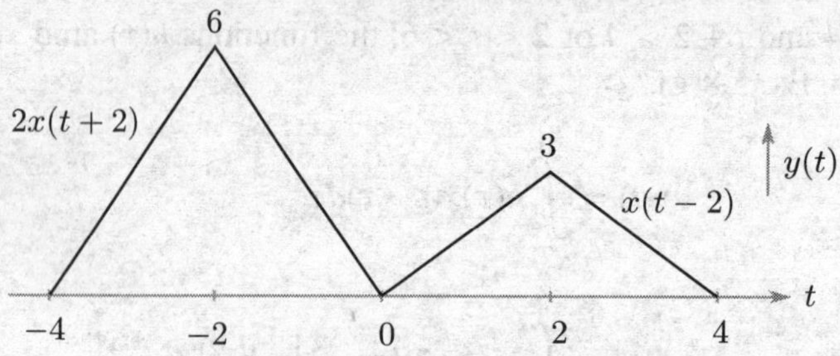

Fig. Ex. 5.9(c)

5.4 Properties of LTI systems

In chapter 4, four important properties: linearity, stability, causality and time-invariance are defined. In this section, we shall investigate stability and causality properties as related to linear time-invariant systems.

The input-output characteristics of an LTI system can be completely described by its impulse response $h(t)$. Hence, all properties of a system can be found using $h(t)$.

5.4.1 Causality

A linear time-invariant continuous-time system is causal if and only if its impulse response, $h(t)$ is zero for $t < 0$. That is, a causal impulse response, $h(t)$ automatically implies a causal system.

The above statement can be proved as follows:

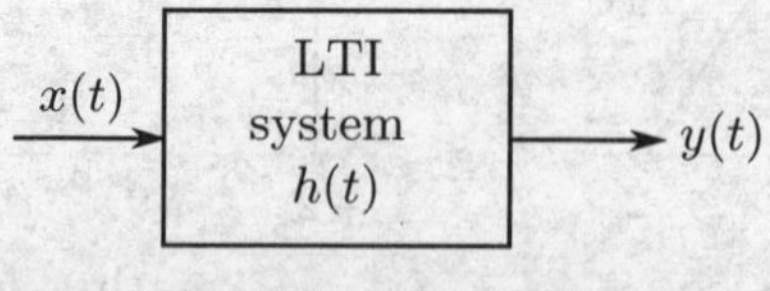

Fig. 5.4: *LTI system*

Let us consider an LTI system characterized by its impulse response, $h(t)$ as shown in Fig. 5.4. The response of this system to an arbitrary input $x(t)$ is given by

$$\begin{aligned} y(t) &= x(t) * h(t) \\ &= \int_{\tau=-\infty}^{\infty} x(\tau)h(t-\tau)d\tau \end{aligned} \tag{5.16}$$

$$\Rightarrow \quad y(t) = \int_{\tau=-\infty}^{t} x(\tau)h(t-\tau)d\tau + \int_{\tau=t}^{\infty} x(\tau)h(t-\tau)d\tau \tag{5.17}$$

If $h(t) = 0$ for $t < 0$, then $h(t-\tau) = 0$ for $t - \tau < 0$, which is equivalent to $h(t-\tau) = 0$ for $\tau > t$. Using this fact in Eq. (5.17), we find that the second integral on the right-hand side of Eq. (5.17) is zero. Thus, we can write $y(t)$ as

$$y(t) = \int_{-\infty}^{t} x(\tau)h(t-\tau)d\tau \tag{5.18}$$

Form Eq. (5.18), it is clear that the output, $y(t)$ at any time t is a weighted integral of the values of the input $x(\tau)$ for $\tau \leq t$, that is, only the present and past values of the input. Hence, the system is causal by the use of definition for Causality. This ends the proof.

5.4.2 Stability

A linear time-invariant continuous-time system is BIBO stable, if and only if its impulse response, $h(t)$ is absolutely integrable. That is, $\int_{-\infty}^{\infty} |h(\tau)|d\tau < \infty$ for BIBO stability.

The above statement can be proved as follows:
Refer Fig. 5.4. The response of the LTI system shown in Fig. 5.4 is

$$\begin{aligned} y(t) &= x(t) * h(t) \\ &= h(t) * x(t) \\ &= \int_{-\infty}^{\infty} h(\tau)x(t-\tau)d\tau \end{aligned} \tag{5.19}$$

Taking magnitudes on both sides of Eq. (5.19) gives

$$|y(t)| = \left| \int_{-\infty}^{\infty} h(\tau)x(t-\tau)d\tau \right| \tag{5.20}$$

Since, magnitude of an integral is less than or equal to the integral of the magnitude, Eq. (5.20) can be written as

$$|y(t)| \leq \int_{-\infty}^{\infty} |h(\tau)x(t-\tau)|d\tau \tag{5.21}$$

But, the magnitude of a product is the product of the magnitudes. Hence, we can write Eq. (5.21) as

$$|y(t)| \leq \int_{-\infty}^{\infty} |h(\tau)||x(t-\tau)|d\tau \tag{5.22}$$

Since $x(t)$ is bounded, we have

$$|x(t)| \leq M_x < \infty$$

By translating a signal $x(t)$, its bound does not change. Hence, we can write

$$|x(t-\tau)| \leq M_x < \infty \tag{5.23}$$

Substituting Eq. (5.23) in Eq. (5.22), we get

$$|y(t)| \leq \int_{-\infty}^{\infty} |h(\tau)|d\tau \times M_x$$

Therefore, if $\int_{-\infty}^{\infty} |h(\tau)|d\tau$ is finite, then $|y(t)|$ for all t is given by

$$|y(t)| \leq M_y < \infty.$$

Thus, we have shown that the system is BIBO stable, since every bounded input produces a bounded output. Hence, the proof of the theorem.

Example 5.10. For each impulse response listed below, determine whether the corresponding system is: (i) Causal (ii) stable

(a) $h(t) = e^{-2|t|}$ (b) $h(t) = e^{2t}u(t-1)$

Solution:

(a) Given: $h(t) = e^{-2|t|}$

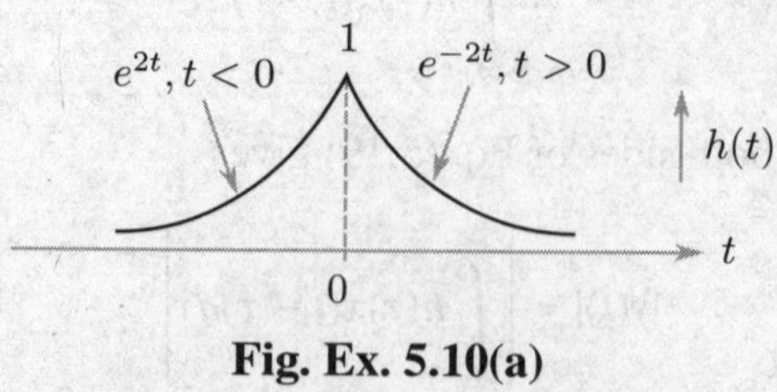

Fig. Ex. 5.10(a)

(i) Since $h(t)$ is not zero for $t < 0$, the system under investigation is anti-causal.

(ii) Let

$$S = \int_{-\infty}^{\infty} |h(t)| dt$$

Then,

$$S = \int_{-\infty}^{0} e^{2t} dt + \int_{0}^{\infty} e^{-2t} dt$$
$$= \frac{1}{2} + \frac{1}{2}$$
$$= 1$$

Since, S is finite, the system is BIBO stable.

(b) Given:

$$h(t) = e^{2t} u(t-1)$$

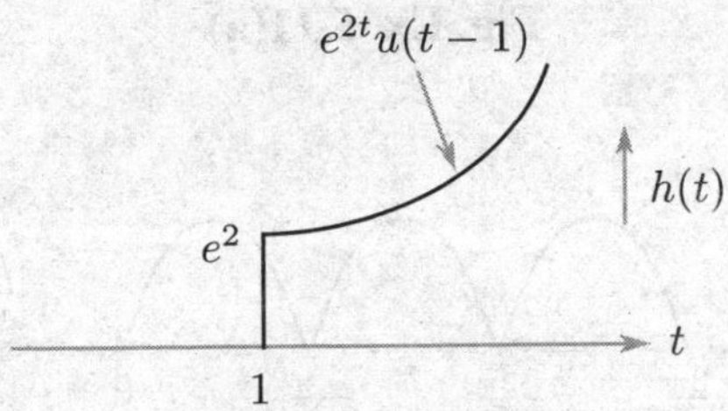

Fig. Ex. 5.10(b)

(i) The system under investigation is causal. This is because, $h(t)$ is zero for $t < 0$.

(ii) Let

$$S = \int_{-\infty}^{\infty} |h(t)| dt$$

Then,

$$S = \int_{1}^{\infty} e^{2t} dt = \infty$$

Since S is infinity, the system given in the problem is not BIBO stable.

Example 5.11. Determine whether the system given below for causality and stability. The systems are described by their respective impulse responses.

(a) $h(t) = \sin 5t u(-t)$ (b) $h(t) = e^{-t} u(-t-1)$

Solution:

(a) Since $u(-t) = \begin{cases} 1, & -t > 0 \text{ or } t < 0 \\ 0, & -t < 0 \text{ or } t > 0, \end{cases}$

We can write

$$h(t) = \sin 5tu(-t)$$
$$= \begin{cases} \sin 5t, & t < 0 \\ 0, & t > 0 \end{cases}$$

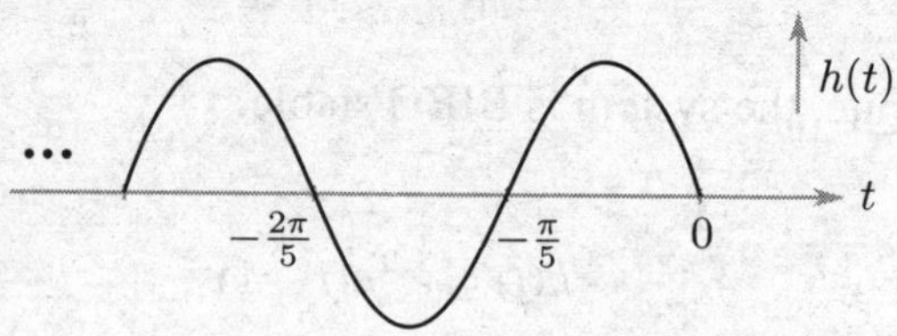

Fig. Ex. 5.11(a)

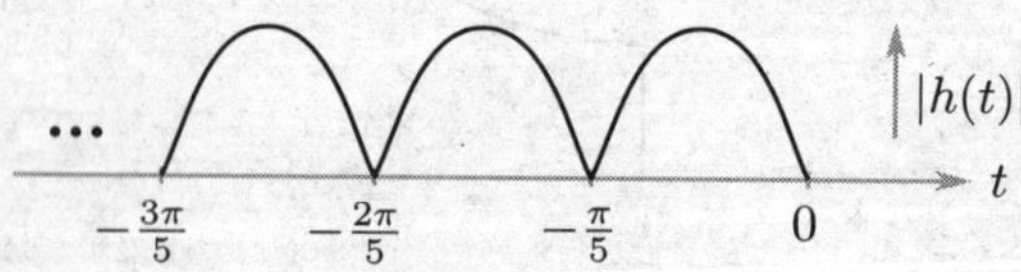

Fig. Ex. 5.11(b)

(i) Referring Fig. Ex. 5.11(a), we find that $h(t) \neq 0$ for $t < 0$. Hence, system is not causal.

(ii)

$$\int_{-\infty}^{\infty} |h(t)|dt = \int_{-\infty}^{0} |\sin 5t|dt = \infty$$

Since impulse response of the system is not absolutely integrable, system is not BIBO stable.

(b) Since,

$$u(-t-1) = \begin{cases} 1, & -t-1 > 0 \text{ or } t < -1 \\ 0, & -t-1 < 0 \text{ or } t > -1 \end{cases}$$

we can write,

$$h(t) = \begin{cases} e^{-t}, & t < -1 \\ 0, & t > -1 \end{cases}$$

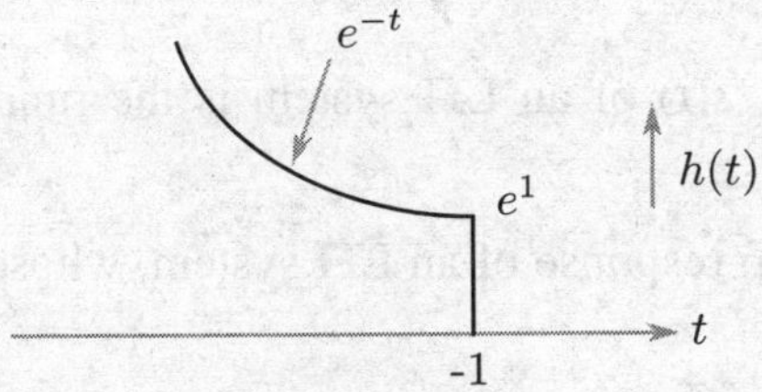

Fig. Ex. 5.11(c)

(i) Since, $h(t) \neq 0$ for $t < 0$, the system is not causal.

(ii) We have,

$$S = \int_{-\infty}^{\infty} |h(t)| dt = \int_{-\infty}^{-1} e^{-t} dt = \infty$$

Since, impulse response of the system is not absolutely integrable, system is not stable.

5.5 Step response of a linear time-invariant system

Let us consider a linear time-invariant system characterized by its impulse response, $h(t)$ as shown in Fig. 5.5.

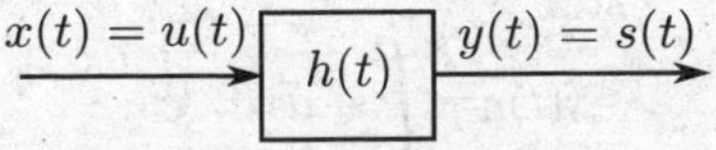

Fig. 5.5 *LTI system*

The step response of the LTI system shown in Fig. 5.5 is

$$\begin{aligned} s(t) &= u(t) * h(t) = h(t) * u(t) \\ &= \int_{-\infty}^{\infty} h(\tau) u(t - \tau) d\tau \end{aligned} \tag{5.24}$$

From the definition of step function, we have

$$u(t - \tau) = \begin{cases} 1, & t - \tau > 0 \text{ or } \tau < t \\ 0, & t - \tau < 0 \text{ or } \tau > t \end{cases} \tag{5.25}$$

Substituting Eq. (5.25) in Eq. (5.24), we get

$$s(t) = \int_{-\infty}^{t} h(\tau)d\tau \tag{5.26}$$

Hence, step response, $s(t)$ of an LTI system is the running integral of its impulse response, $h(t)$.

Example 5.12. Find the step response of an LTI system, whose impulse response is given by $h(t) = e^{-|t|}$.

Solution:

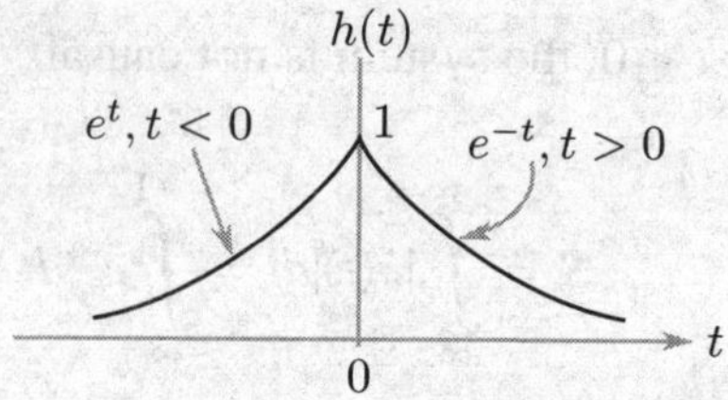

Fig. Ex. 5.12(a)

The step response is given by

$$s(t) = \int_{-\infty}^{t} h(\tau)d\tau$$

Case (i): For $t < 0$,

$$s(t) = \int_{-\infty}^{t} e^{\tau}d\tau = e^{t}$$

Case (ii): For $t > 0$,

$$s(t) = \int_{-\infty}^{0} e^{\tau}d\tau + \int_{0}^{t} e^{-\tau}d\tau$$

$$= 2 - e^{-t}$$

Summarizing the results, we can write

$$s(t) = \begin{cases} e^{t}, & t < 0 \\ 2 - e^{-t}, & t > 0 \end{cases}$$

It may be noted that $s(0) = 1$.

Example 5.13. Find the step response of an LTI system, whose impulse response is given by $h(t) = t^2u(t)$.

Solution:

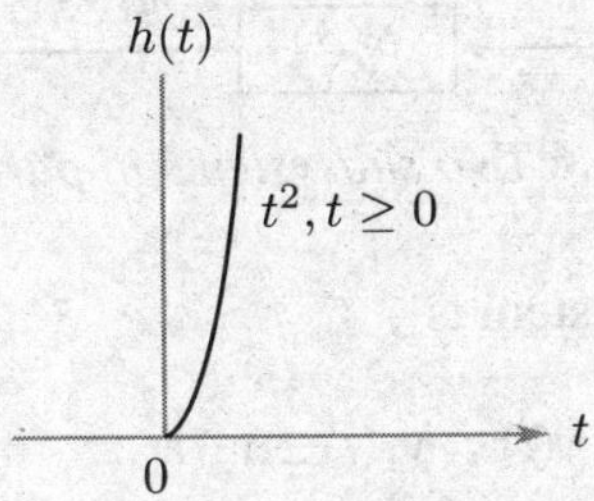

Fig. Ex. 5.13(a)

The step response is given by

$$s(t) = \int_{-\infty}^{t} h(\tau)d\tau$$

Case (i): For $t < 0$, $\quad s(t) = 0$

Case (ii): For $t > 0$,

$$s(t) = \int_{0}^{t} \tau^2 d\tau$$

$$= \frac{t^3}{3}$$

Hence,

$$s(t) = \begin{cases} 0, & t < 0 \\ \frac{t^3}{3}, & t > 0 \end{cases}$$

It may be noted that $s(0) = 0$.

5.6 Representation for LTI systems

5.6.1 Two systems in parallel

Let us consider, two subsystems having impulse responses $h_1(t)$ and $h_2(t)$ connected in parallel as shown in Fig. 5.6.

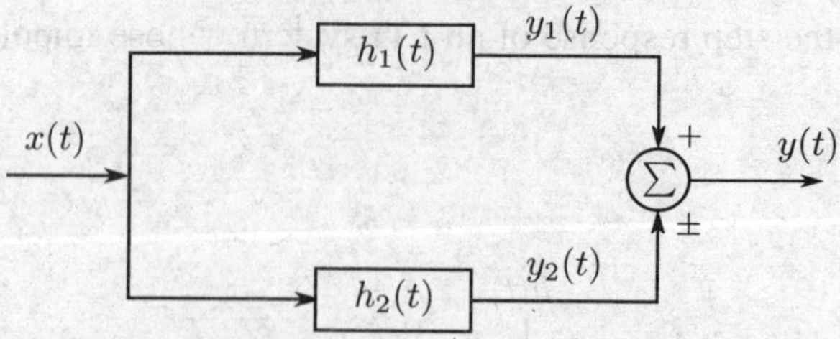

Fig. 5.6 *Two subsystems in parallel*

The response of the overall system is

$$y(t) = y_1(t) \pm y_2(t)$$

$$\Rightarrow \quad y(t) = x(t) * h_1(t) \pm x(t) * h_2(t)$$

Using distributive property of convolution integral, we can write

$$y(t) = x(t) * [h_1(t) \pm h_2(t)] \tag{5.27}$$

Translating Eq. (5.27) into a block diagram, we get the representation shown in Fig. 5.7.

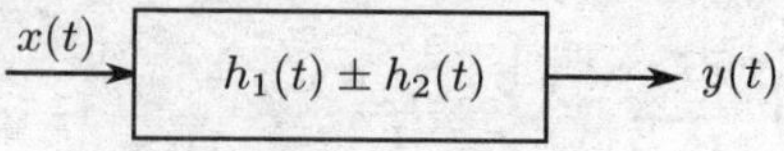

Fig. 5.7 *Overall system*

5.6.2 Two systems in cascade

Consider two subsystems having impulse responses $h_1(t)$ and $h_2(t)$ connected in cascade as shown in Fig. 5.8

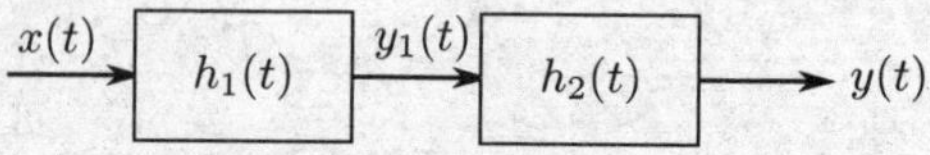

Fig. 5.8 *Two systems in cascade*

From Fig. 5.8, we can write

$$y_1(t) = x(t) * h_1(t)$$

Also,

$$y(t) = y_1(t) * h_2(t)$$

Therefore,

$$y(t) = x(t) * h_1(t) * h_2(t)$$

Using associative property of convolution, we can write

$$y(t) = x(t) * [h_1(t) * h_2(t)] \tag{5.28}$$

Translating Eq. (5.28) into a block diagram, we get block diagram shown in Fig. 5.9.

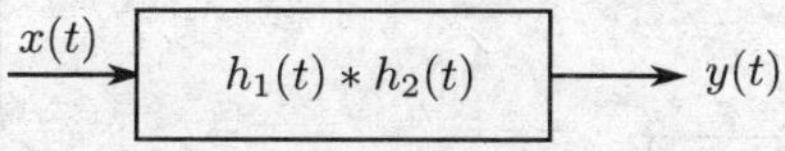

Fig. 5.9 *Overall system*

It follows that for m cascaded systems, impulse response of the total system is given by

$$h(t) = h_1(t) * h_2(t) * \ldots * h_m(t) \tag{5.29}$$

Example 5.14. Find the total response of the system shown in Fig. Ex. 5.14.

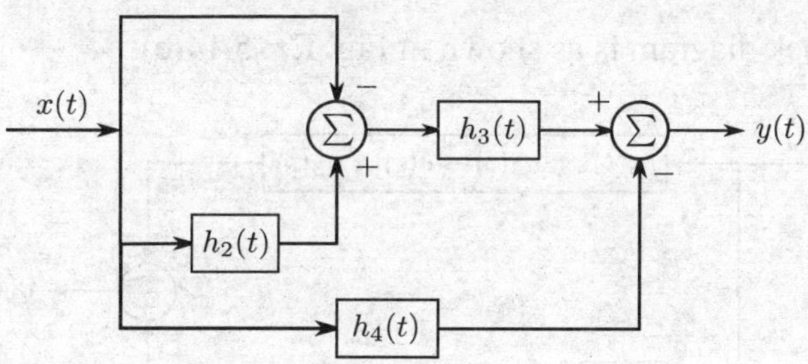

Fig. Ex. 5.14

Solution: For convolution operation, $\delta(t)$ is the identity operator. Hence, the given system can be rewritten as shown in Fig. Ex. 5.14(a).

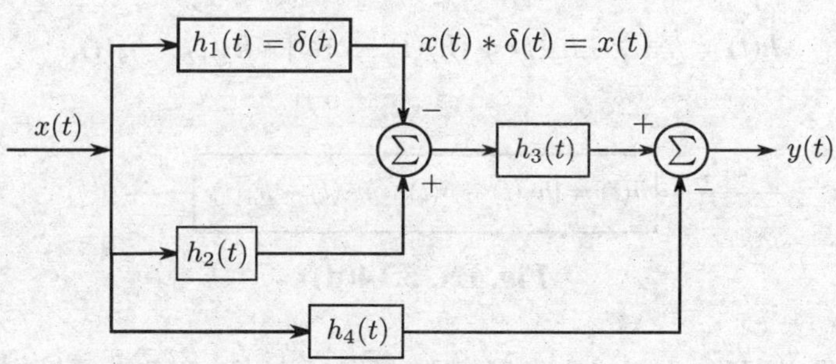

Fig. Ex. 5.14(a)

The impulse response of parallel systems 1 and 2 is given by

$$\begin{aligned} h_a(t) = h_{12}(t) &= h_2(t) - h_1(t) \\ &= h_2(t) - \delta(t) \end{aligned}$$

This results in Fig. Ex. 5.14(b).

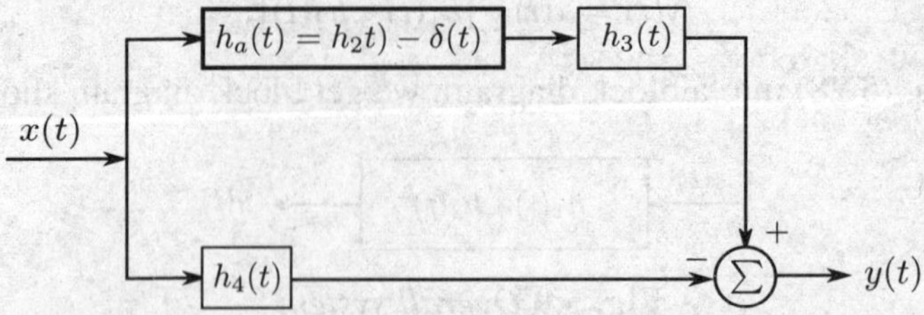

Fig. Ex. 5.14(b)

The cascade connection of system a and system 3 results in the impulse response given by

$$h_b(t) = h_{123}(t) = [h_2(t) - \delta(t)] * h_3(t)$$

The resultant block diagram is as shown in Fig. Ex. 5.14(c).

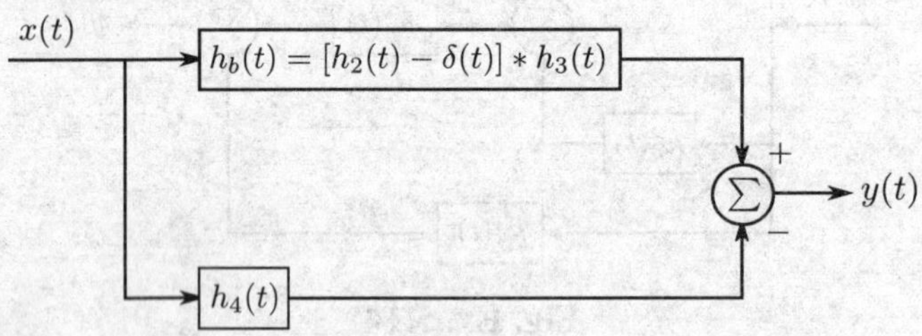

Fig. Ex. 5.14(c)

We add $h_b(t)$ and $h_4(t)$ to give the overall impulse response, as shown in Fig. Ex. 5.14(d).

$$h(t) = h_b(t) + h_4(t) = [h_2(t) - \delta(t)] * h_3(t) - h_4(t)$$

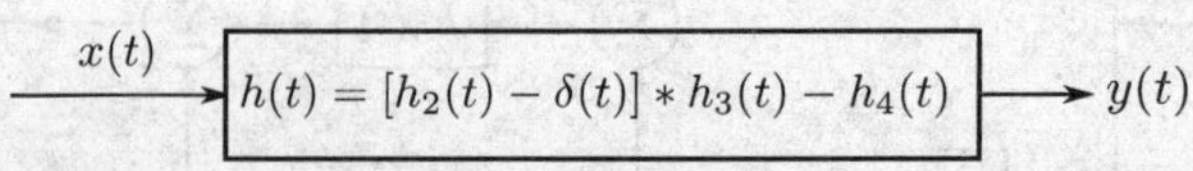

Fig. Ex. 5.14(d)

Example 5.15. The LTI system shown in Fig. Ex. 5.15 is made up of four sub-systems having the following impulse responses:

$$h_1(t) = \delta(t), \qquad h_2(t) = \delta(t-1)$$

$$h_3(t) = \delta(t-2), \quad h_4(t) = \delta(t+1), \quad h_5(t) = \delta(t)$$

Find the impulse response of the system.

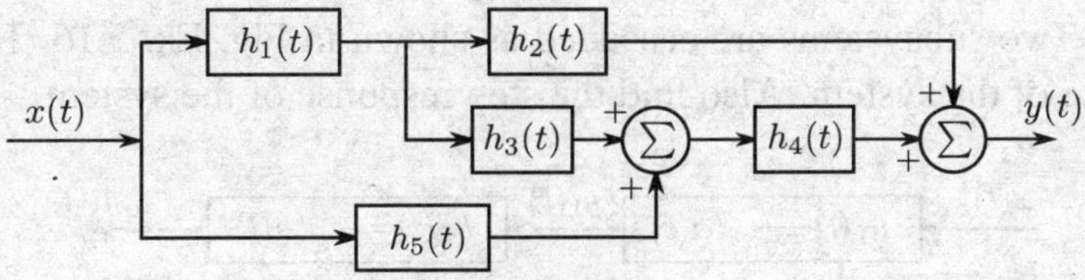

Fig. Ex. 5.15

Solution: Let $y_i(t)$ be the output of the ith subsystem.

We have,

$$\begin{aligned}
y_1(t) &= x(t) * h_1(t) \\
y_2(t) &= y_1(t) * h_2(t) = x(t) * h_1(t) * h_2(t) \\
&= h_1(t) * h_2(t) * x(t) \\
y_3(t) &= y_1(t) * h_3(t) = x(t) * h_1(t) * h_3(t) \\
&= h_1(t) * h_3(t) * x(t) \\
y_5(t) &= x(t) * h_5(t) \\
y_4(t) &= [y_3(t) + y_5(t)] * h_4(t) \\
&= [h_1(t) * h_3(t) * x(t) + x(t) * h_5(t)] * h_4(t) \\
&= [h_1(t) * h_3(t) + h_5(t)] * h_4(t) * x(t)
\end{aligned}$$

Hence,

$$\begin{aligned}
y(t) &= y_2(t) + y_4(t) \\
&= \{h_1(t) * h_2(t) + [h_1(t) * h_3(t) + h_5(t)] * h_4(t)\} * x(t)
\end{aligned}$$

Thus, the impulse response of the system is

$$h(t) = h_1(t) * h_2(t) + h_1(t) * h_3(t) * h_4(t) + h_5(t) * h_4(t)$$

$$\Rightarrow \qquad h(t) = \delta(t) * \delta(t-1) + \delta(t) * \delta(t-2) * \delta(t+1) + \delta(t) * \delta(t+1)$$

Recall:

$$\delta(t-\alpha) * \delta(t-\beta) = \delta(t-\alpha-\beta)$$

Hence,

$$\begin{aligned}
h(t) &= \delta(t-1) + \delta(t-2) * \delta(t+1) + \delta(t+1) \\
&= \delta(t-1) + \delta(t-1) + \delta(t+1) \\
&= 2\delta(t-1) + \delta(t+1)
\end{aligned}$$

Example 5.16. Two subsystems are cascaded as shown in Fig. Ex. 5.16. Find the overall impulse response of the system. Also find the step response of the system.

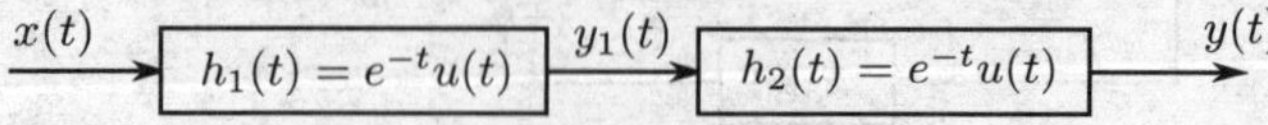

Fig. Ex. 5.16

Solution: The overall impulse response of the system is

$$h(t) = h_1(t) * h_2(t)$$

$$= \int_{-\infty}^{\infty} h_1(\tau)h_2(t-\tau)d\tau$$

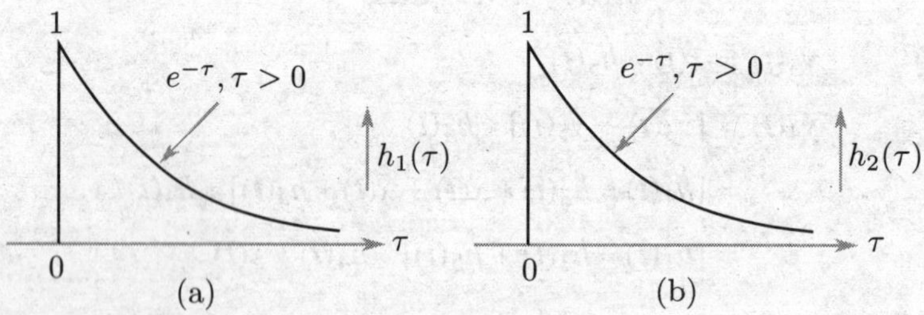

Fig. Ex. 5.16(a): *Signal,* $h_1(\tau)$ **(b):** *Signal,* $h_2(\tau)$

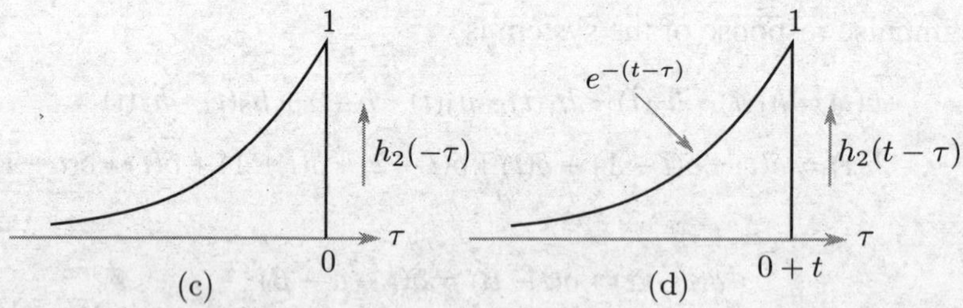

Fig. Ex. 5.16(c): *Folded signal,* $h_2(-\tau)$ **(d):** *Folded and shifted signal,* $h_2(t-\tau)$

(i) For $t < 0$, the signals $h_1(\tau)$ and $h_2(t-\tau)$ do not overlap. Hence, $h(t) = 0$.

(ii) For $t > 0$, the signals $h_1(\tau)$ and $h_2(t-\tau)$ overlap as shown in Fig. Ex. 5.16(e).

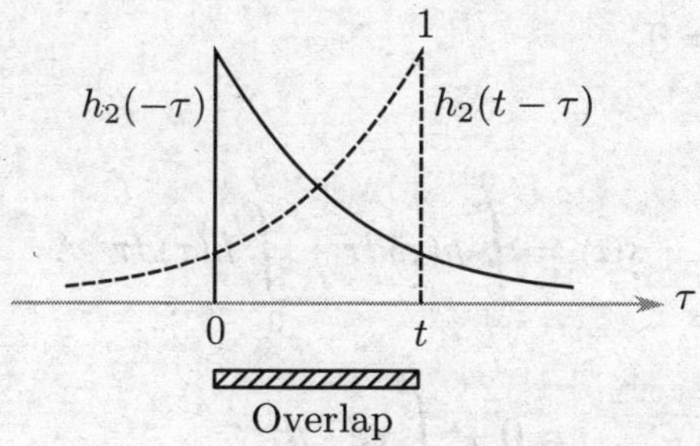

Fig. Ex. 5.16(e)

$$h(t) = \int_0^t e^{-\tau} \times e^{-(t-\tau)}\, d\tau = e^{-t} \int_0^t d\tau = te^{-t}$$

Summarizing, we get

$$h(t) = \begin{cases} te^{-t}, & t > 0 \\ 0, & t < 0 \end{cases} \quad \text{or} \quad h(t) = te^{-t}u(t)$$

Thus,

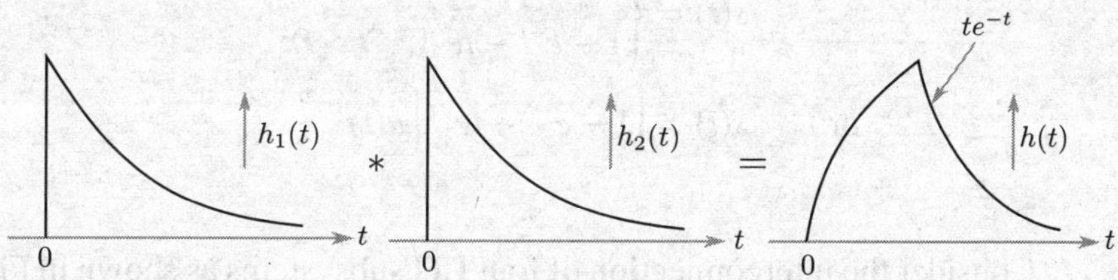

Fig. Ex. 5.16(f)

The step response of the system is

$$y(t) = s(t) = x(t) * h(t) = u(t) * h(t) = \int_{-\infty}^{t} h(\tau)d\tau \text{ (Running integral of } h(t))$$

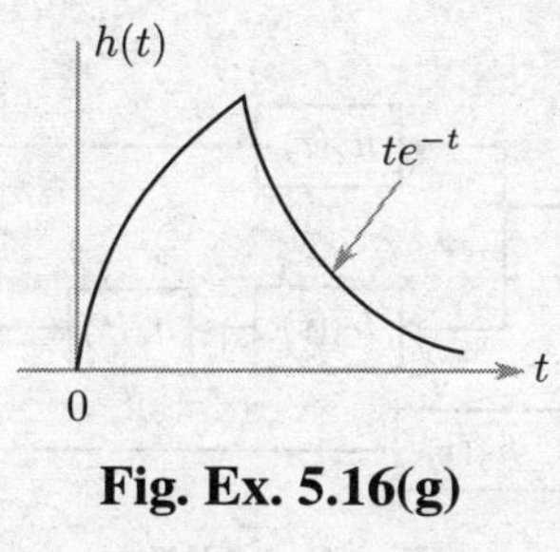

Fig. Ex. 5.16(g)

Refer Fig. Ex. 5.16(g) for the analysis.

Case (i): For $t < 0$, $s(t) = 0$

Case (ii): For $t > 0$,

$$\begin{aligned} s(t) &= \int_{-\infty}^{0} h(\tau)d\tau + \int_{0}^{t} h(\tau)d\tau \\ &= 0 + \int_{0}^{t} \tau e^{-\tau} d\tau \\ &= \tau \frac{e^{-\tau}}{-1}\bigg|_0^t - \int_0^t -e^{-\tau} \times 1 d\tau \\ &= -te^{-t} - e^{-\tau}\big|_0^t \\ &= -te^{-t} - e^{-t} + 1 \\ &= 1 - e^{-t} - te^{-t} \end{aligned}$$

Summarizing,

$$s(t) = \begin{cases} 0, & t < 0 \\ 1 - e^{-t} - te^{-t}, & t > 0 \end{cases}$$

or $$s(t) = \left[1 - e^{-t} - te^{-t}\right] u(t)$$

Example 5.17. Consider the interconnection of four LTI subsystems as shown in Fig. Ex. 5.17. Find the overall impulse response of the system. Also, find $h(t)$ when

$$\begin{aligned} h_1(t) &= 4\delta(t) \\ h_2(t) &= h_3(t) = (t + 1)u(t) \\ h_4(t) &= \delta(t - 1) \\ h_5(t) &= \delta(t) - 4\delta(t - 3) \end{aligned}$$

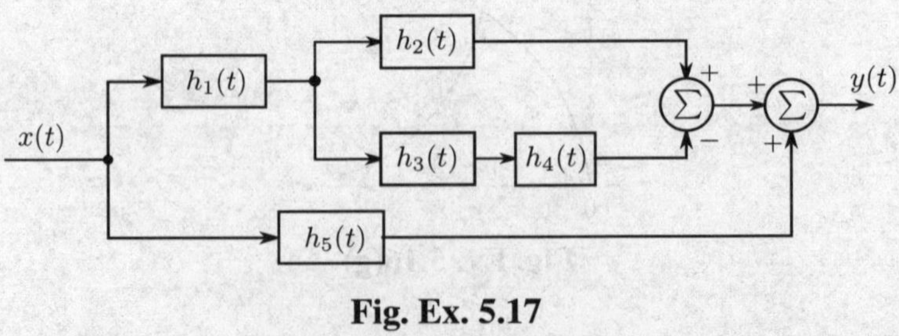

Fig. Ex. 5.17

Solution: Fig. Ex. 5.17 can be reduced to the block diagram shown in Fig. Ex. 5.17(a).

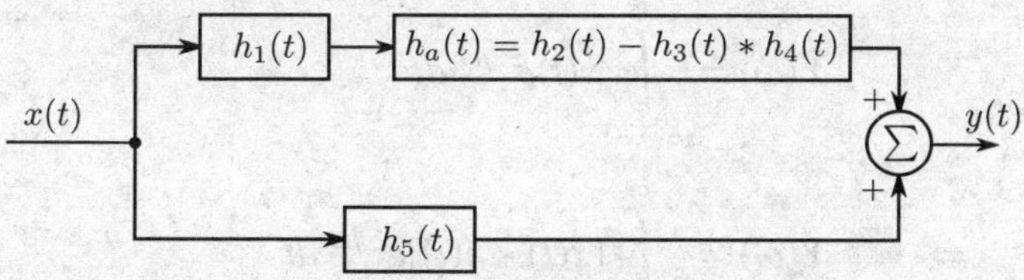

Fig. Ex. 5.17(a)

Hence, overall impulse response of the system is

$$h(t) = h_1(t) * h_a(t) + h_5(t)$$

Our immediate objective is to find $h_a(t)$.

$$\begin{aligned} h_a(t) &= h_2(t) - h_2(t) * h_4(t) \\ &= (t+1)u(t) - (t+1)u(t) * \delta(t-1) \end{aligned}$$

Recall the convolution property of an impulse function:

$$f(t) * \delta(t-\alpha) = f(t-\alpha)$$

Hence,

$$\begin{aligned} h_a(t) &= (t+1)u(t) - (t-1+1)u(t-1) \\ &= (t+1)u(t) - tu(t-1) \end{aligned}$$

Hence,

$$\begin{aligned} h(t) &= h_1(t) * h_a(t) + h_5(t) \\ &= 4\delta(t) * [(t+1)u(t) - tu(t-1)] + \delta(t) - 4\delta(t-3) \\ &= 4(t+1)u(t) - 4tu(t-1) + \delta(t) - 4\delta(t-3) \end{aligned}$$

5.7 Convolution using Fourier transforms

5.7.1 Time-domain convolution

If $x(t) \xleftrightarrow{\textbf{FT}} X(\omega)$ and $h(t) \xleftrightarrow{\textbf{FT}} H(\omega)$, then

$$y(t) = x(t) * h(t) \xleftrightarrow{\textbf{FT}} Y(\omega) = X(\omega)H(\omega)$$

This property can be proved as follows:

By definition,

$$Y(\omega) = \int_{t=-\infty}^{\infty} y(t)e^{-j\omega t}dt$$

$$\Rightarrow \quad Y(\omega) = \int_{t=-\infty}^{\infty} [x(t) * h(t)]e^{-j\omega t}dt$$

$$= \int_{t=-\infty}^{\infty}\int_{\tau=-\infty}^{\infty} x(\tau)h(t-\tau)d\tau\, e^{-j\omega t}dt$$

$$= \int_{\tau=-\infty}^{\infty} x(\tau)\left[\int_{t=-\infty}^{\infty} h(t-\tau)e^{-j\omega t}dt\right]d\tau$$

Put $t - \tau = \lambda$, then $dt = d\lambda$
Accordingly, we get

$$Y(\omega) = \int_{\tau=-\infty}^{\infty} x(\tau)\int_{\lambda=-\infty}^{\infty} h(\lambda)e^{-j\omega(\tau+\lambda)}d\lambda\, d\tau$$

$$= \int_{\tau=-\infty}^{\infty} x(\tau)e^{-j\omega\tau}d\tau \int_{\lambda=-\infty}^{\infty} h(\lambda)e^{-j\omega\lambda}d\lambda$$

$$= X(\omega)H(\omega)$$

Hence, the proof.

5.7.2 Frequency-domain Convolution (Modulation property)

If $x(t) \overset{\textbf{FT}}{\longleftrightarrow} X(\omega)$ and $h(t) \overset{\textbf{FT}}{\longleftrightarrow} H(\omega)$, then

$$y(t) = x(t)h(t) \overset{\textbf{FT}}{\longleftrightarrow} Y(\omega) = \frac{1}{2\pi}[X(\omega) * H(\omega)].$$

The frequency-domain convolution property can be proved as follows:
We know that

$$Y(\omega) = \int_{t=-\infty}^{\infty} y(t)e^{-j\omega t}dt$$

$$= \int_{t=-\infty}^{\infty} x(t)h(t)e^{-j\omega t}dt \qquad (5.28)$$

From the definition of inverse Fourier transform, we have

$$x(t) = \frac{1}{2\pi} \int_{\lambda=-\infty}^{\infty} X(\lambda) e^{j\lambda t} d\lambda \tag{5.29}$$

Substituting Eq. (5.29) in Eq. (5.28), we get

$$Y(\omega) = \int_{t=-\infty}^{\infty} \left[\frac{1}{2\pi} \int_{\lambda=-\infty}^{\infty} X(\lambda) e^{j\lambda t} d\lambda \right] h(t) e^{-j\omega t} dt$$

$$= \frac{1}{2\pi} \int_{\lambda=-\infty}^{\infty} X(\lambda) \int_{t=-\infty}^{\infty} h(t) e^{-j(\omega-\lambda)t}\, dt\, d\lambda$$

$$= \frac{1}{2\pi} \int_{\lambda=-\infty}^{\infty} X(\lambda) H(\omega - \lambda) d\lambda$$

$$= \frac{1}{2\pi} [X(\omega) * H(\omega)]$$

This completes the proof.

Corollary: *If* $x(t) \overset{\mathbf{FT}}{\longleftrightarrow} X(\omega)$, *then*

$$\int_{-\infty}^{t} x(\tau) d\tau \overset{\mathbf{FT}}{\longleftrightarrow} \frac{1}{j\omega} X(\omega) + \pi X(0) \delta(\omega)$$

The above property is known as integration property and is proved in article 3.4.6. This property is very useful for finding **FT** *of signals that are difficult to find using defining equation for* **FT**. *For example, let us find* **FT** *of* $y(t)$ *defined below.*

$$y(t) = \int_{-\infty}^{t} e^{-2|\tau|}\, d\tau \tag{5.30}$$

Let us identify $x(t) = e^{-2|t|}$.

The next step in the analysis is to find **FT** *of* $x(t)$. *Rewrite the double-sided exponential signal as*

$$x(t) = \begin{cases} e^{-2t}, & t > 0 \\ e^{2t}, & t < 0 \end{cases} \tag{5.31}$$

By definition,

$$X(\omega) = \int_{-\infty}^{\infty} x(t)e^{-j\omega t}\,dt$$

$$\Rightarrow X(\omega) = \int_{-\infty}^{0} e^{2t}e^{-j\omega t}\,dt + \int_{0}^{\infty} e^{-2t}e^{-j\omega t}\,dt$$

$$= \frac{4}{4+\omega^2}$$

Hence, $X(0) = 1$.
We know that,

$$y(t) = \int_{-\infty}^{t} x(\tau)d\tau \overset{\mathbf{FT}}{\longleftrightarrow} Y(\omega) = \frac{1}{j\omega}X(\omega) + \pi X(0)\delta(\omega)$$

Hence,

$$Y(\omega) = \frac{1}{j\omega}\left[\frac{4}{4+\omega^2}\right] + \pi \times 1 \times \delta(\omega)$$

$$= \frac{4}{j\omega(4+\omega^2)} + \pi\delta(\omega) \tag{5.32}$$

Thus, this property is very useful for finding **FT** *of signals that are integrals of signals whose* **FT** *exists.*

Example 5.18. Using Convolution theorem, find the inverse Fourier transform of $X(\omega) = \dfrac{1}{(a+j\omega)^2}$

Solution: Rewrite $X(\omega)$ as

$$X(\omega) = \frac{1}{a+j\omega} \times \frac{1}{a+j\omega}$$

$$= X_1(\omega)X_2(\omega) \tag{5.33}$$

where

$$X_1(\omega) = X_2(\omega) = \frac{1}{a+j\omega}$$

Taking inverse Fourier transform on both sides of Eq. (5.33) gives

$$x(t) = x_1(t) * x_2(t) \tag{5.34}$$

Recall:

$$e^{-at}u(t) \overset{\mathbf{FT}}{\longleftrightarrow} \frac{1}{a + j\omega}$$

Hence,

$$x_1(t) = x_2(t) = e^{-at}u(t)$$

Eq. (5.34) can also be written as

$$x(t) = \int_{\tau=-\infty}^{\infty} x_1(\tau)x_2(t-\tau)d\tau$$

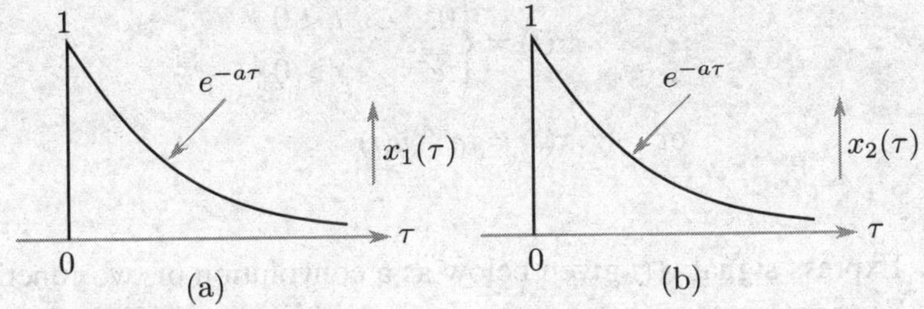

Fig. Ex. 5.18(a): *Signal,* $x_1(\tau)$ **(b):** *Signal,* $x_2(\tau)$

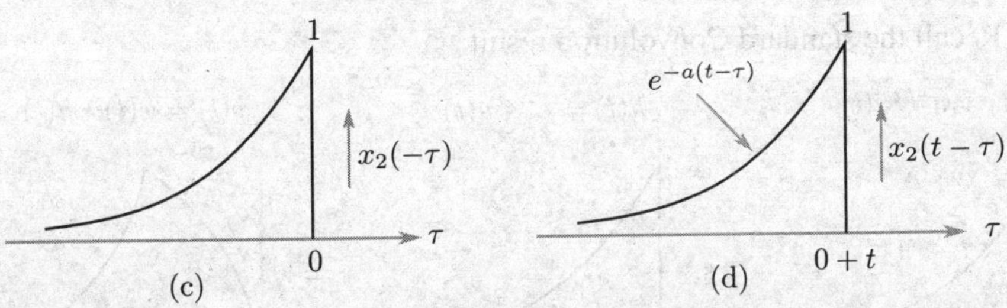

Fig. Ex. 5.18(c): *Folded signal,* $x_2(-\tau)$ **(d):** *Folded and shifted signal,* $x_2(t-\tau)$

Case (i): For $t < 0$, the signals $x_1(\tau)$ and $x_2(t-\tau)$ do not overlap. Hence, $x(t) = 0$.

Case (ii):

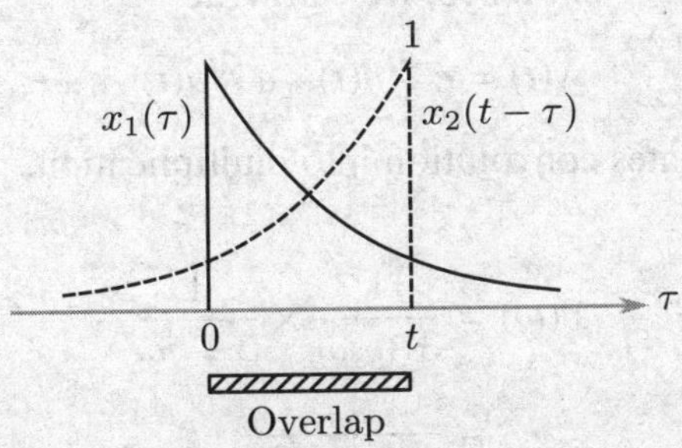

Fig. Ex. 5.18(e)

For $t > 0$, the functions $x_1(\tau)$ and $x_2(t-\tau)$ overlap as shown in Fig. Ex. 5.18(e). Consequently,

$$x(t) = \int_0^t e^{-a\tau} \times e^{-a(t-\tau)} d\tau$$

$$= e^{-at} \int_0^t d\tau = te^{-at}$$

Summarizing the convolution results, we get

$$x(t) = \begin{cases} 0, & t < 0 \\ te^{-at}, & t > 0 \end{cases}$$

or $$x(t) = te^{-at}u(t)$$

Example 5.19. Express signal $y(t)$ given below as a convolution of two functions $x(t)$ and $h(t)$ and find its Fourier transform $Y(\omega)$ as the product of $X(\omega)$ and $H(\omega)$.

$$y(t) = te^{-3t}u(t)$$

Solution: Recall the standard Convolution result:

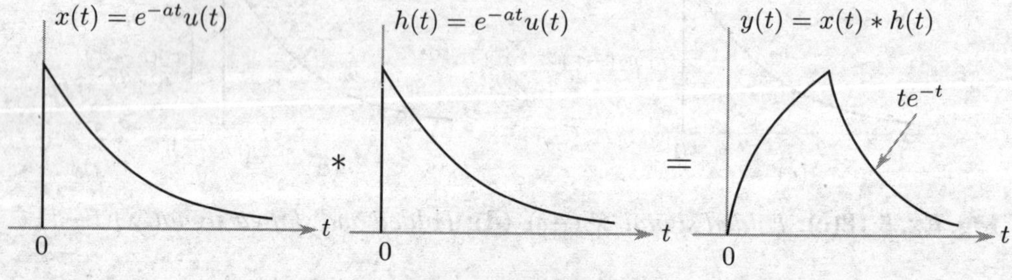

Fig. Ex. 5.19

In the present problem, $a = 3$. Hence, we can write

$$y(t) = e^{-3t}u(t) * e^{-3t}u(t)$$

Fourier transform translates convolution into multiplication.
Hence,

$$Y(\omega) = \frac{1}{3 + j\omega} \times \frac{1}{3 + j\omega}$$

$$= \frac{1}{(3 + j\omega)^2}$$

It may be noted that for finding Fourier transform of $e^{-3t}u(t)$, we have made use of the following Fourier transform pair:

$$e^{-at}u(t) \overset{\mathbf{FT}}{\longleftrightarrow} \frac{1}{a+j\omega}$$

Example 5.20. Let $X_1(\omega)$ and $X_2(\omega)$ be the Fourier transforms of $x_1(t)$ and $x_2(t)$, respectively. Find the Fourier transform $Y(\omega)$ of $y(t) = x_1(t)x_2(t)$. Given:

$$X_1(\omega) = \text{rect}(\omega/4) \quad \text{and} \quad X_2(\omega) = \text{rect}(\omega/4)$$

Solution: The spectra $X_1(\omega)$ and $X_2(\omega)$ are shown in Fig. Ex. 5.20(a).

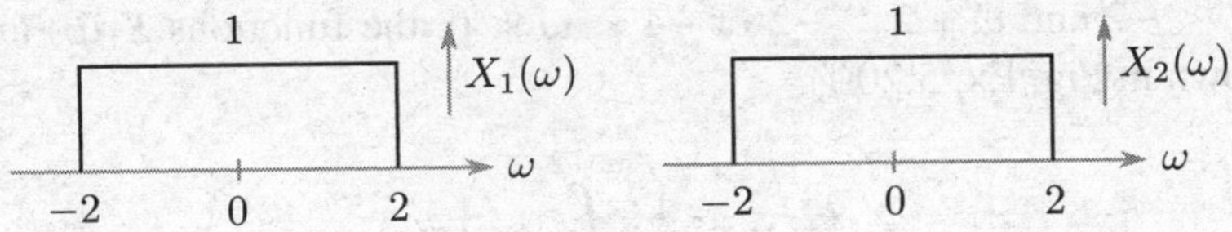

Fig. Ex. 5.20(a)

Given,

$$y(t) = x_1(t)x_2(t)$$

Taking Fourier transform on both sides, we get

$$Y(\omega) = \frac{1}{2\pi}[X_1(\omega) * X_2(\omega)]$$

$$\Rightarrow \quad Y(\omega) = \frac{1}{2\pi}\int_{\beta=-\infty}^{\infty} X_1(\beta)X_2(\omega-\beta)d\beta$$

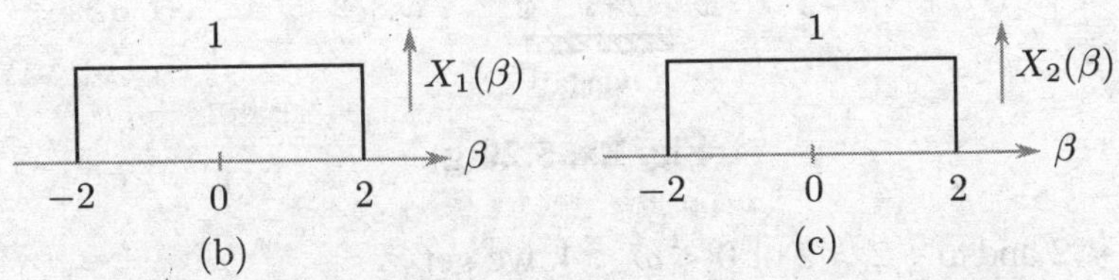

Fig. Ex. 5.20(b): *Function, $X_1(\beta)$* **(c):** *Function, $X_2(\beta)$*

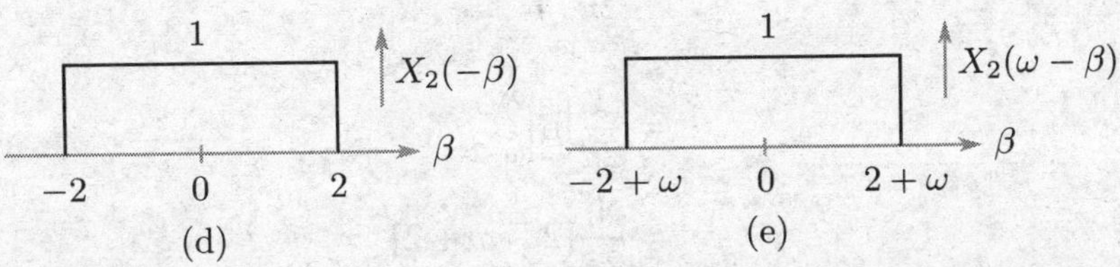

Fig. Ex. 5.20(d): *Folded function, $X_2(-\beta)$* **(e):** *Folded and translated function, $X_2(\omega-\beta)$*

Case (i): For $\omega + 2 < -2$ or $\omega < -4$, the functions $X_1(\beta)$ and $X_2(\omega - \beta)$ do not overlap. Hence, $Y(\omega) = 0$.

Case (ii):

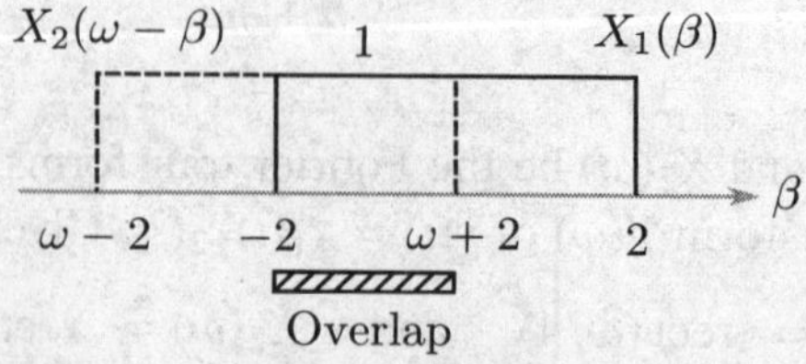

Fig. Ex. 5.20(f)

For $\omega - 2 < -2$ and $\omega + 2 > -2$ or $-4 < \omega < 0$, the functions $X_1(\beta)$ and $X_2(\omega - \beta)$ overlap as shown in Fig. Ex. 5.20(f).

$$Y(\omega) = \frac{1}{2\pi}\int_{-2}^{\omega+2} 1 \times 1 d\beta$$

$$= \frac{1}{2\pi}\left[\beta\right]_{-2}^{\omega+2}$$

$$= \frac{1}{2\pi}[\omega + 4]$$

Case (iii):

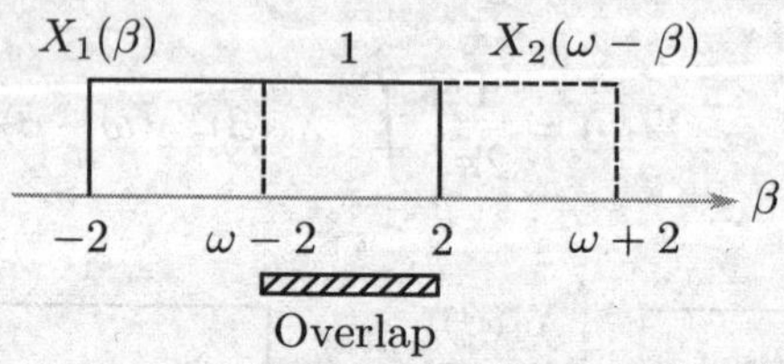

Fig. Ex. 5.20(g)

For $\omega - 2 < 2$ and $\omega + 2 > 2$ or $0 < \omega < 4$, we get

$$Y(\omega) = \frac{1}{2\pi}\int_{\omega-2}^{2} 1 \times 1 d\beta$$

$$= \frac{1}{2\pi}\left[\beta\right]_{\omega-2}^{2}$$

$$= \frac{1}{2\pi}[2 - \omega + 2]$$

$$= \frac{1}{2\pi}(4 - \omega)$$

Case (iv): For $\omega - 2 > 2$ or $\omega > 4$, the functions $X_1(\beta)$ and $X_2(\omega-\beta)$ do not overlap. Hence, $Y(\omega) = 0$.

Summarizing the Convolution results, we get

$$Y(\omega) = \begin{cases} 0, & \omega < -4 \\ \dfrac{1}{2\pi}(4+\omega), & -4 < \omega < 0 \\ \dfrac{1}{2\pi}(4-\omega), & 0 < \omega < 4 \\ 0, & \omega > 4 \end{cases}$$

The spectrum, $Y(\omega)$ is as shown in Fig. Ex. 5.20(h).

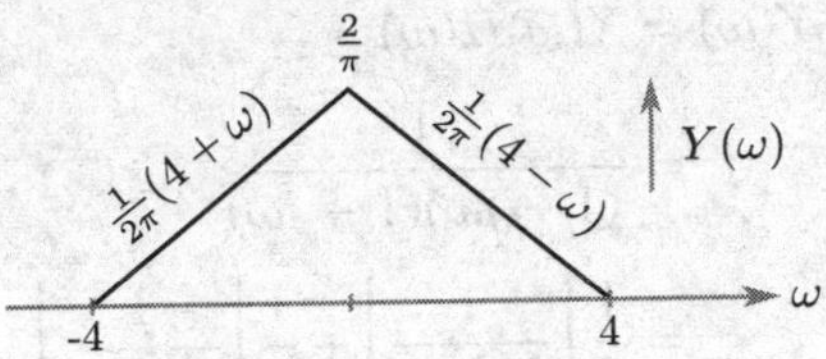

Fig. Ex. 5.20(h)

Example 5.21. Refer the signals $x(t)$ and $h(t)$ given below:

$$x(t) = e^{-t}u(t); \quad h(t) = e^{t}u(-t)$$

Compute, $y(t) = x(t) * h(t)$ by using the convolution property, and inverse transforming.

Solution: Recall:

$$e^{-at}u(t) \overset{\textbf{FT}}{\longleftrightarrow} \frac{1}{a+j\omega}$$

Letting $a = 1$, we get

$$x(t) = e^{-t}u(t) \overset{\textbf{FT}}{\longleftrightarrow} X(\omega) = \frac{1}{1+j\omega}$$

Given:

$$h(t) = e^{t}u(-t)$$

Let us find $H(\omega)$ by using the defining equation.

$$H(\omega) = \int_{-\infty}^{\infty} h(t)e^{-j\omega t}\,dt$$

$$= \int_{-\infty}^{\infty} e^{t}u(-t)e^{-j\omega t}\,dt$$

$$= \int_{-\infty}^{0} e^{t} e^{-j\omega t} dt$$

$$= \left[\frac{e^{(1-j\omega)t}}{1-j\omega} \right]_{-\infty}^{0}$$

$$= \frac{1}{1-j\omega}$$

Also, given in the problem,

$$y(t) = x(t) * h(t)$$

Applying convolution property, we get

$$Y(\omega) = X(\omega)H(\omega)$$

$$= \frac{1}{(1+j\omega)(1-j\omega)}$$

$$= \frac{1}{2}\left[\frac{1}{1+j\omega}\right] + \frac{1}{2}\left[\frac{1}{1-j\omega}\right]$$

Taking inverse Fourier transform on both sides, we get

$$y(t) = \frac{1}{2}e^{-t}u(t) + \frac{1}{2}e^{t}u(-t)$$

$$= \frac{1}{2}e^{-|t|}$$

Example 5.22. Compute the convolution of the following pair of signals $x_1(t)$ and $x_2(t)$ by using convolution property and then inverse transforming.

$$x_1(t) = te^{-2t}u(t); \quad x_2(t) = e^{-4t}u(t)$$

Solution: Recall the standard transform pairs:

(i) $te^{-at}u(t) \overset{\text{FT}}{\longleftrightarrow} \dfrac{1}{(a+j\omega)^2}$

(ii) $e^{-at}u(t) \overset{\text{FT}}{\longleftrightarrow} \dfrac{1}{a+j\omega}$

Hence,

$$X_1(\omega) = \frac{1}{(2+j\omega)^2}$$

and $$X_2(\omega) = \frac{1}{4+j\omega}$$

Let

$$y(t) = x_1(t) * x_2(t)$$

Applying convolution property of Fourier transform, we get

$$\begin{aligned} Y(\omega) &= X_1(\omega)X_2(\omega) \\ &= \frac{1}{(2+j\omega)^2(4+j\omega)} \end{aligned}$$

Letting $j\omega = \nu$, we get

$$Y(\omega) = \frac{1}{(\nu+2)^2(\nu+4)}$$

$$\Rightarrow \quad Y(\omega) = \frac{K_1}{\nu+4} + \frac{K_2}{\nu+2} + \frac{K_3}{(\nu+2)^2}$$

where

$$K_1 = \left.\frac{1}{(\nu+2)^2}\right|_{\nu=-4} = \frac{1}{4}$$

$$K_2 = \frac{d}{d\nu}\left[\frac{1}{\nu+4}\right]_{\nu=-2} = -\left.\frac{1}{(\nu+4)^2}\right|_{\nu=-2} = -\frac{1}{4}$$

$$K_3 = \left.\frac{1}{\nu+4}\right|_{\nu=-2} = \frac{1}{2}$$

Hence,

$$Y(\omega) = \frac{1}{4}\left[\frac{1}{4+j\omega}\right] - \frac{1}{4}\left[\frac{1}{2+j\omega}\right] + \frac{1}{2}\left[\frac{1}{(2+j\omega)^2}\right]$$

Taking inverse Fourier transform on both sides, we get

$$y(t) = \frac{1}{4}e^{-4t}u(t) - \frac{1}{4}e^{-2t}u(t) + \frac{1}{2}te^{-2t}u(t)$$

Example 5.23. Find the Fourier transform of the signal

$$x(t) = \frac{d}{dt}\left[e^{-4t}u(t) * e^{-t}u(t-1)\right]$$

Solution: Let

$$x_1(t) = e^{-4t}u(t)$$

and

$$x_2(t) = e^{-t}u(t-1)$$

Recall:

$$e^{-at}u(t) \overset{\textbf{FT}}{\longleftrightarrow} \frac{1}{a+j\omega}$$

Hence,

$$x_1(t) = e^{-4t}u(t) \overset{\text{FT}}{\longleftrightarrow} X_1(\omega) = \frac{1}{4 + j\omega}$$

Rewrite $x_2(t)$ as

$$\begin{aligned} x_2(t) &= e^{-1}e^{-(t-1)}u(t-1) \\ &= e^{-1}x_3(t-1), \quad \text{where } x_3(t) = e^{-t}u(t) \end{aligned}$$

Hence,

$$x_3(t) = e^{-t}u(t) \overset{\text{FT}}{\longleftrightarrow} X_3(\omega) = \frac{1}{1 + j\omega}$$

Applying time shift and linearity properties, we get

$$\begin{aligned} x_2(t) = e^{-1}x_3(t-1) \overset{\text{FT}}{\longleftrightarrow} X_2(\omega) &= e^{-1}e^{-j\omega}X_3(\omega) \\ &= e^{-1}\frac{e^{-j\omega}}{1 + j\omega} \end{aligned}$$

Applying the time differentiation and convolution properties, we get

$$x(t) = \frac{d}{dt}[x_1(t) * x_2(t)] \overset{\text{FT}}{\longleftrightarrow} X(\omega) = j\omega X_1(\omega)X_2(\omega)$$

Hence,

$$\begin{aligned} X(\omega) &= j\omega\frac{1}{4 + j\omega}e^{-1}\frac{e^{-j\omega}}{(1 + j\omega)} \\ &= e^{-1}\frac{j\omega e^{-j\omega}}{(4 + j\omega)(1 + j\omega)} \end{aligned}$$

Example 5.24. An LTI system is characterized by an impulse response, $h(t) = e^{-2t}u(t)$. Find the response of the system to an input given by $x(t) = e^{-3t}u(t)$. Use convolution property and then inverse Fourier transforming.

Solution: Recall the Fourier transform pair:

$$e^{-at}u(t) \overset{\text{FT}}{\longleftrightarrow} \frac{1}{a + j\omega}$$

Thus, we have

$$X(\omega) = \frac{1}{j\omega + 3}, \quad H(\omega) = \frac{1}{j\omega + 2}$$

Recall:

$$y(t) = x(t) * h(t) \overset{\text{FT}}{\longleftrightarrow} Y(\omega) = X(\omega)H(\omega)$$

Hence,

$$Y(\omega) = \frac{1}{(j\omega + 2)(j\omega + 3)}$$

Letting $j\omega = \nu$, we get

$$Y(\omega) = \frac{1}{(\nu + 2)(\nu + 3)}$$

$$= \frac{K_1}{\nu + 2} + \frac{K_2}{\nu + 3}$$

where, $$K_1 = \left.\frac{1}{\nu + 3}\right|_{\nu=-2} = 1$$

and $$K_2 = \left.\frac{1}{\nu + 2}\right|_{\nu=-3} = -1$$

Thus, $$Y(\omega) = \frac{1}{2 + j\omega} - \frac{1}{3 + j\omega}$$

Taking inverse Fourier transform on both sides we get

$$y(t) = e^{-2t}u(t) - e^{-3t}u(t)$$

5.8 Correlation

Correlation is an operation analogous to convolution. Correlation involves pushing one function against the other and finding the area under the resulting product. While performing, correlation, no folding is performed, unlike in convolution. The correlation of two identical functions *x*(*t*) is called autocorrelation and is denoted by $\phi_{xx}(t)$. The correlation of two different functions *x*(*t*) and *y*(*t*) is called cross-correlation and is denoted by $\phi_{xy}(t)$ or $\phi_{yx}(t)$.

Summarizing, we have

$$\phi_{xx}(t) = \int_{-\infty}^{\infty} x(\tau)x(\tau - t)d\tau \tag{5.35}$$

$$\phi_{xy}(t) = \int_{-\infty}^{\infty} x(\tau)y(\tau - t)d\tau \tag{5.36}$$

$$\phi_{yx}(t) = \int_{-\infty}^{\infty} y(\tau)x(\tau - t)d\tau \tag{5.37}$$

The variable *t* is defined as time lag. It may be noted that the definitions of cross-correlation are not unique, and some others may prefer to interchange the definitions of $\phi_{xy}(t)$ and $\phi_{yx}(t)$.

5.8.1 Correlation as convolution

There is striking similarity between correlation and convolution. This similarity could be understood by a simple mathematical analysis that follows:

Let us consider two continuous-time signals $x(t)$ and $y(t)$. Thus convolution of $x(t)$ and $y(-t)$ is

$$x(t) * y(-t) = \int_{-\infty}^{\infty} x(\tau)y(-(t-\tau))d\tau$$

$$= \int_{-\infty}^{\infty} x(\tau)y(\tau - t)d\tau \tag{5.38}$$

Also, the cross-correlation of $x(t)$ and $y(t)$ is

$$\phi_{xy}(t) = \int_{-\infty}^{\infty} x(\tau)y(\tau - t)d\tau \tag{5.39}$$

Inspection of Eq. (5.38) and Eq. (5.39) implies that

$$\phi_{xy}(t) = x(t) * y(-t)$$

Hence, the absence of folding actually implies that correlation of $x(t)$ and $y(t)$ is equivalent to convolution of $x(t)$ with the folded version $y(-t)$.

The absence of folding implies that the correlation depends on which function is shifted and in general, correlation is not commutative, that is, $\phi_{xy}(t) \neq \phi_{yx}(t)$. Since shifting $y(\tau)$ to right by t seconds is equivalent to shifting $x(\tau)$ to left by t seconds, the correlation $\phi_{xy}(t)$ is related to $\phi_{yx}(t)$ by

$$\phi_{xy}(t) = \phi_{yx}(-t)$$

5.8.2 Important properties of correlation

The correlation function provides the measure of similarity between signals $x(t)$ and $y(t)$. Since the order matters in correlation, we have

$$\phi_{xy}(t) = \int_{-\infty}^{\infty} x(\tau)y(\tau - t)d\tau \tag{5.40}$$

$$\phi_{yx}(t) = \int_{-\infty}^{\infty} y(\tau)x(\tau - t)d\tau \tag{5.41}$$

Letting $t = 0$ in Eq. (5.40) and Eq. (5.41) gives

$$\phi_{xy}(0) = \phi_{yx}(0) = \int_{-\infty}^{\infty} x(\tau)y(\tau)d\tau \tag{5.42}$$

The cross-correlation also satisfies the inequality

$$\begin{aligned} |\phi_{xy}(t)| &\leq \sqrt{\phi_{xx}(0)\phi_{yy}(0)} \\ &= \sqrt{E_x E_y} \end{aligned} \tag{5.43}$$

where E_x and E_y are the energies of signals $x(t)$ and $y(t)$. That is,

$$E_x = \int_{-\infty}^{\infty} x^2(t)\,dt$$

$$\text{and} \qquad E_y = \int_{-\infty}^{\infty} y^2(t)\,dt \tag{5.44}$$

5.8.3 Autocorrelation

The autocorrelation involves correlation of identical signals. Autocorrelation is commutative in nature, that is, it can be performed in any order.

Autocorrelation is a measure of similarity of a given function with itself when shifted by some amount. Therefore, under no shift, the two functions match and under this condition, autocorrelation is maximum. Physically if an individual compares his two photographs taken at the same instant, there is maximum similarity. If one compares his two photographs taken at two different instants of time, the similarity decreases as the dates on which the photographs were taken are widely different. By the same token, with increasing shift, the similarity and hence the correlation between a function and its shifted version decreases. As the shift approaches infinity, the similarity almost vanishes and hence, the autocorrelation function tends to zero.

5.8.4 Properties of autocorrelation

We have, $\phi_{xy}(t) = \phi_{yx}(-t)$. If $x = y$, we get $\phi_{xx}(t) = \phi_{xx}(-t)$. This means that autocorrelation function is an even function of time. For an even signal $x(t)$, we have $x(-t) = x(t)$. This means that autocorrelation of an even function $x(t)$ also equals the convolution of $x(t)$ with $x(t)$ itself.

Autocorrelation function is maximum at the origin. That is,

$$|\phi_{xx}(t)| \leq \phi_{xx}(0) \tag{5.45}$$

The result stated in Eq. (5.45) can be proved as follows:
We know that,

$$[x(\tau) \pm x(\tau - t)]^2 \geq 0$$

Integrating with respect to τ, we get

$$\int_{-\infty}^{\infty} [x(\tau) \pm x(\tau - t)]^2 d\tau \geq 0$$

$$\Rightarrow \quad \int_{-\infty}^{\infty} x^2(\tau) d\tau + \int_{-\infty}^{\infty} x^2(\tau - t) d\tau \pm 2 \int_{-\infty}^{\infty} x(\tau) x(\tau - t) d\tau \geq 0$$

$$\Rightarrow \quad \phi_{xx}(0) + \phi_{xx}(0) \pm 2\phi_{xx}(t) \geq 0$$

$$\Rightarrow \quad \phi_{xx}(0) \geq \pm\phi_{xx}(t)$$

$$\Rightarrow \quad |\phi_{xx}(t)| \leq \phi_{xx}(0)$$

It follows from the above equation that the autocorrelation function is finite and positive for all t.

Example 5.25. Find the autocorrelation of the signal, $x(t) = e^{-at}u(t)$.

Solution:

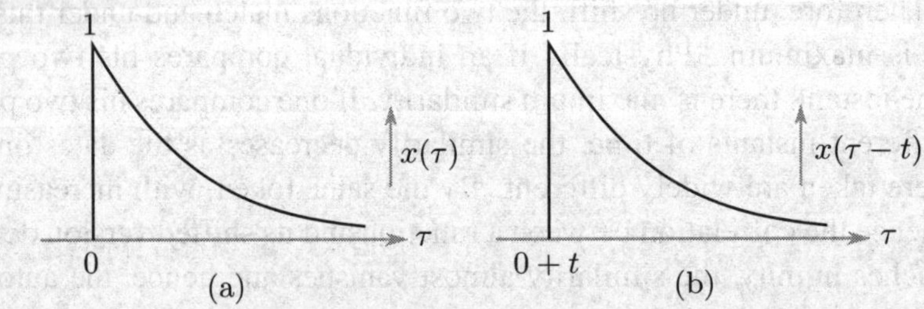

Fig. Ex. 5.25(a): *Signal,* $x(\tau)$ **(b):** *Signal,* $x(\tau)$ *shifted to right by t seconds.*

The autocorrelation of $x(t)$ is defined as follows:

$$\phi_{xx}(t) = \int_{-\infty}^{\infty} x(\tau) x(\tau - t) d\tau$$

The functions $x(\tau)$ and $x(\tau - t)$ are as depicted in Fig. Ex. 5.25(a) and Fig. Ex. 5.25(b), respectively.

Also,

$$x(\tau) = e^{-a\tau}u(\tau)$$

and $$x(\tau - t) = e^{-a(\tau-t)}u(\tau - t)$$

Case (i): For $t < 0$, the functions $x(\tau)$ and $x(\tau - t)$ overlap as shown in Fig. Ex. 5.25(c).

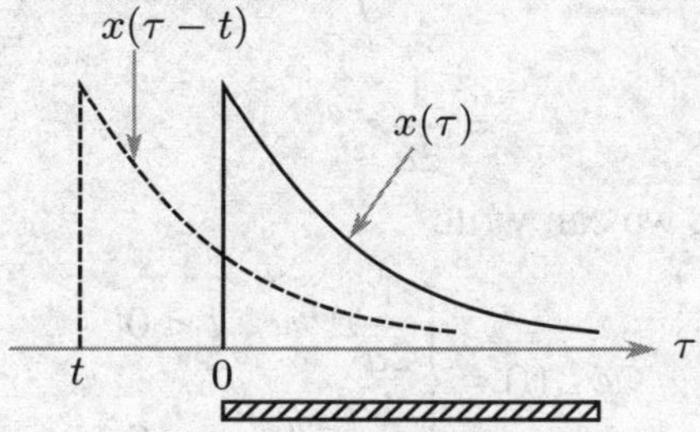

Fig. Ex. 5.25(c)

We have,

$$\phi_{xx}(t) = \int_0^\infty e^{-a\tau}e^{-a(\tau-t)}\,d\tau$$

$$= e^{at}\int_0^\infty e^{-2a\tau}d\tau$$

$$= e^{at}\left[\frac{e^{-2a\tau}}{-2a}\right]_0^\infty$$

$$= \frac{1}{2a}e^{at}$$

Case (ii): For $t > 0$, the functions, $x(\tau)$ and $x(\tau - t)$ overlap as depicted in Fig. Ex. 5.25(d).

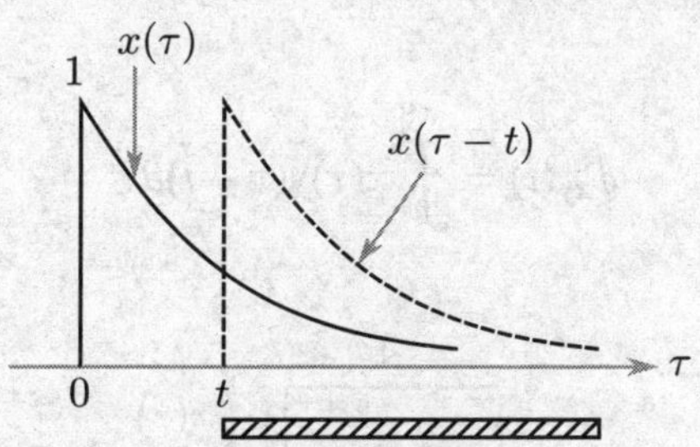

Fig. Ex. 5.25(d)

Referring Fig. Ex. 5.25(d), we can write

$$\phi_{xx}(t) = \int_{t}^{\infty} e^{-a\tau} e^{-a(\tau-t)} d\tau$$

$$= e^{at} \int_{t}^{\infty} e^{-2a\tau} d\tau$$

$$= \frac{1}{2a} e^{-at}$$

Summarizing the results, we can write

$$\phi_{xx}(t) = \begin{cases} \frac{1}{2a} e^{at}, & t < 0 \\ \frac{1}{2a} e^{-at}, & t > 0 \end{cases}$$

$$= \frac{1}{2a} e^{-a|t|}$$

Example 5.26. Evaluate the cross-correlation between the following signals:

$$x(t) = u(t) - u(t-1)$$

and $$y(t) = u(t) + u(t-1) - 2u(t-2)$$

Solution: We have,

$$x(\tau) = \begin{cases} 1, & 0 < \tau < 1 \\ 0, & \text{elsewhere} \end{cases}$$

and $$y(\tau) = \begin{cases} 1, & 0 < \tau < 1 \\ 2, & 1 < \tau < 2 \\ 0, & \tau > 2 \text{ and } \tau < 0 \end{cases}$$

To find $\phi_{xy}(t)$:
By definition,

$$\phi_{xy}(t) = \int_{-\infty}^{\infty} x(\tau) y(\tau - t) d\tau$$

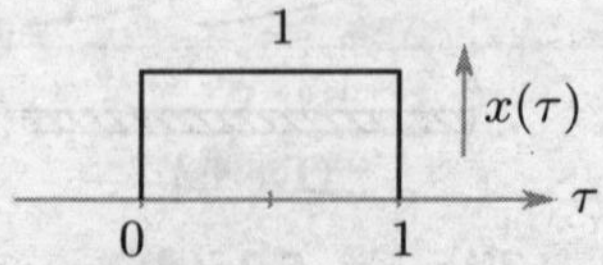

Fig. Ex. 5.26(a): *Signal,* $x(\tau)$

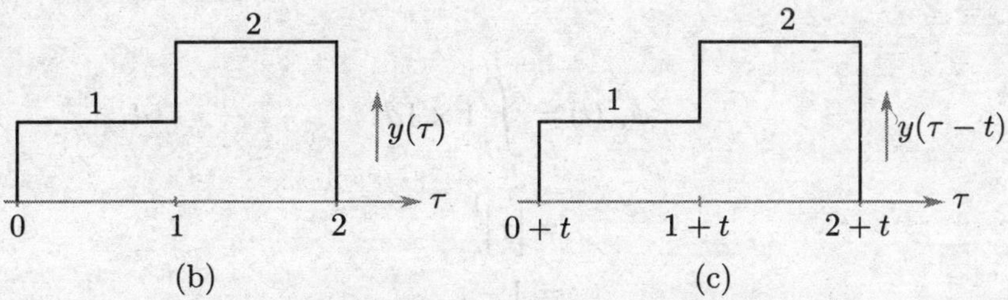

Fig. Ex. 5.26(b): *Signal, $y(\tau)$* **(c):** *Signal, $y(\tau)$ shifted to right by t seconds*

Case (i): For $t + 2 < 0$ or $t < -2$, the functions $x(\tau)$ and $y(\tau - t)$ do not overlap. Hence, $\phi_{xy}(t) = 0$

Case (ii):

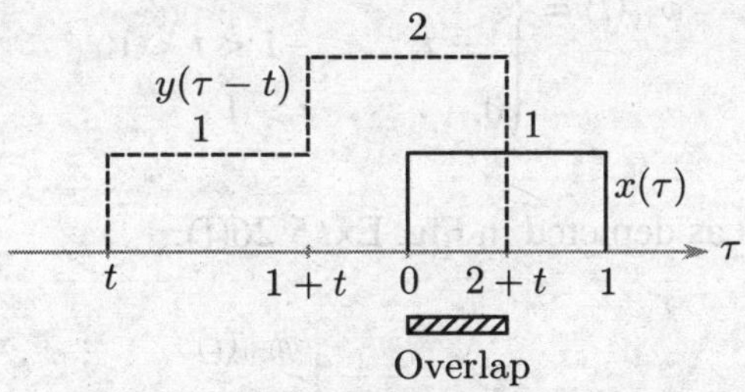

Fig. Ex. 5.26(d)

For $t + 1 < 0$ and $t + 2 > 0$ or $-2 < t < -1$, we have

$$\phi_{xy}(t) = \int_0^{t+2} 1 \times 2 d\tau$$

$$= 2\Big[\tau\Big]_0^{t+2} = 2(t + 2)$$

Case (iii): For $t < 1$ and $t + 2 > 1$ or $-1 < t < 1$, we get

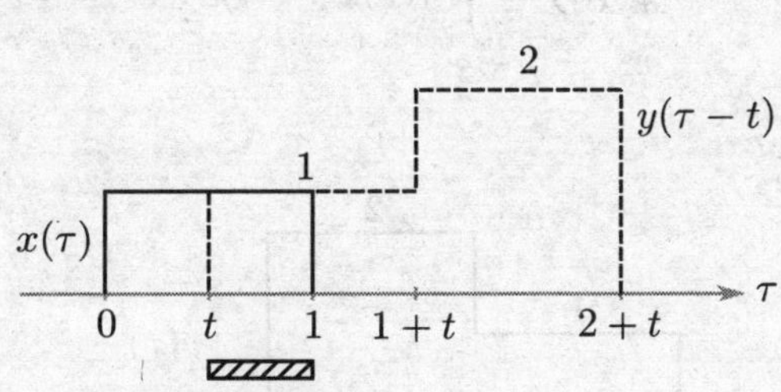

Fig. Ex. 5.26(e)

$$\phi_{xy}(t) = \int_{t}^{1} 1 \times 1 d\tau$$
$$= \left[\tau\right]_t^1$$
$$= 1 - t$$

Case (iv): For $t > 1$, the functions $x(\tau)$ and $y(\tau - t)$ do not overlap. Hence, $\phi_{xy}(t) = 0$.

Summarizing the results, we get

$$\phi_{xy}(t) = \begin{cases} 0, & t < -2 \\ 2(t+2), & -2 < t < -1 \\ 1 - t, & -1 < t < 1 \\ 0, & t > 1 \end{cases}$$

The sketch of $\phi_{xy}(t)$ is as depicted in Fig. Ex. 5.26(f).

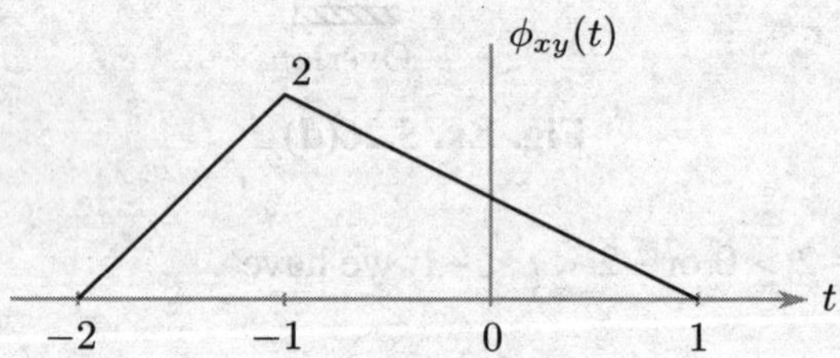

Fig. Ex. 5.26(f)

To find $\phi_{yx}(t)$:

By definition,

$$\phi_{yx}(t) = \int_{-\infty}^{\infty} y(\tau)x(\tau - t)d\tau$$

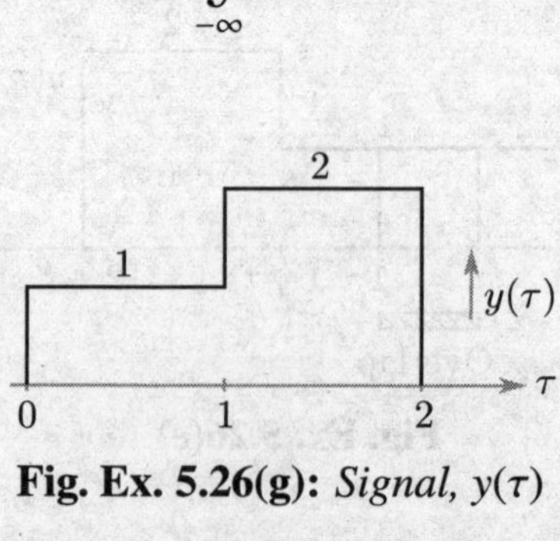

Fig. Ex. 5.26(g): *Signal, $y(\tau)$*

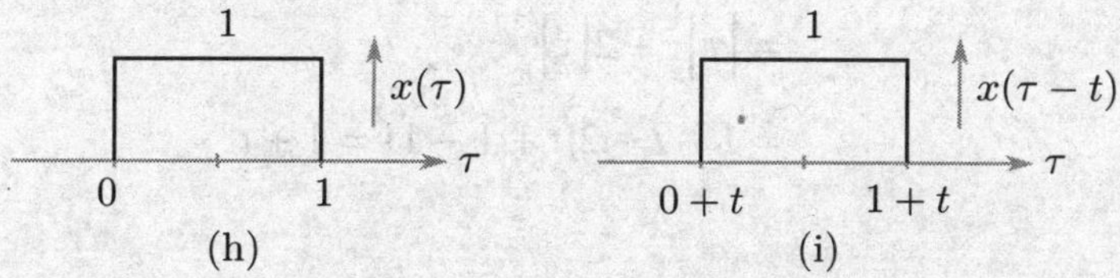

Fig. Ex. 5.26(h): *Signal,* $x(\tau)$ **(i):** *Translated signal,* $x(\tau - t)$

Case (i): For $t + 1 < 0$ or $t < -1$ the functions $y(\tau)$ and $x(\tau - t)$ do not overlap. Hence, $\phi_{yx}(t) = 0$.

Case (ii): For $t < 0$ and $t + 1 > 0$ or $-1 < t < 0$, we get the overlap os shown in Fig. Ex. 5.26(j).

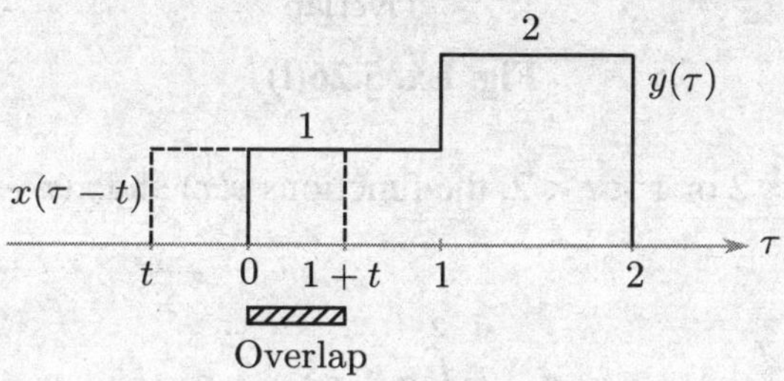

Fig. Ex. 5.26(j)

$$\phi_{yx}(t) = \int_{0}^{t+1} 1 \times 1 d\tau = t + 1$$

Case (iii): For $t < 1$ and $t + 1 > 1$ or $0 < t < 1$, we get the overlap as depicted in Fig. Ex. 5.26(k).

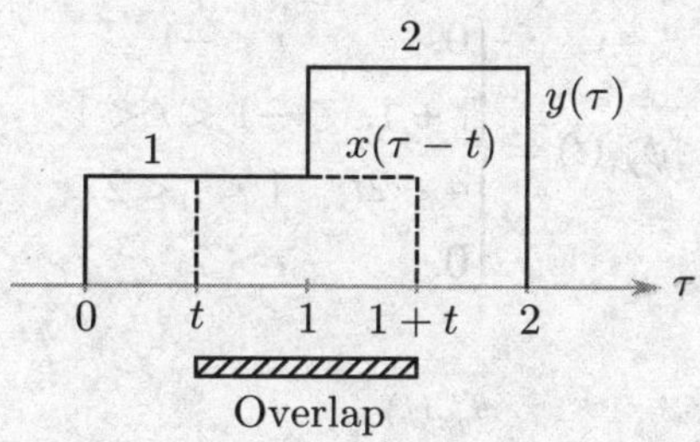

Fig. Ex. 5.26(k)

$$\phi_{yx}(t) = \int_{t}^{t+1} y(\tau)x(\tau - t)d\tau$$

$$= \int_{t}^{1} 1 \times 1 d\tau + \int_{1}^{t+1} 2 \times 1 d\tau$$

$$= \left[\tau\right]_t^1 + 2\left[\tau\right]_1^{t+1}$$
$$= 1 - t + 2[t + 1 - 1] = 1 + t$$

Case (iv):

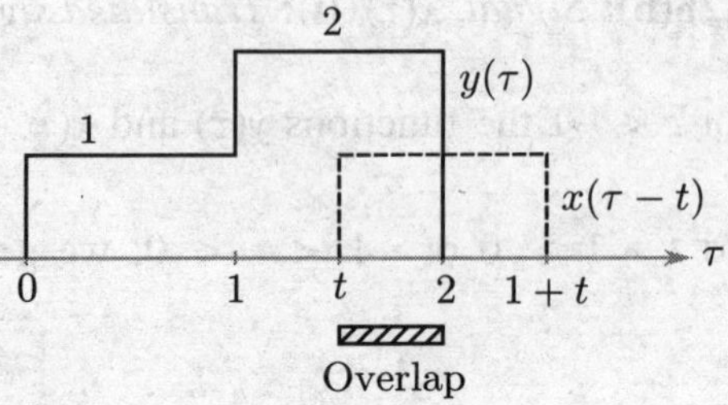

Fig. Ex. 5.26(l)

For $t < 2$ and $t + 1 > 2$ or $1 < t < 2$, the functions $y(\tau)$ and $x(\tau - t)$ overlap as shown in Fig. Ex. 5.26(l).

$$\phi_{yx}(t) = \int_t^2 2 \times 1 d\tau$$
$$= 2\left[\tau\right]_t^2$$
$$= 2[2 - t] = 4 - 2t$$

Case (v): For $t > 2$, $y(\tau)$ and $x(\tau - t)$ do not overlap. Hence, $\phi_{yx}(t) = 0$.

Summarizing, we get

$$\phi_{yx}(t) = \begin{cases} 0, & t < -1 \\ t + 1, & -1 < t < 1 \\ 4 - 2t, & 1 < t < 2 \\ 0, & t > 2 \end{cases}$$

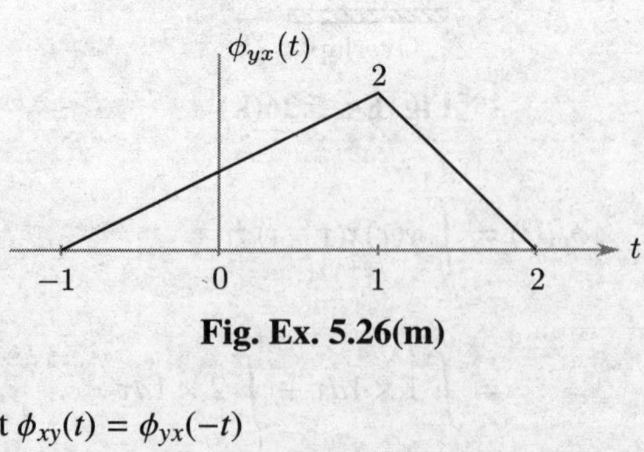

Fig. Ex. 5.26(m)

Comment: We find that $\phi_{xy}(t) = \phi_{yx}(-t)$

Example 5.27. Find the cross-correlation functions, $\phi_{xy}(t)$ and $\phi_{yx}(t)$ for the signals given below:

$$x(t) = e^{-t}u(t); \quad y(t) = e^{t}u(-t)$$

Solution: We know that,

$$u(-t) = \begin{cases} 1, & -t > 0 \text{ or } t < 0 \\ 0, & -t < 0 \text{ or } t > 0 \end{cases}$$

Hence,

$$y(t) = \begin{cases} e^{t}, & t < 0 \\ 0, & t > 0 \end{cases}$$

$$\Rightarrow \quad y(\tau - t) = \begin{cases} e^{\tau - t}, & \tau - t < 0 \text{ or } \tau < t \\ 0, & \tau > t \end{cases}$$

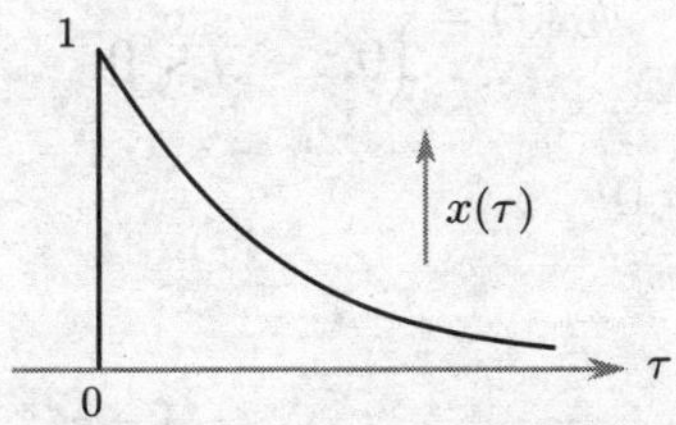

Fig. Ex. 5.27(a): *Signal, $x(\tau)$*

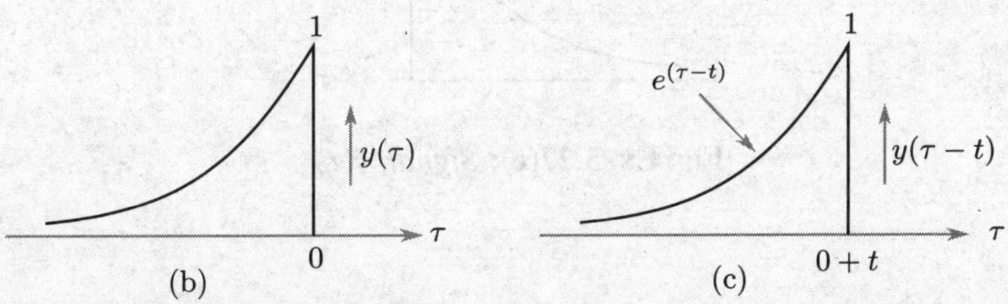

Fig. Ex. 5.27(b): *Signal, $y(\tau)$* **(c):** *Signal, $y(\tau)$ shifted to right by t seconds*

By definition,

$$\phi_{xy}(t) = \int_{-\infty}^{\infty} x(\tau)y(\tau - t)d\tau$$

Case (i): For $t < 0$, the functions $x(\tau)$ and $y(\tau - t)$ do not overlap. Hence, $\phi_{xy}(t) = 0$.

Case (ii): For $t > 0$, the functions $x(\tau)$ and $y(\tau - t)$ overlap as shown in Fig. Ex. 5.27(d).

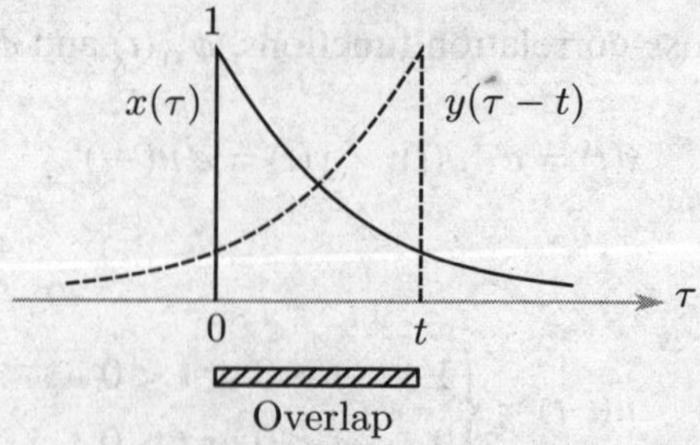

Fig. Ex. 5.27(d)

$$\phi_{xy}(t) = \int_{0}^{t} e^{-\tau} \times e^{(\tau-t)} d\tau = te^{-t}$$

Summarizing, the results, we get

$$\phi_{xy}(t) = \begin{cases} te^{-t}, & t > 0 \\ 0, & t < 0 \end{cases}$$

It may be noted that $\phi_{xy}(0) = 0$

To find $\phi_{yx}(t)$:

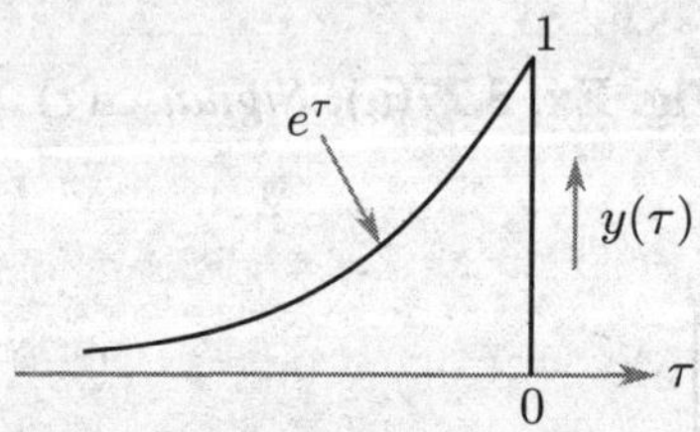

Fig. Ex. 5.27(e): *Signal, y(τ)*

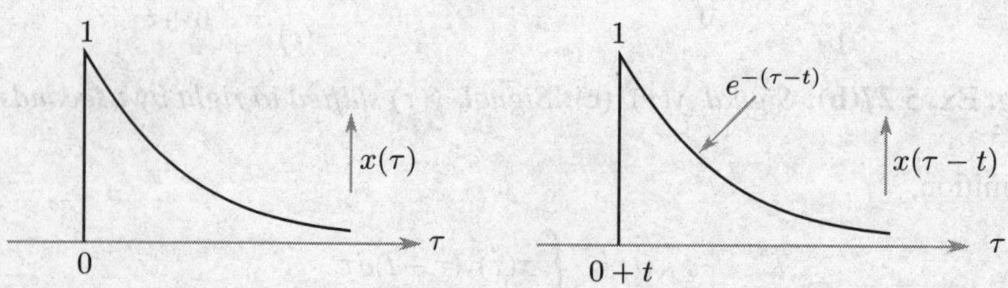

(f) (g)

Fig. Ex. 5.27(f): *Signal, x(τ)* **(g):** *Signal, x(τ) shifted to right by τ seconds*

By definition,

$$\phi_{yx}(t) = \int_{-\infty}^{\infty} y(\tau)x(\tau - t)d\tau$$

Case (i): For $t > 0$, the functions $y(\tau)$ and $x(\tau - t)$ do not overlap. Hence, $\phi_{yx}(t) = 0$.

Case (ii):

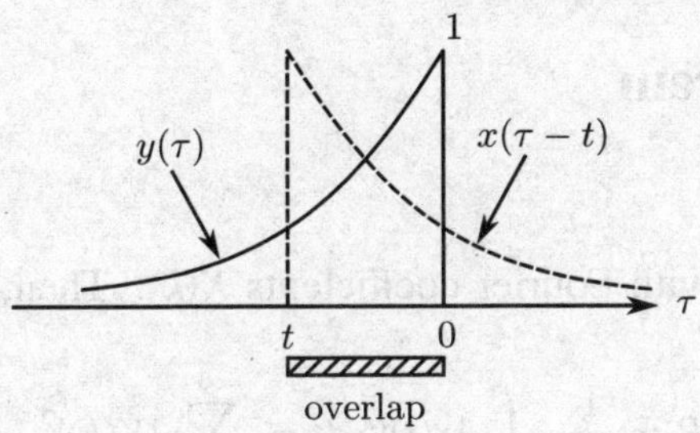

Fig. Ex. 5.27(h)

For $t < 0$, the overlap between $y(\tau)$ and $x(\tau - t)$ is as shown is Fig. Ex. 5.27(h).

$$\phi_{yx}(t) = \int_{t}^{0} e^{\tau} \times e^{-(\tau - t)} d\tau$$

$$= \int_{t}^{0} e^{t} d\tau$$

$$= e^{t}[\tau]_{t}^{0}$$

$$= -te^{t}$$

Summarizing, the results, we get

$$\phi_{yx}(t) = \begin{cases} 0, & t > 0 \\ -te^{t}, & t < 0 \end{cases}$$

Comment: We find that, $\phi_{xy}(t) = \phi_{yx}(-t)$.

Example 5.28. Given below are the autocorrelation functions of a certain signal $x(t)$. State the reason/reasons why these functions cannot qualify as autocorrelation functions.

(i) $\phi_{xx}(t) = e^{-2t}u(t)$

(ii) $\phi_{xx}(t) = 2te^{-\alpha|t|}$

(iii) $\phi_{xx}(t) = -\left[\dfrac{1 - t^2}{6 + t^2}\right]$

Solution:

(i) Not an autocorrelation function because $\phi_{xx}(t)$ is not even symmetric.

(ii) Not an autocorrelation function because $\phi_{xx}(t)$ is not maximum at the origin and is also not even symmetric.

(iii) Not an autocorrelation function because $\phi_{xx}(t)$ is not non-negative.

5.9 Parseval's theorem

5.9.1 Power signals

Let $x(t)$ be a power signal with Fourier coefficients $X(k)$. Then, the average power in the waveform is given by

$$P = \frac{1}{T_0} \int_{<T_0>} |x(t)|^2 \, dt = \sum_{k=-\infty}^{\infty} |X(k)|^2$$

What this theorem tells us is that there are two ways in which the average power in a periodic waveform can be evaluated:

(i) From the analytical definition of $x(t)$ using the integral formulation given in the statement.

(ii) From the series representation for $x(t)$ by summing the squared moduli of the coefficients as shown in the infinite sum.

Proof: Till this section, we have been assuming throughout that $x(t)$ is a real function of t. But as a general case, let us assume that $x(t)$ is complex. Then, $|x(t)|^2 = x(t)x^*(t)$ (definitely this is true even when $x(t)$ is real).

$$P = \frac{1}{T_0} \int_{<T_0>} |x(t)|^2 \, dt$$

$$= \frac{1}{T_0} \int_{<T_0>} x(t)x^*(t) \, dt$$

We have the synthesis equation,

$$x(t) = \sum_{k=-\infty}^{\infty} X(k)e^{jk\omega_0 t}$$

Taking conjugates on both the sides, we get

$$x^*(t) = \sum_{k=-\infty}^{\infty} X^*(k)e^{-jk\omega_0 t}$$

Substituting the expression for $x^*(t)$ in the expression for P, we get

$$P = \frac{1}{T_0} \int_{<T_0>} x(t) \left[\sum_{k=-\infty}^{\infty} X^*(k) e^{-jk\omega_0 t} \right] dt$$

Changing the order of summation and integration, we get,

$$P = \frac{1}{T_0} \sum_{k=-\infty}^{\infty} X^*(k) \int_{<T_0>} x(t) e^{-jk\omega_0 t}\, dt$$

$$= \sum_{k=-\infty}^{\infty} X^*(k) X(k)$$

$$= \sum_{k=-\infty}^{\infty} |X(k)|^2$$

Hence, the proof.

5.9.2 Energy signals and energy density spectrum

Let $x(t)$ be an energy signal with Fourier transform $X(\omega)$. Then the total (1− ohm) energy in $x(t)$ is given by

$$E = \int_{-\infty}^{\infty} |x(t)|^2 dt = \frac{1}{2\pi} \int_{-\infty}^{\infty} |X(\omega)|^2 d\omega \tag{5.46}$$

Parseval's theorem states that the total energy in a pulse can be computed in either one of two ways:

(i) From the time-domain analytical expression for $x(t)$, using the first integral in Eq. (5.46).

(ii) From the frequency-domain spectrum of the pulse, using the second integral in Eq. (5.46).

Proof: The energy of a pulse signal $x(t)$ is given by

$$E = \int_{t=-\infty}^{\infty} |x(t)|^2\, dt$$

$$= \int_{t=-\infty}^{\infty} x(t) x^*(t)\, dt \tag{5.47}$$

From the definition of inverse Fourier transform, we have

$$x(t) = \frac{1}{2\pi} \int_{\omega=-\infty}^{\infty} X(\omega) e^{j\omega t} d\omega$$

Taking conjugates on both sides, we get

$$x^*(t) = \frac{1}{2\pi} \int_{\omega=-\infty}^{\infty} X^*(\omega) e^{-j\omega t} d\omega \tag{5.48}$$

Substituting Eq. (5.48) in Eq. (5.47), we get

$$\begin{aligned} E &= \int_{t=-\infty}^{\infty} x(t) \left[\frac{1}{2\pi} \int_{\omega=-\infty}^{\infty} X^*(\omega) e^{-j\omega t} d\omega \right] dt \\ &= \frac{1}{2\pi} \int_{\omega=-\infty}^{\infty} X^*(\omega) \left[\int_{t=-\infty}^{\infty} x(t) e^{-j\omega t}\, dt \right] d\omega \\ &= \frac{1}{2\pi} \int_{\omega=-\infty}^{\infty} X^*(\omega) X(\omega)\, d\omega \\ &= \frac{1}{2\pi} \int_{\omega=-\infty}^{\infty} |X(\omega)|^2\, d\omega \end{aligned} \tag{5.49}$$

If we use $\omega = 2\pi f$, we obtain

$$E = \int_{-\infty}^{\infty} |X(f)|^2 df \tag{5.50}$$

Because the function $|X(\omega)|^2$ is a real and even function of ω, we can rewrite energy Eq. (5.49) as

$$\begin{aligned} E &= 2 \times \frac{1}{2\pi} \int_{\omega=0}^{\infty} |X(\omega)|^2\, d\omega \\ &= \frac{1}{\pi} \int_{\omega=0}^{\infty} |X(\omega)|^2\, d\omega \end{aligned} \tag{5.51}$$

The energy contained in a pulse in the frequency range $\omega_1 < \omega < \omega_2$ is

$$\Delta E = \frac{1}{\pi}\int_{\omega_1}^{\omega_2} |X(\omega)|^2\, d\omega \qquad (5.52)$$

Using Eq. (5.52), we can find the amount of energy contained in a pulse over any desired range of frequencies.

The energy density spectrum or energy spectral density of a pulse whose Fourier transform is $X(\omega)$ is defined as follows:

$$\phi_{xx}(\omega) = |X(\omega)|^2 \qquad (5.53)$$

The unit of $\phi_{xx}(\omega)$ is Joules per hertz. Some author define ϕ_{xx} as $\frac{1}{2\pi}|X(\omega)|^2$ and in that case, the unit is Joules/radian.

In communication Engineering we prefer $|X(f)|^2$ as the energy spectral density and describes the distribution of energy with frequency. The f-form of energy spectral density is denoted by $\phi_{xx}(f)$. $\phi_{xx}(f)$ has the unit Joules per hertz. **Note that $\phi_{xx}(f)$ is real, non-negative, and even function of f.**

Example 5.29. Find the expression for the energy density spectrum of the pulse

$$x(t) = \begin{cases} t, & -1 < t < 1 \\ 0, & \text{elsewhere} \end{cases}$$

Solution: The triangular pulse is as shown in Fig. Ex. 5.29(a).

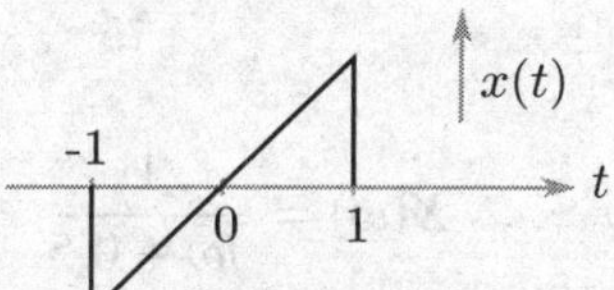

Fig. Ex. 5.29(a)

From Fig. Ex. 5.29(a), we find that $x(t)$ is an odd function of t. Thus, its Fourier transform should be purely imaginary.

By definition,

$$X(\omega) = \int_{-\infty}^{\infty} x(t)e^{-j\omega t}\, dt$$

$$= \int_{-1}^{1} t\cos\omega t\, dt - j\int_{-1}^{1} t\sin\omega t\, dt$$

Since $t\cos\omega t$ is odd, first integral on the right-hand side of the above equation is zero. Thus,

$$X(\omega) = -j\int_{-1}^{1} t\sin\omega t\,dt = -2j\int_{0}^{1} t\sin\omega t\,dt$$

$$\Rightarrow \quad X(\omega) = -2j\left[-t\frac{\cos\omega t}{\omega} + \frac{\sin\omega t}{\omega^2}\right]_0^1$$

$$= 2j\left[\frac{\omega\cos\omega - \sin\omega}{\omega^2}\right]$$

As expected $X(\omega)$ is purely imaginary.

The energy density spectrum of the pulse is

$$\phi_{xx}(\omega) = |X(\omega)|^2 = 4\left|\frac{\omega\cos\omega - \sin\omega}{\omega^2}\right|^2$$

Example 5.30. Find the energy contained by the signal $x(t) = e^{-0.5t}u(t)$ in the frequency band $|\omega| \le 0.5$ rad / sec.

Solution: Recall:

$$e^{-\alpha t}u(t) \overset{\textbf{FT}}{\longleftrightarrow} \frac{1}{j\omega + \alpha}$$

Hence,

$$X(\omega) = \frac{1}{j\omega + 0.5}$$

$$\Rightarrow \quad |X(\omega)|^2 = \frac{1}{\omega^2 + (0.5)^2}$$

Hence,

$$\Delta E = \frac{1}{2\pi}\int_{-0.5}^{0.5} |X(\omega)|^2 d\omega$$

$$= \frac{1}{2\pi} \times 2\int_{0}^{0.5} \frac{1}{\omega^2 + (0.5)^2} d\omega$$

$$= \frac{1}{\pi} \times 2\left[\tan^{-1}\left(\frac{\omega}{0.5}\right)\right]_0^{0.5}$$

$$= \frac{1}{\pi} \times 2\left[\frac{\pi}{4} - 0\right]$$
$$= \frac{2}{4} = \frac{1}{2}\text{ J}$$

Example 5.31. Consider an ideal low-pass filter with frequency response,

$$H(\omega) = \begin{cases} 1, & |\omega| < \omega_C \\ 0, & |\omega| > \omega_C \end{cases}$$

The input to the filter is

$$x(t) = e^{-2t}u(t)$$

Find the value of ω_C such that this filter output possesses exactly one-half of the normalized energy of the input signal $x(t)$.

Solution: Given,

$$x(t) = e^{-2t}u(t)$$
$$\Rightarrow \qquad X(\omega) = \frac{1}{j\omega + 2}$$

The output of the filter is

$$y(t) = x(t) * h(t)$$

Taking Fourier transform on both sides of the above equation, we get

$$Y(\omega) = X(\omega)H(\omega)$$
$$\Rightarrow \qquad Y(\omega) = \begin{cases} \dfrac{1}{2 + j\omega} \times 1, & |\omega| < \omega_C \\ \dfrac{1}{2 + j\omega} \times 0, & |\omega| > \omega_C \end{cases}$$
$$= \begin{cases} \dfrac{1}{1 + 2j\omega}, & |\omega| < \omega_C \\ 0, & |\omega| > \omega_C \end{cases}$$

The normalized energy of the input signal is

$$E_x = \int_{-\infty}^{\infty} |x(t)|^2\, dt = \int_{0}^{\infty} e^{-4t}\, dt = \frac{1}{4}\text{ J}$$

According to Parseval's theorem, the normalized energy of the output signal is

$$E_y = \frac{1}{2\pi}\int_{-\infty}^{\infty} |Y(\omega)|^2 d\omega = \frac{1}{2\pi}\int_{-\omega_C}^{\omega_C}\left[\frac{1}{\sqrt{2^2+\omega^2}}\right]^2 d\omega$$

$$= \frac{1}{2\pi}\int_{-\omega_C}^{\omega_C}\frac{1}{\omega^2+2^2}d\omega$$

$$= \frac{1}{2\pi}\tan^{-1}\left(\frac{\omega_C}{2}\right)\text{ J}$$

Given: $E_y = \frac{1}{2}E_x$

$$\Rightarrow \qquad \frac{1}{2\pi}\tan^{-1}\left(\frac{\omega_C}{2}\right) = \frac{1}{2}\times\frac{1}{4}$$

Hence, $\omega_C = 2$ rad / sec.

Example 5.32. A series RC circuit with $\tau = RC = 1$ sec is excited by a voltage $x(t) = \delta(t)$. The output is the voltage across the capacitor. Find the total energy in the output. Also compute the total signal energy in the output over $|f| \leq \frac{1}{2\pi}$ Hz.

Solution:

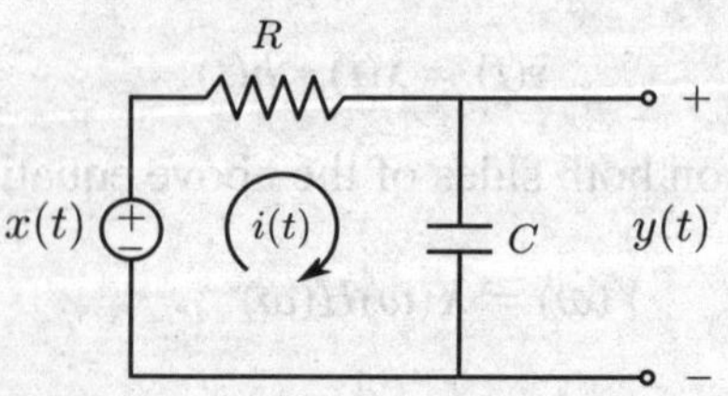

$RC = \tau = 1\text{Sec}$

Fig. Ex. 5.32

Applying KVL to the input loop, we get

$$x(t) = Ri(t) + y(t)$$

But

$$i(t) = C\frac{dy(t)}{dt}$$

Hence,

$$x(t) = RC\frac{dy(t)}{dt} + y(t)$$

Taking Fourier transforms on both sides, we get

$$X(\omega) = RC\,j\omega Y(\omega) + Y(\omega)$$

$$\Rightarrow \qquad H(\omega) = \frac{Y(\omega)}{X(\omega)} = \frac{1}{1 + j\omega RC}$$

Here, $H(\omega)$ is the transfer function or frequency response of the RC circuit. Since $RC = 1$, we get

$$H(\omega) = \frac{1}{1 + j\omega}$$

Given,

$$x(t) = \delta(t) \quad \Rightarrow \quad X(\omega) = 1$$

The response of the *RC* circuit is

$$y(t) = x(t) * h(t)$$

Taking Fourier transform on both sides, we get

$$\begin{aligned} Y(\omega) &= X(\omega)H(\omega) \\ &= 1 \times \frac{1}{1 + j\omega} \\ &= \frac{1}{1 + j\omega} \end{aligned}$$

Recall:

$$e^{-at}u(t) \stackrel{\text{FT}}{\longleftrightarrow} \frac{1}{a + j\omega}$$

Hence,

$$y(t) = e^{-t}u(t)$$

The energy contained in $y(t)$ is

$$\begin{aligned} E_y &= \int_{-\infty}^{\infty} |y(t)|^2\, dt \\ &= \int_{0}^{\infty} e^{-2t}\, dt \\ &= \frac{1}{2}\ \text{J} \end{aligned}$$

According to Parseval's theorem, the energy contained in $y(t)$ is

$$E_y = \frac{1}{2\pi} \int_{-\infty}^{\infty} |Y(\omega)|^2 d\omega$$

Since $\omega = 2\pi f$, we get

$$E_y = \frac{1}{2\pi} \int_{-\infty}^{\infty} |Y(f)|^2 2\pi \, df$$

$$= \int_{-\infty}^{\infty} |Y(f)|^2 \, df$$

Let E_1 be the energy content of the signal $y(t)$ in the frequency band $-\frac{1}{2\pi} \leq f \leq \frac{1}{2\pi}$. Then,

$$E_1 = \int_{-1/2\pi}^{1/2\pi} \left| \frac{1}{1 + j2\pi f} \right|^2 df$$

$$= 2 \int_{0}^{1/2\pi} \frac{1}{1 + 4\pi^2 f^2} df$$

$$= \frac{1}{2\pi^2} \int_{0}^{1/2\pi} \frac{1}{f^2 + \left(\frac{1}{2\pi}\right)^2} df$$

$$= \frac{1}{2\pi^2} \times 2\pi \tan^{-1}(2\pi f) \Big|_{0}^{1/2\pi}$$

$$= 0.25 \text{ J}$$

Example 5.33. Evaluate the integral given below:

$$I = \int_{-\infty}^{\infty} \frac{4}{(\omega^2 + 1)^2} d\omega$$

Solution:

$$I = \int_{-\infty}^{\infty} \frac{4}{(\omega^2 + 1)^2} d\omega$$

$$= \frac{1}{2\pi} \int_{-\infty}^{\infty} \left| \frac{\sqrt{8\pi}}{\omega^2 + 1} \right|^2 d\omega \tag{5.54}$$

Recall Parseval's theorem:

$$\int_{-\infty}^{\infty} |x(t)|^2\, dt = \frac{1}{2\pi}\int_{-\infty}^{\infty} |X(\omega)|^2 d\omega \tag{5.55}$$

Comparing Eq. (5.54) and Eq. (5.55), we get

$$X(\omega) = \frac{\sqrt{8\pi}}{\omega^2 + 1}$$

We know that,

$$e^{-a|t|} \overset{\textbf{FT}}{\longleftrightarrow} \frac{2a}{\omega^2 + a^2}$$

Letting $a = 1$, we get

$$e^{-|t|} \overset{\textbf{FT}}{\longleftrightarrow} \frac{2}{\omega^2 + 1}$$

Hence,

$$x(t) = \sqrt{2\pi}e^{-|t|} \overset{\textbf{FT}}{\longleftrightarrow} X(\omega) = \frac{2\sqrt{2\pi}}{\omega^2 + 1}$$

Therefore,

$$I = \int_{-\infty}^{\infty} |x(t)|^2\, dt$$

$$= 2\int_{0}^{\infty} 2\pi e^{-2t}\, dt = 2\pi$$

Example 5.34. Evaluate the integral given below:

$$I = \int_{-\infty}^{\infty} \frac{2}{|j\omega + 2|^2} d\omega$$

Solution: Rewrite the given integral as

$$I = \frac{1}{2\pi}\int_{-\infty}^{\infty} \frac{4\pi}{|j\omega + 2|^2} d\omega$$

$$= \frac{1}{2\pi}\int_{-\infty}^{\infty} \left|\frac{2\sqrt{\pi}}{j\omega + 2}\right|^2 d\omega \tag{5.56}$$

Recall parseval's theorem:

$$\int_{-\infty}^{\infty} |x(t)|^2\, dt = \frac{1}{2\pi} \int_{-\infty}^{\infty} |X(\omega)|^2 d\omega \tag{5.57}$$

Comparing Eq. (5.56) and Eq. (5.57), we get

$$X(\omega) = \frac{2\sqrt{\pi}}{j\omega + 2} \tag{5.58}$$

Recall the standard Fourier transform pair:

$$e^{-at}u(t) \xleftrightarrow{\textbf{FT}} \frac{1}{j\omega + a}$$

Taking inverse Fourier transform on both sides of Eq. (5.58), we get

$$x(t) = 2\sqrt{\pi}e^{-2t}u(t)$$

Hence,

$$I = \int_{-\infty}^{\infty} |x(t)|^2\, dt = \int_{0}^{\infty} 4\pi e^{-4t}\, dt = \pi$$

Example 5.35. If $x(t) = e^{-a|(t)|}$, verify Parseval's theorem for energy signals.

Solution: The Parseval's theorem for pulses is

$$\int_{-\infty}^{\infty} |x(t)|^2\, dt = \frac{1}{2\pi} \int_{-\infty}^{\infty} |X(\omega)|^2\, d\omega$$

$$= \frac{1}{2\pi} \times 2 \int_{0}^{\infty} |X(\omega)|^2\, d\omega$$

To find $\int_{-\infty}^{\infty} |x(t)|^2\, dt$:

$$\int_{-\infty}^{\infty} e^{-2a|t|}\, dt = \int_{-\infty}^{0} e^{2at}\, dt + \int_{0}^{\infty} e^{-2at} dt$$

$$= \left.\frac{e^{2at}}{2a}\right|_{-\infty}^{0} + \left.\frac{e^{-2at}}{-2a}\right|_{0}^{\infty}$$

$$= \frac{1}{2a} + \frac{1}{2a} = \frac{1}{a}$$

To find $\frac{1}{\pi}\int_0^\infty |X(\omega)|^2 d\omega$:

The Fourier transform of $x(t)$ is

$$X(\omega) = \frac{2a}{a^2 + \omega^2}$$

Hence,

$$\frac{1}{\pi}\int_0^\infty |X(\omega)|^2 d\omega = \frac{1}{\pi}\int_0^\infty \frac{4a^2}{(a^2+\omega^2)^2}\, d\omega$$

$$= \frac{4a^2}{\pi}\frac{1}{2a^2}\left[\frac{\omega}{\omega^2+a^2} + \frac{1}{a}\tan^{-1}\frac{\omega}{a}\right]\Bigg|_0^\infty$$

$$= \frac{2}{\pi}\left[0 + \frac{\pi}{2a} - 0 - 0\right] = \frac{1}{a}$$

Hence,

$$\int_{-\infty}^{\infty} |x(t)|^2\, dt = \frac{1}{\pi}\int_0^\infty |X(\omega)|^2\, d\omega$$

Example 5.36. The current in a 1Ω resistor is

$$i = 20e^{-2t}u(t)\, A.$$

What percentage of total energy dissipated in the resistor can be associated with the frequency band $0 \le \omega \le 2\sqrt{3}$ rad/sec?

Solution: The total energy dissipated in the 1Ω resistor is

$$E_{1\Omega} = \int_0^\infty 400e^{-4t}\, dt$$

$$= 400\left[\frac{e^{-4t}}{4}\right]\Bigg|_0^\infty$$

$$= 100 \text{ J}$$

We can check this total energy calculation with Parseval's theorem:

$$I(\omega) = \frac{20}{2 + j\omega}$$

$$\Rightarrow \qquad |I(\omega)| = \frac{20}{\sqrt{4+\omega^2}}$$

Therefore,

$$E_{1\Omega} = \frac{1}{\pi}\int_0^\infty |I(\omega)|^2\, d\omega$$

$$= \frac{1}{\pi}\int_0^\infty \frac{400}{4+\omega^2}\, d\omega$$

$$= \frac{400}{\pi}\left[\frac{1}{2}\tan^{-1}\left(\frac{\omega}{2}\right)\right]_0^\infty = 100 \text{ J}$$

The energy associated with the frequency band $0 \le \omega \le 2\sqrt{3}$ rad/sec is

$$\Delta E = \frac{1}{\pi}\int_0^{2\sqrt{3}} \frac{400}{4+\omega^2}\, d\omega$$

$$= \frac{400}{\pi}\left[\frac{1}{2}\tan^{-1}\left(\frac{\omega}{2}\right)\right]_0^{2\sqrt{3}} = \frac{200}{3} \text{ J}$$

Hence, the percentage of the total energy associated with this range of frequencies is

$$\eta = \frac{200/3}{100} \times 100 = 66.67\%$$

5.10 Relation between autocorrelation function and energy spectral density

The Parseval's theorem for energy signals in f-form is

$$E = \int_{-\infty}^{\infty} |x(t)|^2\, dt = \int_{-\infty}^{\infty} |X(f)|^2\, df \tag{5.59}$$

From Eq. (5.59), we find that the total energy in the signal $x(t)$ equals the area under the squared magnitude spectrum, $|X(f)|^2$. The function $\phi_{xx}(f) = |X(f)|^2$ is called energy density spectrum or energy spectral density and explains the distribution of energy with frequency. Let us rewrite $\phi_{xx}(f)$ as

$$\phi_{xx}(f) = X(f)X^*(f) \tag{5.60}$$

The Fourier transform of a real signal is

$$X(f) = \int_{-\infty}^{\infty} x(t)e^{-j2\pi ft}df \tag{5.61}$$

Taking complex conjugates on both sides of Eq. (5.61) gives

$$X^*(f) = \int_{-\infty}^{\infty} x^*(t)e^{j2\pi ft}df \tag{5.62}$$

For a real signal, $x^*(t) = x(t)$. Making use of this fact in Eq. (5.62) gives

$$X^*(f) = \int_{-\infty}^{\infty} x(t)e^{-j2\pi(-f)t} = X(-f) \tag{5.63}$$

Making use of Eq. (5.63) in Eq. (5.60) gives

$$\phi_{xx}(f) = X(f)X(-f) \tag{5.64}$$

Recall: If $x(t) \overset{\textbf{FT}}{\longleftrightarrow} X(f)$, then $x(-t) \overset{\textbf{FT}}{\longleftrightarrow} X(-f)$
Hence, taking inverse Fourier transform on both sides of Eq. (5.64) gives

$$\phi_{xx}(t) = x(t) * x(-t)$$

In other words, the energy spectral density and autocorrelation function form a Fourier transform pair. That is,

$$\text{(Autocorrelation)}\ \phi_{xx}(t) \overset{\textbf{FT}}{\longleftrightarrow} \phi_{xx}(f)\ \text{(energy spectral density)} \tag{5.65}$$

Eq. (5.65) is known as Weiner-Khincthine theorem.

5.11 Power spectral density

A signal $x(t)$ is said to be a power signal, if the average power of the signal is a finite quantity. That is,

$$P = \lim_{T\to\infty} \frac{1}{T} \int_{-T/2}^{T/2} |x(t)|^2\, dt = M < \infty,$$

where M is any finite constant. Usually, periodic signals and random signals have finite average power and are viewed as power signals.

The parseval's theorem for periodic signals is dealt in article 5.9.1. However, we present the result of parseval's theorem for periodic signals for explaining power spectral density.

$$P = \frac{1}{T_0} \int_{<T_0>} |x(t)|^2 \, dt$$

$$= \sum_{k=-\infty}^{\infty} |x(k)|^2 \qquad (5.66)$$

What parseval's theorem tells us is that there are two ways in which the average power in a periodic waveform can be computed:

(i) From the analytical definition of $x(t)$ using the integral formulation given in Eq. (5.66).

(ii) From the Fourier series representation of $x(t)$ by summing the squared moduli of the coefficients as shown in the infinite summation of Eq. (5.66).

In other words, Eq. (5.66), links the time-domain representation of power (the integral) with the frequency-domain representation (the infinite sum).

Parseval's theorem is a statement that can be considered term by term, for example if the average power in a periodic waveform is given by Eq. (5.66), then the power in, say, the second term in the complex Fourier series must be

$$P(2) = |X(2)|^2$$

and hence we are able to find how much power, any term or group of terms in the Fourier series contributes to the total average power. Thus, if we plot the values of $|X(k)|^2$ versus k, we would obtain a line spectrum of the power in the waveform. This line spectrum of power is also known as power spectral density.

For power signals, autocorrelation function, $\phi_{xx}(t)$ and power spectral density, $\phi_{xx}(f)$ form a Fourier transform pair and play a role similar to the signal $x(t)$ and its Fourier transform $X(f)$. It may be noted that $\phi_{xx}(0)$ gives the average power in $x(t)$. The power spectral density (PSD) is often used to characterize random signals. It is important to note that PSD of a signal is not unique because many time signals can yield the same PSD.

Example 5.37. Consider the periodic waveform $x(t) = 4 + 2\cos 3t + 3\sin 4t$.

(a) What is the value of T_0?

(b) What is the total average power?

(c) Find the Complex Fourier Coefficients.

(d) Using Parseval's theorem, find the power spectrum.

(e) Show that the total average power from part (d) agrees with what is obtained in part (b).

Solution:

(a)

$$x(t) = 4 + 2\cos 3t + 3\sin 4t$$

$$\omega_{0_1} = 3, \ \omega_{0_2} = 4$$

$$\omega_0 = \text{GCD}(\omega_{0_1}, \omega_{0_2}) = 1.$$

Then,

$$T_0 = \frac{2\pi}{\omega_0} = 2\pi$$

(b)

$$P = \frac{1}{T_0}\int_{<T_0>} |x(t)|^2\, dt$$

$$= \frac{1}{2\pi}\int_{-\pi}^{\pi} [4 + 2\cos 3t + 3\sin 4t]^2\, dt$$

$$= \frac{1}{2\pi}\left[\int_{-\pi}^{\pi} 4^2\, dt + \int_{-\pi}^{\pi} (2\cos 3t)^2\, dt + \int_{-\pi}^{\pi} (3\sin 4t)^2\, dt\right]$$

All the remaining products after integration become zero (because sines and cosines form an orthogonal set).

Therefore, $P = 16 + 2 + \dfrac{9}{2} = \dfrac{45}{2}$ watts.

(c) Given

$$x(t) = 4 + 2\cos 3t + 3\sin 4t$$

$$= 4 + 2\left[\frac{e^{j3t} + e^{-j3t}}{2}\right] + 3\left[\frac{e^{j4t} - e^{-j4t}}{2j}\right]$$

Since, $\omega_0 = 1$, we get

$$x(t) = 4 + e^{-j3\omega_0 t} + e^{j3\omega_0 t} + \frac{3}{2j}e^{j4\omega_0 t} - \frac{3}{2j}e^{-j4\omega_0 t} \tag{5.67}$$

Comparing Eq. (5.67) with

$$x(t) = \sum_{k=-\infty}^{\infty} X(k)e^{jk\omega_0 t}$$

we get,

$$X(-4) = \frac{-3}{2j}, \quad X(-3) = 1, \quad X(0) = 4$$

$$X(3) = 1, \quad X(4) = \frac{3}{2j}$$

(d) By Parseval's theorem, the power contributed by each of the terms in the series is

$$P(-4) = \left|\frac{3j}{2}\right|^2 = \frac{9}{4}$$

$$P(-3) = |1|^2 = 1$$

$$P(0) = |4|^2 = 16$$

$$P(3) = 1, \quad P(4) = \frac{9}{4}$$

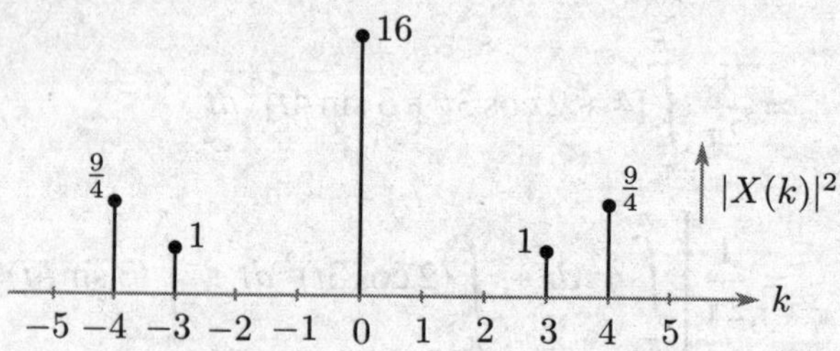

Fig. Ex. 5.37 : *Power spectrum.*

$$P = \sum_{k=-\infty}^{\infty} |X(k)|^2$$

$$= 2 \times \frac{9}{4} + 2 \times 1 + 16$$

$$= \frac{45}{2} \text{ Watts.}$$

Hence, the Parseval's theorem.

Example 5.38. Find the Complex Fourier Coefficient, the power spectrum and the total power for $x(t)$ defined below.

$$x(t) = 5\sin(5t) + 7\cos(6t) + 3$$

Also, find the total power,

(a) if $x(t)$ were a voltage across 10,000 ohms and

(b) if $x(t)$ were a current through 0.01 ohm.

Solution: Given

$$x(t) = 5\sin(5t) + 7\cos(6t) + 3$$

$$\Rightarrow \quad x(t) = \frac{5}{2j}e^{j5t} - \frac{5}{2j}e^{-j5t} + \frac{7}{2}e^{j6t} + \frac{7}{2}e^{-j6t} + 3 \tag{5.68}$$

Also,

$$\begin{aligned} \omega_0 &= \text{GCD}[\omega_{0_1}, \omega_{0_2}] \\ &= \text{GCD}[5, 6] = 1 \\ T_0 &= \frac{2\pi}{\omega_0} = 2\pi \end{aligned}$$

Making use of ω_0 in Eq. (5.68), we get

$$x(t) = \frac{5}{2j}e^{j5\omega_0 t} - \frac{5}{2j}e^{-j5\omega_0 t} + \frac{7}{2}e^{j6\omega_0 t} + \frac{7}{2}e^{-j6\omega_0 t} + 3 \tag{5.69}$$

Comparing Eq. (5.69) with

$$x(t) = \sum_{k=-\infty}^{\infty} X(k)e^{jk\omega_0 t}$$

we get,

$$X(-6) = \frac{7}{2}, \quad X(-5) = \frac{5j}{2}, \quad X(0) = 3$$

$$X(5) = \frac{-5j}{2}, \quad X(6) = \frac{7}{2}$$

We know that,

$$P(k) = |X(k)|^2 \tag{5.70}$$

Therefore,

$$P(-6) = \frac{49}{4}, \quad P(-5) = \frac{25}{4}, \quad P(0) = 9, \quad P(5) = \frac{25}{4}, \quad P(6) = \frac{49}{4}$$

From Parseval's theorem, the total power expended in a 1 ohm resistor is given by

$$\begin{aligned} P &= \sum_{k=-\infty}^{\infty} |X(k)|^2 \\ &= 2 \times \frac{49}{4} + 2 \times \frac{25}{4} + 9 \\ &= 46 \text{ W} \end{aligned}$$

(a) If $x(t)$ were a voltage waveform impressed across 10,000 Ω, then

$$P_{\text{total}} = \frac{P}{R} = \frac{46}{10000}$$
$$= 4.6 \times 10^{-3}\ \text{W}$$

(b) On the other hand, if $x(t)$ were a current waveform impressed across 0.01Ω, then

$$P_{\text{total}} = P\ R = 46 \times 0.01$$
$$= 0.46\ \text{W}$$

5.12 Detection of periodic signals in the presence of noise by correlation

Let $s(t)$ be a periodic transmitted signal corrupted by a noise signal $n(t)$. Assuming the noise to be additive, the received signal is given by

$$r(t) = s(t) + n(t) \tag{5.71}$$

In Eq. (5.71), $n(t)$ is a signal with random amplitude variations.

Our objective is to identify the transmitted signal $s(t)$ mixed with noise by making use of the received signal $r(t)$.

The detection of $s(t)$ can be carried out by cross-correlating the received signal $r(t)$ with another periodic signal $c(t)$. The signal $c(t)$ must have the same frequency as the transmitted signal $s(t)$. The only disadvantage of this technique is that we must know the frequency of signal $s(t)$ (to be detected) beforehand. In majority of the cases, the frequency of $s(t)$ is known beforehand.

Refer the block diagram of a typical communication system shown in Fig. 5.9. In Fig. 5.9, $c(t)$ is a signal generated locally at the receiver and has the same frequency as that of $s(t)$.

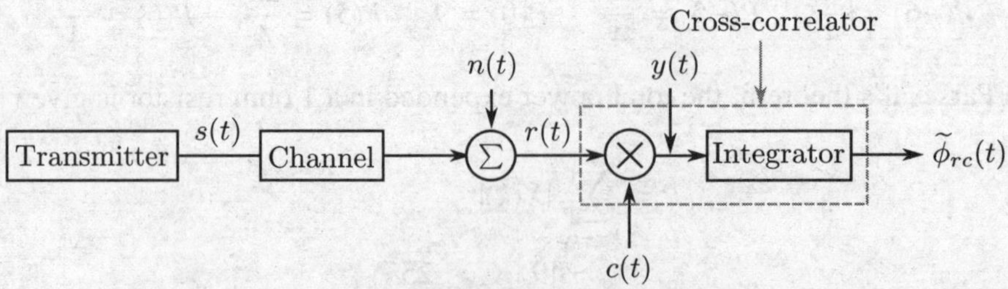

Fig. 5.9: *Typical block diagram of a communication system*

Since $r(t)$ and $c(t)$ are periodic signals, we define the average cross-correlation of these two signals as (output of cross-correlator shown in Fig. 5.9)

$$\begin{aligned}\widetilde{\phi}_{rc}(t) &= \lim_{T\to\infty} \frac{1}{T} \int_{-T/2}^{T/2} r(\tau)c(\tau - t)d\tau \\ &= \lim_{T\to\infty} \frac{1}{T} \int_{-T/2}^{T/2} [s(\tau) + n(\tau)]c(\tau - t)d\tau \\ &= \widetilde{\phi}_{sc}(t) + \widetilde{\phi}_{nc}(t) \qquad (5.72)\end{aligned}$$

It may be noted that $c(t)$ and $n(t)$ are uncorrelated. Hence, $\widetilde{\phi}_{nc}(t) = 0$. Consequently, we get

$$\widetilde{\phi}_{rc}(t) = \widetilde{\phi}_{sc}(t) \qquad (5.73)$$

Since $s(t)$ and $c(t)$ are signals of same frequency, it implies that $\widetilde{\phi}_{sc}(t)$ is also a periodic function of the same frequency. It is very clear that if the cross-correlation of a corruputed signal $r(t)$ with $c(t)$ yields a periodic signal, $r(t)$ must have a periodic component of same frequency as that of $c(t)$. Thus, the signal $s(t)$ is detected in the presence of additive noise $n(t)$.

5.13 Extraction of a signal from noise by filtering

A signal corrupted by noise can be detected by either cross-correlation technique or by filtering. Both the techniques are equivalent. However, cross-correlation is a time-domain technique whereas filtering is a frequency-domain technique.

Consider the cross-correlation function $\phi_{12}(t)$ of the signals $r_1(t)$ and $r_2(t)$.

If $r_1(t) \overset{\textbf{FT}}{\longleftrightarrow} R_1(\omega)$ and $r_2(t) \overset{\textbf{FT}}{\longleftrightarrow} R_2(\omega)$, then

$$\phi_{12}(t) = r_1(t) * r_2(-t) \overset{\textbf{FT}}{\longleftrightarrow} \phi_{12}(\omega) = R_1(\omega)R_2(-\omega) \qquad (5.74)$$

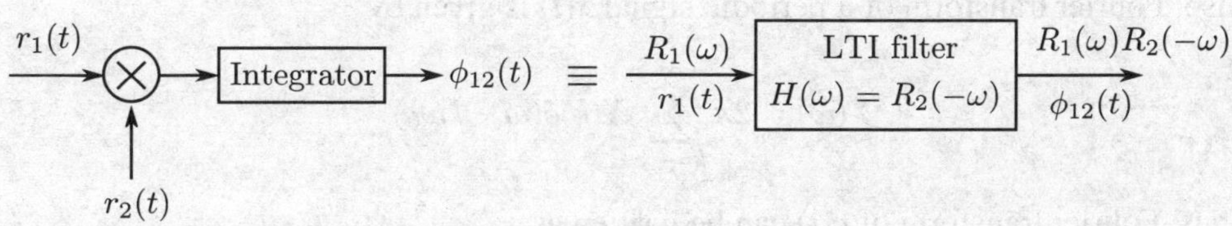

Fig. 5.10: *Equivalence between cross-correlation and filtering*

The output of a linear time-invariant filter is the convolution of its input $x(t)$ with the impulse response $h(t)$ of the filter. In this case $x(t) = r_1(t)$ and $h(t) = r_2(-t)$. The Fourier transform of the impulse response is known as the transfer function, $H(\omega)$ of the filter. Hence, in this case, $H(\omega) = R_2(-\omega)$. The technique of obtaining $\phi_{12}(t)$ by filtering is depicted in Fig. 5.10.

Summarizing, the cross-correlation function of signals $r_1(t)$ and $r_2(t)$ is the response of an LTI system with transfer function $R_2(-\omega)$.

The concept developed above may now be applied to detect a periodic signal $s(t)$ corrupted by a random noise signal, $n(t)$. Let us denote the received signal by $r(t)$. Then, we have

$$r(t) = s(t) + n(t) \tag{5.75}$$

The periodic component $s(t)$ present in $r(t)$ can be recoverd by cross-correlating $r(t)$ with another periodic signal $c(t)$ having same frequency as that of $c(t)$. Cross-correlation of $r(t)$ and $c(t)$ is equivalent to passing $r(t)$ through a linear time-invariant filter having the transfer function of $C(-\omega)$. To understand the filtering operation it is required to find the Fourier transform, $C(-\omega)$.

We have seen in chapter 3, that any periodic signal $x(t)$ has the following exponential Fourier series representation.

$$x(t) = \sum_{k=-\infty}^{\infty} X(k)e^{jk\omega_0 t}, \quad \omega_0 = \frac{2\pi}{T_0} \tag{5.76}$$

where $X(k)$ is the Fourier coefficient and is given by

$$X(k) = \frac{1}{T_0} \int_{<T_0>} x(t)e^{-jk\omega_0 t}\, dt$$

By the same token, we can represent $c(t)$ by

$$c(t) = \sum_{k=-\infty}^{\infty} C(k)e^{jk\omega_0 t} \tag{5.77}$$

$$\text{where} \qquad C(k) = \frac{1}{T_0} \int_{<T_0>} c(t)e^{-jk\omega_0 t}\, dt$$

Also, Fourier transform of a periodic signal $x(t)$ is given by

$$X(\omega) = 2\pi \sum_{k=-\infty}^{\infty} X(k)\delta(\omega - k\omega_0) \tag{5.78}$$

Thus, Fourier transform of $c(t)$ can be written as

$$C(\omega) = 2\pi \sum_{k=-\infty}^{\infty} C(k)\delta(\omega - k\omega_0) \tag{5.79}$$

Since $C(-\omega) = C^*(\omega)$, we get

$$C(-\omega) = 2\pi \sum_{k=-\infty}^{\infty} C^*(k)\delta(\omega - k\omega_0) \tag{5.80}$$

Thus, $C(-\omega)$ has impulses located at $\omega = 0, \pm\omega_0, \pm 2\omega_0 \cdots$ etc and each impulse has a strength equal to $C^*(k)$. The output of the filter is the input $R(\omega)$ multiplied by the frequency response, $C(-\omega)$. Hence, the output of the filter consists of a signal with frequency components $\omega = 0, \pm\omega_0, \pm 2\omega_0 \ldots$ etc. Thus, the filter allows to pass through only the frequency components (fundamental frequency and harmonics) of $s(t)$ and also filters out all the noise components. The various frequency components of $s(t)$ are attenuated by $C^*(k)$ and hence, the output of this filter has the same fundamental frequency (ω_0) as that of $s(t)$ but, its waveform has a shape different from that of $s(t)$.

If the output waveform at the filter output must have same shape as that of $s(t)$, it is required that the frequency response of the filter ($C(-\omega)$) must have impulses at $\omega = 0$, $\pm\omega_0, \pm 2\omega_0 \cdots$. etc and these impulses must have a uniform weight, that is, $C^*(k)$ must be identical for all impulses. We know that, Fourier transform of a Dirac comb is a Dirac comb. A Dirac comb is the name given to a periodic impulse train having identical weights for all of its impulses. Thus, for a distortionless recovery of $s(t)$ at the output of the filter, it implies that $C(-\omega)$ must be a Dirac comb. In time-domain, this filtering operation tantamounts to convolution of $r(t)$ with a Dirac comb (a train of impulses defined by $\sum_{k=-\infty}^{\infty} \delta(t - kT_0)$).

Summary of Important points

- The response of an LTI system is found by convoluting input with impulse response of the system.
- We use $x(t) * h(t)$ as a shorthand notation for $\int_{-\infty}^{\infty} x(\tau)h(t-\tau)d\tau$.
- Convolution yields the zero-state response of an LTI system.
- Convolution may be performed in any order.
- Convolution is generally a smoothing and stretching operation, unless the convolved signals contain impulses.
- For stability: $h(t)$ must be absolutely integrable.
- For Causality: $h(t)$ must be zero for negative-time.
- Correlation is an operation similar to convolution. It involves sliding one function past the other and finding the area under the resulting product.
- Correlation is the convolution of one signal with a folded version of the other.

- The autocorrelation involves identical function and hence is commutative.
- Autocorrelation is finite and nonnegative for all t.
- The autocorrelation function is always even symmetric with a maximum at the origin.
- The energy-density spectrum plot of a signal is the squared magnitude of its spectrum.
- Parseval's relation implies that the energy can be found from its spectrum $|X(\omega)|$.
- Convolution and multiplication are duals.

Exercise Problems

EP. 5.1 Convolute the two continuous-time signals given below:

$$x_1(t) = 2u(t+2) - 2u(t-2)$$

$$x_2(t) = \begin{cases} 0, & t < -4 \\ e^{-|t|}, & -4 < t < 4 \\ 0, & t > 4 \end{cases}$$

Ans:

$$x_1(t) * x_2(t) = \begin{cases} 0, & t < -6 \\ 2[e^{t+2} - e^{-4}], & -6 < t < -2 \\ 2[1 - e^{t-2}] + 2[1 - e^{-(t+2)}], & -2 < t < 2 \\ 2[e^{-(t-2)} - e^{-4}], & 2 < t < 6 \\ 0, & t > 6 \end{cases}$$

EP. 5.2 Find $x_1(t) * x_2(t)$, where

$$x_1(t) = u(t-2) - u(t-4)$$
$$x_2(t) = u(t-2) - u(t-4)$$

Ans:

$$x_1(t) * x_2(t) = (t-4)u(t-4) - 2(t-6)u(t-6) + (t-8)u(t-8)$$

EP. 5.3 Find $x_1(t) * x_2(t)$, where

$$x_1(t) = \delta(t-1)$$
$$x_2(t) = \delta(t-2)$$

Ans: $x_1(t) * x_2(t) = \delta(t-3)$

EP. 5.4 An LTI system is characterized by an impulse response given by

$$h(t) = u(t-1) - u(t-3)$$

Find the response of the system to an input given by

$$x(t) = e^{-t}u(t)$$

Ans:

$$y(t) = \begin{cases} 0, & t < 1 \\ 1 - e^{-(t-1)}, & 1 < t < 3 \\ e^{-(t-3)} - e^{-(t-1)}, & t > 3 \end{cases}$$

EP. 5.5 An LTI system is characterized by an impulse response given by

$$h(t) = 2u(1-t)$$

Find the response of the system to an input given by

$$x(t) = e^{t}u(-t)$$

Ans: $y(t) = 2[1 - e^{-(1-t)}]u(1-t)$

EP. 5.6 Find $x(t) * h(t)$, where $h(t)$ and $x(t)$ are as shown in Fig. Ep. 5.6

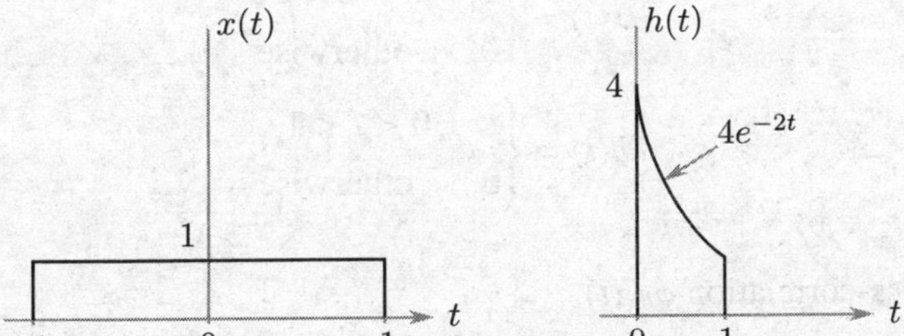

Fig. EP. 5.6

Ans:

$$x(t) * h(t) = \begin{cases} 0, & t < -2 \\ 4[1 - e^{-(t+2)}], & -2 < t < -1 \\ 2.53, & -1 < t < 2 \\ 4[e^{-(t-2)} - e^{-1}], & 2 < t < 3 \\ 0, & t > 3 \end{cases}$$

EP. 5.7 Given:

$$x(t) = \begin{cases} 1, & -1 < t < 0 \\ 0, & \text{elsewhere} \end{cases}$$

Find the autocorrelation, $\phi_{xx}(t)$.

Ans:

$$\phi_{xx}(t) = \begin{cases} 1 + t, & -1 < t < 0 \\ 1 - t, & 0 < t < 1 \\ 0, & \text{elsewhere} \end{cases}$$

EP. 5.8 Given:

$$x(t) = \begin{cases} t, & 0 < t < 1 \\ 0, & \text{elsewhere} \end{cases}$$

Find the autocorrelation, $\phi_{xx}(t)$.

Ans:

$$\phi_{xx}(t) = \begin{cases} -\frac{1}{2}t(t^2 - 1) + \frac{1}{3}(t^3 + 1), & -1 < t < 0 \\ \frac{1}{6}(t - 1)^2(t + 2), & 0 < t < 1 \\ 0, & \text{elsewhere} \end{cases}$$

EP. 5.9 Given:

$$x(t) = \begin{cases} 1, & -1 < t < 0 \\ 0, & \text{otherwise} \end{cases}$$

$$h(t) = \begin{cases} t, & 0 < t < 1 \\ 0, & \text{otherwise} \end{cases}$$

Find the cross-correlation $\phi_{hx}(t)$.

Ans:

$$\phi_{hx}(t) = \begin{cases} \frac{1}{2}t^2, & 0 < t < 1 \\ \frac{1}{2}[1 - (t - 1)^2], & 1 < t < 2 \\ 0, & \text{otherwise} \end{cases}$$

EP. 5.10 Find the energy stored in the signal $x(t) = e^{-t}u(t)$. Find the value of a such that 99% of the signal energy lies in the range $0 \leq t \leq a$. What is the corresponding signal bandwidth B, where B is such that 99% of the signal energy lies in the band $0 \leq \omega \leq B$.

Ans: $a = 0.495, \quad B = 63.66$

EP. 5.11 Find the total energy of the signal $x(t) = e^{-t}u(t)$. What percentage of total energy is contained in the frequency band, $-7 \leq \omega \leq 7$?

Ans: $E_T = \frac{1}{2}$ Joule, 91%.

EP. 5.12 Let $x(t) = e^{-2t}u(t)$. What is ω_B if the frequency band $(-\omega_B \leq \omega \leq \omega_B)$ is to contain 50% of the total signal energy ?

Ans: $|\omega| \leq 2$ rad/second

Chapter 6

Laplace transforms

Fourier transform enables us to do a lot of things. For example, analyze frequency response of LTI systems, sampling, modulation etc. Then why do we need yet another transform ? one view is that Laplace transform is an extension of Fourier transform to allow analysis of broader class of signals and systems. In particular, Fourier transform cannot handle unbounded signals and unstable systems, that is, when

$$\int_{-\infty}^{\infty} |x(t)|\, dt = \infty$$

In many applications, we do need to deal with unstable systems, for example oscillators or stabilizing an inverted pendulum. All such systems are analyzed using Laplace transforms.

The Laplace transform was discovered originally by L. Euler (1707-1783) the eminent swiss mathematician and physicist of 18th-century, but is named in the honour of French mathematician and astronomer P.S. Laplace (1749-1827), who used the transform in his work on astrophysics.

The Laplace transform is a very powerful tool in solving differential equations and it also plays a very significant role in dealing with linear time-invariant systems.

6.1 Definitions of Laplace transforms

A transform is a change in the mathematical description of a physical variable to facilitate computation.

Peirre-Simon Laplace is credited with a transform that bears his name. By definition, Laplace transform $X(s)$ of a function $x(t)$ is given by

$$\mathscr{L}\{x(t)\} = X(s) = \int_{-\infty}^{\infty} x(t)e^{-st}\, dt \tag{6.1}$$

where $\mathscr{L}\{\cdot\}$ indicates the operator for Laplace transform. Definition given by Eq. (6.1) is called bilateral or two-sided Laplace transform. The integral represented by Eq. (6.1) only converges for a bounded region of the complex s-plane. This region is called the region of convergence (or ROC). We shall determine the ROCs with some examples, once we are more comfortable with the Laplace transform.

In order to reconstruct $x(t)$ from $X(s)$, the reverse operation to the Laplace transform must be used. This is the inverse Laplace transform;

$$x(t) = \mathscr{L}^{-1}\{X(s)\} = \frac{1}{2\pi j}\int_{C-j\infty}^{C+j\infty} X(s)e^{st}ds \tag{6.2}$$

The real number C determines the location of the path of integration within the complex s-plane. This path must lie within the ROC, but its exact location and shape does not influence the result.

We now modify Eq. (6.1) to obtain a form of Laplace transform that is useful in many applications. Let us rewrite Eq. (6.1) as

$$X(s) = \int_{-\infty}^{0} x(t)e^{-st}\,dt + \int_{0}^{\infty} x(t)e^{-st}\,dt \tag{6.3}$$

Next, we define $x(t) = 0$ for $t < 0$. Then, the first integral in Eq. (6.3) becomes zero. The resulting transform, called the unilateral, or single-sided Laplace transform is given by

$$X(s) = \int_{0}^{\infty} x(t)e^{-st}\,dt \tag{6.4}$$

Eq. (6.4) is well suited to causal signals that is, $x(t) = 0$ for $t < 0$ and problems with intial conditions.

The equation for the inverse Laplace transform, Eq. (6.2), is the same for both bilateral and unilateral Laplace transforms.

The pairing of a function of time $x(t)$ and its Laplace transform $X(s)$ can also be identified by the symbol $\overset{\textbf{LT}}{\longleftrightarrow}$, that is,

$$x(t) \overset{\textbf{LT}}{\longleftrightarrow} X(s)$$

Unless otherwise specified Laplace transform means bilateral Laplace transform.

6.1.1 The Existance of Laplace transform: Region of Convergence

The Laplace transform of a function $x(t)$ is reiterated below:

$$\mathscr{L}\{x(t)\} = X(s) = \int_{-\infty}^{\infty} x(t)e^{-st}\,dt \tag{6.5}$$

We should stop and ask, under what conditions the integral of Eq. (6.5) converges to a finite value. Naturally, Laplace transform will exist if

$$\int_{-\infty}^{\infty} |x(t)e^{-st}|\, dt < \infty \tag{6.6}$$

Letting $s = \sigma + j\omega$ in Eq. (6.6) gives

$$\int_{-\infty}^{\infty} |x(t)e^{-\sigma t}||e^{-j\omega t}|\, dt < \infty$$

Since,

$$\begin{aligned} |e^{-j\omega t}| &= |\cos\omega t - j\sin\omega t| \\ &= \sqrt{\cos^2\omega t + \sin^2\omega t} \\ &= 1 \end{aligned}$$

Eq. (6.6) reduces to

$$\int_{-\infty}^{\infty} |x(t)e^{-\sigma t}|\, dt < \infty \tag{6.7}$$

The condition imposed in Eq. (6.7) is only a sufficient condition for the existance of Laplace transform. The range of values of σ for which Eq. (6.7) converges is called the Region Of Convergence (ROC).

The examples to follow should help as to know the connection between unilateral and bilateral Laplace transforms and the determination of ROC.

6.1.2 Poles and Zeros of rational Laplace transforms

Complex or real exponential signals have rational Laplace transforms. That is, they are ratios of a numerator polynomial and a denominator polynomial.

$$X(s) = \frac{N(s)}{D(s)}$$

The roots of $N(s) = 0$ are called the zeros of the Laplace transform. If z_1 is a zero, then $X(z_1) = 0$. The roots of $D(s) = 0$ are called the poles of Laplace transform. If p_1 is a pole, then $X(p_1) = \infty$.

For example,

$$X(s) = \frac{2s+1}{s^2+s-2}, \qquad -2 < \sigma < 1$$

has poles at $p_1 = 1$, $p_2 = -2$ and zero at $z_1 = -\frac{1}{2}$.

Example 6.1. Find Laplace transform of an exponential signal defined by $x(t) = e^{-at}u(t)$.

Solution: In Fig. Ex. 6.1(a), we have shown the sketch of a real decaying exponential signal.

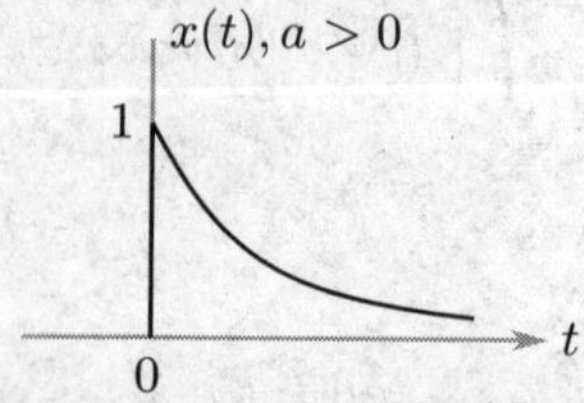

Fig. Ex. 6.1(a)

By definition,

$$\mathscr{L}\{x(t)\} = X(s) = \int_{-\infty}^{\infty} x(t)e^{-st}\,dt$$

$$\Rightarrow \quad X(s) = \int_{-\infty}^{\infty} e^{-at}u(t)e^{-st}\,dt$$

Since,

$$u(t) = \begin{cases} 1, & t > 0 \\ 0, & t < 0 \end{cases}$$

we get

$$\begin{aligned} X(s) &= \int_{0}^{\infty} e^{-at}e^{-st}\,dt \\ &= \int_{0}^{\infty} e^{-(a+s)t}\,dt \\ &= \underbrace{\frac{e^{-(a+s)\infty}}{a+s}}_{\text{Term-I}} + \underbrace{\frac{e^{0}}{a+s}}_{\text{Term-II}} \end{aligned}$$

The first term on the right-hand side of the above equation will be zero if and only if the power of the exponent is negative. Hence, real part of $(a + s)$ is greater than zero, that is,

$$\begin{aligned} & \text{Re}\{s + a\} > 0 \\ \Rightarrow \quad & \text{Re}\{\sigma + j\omega + a\} > 0 \\ \Rightarrow \quad & \sigma + a > 0 \\ \Rightarrow \quad & \sigma > -a \end{aligned}$$

If a is complex, then the ROC will be $\sigma > \mathrm{Re}\{-a\}$.

Summarizing, we can write

$$x(t) = e^{-at}u(t) \overset{\mathbf{LT}}{\longleftrightarrow} X(s) = \frac{1}{s+a}; \quad \mathrm{ROC}: \sigma > -a$$

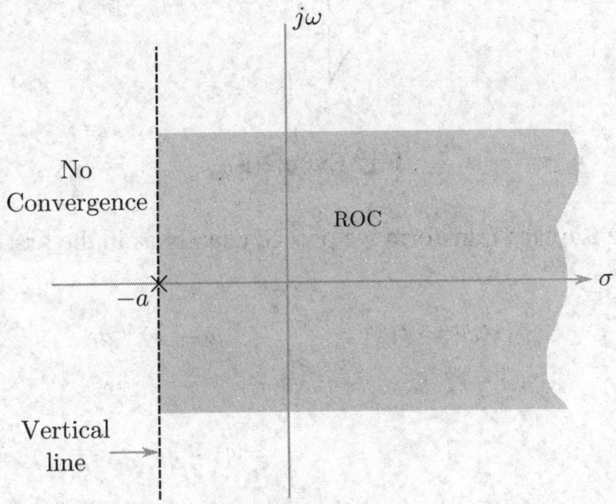

Fig. Ex. 6.1(b)

Since $x(t) = 0$ for $t < 0$, the bilateral and unilateral transforms give identical values.

The ROC in the s-plane is as shown in Fig. Ex. 6.1(b). It contains all the values of s for which the real part of s, namely σ is greater than $-a$. We can express this in this way: a vertical line through the pole divides the s-plane into two halves. The Laplace transform exists for those values of s that are in the right-half.

For $a = 0$, we obtain an important special case:

$$\mathscr{L}\{u(t)\} = \frac{1}{s}; \quad \mathrm{ROC}: \sigma > 0.$$

Example 6.2. Find the Laplace transform of a function $x(t)$ that disappears for $t > 0$. That is,

$$x(t) = -e^{-at}u(-t)$$

Solution: We know that,

$$u(-t) = \begin{cases} 1, & -t > 0 \text{ or } t < 0 \\ 0, & -t < 0 \text{ or } t > 0 \end{cases}$$

Hence,
$$x(t) = \begin{cases} -e^{-at}, & t < 0 \\ 0, & t > 0 \end{cases}$$

In Fig. Ex. 6.2(a), the sketch of $x(t)$ is shown for $a < 0$ and real. From this figure, we find that $x(t)$ is absent for $t > 0$.

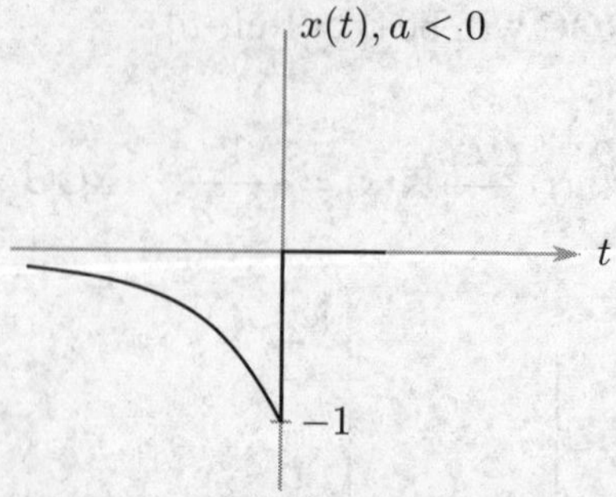

Fig. Ex. 6.2(a)

To determine the Laplace transform we proced exactly as in the first example:

$$\begin{aligned}
\mathscr{L}\{x(t)\} = X(s) &= -\int_{-\infty}^{\infty} e^{-at}u(-t)e^{-st}\,dt \\
&= -\int_{-\infty}^{0} e^{-(a+s)t}\,dt \\
&= \left.\frac{e^{-(a+s)t}}{(a+s)}\right|_{-\infty}^{0} \\
&= \underbrace{\frac{e^{0}}{a+s}}_{\text{Term-I}} - \underbrace{\frac{e^{(a+s)\infty}}{a+s}}_{\text{Term-II}}
\end{aligned}$$

Term II will be zero if and only if real part of the exponent is negative. That is,

$$\begin{aligned}
& \text{Re}\{a+s\} < 0 \\
\Rightarrow\quad & \text{Re}\{a+\sigma+j\omega\} < 0 \\
\Rightarrow\quad & a+\sigma < 0 \\
\Rightarrow\quad & \sigma < -a
\end{aligned}$$

Summarizing, we have

$$X(s) = \frac{1}{s+a}; \quad \text{ROC} : \sigma < -a$$

If a is complex, then the ROC will be $\sigma < \text{Re}\{-a\}$.

The Laplace transform $X(s)$ has the same form as in Example 6.1. The difference lies in the ROC, which in this case is the half of the plane on the left-side of the vertical line. Now, it is clear that different time functions can have the same Laplace transform but with different ROCs. Hence to find inverse Laplace transform, the ROC must be known.

The ROC of $X(s)$ for this example is represented in Fig. Ex. 6.2(b).

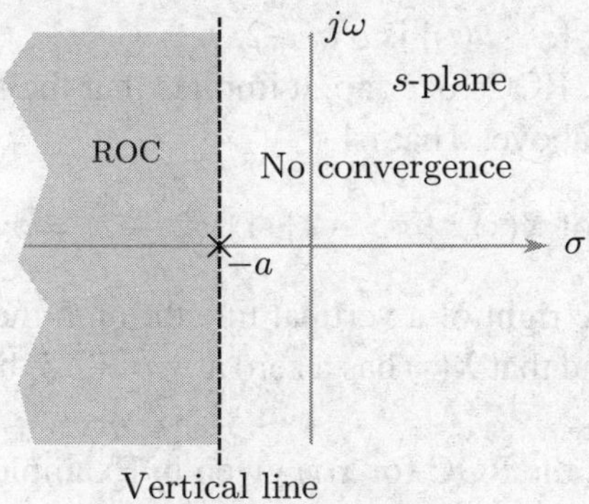

Fig. Ex. 6.2(b)

Example 6.3. Find Laplace transform of the signal given below:

$$x(t) = e^{-t}u(t) + e^{-2t}u(t)$$

Solution: By definition,

$$\begin{aligned}
\mathscr{L}\{x(t)\} &= X(s) \\
&= \int_{-\infty}^{\infty} x(t)e^{-st}\,dt \\
&= \int_{-\infty}^{\infty} [e^{-t} + e^{-2t}]u(t)e^{-st}\,dt \\
&= \int_{0}^{\infty} e^{-(s+1)t}\,dt + \int_{0}^{\infty} e^{-(s+2)t}\,dt \\
&= \left.\frac{e^{-(s+1)t}}{-(s+1)}\right|_0^{\infty} + \left.\frac{e^{-(s+2)t}}{-(s+2)}\right|_0^{\infty} \\
&= -\underbrace{\frac{e^{-(s+1)\infty}}{(s+1)}}_{\text{Term-I}} + \underbrace{\frac{e^{0}}{s+1}}_{\text{Term-II}} - \underbrace{\frac{e^{-(s+2)\infty}}{(s+2)}}_{\text{Term-III}} + \underbrace{\frac{e^{0}}{s+2}}_{\text{Term-IV}} \\
&= \frac{1}{s+1} + \frac{1}{s+2} \\
&= \frac{2s+3}{s^2+3s+2}
\end{aligned}$$

Term I will be zero if and only if $\text{Re}\{s + 1\} > 0$ or $\sigma > -1$. Similarly, Term III will be zero if and only if $\text{Re}\{s + 2\} > 0$ or $\sigma > -2$. Summarizing, we have the ROC of $\mathscr{L}\{e^{-t}u(t)\}$ is $\sigma > -1$ and the ROC of $\mathscr{L}\{e^{-2t}u(t)\}$ is $\sigma > -2$.

As $X(s)$ exists if both the ROCs overlap, it implies that the ROC of $X(s)$ is the intersection of the two ROCs found above. That is,

$$\text{ROC of } X(s) : \{\sigma > -1\} \cap \{\sigma > -2\} = \sigma > -1$$

Thus, the ROC lies to the right of a vertical line through the rightmost pole. Also from the expression of $X(s)$, we find that $X(s)$ has a zero at $s = -1.5$, but this zero has no influence on the ROC.

Fig. Ex. 6.3(a) illustrates the ROC for $x(t)$ given in Example 6.3.

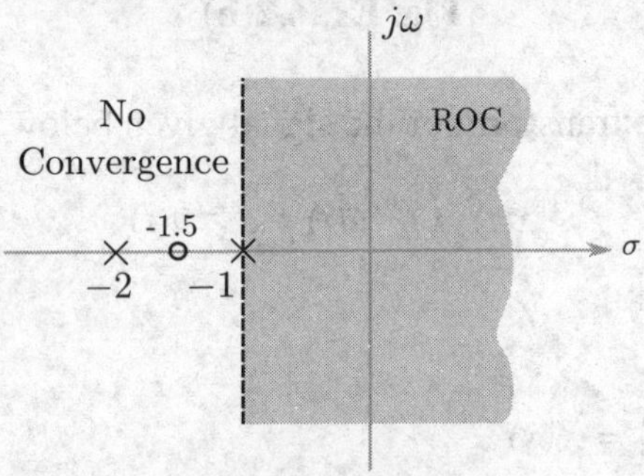

Fig. Ex. 6.3(a)

Example 6.4. Find Laplace transform of the signal $x(t)$ given below:

$$x(t) = \begin{cases} e^{-2t}, & t > 0 \\ -e^{2t}, & t < 0 \end{cases}$$

Solution: The signal given in Ex 6.4 is known as double-sided exponential signal and is sketched in Fig. Ex. 6.4(a).

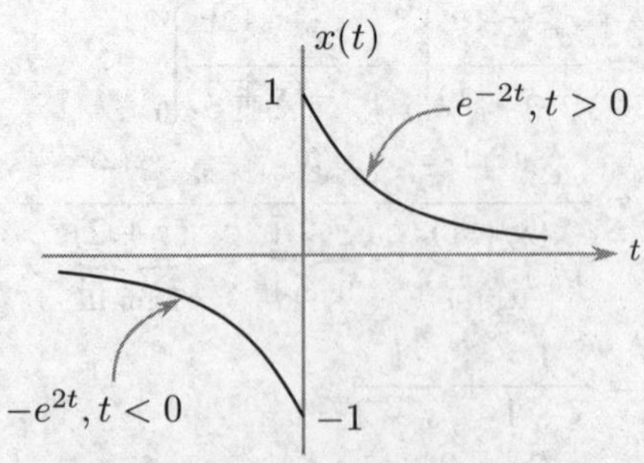

Fig. Ex. 6.4(a)

By definition,

$$\mathscr{L}\{x(t)\} = X(s) = \int_{-\infty}^{\infty} x(t)e^{-st}\,dt$$

$$= \int_{-\infty}^{\infty} [e^{-2t}u(t) - e^{2t}u(-t)]e^{-st}\,dt$$

$$= \int_{0}^{\infty} e^{-(s+2)t}\,dt - \int_{-\infty}^{0} e^{-(s-2)t}\,dt$$

$$= \left.\frac{e^{-(s+2)t}}{-(s+2)}\right|_{0}^{\infty} - \left.\frac{e^{-(s-2)t}}{-(s-2)}\right|_{-\infty}^{0}$$

$$= \underbrace{-\frac{e^{-(s+2)\infty}}{(s+2)}}_{\text{Term-I}} + \underbrace{\frac{1}{s+2}}_{\text{Term-II}} - \underbrace{\frac{e^{(s-2)\infty}}{(s-2)}}_{\text{Term-III}} + \underbrace{\frac{1}{s-2}}_{\text{Term-IV}}$$

Term-I will be zero if and only if, $\text{Re}\{(s+2)\} > 0$ or $\sigma > -2$. Similarly Term-III will be zero if and only if, $\text{Re}\{(s-2)\} < 0$ or $\sigma < 2$.

Summarizing, ROC of $\mathscr{L}\{e^{-2t}u(t)\}$ is $\sigma > -2$ and the ROC of $\mathscr{L}\{-e^{2t}u(-t)\}$ is $\sigma < 2$. Hence, the ROC of $X(s)$ is the intersection of the two ROCs found above. Thus,

$$\text{ROC of } X(s): \; \{\sigma > -2\} \cap \{\sigma < 2\} = -2 < \sigma < 2$$

The ROC of $X(s)$ for the Example 6.4 lies in the strip between the two poles as depicted in Fig. Ex. 6.4(b).

Summarizing, we have

$$X(s) = \frac{1}{s+2} + \frac{1}{s-2} = \frac{2s}{(s+2)(s-2)}; \quad \text{ROC}: -2 < \sigma < 2$$

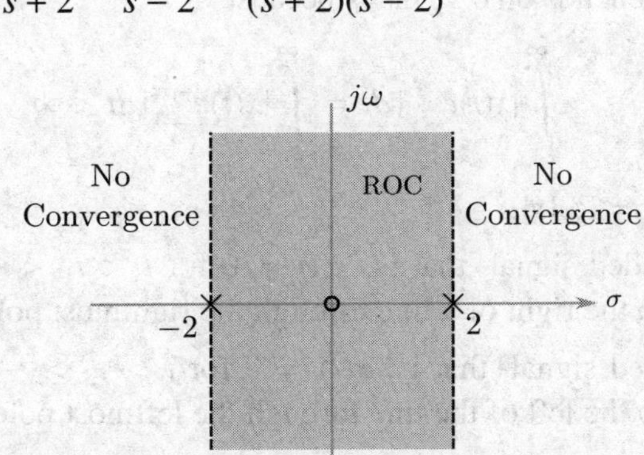

Fig. Ex. 6.4(b)

Example 6.5. Find Laplace transform of the signal given below:

$$x(t) = e^{-t}u(t) - e^{-2t}u(-t)$$

Solution: By definition,

$$X(s) = \int_{-\infty}^{\infty} x(t)e^{-st}\,dt$$

$$\Rightarrow \qquad X(s) = \int_{0}^{\infty} e^{-t}e^{-st}\,dt - \int_{-\infty}^{0} e^{-2t}e^{-st}\,dt$$

$$= \left.\frac{e^{-(s+1)t}}{-(s+1)}\right|_0^{\infty} + \left.\frac{e^{-(s+2)t}}{s+2}\right|_{-\infty}^{0}$$

$$= \underbrace{-\frac{e^{-(s+1)\infty}}{s+1}}_{\text{Term-I}} + \underbrace{\frac{1}{s+1}}_{\text{Term-II}} + \underbrace{\frac{1}{s+2}}_{\text{Term-III}} - \underbrace{\frac{e^{(s+2)\infty}}{s+2}}_{\text{Term-IV}}$$

Term-I will converge to zero when $\text{Re}\{(s+1)\} > 0$ or $\sigma > -1$. Similarly, Term-IV will converge to zero only when $\text{Re}\{s+2\} < 0$ or $\sigma < -2$. Since the ROC of $X(s)$ is $\{\sigma > -1\} \cap \{\sigma < -2\} = \phi$ where ϕ is the null set, it implies the two ROCs do not overlop. Hence, the signal given in Example 6.5 has no Laplace transform.

6.2 Properties of Region of Convergence

The properties that we observed in Example 6.1 through Example 6.5 can be generalized for more complicated signals.

(1) The region of convergence is a strip parallel to the imaginary axis in the s-plane. That is, the ROC only depends on σ. This is because

$$\int_{-\infty}^{\infty} |x(t)e^{-st}|\,dt = \int_{-\infty}^{\infty} |x(t)e^{-\sigma t}|\,dt < \infty$$

depends only on $\sigma = \text{Re}\{s\}$.

(2) If $x(t)$ is a right-sided signal, that is, $x(t) = 0$ for $t < t_1 < \infty$ then, the region of convergence lies to the right of a line through the rightmost pole.

(3) If $x(t)$ is a left-sided signal, that is, $x(t) = 0$ for $t > t_2 > -\infty$, then the region of convergence lies to the left of the line through the leftmost pole.

(4) When $x(t)$ is bilateral—the sum of right-sided and left-sided signals, the region of convergence is a strip between two poles, as long as the right-sided and left-sided regions of convergence overlap.

(5) The region of convergence does not contain any poles.

This property can be proved as follows:

Let

$$X(s) = \frac{N(s)}{D(s)}$$

Poles are places where $D(s) = 0$.

$$\Rightarrow \quad X(s) = \frac{N(s)}{D(s)} = \infty$$

$$\Rightarrow \quad X(s) \text{ is not convergent.}$$

(6) If $x(t)$ is a finite duration signal, that is, $x(t) = 0$ except in a finite interival $t_1 < t < t_2$, then the region of convergence is the entire s-plane except possibly $s = 0$ or $s = \infty$.

(7) If ROC of $X(s)$ includes the $j\omega$-axis, then Fourier transform of $x(t)$ exists.

6.3 Laplace transforms for common signals

When dealing with LTI systems, it is sufficient just to know some of the more frequently occurring transform pairs. The most commonly used are summarized is Table 6.1. Instead of having to re-evaluate the Laplace transform of a given signal, we can simply refer to such a table and read out the desired transform.

Table 6.1: Some important Laplace transform pairs; Re{s} is denoted by σ.

$x(t)$	$X(s)$	ROC
$u(t)$	$\frac{1}{s}$	$\sigma > 0$
$e^{-at}u(t)$	$\frac{1}{s+a}$	$\sigma > \text{Re}\{-a\}$
$-e^{-at}u(-t)$	$\frac{1}{s+a}$	$\sigma < \text{Re}\{-a\}$
$te^{-at}u(t)$	$\frac{1}{(s+a)^2}$	$\sigma > \text{Re}\{-a\}$
$t^n e^{-at}u(t)$	$\frac{n!}{(s+a)^{n+1}}$	$\sigma > \text{Re}\{-a\}$
$tu(t)$	$\frac{1}{s^2}$	$\sigma > 0$
$t^n u(t)$	$\frac{n!}{s^{n+1}}$	$\sigma > 0$
$\sin \omega_0 t\, u(t)$	$\frac{\omega_0}{s^2+\omega_0^2}$	$\sigma > 0$
$\cos \omega_0 t\, u(t)$	$\frac{s}{s^2+\omega_0^2}$	$\sigma > 0$
$e^{-at} \cos \omega_0 t\, u(t)$	$\frac{s+a}{(s+a)^2+\omega_0^2}$	$\sigma > \text{Re}\{-a\}$
$e^{-at} \sin \omega_0 t\, u(t)$	$\frac{\omega_0}{(s+a)^2+\omega_0^2}$	$\sigma > \text{Re}\{-a\}$

Example 6.6. Find the Laplace transform of the function given below:

$$x(t) = t\,u(t)$$

Solution: Rewrite $x(t)$ as

$$x(t) = \begin{cases} t, & t > 0 \\ 0, & t < 0 \end{cases}$$

By definition, we have

$$\mathscr{L}\{x(t)\} = X(s) = \int_{-\infty}^{\infty} x(t)e^{-st}\,dt$$

$$\Rightarrow \qquad X(s) = \int_{0}^{\infty} te^{-st}\,dt$$

Recall the standard integral:

$$\int te^{at}\,dt = \frac{e^{at}}{a^2}[at - 1]$$

Hence,

$$X(s) = -\frac{e^{-st}}{s^2}(st + 1)\Big|_0^{\infty}$$

$$= \frac{e^{-s\infty}}{s^2}(\infty t + 1) + \frac{1}{s^2}$$

The first term on the right-hand side of the above equation will be zero only when $\text{Re}\{s\} > 0$. That is, $\text{Re}\{\sigma + j\omega\} > 0$. Hence, the ROC of $X(s)$ is $\sigma > 0$.

Thus, $tu(t) \overset{\text{LT}}{\longleftrightarrow} \frac{1}{s^2}$; ROC : $\sigma > 0$.

The sketch of ROC is as shown in Fig. Ex. 6.6(a).

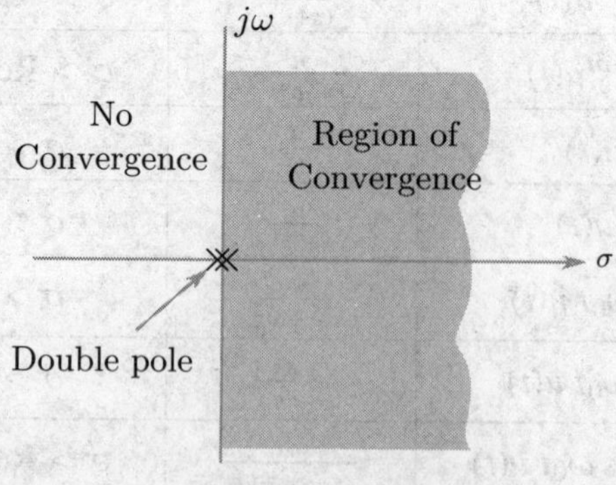

Fig. Ex. 6.6(a)

Example 6.7. Find Laplace transform of

$$x(t) = \cos \omega_0 t\, u(t)$$

Solution: By definition,

$$\begin{aligned}
X(s) &= \int_{-\infty}^{\infty} x(t)e^{-st}\, dt \\
&= \int_{-\infty}^{\infty} \cos \omega_0 t\, u(t)e^{-st}\, dt \\
&= \int_{0}^{\infty} \cos \omega_0 t e^{-st}\, dt \\
&= \int_{0}^{\infty} \left[\frac{e^{j\omega_0 t} - e^{-j\omega_0 t}}{2}\right] e^{-st}\, dt \\
&= \frac{1}{2}\int_{0}^{\infty} e^{-(s-j\omega_0)t}\, dt + \frac{1}{2}\int_{0}^{\infty} e^{-(s+j\omega_0)t}\, dt \\
&= \frac{1}{2}\left[\frac{-e^{-(s-j\omega_0)t}}{(s-j\omega_0)}\right]_0^{\infty} + \frac{1}{2}\left[\frac{-e^{-(s+j\omega_0)t}}{(s+j\omega_0)}\right]_0^{\infty} \\
&= -\underbrace{\frac{1}{2}\frac{e^{-(s-j\omega_0)\infty}}{(s-j\omega_0)}}_{\text{Term-I}} + \underbrace{\frac{1}{2(s-j\omega_0)}}_{\text{Term-II}} - \underbrace{\frac{1}{2}\frac{e^{-(s+j\omega_0)\infty}}{s+j\omega_0}}_{\text{Term-III}} + \underbrace{\frac{1}{2(s+j\omega_0)}}_{\text{Term-IV}}
\end{aligned}$$

Term-I will be zero if and only if $\text{Re}\{s - j\omega_0\} > 0$ or $\text{Re}\{\sigma + j\omega - j\omega_0\} > 0 \Rightarrow \sigma > 0$. Hence, the region of convergence of $\mathscr{L}\left\{\frac{1}{2}e^{j\omega_0 t}u(t)\right\}$ is $\text{ROC}_1 : \sigma > 0$. Similarly, Term-III will be zero only when $\text{Re}\{s + j\omega_0\} > 0 \Rightarrow \text{Re}\{\sigma + j\omega + j\omega_0\} > 0 \Rightarrow \sigma > 0$. Thus, the region of convergence of $\mathscr{L}\left\{\frac{1}{2}e^{-j\omega_0 t}u(t)\right\}$ is $\text{ROC}_2 : \sigma > 0$. Hence, the ROC of $X(s) = \text{ROC}_1 \cap \text{ROC}_2 = \sigma > 0$.

Summarizing, we have

$$\begin{aligned}
X(s) &= \underbrace{\frac{1}{2(s-j\omega_0)}}_{\text{ROC}_1:\sigma>0} + \underbrace{\frac{1}{2(s+j\omega_0)}}_{\text{ROC}_2:\sigma>0} \\
&= \frac{s}{s^2+\omega_0^2}; \quad \text{ROC} : \sigma > 0
\end{aligned}$$

Example 6.8. By using the defining equation for Bilateral Laplace transforms, find $X(s)$ for the signals given below:

(a) $x(t) = u(t) - u(t-1)$ (b) $x(t) = te^{-t}u(t)$

Solution:

(a) Given:

$$x(t) = u(t) - u(t-1)$$

$$\Rightarrow \quad x(t) = \begin{cases} 1, & 0 < t < 1 \\ 0, & \text{otherwise} \end{cases}$$

Hence,

$$X(s) = \int_0^1 1 \cdot e^{-st}\, dt$$

$$= -\frac{1}{s}\left[e^{-s} - 1\right]$$

$$= \frac{1}{s}\left[1 - e^{-s}\right]$$

The above result is valid for all values of s. Hence, the ROC is the entire s-plane.

(b) Given:

$$x(t) = te^{-t}u(t)$$

$$\Rightarrow \quad x(t) = \begin{cases} te^{-t}, & t > 0 \\ 0, & t < 0 \end{cases}$$

Hence,

$$X(s) = \int_{-\infty}^{\infty} x(t)e^{-st}\, dt$$

$$= \int_0^{\infty} te^{-t}e^{-st}\, dt$$

$$= \int_0^{\infty} te^{-(s+1)t}\, dt$$

Recall:

$$\int te^{at}\, dt = \frac{e^{at}}{a^2}[at - 1]$$

Thus,

$$X(s) = \frac{e^{-(s+1)t}}{(s+1)^2}[-(s+1)t - 1]\bigg|_0^\infty$$

The term, $\dfrac{e^{-(s+1)\infty}}{(s+1)^2}$ will become zero if and only if

$$\text{Re}\{s+1\} > 0 \Rightarrow \text{Re}\{\sigma + j\omega + 1\} > 0 \Rightarrow \sigma + 1 > 0 \Rightarrow \sigma > -1$$

Hence,

$$X(s) = \frac{1}{(s+1)^2}; \quad \text{ROC} : \sigma > -1$$

Example 6.9. By direct integration, find Laplace transform of the signal given below:

$$x(t) = t \cos \omega_0 t \, u(t)$$

Solution: By definition,

$$X(s) = \int_{-\infty}^{\infty} x(t) e^{-st} \, dt$$

$$\Rightarrow \quad X(s) = \int_{-\infty}^{\infty} t \cos \omega_0 t \, u(t) e^{-st} \, dt$$

$$= \int_0^\infty t \times \frac{1}{2}\left[e^{j\omega_0 t} + e^{-j\omega_0 t}\right] e^{-st} \, dt$$

$$= \frac{1}{2}\int_0^\infty t e^{-(s-j\omega_0)t} \, dt + \frac{1}{2}\int_0^\infty t e^{-(j\omega_0+s)t} \, dt$$

$$= \frac{1}{2}\left[\frac{e^{-(s-j\omega_0)t}}{(s-j\omega_0)^2}\{-(s-j\omega_0)t - 1\}\right]_0^\infty$$

$$+ \frac{1}{2}\left[\frac{e^{-(s+j\omega_0)t}}{(s+j\omega_0)^2}\{-(s+j\omega_0)t - 1\}\right]_0^\infty$$

It may be noted that $e^{-(s\pm j\omega_0)\infty}$ will be zero only when

$$\text{Re}\{(s \pm j\omega_0)\} > 0$$

$$\Rightarrow \text{Re}\{(\sigma + j\omega \pm j\omega_0)\} > 0 \Rightarrow \sigma > 0$$

Hence,

$$X(s) = \frac{1}{2}\left[\frac{1}{(s - j\omega_0)^2} + \frac{1}{(s + j\omega_0)^2}\right]$$
$$= \frac{s^2 - \omega_0^2}{(s^2 + \omega_0^2)^2}; \quad \text{ROC} : \sigma > 0$$

Example 6.10. Find $X(s)$ by direct integration for the signal given below:

$$x(t) = \cosh at \;\; u(t)$$

Solution: By definition,

$$X(s) = \int_{-\infty}^{\infty} x(t)e^{-st}\, dt$$
$$= \int_{-\infty}^{\infty} \cosh at \;\; u(t)e^{-st}\, dt$$
$$= \int_{0}^{\infty} \left[\frac{e^{at} + e^{-at}}{2}\right] e^{-st} dt$$
$$= \frac{1}{2}\int_{0}^{\infty} e^{-(s-a)t}\, dt + \frac{1}{2}\int_{0}^{\infty} e^{-(s+a)t}\, dt$$
$$= \frac{1}{2}\left[\frac{-e^{-(s-a)t}}{(s-a)}\right]_0^{\infty} + \frac{1}{2}\left[\frac{-e^{-(s+a)t}}{(s+a)}\right]_0^{\infty}$$
$$= \underbrace{-\frac{1}{2}\frac{e^{-(s-a)\infty}}{(s-a)}}_{\text{Term-I}} + \underbrace{\frac{1}{2(s-a)}}_{\text{Term-II}} \underbrace{- \frac{1}{2}\frac{e^{-(s+a)\infty}}{(s+a)}}_{\text{Term-III}} + \underbrace{\frac{1}{2(s+a)}}_{\text{Term-IV}}$$

Term-I will be zero only when $\text{Re}\{(s-a)\} > 0 \Rightarrow \text{Re}\{(\sigma + j\omega - a)\} > 0 \Rightarrow \sigma - a > 0 \Rightarrow \sigma > a$. Hence, the ROC of $\mathscr{L}\left\{\frac{1}{2}e^{at}u(t)\right\}$ is $\text{ROC}_1 = \sigma > a$.

Term-III will be zero only when $\text{Re}\{s + a\} > 0 \Rightarrow \text{Re}\{\sigma + j\omega + a\} > 0 \Rightarrow \sigma > -a$. Hence, the ROC of $\mathscr{L}\left\{\frac{1}{2}e^{-at}u(t)\right\}$ is $\text{ROC}_2 = \sigma > -a$.

Thus, the region of convergence of $X(s)$ is

$$\begin{aligned}\text{ROC} &= \text{ROC}_1 \cap \text{ROC}_2\\ &= \{\sigma > a\} \cap \{\sigma > -a\}\\ &= \begin{cases} a, & a > 0\\ -a, & a < 0\end{cases}\\ &= |a|\end{aligned}$$

Thus,

$$\begin{aligned}X(s) &= \underbrace{\frac{1}{2(s-a)}}_{\text{ROC}_1:\sigma>a} + \underbrace{\frac{1}{2(s+a)}}_{\text{ROC}_2:\sigma>-a}\\ &= \frac{s}{s^2-a^2}; \quad \text{ROC}: \sigma > |a|\end{aligned}$$

Example 6.11. Find Laplace transform of the waveform shown in Fig. Ex. 6.11 by direct integration.

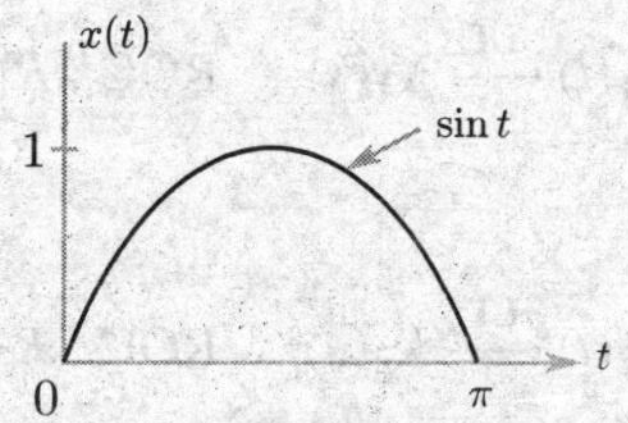

Fig. Ex. 6.11

Solution: From Fig. Ex. 6.11, we can write

$$x(t) = \begin{cases}\sin t, & 0 < t < \pi\\ 0, & \text{otherwise}\end{cases}$$

Hence,

$$\begin{aligned}\mathscr{L}\{x(t)\} &= X(s)\\ &= \int_0^{\pi} \sin t\, e^{-st}\, dt\end{aligned}$$

Recall the standard integral:

$$\int a^{ax} \sin bx\, dx = \frac{e^{ax}}{a^2+b^2}[a \sin bx - b\cos bx]$$

Therefore,

$$X(s) = \frac{e^{-st}}{s^2+1}[-s\sin t - \cos t]\Big|_{t=0}^{\pi}$$

$$= \frac{1+e^{-\pi s}}{s^2+1}; \quad \text{ROC: All } s$$

6.4 Properties of the Laplace transform

Properties present the frequency-domain effect of time-domain characteristics and operations on signals and vice versa. In addition, they are used to find new transform pairs more easily.

6.4.1 Linearity

The Laplace transform of a linear combination of signals is the same linear combination of their individual Laplace transforms. Mathematically, this property can be stated as follows:

If

$$x_1(t) \overset{\textbf{LT}}{\longleftrightarrow} X_1(s); \quad \text{ROC}: R_1$$

and

$$x_2(t) \overset{\textbf{LT}}{\longleftrightarrow} X_2(s); \quad \text{ROC}: R_2$$

then

$$y(t) = a_1x_1(t) + a_2x_2(t) \overset{\textbf{LT}}{\longleftrightarrow} Y(s) = a_1X_1(s) + a_2X_2(s); \quad \text{ROC}: R_y \supset R_1 \cap R_2$$

The set notation $A \supset B$ means that set A contains set B, while $A \cap B$ denotes the intersection of sets A and B, that is, the set containing all elements common to A and B. Thus, R_y is atleast as large as the intersection of the ROCs R_1 and R_2. ROC: R_y can be bigger due to pole-zero cancellation.

Proof: From the definition of Laplace transform we have

$$Y(s) = \int_{-\infty}^{\infty} y(t)e^{-st}\,dt$$

$$= \int_{-\infty}^{\infty} [a_1x_1(t) + a_2x_2(t)]e^{-st}\,dt$$

$$= a_1 \int_{-\infty}^{\infty} x_1(t)e^{-st}\, dt + a_2 \int_{-\infty}^{\infty} x_2(t)e^{-st}\, dt$$

$$= a_1X_1(s) + a_2X_2(s)$$

ROC: R_y can be greater than or equal to $R_1 \cap R_2$.

6.4.2 Time shifting

If

$$x(t) \overset{\textbf{LT}}{\longleftrightarrow} X(s); \quad \text{ROC} : R_x$$

then

$$y(t) = x(t - t_0) \overset{\textbf{LT}}{\longleftrightarrow} Y(s) = e^{-t_0 s}X(s); \quad \text{ROC} : R_y = R_x.$$

Proof: By definition, we have

$$Y(s) = \int_{-\infty}^{\infty} y(t)e^{-st}\, dt$$

$$= \int_{-\infty}^{\infty} x(t - t_0)e^{-st}\, dt$$

Letting $t - t_0 = \lambda$, we get

$$Y(s) = \int_{\lambda=-\infty}^{\infty} x(\lambda)e^{-s(t_0+\lambda)} d\lambda$$

$$= e^{-t_0 s} \int_{-\infty}^{\infty} x(\lambda)e^{-s\lambda} d\lambda$$

$$= e^{-t_0 s}X(s)$$

ROC: $R_y = R_x$

6.4.3 Shifting in s-domain

If

$$x(t) \overset{\textbf{LT}}{\longleftrightarrow} X(s); \quad \text{ROC} : R_x$$

then

$$y(t) = e^{s_0 t}x(t) \overset{\textbf{LT}}{\longleftrightarrow} Y(s) = X(s - s_0); \quad \text{ROC} : R_y = R_x + \text{Re}\{s_0\} = R_x + \sigma_0$$

Proof: By definition, we have

$$Y(s) = \int_{-\infty}^{\infty} y(t)e^{-st}\,dt$$

$$= \int_{-\infty}^{\infty} e^{s_0 t}x(t)e^{-st}\,dt$$

$$= \int_{-\infty}^{\infty} x(t)e^{-(s-s_0)t}\,dt$$

$$= X(s-s_0)$$

ROC: $R_y = R_x + \text{Re}\{s_0\}$

6.4.4 Time-scaling

If

$$x(t) \overset{\textbf{LT}}{\longleftrightarrow} X(s); \quad \text{ROC} : R_x$$

then

$$y(t) = x(\alpha t) \overset{\textbf{LT}}{\longleftrightarrow} Y(s) = \frac{1}{|\alpha|}X\left(\frac{s}{\alpha}\right); \quad \text{ROC} : R_y = \frac{R_x}{\alpha}$$

Proof: By using the defining equation for Laplace transform, we have

$$Y(s) = \int_{t=-\infty}^{\infty} y(t)e^{-st}\,dt = \int_{-\infty}^{\infty} x(\alpha t)e^{-st}\,dt$$

Let $\alpha t = \lambda$ and $\alpha > 0 \Rightarrow \alpha dt = d\lambda$.

Hence,

$$Y(s) = \int_{\lambda=-\infty}^{\infty} x(\lambda)e^{-(\frac{s}{\alpha})\lambda}\frac{d\lambda}{\alpha}$$

$$= \frac{1}{\alpha}\int_{-\infty}^{\infty} x(\lambda)e^{-(\frac{s}{\alpha})\lambda}d\lambda$$

$$= \frac{1}{\alpha}X\left(\frac{s}{\alpha}\right)$$

Similarly, when $\alpha < 0$, we can show that

$$Y(s) = -\frac{1}{\alpha}X\left(\frac{s}{\alpha}\right)$$

Hence, summarizing the results, we can write

$$Y(s) = \frac{1}{|\alpha|}X\left(\frac{s}{\alpha}\right); \text{ ROC} : R_y = \frac{R_x}{\alpha}$$

where the new ROC is the original one, expanded or contracted by $\frac{1}{|\alpha|}$ and flipped around imaginary axis if $\alpha < 0$.

As an example, consider a signal $x(t)$ whose Laplace transform has the ROC: R_x shown in Fig. 6.1(a). After the time expansion and reversal $x(-0.5t)$, the resulting Laplace transform has the ROC shown in Fig. 6.1(b).

$$y(t) = x(-0.5t) \overset{\text{LT}}{\longleftrightarrow} Y(s) = 2X(-2s); \quad \text{ROC} : R_y = -2R_x$$

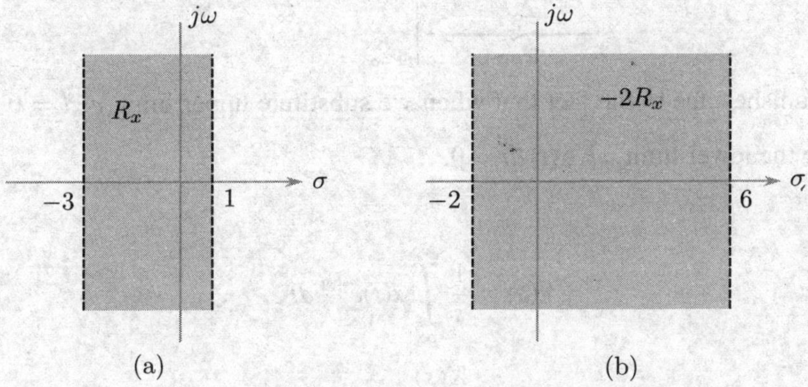

Fig 6.1: *Expansion of* ROC *after time-scaling*

6.4.5 Time-domain integration

If

$$x(t) \overset{\text{LT}}{\longleftrightarrow} X(s); \quad \text{ROC} : R_x,$$

then

$$y(t) = \int_{-\infty}^{t} x(t)\,dt \overset{\text{LT}}{\longleftrightarrow} Y(s) = \frac{1}{s}X(s); \quad \text{ROC} : R_y = R_x \cap \{\sigma > 0\}$$

Proof: By definition of Laplace transform, we have

$$Y(s) = \int_{-\infty}^{\infty} y(t)e^{-st}\,dt$$

$$\text{But} \qquad y(t) = \int_{-\infty}^{t} x(t)\,dt$$

Hence,

$$Y(s) = \int_{-\infty}^{\infty} \left\{ \int_{-\infty}^{t} x(t)\,dt \right\} e^{-st}\,dt$$

Carrying out integration by parts, we get

$$Y(s) = \underbrace{\frac{-e^{-st}}{s} \int_{-\infty}^{t} x(t)\,dt}_{\text{Term-I}} \Bigg|_{t=-\infty}^{\infty} + \frac{1}{s}\int_{-\infty}^{\infty} x(t)e^{-st}\,dt$$

Term-I vanishes due to the fact that when we substitute upper limit, $e^{-st} = 0$ and when we substitute the lower limit, $\int_{-\infty}^{t} x(t)\,dt = 0$.

Hence,

$$Y(s) = \frac{1}{s}\int_{-\infty}^{\infty} x(t)e^{-st}\,dt$$

$$= \frac{X(s)}{s}$$

ROC: $R_y = R_x \cap \{\sigma > 0\}$

The form of the ROC : R_y follows from the possible introduction of an additional pole at $s = 0$ due to the multiplication by $\frac{1}{s}$.

6.4.6 Differentiation in time-domain

If

$$x(t) \xleftrightarrow{\text{LT}} X(s); \quad \text{ROC} : R_x$$

then

$$y(t) = \frac{dx(t)}{dt} \xleftrightarrow{\text{LT}} Y(s) = sX(s) - x(0); \quad \text{ROC} : R_y \supset R_x$$

It may be noted that $R_y = R_x$ unless a pole-zero cancellation exists at $s = 0$.

Proof: By definition,

$$x(t) = \frac{1}{2\pi j} \int_{c-j\infty}^{c+j\infty} X(s)e^{st} ds \tag{6.8}$$

Differentiating both sides of the above equation with respect to t, we obtain

$$\frac{dx(t)}{dt} = \frac{1}{2\pi j} \int_{c-j\infty}^{c+j\infty} sX(s)e^{st} ds \tag{6.9}$$

Comparing Eq. (6.9) with Eq. (6.8), we conclude that $\frac{dx(t)}{dt}$ is the inverse Laplace transform of $sX(s)$.

Hence,

$$y(t) = \frac{dx(t)}{dt} \overset{\mathbf{LT}}{\longleftrightarrow} Y(s) = sX(s); \quad \text{ROC} : R_y \supset R_x$$

6.4.7 Differentiation in s-domain

If

$$x(t) \overset{\mathbf{LT}}{\longleftrightarrow} X(s); \quad \text{ROC} : R_x$$

then

$$y(t) = -tx(t) \overset{\mathbf{LT}}{\longleftrightarrow} Y(s) = \frac{dX(s)}{ds}; \quad \text{ROC} : R_y = R_x$$

Proof: By definition,

$$X(s) = \int_{-\infty}^{\infty} x(t)e^{-st}\, dt \tag{6.10}$$

Differentiating both sides of the above equation with respect to s, we get

$$\frac{dX(s)}{ds} = \int_{-\infty}^{\infty} [-tx(t)]e^{-st}\, dt \tag{6.11}$$

Comparing Eq. (6.11) with Eq. (6.10), we conclude that $\frac{dX(s)}{ds}$ is the Laplace transform of $[-tx(t)]$.

Hence,

$$y(t) = -tx(t) \overset{\mathbf{LT}}{\longleftrightarrow} Y(s) = \frac{dX(s)}{ds}; \quad \text{ROC} : R_y = R_x$$

6.4.8 Convolution

If

$$x(t) \overset{\text{LT}}{\longleftrightarrow} X(s); \quad \text{ROC} : R_x$$

and

$$h(t) \overset{\text{LT}}{\longleftrightarrow} H(s); \quad \text{ROC} : R_h$$

then

$$y(t) = x(t) * h(t) \overset{\text{LT}}{\longleftrightarrow} Y(s) = X(s)H(s); \quad \text{ROC} : R_y \supset R_x \cap R_h$$

Thus, convolution in time-domain corresponds to multiplication in the complex frequency-domain. This is an extremely useful property for LTI system analysis. Note that the resulting ROC(R_y) includes the intersection of two original ROCs (R_x and R_y), but it may be larger, for example, when pole-zero concellation occurs.

Proof: From the definition of Laplace transform, we have

$$Y(s) = \int_{t=-\infty}^{\infty} y(t)e^{-st}\,dt$$

$$= \int_{t=-\infty}^{\infty} [x(t) * h(t)]e^{-st}\,dt \tag{6.12}$$

From the definition of convolution, we have

$$x(t) * h(t) = \int_{\tau=-\infty}^{\infty} x(\tau)h(t-\tau)\,d\tau \tag{6.13}$$

Substituting Eq. (6.13) in Eq. (6.12), we get

$$Y(s) = \int_{t=-\infty}^{\infty} \int_{\tau=-\infty}^{\infty} x(\tau)h(t-\tau)d\tau\, e^{-st}\,dt \tag{6.14}$$

Let $t - \tau = \lambda \quad \Rightarrow dt = d\lambda$
Then, Eq. (6.14) becomes

$$Y(s) = \int_{\lambda=-\infty}^{\infty} \int_{\tau=-\infty}^{\infty} x(\tau)h(\lambda)\,d\tau e^{-s(\lambda+\tau)} d\lambda$$

$$= \int_{\tau=-\infty}^{\infty} x(\tau)e^{-s\tau} d\tau \int_{\lambda=-\infty}^{\infty} h(\lambda)e^{-s\lambda} d\lambda$$

$$= X(s)H(s)$$

ROC: $R_y \supset R_x \cap R_h$.

Table 6.2 summarizes the properties of Laplace transform presented in this section.

Table 6.2: Properties of the Laplace transform

Property	Signal	Transform	ROC
	$x(t)$	$X(s)$	R_x
	$x_1(t)$	$X_1(s)$	R_1
	$x_2(t)$	$X_2(s)$	R_2
	$y(t)$	$Y(s)$	R_y
Linearity	$y(t) = a_1x_1(t) + a_2x_2(t)$	$Y(s) = a_1X_1(s) + a_2X_2(s)$	$R_y \supset R_1 \cap R_2$
Time shifting	$y(t) = x(t - t_0)$	$Y(s) = e^{-t_0 s}X(s)$	$R_y = R_x$
Shifting in s	$y(t) = e^{s_0 t}x(t)$	$Y(s) = X(s - s_0)$	$R_y = R_x + \mathrm{Re}\{s_0\}$
Time-scaling	$y(t) = x(\alpha t)$	$Y(s) = \frac{1}{\lvert\alpha\rvert}X(\frac{s}{\alpha})$	$R_y = \frac{R_x}{\alpha}$
Differentiation in t	$y(t) = \frac{dx(t)}{dt}$	$Y(s) = sX(s)$	$R_y \supset R_x$
Differentiation in s	$y(t) = -tx(t)$	$Y(s) = \frac{dX(s)}{ds}$	$R_y = R_x$
Integration	$y(t) = \int_{-\infty}^{t} x(t)\,dt$	$Y(s) = \frac{1}{s}X(s)$	$R_y \supset R_x \cap \{\sigma > 0\}$
Convolution	$y(t) = x(t) * h(t)$	$Y(s) = X(s)H(s)$	$R_y \supset R_1 \cap R_2$

6.5 Partial fraction expansion

The partial fraction expansion of $X(s)$ requires the following two conditions:

(i) $X(s)$ must be a strictly proper rational function and

(ii) a denominator in factored form. The structure of expansion depends on the nature of the factors in $Q(s)$. The constants in the numerator of partial fraction expansion are called residues.

6.5.1 Distinct factors

Let

$$X(s) = \frac{N(s)}{D(s)}$$

If $D(s)$ contains only distinct factors, we can write $X(s)$ in the form

$$X(s) = \frac{N(s)}{(s + p_1)(s + p_2)\cdots(s + P_N)}$$

$$= \frac{K_1}{s + p_1} + \frac{K_2}{s + p_2} + \cdots + \frac{K_N}{s + p_N}$$

To find i^{th} coefficient, K_i, we multiply both sides by $(s + p_i)$ to get

$$(s + p_i)X(s) = \frac{K_1(s + p_i)}{(s + p_1)} + \cdots + K_i + \cdots + \frac{K_N(s + p_i)}{(s + p_N)}$$

Let us put $s = -p_i$ on both sides. This gives

$$K_i = (s + p_i)X(s)\big|_{s=-p_i}$$

Let $D(s)$ contain some complex factors: $(s+C_1), (s+C_1^*), (s+C_2), (s+C_2^*)\ldots$ etc. Then, we can write $X(s)$ in the form

$$X(s) = \frac{K_1}{s + p_1} + \frac{K_2}{s + p_2} + \cdots + \frac{A_1}{s + C_1} + \frac{A_2}{s + C_1^*} + \frac{A_3}{s + C_2} + \frac{A_4}{s + C_2^*} + \cdots$$

For a real root, the coefficient (residue) will also be real. For each pair of complex conjugate roots, the residues will also be complex conjugates, that is, $A_i = A_i^*$, and we need to compute only one of these.

6.5.2 Repeated factors

If a factor is repeated m times, it is said to have a multiplicity of m. Let $X(s)$ be of the form

$$\begin{aligned} X(s) &= \frac{N(s)}{(s + p_1)(s + \alpha)^m} \\ &= \frac{K_1}{s + p_1} + \frac{A_0}{(s + \alpha)^m} + \frac{A_1}{(s + \alpha)^{m-1}} + \cdots + \frac{A_{m-1}}{s + \alpha} \end{aligned}$$

The coefficient K_1 associated with $(s + p_1)$ is found as follows:

$$K_1 = X(s)(s + p_1)\big|_{s=-p_1}$$

The coefficients A_i for the repeated factor require $(s + \alpha)^m X(s)$ and its derivatives for their evaluation.

$$\begin{aligned} A_0 &= (s + \alpha)^m X(s)\big|_{s=-\alpha} \\ A_1 &= \frac{1}{1!}\frac{d}{ds}\left[(s + \alpha)^m X(s)\right]\bigg|_{s=-\alpha} \\ &\vdots \\ A_{m-1} &= \frac{1}{(m-1)!}\frac{d^{m-1}}{ds^{m-1}}\left[(s + \alpha)^m X(s)\right]\bigg|_{s=-\alpha} \end{aligned}$$

Example 6.12. An LTI system is characterized by an impulse response given by $h(t) = [e^{-2t} + e^{-t}]u(t)$. Find the response of the system for an input

$$x(t) = -e^{-2t}u(t) + \delta(t)$$

Solution: Recall:

$$e^{-at}u(t) \overset{\mathbf{LT}}{\longleftrightarrow} \frac{1}{s+a}; \quad \text{ROC} : \sigma > -a$$

The Laplace transform of $\delta(t)$ is found as follows:

$$\mathscr{L}\{\delta(t)\} = \int_{-\infty}^{\infty} \delta(t)e^{-st}\, dt$$

Using sifting property of an impulse function, we get

$$\mathscr{L}\{\delta(t)\} = e^{-st}\Big|_{t=0} = 1; \quad \text{ROC}: \text{ All } s$$

Hence,

$$h(t) = \left[e^{-2t} + e^{-t}\right]u(t) \overset{\mathbf{LT}}{\longleftrightarrow} H(s) = \frac{2s+3}{(s+2)(s+1)}; \quad \text{ROC} : \sigma > -1$$

$$x(t) = -e^{-2t}u(t) + \delta(t) \overset{\mathbf{LT}}{\longleftrightarrow} X(s) = -\frac{1}{s+2} + 1 = \frac{s+1}{s+2}; \quad \text{ROC} : \sigma > -2$$

Note that ROC does not contain any poles. Also $x(t)$ and $h(t)$ are right-sided signals. Hence, their ROCs lie to the right of a vertical line through the rightmost pole.

Also,

$$y(t) = x(t) * h(t) \overset{\mathbf{LT}}{\longleftrightarrow} Y(s) = X(s)H(s); \quad \text{ROC} : R_y = R_x \supset R_h$$

Thus,

$$Y(s) = H(s)X(s) = \frac{(2s+3)(s+1)}{(s+2)(s+1)(s+2)}$$

$$\text{ROC} : R_y = \{\sigma > -2\} \cap \{\sigma > -1\} = \sigma > -1$$

$$\Rightarrow \qquad Y(s) = \frac{(2s+3)}{(s+2)^2}; \quad \text{ROC} : R_y = \sigma > -2$$

(after pole-zero Concellation).

Expanding this transform into partial fractions, we get

$$\begin{aligned} Y(s) &= \frac{2s+3}{(s+2)^2} \\ &= \frac{A}{s+2} + \frac{B}{(s+2)^2}; \quad \text{ROC} : R_y = \sigma > -2 \end{aligned}$$

where,

$$A = \frac{d}{ds}[2s+3]\Big|_{s=-2} = 2$$

and

$$B = 2s + 3|_{s=-2} = -1$$

Thus,

$$Y(s) = \frac{2}{s+2} - \frac{1}{(s+2)^2}$$

Recall: $te^{-at}u(t) \overset{\text{LT}}{\longleftrightarrow} \frac{1}{(s+a)^2}$; ROC : $R = \sigma > -a$

Hence,

$$y(t) = 2e^{-2t}u(t) - te^{-2t}u(t)$$

Example 6.13. Given

$$X(s) = \frac{s+3}{(s+1)(s-2)}$$

Mark all possible ROCs. Comment on the nature of $x(t)$. Also, check if Fourier transform exists ?

Solution:

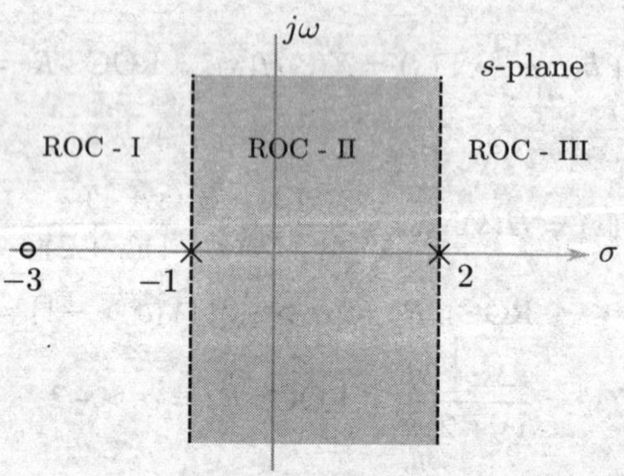

Fig. Ex. 6.13

Three possible ROCs are sketched in Fig. Ex. 6.13.
The nature of $x(t)$ is tabulated below:

$x(t)$ is right-sided	ROC-III
$x(t)$ is left-sided	ROC-I
$x(t)$ extends for all time	ROC-II

If ROC of $X(s)$ includes the $j\omega$ axis, then Fourier transform of $x(t)$ exists. Only ROC-II includes $j\omega$ axis. Hence Fourier transform of $x(t)$ exists if and only if $x(t)$ extends for all t.

Example 6.14. Find inverse Laplace transform of $X(s)$ given below:

$$X(s) = \frac{s+3}{s(s+1)(s-2)}; \quad 0 < \text{Re}\{s\} < 2$$

Solution: Using the theory of partial fractions, we can write

$$X(s) = \underbrace{\frac{K_1}{s+1}}_{\text{ROC}:R_1} + \underbrace{\frac{K_2}{s}}_{\text{ROC}:R_2} + \underbrace{\frac{K_3}{s-2}}_{\text{ROC}:R_3}$$

In order to have ROC $\supset R_1 \cap R_2 \cap R_3$, the only possibility for the individual fractions ROCs is the following:

$$\text{ROC} : R_1 = \sigma > -1, \quad \text{ROC} : R_2 = \sigma > 0, \quad \text{ROC} : R_3 = \sigma < 2$$

Thus,

$$X(s) = \underbrace{\frac{K_1}{s+1}}_{R_1=\sigma>-1} + \underbrace{\frac{K_1}{s}}_{R_2=\sigma>0} + \underbrace{\frac{K_3}{s-2}}_{R_3=\sigma<2}$$

The values of K_1, K_2 and K_3 are found as follows:

$$K_1 = \left.\frac{s+3}{s(s-2)}\right|_{s=-1} = \frac{2}{3}$$

$$K_2 = \left.\frac{(s+3)}{(s+1)(s-2)}\right|_{s=0} = -\frac{3}{2}$$

$$K_3 = \left.\frac{(s+3)}{s(s+1)}\right|_{s=2} = \frac{5}{6}$$

Hence,

$$X(s) = \underbrace{\frac{2}{3}\left[\frac{1}{s+1}\right]}_{R_1=\sigma>-1} - \underbrace{\frac{3}{2}\left[\frac{1}{s}\right]}_{R_2=\sigma>0} + \underbrace{\frac{5}{6}\left[\frac{1}{s-2}\right]}_{R_3=\sigma<2}$$

Recall:

(1) $e^{-at}u(t) \overset{\text{LT}}{\longleftrightarrow} \dfrac{1}{s+a}$; ROC : $\sigma > -a$

(2) $-e^{-at}u(-t) \overset{\text{LT}}{\longleftrightarrow} \dfrac{1}{s+a}$; ROC : $\sigma < -a$

(3) $u(t) \overset{\text{LT}}{\longleftrightarrow} \dfrac{1}{s}$; ROC : $\sigma > 0$

Hence,

$$x(t) = \frac{2}{3}e^{-t}u(t) - \frac{3}{2}u(t) - \frac{5}{6}e^{2t}u(-t)$$

Example 6.15. It is known that

$$X_1(s) = \frac{2s+3}{s^2+3s+2}$$

and

$$X_2(s) = \frac{3s+1}{s^2+4s+3}$$

are the Laplace transforms of two right-sided signals.

(a) Find the poles of $X_1(s)$ and give the region of convergence.

(b) Find the poles of $X_2(s)$ and give the region of convergence.

(c) Find the poles and zeros of $X_1(s) + X_2(s)$ and give the region of convergence.

Solution:

(a) Rewrite $X_1(s)$ as

$$X_1(s) = \frac{2s+3}{(s+1)(s+2)}$$

Poles of $X_1(s)$: $s = -1, -2$

The poles and ROC of $X_1(s)$ are as shown in Fig. Ex. 6.15(a).

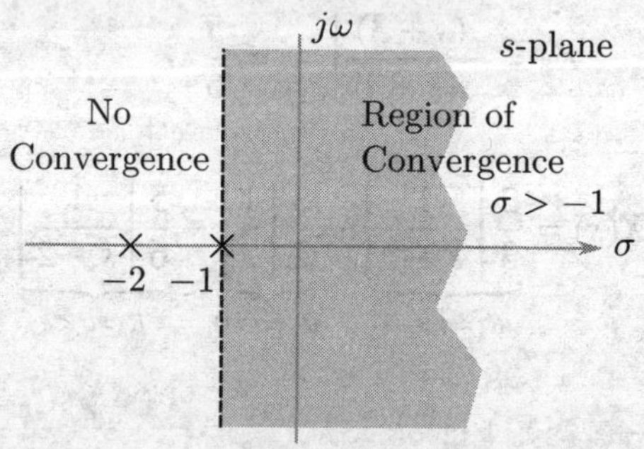

Fig. Ex. 6.15(a)

(b) Factorising the denominator of $X_2(s)$, we get

$$X_2(s) = \frac{(3s+1)}{(s+3)(s+1)}$$

Poles of $X_2(s)$: $s = -1, -3$

The poles and the ROC of $X_2(s)$ are sketched in Fig. Ex. 6.15(b).

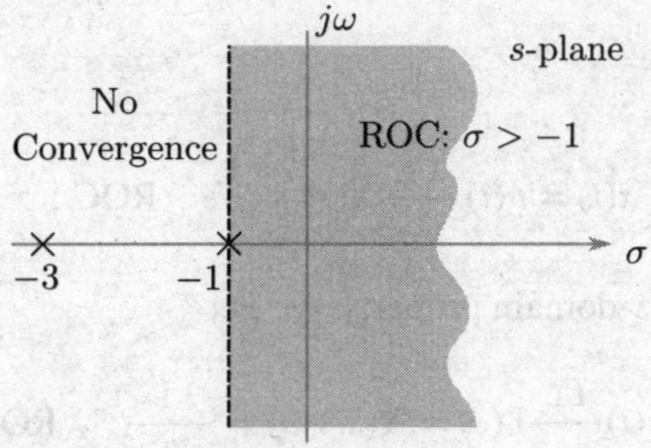

Fig. Ex. 6.15(b)

(c) $$X_1(s) + X_2(s) = \frac{5s^2 + 16s + 11}{(s+3)(s+2)(s+1)}$$
$$= \frac{5s + 11}{(s+2)(s+3)}$$

Note that pole-zero cancellation takes place.

The poles, zero and ROC of $X_1(s) + X_2(s)$ are sketched in Fig. Ex. 6.15(c).

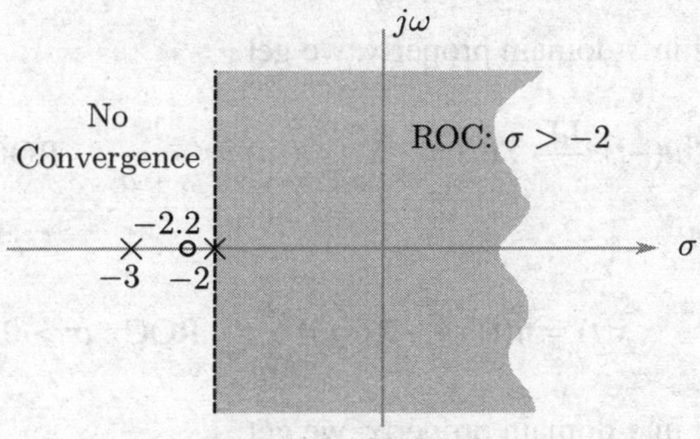

Fig. Ex. 6.15(c)

Example 6.16. Using properties, find Laplace transform of the signals given below:

(a) $y(t) = e^{-at}u(t)$

(b) $y(t) = -e^{-at}u(-t)$

(c) $y(t) = te^{-at}u(t)$

(d) $y(t) = \frac{d^2}{dt^2}[e^{-3(t-2)}u(t-2)]$

Solution:

(a) Recall:

$$x(t) = u(t) \overset{\text{LT}}{\longleftrightarrow} X(s) = \frac{1}{s}; \quad \text{ROC} : \sigma > 0$$

Applying shifting in s-domain property, we get

$$y(t) = e^{-at}u(t) \overset{\text{LT}}{\longleftrightarrow} Y(s) = X(s+a) = \frac{1}{s+a}; \quad \text{ROC} : \sigma > \text{Re}\{-a\}$$

It may be noted that if a is real, then ROC is $\sigma > -a$.

(b) Recall:

$$x(t) = u(t) \overset{\text{LT}}{\longleftrightarrow} X(s) = \frac{1}{s}; \quad \text{ROC} : \sigma > 0$$

Applying scaling property with $\alpha = -1$, we get

$$x_1(t) = u(-t) \overset{\text{LT}}{\longleftrightarrow} X_1(s) = \frac{1}{|-1|}X(-s) = \frac{-1}{s}; \quad \text{ROC} : \sigma < 0$$

Applying shifting in s-domain property, we get

$$y(t) = -e^{-at}u(-t) \overset{\text{LT}}{\longleftrightarrow} Y(s) = -X_1(s+a) = \frac{1}{s+a}; \quad \text{ROC} : \sigma < \text{Re}\{-a\}$$

(c) Recall:

$$x(t) = u(t) \overset{\text{LT}}{\longleftrightarrow} X(s) = \frac{1}{s}; \quad \text{ROC} : \sigma > 0$$

Applying shifting in s-domain property, we get

$$x_1(t) = e^{-at}u(t) \overset{\text{LT}}{\longleftrightarrow} X_1(s) = X(s+a) = \frac{1}{s+a}; \quad \text{ROC} : \sigma > \text{Re}\{-a\}$$

Applying differentiation in s-domain property, we get

$$\begin{aligned} y(t) = te^{-at}u(t) \overset{\text{LT}}{\longleftrightarrow} Y(s) &= -\frac{d}{ds}X_1(s) \\ &= -\frac{d}{ds}\left[\frac{1}{s+a}\right] \\ &= \frac{1}{(s+a)^2}; \quad \text{ROC} : \sigma > \text{Re}\{-a\} \end{aligned}$$

(d) We know that,

$$e^{-3t}u(t) \overset{\text{LT}}{\longleftrightarrow} \frac{1}{s+3}; \quad \text{ROC} : \sigma > -3$$

The time-shift property implies

$$e^{-3(t-2)}u(t-2) \overset{\text{LT}}{\longleftrightarrow} \frac{1}{s+3}e^{-2s}; \quad \text{ROC} : \sigma > -3$$

Finally, applying time differentiation property twice, we get

$$y(t) = \frac{d^2}{dt^2}\left[e^{-3(t-2)}u(t-2)\right] \overset{\text{LT}}{\longleftrightarrow} Y(s) = \frac{s^2 e^{-2s}}{s+3}; \quad \text{ROC} : \sigma > -3$$

Example 6.17. Find the inverse Laplace transform of $X(s)$ given below:

$$X(s) = \frac{2-2s}{(s+1)(s+2)(s+5)}; \quad \text{ROC} : \text{Re}\{s\} > -1$$

Solution: Using the theory of partial fractions, we can write

$$X(s) = \frac{K_1}{s+1} + \frac{K_2}{s+2} + \frac{K_3}{s+5}$$

The constants associated with partial fraction expansion of $X(s)$ can be found as follows:

$$\begin{aligned}
K_1 &= X(s)(s+1)\big|_{s=-1} \\
&= \left.\frac{2-2s}{(s+2)(s+5)}\right|_{s=-1} = 1 \\
K_2 &= X(s)(s+2)\big|_{s=-2} \\
&= \left.\frac{(2-2s)}{(s+1)(s+5)}\right|_{s=-2} = -2 \\
K_3 &= X(s)(s+5) \\
&= \left.\frac{(2-2s)}{(s+1)(s+2)}\right|_{s=-5} = 1
\end{aligned}$$

Hence,

$$X(s) = \frac{1}{s+1} - \frac{2}{s+2} + \frac{1}{s+5}$$

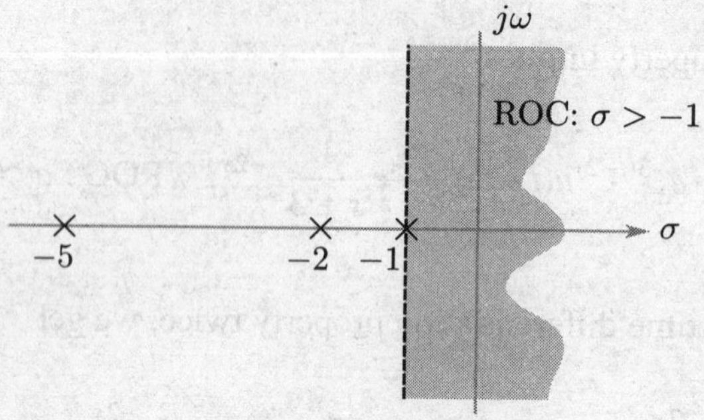

Fig. Ex. 6.17(a)

By looking at the ROC of $X(s)$ sketched in Fig. Ex. 6.17(a), it implies that $x(t)$ is a right-sided signal.

Recall:

$$e^{-at}u(t) \overset{\textbf{LT}}{\longleftrightarrow} \frac{1}{s+a}; \quad \text{ROC} : \sigma > -a$$

Hence,

$$x(t) = e^{-t}u(t) - 2e^{-2t}u(t) + e^{-5t}u(t)$$

Example 6.18. Repeat the previous problem for the ROCs given below:

(i) $\text{ROC}_1 : \text{Re}\{s\} < -5$ (ii) $\text{ROC}_2 : -2 < \text{Re}\{s\} < -1$

Solution: From Example 6.17, we have

$$X(s) = \frac{1}{s+1} - \frac{2}{s+2} + \frac{1}{s+5}$$

(i)

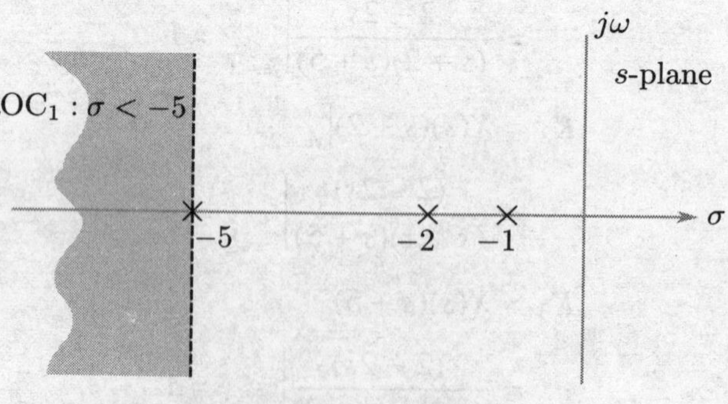

Fig. Ex. 6.18(a)

From Fig. Ex. 6.18(a), we observe that ROC_1 lies to the left of a line through the left-most pole. Hence, $x(t)$ is a left-sided signal.

Recall:

$$-e^{-at}u(-t) \overset{\textbf{LT}}{\longleftrightarrow} \frac{1}{s+a}; \quad \text{ROC}: \sigma < -a$$

Hence, inverse Laplace transform of $X(s)$ yields

$$x(t) = -e^{-t}u(-t) + 2e^{-2t}u(-t) - e^{-5t}u(-t)$$

(ii)

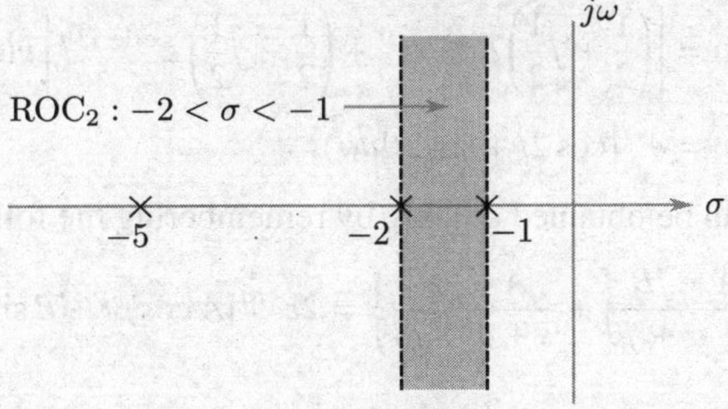

Fig. Ex. 6.18(b)

From Fig. Ex. 6.18(b), we observe that the region of convergence is a strip between two poles. Hence, $x(t)$ is bilateral; the sum of right-sided and left-sided signal. The poles at -2 and -5 give right-sided signals whereas the pole at -1 gives a left-sided signal.

Hence,

$$x(t) = -e^{-t}u(-t) - 2e^{-2t}u(t) + e^{-5t}u(t)$$

Example 6.19. Find inverse Laplace transform of $X(s) = \dfrac{s+3}{s^2+2s+5}$. Assume a right-sided region of convergence.

Solution: Let $s^2 + 2s + 5 = 0$.

Then $s = -1 \pm j2$

Rewriting $X(s)$ in the factored from, we get

$$X(s) = \frac{s+3}{(s+1+j2)(s+1-j2)}$$
$$= \frac{K_1}{s+1+j2} + \frac{K_2}{s+1-j2}$$

where

$$K_1 = \left.\frac{s+3}{s+1-j2}\right|_{s=-1+j2} = \frac{1}{2} + j\frac{1}{2}$$

and

$$K_2 = K_1^* = \frac{1}{2} - j\frac{1}{2}$$

Hence,

$$X(s) = \left(\frac{1}{2} + j\frac{1}{2}\right)\left[\frac{1}{s+1+j2}\right] + \left(\frac{1}{2} - j\frac{1}{2}\right)\left[\frac{1}{s+1-j2}\right]$$

Taking inverse Laplace transforms on both sides of the above equation, we get

$$x(t) = \left[\left(\frac{1}{2} + j\frac{1}{2}\right)e^{-(1+j2)t} + \left(\frac{1}{2} - j\frac{1}{2}\right)e^{-(1-j2)t}\right]u(t)$$

$$= e^{-t}[\cos 2t + \sin 2t]u(t)$$

The above result can be obtained quickly by remembering the following result:

$$\mathscr{L}^{-1}\left\{\frac{A+jB}{s+\alpha+j\beta} + \frac{A-jB}{s+\alpha-j\beta}\right\} = 2e^{-\alpha t}[A\cos\beta t + B\sin\beta t]u(t)$$

Alternate method:

Rewrite $X(s)$ as

$$X(s) = \frac{(s+1)+2}{(s^2+2s+1)+4}$$

$$= \frac{(s+1)+2}{(s+1)^2+2^2}$$

$$= \frac{s+1}{(s+1)^2+2^2} + \frac{2}{(s+1)^2+2^2}$$

Recall:

(1) $e^{-at}\cos bt\, u(t) \overset{\text{LT}}{\longleftrightarrow} \dfrac{s+a}{(s+a)^2+b^2}$; ROC : $\sigma > -a$

(2) $e^{-at}\sin bt\, u(t) \overset{\text{LT}}{\longleftrightarrow} \dfrac{b}{(s+a)^2+b^2}$; ROC : $\sigma > -a$

Hence,

$$x(t) = e^{-t}\cos 2t\, u(t) + e^{-t}\sin 2t\, u(t)$$

Example 6.20. Find inverse Laplace transform of the function given below:

$$X(s) = \frac{4}{(s+1)(s+2)^3}; \quad \text{ROC} : \sigma > -1$$

Solution: The partial fraction expansion of $X(s)$ is

$$X(s) = \frac{K_1}{s+1} + \frac{K_2}{(s+2)^3} + \frac{K_3}{(s+2)^2} + \frac{K_4}{(s+2)}$$

where,

$$K_1 = X(s)(s+1)\big|_{s=-1}$$

$$= \frac{4}{(s+2)^3}\bigg|_{s=-1} = 4$$

$$K_2 = X(s)(s+2)^3\big|_{s=-2}$$

$$= \frac{4}{s+1}\bigg|_{s=-2} = -4$$

$$K_2 = \frac{d}{ds}[X(s)(s+2)^3]\bigg|_{s=-2}$$

$$= \frac{d}{ds}\left[\frac{4}{s+1}\right]\bigg|_{s=-2} = -4$$

$$K_3 = \frac{1}{2!}\frac{d^2}{ds^2}[X(s)(s+2)^3]\bigg|_{s=-2}$$

$$= \frac{1}{2!}\frac{d^2}{ds^2}\left[\frac{4}{s+1}\right]\bigg|_{s=-2}$$

$$= -4$$

This gives the result

$$X(s) = \frac{4}{s+1} - \frac{4}{(s+2)^3} - \frac{4}{(s+2)^2} - \frac{4}{s+2}$$

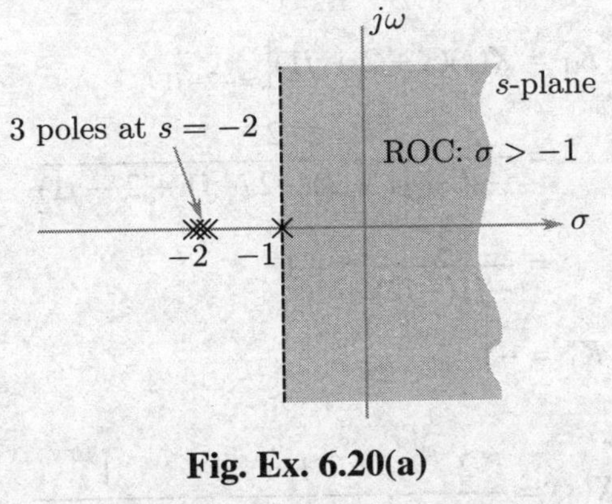

Fig. Ex. 6.20(a)

Referring Fig. Ex. 6.20(a), we find that the ROC lies to the right of the line through the rightmost pole. This means $x(t)$ is a right-sided signal.

Remember:

$$\mathscr{L}^{-1}\left\{\frac{K}{(s+\alpha)^n}\right\} = \frac{K}{(n-1)!}t^{n-1}e^{-\alpha t}u(t)$$

Hence,

$$x(t) = 4e^{-t}u(t) - 2t^2e^{-2t}u(t) - 4te^{-2t}u(t) - 4e^{-2t}u(t)$$

Example 6.21. Find inverse Laplace transform of the function given below:

$$X(s) = \frac{2}{(s+2)(s^2+4s+5)}$$

Assume a right-sided region of convergence.

Solution: The poles of $X(s)$ are

(i) $s+2=0 \Rightarrow s=-2$

(ii) $s^2+4s+5=0 \Rightarrow s=-2\pm j1$

Let us rewrite $X(s)$ in the factored form as

$$X(s) = \frac{2}{(s+2)(s+2+j1)(s+2-j1)}$$

The partial fraction expansion of $X(s)$ is

$$X(s) = \frac{A}{s+2} + \frac{K_1}{s+2+j1} + \frac{K_1^*}{s+2-j1}$$

where,

$$A = X(s)(s+2)\big|_{s=-2} = 2$$

$$K_1 = X(s)(s+2+j1)\big|_{s=-2-j1}$$

$$= \frac{2}{(-2-j1+2)(-2-j1+2-j1)}$$

$$= \frac{2}{-j1(-j2)} = -1$$

$$K_1^* = -1$$

Hence,

$$X(s) = \frac{2}{s+2} - \frac{1}{s+2+j1} - \frac{1}{s+2-j1}$$

It is stated in the problem that the region of convergence is right-sided. Hence, inverse Laplace transform of $X(s)$ yields,

$$x(t) = \left[2e^{-2t} - e^{-(2+j1)t} - e^{-(2-j1)t}\right]u(t)$$
$$= \left[2e^{-2t} - 2e^{-2t}\cos t\right]u(t)$$

Example 6.22. Find inverse Laplace transform of $X(s)$ given below:

$$X(s) = \frac{2[s - 2e^{-s}]}{(s+1)(s+2)}; \quad \text{ROC} : \sigma > -1$$

Solution: Rewrite $X(s)$ in the following manner:

$$X(s) = \frac{2s}{(s+1)(s+2)} - \frac{4e^{-s}}{(s+1)(s+2)}$$
$$= 2\left[\frac{1}{s+1} - \frac{1}{s+2}\right] - 4\left[\frac{1}{s+1} - \frac{1}{s+2}\right]e^{-s}$$

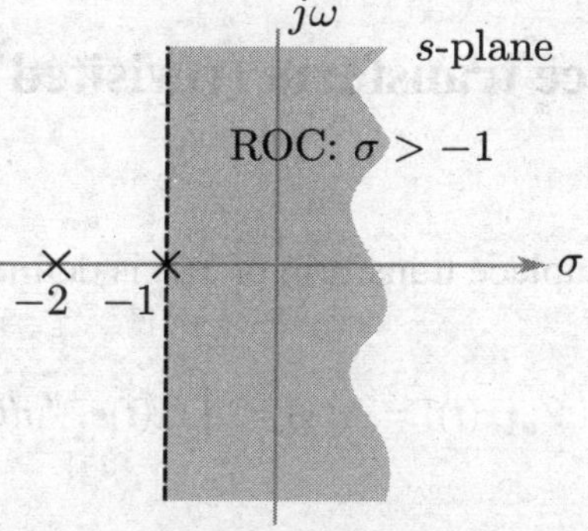

Fig. Ex. 6.22(a)

From Fig. Ex. 6.22(a), we find that ROC lies to the right of the line through the rightmost pole. Hence, $x(t)$ is a right-sided signal.

Inverse Laplace transform of $X(s)$ yields

$$x(t) = 2\left[e^{-t} - e^{-2t}\right]u(t) - 4\left[(e^{-t} - e^{-2t})u(t)\right]_{t \to t-1}$$
$$= 2\left[e^{-t} - e^{-2t}\right]u(t) - 4\left[e^{-(t-1)} - e^{-2(t-1)}\right]u(t-1)$$

Example 6.23. Find the time signal $x(t)$ corresponding to the transformed function given below:

$$X(s) = \frac{4\left[s^2 - e^{-s}\right]}{(s+1)(s+2)}; \quad \text{ROC} : \sigma > -1$$

Solution: Let us seperate terms in $X(s)$ and use e^{-s} as a delay. This gives

$$X(s) = \frac{4s^2}{(s+1)(s+2)} - \frac{4e^{-s}}{(s+1)(s+2)}$$

The first term on the right-hand side is not a proper fraction. It is made a proper fraction by using division. Thus, we get

$$X(s) = 4 + \frac{4}{s+1} - \frac{16}{s+2} - \left[\frac{4}{s+1} - \frac{4}{s+2}\right]e^{-s}$$

The ROC given in the problem implies that $\mathscr{L}^{-1}\{X(s)\}$ is a right-sided signal. Hence,

$$x(t) = 4\delta(t) + 4e^{-t}u(t) - 16e^{-2t}u(t) - \left[4e^{-t}u(t) - 4e^{-2t}u(t)\right]_{t\to t-1}$$

$$\Rightarrow \quad x(t) = 4\delta(t) + \left[4e^{-t} - 16e^{-2t}\right]u(t) - [4e^{-(t-1)} - 4e^{-2(t-1)}]u(t-1)$$

6.6 Unilateral Laplace transform (revisited)

6.6.1 Definitions

The one-sided or unilateral Laplace transform of $x(t)$ is defined as follows:

$$\mathscr{L}_u\{x(t)\} = X(s) = \int_{0^-}^{\infty} x(t)e^{-st}\,dt \tag{6.15}$$

It may be noted that s is a complex variable and is given by $s = \sigma + j\omega$. Notice that for this transformation $x(t) = 0$ for $t < 0$. The lower limit of integration is chosen to be 0^- (rather than 0 or 0^+) to include any discontinuity that may be present at $t = 0$. If it is known that no such discontinuity exists at $t = 0^-$, we simply interpret $x(0^-)$ as $x(0)$. Some authors denote $\mathscr{L}_u\{x(t)\}$ by $X_1(s)$, but we shall denote both $\mathscr{L}\{x(t)\}$ and $\mathscr{L}_u\{x(t)\}$ by $X(s)$.

6.6.2 Basic properties

If may be noted that two signals that differ for $t < 0$ but are equal for $t \geq 0$ will have identical unilateral Laplace transform. Also, it is crucial to note that unilateral Laplace transform of $x(t)$ and bilateral Laplace transform of $x(t)\,u(t)$ are identical, assuming there is no impulse at $t = 0$. This consequently means that Laplace transform properties for causal, right-sided signals also apply to unilateral transform as well. The basic properties of the unilateral Laplace transform that are different from those of bilateral Laplace transform are time-shift, time-differentiation and time-integration properties. They are presented below:

(1) **Time-shift**

For $x(t) = 0,\ t < 0$, if $x(t) \overset{\text{UL}}{\longleftrightarrow} X(s)$, then for a time delay of $t_0 > 0$,

$$x(t - t_0) \overset{\text{UL}}{\longleftrightarrow} e^{-st_0}X(s);\ t_0 > 0$$

Hence, for a time delay, this time-shifting property is exactly the same as that for a bilateral Laplace transform. In case, where $x(t) \neq 0$ for $t < 0$, a time delay can make a part of the signal appear for $t > 0$, which could change the unilateral Laplace transform. For a time lead, the part of the signal shifted to negative-times is lost. In both of these cases, there is no simple relationship between the resulting (obtained using time shift property) and the original (found using defining equation) unilateral Laplace transforms.

(2) **Differentiation in the time-domain**

$$\text{If} \quad x(t) \overset{\text{UL}}{\longleftrightarrow} X(s), \quad \text{then} \quad \frac{dx(t)}{dt} \overset{\text{UL}}{\longleftrightarrow} sX(s) - x(0^-),$$

provided that the $\lim_{t\to\infty} x(t)e^{-st} = 0$.

Repeated application of this property gives

$$\frac{d^2x(t)}{dt^2} \overset{\text{UL}}{\longleftrightarrow} s^2X(s) - sx(0^-) - x'(0^-)$$

$$\frac{d^nx(t)}{dt^n} \overset{\text{UL}}{\longleftrightarrow} s^nX(s) - s^{n-1}x(0^-) - s^{n-2}x'(0^-) - x^{n-1}(0^-)$$

$$\text{where} \quad x^i(0^-) = \left.\frac{d^ix(t)}{dt^i}\right|_{t=0^-}$$

(3) **Integration in the time-domain**

$$\text{If} \quad x(t) \overset{\text{UL}}{\longleftrightarrow} X(s), \quad \text{then} \quad \int_{0^-}^{t} x(\tau)\,d\tau \overset{\text{UL}}{\longleftrightarrow} \frac{1}{s}X(s)$$

$$\text{and} \quad \int_{-\infty}^{t} x(\tau)d\tau \overset{\text{UL}}{\longleftrightarrow} \frac{1}{s}X(s) + \frac{1}{s}\int_{-\infty}^{0^-} x(\tau)\,d\tau$$

6.6.3 System function

System function, $H(s)$ of a linear time-invariant system is defined as the ratio of unilateral Laplace transform of the output to unilateral Laplace transform of the input under the assumption that all initial conditions are zero. That is,

$$H(s) = \frac{Y(s)}{X(s)}$$

The system function characterizes the system. That is, system properties correspond to properties of $H(s)$ and its ROC.

The inverse Laplace transform of $H(s)$, namely $h(t)$ is known as impulse response of the system.

A system is stable, if its impulse response is absolutely integrable. That is,

$$\int_{-\infty}^{\infty} |h(t)|\, dt < \infty$$

It is not necessary to find $\mathscr{L}^{-1}\{H(s)\}$ to find stability. Equivalently, an LTI system is BIBO stable if ROC of $H(s)$ includes $j\omega$ axis.

Recall that $h(t) = 0$ for $t < 0$ for a causal system and thus is right-sided. Therefore, the ROC associated with the transfer function of a causal system is right-half plane.

The stability condition for a causal LTI system with a proper, rational transfer function is stated as follows:

A causal system with a proper rational transfer function $H(s)$ is stable if and only if all of its poles are in the left-half of the s-plane; that is, if all of the poles have negative real parts.

Example 6.24. Find unilateral Laplace transform of

$$y(t) = e^{-4t}\sin(2t)u(2t)$$

starting from the unilateral Laplace transform of $x(t) = e^{-2t}\sin tu(t)$. Also find the poles of $X(s)$ and $Y(s)$.

Solution: Recall the Laplace transform pair:

$$e^{-at}\sin bt\, u(t) \overset{\text{UL}}{\longleftrightarrow} \frac{b}{(s+a)^2 + b^2}$$

Hence,

$$x(t) = e^{-2t}\sin t\, u(t) \overset{\text{UL}}{\longleftrightarrow} X(s) = \frac{1}{(s+2)^2 + 1^2}$$

$$= \frac{1}{s^2 + 4s + 5}$$

The poles of $X(s)$ are found as the roots of the equation $s^2 + 4s + 5 = 0$. The two poles are located at $-2 + j1$ and $-2 - j1$.

Recall the time-scaling property:

$$\text{If} \quad x(t) \overset{\text{UL}}{\longleftrightarrow} X(s), \quad \text{then} \quad x(at) \overset{\text{UL}}{\longleftrightarrow} \frac{1}{a}X\left(\frac{s}{a}\right);\ a > 0$$

Hence,

$$y(t) = x(2t) \stackrel{\textbf{UL}}{\longleftrightarrow} Y(s) = \frac{1}{2}\left[\frac{1}{\left(\frac{s}{2}\right)^2 + 4\left(\frac{s}{2}\right) + 5}\right]$$

$$= \frac{2}{(s+4-j2)(s+4+j2)}$$

The two poles are located at $-4 + j2$ and $-4 - j2$.

Example 6.25. Perform, $y(t) = e^{2t}u(t) * e^{-2t}u(t)$ by using convolution property of unilateral Laplace transforms.

Solution: We know that,

$$e^{-at}u(t) \stackrel{\textbf{UL}}{\longleftrightarrow} \frac{1}{s+a}$$

Hence,

$$x_1(t) = e^{2t}u(t) \stackrel{\textbf{UL}}{\longleftrightarrow} X_1(s) = \frac{1}{s-2}$$

$$x_2(t) = e^{-2t}u(t) \stackrel{\textbf{UL}}{\longleftrightarrow} X_2(s) = \frac{1}{s+2}$$

Recall:

$$y(t) = x_1(t) * x_2(t) \stackrel{\textbf{UL}}{\longleftrightarrow} Y(s) = X_1(s)X_2(s)$$

Hence,

$$Y(s) = \frac{1}{(s+2)(s-2)}$$

$$= \frac{1}{4}\left[\frac{1}{s-2} - \frac{1}{s+2}\right]$$

Taking inverse Laplace transform on both sides, we get

$$y(t) = \frac{1}{4}\left[e^{2t}u(t) - e^{-2t}u(t)\right]$$

Example 6.26. Find zero-input, zero-state and total response of the system governed by the differential equation given below:

$$\frac{d^2y(t)}{dt^2} + 4\frac{dy(t)}{dt} + 4y(t) = \frac{d^2x(t)}{dt^2} + \frac{dx(t)}{dt} + 2x(t)$$

with initial conditions $y(0^-) = 2$ and $\left.\frac{dy(t)}{dt}\right|_{t=0^-} = 3$ and input $x(t) = u(t)$.

Solution: Taking unilateral Laplace transform on both sides of the given differential equation, we get

$$s^2Y(s) - y(0^-)s - y'(0^-) + 4[sY(s) - y(0^-)] + 4Y(s) = s^2X(s) + sX(s) + 2X(s)$$

To find zero-input response:

Letting $X(s) = 0$, we get

$$s^2Y(s) - 2s - 3 + 4[sY(s) - 2] + 4Y(s) = 0$$

$$\Rightarrow \quad Y(s) = \frac{2s+11}{(s+2)^2}$$

$$= \frac{K_1}{s+2} + \frac{K_2}{(s+2)^2}$$

where,

$$K_1 = \frac{1}{1!}\frac{d}{ds}\left[(s+2)^2Y(s)\right]\Big|_{s=-2}$$

$$= \frac{d}{ds}(2s+11)\Big|_{s=-2} = 2$$

$$K_2 = (s+2)^2Y(s)\Big|_{s=-2}$$

$$= 2s + 11\Big|_{s=-2} = 7$$

Hence,

$$Y(s) = \frac{2}{s+2} + \frac{7}{(s+2)^2}$$

Taking inverse Laplace transform on both sides, we get

$$y_{ZI}(t) = 2e^{-2t}u(t) + 7te^{-2t}u(t)$$

To find zero-state response:

Assuming all initial conditions to be zero, we get

$$s^2Y(s) + 4sY(s) + 4Y(s) = s^2X(s) + sX(s) + 2X(s)$$

Since,

$$x(t) = u(t) \overset{\text{UL}}{\longleftrightarrow} X(s) = \frac{1}{s},$$

we get

$$s^2Y(s) + 4sY(s) + 4Y(s) = s + 1 + \frac{2}{s} \Rightarrow Y(s) = \frac{s^2+s+2}{s(s^2+4s+4)}$$

Expanding into partial fractions, we get

$$Y(s) = \frac{0.5}{s} + \frac{0.5}{s+2} - \frac{2}{(s+2)^2}$$

Taking inverse Laplace transform on both sides, we get

$$y_{ZS}(t) = [0.5 + 0.5e^{-2t} - 2te^{-2t}]u(t)$$

Total response:

$$\begin{aligned} y(t) &= y_{ZI}(t) + y_{ZS}(t) \\ &= \left[0.5 + 2.5e^{-2t} + 5te^{-2t}\right] u(t) \end{aligned}$$

The terms that tend to zero in $y(t)$ as $t \to \infty$ gives the transient response of the system. The terms that remain in $y(t)$ as $t \to \infty$ defines the steady-state response of the system. Hence, in the present problem, the steady-state response is $0.5u(t)$ and transient response is

$$[2.5e^{-2t} + 5te^{-2t}]u(t).$$

Example 6.27. Using properties, find the unilateral Laplace transforms of the signals given below:

(a) $x(t) = \frac{d}{dt}[tu(t)]$

(b) $x(t) = e^{-t}u(t) * \sin 4\pi t\, u(t)$

(c) $x(t) = \int_0^t e^{-3\tau} \cos 4\tau\, u(\tau)\, d\tau$

Solution:

(a) Let $y(t) = t\,u(t)$. Then, $Y(s) = \frac{1}{s^2}$.

Applying differentiation in time-domain property, we get

$$x(t) = \frac{dy(t)}{dt} \overset{\text{UL}}{\longleftrightarrow} X(s) = sY(s) - y(0^-)$$

Hence,

$$X(s) = s\frac{1}{s^2} - 0 = \frac{1}{s}$$

(b) Let $x_1(t) = e^{-t}u(t)$ and $x_2(t) = \sin 4\pi t\, u(t)$

Recall:

$$e^{-at}u(t) \overset{\textbf{UL}}{\longleftrightarrow} \frac{1}{s+a}$$

Hence,

$$x_1(t) = e^{-t}u(t) \overset{\textbf{UL}}{\longleftrightarrow} X_1(s) = \frac{1}{s+1}$$

Recall:

$$\sin \omega_0 t\, u(t) \overset{\textbf{UL}}{\longleftrightarrow} \frac{\omega_0}{s^2 + \omega_0^2}$$

Hence,

$$x_2(t) = \sin 4\pi t\, u(t) \overset{\textbf{UL}}{\longleftrightarrow} X_2(s) = \frac{4\pi}{s^2 + (4\pi)^2}$$

Applying time-domain convolution property, we get

$$x(t) = x_1(t) * x_2(t) \overset{\textbf{UL}}{\longleftrightarrow} X(s) = X_1(s)X_2(s)$$

Hence,

$$X(s) = X_1(s)X_2(s)$$

$$= \frac{4\pi}{(s+1)\left[s^2 + (4\pi)^2\right]}$$

(c) Given:

$$x(t) = \int_0^t e^{-3\tau} \cos 4\tau\, d\tau$$

$$\Rightarrow \qquad x(t) = \int_{-\infty}^t e^{-3\tau} \cos 4\tau u(\tau) d\tau$$

Let

$$y(t) = e^{-3t} \cos 4t\, u(t)$$

Then,

$$Y(s) = \frac{s+3}{(s+3)^2 + 4^2}$$

Recall:

$$x(t) = \int_{-\infty}^t y(\tau) d\tau \overset{\textbf{UL}}{\longleftrightarrow} X(s) = \frac{1}{s}\int_{-\infty}^{0^-} y(\tau) d\tau + \frac{Y(s)}{s}$$

Since $y(t) = 0$ for $t < 0$, it implies that the integral, $\frac{1}{s}\int_{-\infty}^{0^-} y(\tau)\,d\tau = 0$.

Hence,

$$X(s) = 0 + \frac{1}{s}\left[\frac{s+3}{(s+3)^2+4^2}\right]$$

$$= \frac{(s+3)}{s[(s+3)^2+4^2]}$$

Example 6.28. Given the transform pair $x(t) \overset{\text{UL}}{\longleftrightarrow} \frac{s}{s+1}$ determine the unilateral Laplace transform of each of the following signals:

(a) $x(3t)$

(b) $x(t) * 2tx(t)$

(c) $e^{-3t}x(t)$

(d) $\int_0^t x(2\tau)\,d\tau$

Solution:

(a) Recall:

$$x(at) \overset{\text{UL}}{\longleftrightarrow} \frac{1}{a}X\left(\frac{s}{a}\right);\ a > 0$$

Hence,

$$y(t) = x(3t) \overset{\text{UL}}{\longleftrightarrow} Y(s) = \frac{1}{3}X\left(\frac{s}{3}\right)$$

$$\Rightarrow \qquad Y(s) = \frac{1}{3}\left[\frac{\frac{s}{3}}{\frac{s}{3}+1}\right]$$

$$= \frac{1}{3}\left[\frac{s}{s+3}\right]$$

(b) Let

$$y(t) = x(t) * 2tx(t)$$

$$= x(t) * x_1(t)$$

where $x_1(t) = 2tx(t)$.

Recall:

$$-tx(t) \overset{\text{UL}}{\longleftrightarrow} \frac{dX(s)}{ds}$$

Hence,

$$x_1(t) = 2t\,u(t) \overset{\textbf{UL}}{\longleftrightarrow} X_1(s) = -2\frac{d}{ds}X(s)$$

Therefore,

$$\begin{aligned} X_1(s) &= -2\frac{d}{ds}\left(\frac{s}{s+1}\right) \\ &= -2\left[\frac{(s+1)\times 1 - s\times 1}{(s+1)^2}\right] \\ &= \frac{-2}{(s+1)^2} \end{aligned}$$

Recall:

$$x_1(t) * x_2(t) \overset{\textbf{UL}}{\longleftrightarrow} X_1(s)X_2(s)$$

Hence,

$$y(t) = x(t) * x_1(t) \overset{\textbf{UL}}{\longleftrightarrow} Y(s) = X(s)X_1(s)$$

Thus,

$$Y(s) = \frac{s}{s+1} \times \frac{-2}{(s+1)^2} = \frac{-2s}{(s+1)^3}$$

(c) Recall:

$$e^{s_0 t}x(t) \overset{\textbf{UL}}{\longleftrightarrow} X(s - s_0)$$

Hence,

$$y(t) = e^{-3t}x(t) \overset{\textbf{UL}}{\longleftrightarrow} Y(s) = X(s+3)$$

Thus,

$$\begin{aligned} Y(s) &= \left.\frac{s}{s+1}\right|_{s\to s+3} \\ &= \frac{s+3}{s+4} \end{aligned}$$

(d) Let

$$y(t) = \int_0^t x(2\tau)d\tau$$

If we assume $x(t) = 0$ for $t < 0$, then

$$y(t) = \int_{-\infty}^{t} x(2\tau)d\tau = \int_{-\infty}^{t} x_1(\tau)d\tau \quad \text{where} \quad x_1(t) = x(2t)$$

Recall:

$$x(at) \overset{\text{UL}}{\longleftrightarrow} \frac{1}{a}X\left(\frac{s}{a}\right); a > 0$$

Hence,

$$x_1(t) = x(2t) \overset{\text{UL}}{\longleftrightarrow} X_1(s) = \frac{1}{2}X\left(\frac{s}{2}\right)$$

Recall:

$$y(t) = \int_{-\infty}^{t} x_1(\tau)d\tau \overset{\text{UL}}{\longleftrightarrow} Y(s) = \frac{1}{s}X_1(s) + \frac{1}{s}\int_{-\infty}^{0^-} x_1(\tau)d\tau$$

The integral $\frac{1}{s}\int_{-\infty}^{0^-} x_1(\tau)\,d\tau = 0$ because $x_1(t) = 0$ for $t < 0$.

Hence,

$$\begin{aligned} Y(s) &= \frac{1}{s}\left[\frac{1}{2}X\left(\frac{s}{2}\right)\right] + 0 \\ &= \frac{1}{2s}\left[\frac{\frac{s}{2}}{\frac{s}{2}+1}\right] \\ &= \frac{1}{2}\left[\frac{1}{s+2}\right] \end{aligned}$$

Example 6.29. Using the method of partial fractions, find the inverse Laplace transform of the unilateral Laplace transforms given below:

(a) $X(s) = -\frac{9}{(s+1)(s^2+2s+10)}$

(b) $X(s) = \frac{s+4+e^{-4s}}{s^2+5s+6}$

Solution:

(a) The partial fraction expansion of $X(s)$ is

$$X(s) = \frac{A}{s+1} + \frac{Bs+C}{s^2+2s+10}$$

Let us now proceed to find the values of A, B and C

$$A(s^2+2s+10) + (Bs+C)(s+1) = -9$$

(1) Letting $s = 0$, we get

$$10A + C = -9$$

(2) Letting $s = 1$, we get

$$13A + 2B + 2C = -9$$

(3) Letting $s = -1$, we get

$$9A + 0 = -9$$

Solving, we get

$$A = -1, \quad B = 1 \quad \text{and} \quad C = 1$$

Hence,

$$X(s) = \frac{-1}{s+1} + \frac{s+1}{s^2+2s+1+9}$$
$$= \frac{-1}{(s+1)} + \frac{(s+1)}{(s+1)^2+3^2}$$

Taking inverse transforms, we get

$$x(t) = \left[-e^{-t} + e^{-t}\cos 3t\right]u(t)$$

(b) Rewrite $X(s)$ as

$$X(s) = \frac{s+4}{s^2+5s+6} + \frac{e^{-4s}}{s^2+5s+6}$$
$$= \frac{s+4}{(s+2)(s+3)} + \frac{e^{-4s}}{(s+2)(s+3)}$$
$$\Rightarrow \qquad X(s) = \left[\frac{K_1}{s+2} + \frac{K_2}{s+3}\right] + \left[\frac{1}{s+2} - \frac{1}{s+3}\right]e^{-4s}$$

where,

$$K_1 = \left.\frac{s+4}{s+3}\right|_{s=-2} = 2$$
$$K_2 = \left.\frac{s+4}{s+2}\right|_{s=-3} = -1$$

Hence,

$$X(s) = \left[\frac{2}{s+2} - \frac{1}{s+3}\right] + \left[\frac{1}{s+2} - \frac{1}{s+3}\right]e^{-4s}$$

Taking inverse transforms, we get

$$x(t) = \left[2e^{-2t} - e^{-3t}\right]u(t) + \left[e^{-2t}u(t) - e^{-3t}u(t)\right]_{t\to t-4}$$
$$\Rightarrow \qquad x(t) = \left[2e^{-2t} - e^{-3t}\right]u(t) + \left[e^{-2(t-4)} - e^{-3(t-4)}\right]u(t-4)$$

Example 6.30. An LTI system is described by the differential equation given below:

$$\frac{d^2y(t)}{dt^2} + 5\frac{dy(t)}{dt} + 6y(t) = x(t) + \frac{dx(t)}{dt}$$

Determine the transfer function and impulse response of the system.

Solution: Taking unilateral Laplaces transform on both sides of the given differential equation with all intial conditions equal to zero, we get

$$s^2Y(s) + 5sY(s) + 6Y(s) = X(s) + sX(s)$$

$$\Rightarrow \qquad H(s) = \frac{Y(s)}{X(s)}$$

$$= \frac{s+1}{s^2+5s+6}$$

$$= \frac{s+1}{(s+2)(s+3)}$$

The inverse Laplace transform of $H(s)$ is known as impulse response of the system. The partial fraction expansion of $H(s)$ gives

$$H(s) = \frac{K_1}{s+2} + \frac{K_2}{s+3}$$

where,

$$K_1 = \left.\frac{s+1}{s+3}\right|_{s=-2} = -1$$

$$K_2 = \left.\frac{s+1}{s+2}\right|_{s=-3} = 2$$

Hence,

$$H(s) = \frac{-1}{s+2} + \frac{2}{s+3}$$

Taking inverse transforms, we get

$$h(t) = -e^{-2t}u(t) + 2e^{-3t}u(t)$$

Example 6.31. Determine the zero-input response and zero-state response of an LTI system described by the differential equation:

$$5\frac{dy(t)}{dt} + 10y(t) = 2x(t), \;\; y(0^-) = 2, \;\; x(t) = u(t)$$

Also, find the impulse response of the system.

Solution: Taking unilateral Laplace transform on both sides of the given differential equation, we get

$$5[sY(s) - y(0^-)] + 10Y(s) = 2X(s) \quad \text{where} \quad X(s) = \frac{1}{s}$$

To find zero-input response:

The zero-input is found by letting $X(s) = 0$. This gives

$$5[sY(s) - 2] + 10Y(s) = 0$$

$$\Rightarrow \qquad Y(s) = \frac{10}{5s + 10}$$
$$= \frac{2}{s + 2}$$

Hence,

$$y_{ZI}(t) = 2e^{-2t}u(t)$$

To find zero-state response:

The zero-state response is found by assuming all initial conditions to be zero. This leads to

$$5sY(s) + 10Y(s) = \frac{2}{s}$$

$$\Rightarrow \qquad Y(s) = \frac{2}{s(5s + 10)}$$
$$= \frac{2}{5} \times \frac{1}{s(s + 2)}$$
$$= \frac{1}{5}\left[\frac{1}{s} - \frac{1}{s + 2}\right]$$

Hence,

$$y_{ZS}(t) = \frac{1}{5}\left[1 - e^{-2t}\right]u(t)$$

To find total response:

$$y(t) = y_{ZI}(t) + y_{ZS}(t)$$
$$= 2e^{-2t}u(t) + \frac{1}{5}u(t) - \frac{1}{5}e^{-2t}u(t)$$
$$= \frac{9}{5}e^{-2t}u(t) + \frac{1}{5}u(t)$$

To find impulse response of the system:

Taking unilateral Laplace transform on both sides of the given differential equation with initial condition equal to zero, we get

$$5sY(s) + 10Y(s) = 2X(s)$$

Hence,

$$H(s) = \frac{2}{5s+10} = \frac{2}{5}\left[\frac{1}{s+2}\right]$$

Therefore,

$$h(t) = \frac{2}{5}e^{-2t}u(t)$$

Example 6.32. Find the convolution of two indentical rectangular pulses. Each rectangular pulse has unit amplitude and duration equal to $2T$ seconds. Also, the pulse is centered at $t = T$.

Solution: The pulse given in the problem is as shown in Fig. Ex. 6.32.

Referring Fig. Ex. 6.32, we can write

$$x(t) = u(t) - u(t - 2T)$$

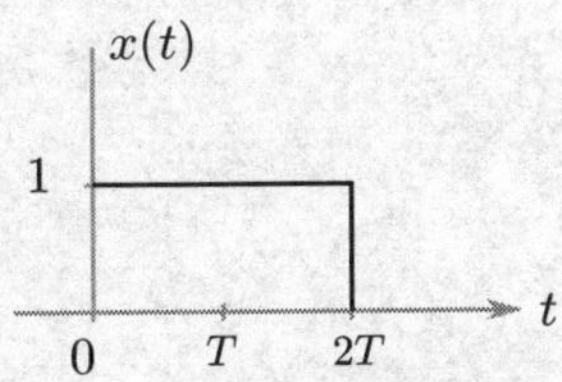

Fig. Ex. 6.32

Taking Laplace transform, we get

$$X(s) = \frac{1}{s} - \frac{1}{s}e^{-2Ts}$$
$$= \frac{1}{s}\left(1 - e^{-2Ts}\right)$$

Let

$$y(t) = x(t) * x(t)$$

Then,

$$Y(s) = X^2(s)$$

$$= \left[\frac{1 - e^{-2Ts}}{s}\right]^2$$

$$\Rightarrow \qquad Y(s) = \frac{1}{s^2} - \frac{2}{s^2}e^{-2Ts} + \frac{1}{s^2}e^{-4Ts}$$

Taking inverse Laplace transform, we get

$$y(t) = tu(t) - 2(t - 2T)u(t - 2T) + (t - 4T)u(t - 4T)$$
$$= r(t) - 2r(t - 2T) + r(t - 4T)$$

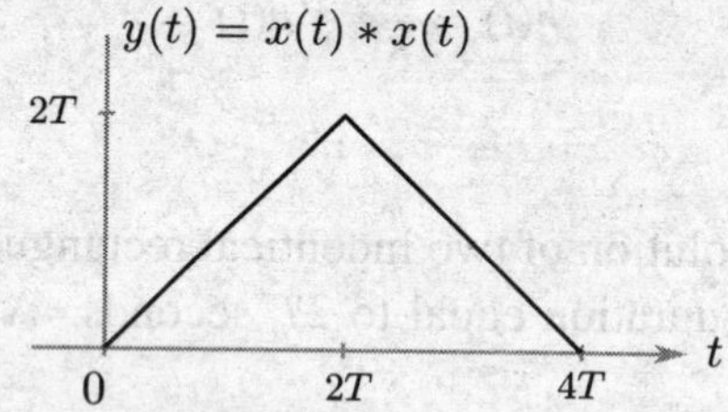

Fig. Ex. 6.32(a)

Fig. Ex. 6.32(a) shows $y(t)$.

Example 6.33. Find the unilateral Laplace transform of each of the following functions:

(a) $x(t) = t^2 \cos(2t + 30°)u(t)$

(b) $x(t) = 2tu(t) - 4\frac{d}{dt}\delta(t)$

(c) $x(t) = 5u\left(\frac{t}{3}\right)$

(d) $x(t) = 5e^{-\frac{t}{2}}u(t)$

Solution:

(a) Let us first find the unilateral Laplace transform of $\cos(2t + 30°)u(t)$

$$\mathscr{L}_u\{\cos(2t + 30°)u(t)\} = \mathscr{L}_u\{\cos 30° \cos 2tu(t) - \sin 30° \sin 2t\, u(t)\}$$
$$= \cos 30° \mathscr{L}_u\{\cos 2t\, u(t)\} - \sin 30° \mathscr{L}_u\{\sin 2tu(t)\}$$
$$= \cos 30° \left[\frac{s}{s^2 + 4}\right] - \sin 30° \left[\frac{2}{s^2 + 4}\right]$$
$$= \frac{s \cos 30° - 2 \sin 30°}{s^2 + 4}$$

The Laplace transform of $x(t)$ is now found by using differentiation in s-domain property. The statement of this property is given below:
If

$$\mathscr{L}_u\{f(t)\} = F(s), \quad \text{then}$$
$$\mathscr{L}_u\{t^n f(t)\} = (-1)^n \frac{d^n F(s)}{ds^n}$$

Hence,

$$\begin{aligned}
\mathscr{L}_u\{t^2 \cos(2t + 30°)\} &= \frac{d^2}{ds^2}[\mathscr{L}_u\{\cos(2t + 30°)u(t)\}] \\
&= \frac{d^2}{ds^2}\left[\frac{s\cos 30° - 2\sin 30°}{s^2 + 4}\right] \\
&= \frac{d^2}{ds^2}\left[\frac{\frac{\sqrt{3}}{2}s - 1}{s^2 + 4}\right] \\
&= \frac{d}{ds}\frac{d}{ds}\left[\frac{\frac{\sqrt{3}}{s}s - 1}{s^2 + 4}\right] \\
&= \frac{d}{ds}\frac{d}{ds}\left[\left(\frac{\sqrt{3}}{2}s - 1\right)(s^2 + 4)^{-1}\right] \\
&= \frac{d}{ds}\left[\left(\frac{\sqrt{3}}{2}(s^2 + 4)^{-1}\right) - 2s\left(\frac{\sqrt{3}}{2}s - 1\right)(s^2 + 4)^{-2}\right] \\
&= \frac{\frac{\sqrt{3}}{2}(-2s)}{(s^2 + 4)^2} - 2\frac{\frac{\sqrt{3}}{2}s - 1}{(s^2 + 4)^2} - \frac{2s\frac{\sqrt{3}}{2}}{(s^2 + 4)^2} + \frac{8s^2\left(\frac{\sqrt{3}}{2}s - 1\right)}{(s^2 + 4)^3} \\
&= \frac{8 - 12\sqrt{3}s - 6s^2 + \sqrt{3}s^2}{(s^2 + 4)^3}
\end{aligned}$$

(b) $x(t) = 2tu(t) - 4\frac{d}{dt}\delta(t)$

$$\mathscr{L}_u\{x(t)\} = X(s) = 2\mathscr{L}_u\{tu(t)\} - 4\mathscr{L}_u\left\{\frac{d}{dt}\delta(t)\right\}$$

We know that whenever a function $f(t)$ is discountinuous at the origin, we have $\mathscr{L}_u\left\{\frac{d}{dt}f(t)\right\} = sF(s) - f(0^-)$. Applying this relation to the second term on the right side of the above equation, we get

$$\begin{aligned} X(s) &= 2\frac{1}{s^2} - 4[s \times 1 - \delta(0^-)] \\ &= \frac{2}{s^2} - 4[s - 0] \\ &= \frac{2}{s^2} - 4s \end{aligned}$$

(c) $x(t) = 5u\left(\frac{t}{3}\right)$

Using scaling property,

$$\mathscr{L}_u\{f(at)\} = \frac{1}{a}F\left(\frac{s}{a}\right)$$

we get,

$$\begin{aligned} \mathscr{L}_u\{x(t)\} = X(s) &= 5 \times \frac{1}{1/3}\mathscr{L}_u\{u(t)\}_{s\to\left(\frac{s}{\frac{1}{3}}\right)} \\ &= 5 \times \frac{1}{1/3} \times \left[\frac{1}{s}\right]_{s\to\left(\frac{s}{\frac{1}{3}}\right)} \\ &= \frac{5}{s} \end{aligned}$$

(d) $x(t) = 5e^{-\frac{t}{2}}u(t)$

We know the Laplace transform pair:

$$\mathscr{L}_u\{e^{-at}u(t)\} = \frac{1}{s+a}$$

Hence,

$$\begin{aligned} \mathscr{L}_u\{x(t)\} = X(s) &= 5\mathscr{L}_u\left\{e^{-\frac{1}{2}t}u(t)\right\} \\ &= 5\frac{1}{s+\frac{1}{2}} = \frac{10}{2s+1} \end{aligned}$$

Example 6.34. Find the unilateral Laplace transform of the following functions:

(a) $x(t) = t\cos at$

(b) $x(t) = \dfrac{1}{2a^2}\sin at \sinh(at)$

(c) $x(t) = \dfrac{\sin^2 \omega t}{t}$

Solution:

(a) $x(t) = t\cos at$

We know that

$$\mathscr{L}_u\{tf(t)\} = -\frac{d}{ds}F(s)$$

Let

$$f(t) = \cos at$$

$$\Rightarrow \quad F(s) = \frac{s}{s^2 + a^2}$$

Hence,

$$\mathscr{L}_u\{t\cos at\} = \mathscr{L}_u\{tf(t)\} = -\frac{d}{ds}\left[\frac{s}{s^2+a^2}\right]$$
$$= \frac{s^2 - a^2}{(s^2 + a^2)^2}$$

(b)

$$x(t) = \frac{1}{2a^2}\sin at \sinh at$$
$$= \frac{1}{2a^2}\left[\frac{1}{2}e^{at}\sin at - \frac{1}{2}e^{-at}\sin at\right]$$
$$= \frac{1}{4a^2}\left[e^{at}\sin at - e^{-at}\sin at\right]$$

We know the shifting in s-domain property:

$$\mathscr{L}_u\{e^{s_0 t}f(t)\} = F(s)\Big|_{s\to(s-s_0)}$$

Applying this property along with linearity property, we get

$$\begin{aligned}
\mathscr{L}_u\{x(t)\} &= X(s) \\
&= \frac{1}{4a^2}[\mathscr{L}_u\{e^{at}\sin at\} - \mathscr{L}_u\{e^{-at}\sin at\}] \\
&= \frac{1}{4a^2}\left[\left.\frac{a}{s^2+a^2}\right|_{s\to s-a} - \left.\frac{a}{s^2+a^2}\right|_{s\to s+a}\right] \\
&= \frac{1}{4a^2}\left[\frac{a}{(s-a)^2+a^2} - \frac{a}{(s+a)^2+a^2}\right] \\
&= \frac{s}{[(s-a)^2+a^2][(s+a)^2+a^2]}
\end{aligned}$$

(c) $x(t) = \dfrac{1}{t}\sin^2 \omega t$

We know that

$$\mathscr{L}_u\{f(t)\} = F(s) = \int_0^\infty f(t)e^{-st}\,dt$$

Hence,

$$\begin{aligned}
\int_s^\infty F(s)ds &= \int_0^\infty f(t)\int_s^\infty e^{-st}ds\,dt \\
&= \int_0^\infty f(t)\left[\frac{e^{-st}}{-t}\right]_s^\infty dt \\
&= \int_0^\infty \frac{f(t)}{t}e^{-st}\,dt \\
&= \mathscr{L}_u\left[\frac{f(t)}{t}\right]
\end{aligned}$$

In the present case,

$$\begin{aligned}
f(t) &= \sin^2 \omega t \\
&= \left[\frac{1}{j2}e^{j\omega t} - \frac{1}{j2}e^{-j\omega t}\right]^2 \\
&= \frac{e^{j2\omega t} - 2 + e^{-j2\omega t}}{-4}
\end{aligned}$$

Hence,

$$F(s) = -\frac{1}{4}\left[\frac{1}{s - j2\omega}\right] + \frac{1}{2}\left(\frac{1}{s}\right) - \frac{1}{4}\left[\frac{1}{s + j2\omega}\right]$$

Finally,

$$\begin{aligned}
X(s) &= \mathscr{L}_u\left\{\frac{1}{t}\sin^2 \omega t\right\} \\
&= \mathscr{L}_u\left\{\frac{1}{t}f(t)\right\} \\
&= \int_s^\infty F(s)\,ds = \lim_{x\to\infty}\int_s^x f(x)\,dx \\
&= \lim_{x\to\infty}\left[\frac{\log_e(x - j2\omega) - \log_e(s - j2\omega) - 2\log_e x + 2\log_e s + \log_e(x + j2\omega) - \log_e(s + j2\omega)}{-4}\right] \\
&= -\frac{1}{4}\log_e\left(\frac{x^2 + 4\omega^2}{x^2}\right)_{x\to\infty} + \frac{1}{4}\log_e\left[\frac{s^2 + 4\omega^2}{s^2}\right] \\
&= \frac{1}{4}\log_e\left[\frac{s^2 + 4\omega^2}{s^2}\right]
\end{aligned}$$

6.7 Initial-value theorem

The initial-value theorem allows us to find the initial value $x(0)$ directly from its Laplace transform $X(s)$.

If $x(t)$ is a causal signal, then

$$x(0) = \lim_{s\to\infty} sX(s)$$

When $x(t)$ is discontinuous at the origin, the argument 0 on the left-side of the above equation should be read as 0^-. Accordingly, For a discontinuous function $x(t)$ at the origin, we get

$$x(0^-) = \lim_{s\to\infty} sX(s)$$

Proof: To prove this property, we use the time differentiation property.

$$\mathscr{L}_u\left\{\frac{dx(t)}{dt}\right\} = sX(s) - x(0)$$

The unilateral Laplace transform of $\dfrac{dx(t)}{dt}$ is given by

$$\mathscr{L}_u\left\{\frac{dx(t)}{dt}\right\} = \int_0^\infty \frac{dx(t)}{dt}e^{-st}\,dt$$

Hence,

$$\int_0^\infty \frac{dx(t)}{dt} e^{-st}\, dt = sX(s) - x(0)$$

If we let $s \to \infty$, then the integral on the left side of the above equation tends to zero. Thus,

$$\lim_{s\to\infty} [sX(s) - x(0)] = 0$$

$$\Rightarrow \qquad x(0) = \lim_{s\to\infty} sX(s)$$

Example 6.35. Find the initial value of

$$F(s) = \frac{s+1}{(s+1)^2 + 3^2}$$

Solution:

$$f(0) = \lim_{s\to\infty} sX(s) = \lim_{s\to\infty} s\left[\frac{s+1}{(s+1)^2+3^2}\right]$$

$$= \lim_{s\to\infty} \frac{s^2+s}{(s+1)^2+3^2}$$

$$= \lim_{s\to\infty} \frac{s^2\left[1+\frac{1}{s}\right]}{s^2\left[1+\frac{2}{s}+\frac{10}{s^2}\right]} = 1$$

We know the transform pair:

$$\mathscr{L}_u\{e^{-bt}\cos at\} = \frac{s+b}{(s+b)^2+a^2}$$

Hence, inverse Laplace transform of $F(s)$ yields

$$f(t) = e^{-t}\cos 3t$$

At $t = 0$, we get $f(0) = 1$.
This verifies the theorem.

6.8 Final-value theorem

The final-value theorem allows us to find the final value of $x(t)$ directly from its Laplace transform $X(s)$.

It $x(t)$ is a causal signal, then

$$\lim_{t\to\infty} x(t) = \lim_{s\to 0} sX(s)$$

Proof: The unilateral Laplace transform of $\dfrac{dx(t)}{dt}$ is given by

$$sX(s) - x(0) = \int_0^{\infty} \frac{dx(t)}{dt} e^{-st}\, dt$$

Taking the limit $s \to 0$ on both sides, we get

$$\begin{aligned}\lim_{s\to 0}[sX(s) - x(0)] &= \lim_{s\to 0} \int_0^{\infty} \frac{dx(t)}{dt} e^{-st}\, dt \\ &= \int_0^{\infty} \frac{dx(t)}{dt} \left[\lim_{s\to 0} e^{-st}\right] dt \\ &= \int_0^{\infty} \frac{dx(t)}{dt}\, dt \\ &= x(t)\Big|_0^{\infty} = x(\infty) - x(0)\end{aligned}$$

Since

$$\lim_{s\to 0}[sX(s) - x(0)] = \lim_{s\to 0} sX(s) - x(0)$$

we get,

$$\lim_{s\to 0} sX(s) - x(0) = x(\infty) - x(0)$$

Therefore,

$$x(\infty) = \lim_{s\to 0} sX(s)$$

This proves the final value theorem.

The final value theorem may be applied if, and only if all the poles of $X(s)$ have a real part that is negative.

The final value theorem is very useful since we can find $x(\infty)$ from $X(s)$. However, one must be careful in using the final value theorem since the function $x(t)$ may not have a final value as $t \to \infty$. For example, consider $x(t) = \sin at$ having $X(s) = \dfrac{a}{s^2 + a^2}$. Now we know $\lim_{t\to\infty} \sin at$ does not exist. However, if we uncarefully use the final value theorem in this case, we would obtain:

$$\lim_{s\to 0} sX(s) = \lim_{s\to 0} s\left[\frac{a}{s^2 + a^2}\right] = 0$$

The final value theorem has failed because the poles of $X(s)$ lie on $j\omega$- axis. Therefore, we conclude that for final value theorem to give a valid result, poles of $X(s)$ should not lie to the right-side of the s-plane or on the $j\omega$-axis.

Example 6.36. Find the final value of

$$X(s) = \frac{10}{(s+1)^2 + 10^2}$$

Solution:

$$\lim_{t\to\infty} x(t) = x(\infty) = \lim_{s\to 0}[sX(s)] = \lim_{s\to 0} \frac{s10}{(s+1)^2 + 10^2} = 0$$

We know the Laplace transform pair

$$\mathscr{L}_u\left[e^{-at}\sin bt\right] = \frac{b}{(s+a)^2 + b^2}$$

Hence,

$$x(t) = \mathscr{L}_u^{-1}\{X(s)\} = \mathscr{L}_u^{-1}\left\{\frac{10}{(s+1)^2 + 10^2}\right\} = e^{-t}\sin 10t$$

Thus, $x(\infty) = 0$.
This verifies the result obtained from final-value theorem.

Example 6.37. Apply the initial and final value theorems to each of the functions given below:

(a) $F(s) = \dfrac{s^2 + 5s + 10}{s+6}$ (b) $F(s) = \dfrac{s^2 + 5s + 10}{5(s^2 + 6s + 8)}$

Solution: Since $F(s)$ referred in (a) and (b) are improper [1] fractions, the corresponding time-domain counterparts, $f(t)$ contain impulses.

Thus, neither the initial value theorem nor the final value theorem may be applied to these transformed functions.

Example 6.38. Find the initial and final values of $f(t)$ when $F(s) = \dfrac{60}{s^2 - 2s + 1}$

Solution: Initial value theorem:

$$f(0) = \lim_{s\to\infty} sF(s)$$

$$= \lim_{s\to\infty} s\frac{60}{s^2 - 2s + 1} = 0$$

Final value theorem:

The poles of $F(s)$ are given by finding the roots of the denominator polynomial. That is,

$$s^2 - 2s + 1 = 0 \quad \Rightarrow \quad (s-1)^2 = 0 \quad \Rightarrow \quad s = 1, 1$$

Since both the poles of $F(s)$ lie to the right of the s plane, final value theorem cannot be used to find $f(\infty)$.

[1] If the degree of the numerator polynomial is greater than or equal to the degree of the denominator polynomial, the fraction is said to be improper.

6.9 Three important singularity functions

The three important singularity functions employed in circuit and system analysis are:

(i) unit step function, $u(t)$

(ii) delta function, $\delta(t)$

(iii) ramp function, $r(t)$.

They are called singularity functions because they are either not finite or they do not possess finite derivatives everywhere.

The mathematical definition of unit step function is

$$u(t) = \begin{cases} 0, & t < 0 \\ 1, & t > 0 \end{cases} \tag{6.16}$$

The step function is not defined at $t = 0$. Thus, the unit step function $u(t)$ is 0 for negative values of t, and 1 for positive values of t. Often it is advantageous to define the unit step function as follows:

$$u(t) = \begin{cases} 1, & t \geq 0^+ \\ 0, & t \leq 0^- \end{cases}$$

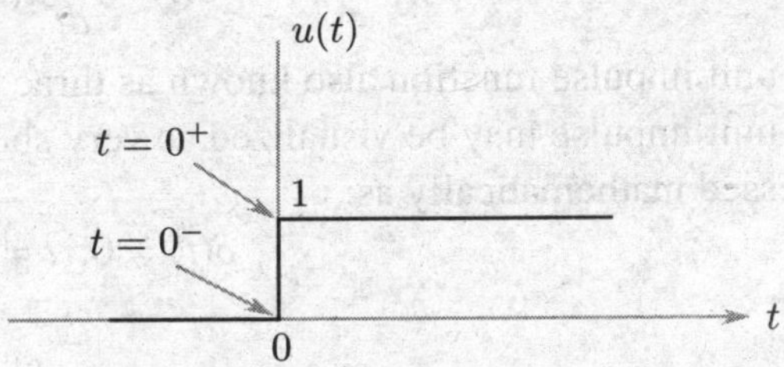

Fig. 6.2: *The unit step function*

A discontinuity may occur at time other than $t = 0$; for example, in sequential switching, the unit step function that occurs at $t = a$ is expressed as $u(t - a)$.

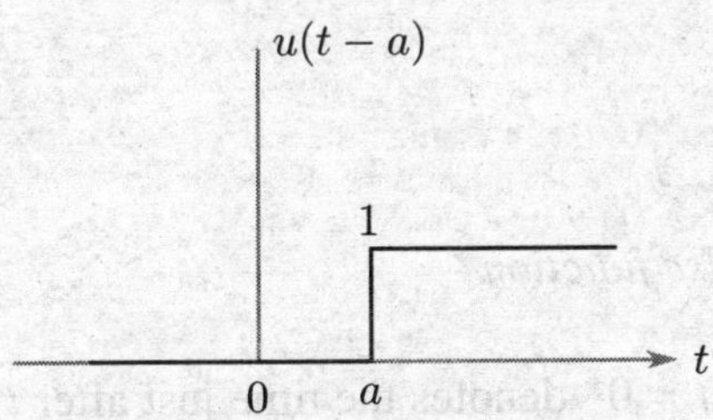

Fig. 6.3: *The step function occuring at t = a*

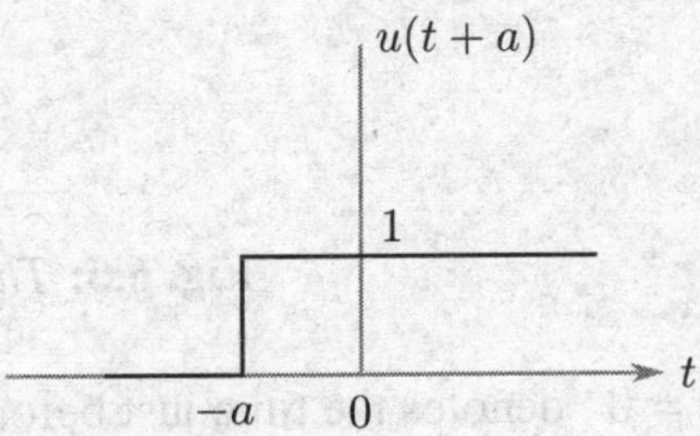

Fig. 6.4: *The step function occuring at t = −a*

Thus,

$$u(t-a) = \begin{cases} 0, & t-a<0 \text{ or } t<a \\ 1, & t-a>0 \text{ or } t>a \end{cases}$$

Similarly, the unit step function that occurs at $t = -a$ is expressed as $u(t+a)$. Thus,

$$u(t+a) = \begin{cases} 0, & t+a<0 \text{ or } t<-a \\ 1, & t+a>0 \text{ or } t>-a \end{cases}$$

We use step function to represent an abrupt change in voltage or current, like the change that occur in the circuits of control engineering and digital systems. For example, the voltage

$$v(t) = \begin{cases} 0, & t<a \\ K, & t>a \end{cases}$$

may be expressed in terms of the unit step function as

$$v(t) = Ku(t-a) \tag{6.17}$$

The derivative of the unit step function $u(t)$ is the unit impulse function $\delta(t)$. That is,

$$\delta(t) = \frac{d}{dt}u(t) \tag{6.18}$$

The unit impulse function also known as dirac delta function is shown in Fig. 6.5.

The unit impulse may be visualized as very short duration pulse of unit area. This may be expressed mathematically as:

$$\delta(t) = 0; \ t \neq 0$$

and

$$\int_{-\infty}^{\infty} \delta(t)\,dt = 1 \Rightarrow \int_{0^-}^{0^+} \delta(t)\,dt = 1 \tag{6.19}$$

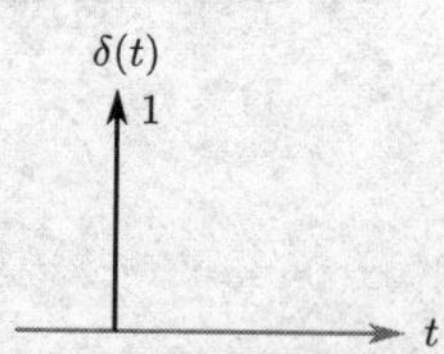

Fig. 6.5: *The unit impulse function*

where $t = 0^-$ denotes the time just before $t = 0$ and $t = 0^+$ denotes the time just after $t = 0$. Since the area under the unit impulse is unity, it is a practice to write '1' beside the arrow that is used to symbolize the unit impulse function as shown in Fig. 6.5. When the impulse has a strength other than unity, the area of the impulse function is equal to its strength. For example, an impulse function $5\delta(t)$ has an area of 5 units. Fig. 6.6 shows impulse functions, $2\delta(t+2)$, $5\delta(t)$ and $-2\delta(t-3)$.

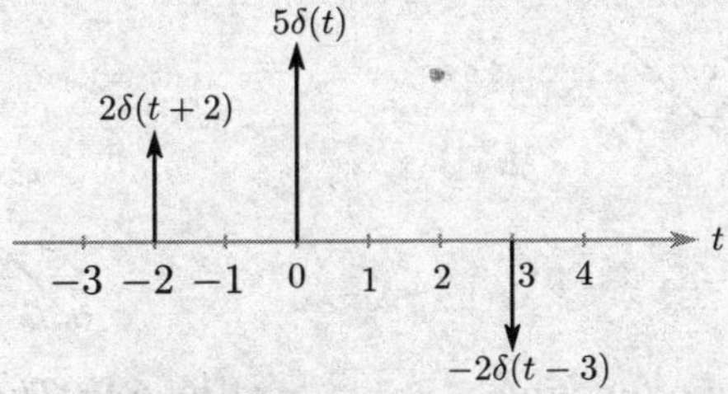

Fig. 6.6: *Three impulse functions*

An important property of the unit impulse function is what is often called the sifting property; which is exhibited by the following integral:

$$\int_{t_1}^{t_2} f(t)\delta(t-t_0)\,dt = \begin{cases} f(t_0), & t_1 < t_0 < t_2 \\ 0, & \text{otherwise} \end{cases}$$

for a finite t_0 and any $f(t)$ continuous at t_0.

Integrating the unit step function results in the unit ramp function $r(t)$.

$$r(t) = \int_{-\infty}^{t} u(\tau)d\tau = tu(t) \tag{6.20}$$

$$\text{or} \qquad r(t) = \begin{cases} 0, & t \le 0 \\ t, & t \ge 0 \end{cases}$$

Fig. 6.7 shows the ramp function.

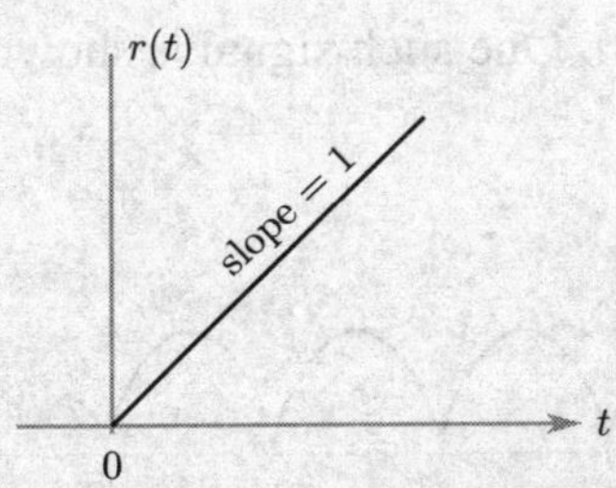

Fig. 6.7: *The unit ramp function*

In general, a ramp is a function that changes at a constant rate.

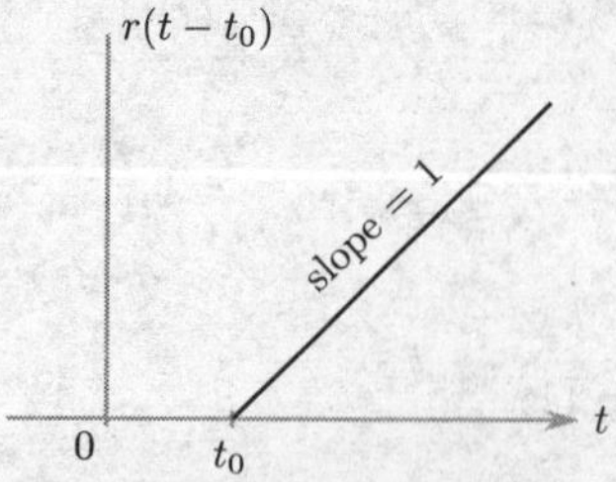

Fig. 6.8: *The unit ramp function delayed by t_0*

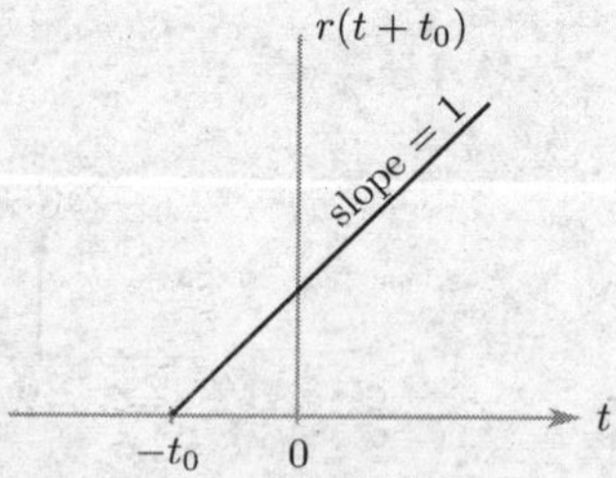

Fig. 6.9: *The unit ramp function advanced by t_0*

A delayed ramp function is shown in Fig. 6.8. Mathematically, it is described as follows:

$$r(t-t_0) = \begin{cases} 0, & t \leq t_0 \\ t-t_0, & t \geq t_0 \end{cases}$$

An advanced ramp function is shown in Fig. 6.9. Mathematically, it is described as follows:

$$r(t+t_0) = \begin{cases} 0, & t \leq -t_0 \\ t+t_0, & t \geq -t_0 \end{cases}$$

It is very important to note that the three sigularity functions are related by differentiation as

$$\delta(t) = \frac{du(t)}{dt}, \quad u(t) = \frac{dr(t)}{dt}$$

or by integration as

$$u(t) = \int_{-\infty}^{t} \delta(t)\, dt, \quad r(t) = \int_{-\infty}^{t} u(\tau) d\tau$$

6.10 Switched time periodicity

A periodic signal $x(t)$ turned on at $t = 0$ may be expressed as $x(t) = x_p(t)u(t)$. This means that $x(t)$ is periodic only for $t \geq 0$. One such signal is shown in Fig. 6.10.

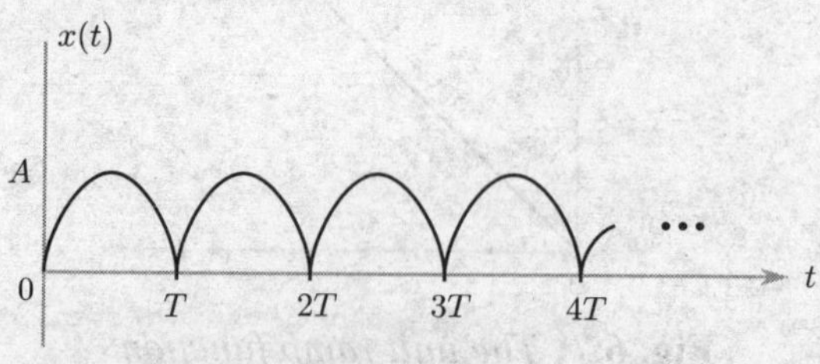

Fig. 6.10: *A switched periodic function*

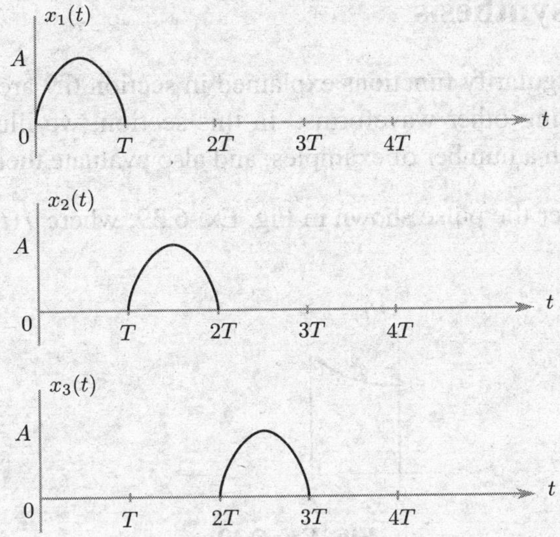

Fig. 6.11: *Decomposition of switched periodic function*

The switched periodic function $x(t)$ can be represented as the sum of time-shifted functions as shown in Fig. 6.11. That is,

$$\begin{aligned} x(t) &= x_1(t) + x_2(t) + x_3(t) + \cdots \\ &= x_1(t) + x_1(t-T) + x_1(t-2T) + \cdots \end{aligned} \tag{6.21}$$

where $x_1(t)$ is the waveform described over the first period of $x(t)$.

Taking unilateral Laplace transform on both sides of Eq. (6.21) with time-shift property applied, we get

$$\begin{aligned} X(s) &= X_1(s) + X_1(s)e^{-Ts} + X_1(s)e^{-2Ts} + \cdots \\ &= X_1(s)[1 + e^{-Ts} + e^{-2Ts} + \cdots] \end{aligned}$$

Recall:

$$\sum_{n=0}^{\infty} a^n = \frac{1}{1-a}; \ |a| < 1$$

Hence,

$$X(s) = \frac{X_1(s)}{1 - e^{-sT}} \tag{6.22}$$

In Eq. (6.22), $X_1(s)$ is the unilateral Laplace transform of $x(t)$ defined over first period only. Hence, we have shown that unilateral Laplace transform of a periodic function is the unilateral Laplace transform evaluated over its first period divided by $1 - e^{-Ts}$.

6.11 Waveform synthesis

The three important singularity functions explained in section 6.9 are very useful as building blocks in constructing other waveforms. In this section, we illustrate the concept of waveform synthesis with a number of examples, and also evaluate their Laplace transforms.

Example 6.39. Consider the pulse shown in Fig. Ex. 6.39, where $f(t) = e^{2t}$ for $0 < t < T$. Find $F(s)$ for the pulse.

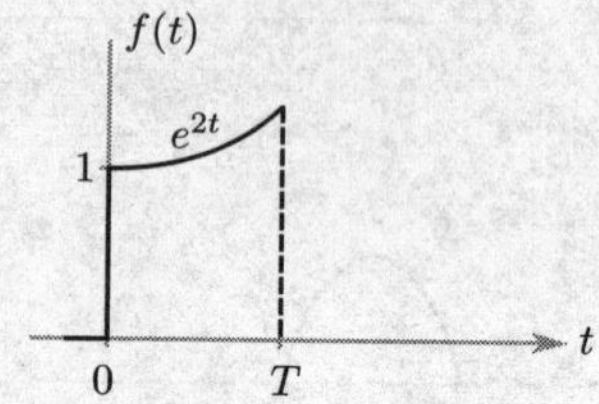

Fig. Ex. 6.39

Solution: The discrete pulse $f(t)$ could be imagined as the product of signal $x(t)$ and $g(t)$ as shown in Fig. Ex. 6.39(a) and (b) respectively.

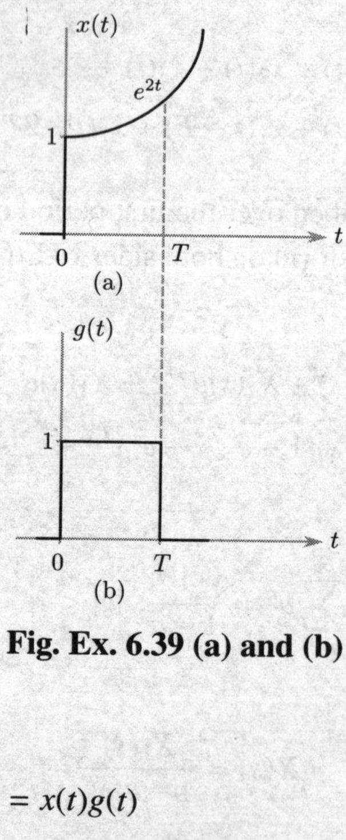

Fig. Ex. 6.39 (a) and (b)

That is,

$$f(t) = x(t)g(t)$$
$$= e^{2t}[u(t) - u(t - T)]$$
$$= e^{2t}u(t) - e^{2t}u(t - T)$$

$$= e^{2t}u(t) - e^{2(t-T+T)}u(t-T)$$
$$= e^{2t}u(t) - e^{2T}e^{2(t-T)}u(t-T)$$

Hence,

$$\mathscr{L}_u\{f(t)\} = F(s) = \frac{1}{s-2} - \frac{e^{2T}e^{-Ts}}{s-2} = \frac{1}{s-2} - \frac{e^{-T(s-2)}}{s-2} = \frac{1-e^{-T(s-2)}}{(s-2)}$$

Alternate method:

$$\mathscr{L}_u\{f(t)\} = F(s) \triangleq \int_0^{\infty} f(t)e^{-st}\,dt$$
$$= \int_0^{T} e^{2t}e^{-st}\,dt$$
$$= \frac{1-e^{-T(s-2)}}{(s-2)}$$

Example 6.40. Find the unilateral Laplace transform of $f(t)$ shown in Fig. Ex. 6.40.

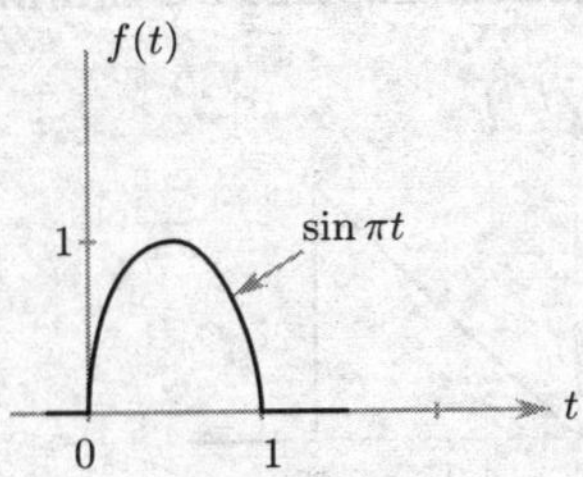

Fig. Ex. 6.40

Solution: $f(t)$ is a discrete pulse and can be expressed mathematically as (refer Fig. Ex. 6.40(a)):

$$f(t) = x(t)g(t)$$
$$= \sin \pi t[u(t) - u(t-1)]$$
$$= \sin \pi t\, u(t) - \sin \pi t\, u(t-1)$$
$$= \sin \pi t\, u(t) - \sin[\pi(t-1+1)]u(t-1)$$
$$\Rightarrow \quad f(t) = \sin \pi t\, u(t) - \{\sin[\pi(t-1)]\cos \pi + \cos[\pi(t-1)]\sin \pi\}\, u(t-1)$$

Hence,

$$F(s) = \mathscr{L}_u\{f(t)\} = \frac{\pi}{s^2+\pi^2} + \frac{e^{-s}\pi}{s^2+\pi^2}$$
$$= \frac{\pi}{s^2+\pi^2}[1+e^{-s}]$$

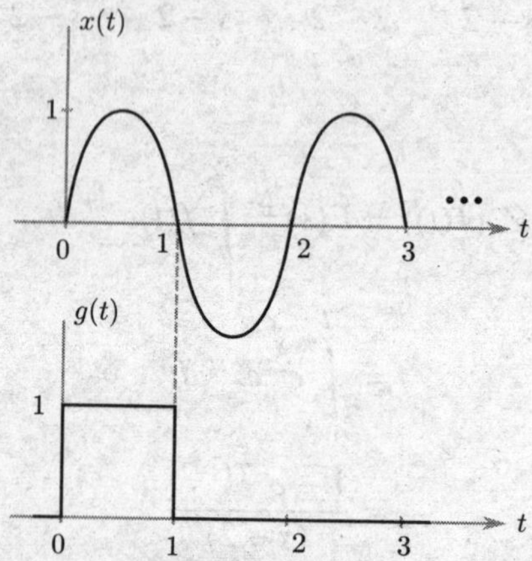

Fig. Ex. 6.40(a)

Example 6.41. Determine the unilateral Laplace transform of $f(t)$ shown in Fig. Ex. 6.41.

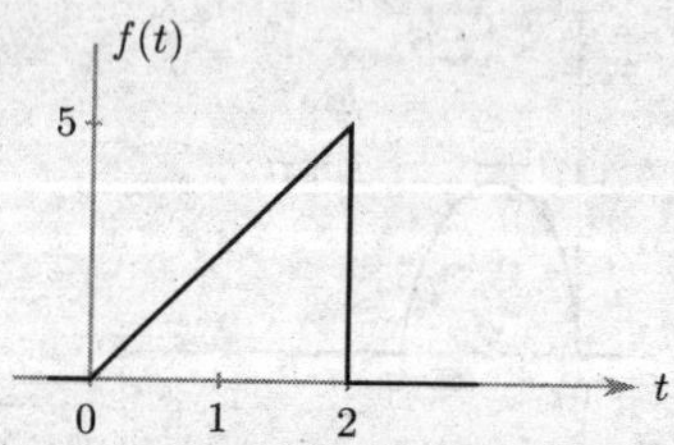

Fig. Ex. 6.41

Solution: We can write,

$$f(t) = x(t)g(t)$$
$$= \left[\frac{5}{2}t\right][u(t) - u(t-2)]$$
$$= \frac{5}{2}\,tu(t) - \frac{5}{2}\,tu(t-2)$$
$$= \frac{5}{2}\,tu(t) - \frac{5}{2}(t-2+2)u(t-2)$$
$$= \frac{5}{2}\,tu(t) - \frac{5}{2}(t-2)u(t-2) - 5u(t-2)$$

Hence,

$$\mathscr{L}_u\{f(t)\} = F(s) = \frac{5}{2}\left(\frac{1}{s^2}\right) - \frac{5}{2}\left(\frac{1}{s^2}\right)e^{-2s} - \frac{5}{s}e^{-2s}$$

$$= \frac{5}{2s^2}[1 - e^{-2s} - 2se^{-2s}]$$

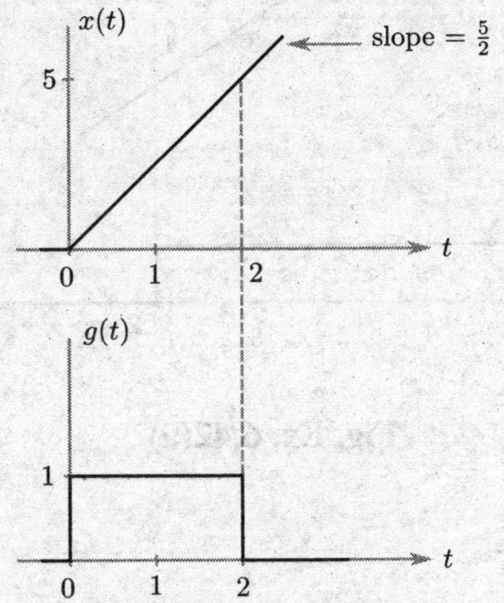

Fig. Ex. 6.41(a)

Example 6.42. Find the Laplace transform of $f(t)$ shown in Fig. Ex. 6.42

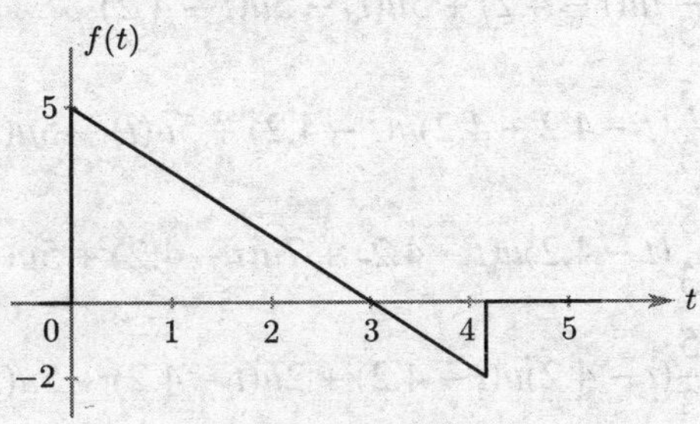

Fig. Ex. 6.42

Solution: The equation of a straight line is $y = mx + c$, where m = slope of the line and c = intercept on y-axis.

Hence, $f(t) = \frac{-5}{3}t + 5$

When $f(t) = -2$, let us find t.

That is,

$$-2 = \frac{-5}{3}t + 5 \quad \Rightarrow \quad t = 4.2 \text{ Seconds}$$

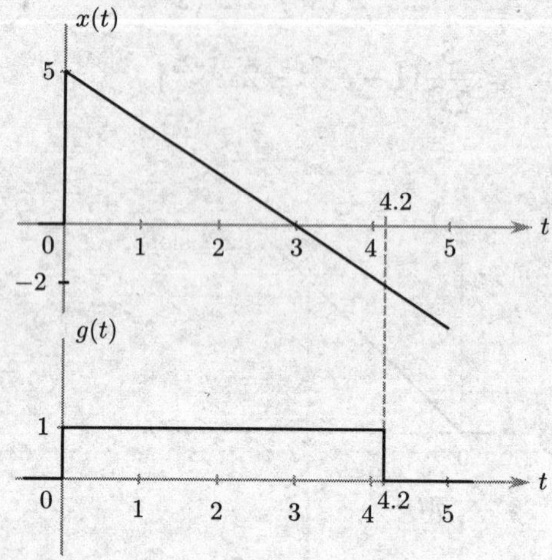

Fig. Ex. 6.42(a)

Mathematically,

$$\begin{aligned}
f(t) &= x(t)g(t) \\
&= \left[\frac{-5}{3}t + 5\right][u(t) - u(t-4.2)] \\
&= \frac{-5}{3}\,tu(t) + \frac{5}{3}\,tu(t-4.2) + 5u(t) - 5u(t-4.2) \\
&= \frac{-5}{3}\,tu(t) + \frac{5}{3}(t - 4.2 + 4.2)u(t-4.2) + 5u(t) - 5u(t-4.2) \\
&= \frac{-5}{3}\,tu(t) + \frac{5}{3}(t-4.2)u(t-4.2) + 7u(t-4.2) + 5u(t) - 5u(t-4.2) \\
&= \frac{-5}{3}\,tu(t) + \frac{5}{3}(t-4.2)u(t-4.2) + 2u(t-4.2) + 5u(t)
\end{aligned}$$

Hence,

$$\begin{aligned}
F(s) &= \mathscr{L}_u\{f(t)\} \\
&= \frac{-5}{3s^2} + \frac{5}{3s^2}e^{-4.2s} + \frac{2}{s}e^{-4.2s} + \frac{5}{s} \\
&= \frac{-5 + 5e^{-4.2s} + 6se^{-4.2s} + 15s}{3s^2}
\end{aligned}$$

Example 6.43. Express the voltage pulse shown in Fig. Ex. 6.43 in terms of unit step function and then find $V(s)$. Also, find $\mathscr{L}_u\left\{\frac{dv(t)}{dt}\right\}$.

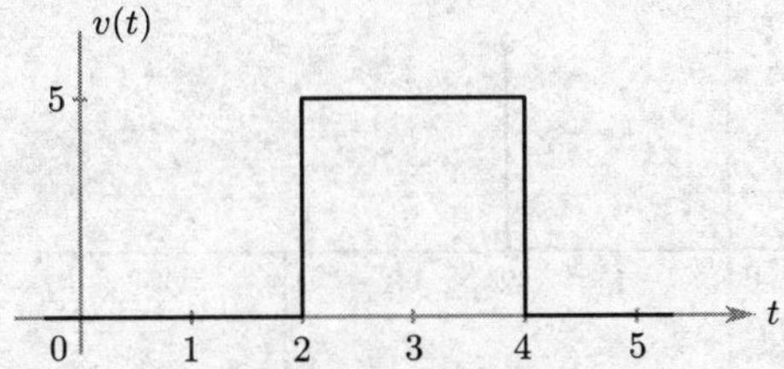

Fig. Ex. 6.43

Solution: The pulse shown in Fig. Ex. 6.43 is the gate function. This function may be regarded as a step function that switches on at $t = 2$ secs and switches off at $t = 4$ secs.

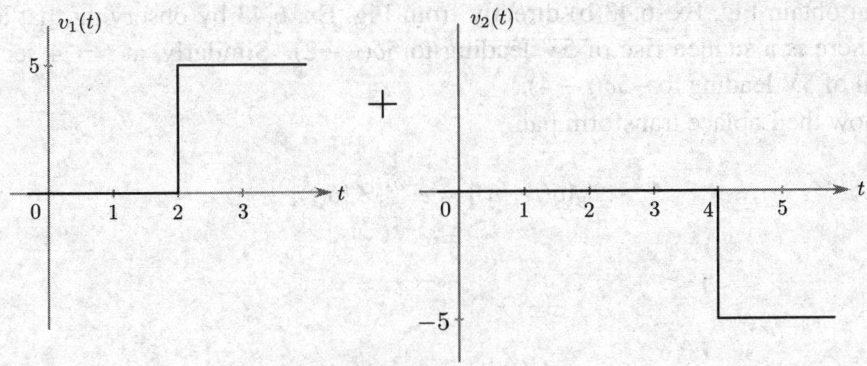

Fig. Ex. 6.43(a)

Referring to Figs. Ex. 6.43 and Ex. 6.43(a), we can write

$$v(t) = v_1(t) + v_2(t)$$

$$\Rightarrow \quad v(t) = 5u(t-2) - 5u(t-4)$$

Hence,

$$V(s) = \frac{5}{s}e^{-2s} - \frac{5}{s}e^{-4s} = \frac{5}{s}[e^{-2s} - e^{-4s}]$$

Taking the derivative of $v(t)$, we get

$$\frac{dv(t)}{dt} = 5[\delta(t-2) - \delta(t-4)]$$

Fig. Ex. 6.43(b) shows the graph of $\dfrac{dv(t)}{dt}$.

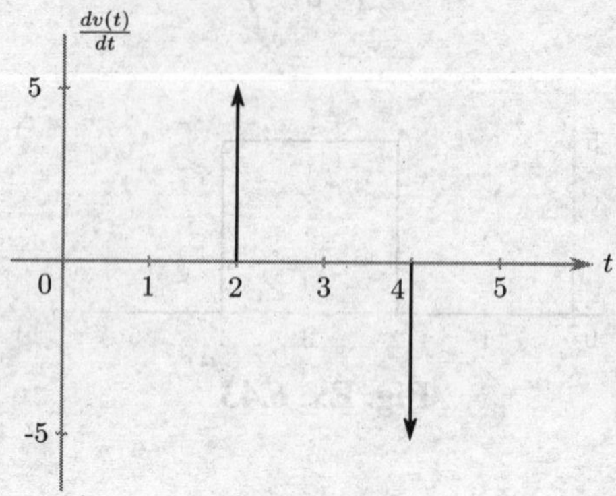

Fig. Ex. 6.43(b)

We can obtain Fig. Ex. 6.43(b) directly from Fig. Ex. 6.43 by observing that at $t = 2$ seconds, there is a sudden rise of 5V leading to $5\delta(t-2)$. Similarly, at $t = 4$ seconds, a sudden fall of 5V leading to $-5\delta(t-4)$.

We know the Laplace transform pair

$$\begin{aligned}\mathscr{L}_u\{\delta(t-a)\} &= e^{-as}\mathscr{L}\{\delta(t)\}\\ &= e^{-as}\end{aligned}$$

Hence,

$$\mathscr{L}_u\left\{\frac{dv(t)}{dt}\right\} = 5[e^{-2s} - e^{-4s}]$$

Example 6.44. Express the current pulse in Fig. Ex. 6.44 in terms of the unit step.

Find: (i) $\mathscr{L}_u\{i(t)\}$ (ii) $\mathscr{L}_u\{\int i(t)dt\}$

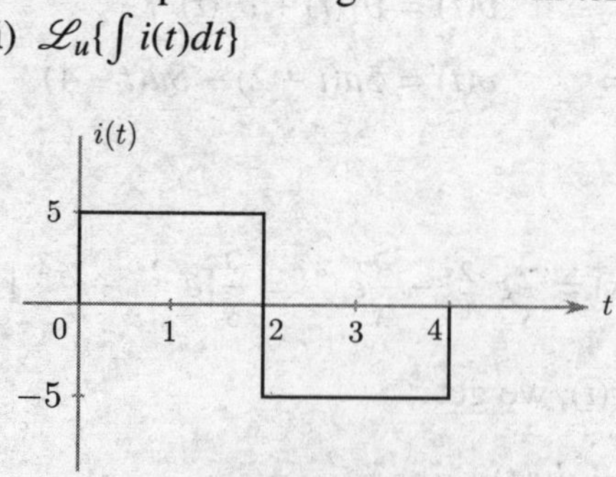

Fig. Ex. 6.44

Solution:

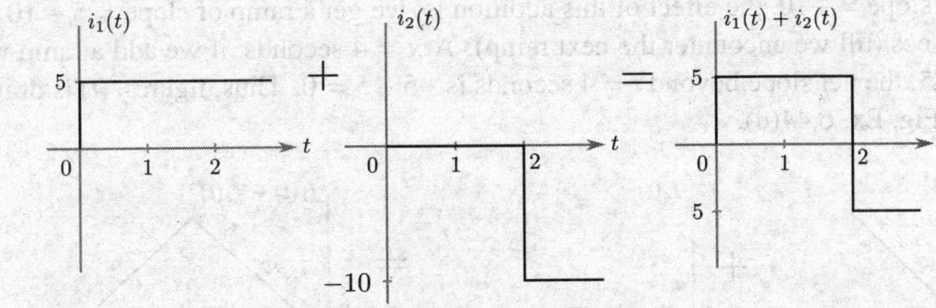

Fig. Ex. 6.44(a)

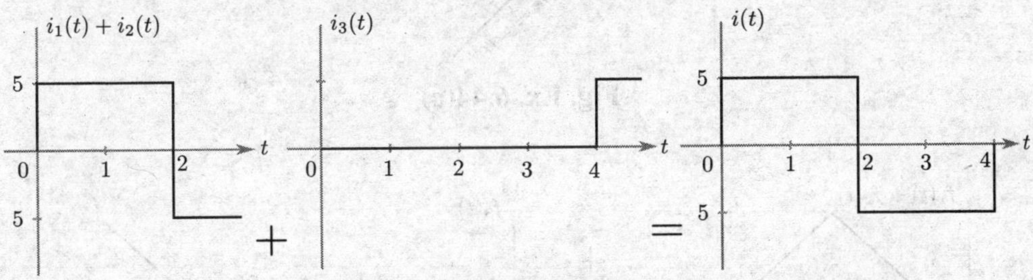

Fig. Ex. 6.44(b)

Referring to Figs. Ex. 6.44 (a) and (b) and using the principle of synthesis, we can write

$$\begin{aligned} i(t) &= i_1(t) + i_2(t) + i_3(t) \\ &= 5u(t) - 10u(t-2) + 5u(t-4) \end{aligned}$$

The unilateral Laplace transform of the above equation yields

$$\begin{aligned} I(s) &= \frac{5}{s} - \frac{10}{s}e^{-2s} + \frac{5}{s}e^{-4s} \\ &= \frac{5}{s}[1 - 2e^{-2s} + e^{-4s}] \\ &= \frac{5}{s}[1 - e^{-2s}]^2 \end{aligned}$$

Let

$$f(t) = \int i(t)\,dt$$

then,

$$\begin{aligned} f(t) &= \int [5u(t) - 10u(t-2) + 5u(t-4)]\,dt \\ &= 5r(t) - 10r(t-2) + 5r(t-4) \\ &= f_1(t) + f_2(t) + f_3(t) \end{aligned}$$

The function $f_1(t)$ is a ramp of slope = 5 as shown in Fig. Ex. 6.44(c). To this, if we add a ramp of slope = -10, the effect of this addition is, we get a ramp of slope = 5 – 10 = –5 for $t \geq 2$ secs (till we encounter the next ramp). At $t = 4$ seconds, if we add a ramp with a slope of +5, the net slope beyond $t = 4$ seconds is $-5 + 5 = 0$. Thus, figure $f(t)$ is drawn as shown in Fig. Ex. 6.44(d).

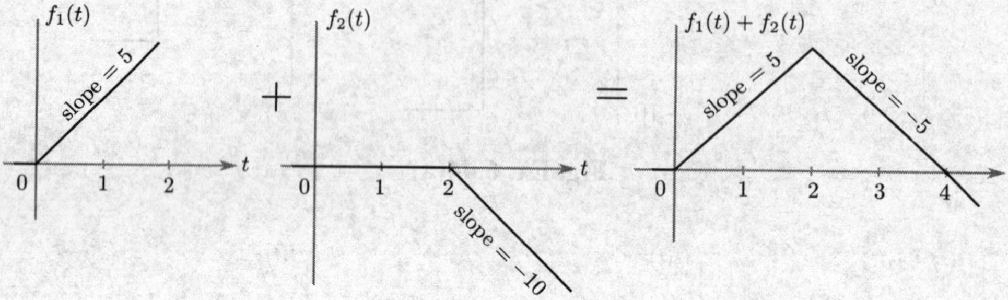

Fig. Ex. 6.44(c)

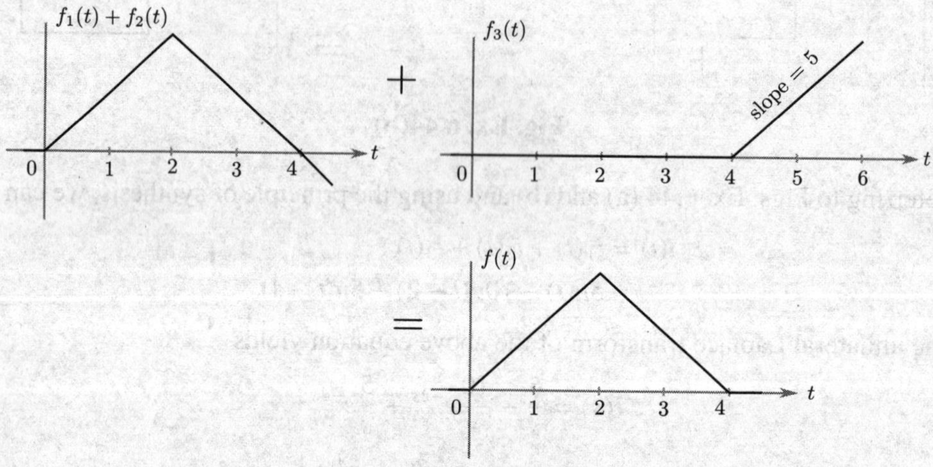

Fig. Ex. 6.44(d)

$$\begin{aligned}
\mathscr{L}_u\{f(t)\} &= F(s) \\
&= \mathscr{L}_u\{5r(t) - 10r(t-2) + 5r(t-4)\} \\
&= \frac{5}{s^2} - \frac{10}{s^2}e^{-2s} + \frac{4}{s^2}e^{-4s} \\
&= \frac{5}{s^2}[1 - 2e^{-2s} + e^{-4s}]
\end{aligned}$$

Example 6.45. Express the sawtooth function shown in Fig. Ex. 6.45 in terms of singularity functions. Then, find $\mathscr{L}_u\{v(t)\}$.

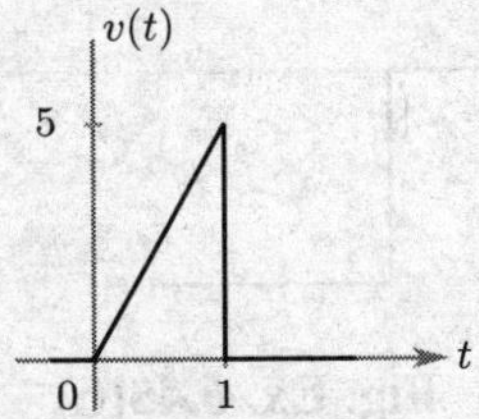

Fig. Ex. 6.45

Solution: There are three methods to solve this problem.

Method 1:

The function $v_1(t)$ is a ramp function of slope = +5. This slope +5 should continue till $t = 1$ second. Hence, at $t = 1$ second, a ramp of slope $t = -5$ is added to $v_1(t)$. The graph of $v_1(t) + v_2(t)$ is shown in Fig. Ex. 6.45(a). Next, to $v_1(t) + v_2(t)$, a step of −5V is added at $t = 1$ second.

Hence,

$$\begin{aligned} v(t) &= v_1(t) + v_2(t) + v_3(t) \\ &= 5r(t) - 5r(t-1) - 5u(t-1) \\ V(s) = \mathscr{L}_u\{f(t)\} &= \frac{5}{s^2} - \frac{5}{s^2}e^{-s} - \frac{5}{s}e^{-s} \\ &= \frac{5}{s^2}[1 - e^{-s} - se^{-s}] \end{aligned}$$

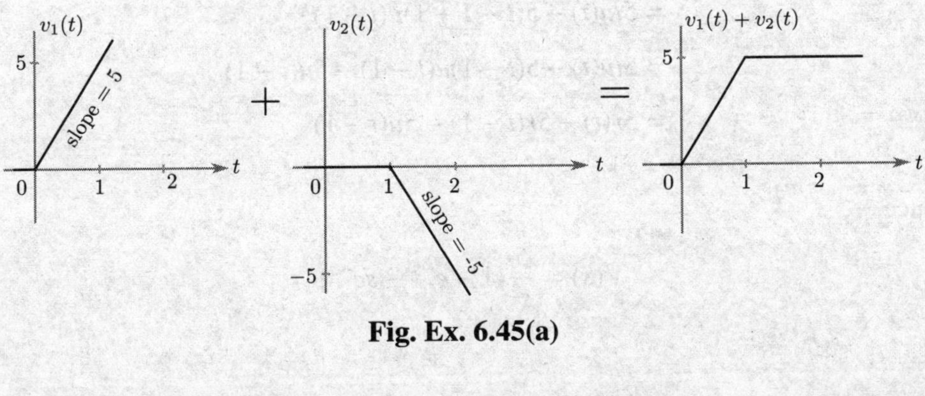

Fig. Ex. 6.45(a)

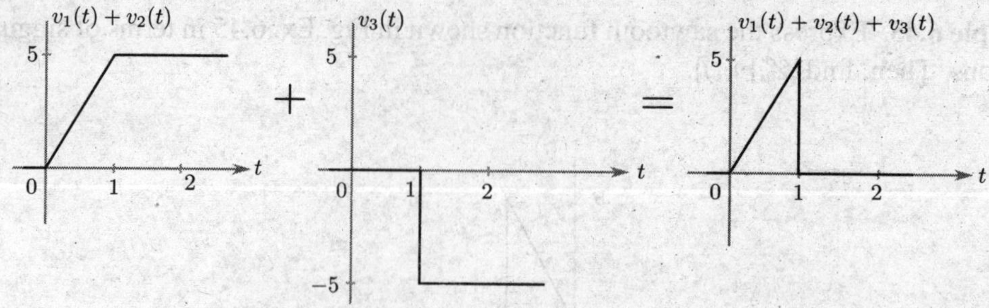

Fig. Ex. 6.45(b)

Method 2:

This method involves graphical manipulation.

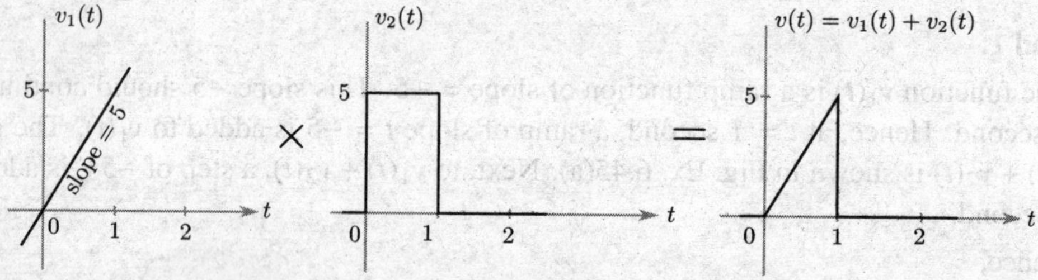

Fig. Ex. 6.45(c)

The equation of a straight line passing through the origin is $y = mx$, where m = slope of the line. This allows us to write $v_1(t) = 5t$. From Fig. Ex. 6.45(c), we can write

$$\begin{aligned}
v(t) &= v_1(t)v_2(t) \\
&= 5t[u(t) - u(t-1)] \\
&= 5tu(t) - 5tu(t-1) \\
&= 5tu(t) - 5(t-1+1)u(t-1) \\
&= 5tu(t) - 5(t-1)u(t-1) - 5u(t-1) \\
&= 5r(t) - 5r(t-1) - 5u(t-1)
\end{aligned}$$

Hence,

$$V(s) = \frac{5}{s^2}[1 - e^{-s} - se^{-s}]$$

Method 3:

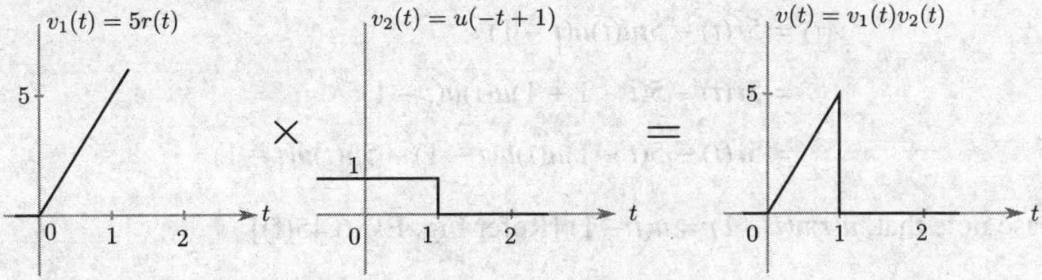

Fig. Ex. 6.45(d)

This method also involves graphical manipulation. We observe from Fig. Ex. 6.45(d) that $v(t)$ is equal to multiplication of a ramp function and a unit step function.

Thus,

$$\begin{aligned} v(t) &= v_1(t)v_2(t) \\ &= 5r(t)[u(-t+1)] \end{aligned}$$

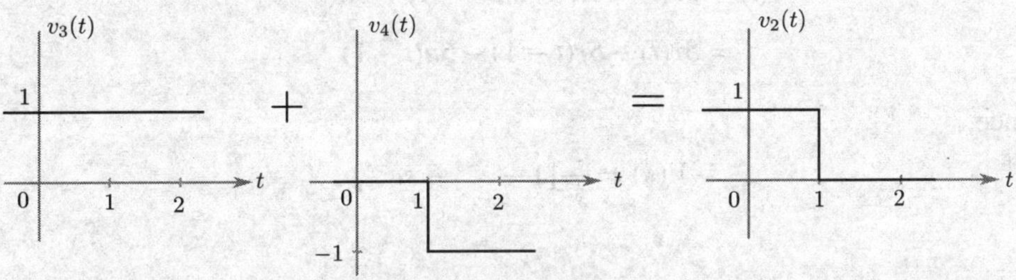

Fig. Ex. 6.45(e)

From Fig. Ex. 6.45(e), we can write

$$\begin{aligned} v_2(t) &= v_3(t) + v_4(t) \\ \Rightarrow \quad v_2(t) &= 1 - u(t-1) \\ \Rightarrow \quad u(-t+1) &= 1 - u(t-1) \end{aligned}$$

Hence,

$$\begin{aligned} v(t) &= 5r(t)[1 - u(t-1)] \\ &= 5r(t) - 5r(t)u(t-1) \end{aligned}$$

We know that,

$$r(t) = tu(t)$$

Hence,

$$\begin{aligned} v(t) &= 5r(t) - 5tu(t)u(t-1) \\ &= 5r(t) - 5(t-1+1)u(t)u(t-1) \\ &= 5r(t) - 5(t-1)u(t)u(t-1) - 5u(t)u(t-1) \end{aligned}$$

Please note that, $u(t)u(t-1) = u(t-1)$ [Refer Fig. Ex. 6.45(f)]

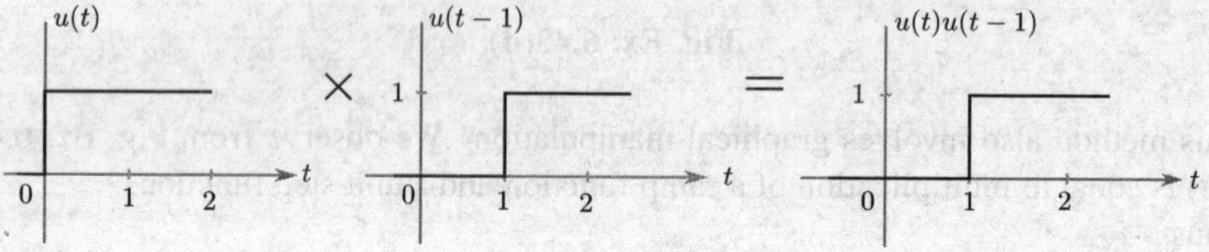

Fig. Ex. 6.45(f)

Thus,

$$\begin{aligned} v(t) &= 5r(t) - 5(t-1)u(t-1) - 5u(t-1) \\ &= 5r(t) - 5r(t-1) - 5u(t-1) \end{aligned}$$

Hence,

$$V(s) = \frac{5}{s^2}[1 - e^{-s} - se^{-s}]$$

Example 6.46. Given the signal

$$x(t) = \begin{cases} 3, & t < 0 \\ -2, & 0 < t < 1 \\ 2t - 4, & t > 1 \end{cases}$$

Express $x(t)$ in terms of singularity functions. Also, find $\mathscr{L}_u\{x(t)\}$.

Solution: The signal $x(t)$ may be viewed as follows:

(i) In the interval, $t < 0$, $x(t)$ may be regarded as $3u(-t)$

(ii) In the interval, $0 < t < 1$, $x(t)$ may be viewed as $-2[u(t) - u(t-1)]$ and

(iii) for $t > 1$, $x(t)$ may be viewed as $(2t - 4)u(t - 1)$

Thus,

$$x(t) = 3u(-t) - 2[u(t) - u(t-1)] + (2t-4)u(t-1)$$

$$\Rightarrow \quad x(t) = 3[1 - u(t)] - 2u(t) + 2u(t-1) + 2tu(t-1) - 4u(t-1)$$

$$= 3 - 5u(t) - 2u(t-1) + 2(t-1+1)u(t-1)$$

$$= 3 - 5u(t) - 2u(t-1) + 2(t-1)u(t-1) + 2u(t-1)$$

$$= 3 - 5u(t) + 2r(t-1)$$

$\mathscr{L}_u\{x(t)\}$ cannot be found because $x(t)$ contains a constant 3 for $-\infty < t < 0$ (a noncausal signal).

Example 6.47. Express $f(t)$ in terms of singularity functions and then find $\mathscr{L}_u\{f(t)\}$.

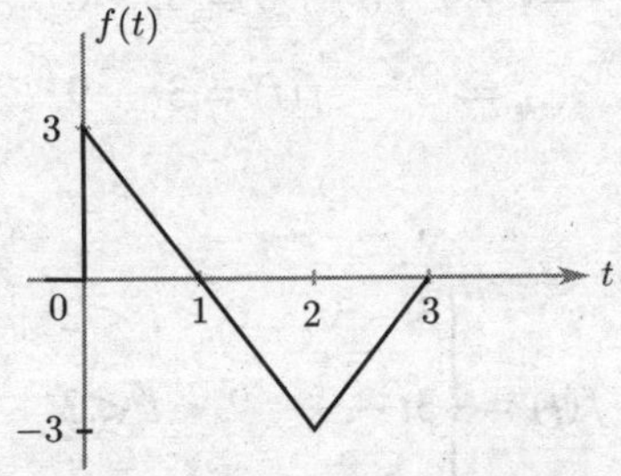

Fig. Ex. 6.47

Solution: To find $f(t)$ for $0 < t < 2$:

Equation of the straight line 1 is

$$\frac{y - y_1}{x - x_1} = \frac{y_2 - y_1}{x_2 - x_1}$$

Here, y is $f(t)$ and x is t.

Hence,

$$\frac{f(t) - 3}{t - 0} = \frac{-3 - 3}{2 - 0}$$

$$\Rightarrow \quad 2f(t) - 6 = -6t$$

$$\Rightarrow \quad f(t) = 3 - 3t$$

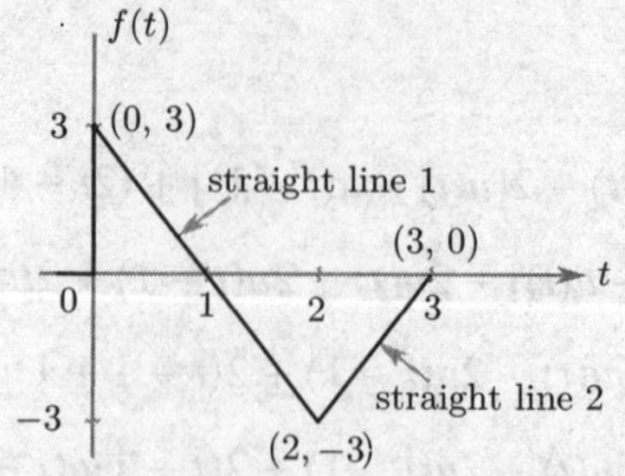

Fig. Ex. 6.47(a)

To find $f(t)$ for $2 < t < 3$:
The equation of straight line 2 is

$$\frac{f(t)+3}{t-2} = \frac{0+3}{3-2}$$

$$\Rightarrow \quad f(t)+3 = 3t-6$$

$$\Rightarrow \quad f(t) = 3t-9$$

Hence

$$f(t) = \begin{cases} 3-3t, & 0 < t < 2 \\ 3t-9, & 2 < t < 3 \\ 0, & \text{otherwise} \end{cases}$$

The above equation may also be written as:

$$\begin{aligned} f(t) &= [3-3t][u(t)-u(t-2)] + [3t-9][u(t-2)-u(t-3)] \\ &= 3u(t) - 3u(t-2) - 3tu(t) + 3tu(t-2) + 3tu(t-2) \\ &\quad - 3tu(t-3) - 9u(t-2) + 9u(t-3 \end{aligned}$$

$$\begin{aligned} \Rightarrow \quad f(t) &= 3u(t) - 12u(t-2) - 3tu(t) + 6tu(t-2) - 3tu(t-3) + 9u(t-3) \\ &= 3u(t) - 12u(t-2) - 3tu(t) + 6(t-2+2)u(t-2) \\ &\quad - 3(t-3+3)u(t-3) + 9u(t-3) \\ &= 3u(t) - 12u(t-2) - 3tu(t) + 6(t-2)u(t-2) \\ &\quad + 12u(t-2) - 3(t-3)u(t-3) - 9u(t-3) + 9u(t-3) \end{aligned}$$

$$f(t) = 3u(t) - 3tu(t) + 6(t-2)u(t-2) - 3(t-3)u(t-3)$$

Hence,

$$F(s) = \mathscr{L}\{f(t)\}$$
$$= \frac{3}{s} - \frac{3}{s^2} + \frac{6}{s^2}e^{-2s} - \frac{3}{s^2}e^{-3s}$$

Example 6.48. Express the function $f(t)$ shown in Fig. Ex. 6.48 using singularity functions and then find $\mathscr{L}_u\{f(t)\}$.

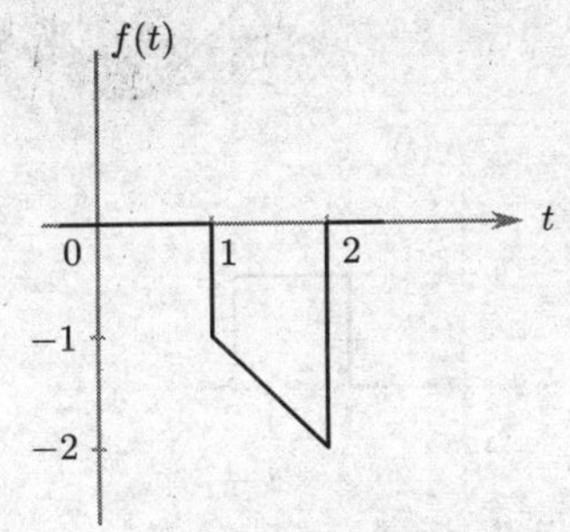

Fig. Ex. 6.48

Solution:

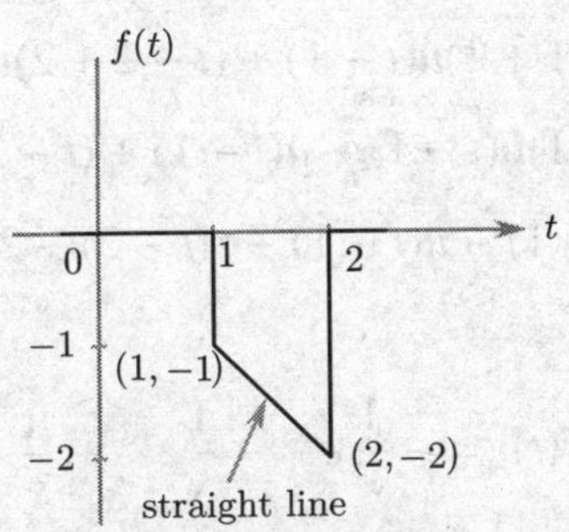

Fig. Ex. 6.48(a)

Equation of the straight line shown in Fig. Ex. 6.48(b) is

$$\frac{f_1(t) + 1}{t - 1} = \frac{-2 + 1}{s - 1}$$

$$\Rightarrow \qquad f_1(t) + 1 = -t + 1$$

$$\Rightarrow \qquad f_1(t) = -t$$

Referring Figs. 6.48(a) and 6.48(b), we can write

$$f(t) = f_1(t)g(t)$$

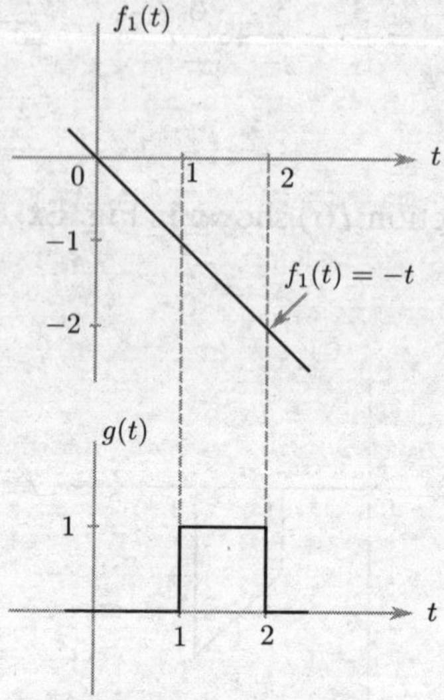

Fig. Ex. 6.48(b)

$$\begin{aligned}
\Rightarrow \qquad f(t) &= -t[u(t-1) - u(t-2)] \\
&= -(t-1+1)u(t-1) + (t-2+2)u(t-2) \\
&= -(t-1)u(t-1) - u(t-1) + (t-2)u(t-2) + 2u(t-2) \\
&= -r(t-1) - u(t-1) + r(t-2) + 2u(t-2)
\end{aligned}$$

Hence,

$$F(s) = \mathscr{L}_u\{f(t)\} = -\frac{1}{s^2}e^{-s} - \frac{1}{s}e^{-s} + \frac{1}{s^2}e^{-2s} + \frac{2}{s}e^{-2s}$$

Example 6.49. Find the Laplace transform of the function $f(t)$ shown in Fig. Ex. 6.49.

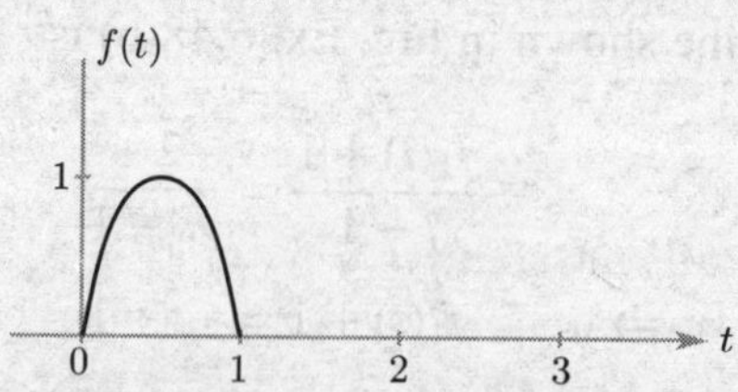

Fig. Ex. 6.49

Solution: **Method 1:**

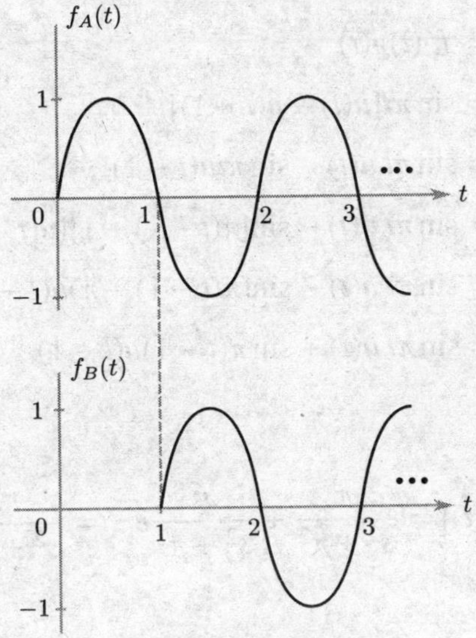

Fig. Ex. 6.49(a)

We can write,

$$f(t) = f_A(t) + f_B(t) = \sin \pi t \, u(t) + \sin \pi(t-1)u(t-1)$$

Hence,

$$F(s) = \mathscr{L}_u\{f(t)\} = \frac{\pi}{s^2+\pi^2} + \frac{\pi}{s^2+\pi^2}e^{-s} = \frac{\pi}{s^2+\pi^2}[1+e^{-s}]$$

Method 2:

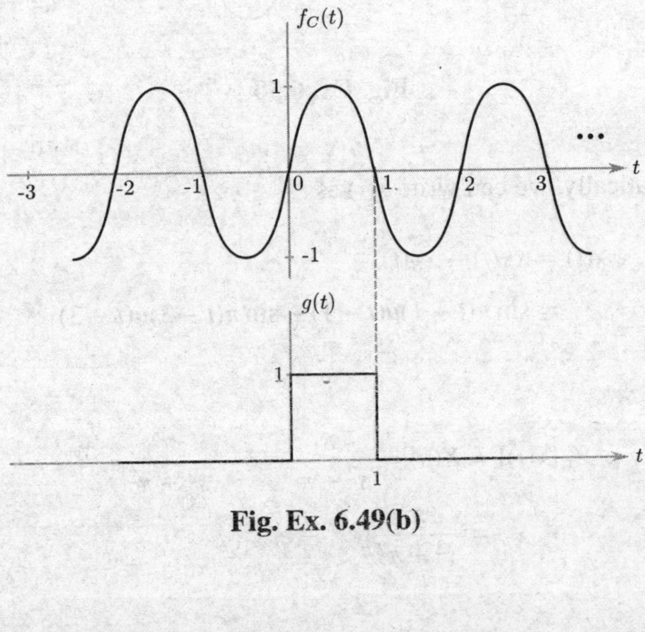

Fig. Ex. 6.49(b)

Graphically, we can manipulate $f(t)$ as

$$\begin{aligned} f(t) &= f_C(t)g(t) \\ &= \sin \pi t[u(t) - u(t-1)] \\ &= \sin \pi t\, u(t) - \sin \pi t u(t-1) \\ &= \sin \pi t\, u(t) - \sin[\pi(t-1) + 1][u(t-1)] \\ &= \sin \pi t\, u(t) - \sin(\pi(t-1) + \pi)u(t-1) \\ &= \sin \pi t\, u(t) + \sin \pi(t-1)u(t-1) \end{aligned}$$

Hence,

$$F(s) = \mathscr{L}_u\{f(t)\} = \frac{\pi}{s^2+\pi^2} + \frac{\pi}{s^2+\pi^2}e^{-s} = \frac{\pi}{s^2+\pi^2}[1 + e^{-s}]$$

Example 6.50. Find the Laplace transform of the signal $x(t)$ shown in Fig. Ex. 6.50.

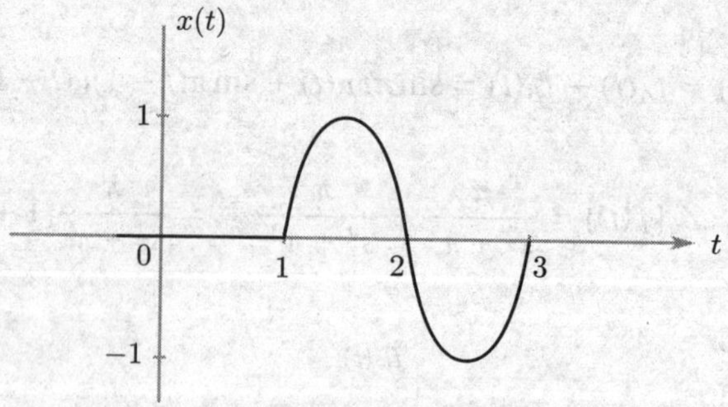

Fig. Ex. 6.50

Solution: Mathematically, we can write $x(t)$ as

$$\begin{aligned} x(t) &= x_A(t) - x_B(t) \\ &= \sin \pi(t-1)u(t-1) - \sin \pi(t-3)u(t-3) \end{aligned}$$

Hence,

$$\begin{aligned} \mathscr{L}_u\{x(t)\} = X(s) &= \frac{\pi}{s^2+\pi^2}e^{-s} - \frac{\pi}{s^2+\pi^2}e^{-3s} \\ &= \frac{\pi}{s^2+\pi^2}[e^{-s} - e^{-3s}] \end{aligned}$$

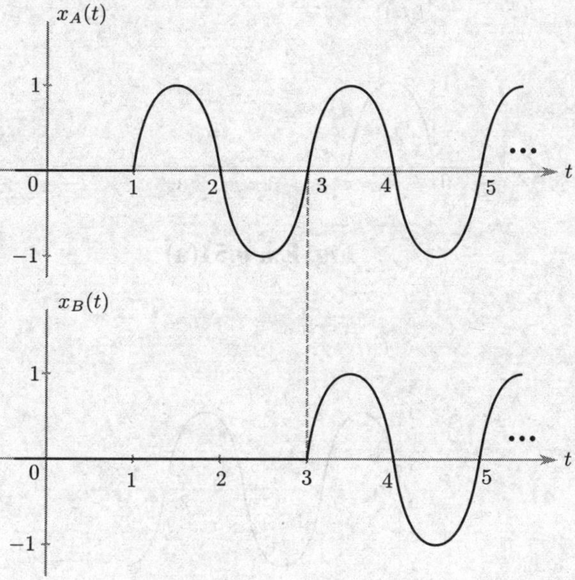

Fig. Ex. 6.50(a)

Example 6.51. Refer the waveform shown in Fig. Ex. 6.51. The equation for the waveform is $\sin t$ from 0 to π, $-\sin t$ from π to 2π. Show that the Laplace transform of this waveform is $F(s) = \dfrac{1}{s^2+1}\coth\left(\dfrac{\pi s}{2}\right)$.

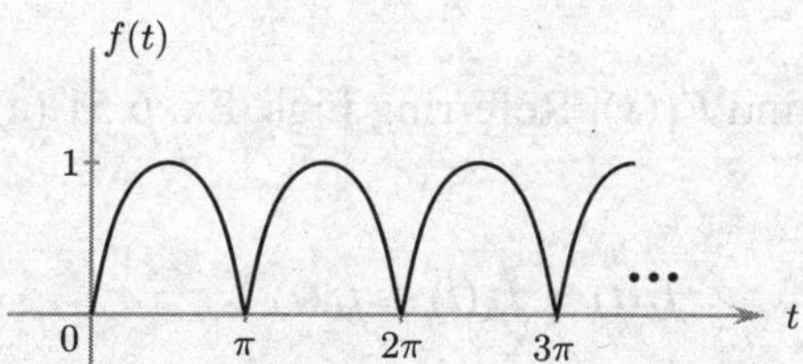

Fig. Ex. 6.51

Solution: $f(t)$ is a periodic waveform with a period $T = \pi$ seconds. Let $f_1(t)$ be the waveform $f(t)$ described over only one period. The unilateral Laplace transforms of $f(t)$ and $f_1(t)$ are related as

$$F(s) = \frac{F_1(s)}{1 - e^{-sT}}$$

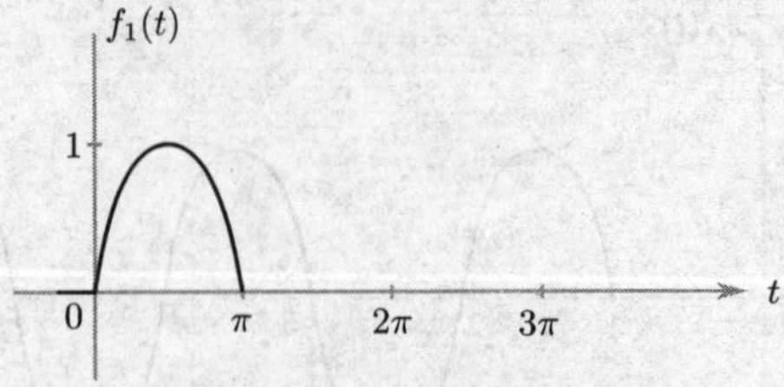

Fig. Ex. 6.51(a)

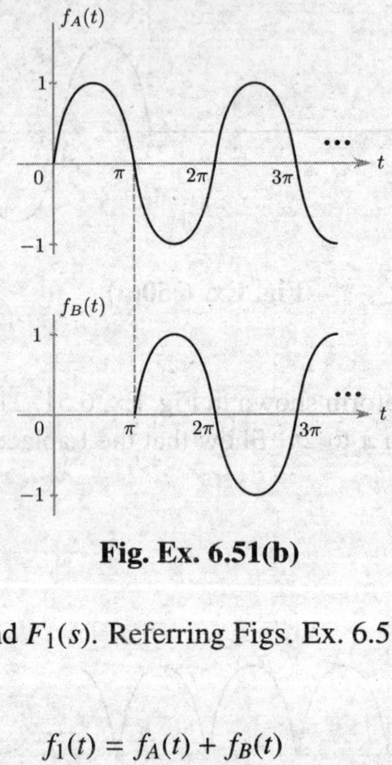

Fig. Ex. 6.51(b)

Let us now proceed to find $F_1(s)$. Referring Figs. Ex. 6.51 (a) and (b), we can write

$$\begin{aligned} f_1(t) &= f_A(t) + f_B(t) \\ &= \sin tu(t) + \sin(t-\pi)u(t-\pi) \\ \Rightarrow \qquad F_1(s) &= \frac{1}{s^2+1} + \frac{1}{s^2+1}e^{-\pi s} \\ &= \frac{(1+e^{-\pi s})}{s^2+1} \end{aligned}$$

Hence,

$$\begin{aligned} F(s) &= \frac{F_1(s)}{1-e^{-sT}} = \frac{F_1(s)}{1-e^{-\pi s}} \\ \Rightarrow \qquad F(s) &= \frac{(1+e^{-\pi s})}{(s^2+1)(1-e^{-\pi s})} \end{aligned}$$

$$= \frac{1}{s^2+1}\,\frac{\dfrac{\cancel{e^{-\pi s/2}}[e^{\pi s/2}+e^{-\pi s/2}]}{2}}{\dfrac{\cancel{e^{-\pi s/2}}[e^{\pi s/2}-e^{-\pi s/2}]}{2}}$$

$$\Rightarrow \qquad F(s) = \frac{1}{s^2+1}\,\frac{\cosh\left(\frac{\pi s}{2}\right)}{\sinh\left(\frac{\pi s}{2}\right)}$$

$$= \frac{1}{s^2+1}\coth\left(\frac{\pi s}{2}\right)$$

Example 6.52. Find the Laplace transform of the pulse shown in Fig. Ex. 6.52.

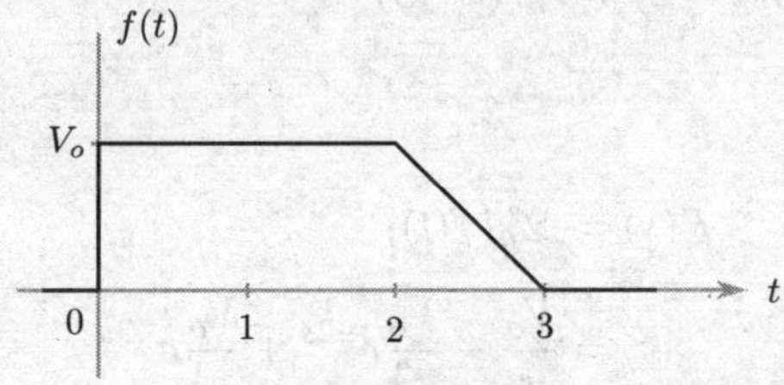

Fig. Ex. 6.52

Solution: We can describe Fig. Ex. 6.52 mathematically as

$$f(t) = \begin{cases} V_o, & 0 < t < 2 \\ -V_o t + 3V_o, & 2 < t < 3 \\ 0, & \text{otherwise} \end{cases}$$

The expression for $f(t)$ for $2 < t < 3$ is obtained as follows:
Equation of a straight line between two points is given by

$$\frac{y-y_1}{x-x_1} = \frac{y_2-y_1}{x_2-x_1}$$

In the present context, $y = f(t)$, $x = t$, $(x_1, y_1) = (2, V_o)$ and $(x_2, y_2) = (3, 0)$
Hence,

$$\frac{f(t)-V_o}{t-2} = \frac{0-V_o}{3-2}$$

$$\Rightarrow \qquad f(t) = -V_o t + 3V_o$$

The time-domain expression for $f(t)$ between $t = 0$ and 3 could be written using graphical manipulation as

$$f(t) = V_o[u(t)-u(t-2)] + [-V_o t + 3V_o][u(t-2)-u(t-3)]$$

The first term on the right-side of the above equation defines $f(t)$ for $0 < t < 2$ and the second term on the right-side defines $f(t)$ for $2 < t < 3$.

$$\begin{aligned} f(t) &= V_o u(t) - V_o u(t-2) - V_o t u(t-2) + V_o t\, u(t-3) + 3V_o u(t-2) - 3V_o u(t-3) \\ &= V_o u(t) - V_o u(t-2) - V_o(t-2+2)u(t-2) \\ &\quad + V_o(t-3+3)u(t-3) + 3V_o u(t-2) - 3V_o u(t-3) \\ &= V_o u(t) - V_o u(t-2) - V_o(t-2)u(t-2) - 2V_o u(t-2) + V_o(t-3)u(t-3) \\ &\quad + 3V_o u(t-3) + 3V_o u(t-2) - 3V_o u(t-3) \\ &= V_o u(t) - V_o(t-2)u(t-2) + V_o(t-3)u(t-3) \end{aligned}$$

$$\Rightarrow \quad f(t) = V_o u(t) - V_o r(t-2) + V_o r(t-3)$$

Hence,

$$\begin{aligned} F(s) &= \mathscr{L}_u\{f(t)\} \\ &= \frac{V_o}{s} - \frac{V_o}{s^2}e^{-2s} + \frac{V_o}{s^2}e^{-3s} \end{aligned}$$

Example 6.53. Consider a staircase waveform which extends to infinity and at $t = nt_0$ jumps to the value $n+1$, being a superposition of unit step functions. Determine the Laplace transform of this waveform.

Solution:

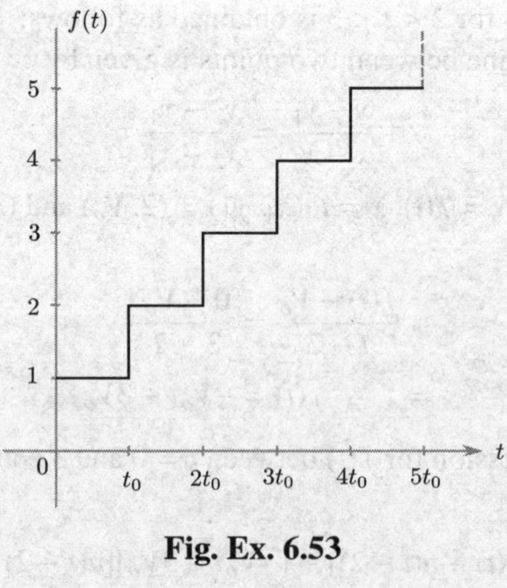

Fig. Ex. 6.53

We can write,

$$f(t) = u(t) + u(t - t_0) + u(t - 2t_0) + \cdots$$

$$\Rightarrow \quad F(s) = \mathscr{L}_u\{f(t)\} = \frac{1}{s} + \frac{1}{s}e^{-t_0 s} + \frac{1}{s}e^{-2t_0 s} + \cdots$$

$$= \frac{1}{s}[1 + e^{-t_0 s} + e^{-2t_0 s} + \cdots]$$

Let $\quad e^{-t_0 s} = x$

then $\quad F(s) = \frac{1}{s}[1 + x + x^2 + \cdots]$

From Binomial theorem, we have

$$(1 - x)^{-1} = 1 + x + x^2 + \cdots$$

Hence,

$$F(s) = \frac{1}{s(1 - x)}$$

$$= \frac{1}{s(1 - e^{-t_0 s})}$$

Example 6.54. Find the Laplace transform of the waveform shown in Fig. Ex. 6.52.

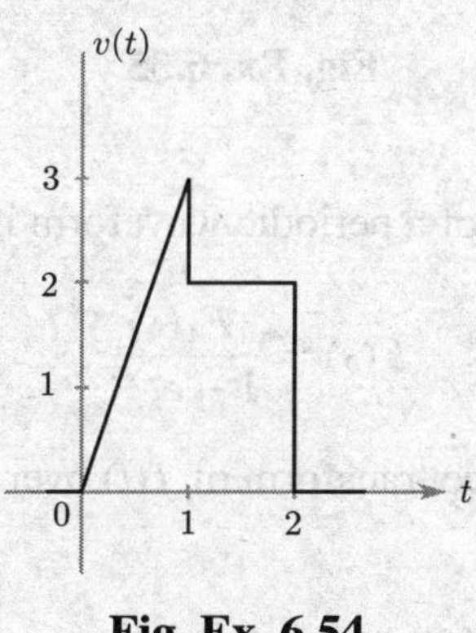

Fig. Ex. 6.54

Solution:

$$v(t) = \begin{cases} 3t, & 0 < t < 1 \\ 2, & 1 < t < 2 \\ 0, & \text{otherwise} \end{cases}$$

or
$$\begin{aligned} v(t) &= 3t[u(t) - u(t-1)] + 2[u(t-1) - u(t-2)] \\ &= 3tu(t) - 3tu(t-1) + 2u(t-1) - 2u(t-2) \\ &= 3tu(t) - 3(t-1+1)u(t-1) + 2u(t-1) - 2u(t-2) \\ &= 3tu(t) - 3(t-1)u(t-1) - 3u(t-1) + 2u(t-1) - 2u(t-2) \\ v(t) &= 3tu(t) - 3(t-1)u(t-1) - u(t-1) - 2u(t-2) \\ &= 3r(t) - 3r(t-1) - u(t-1) - 2u(t-2) \end{aligned}$$

Hence,

$$\begin{aligned} V(s) &= \mathscr{L}_u\{v(t)\} \\ &= \frac{3}{s^2} - \frac{3}{s^2}e^{-s} - \frac{1}{s}e^{-s} - \frac{2}{s}e^{-2s} \end{aligned}$$

Example 6.55. Find the Laplace transform of the periodic waveform shown in Fig. Ex. 6.55.

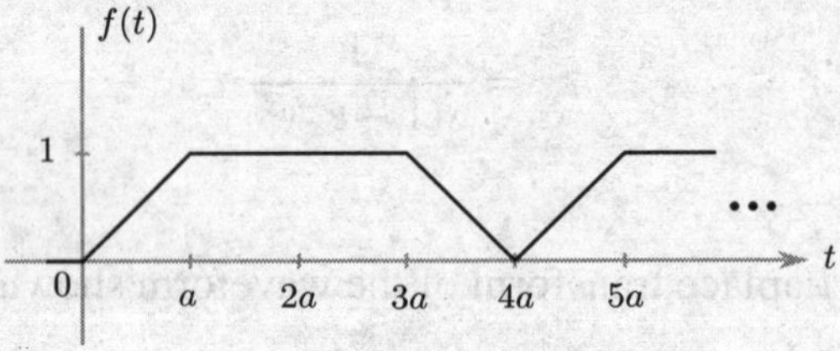

Fig. Ex. 6.55

Solution: The Laplace transform of a periodic waveform is found using the relation

$$F(s) = \frac{F_1(s)}{1 - e^{-sT}}$$

where $F_1(s) = \mathscr{L}_u\{f_1(t)\}$ = Laplace transform of $f(t)$ over $0 < t < T$ and T = fundamental period of $f(t)$.

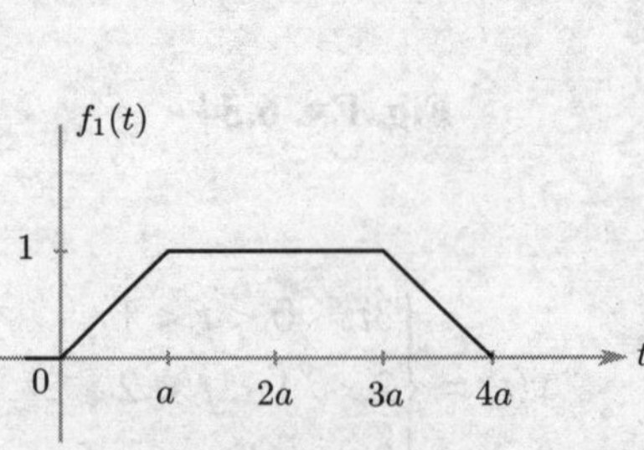

Fig. Ex. 6.55(a)

Referring to Fig. Ex. 6.55(a), we can write:

$$f_1(t) = \begin{cases} \dfrac{t}{a}, & 0 < t < a \\ 1, & a < t < 3a \\ \dfrac{-1}{a}t + 4, & 3a < t < 4a \\ 0, & \text{otherwise} \end{cases}$$

$$\begin{aligned}
\Rightarrow \quad f_1(t) &= \frac{1}{a}t[u(t) - u(t-a)] + [u(t-a) - u(t-3a)] + \left[-\frac{1}{a}t + 4\right][u(t-3a) - u(t-4a)] \\
&= \frac{1}{a}tu(t) - \frac{1}{a}tu(t-a) + u(t-a) - u(t-3a) - \frac{1}{a}tu(t-3a) \\
&\quad + \frac{1}{a}tu(t-4a) + 4u(t-3a) - 4u(t-4a) \\
\Rightarrow \quad f_1(t) &= \frac{1}{a}tu(t) - \frac{1}{a}(t-a+a)u(t-a) + u(t-a) - u(t-3a) \\
&\quad - \frac{1}{a}(t-3a+3a)u(t-3a) + \frac{1}{a}(t-4a+4a)u(t-4a) \\
&\quad + 4u(t-3a) - 4u(t-4a) \\
&= \frac{1}{a}tu(t) - \frac{1}{a}(t-a)u(t-a) - u(t-a) + u(t-a) - u(t-3a) \\
&\quad - \frac{1}{a}(t-3a)u(t-3a) - 3u(t-3a) + \frac{1}{a}(t-4a)u(t-4a) + 4u(t-4a) \\
&\quad + 4u(t-3a) - 4u(t-4a) \\
&= \frac{1}{a}tu(t) - \frac{1}{a}(t-a)u(t-a) - \frac{1}{a}(t-3a)u(t-3a) + \frac{1}{a}(t-4a)u(t-4a) \\
&= \frac{1}{a}r(t) - \frac{1}{a}r(t-a) - \frac{1}{a}r(t-3a) + \frac{1}{a}r(t-4a)
\end{aligned}$$

Hence,

$$\begin{aligned}
F_1(s) &= \mathscr{L}_u\{f_1(t)\} \\
&= \frac{1}{as^2} - \frac{1}{as^2}e^{-as} - \frac{1}{as^2}e^{-3as} + \frac{1}{as^2}e^{-4as} \\
&= \frac{1}{as^2}(1 - e^{as} - e^{-3as} + e^{-4as})
\end{aligned}$$

Alternate method for finding $F_1(s)$:

From Figs. Ex. 6.55(b), (c), (d), we can write

$$\begin{aligned}
f_1(t) &= f_A(t) + f_B(t) + f_C(t) + f_D(t) \\
&= \frac{1}{a}tu(t) - \frac{1}{a}(t-a)u(t-a) - \frac{1}{a}(t-3a)u(t-3a) + \frac{1}{a}(t-4a)u(t-4a)
\end{aligned}$$

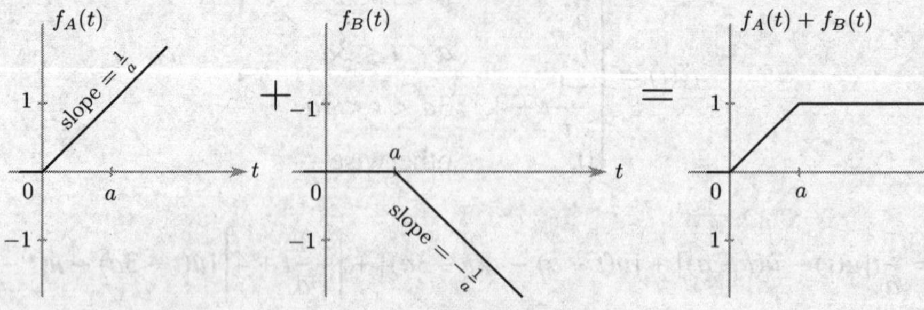

Fig. Ex. 6.55(b)

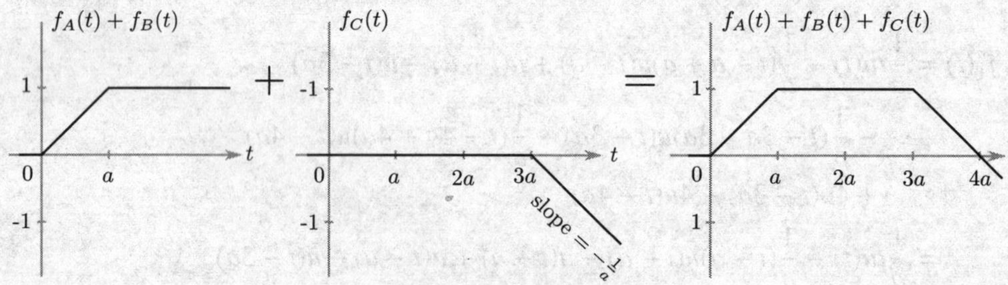

Fig. Ex. 6.55(c)

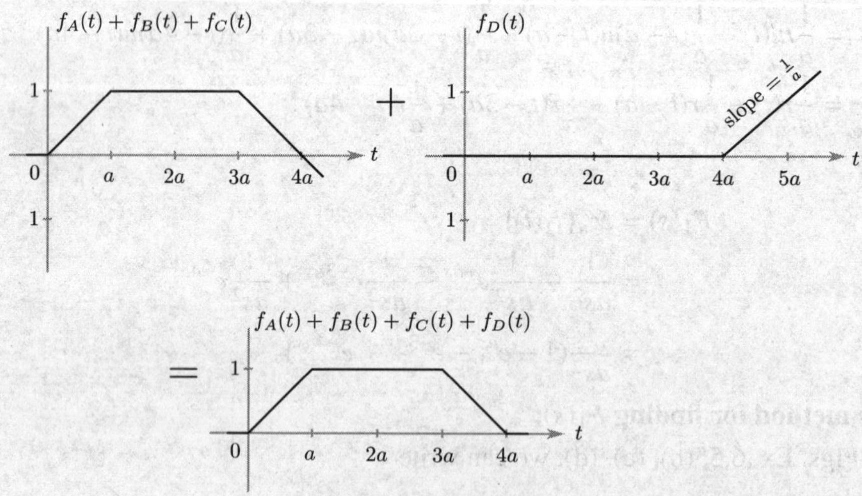

Fig. Ex. 6.55(d)

Hence,

$$F_1(s) = \mathscr{L}_u\{f_1(t)\}$$
$$= \frac{1}{as^2} - \frac{1}{as^2}e^{-as} - \frac{1}{as^2}e^{-3as} + \frac{1}{as^2}e^{-4as}$$

Finally,

$$F(s) = \mathscr{L}_u\{f(t)\}$$
$$= \frac{F_1(s)}{1 - e^{-sT}}$$

Substituting $T = 4a$, we get

$$F(s) = \frac{1}{as^2}\frac{(1 - e^{-as} - e^{-3as} + e^{-4as})}{(1 - e^{-4as})}$$

Example 6.56. Find the Laplace transform of the function $f(t)$ shown in Fig. Ex. 6.56.

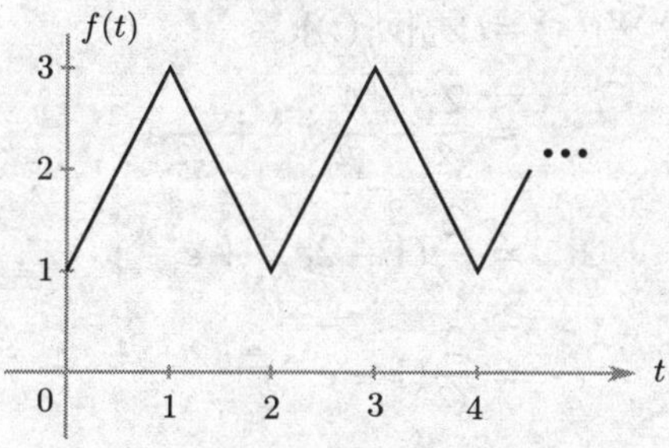

Fig. Ex. 6.56

Solution: Let $f(t) = x(t) + u(t)$, where $x(t)$ is a periodic triangular wave and is as shown in Fig. Ex. 6.56(a).

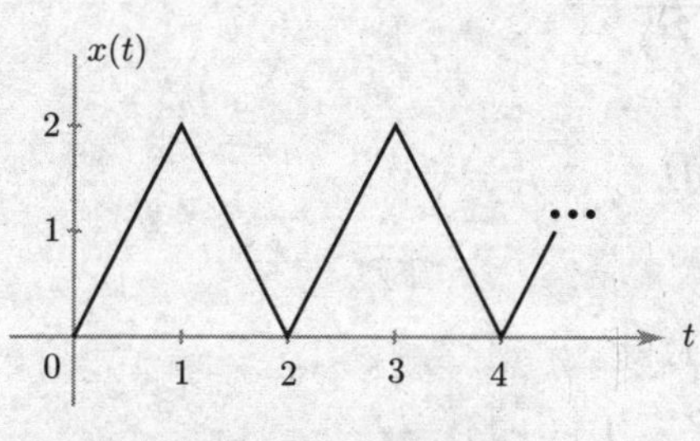

Fig. Ex. 6.56(a)

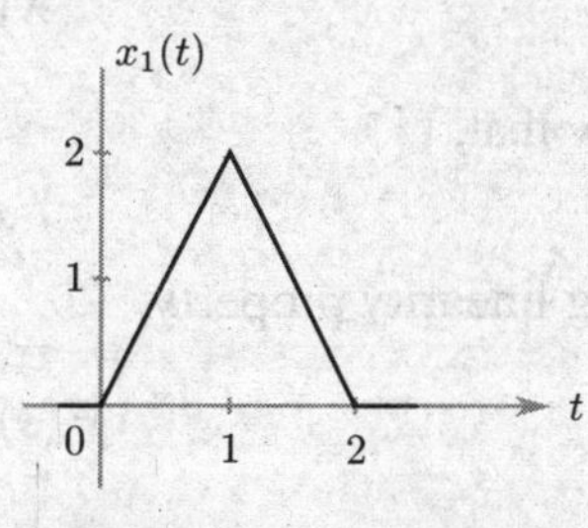

Fig. Ex. 6.56(b)

Let $x_1(t)$ be $x(t)$ within its first period as shown in Fig. Ex. 6.56(b).

Referring to Fig. Ex. 6.56(b), we can write

$$x_1(t) = \begin{cases} 2t, & 0 < t < 1 \\ 4 - 2t, & 1 < t < 2 \end{cases}$$

$$\begin{aligned}
\Rightarrow \qquad x_1(t) &= 2t[u(t) - u(t-1)] + (4-2t)[u(t-1) - u(t-2)] \\
&= 2tu(t) - 2tu(t-1) + 4u(t-1) - 4u(t-2) - 2tu(t-1) + 2tu(t-2) \\
&= 2tu(t) - 2(t-1+1)u(t-1) + 4u(t-1) - 4u(t-2) \\
&\quad - 2(t-1+1)u(t-1) + 2(t-2+2)u(t-2) \\
&= 2tu(t) - 2(t-1)u(t-1) - 2u(t-1) + 4u(t-1) - 4u(t-2) \\
&\quad - 2(t-1)u(t-1) - 2u(t-1) + 2(t-2)u(t-2) + 4u(t-2) \\
\Rightarrow \qquad x_1(t) &= 2tu(t) - 4(t-1)u(t-1) + 2(t-2)u(t-2) \\
\Rightarrow \qquad x_1(t) &= 2r(t) - 4r(t-1) + 2r(t-2)
\end{aligned}$$

Hence,

$$\begin{aligned}
X_1(s) &= \mathscr{L}_u\{x_1(t)\} \\
&= \frac{2}{s^2} - \frac{4}{s^2}e^{-s} + \frac{2}{s^2}e^{-2s} \\
&= \frac{2}{s^2}(1 - 2e^{-s} + e^{-2s}) \\
&= \frac{2}{s^2}(1 - e^{-s})^2
\end{aligned}$$

Since $x(t)$ is periodic,

$$X(s) = \mathscr{L}_u\{x(t)\} = \frac{X_1(s)}{1 - e^{-sT}}$$

where $T = 2$ seconds

Hence,

$$X(s) = \frac{2(1 - e^{-s})^2}{s^2(1 - e^{-2s})}$$

We know that,

$$f(t) = x(t) + u(t)$$

Applying linearity property,

$$\begin{aligned}
F(s) &= X(s) + U(s) \\
&= \frac{2}{s}\frac{(1 - e^{-s})^2}{(1 - e^{-2s})} + \frac{1}{s}
\end{aligned}$$

Example 6.57. Find the Laplace transform of the periodic signal $x(t)$ shown in Fig. Ex. 6.57.

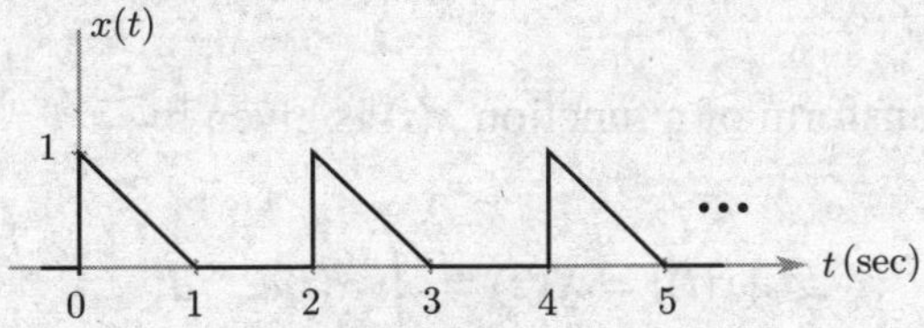

Fig. Ex. 6.57

Solution: From Fig. Ex. 6.57, we find that $T = 2$ Seconds.

The signal $x(t)$ considered over one period is denoted as $x_1(t)$ and is shown in Fig. Ex. 6.57(a).

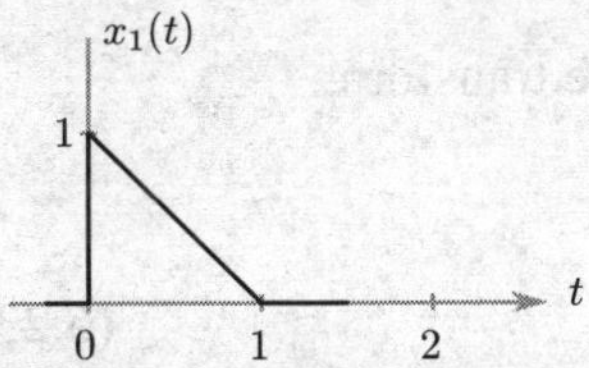

Fig. Ex. 6.57(a)

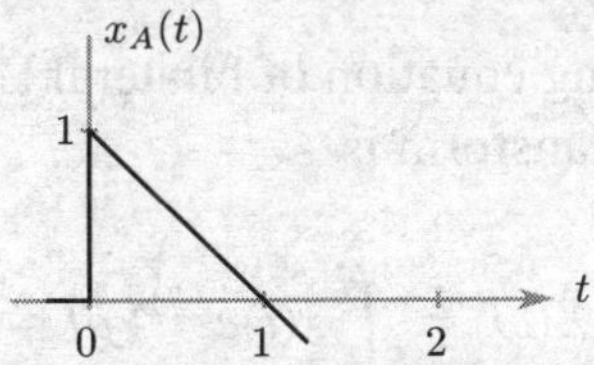

Fig. Ex. 6.57(b)

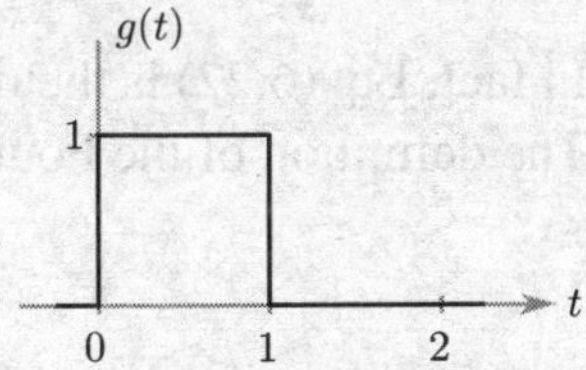

Fig. Ex. 6.57(c)

The signal $x_1(t)$ may be viewed as the multiplication of $x_A(t)$ and $g(t)$.

That is,

$$\begin{aligned}
x_1(t) &= x_A(t)g(t) \\
&= [-t+1][u(t)-u(t-1)] \\
\Rightarrow \quad x_1(t) &= -tu(t)+tu(t-1)+u(t)-u(t-1) \\
&= -tu(t)+(t-1+1)u(t-1)+u(t)-u(t-1) \\
&= -tu(t)+(t-1)u(t-1)+u(t-1)+u(t)-u(t-1) \\
&= u(t)-tu(t)+(t-1)u(t-1) \\
&= u(t)-r(t)+r(t-1)
\end{aligned}$$

Taking Unilateral Laplace Transform, we get

$$\begin{aligned}
X_1(s) &= \frac{1}{s}-\frac{1}{s^2}+\frac{1}{s^2}e^{-s} \\
&= \frac{s-1+e^{-s}}{s^2}
\end{aligned}$$

Hence,

$$X(s) = \frac{X_1(s)}{1-e^{-sT}} = \frac{(s-1+e^{-s})}{s^2(1-e^{-2s})}$$

6.12 Relationship between Laplace transform and Fourier transform

The unilateral Laplace transform of a function $x(t)$ is given by

$$\mathscr{L}_u\{x(t)\} = X(s) = \int_0^\infty x(t)e^{-st}\,dt$$

If we assume $x(t)$ equal zero for $t < 0$, then we can write $x(t) = x(t)u(t)$. Consequently, the above equation becomes

$$X(s) = \int_{-\infty}^{\infty} x(t)e^{-st}\,dt \tag{6.22}$$

In fact, Eq. (6.22) is the defining equation of bilateral Laplace transform.

The definition of the Fourier transform is

$$X(\omega) = \int_{-\infty}^{\infty} x(t)e^{-j\omega t}\,dt \tag{6.23}$$

Letting $s = j\omega$ in Eq. (6.22) gives

$$X(j\omega) = \int_{-\infty}^{\infty} x(t)e^{-j\omega t}\,dt \tag{6.24}$$

The integrals in Eqs. (6.23) and (6.24) are equal and we see an inconsistency in the notation:

$$X(\omega) \stackrel{?}{=} X(j\omega)$$

Since $X(j\omega)$ is a function of ω, it is simply written as $X(\omega)$ and this conflict is resolved.

Thus, the Fourier transform of a function of time is same as its Laplace transform on $j\omega$ axis of the complex s-plane. That is of course only true if the Laplace transform exists as well. The connection between Laplace transform and Fourier transform is that if the region of convergence of $\mathscr{L}\{x(t)\}$ contains $j\omega$ axis, then

$$F\{x(t)\} = \mathscr{L}\{x(t)\}\Big|_{s=j\omega}$$

The function $x(t) = e^{-\alpha t}u(t)$, $\alpha > 0$ has the Laplace transform

$$X(s) = \frac{1}{s+\alpha}; \quad \text{ROC}: \sigma > -\alpha \tag{6.25}$$

The region of convergence lies to right of vertical line passing through the pole at $s = -\alpha$ and is shown in Fig. 6.12.

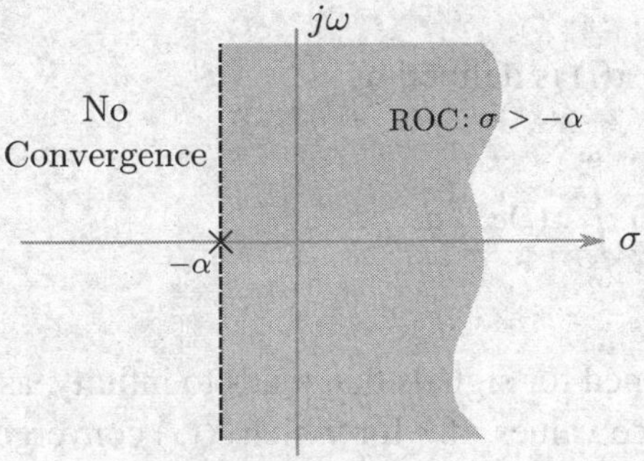

Fig. 6.12: ROC *for* $\alpha > 0$

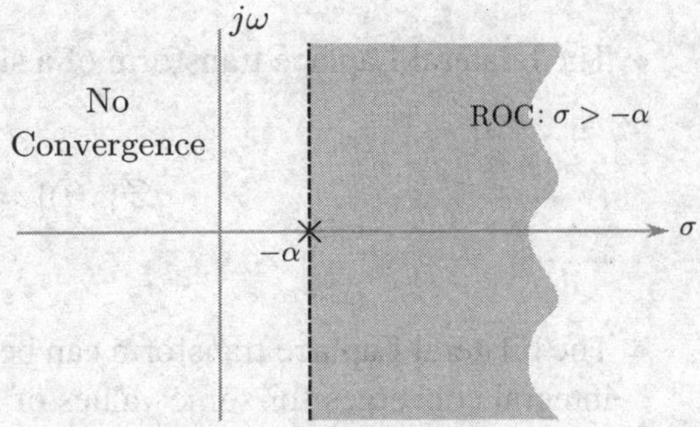

Fig. 6.13: ROC *for* $\alpha < 0$

By substituting $x(t) = e^{-\alpha t}u(t)$ in Eq. (6.23) and evaluating the integral gives

$$X(\omega) = \frac{1}{j\omega + \alpha} \tag{6.26}$$

Eq. (6.26) clearly shows the relation between Fourier and Laplace transforms.

If, however, $\alpha < 0$, then Fourier transform does not exist, because $\int_{-\infty}^{\infty} e^{-\alpha t}u(t)e^{-j\omega t}\, dt$ does not converge. On the other hand, the Laplace transform, of $x(t) = e^{-\alpha t}u(t)$, for $\alpha < 0$ is also $\dfrac{1}{s+\alpha}$ with ROC now lying to the right of $-\alpha$ as depicted in Fig. 6.13. Since ROC does not contain $j\omega$ axis, Fourier transform does not exist.

Example 6.58. Find Laplace transform of the signal $x(t)$ whose Fourier transform is

$$X(\omega) = \frac{1}{4 + \omega^2}$$

Solution: Rewrite $X(\omega)$ as

$$X(\omega) = \frac{1}{4 - (-j\omega)^2}$$

$$\Rightarrow \qquad X(s) = \frac{1}{4 - s^2}$$

$$= \frac{-1}{(s+2)(s-2)}$$

$X(s)$ has poles at $s = -2$ and $s = +2$. The ROC of $X(s)$ has to include $j\omega$ axis. Thus, the ROC of $X(s)$ is $-2 < \sigma < 2$.

Summary of Important points

- The bilateral Laplace transform of a signal $x(t)$ is defined by

$$\mathscr{L}\{x(t)\} = X(s) \int_{-\infty}^{\infty} x(t)e^{-st}dt$$

- The bilateral Laplace transform can be defined for signals that tends to infinity, as the integral converges for some values of s. The values of s for which $X(s)$ converges is called ROC. The ROC is a strip parallel to $j\omega$ axis in the s-plane. The ROC does not contain any poles.

- The bilateral Laplace transform of a signal and its inverse are unique for a specified ROC.

- The Laplace transform of causal signals is

$$\mathscr{L}_u\{x(t)\} = X(s) = \int_{0}^{\infty} x(t)e^{-st}\,dt$$

  The lower of limit of integration is 0^- if $x(t)$ contains impulses. The Laplace transform defined above is known by the name, unilateral Laplace transform.

- For causal signals, unilateral and bilateral Laplace transforms give identical results.

- The inverse Laplace transform is obtained by evaluating a complex integral in the complex plane. However, the transformed functions that we have to deal in engineering are rational functions, and these can be inverted by using partial fraction expansion and then by standard transform pairs with appropriate ROCs.

- The Fourier transform of a signal is simply equal to its Laplace transform of a signal evaluated on $j\omega$ axis, provided ROC of Laplace transform contains $j\omega$ axis.

- The unilateral transform was introduced mainly because of its time differentiation property, which proves particularly useful in finding zero-input and zero-state responses of the system.

- The most important property of Laplace transform is the convolution property, which states that the Laplace transform of convolution of two signals is equal to the product of their Laplace transforms.

Exercise Problems

EP. 6.1 Find Bilateral Laplace transform of the signals given below. Also give their respective ROCs.

(a) $x(t) = e^{-2t}u(t) + e^{-3t}u(t)$ (b) $x(t) = e^{-3t}u(t) + e^{2t}u(-t)$

(c) $x(t) = e^{2t}u(t) + e^{-3t}u(-t)$

Ans:

(a) $X(s) = \dfrac{2\left(s+\frac{5}{2}\right)}{(s+2)(s+3)}$; ROC: $\sigma > -2$

(b) $X(s) = \dfrac{-5}{(s-2)(s+3)}$; ROC: $-3 < \sigma < 2$

(c) $x(t)$ has no transform $X(s)$

EP. 6.2 Find Laplace transform of the signals given below. Also give their respective ROCs.

(a) $x(t) = u(t - t_0)$ (b) $x(t) = e^{-2t}[u(t) - u(t-5)]$

(c) $x(t) = \sum\limits_{k=0}^{\infty} \delta(t - kT)$

Ans:

(a) $X(s) = \dfrac{e^{-t_0 s}}{s}$; ROC: $\sigma > 0$

(b) $X(s) = \dfrac{1}{s+2}[1 - e^{-5(s+2)}]$; ROC: $\sigma > -2$

(c) $X(s) = \dfrac{1}{1 - e^{-sT}}$, ROC: $\sigma > 0$

EP. 6.3 Using properties, find Laplace transform of the signals given below:

(a) $x(t) = \cos\omega_0 t\, u(t)$ (b) $x(t) = e^{-at}\cos\omega_0 t\, u(t)$

(c) $x(t) = t\, u(t)$

Ans:

(a) $X(s) = \dfrac{s}{s^2 + \omega_0^2}$; ROC: $\sigma > 0$

(b) $X(s) = \dfrac{s+a}{(s+a)^2 + \omega_0^2}$; ROC: $\sigma > -a$

(c) $X(s) = \dfrac{1}{s^2}$; ROC: $\sigma > 0$

EP. 6.4 Using properties, find unilateral Laplace transform of the signals given below.

(a) $x(t) = t^2e^{-2t}u(t)$ (b) $x(t) = 2t\dfrac{d}{dt}[e^{-2t}\sin t\, u(t)]$

Ans:

(a) $X(s) = \dfrac{2}{(s+2)^3}$ (b) $X(s) = \dfrac{2(s^2-2)}{(s^2+2s+2)^2}$

EP. 6.5 Find the bilateral Laplace transform and ROC for each of the following signals:

(a) $x(t) = e^{2t}u(-t+2)$ (b) $x(t) = \delta(t-t_0)$

Ans:

(a) $X(s) = -\dfrac{e^{-2(s-2)}}{(s-2)}$; ROC: $\sigma < 2$ (b) $X(s) = e^{-st_0}$; ROC: All s

EP. 6.6 Find inverse Laplace transform of the transformed function given below:

$$X(s) = \frac{s+3}{s^2+3s+2}$$

(i) ROC: $\sigma < -2$

(ii) ROC: $\sigma > -1$

(iii) ROC: $-2 < \sigma < -1$

Ans:

(i) $x(t) = \left[-2e^{-t} + e^{-2t}\right]u(-t)$

(ii) $x(t) = \left[2e^{-t} - e^{-2t}\right]u(t)$

(iii) $x(t) = -e^{-2t}u(t) - 2e^{-t}u(-t)$

EP. 6.7 Find inverse Laplace transform of $X(s)$ given below:

$$X(s) = \frac{s^2+3s+4}{s^2+2s+1};\ \text{ROC}: \sigma < -1$$

Ans:

$$x(t) = \delta(t) - \left[e^{-t} + 2te^{-t}\right]u(-t)$$

EP. 6.8 Find the bilateral Laplace transform and the corresponding ROC for each of the following signals:

(a) $x(t) = e^{-2t}u(t) + e^{-t}u(t) + e^{t}u(-t)$

(b) $x(t) = e^{2t+4}u(t+2)$

(c) $x(t) = \cos 3t\, u(-t) * e^{-t}u(t)$

Ans:

(a) $X(s) = \dfrac{1}{s+2} + \dfrac{1}{s+1} - \dfrac{1}{s-1}$; ROC: $-1 < \sigma < 1$

(b) $X(s) = \dfrac{e^{2s}}{s-2}$; ROC: $\sigma > -2$

(c) $X(s) = \left[-\dfrac{s}{s^2+9}\right]\left[\dfrac{1}{s+1}\right]$; ROC: $-1 < \sigma < 0$.

EP. 6.9 Find the Laplace transform for the half rectified sine wave $x(t)$ shown in Fig. EP. 6.9.

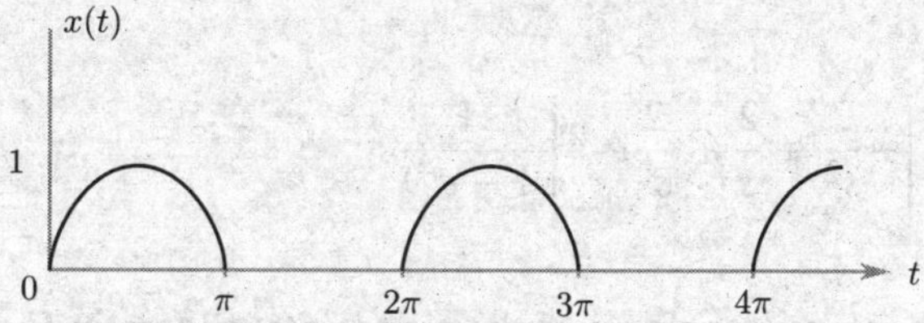

Fig. EP. 6.9

Ans: $X(s) = \dfrac{1}{(s^2+1)(1-e^{-\pi s})}$

EP. 6.10 Find the Laplace transform of the waveform shown in Fig. EP. 6.10.

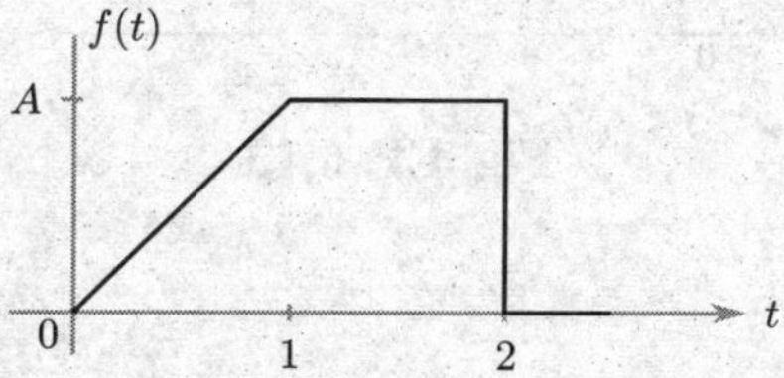

Fig. EP. 6.10

Ans: $F(s) = \frac{A(1-e^{-s})}{s^2} - \frac{A}{s}e^{-2s}$

EP. 6.11 Find the Laplace transform of the periodic waveform shown in Fig. EP. 6.11.

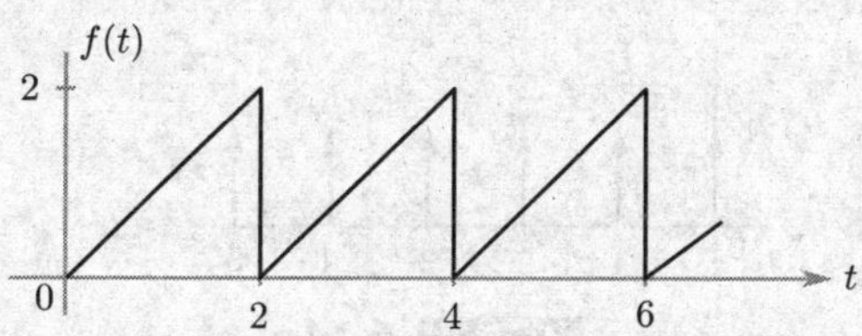

Fig. EP. 6.11

Ans: $F(s) = \left[\dfrac{1}{s^2} - \dfrac{1}{s^2}e^{-2s} - \dfrac{2}{s}e^{-2s}\right]\dfrac{1}{1-e^{-2s}}$

EP. 6.12 Find the Laplace transform of the waveform shown in Fig. EP. 6.12.

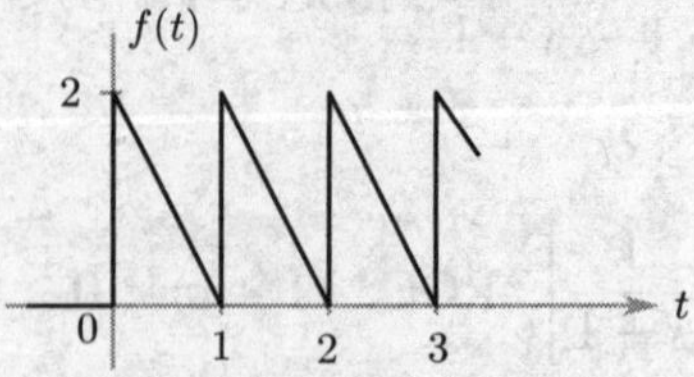

Fig. EP. 6.12

Ans: $$F(s) = \left[-\frac{2}{s^2} + \frac{2}{s} + \frac{2}{s^2}e^{-2}\right]\frac{1}{1-e^{-s}}$$

EP. 6.13 Obtain the Laplace transform of the signal $f(t)$ shown in Fig. EP. 6.13.

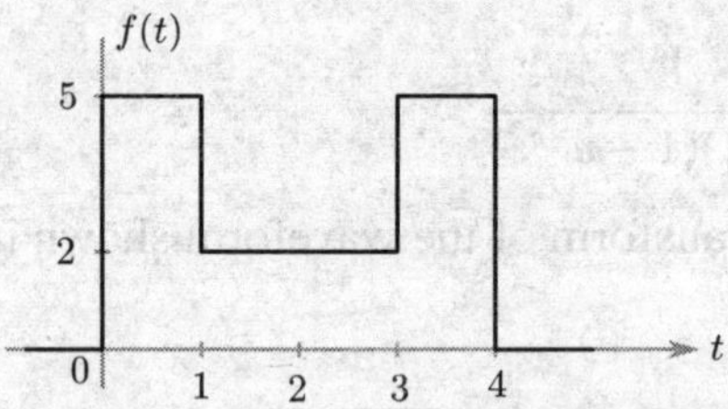

Fig. EP. 6.13

Ans: $$F(s) = \frac{1}{s}[5 - 3e^{-s} + 3e^{-3s} - 5e^{-4s}]$$

EP. 6.14 Obtain the Laplace transform of the unit impulses shown in Fig. EP. 6.14.

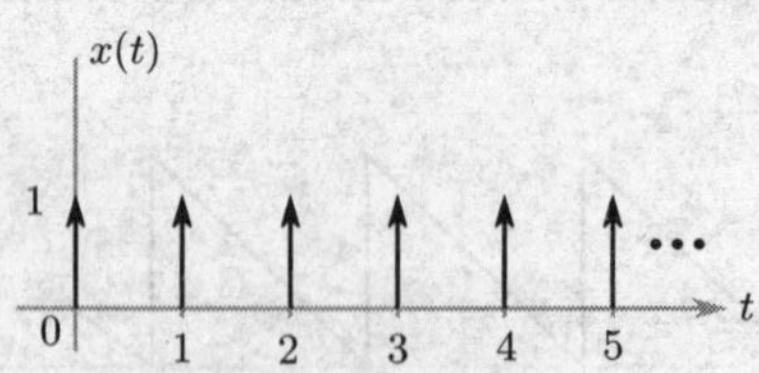

Fig. EP. 6.14

Ans: $$X(s) = \frac{1}{1-e^{-s}}$$

EP. 6.15 Synthesize the wave form shown in Fig. EP. 6.15 using ramp function and obtain the Laplace transform of $f(t)$.

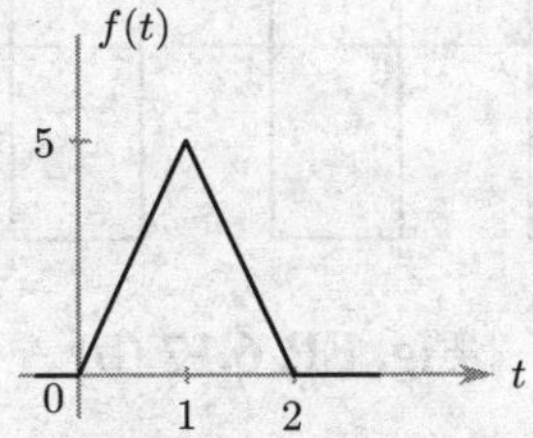

Fig. EP. 6.15

Ans: $F(s) = \dfrac{5}{s^2}[1 - 2e^{-s} + e^{-2s}]$

EP. 6.16 Find the Laplace transform of the voltage wave form as shown in Fig. EP. 6.16.

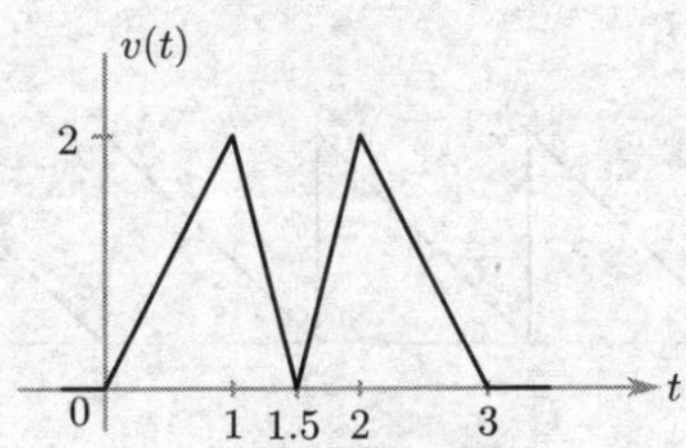

Fig. EP. 6.16

Ans: $V(s) = \dfrac{2}{s^2}\left[1 - 3e^{-s} + 5e^{-1.5s} - 6e^{-2s} + 6e^{-3s}\right]$

EP. 6.17 Find the Laplace transform of the perodic wave forms shown in Figs. EP. 6.17(a) and (b).

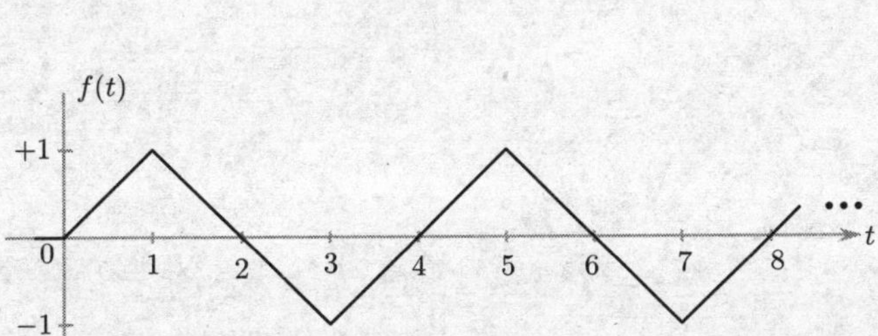

Fig. EP. 6.17 (a)

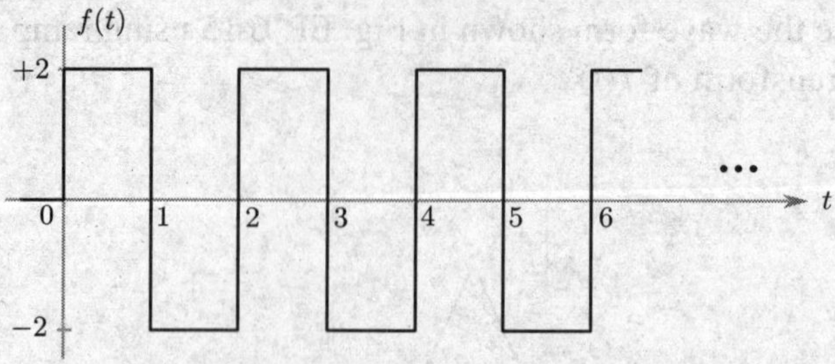

Fig. EP. 6.17 (b)

Ans:

(i) $F(s) = \dfrac{1}{1-e^{-4s}}\left[\dfrac{1}{s^2} - \dfrac{2}{s^2}e^{-s} + \dfrac{1}{s^2}e^{-2s} - \dfrac{1}{s}e^{-2s} + \dfrac{2}{s^2}e^{-3s} - \dfrac{1}{s^2}e^{-4s}\right]$

(ii) $F(s) = \dfrac{1}{1-e^{-2s}}\left[\dfrac{2}{s} - \dfrac{4e^{-s}}{s} + \dfrac{2}{s}e^{-2s}\right]$

EP. 6.18 Find the Laplace transform of the saw tooth wave as shown in Fig. EP. 6.18.

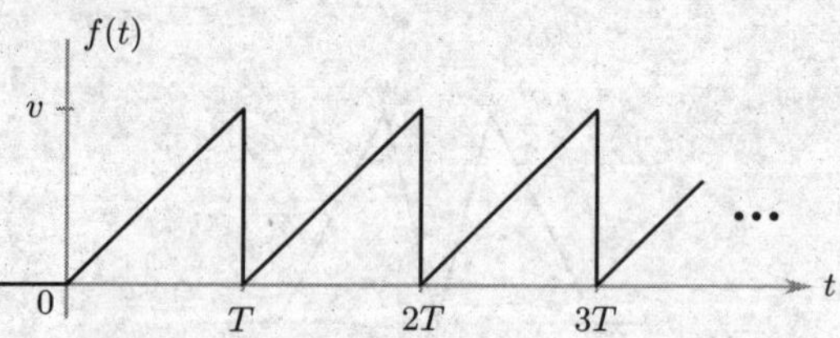

Fig. EP. 6.18

Ans: $F(s) = \dfrac{v(1-e^{-Ts} - Tse^{-Ts})}{Ts^2(1-e^{-Ts})}$

Chapter 7

Sampling

Sampling theorem provides the link between analog world and digital world. The analysis and design of digital communication systems, digital controllers etc, requires an understanding of the sampling process and its implications.

Consider a continuous-time signal $x(t)$. We define sampling as the generation of an ordered number sequence by taking the values of $x(t)$ at specified instants of time. Hence, sampling generates a number sequence $x(T_s), x(2T_s), \ldots\ x(nT_s)$ where, nT_s are the instants at which sampling occurs.

7.1 Brief review of Fourier transform

The f-form definition of Fourier transform and its inverse are used in this chapter. This form gives a better understanding of the process of sampling.

Fourier transform:

$$\mathscr{F}\{x(t)\} = X(f) = \int_{-\infty}^{\infty} x(t)e^{-j2\pi ft}\,dt \tag{7.1}$$

Inverse Fourier transform:

$$\mathscr{F}^{-1}\{X(f)\} = \int_{-\infty}^{\infty} X(f)e^{j2\pi ft}\,df \tag{7.2}$$

The signal $x(t)$ and its inverse $X(f)$ form a unique transform pair, and their relationship is shown symbolically by using a double arrow:

$$x(t) \xleftrightarrow{\textbf{FT}} X(f)$$

The Fourier transform (**FT**)$X(f)$ is in general, complex and may be represented in the following exponential form:

$$X(f) = |X(f)|e^{j\theta(f)}$$

For real signals, $X(f)$ is conjugate symmetric with $X(-f) = X^*(f)$. This implies that $|X(f)|$ displays even symmetry and $\theta(f)$ exhibits odd symmetry.

7.1.1 Effect of signal symmetry on the FT of real-valued signals

- If $x(t)$ is real and even symmetric, then $X(f)$ is real and even symmetric.
- If $x(t)$ is real and odd symmetric, then $X(f)$ is imaginary and odd symmetric.
- If $x(t)$ is real and has no symmetry, then $\text{Re}\{X(f)\}$ is even symmetric and $\text{Im}\{X(f)\}$ is odd symmetric.

7.1.2 Fourier transform pairs and properties

Table 7.1 gives the Fourier transform of some useful signals. It is to be noted that Fourier transform of signals that grow exponentially or faster, does not exist. The Fourier transform is a linear operation and obeys the principle of superposition. Most cmmonly used properties of **FT** are summarized in Table 7.2 as a ready reckoner.

Table 7.1: Some useful transform pairs

Sl.no	$x(t)$	$X(f)$
1	$\delta(t)$	1
2	$\text{rect}(t)$	$\text{sinc}(f)$
3	$\cos 2\pi f_0 t$	$\frac{1}{2}[\delta(f+f_0) + \delta(f-f_0)]$
4	$\sin 2\pi f_0 t$	$\frac{1}{2j}[\delta(f+f_0) - \delta(f-f_0)]$
5	$e^{-at}u(t)$	$\frac{1}{a+j2\pi f}$
6	$te^{-at}u(t)$	$\frac{1}{(a+j2\pi f)^2}$
7	A	$A\delta(f)$
8	$\text{sgn}(t)$	$\frac{1}{j\pi f}$
9	$u(t)$	$\frac{1}{2}\delta(f) + \frac{1}{j2\pi f}$
10	$\sum_{n=-\infty}^{\infty} \delta(t-nT)$	$\frac{1}{T}\sum_{n=-\infty}^{\infty} \delta\left(f - \frac{n}{T}\right)$

Table 7.2: Properties of FT

Property	$x(t)$	$X(f)$
Duality	$X(t)$	$x(-f)$
Time-scaling	$x(at)$	$\frac{1}{\lvert a\rvert}X\left(\frac{f}{a}\right)$
Time shift	$x(t-t_0)$	$e^{-j2\pi f t_0}X(f)$
Frequency shift	$e^{j2\pi\beta t}x(t)$	$X(f-\beta)$
Convolution	$x(t)*h(t)$	$X(f)H(f)$
Multiplication	$x(t)h(t)$	$X(f)*H(f)$
Derivative	$\frac{dx(t)}{dt}$	$j2\pi f\, X(f)$
Modulation	$x(t)\cos 2\pi f_0 t$	$\frac{1}{2}[X(f+f_0)+X(f-f_0)]$
Frequency differentiation	$-j2\pi t x(t)$	$\frac{dX(f)}{df}$
Integration	$\int\limits_{-\infty}^{t} x(\tau)\,d\tau$	$\frac{1}{2}X(0)\delta(f)+\frac{1}{j2\pi f}X(f)$

7.2 Impulse sampling/Uniform sampling theorem for low-pass signals

Statement: If a signal $x(t)$ contains no frequency components for $|f| > f_x$, then it is completely described by instantaneous values $x(nT_s)$ uniformly spaced in time with period $T_s \leq \frac{1}{2f_x}$. If the sampling rate $f_s \geq 2f_x$, and if the sampled values are represented by weighted impulses, then the signal can be exactly reconstructed from its samples by an ideal low-pass filter of bandwidth B, where $f_x \leq B < f_s - f_x$ and $f_s = \frac{1}{T_s}$. The sampling rate, $f_s = 2f_x$ is known as the Nyquist rate and is a theoretically sufficient condition that allows an analog signal to be reconstructed completely from a set of uniformly spaced discrete-time samples.

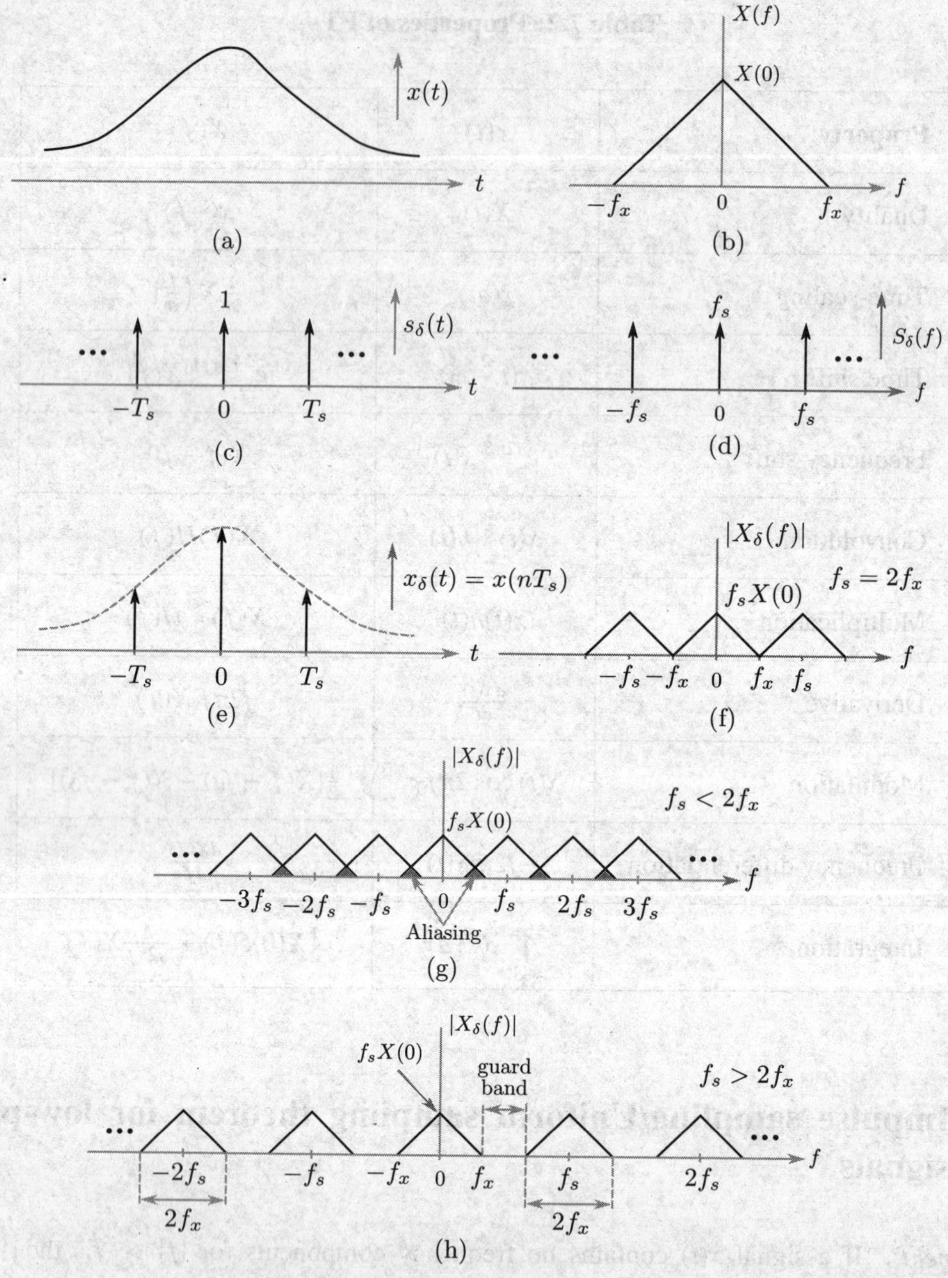

Fig. 7.1: (a) *Low-pass signal* $x(t)$ (b) *Spectrum of* $x(t)$ (c) *Dirac comb* (d) **FT** *of Dirac comb* (e) *Sampled signal* $x(nT_s)$ (f) *Spectrum of sampled signal for* $f_s = 2f_x$ (g) *Spectrum of sampled signal for* $f_s < 2f_x$ (h) *Spectrum of sampled signal for* $f_s > 2f_x$.

Consider an analog signal $x(t)$ as shown in Fig. 7.1(a). The amplitude spectrum of the baseband signal $x(t)$ is as shown in Fig. 7.1(b). The process of sampling known as ideal sampling is accomplished by using a train of impulses known as Dirac comb (Fig. 7.1(c)). Mathematically, a Dirac comb can be represented as follows:

$$s_\delta(t) = \sum_{n=-\infty}^{\infty} \delta(t - nT_s)$$

The Fourier transform of a Dirac comb is also a Dirac comb (Fig. 7.1(d)).

That is,

$$S_\delta(f) = f_s \sum_{n=-\infty}^{\infty} \delta(f - nf_s)$$

The process of sampling is described mathematically as follows:

$$x_\delta(t) = x(t)s_\delta(t)$$

Taking Fourier transform on both sides, we get

$$\begin{aligned} X_\delta(f) &= X(f) * S_\delta(f) \\ &= X(f) * f_s \sum_{n=-\infty}^{\infty} \delta(f - nf_s) \end{aligned}$$

Applying convolution property of an impulse function, we get

$$X_\delta(f) = f_s \sum_{n=-\infty}^{\infty} X(f - nf_s)$$

Using the expression for $X_\delta(f)$, the amplitude spectrum of $x_\delta(t)$ is drawn for $f_s = 2f_x$, $f_s < 2f_x$ and $f_s > 2f_x$. These are shown in Figs. 7.1(f) through 7.1(h).

When $f_s = 2f_x$, the adjoining spectral replicates just touch each other and by using an ideal low-pass filter having a bandwidth, $B = f_x$ Hz, it is possible to recover the continuous-time signal $x(t)$. When $f_s > 2f_x$, the spectral replicates have a larger separation between them, known as the *guard band*, which makes the process of filtering much easier and effective.

When we do undersampling, that is, $f_s < 2f_x$, the spectral replicates overlap each other and the filtered signal will thus be distorted. This type of distortion is called *aliasing*. Aliasing can be avoided if $f_s \geq 2f_x$ or $T_s \leq \frac{1}{2f_x}$.

In summary, we can recover the message signal $x(t)$ from the sampled signal $x_\delta(t)$ if $f_s \geq 2f_x$ and the bandwidth B of the ideal low-pass filter lies between f_x and $f_s - f_x$.

7.2.1 Signal reconstruction

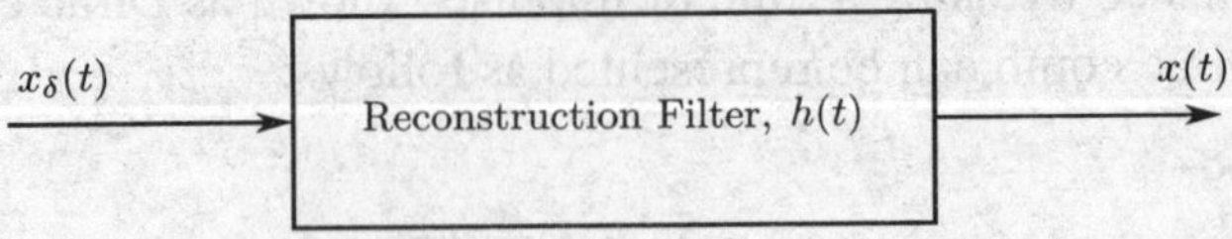

Fig. 7.2 : *Block diagram representing signal reconstruction.*

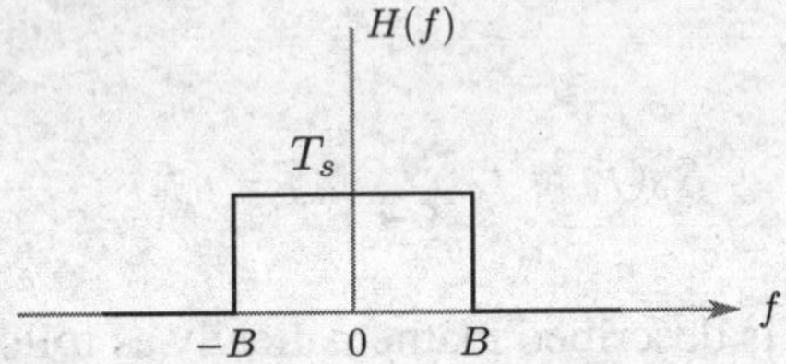

Fig. 7.3 : *Ideal magnitude response of a low-pass filter.*

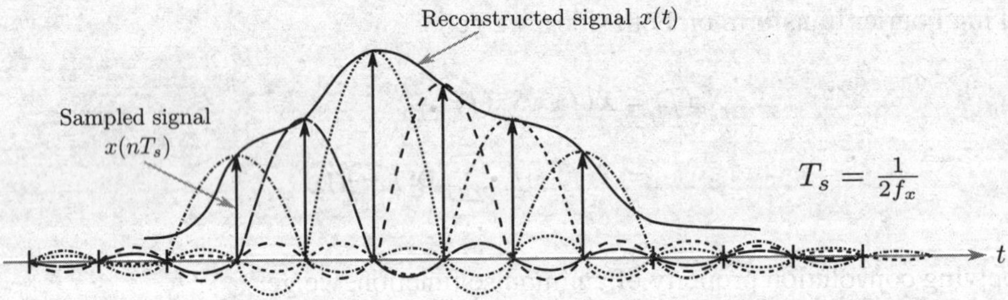

Fig. 7.4 : *Ideal interpolation.*

The concept of signal reconstruction is depicted in Fig. 7.2. Let us consider a reconstruction filter (low-pass filter) having a frequency response as shown in Fig. 7.3. Since the spectral replicates in $X_\delta(f)$ have an amplitude $f_s X(0)$, the gain of the ideal low-pass filter in the pass-band is $\frac{1}{f_s} = T_s$. From the definition of inverse Fourier transform, we have

$$h(t) = \int_{-\infty}^{\infty} H(f) e^{j2\pi f t}\, df$$

$$= \int_{-B}^{B} T_s e^{j2\pi f t}\, df$$

$$= T_s \left[\frac{e^{j2\pi f t}}{j2\pi t} \right]_{-B}^{B}$$

$$= 2BT_s \frac{\sin(2\pi Bt)}{2\pi Bt}$$

$$= 2BT_s \operatorname{sinc}(2Bt) \tag{7.3}$$

It may noted that $\operatorname{sinc}(x) = \frac{\sin \pi x}{\pi x}$. Let us assume that the filter to be an LTI system. Then, the response of the filter is computed using convolution theorem. That is

$$x(t) = x_\delta(t) * h(t)$$

But,

$$x_\delta(t) = x(t)s_\delta(t)$$

Hence,

$$x(t) = x(t)s_\delta(t) * h(t)$$

$$\Rightarrow \qquad x(t) = x(t) \sum_{n=-\infty}^{\infty} \delta(t - nT_s) * h(t) \tag{7.4}$$

Recall the product property of an impulse function:

$$f(t)\delta(t - t_0) = f(t_0)\delta(t - t_0)$$

The above property allows us to write Eq. (7.4) as

$$x(t) = \sum_{n=-\infty}^{\infty} x(nT_s)\delta(t - nT_s) * h(t) \tag{7.5}$$

Substituting Eq. (7.3) in Eq. (7.5) gives

$$x(t) = \sum_{n=-\infty}^{\infty} x(nT_s)\delta(t - nT_s) * 2BT_s \operatorname{sinc}(2Bt) \tag{7.6}$$

Recall the convolution property of an impulse function:

$$f(t) * \delta(t - t_0) = f(t - t_0)$$

Applying the above property to Eq. (7.6) gives

$$x(t) = 2BT_s \sum_{n=-\infty}^{\infty} x(nT_s) \operatorname{sinc}[2B(t - nT_s)] \tag{7.7}$$

Eq. (7.7) is known as the interpolation formula for reconstructing the original signal $x(t)$ from the sequence of sample values $\{x(nT_s)\}$, with sinc function playing the role of an interpolation function. Each sample $x(nT_s)$ is multiplied by a delayed sinc function as shown in Fig. 7.4, and all the resulting delayed sinc pulses are added to obtain $x(t)$.

7.2.2 Effect of undersampling: Aliasing

The low-pass sampling theorem is derived by assuming that the signal $x(t)$ is strictly band-limited, that is, its spectrum is of finite bandwidth. It can be shown that a signal cannot be time-limited and band-limited simultaneously. If a signal is time-limited, it cannot be band-limited and vice-versa (but it can be simultaneously non-time-limited and non-band-limited). Thus, it implies that for $X(f)$ to be of finite bandwidth, the signal $x(t)$ must be of infinite duration. In practice, we have to work with signals that are of finite duration in time. This automatically means that the spectrum of such signals are not of finite bandwidth. This implies that, when a signal of finite duration in time is sampled, an error in the reconstruction process occurs.

Let us consider a finite duration signal $x(t)$ having spectrum $X(f)$, which is of infinite duration as shown in Fig. 7.5(a).If this signal is sampled using a Dirac comb, then the spectrum of the sampled signal is given by

$$X_\delta(f) = f_s \sum_{n=-\infty}^{\infty} X(f - nf_s) \tag{7.8}$$

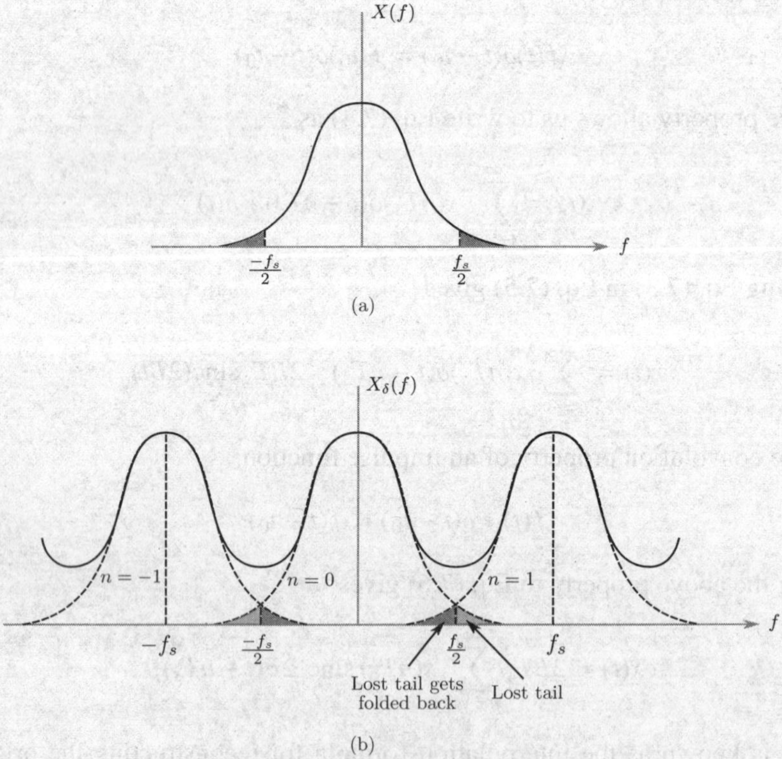

Fig. 7.5 : (a) *Spectrum of a finite duration signal $x(t)$.* (b) *The spectrum of the sampled signal; the dashed curves represent the $X(f - f_s)$ and $X(f + f_s)$ scaled by a factor, f_s.*

The spectrum, $X_\delta(f)$ is shown in Fig. 7.5(b) for $n = -1, 0$ and 1. If we use a low-pass filter(reconstruction filter), with its pass-band extending from $-\frac{fs}{2}$ to $+\frac{fs}{2}$, it no longer gives an undistorted version of the original signal $x(t)$ at its output. This is because of two seperate reasons:

(1) The loss of tail of $X(f)$ beyond $|f| > \frac{fs}{2}$ Hz.

(2) The reappearance of this tail inverted or folded onto the spectrum. It may be noted that the spectral components cross at $\frac{fs}{2} = \frac{1}{2T_s}$ Hz. This frequency is called folding frequency. Rightly called so, because the spectrum folds onto itself at this frequency. For example, a component $\frac{f_s}{2} + f_1$ shows up as a component of lower frequency $\frac{f_s}{2} - f_1$ in the reconstructed signal. Thus the components of frequency above $\frac{f_s}{2}$ appear as components of frequency below $\frac{fs}{2}$. This tail inversion is called spectral folding or aliasing. The phenomenon of aliasing also takes place when a band-limited signal $x(t)$ is undersampled ($f_s < 2f_x$) as illustrated in Fig. 7.1g.

7.3 Natural sampling

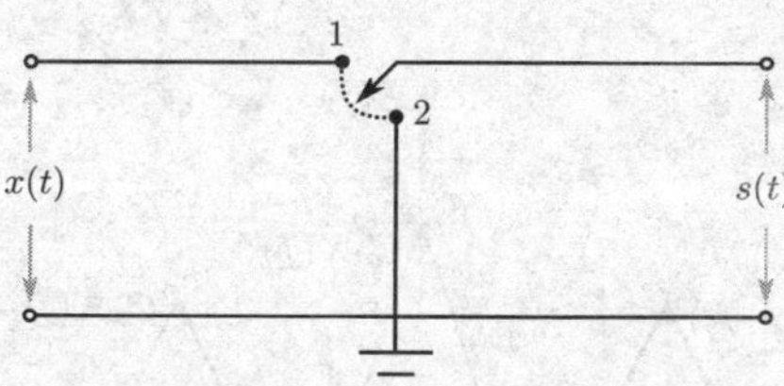

Fig. 7.6 : *Natural sampler*

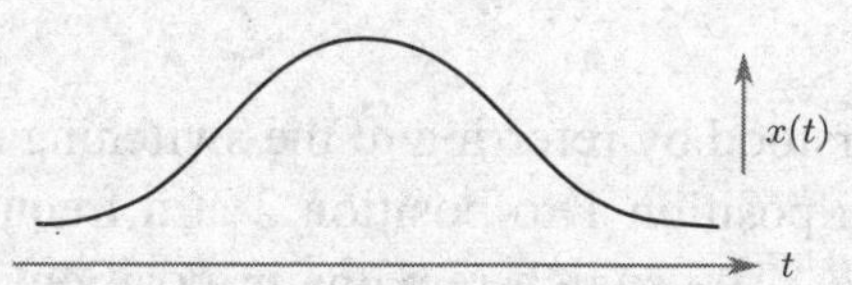

Fig. 7.7 : *Continuous-time signal, x(t)*

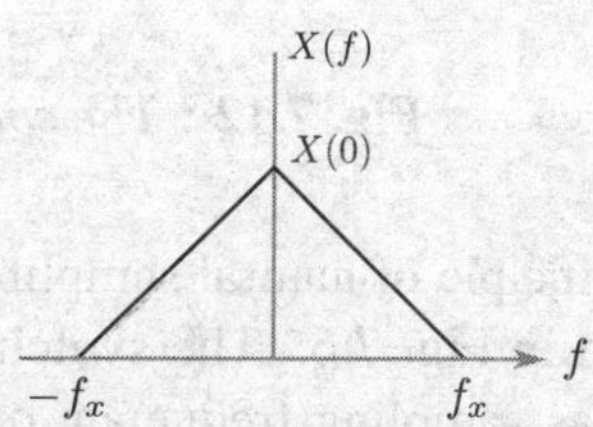

Fig. 7.8 : *Spectrum of x(t)*

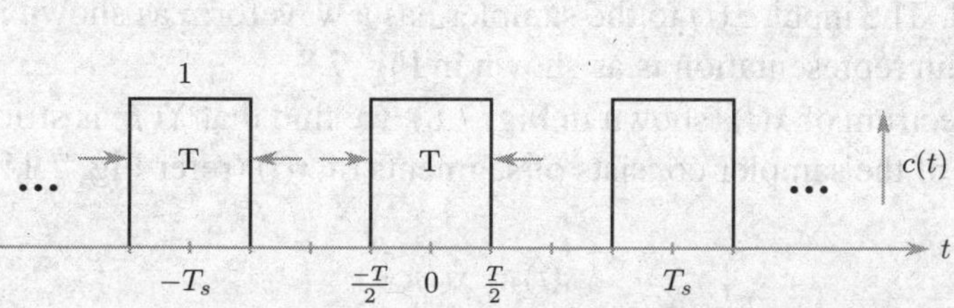

Fig. 7.9 : *Sampling waveform (Period = T_s)*

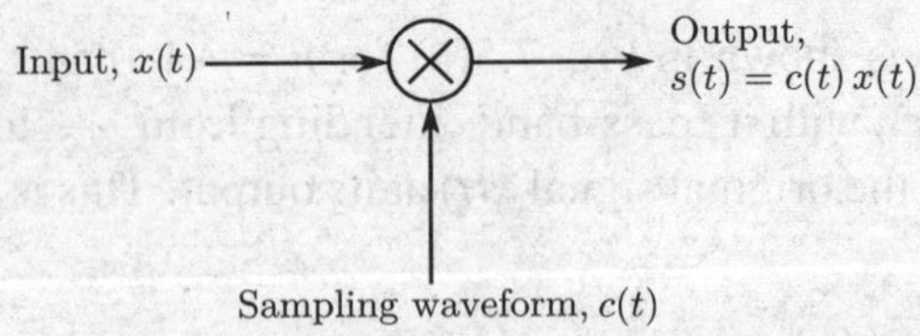

Fig. 7.10 : *Sampling viewed as multiplication*

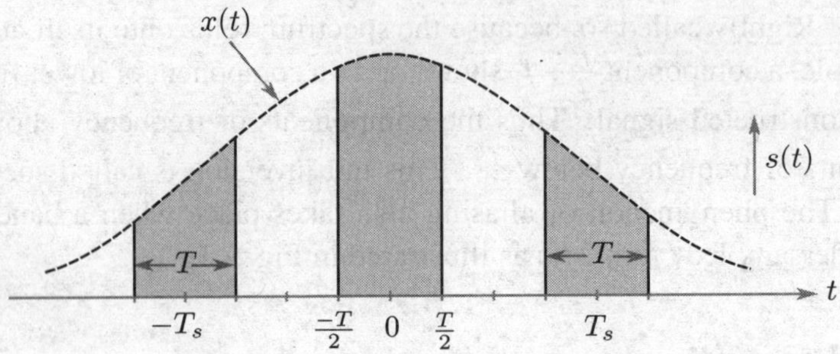

Fig. 7.11 : *Natural samples of x(t)*

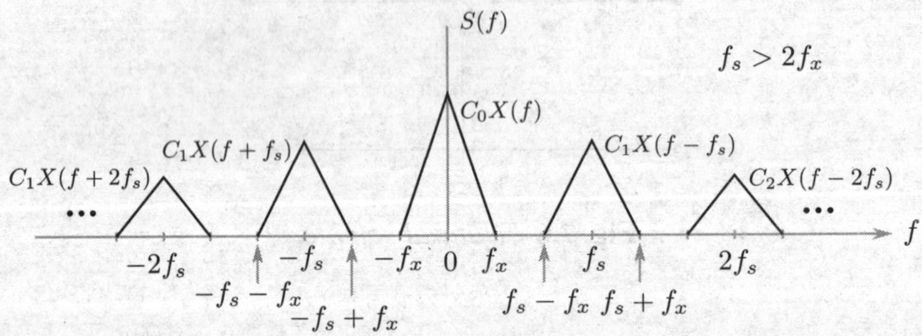

Fig. 7.12 : *The spectrum of natural sampled signal s(t)*

The principle of natural sampling is best understood by referring to the switching sampler shown in Fig. 7.6. The switch changes from position 1 to position 2 at a frequency f_s known as sampling frequency or sampling rate. The switch remains in position 1 for a breif interval equal to T seconds and on the grounded contact for the remainder of each sampling period. The input, $x(t)$ to the sampler has a waveform as shown in Fig. 7.7 and its frequency-domain representation is as shown in Fig. 7.8.

From the spectrum of $x(t)$ (shown in Fig. 7.8), we find that $X(f)$ is strictly band-limited. The output, $s(t)$ of the sampler consists of segments of $x(t)$ (refer Fig. 7.11), and $s(t)$ can be represented as

$$s(t) = x(t)c(t) \tag{7.9}$$

where $c(t)$ is the sampling waveform and is as shown in Fig. 7.9.

Eq. (7.9) implies natural sampling is the product of input, $x(t)$ and the sampling waveform, $c(t)$ and the same has been shown conceptually in Fig. 7.10. Since the samples in $s(t)$ have a top that follows the slope of the input signal $x(t)$, these samples are called natural samples.

The sampling waveform $c(t)$ is periodic and hence may be represented by a trignometric Fourier series. The result is given below

$$c(t) = C_0 + \sum_{n=1}^{\infty} 2C_n \cos 2\pi n f_s t \tag{7.10}$$

$$\text{where,} \quad C_0 = \frac{T}{T_s} \quad \text{and} \quad C_n = f_s T \operatorname{sinc}(n f_s T)$$

Substituting Eq. (7.10) in Eq. (7.9), we get

$$s(t) = C_0 x(t) + 2C_1 x(t) \cos 2\pi f_s t + 2C_2 x(t) \cos 4\pi f_s t + \cdots \tag{7.11}$$

Recall the modulation property: If $g(t) \overset{\mathbf{FT}}{\longleftrightarrow} G(f)$, then

$$g(t) \cos[2\pi(n f_s)t] \overset{\mathbf{FT}}{\longleftrightarrow} \frac{1}{2}[G(f - n f_s) + G(f + n f_s)]$$

Hence, the Fourier transform of Eq. (7.11) yields

$$S(f) = C_0 X(f) + C_1[X(f - f_s) + X(f + f_s)] + C_2[X(f - 2f_s) + X(f + 2f_s)] + \cdots \tag{7.12}$$

Using Eq. (7.12), the spectrum of the sampled signal, $s(t)$ is drawn for $f_s > 2f_x$ and is shown in Fig. 7.12. It is evident from Fig. 7.12, that the sampling operation leaves the message spectrum intact, but it simply repeats periodically in frequency with a fundamental period equal to f_s. We also observe that the message spectrum centred at $f = 0$ is attenuated by a factor C_0. Also, the spectal components at integer multiples of f_s are attenuated by C_n. Since the sampling operation has not altered the message spectrum, it should be possible to reconstruct $x(t)$ from the sampled waveform $s(t)$ by using an ideal low-pass filter having a bandwidth B bounded by the following constraints:

$$f_x < B < f_s - f_x$$

If $A = \frac{1}{T}$ and $T \to 0$, the train of rectangular pulses becomes a Dirac comb. Consequently, natural sampling becomes impulse or ideal sampling.

7.4 Flat-top Sampling

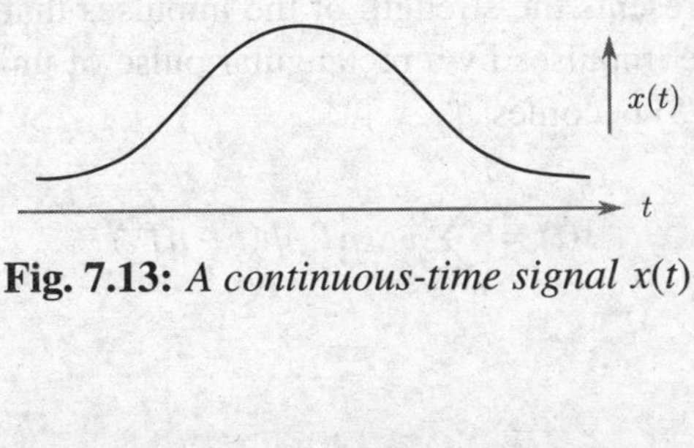

Fig. 7.13: *A continuous-time signal* $x(t)$

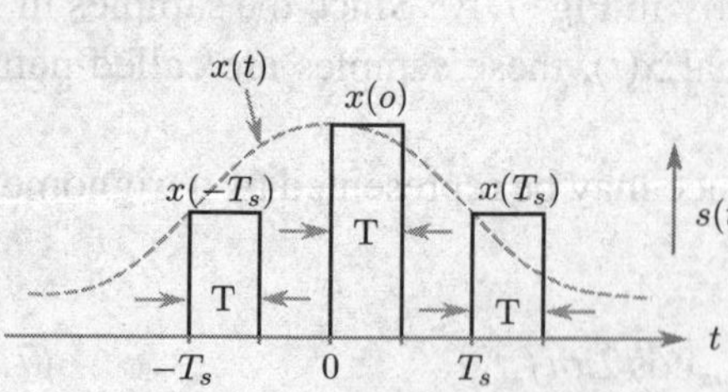

Fig. 7.14 : *Flat-top samples*

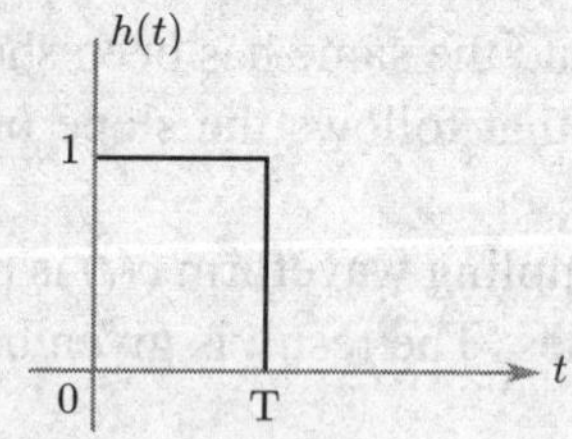

Fig. 7.15: *Rectangular pulse h(t)*

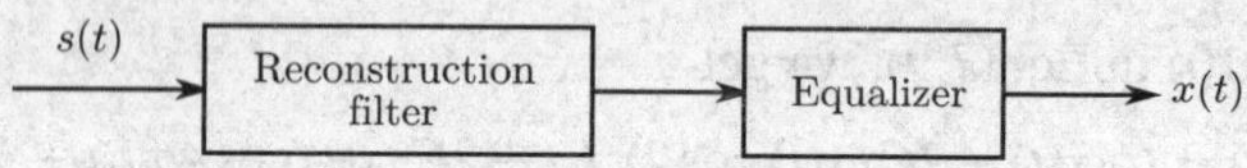

Fig. 7.16: *Conceptual block diagram for signal reconstruction*

Ideal sampling is done by multiplying an analog signal $x(t)$ by a train of impulses that are seperated by T_s seconds. That is,

$$x_\delta(t) = x(t)s_\delta(t) \tag{7.13}$$

where $s_\delta(t)$ is the Dirac comb and has the following time-domain representation:

$$s_\delta(t) = \sum_{n=-\infty}^{\infty} \delta(t - nT_s) \tag{7.14}$$

Substituting Eq. (7.14) in Eq. (7.13) gives

$$x_\delta(t) = x(t) \sum_{n=-\infty}^{\infty} \delta(t - nT_s)$$

Applying the product property of impulse function, we get

$$x_\delta(t) = \sum_{n=-\infty}^{\infty} x(nT_s)\delta(t - nT_s) \tag{7.15}$$

In Eq. (7.15), $x(nT_s)$, represents the strength of the impulses that are seperated by T_s seconds. Let us replace these impulses by a rectangular pulse of unit amplitude and duration equal to T. Then, Eq. (7.15) becomes

$$s(t) = \sum_{n=-\infty}^{\infty} x(nT_s)h(t - nT_s) \tag{7.16}$$

Eq. (7.16) is the time-domain expression for the flat-top samples shown in Fig. 7.14. Each sample shown in Fig. 7.14 has a width equal to T seconds and amplitude equal to $x(nT_s)$.

The next step in the analysis is to perform the convolution of $x_\delta(t)$ and $h(t)$.

$$x_\delta(t) * h(t) = \int_{-\infty}^{\infty} x_\delta(\tau)h(t-\tau)\,d\tau \tag{7.17}$$

Replacing t by τ in Eq. (7.15) and then using that result in Eq. (7.17) gives

$$\begin{aligned} x_\delta(t) * h(t) &= \int_{-\infty}^{\infty} \sum_{n=-\infty}^{\infty} x(nT_s)\delta(\tau - nT_s)h(t-\tau)\,d\tau \\ &= \sum_{n=-\infty}^{\infty} x(nT_s) \int_{-\infty}^{\infty} h(t-\tau)\delta(\tau - nT_s)\,d\tau \end{aligned} \tag{7.18}$$

Recall the sifting property:

$$\int_{-\infty}^{\infty} f(t)\delta(t-t_0)\,dt = f(t)\big|_{t=t_0} = f(t_0)$$

Applying sifting property to Eq. (7.18) gives

$$\begin{aligned} x_\delta(t) * h(t) &= \sum_{n=-\infty}^{\infty} x(nT_s)h(t-\tau)\big|_{\tau=nT_s} \\ &= \sum_{n=-\infty}^{\infty} x(nT_s)h(t-nT_s) \end{aligned} \tag{7.19}$$

Comparing Eq. (7.16) and Eq. (7.19), we find that

$$s(t) = x_\delta(t) * h(t) \tag{7.19}$$

Eq. (7.19) means that flat-top samples can be obtained by convoluting the output of an ideal sampler with a rectangular pulse, $h(t)$ (shown in Fig. (7.15)).

Taking Fourier transform on both sides of Eq. (7.19) gives

$$S(f) = X_\delta(f)H(f) \tag{7.20}$$

To find an expression for the spectrum of $s(t)$, it is necessary to find Fourier transform of $h(t)$.

By definition,

$$H(f) = \int_{-\infty}^{\infty} h(t)e^{-j2\pi ft}\, dt$$

$$\Rightarrow \qquad H(f) = \int_{0}^{T} 1 \cdot e^{-j2\pi ft}\, dt$$

$$= \left. \frac{e^{-j2\pi ft}}{-j2\pi f} \right|_0^T$$

$$= \frac{1 - e^{-j2\pi fT}}{j2\pi f}$$

$$= Te^{-j\pi fT}\left[\frac{e^{j\pi fT} - e^{-j\pi fT}}{(j2)(\pi fT)}\right]$$

$$= Te^{-j\pi fT}\frac{\sin(\pi fT)}{\pi fT}$$

$$= Te^{-j\pi fT}\,\text{sinc}(fT) \tag{7.21}$$

Subsituting Eq. (7.21) in Eq. (7.20), we get

$$S(f) = X_\delta(f)Te^{-j\pi fT}\,\text{sinc}(fT)$$

$$= \left[f_s \sum_{n=-\infty}^{\infty} X(f - nf_s)\right] Te^{-j\pi fT}\,\text{sinc}(fT) \tag{7.22}$$

From Eq. (7.21), we find that $H(f)$ behaves like a low-pass filter since sinc(fT) decreases as f increases. The low-pass action of sinc(fT) attenuates the spectral components of $X_\delta(f)$ present at $\pm fs, \pm 2fs \ldots$ etc., Also, the message spectrum centered at $f = 0$ in $X_\delta(f)$ is multiplied by sinc(fT). Consequently, the high frequency portion of the message spectrum gets attenuated. This loss of high frequency content is called aperture effect. The larger the pulse duration, larger is the aperture effect. Aperture effect can be compensated by using an equalizer having an amplitude response, $\frac{1}{|H(f)|}$ in cascade with the reconstruction filter (shown in Fig. 7.16).

7.5 Practical sample-and-Hold Circuit

A more popular method of obtaining flat-top samples from a continuous-time signal $x(t)$ is known as sample-and-hold technique. An elementary sample-and-hold circuit consists of two FET switches and a capacitor interconnected as shown in Fig. 7.17.

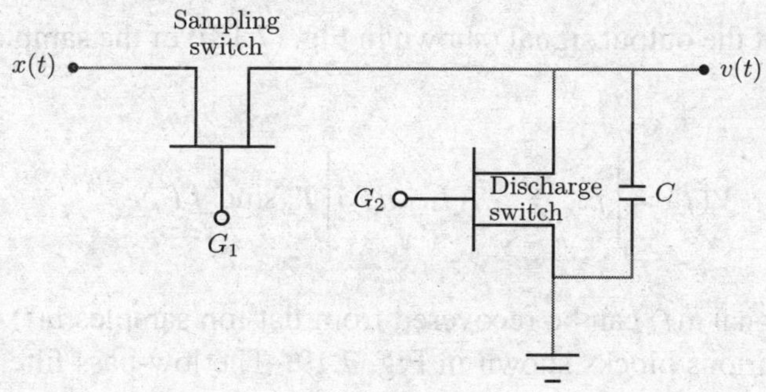

Fig. 7.17: *Sample-and-hold circuit.*

A gate pulse applied at G_1 breifly closes the sampling switch. The capacitor holds the sampled voltage until discharged by a pulse applied to G_2. Periodic gating of the sample-and-hold circuit generates the sampled wave,

$$v(t) = \sum_{n=-\infty}^{\infty} x(nT_s)h(t - nT_s) \tag{7.23}$$

The above equation is nothing but Eq. (7.16). In Eq. (7.23), $h(t)$ represents the impulse response of the sample-and-hold circuit; that is

$$h(t) = \begin{cases} 1, & 0 < t < T_s \\ 0, & \text{otherwise} \end{cases} \tag{7.24}$$

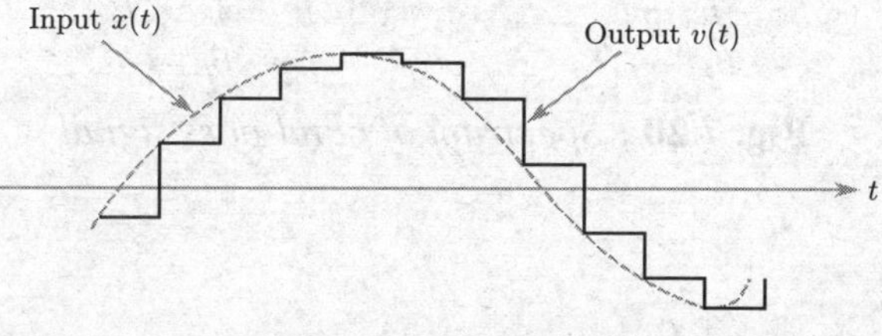

Fig. 7.18 : *Stair-case like output waveform of sample-and-hold circuit.*

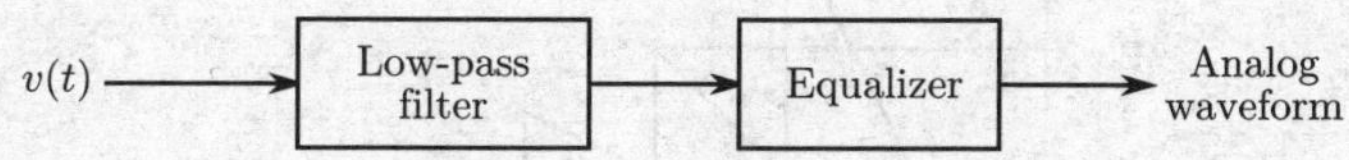

Fig. 7.19 : *Conceptual block diagram for signal reconstruction.*

Accordingly, the transfer function of the sample-and-hold circuit is

$$H(f) \triangleq \int_{-\infty}^{\infty} h(t)e^{-2\pi ft}\,dt \;\Rightarrow H(f) = \int_{0}^{T_s} 1 \times e^{-j2\pi ft}\,dt = T_s\,\text{sinc}(fT_s)e^{-j\pi fT_s} \tag{7.25}$$

where $\text{sinc}(x) = \frac{\sin \pi x}{\pi x}$.

The spectrum of the output signal (shown in Fig. (7.18)) of the samples-and-hold circuit (Refer Eq. (7.22)) is

$$V(f) = \left[fs \sum_{n=-\infty}^{\infty} X(f - nf_s) \right] T_s \operatorname{sinc}(fT_s) e^{-j\pi f T_s} \tag{7.26}$$

The original signal $x(t)$ can be recovered from flat-top samples $v(t)$ by passing flat-top samples through various blocks shown in Fig. 7.19. The low-pass filter has the main aim of suppressing the spectra of flat-top samples present at integer multiplies of f_s and allow the message spectrum centered at $f = 0$ freely. The aperture effect is removed by using an equalizer whose amplitude response equals $\frac{1}{|H(f)|}$.

7.6 Band-pass Sampling theorem

Consider a signal $x(t)$ having a spectrum as shown in Fig. 7.20. The following sampling theorem gives the condition for representing $x(t)$ by its sampled values.

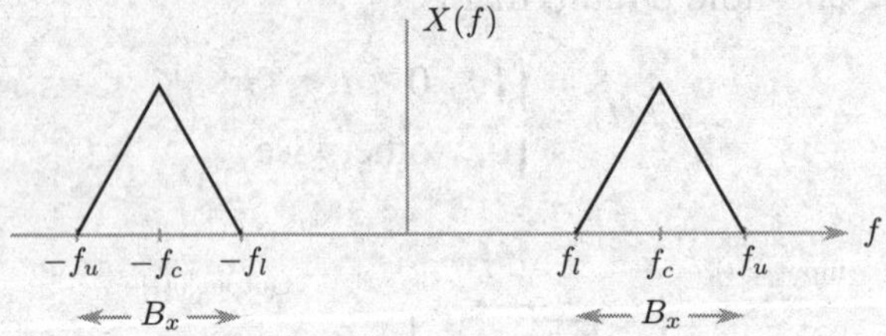

Fig. 7.20 : *Spectrum of band-pass signal*

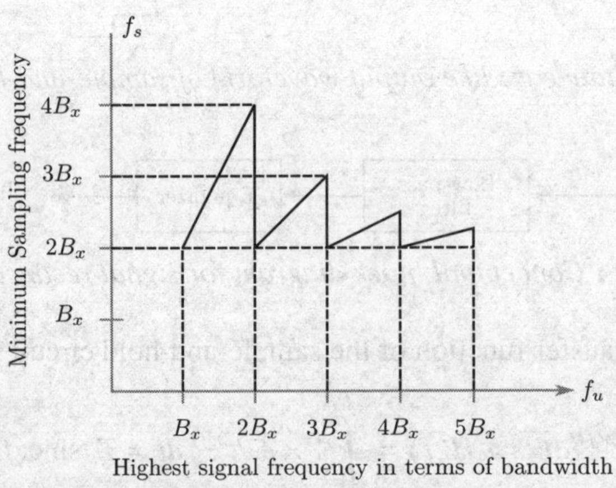

Fig. 7.21: *Minimum sampling frequency for a signal occupying a bandwidth B_x*

The band-pass sampling theorem states that if a band-pass signal $x(t)$ has a spectrum of bandwidth B_x and an upper frequency f_u, then $x(t)$ can be recoved from $x(nT_s)$ by band-pass filtering if $f_s = \frac{2f_u}{m}$, where m is an integer not exceeding $\frac{f_u}{B_x}$. All higher rates are not necessarily usable unless they exceed $2f_u$.

The frequency response of the band-pass filter used for signal reconstruction is given below:

$$H(f) = \begin{cases} 1, & f_l < |f| < f_u \\ 0, & \text{elsewhere} \end{cases}$$

The sampling rate for a band-pass signal depends on the ratio $\frac{f_u}{B_x}$. If $\frac{f_u}{B_x} >> 1$, then the minimum sampling rate approaches $2B_x$. A sketch of f_u versus f_s is shown in Fig. 7.21. One can easily verify that $f_s > 2f_u$ will result in exact signal reconstruction.

Example 7.1. A signal $x(t) = 2\cos 400\pi t + 6\cos 640\pi t$ is ideally sampled at $f_s = 500\ Hz$. If the sampled signal is passed through an ideal low-pass filter with a cutoff frequency of 400 Hz, what frequency components will appear in the filter output?

Solution: Given $x(t) = 2\cos 400\pi t + 6\cos 640\pi t$. Taking **FT** on both sides of the above equation, we get

$$X(f) = [\delta(f-200) + \delta(f+200)] + 3[\delta(f-320) + \delta(f+320)]$$

Using the above equation, the spectrum of the signal $x(t)$ is drawn and it appears as shown in Fig. Ex. 7.1.

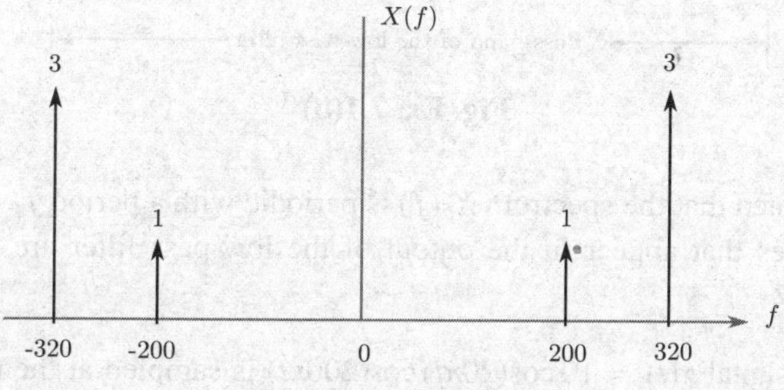

Fig. Ex. 7.1(a)

The **FT** of the sampled signal $x_\delta(t)$ is drawn using the expression given below.

$$X_\delta(f) = f_s \sum_{n=-\infty}^{\infty} X(f - nf_s)$$

$$= f_s \sum_{n=-\infty}^{\infty} X(f - 500n)$$

$$\Rightarrow X_\delta(f) = f_s \sum_{n=-\infty}^{\infty} [\delta(f - 500n - 200) + \delta(f - 500n + 200)]$$
$$+ 3f_s \sum_{n=-\infty}^{\infty} [\delta(f - 500n - 320) + \delta(f - 500n + 320)]$$

The location of impulses in $X_\delta(f)$ are obtained by equating argument of each impulse function to zero. Thus, the location of various impulses in $X_\delta(f)$ are as follows:

(i) $f = 500n + 200$, (ii) $f = 500n - 200$
(iii) $f = 500n + 320$ and (iv) $f = 500n - 320$

The spectrum of $x_\delta(t)$ is as shown in Fig. Ex. 7.1(b).

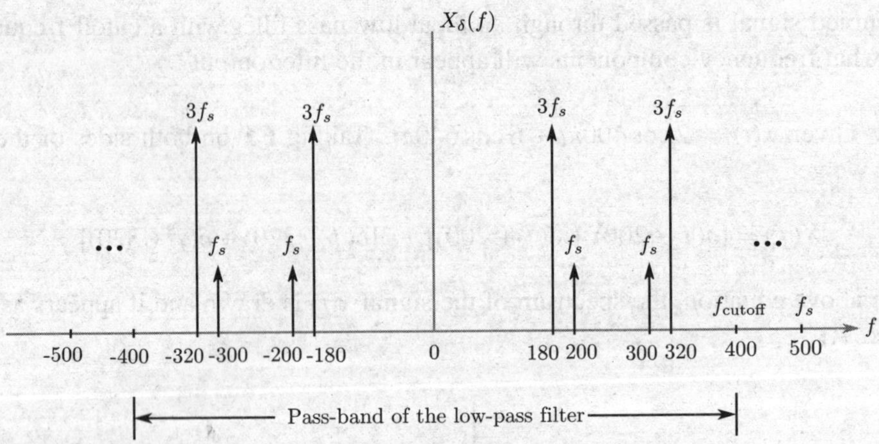

Fig. Ex. 7.1(b)

It should be noted that the spectrum $X_\delta(f)$ is periodic with a period f_s.

The frequencies that appear in the output of the low-pass filter are 180 Hz, 200 Hz, 300 Hz and 320 Hz.

Example 7.2. A signal $x(t) = 10\cos(20\pi t)\cos(200\pi t)$ is sampled at the rate of 250 samples/second.

(a) Sketch the spectrum of the sampled signal.

(b) Specify the cutoff frequency for the ideal reconstruction filter so as to recover $x(t)$ from $x_\delta(t)$.

(c) Specify the Nyquist rate for the signal $x(t)$.

Solution:

(a) Given $x(t) = 10\cos(20\pi t)\cos(200\pi t)$

$\Rightarrow \quad x(t) = 5\cos(220\pi t) + 5\cos(180\pi t)$

Taking **FT** on both sides of the above equation, we get

$$X(f) = 2.5[\delta(f-110) + \delta(f+110) + \delta(f-90) + \delta(f+90)]$$

The spectrum of the signal $x(t)$ is drawn using the above expression and it appears as shown in Fig. Ex. 7.2(a).

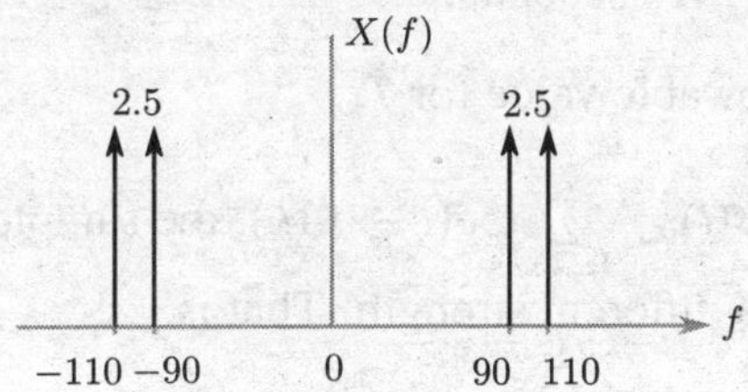

Fig. Ex. 7.2(a)

Recall: $X_\delta(f) = f_s \Sigma_{n=-\infty}^{\infty} X(f - nf_s)$

$$\Rightarrow \quad X_\delta(f) = f_s \sum_{n=-\infty}^{\infty} X(f - 250n)$$

$$= 2.5 f_s \sum_{n=-\infty}^{\infty} \{\delta(f - 250n - 110) + \delta(f - 250n + 110)$$

$$+\delta(f - 250n - 90) + \delta(f - 250n + 90)\}$$

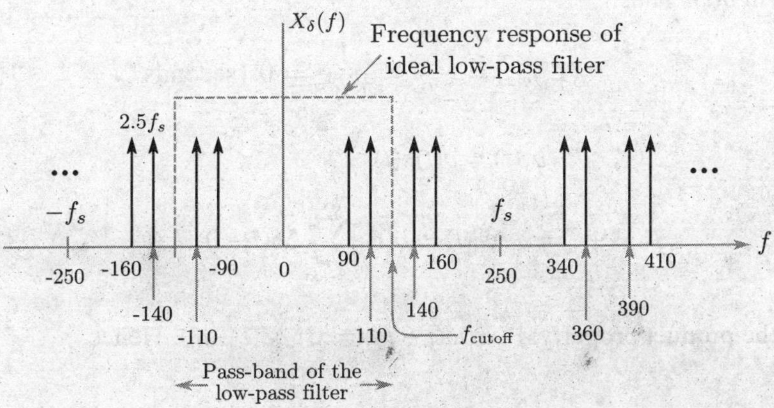

Fig. Ex. 7.2(b)

Using the above expression, the spectrum of the sampled signal is drawn and it appears as shown in Fig. Ex. 7.2(b). It should be noted that $X_\delta(f)$ is symmetric about vertical axis and is periodic with a period equal to f_s.

(b) The cutoff frequency of the ideal Low-pass filter (LPF) should be greater than 110 Hz and less than 140 Hz for recovering $x(t)$ from $x_\delta(t)$.

(c) Nyquist rate, $f_s = 2f_x$, where f_x is the highest frequency present in $x(t)$. Thus, $f_s = 2 \times 110 = 220$Hz.

Example 7.3. The signal $\upsilon(t) = \cos(5\pi t) + 0.5\cos(10\pi t)$ is instantaneously sampled. The interval between the samples is T_s seconds.

(a) Find the maximum allowable value for T_s.

(b) If the sampling signal $s(t) = \sum_{k=-\infty}^{\infty} 5\delta(t - 0.1k)$, the sampled signal consists of a train of impulses each with a different strength. That is,

$$\upsilon_s(t) = \sum_{k=-\infty}^{\infty} I_k\delta(t - 0.1k)$$

Find the values of I_0, I_1, I_2 and show that $I_{k+4} = I_k$.

(c) To reconstruct the signal $\upsilon(t)$, the sampled signal $\upsilon_s(t)$ is passed through a rectangular LPF. Find the minimum filter bandwidth required to reconstruct the signal without distortion.

Solution:

(a) The highest frequency component present in the message signal $\upsilon(t)$ is 5 Hz. According to low-pass sampling theorem: $T_s \leq 1/2f_x$, where f_x is the highest frequency present in $\upsilon(t)$. Thus,

$$(T_s)_{max} = \frac{1}{2f_x} = \frac{1}{2 \times 5} = 0.1\text{seconds}$$

(b)

$$\upsilon_s(t) = \upsilon(t)s(t)$$

$$\Rightarrow \quad \upsilon_s(t) = \upsilon(t)\sum_{k=-\infty}^{\infty} 5\delta(t - 0.1k)$$

Recall the product property: $x(t)\delta(t - t_0) = x(t_0)\delta(t - t_0)$. Hence,

$$\upsilon_s(t) = \sum_{k=-\infty}^{\infty} 5\upsilon(0.1k)\delta(t - 0.1k)$$

$$\Rightarrow \quad v_s(t) = \sum_{k=-\infty}^{\infty} 5[\cos(0.5\pi k) + 0.5\cos(\pi k)]\delta(t - 0.1k)$$

$$= \sum_{k=-\infty}^{\infty} I_k \delta(t - 0.1k)$$

Hence, $I_k = 5[\cos(0.5\pi k) + 0.5\cos(\pi k)]$. The values of I_k for different values of k are given below:

k	I_k
0	7.5
1	−2.5
2	−2.5

Also,

$$\begin{aligned} I_{k+4} &= 5[\cos(0.5\pi(k+4)) + 0.5\cos(\pi(k+4))] \\ &= 5[\cos(0.5\pi k + 2\pi) + 0.5\cos(\pi k + 4\pi)] \\ &= 5[\cos(0.5\pi k) + 0.5\cos(\pi k)] = I_k \end{aligned}$$

(c) Since the highest frequency component of $v(t)$ is 5 Hz, the minimum filter bandwidth is 5 Hz to reconstruct the signal without distortion.

Example 7.4. Fig. Ex. 7.4 shows the Fourier spectra of the signals $x_1(t)$ and $x_2(t)$. Determine the Nyquist interval and sampling rate for signals $x_1(t), x_2(t), x_1^2(t), x_2^2(t), x_2^3(t)$, and $x_1(t)x_2(t)$.

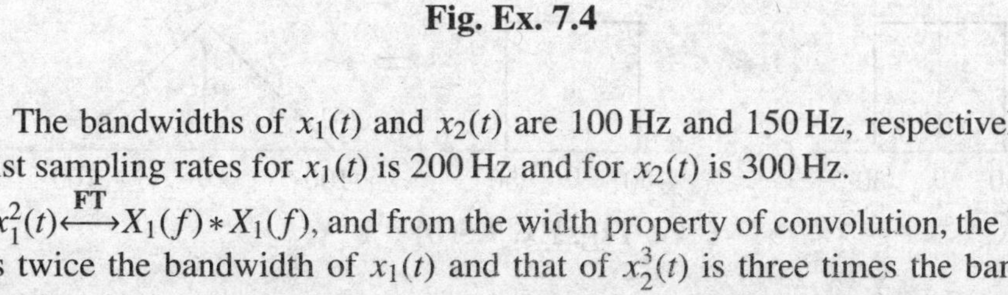

Fig. Ex. 7.4

Solution: The bandwidths of $x_1(t)$ and $x_2(t)$ are 100 Hz and 150 Hz, respectively. Hence, the Nyquist sampling rates for $x_1(t)$ is 200 Hz and for $x_2(t)$ is 300 Hz.

Also $x_1^2(t) \overset{\text{FT}}{\longleftrightarrow} X_1(f) * X_1(f)$, and from the width property of convolution, the bandwidth of $x_1^2(t)$ is twice the bandwidth of $x_1(t)$ and that of $x_2^3(t)$ is three times the bandwidth of $x_2(t)$. Likewise, the bandwidth of $x_1(t)x_2(t)$ is the sum of the bandwidth of $x_1(t)$ and $x_2(t)$. Hence, the Nyquist rate for $x_1^2(t)$ is 400 Hz, for $x_2^2(t)$ is 600 Hz, for $x_2^3(t)$ is 900 Hz and for $x_1(t)x_2(t)$ is 500 Hz.

Example 7.5. Determine the Nyquist rate and the Nyquist sampling interval for the following signals:

(a) $x_1(t) = \text{sinc}(100\pi t)$

(b) $x_2(t) = \text{sinc}^2(100\pi t)$

(c) $x_3(t) = \text{sinc}(100\pi t) + \text{sinc}(50\pi t)$

Solution:

(a) Given $x_1(t) = \text{sinc}(100\pi t)$.

Recall:

(1) The spectrum of a sinc function is rectangular and is of finite width.

(2) The spectrum of a rectangular pulse is sinc function and is of infinite duration.

From the spectrum of $x_1(t)$ shown in Fig. Ex. 7.5(a), we find that the bandwidth of the signal $x_1(t)$ is $f_x = 50\,\text{Hz}$. Hence, the Nyquist rate= $2f_x = 100\,\text{Hz}$ and Nyquist interval = $1/100 = 0.01\,\text{sec}$.

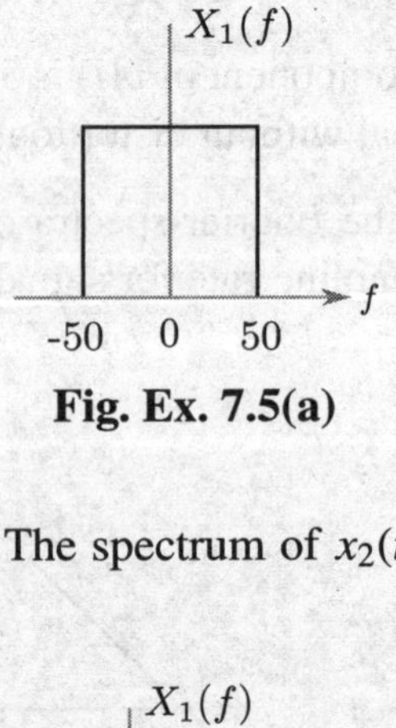

Fig. Ex. 7.5(a)

(b) Given $x_2(t) = \text{sinc}^2(100\pi t)$. The spectrum of $x_2(t)$ is a triangular pulse as shown in Fig. Ex. 7.5(b).

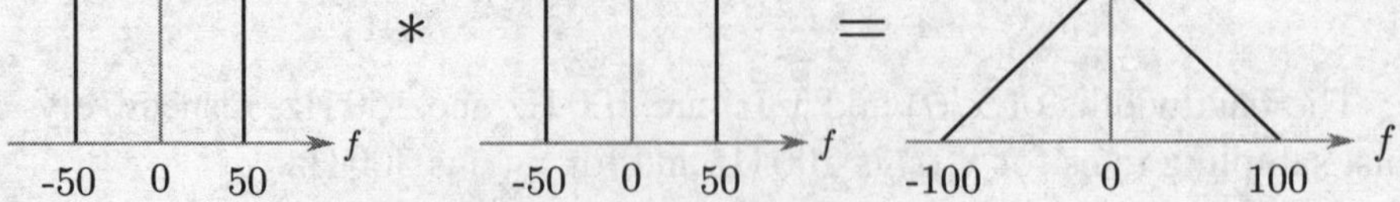

Fig. Ex. 7.5(b)

Hence, the bandwidth of the signal $x_2(t)$ is $100\,\text{Hz}$.

Thus, Nyquist rate = $2 \times 100 = 200\,\text{Hz}$ and Nyquist interval = $1/200$seconds.

(c) Given $x_3(t) = \text{sinc}(100\pi t) + \text{sinc}(50\pi t) = x_A(t) + x_B(t)$.

The spectrum of $x_3(t)$ is as shown in Fig. Ex. 7.5(c).

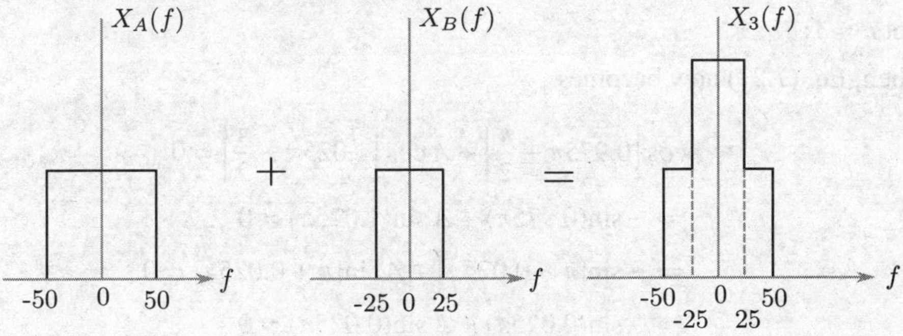

Fig. Ex. 7.5(c)

From Fig. Ex. 7.5(c), we find that the bandwidth of the signal $x_3(t)$ is 50 Hz. Hence, Nyquist rate= $2 \times 50 = 100$Hz and Nyquist interval $= 1/100 = 0.1$ seconds.

Example 7.6. A signal $x(t)$ consists of two frequency components $f_1 = 3.9$ kHz and $f_2 = 4.1$ kHz in such a relationship they cancel out each other when $x(t)$ is sampled at the instants $t = 0, T, 2T, \ldots$ where $T = 125$ micro seconds. The signal $x(t)$ is defined by

$$x(t) = \cos\left[2\pi f_1 t + \frac{\pi}{2}\right] + A\cos[2\pi f_2 t + \phi]$$

Find the values of A and ϕ of the second frequency component.

Solution: Given

$$x(t) = \cos\left[2\pi f_1 t + \frac{\pi}{2}\right] + A\cos[2\pi f_2 t + \phi]$$

Also given in the problem, that,

$$x(nT) = 0$$

Hence, $\cos[2\pi f_1 nT + \pi/2] + A\cos[2\pi f_2 nT + \phi] = 0$

$$\Rightarrow \quad \cos\left[2\pi \times 3.9 \times 10^3 \times 125 \times 10^{-6} \times n + \frac{\pi}{2}\right]$$

$$+ A\cos[2\pi \times 4.1 \times 10^3 \times 125 \times 10^{-6} \times n + \phi] = 0$$

$$\Rightarrow \quad \cos\left[0.975\pi n + \frac{\pi}{2}\right] + A\cos[1.025\pi n + \phi] = 0 \tag{7.27}$$

(i) Let $n = 0$:

Then, Eq. (7.27) becomes

$$\cos\left(\frac{\pi}{2}\right) + A\cos\phi = 0$$

Since,

$$A \neq 0, \quad \text{we get} \quad \phi = \pm\frac{\pi}{2}$$

(ii) Let $n = 1$:

Then, Eq. (7.27) now becomes

$$\cos\left[0.975\pi + \frac{\pi}{2}\right] + A\cos\left[1.025\pi \pm \frac{\pi}{2}\right] = 0$$

$$\Rightarrow -\sin(0.975\pi) \mp A\sin(1.025\pi) = 0$$

$$\Rightarrow -\sin[\pi - 0.025\pi] \mp A\sin[\pi + 0.025\pi] = 0$$

$$\Rightarrow -\sin(0.025\pi) \pm A\sin(0.025\pi) = 0$$

The above equation is satisfied only when $A = 1$ and +ve sign for the second term is taken. The +ve sign can appear in the above equation, only when $\phi = +\pi/2$ in Eq. (7.27). Thus, $A = 1$ and $\phi = +\pi/2$.

Example 7.7. The signals $x_1(t) = 10\cos 100\pi t$ and $x_2(t) = 10\cos 50\pi t$ are both sampled with $f_s = 75$ Hz. Show that two sequences so obtained are identical.

Solution: Given $x_1(t) = 10\cos(100\pi t)$. Taking **FT** on the both sides of the above equation, we get

$$X_1(f) = 5[\delta(f - 50) + \delta(f + 50)]$$

With $f_s = 1/T_s = 75$ Hz, the spectrum of the sampled version of $x_1(t)$ is

$$X_{1\delta}(f) = f_s \sum_{m=-\infty}^{\infty} X_1(f - mf_s)$$

$$= 375 \sum_{m=-\infty}^{\infty} [\delta(f - 50 - 75m) + \delta(f + 50 - 75m)] \tag{7.28}$$

Similarly, the **FT** of $x_2(t)$ is

$$X_2(f) = 5[\delta(f - 25) + \delta(f + 25)]$$

The spectrum of the sampled version of $x_2(t)$ is

$$X_{2\delta}(f) = f_s \sum_{m=-\infty}^{\infty} X_2(f - mf_s)$$

$$= 375 \sum_{m=-\infty}^{\infty} [\delta(f - 25 - 75m) + \delta(f + 25 - 75m)] \tag{7.29}$$

Letting $m = l - 1$ in the first term of Eq. (7.29) and $m = k + 1$ for the second term, we get

$$X_{2\delta}(f) = 375 \sum_{l=-\infty}^{\infty} \delta(f + 50 - 75l) + 375 \sum_{k=-\infty}^{\infty} \delta(f - 50 - 75k) \tag{7.30}$$

Since l and k are dummy variables, they can be replaced by m. Consequently, Eq. (7.30) becomes

$$X_{2\delta}(f) = 375 \sum_{m=-\infty}^{\infty} [\delta(f - 50 - 75m) + \delta(f + 50 - 75m)] \tag{7.31}$$

Comparing Eq. (7.28) and (7.31), we find that the sampled version of $x_1(t)$ and $x_2(t)$ are identical.

Example 7.8. The spectrum of a band-pass signal $x(t)$ has a bandwidth of 0.6 kHz centered around ±12 kHz. Find the Nyquist rate for quadrature sampling of in-phase and quadrature components of the signal $x(t)$.

Solution: The signal $x(t)$ can be expressed interms of inphase and quadrature components as

$$\begin{aligned} x(t) &= x_I(t)\cos 2\pi f_c t - x_Q(t)\sin 2\pi f_c t \\ &= x_I(t)\cos[2\pi \times 12 \times 10^3 t] - x_Q(t)\sin[2\pi \times 12 \times 10^3 t] \end{aligned}$$

where $x_I(t)$ and $x_Q(t)$ are low-pass signals with a bandwidth of

$$f_x = \frac{1}{2} \times 0.6 = 0.3\,\text{kHz}$$

The Nyquist rate for $x_I(t)$ and $x_Q(t)$ is therefore

$$f_s = 2f_x = 2 \times 0.3 = 0.6\,\text{kHz}$$

Example 7.9. A band-pass signal $x(t)$ with a spectrum shown in Fig. Ex. 7.9(a) is ideally sampled. Sketch the spectrum of the sampled signal at $f_s = 25, 45$ and 50 Hz. Indicate if and how the signal can be recovered.

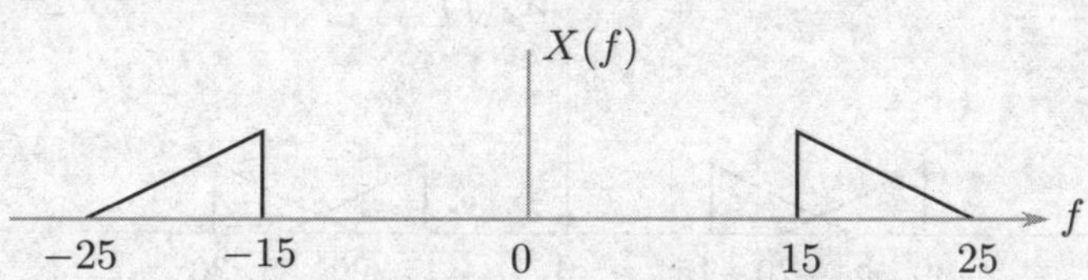

Fig. Ex. 7.9(a)

Solution: From Fig. 7.9(a), we find that $f_u = 25\,\text{Hz}$, $B_x = 25 - 15 = 10\,\text{Hz}$. Then $f_u/B_x = 2.5$ and $m = 2$.

Hence, according to band-pass sampling theorem, it is possible to reconstruct the analog signal $x(t)$ from sample values taken at a rate

$$f_s = \frac{2f_u}{m}$$
$$= \frac{2 \times 25}{2} = 25\,Hz$$

The spectrum of the sampled signal is sketched using the expression,

$$X_\delta(f) = f_s \sum_{n=-\infty}^{\infty} X(f - nf_s)$$
$$\Rightarrow X_\delta(f) = 25 \sum_{n=-\infty}^{\infty} X(f - 25n) \quad (7.32)$$

For $f_s = 25\,\text{Hz}$, the spectrum $X_\delta(f)$ is as shown in Fig. Ex. 7.9(b). Similalry Figs. Ex. 7.9(c) and Ex. 7.9(d) depict the spectra of $x_\delta(t)$ for $f_s = 45$ and $50\,\text{Hz}$, respectively.

From Fig. Ex. 7.9(b), we see that $x(t)$ can be recovered from $x_\delta(t)$ by using an ideal band-pas filter (BPF) with lower cutoff frequency f_{c1}, greater than or equal to 10 Hz and less than 15 Hz, and upper cutoff frequency f_{c2} equal to 25 Hz.

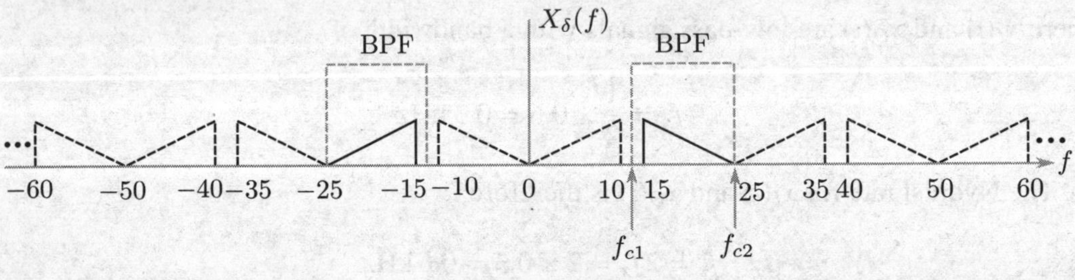

Fig. Ex. 7.9(b)

For $f_s = 45\,\text{Hz}$: From Fig. Ex. 7.9(c), it is not possible to recover $x(t)$ by filtering, due to aliasing.

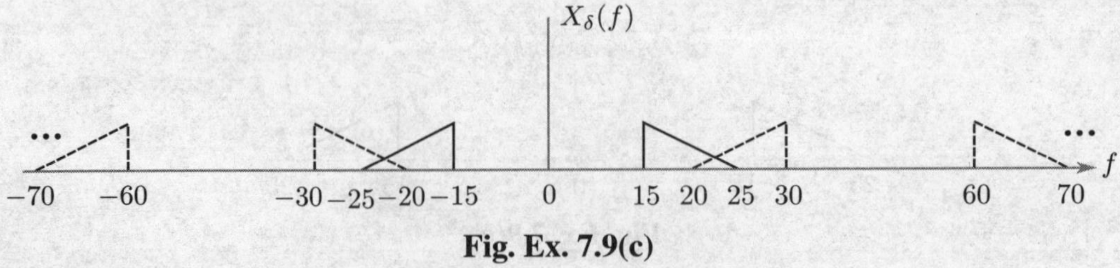

Fig. Ex. 7.9(c)

For $f_s = 50$ Hz: From Fig. Ex. 7.9(d), it is seen that $x(t)$ can be recovered by using an ideal band-pass filter having a lower cutoff frequency = 15 Hz and an upper cutoff frequency = 25 Hz.

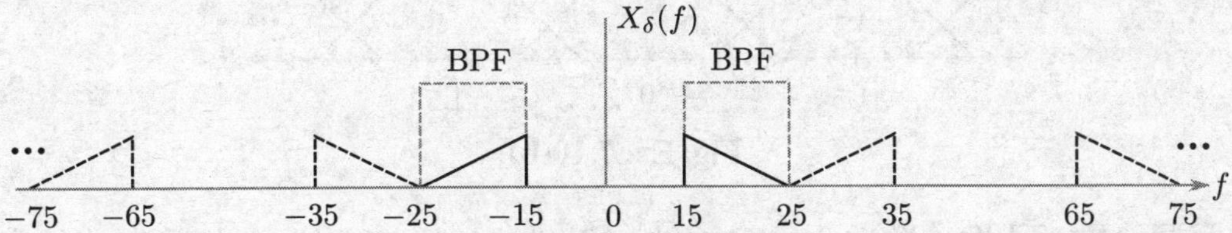

Fig. Ex. 7.9(d)

Example 7.10. A low-pass signal has the spectrum given by

$$X(f) = \begin{cases} 1 - |f|, & |f| < 1 \\ 0, & |f| > 1 \end{cases}$$

Assume that $x(t)$ is sampled at 1.5 Hz, and then applied to a low-pass reconstruction filter with cutoff frequency at 1 Hz. Plot the spectrum of the resulting signal.

Solution: Given

$$X(f) = \begin{cases} 1 - |f|, & |f| < 1 \\ 0, & |f| > 1 \end{cases}$$

Using the above equation, the spectrum of $x(t)$ is drawn and is as shown in Fig. Ex. 7.10(a).

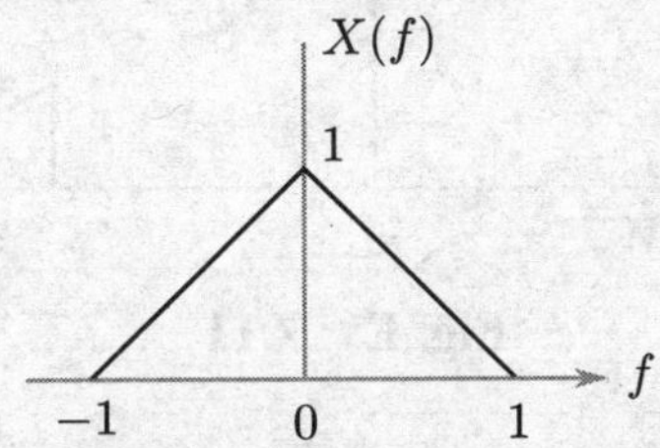

Fig. Ex. 7.10(a)

The spectrum of the sampled signal is given by the expression

$$X_\delta(f) = f_s \sum_{n=-\infty}^{\infty} X(f - nf_s)$$

$$\Rightarrow \quad X_\delta(f) = 1.5 \sum_{n=-\infty}^{\infty} X(f - 1.5n) \tag{7.33}$$

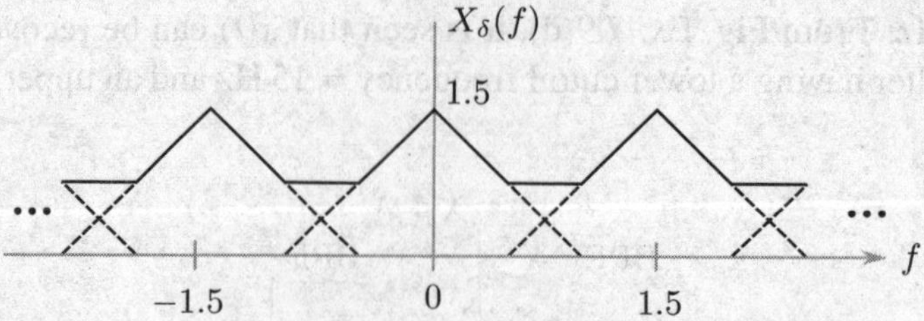

Fig. Ex. 7.10(b)

In Fig. Ex. 7.10(c), $X'(f)$ represents the **FT** of the output $x'(t)$ of the reconstruction filter. The spectrum of $x'(t)$ is significantly different from that of $x(t)$, indicating distortion due to aliasing.

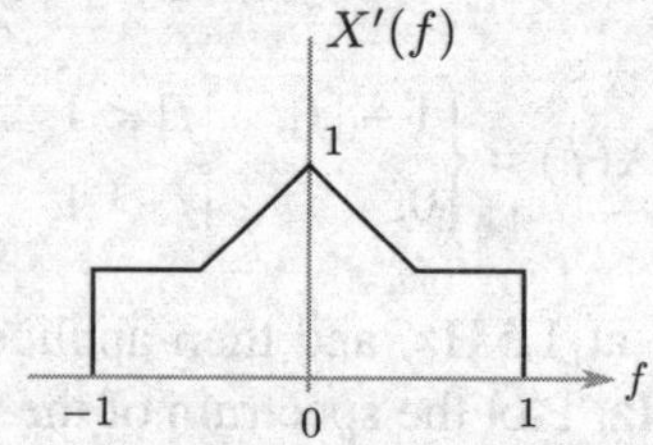

Fig. Ex. 7.10(c)

Example 7.11. Suppose $x(t)$ has the spectrum shown in Fig. Ex. 7.11 with $f_u = 25$ kHz and $W = 10$ kHz. Sketch $X_\delta(f)$ for $f_s = 60$, 45 and 25 kHz. Comment in each case on the possible reconstruction of $x(t)$ from $x_\delta(t)$.

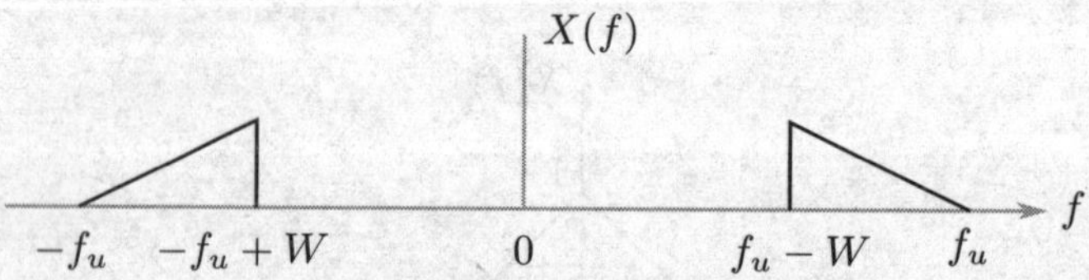

Fig. Ex. 7.11

Solution: According to band-pass sampling theorem, sampling frequency $f_s = 2f_u/m$, where m is an integer not exceeding f_u/B_x. In the present problem, $f_u/B_x = 25/W = 2.5$. Hence, $m = 2$. Thus, $f_s = 2 \times \frac{25}{2} = 25$ kHz. The spectrum of $x_\delta(t)$ is plotted using the expression given below:

$$X_\delta(f) = f_s \sum_{n=-\infty}^{\infty} X(f - nf_s)$$

$$= 25 \sum_{n=-\infty}^{\infty} X(f - 25n) \tag{7.34}$$

It may be noted that in Eq. (7.34) all the frequencies are in kHz.

(i) $f_s = 60\,\text{kHz}$

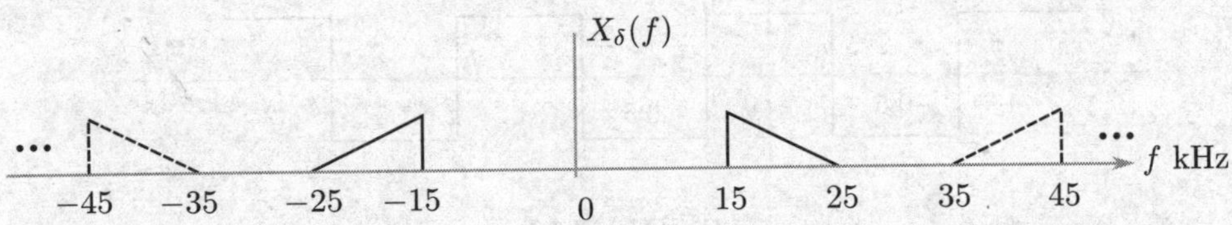

The continuous-time signal $x(t)$ can be recovered by passing $x_\delta(t)$ through an ideal band-pass filter whose upper cutoff frequency, $f_{c_2} = 25$ kHz and lower cutoff frequency, $f_{c_1} = 15$ kHz.

(ii) $f_s = 45\,\text{kHz}$

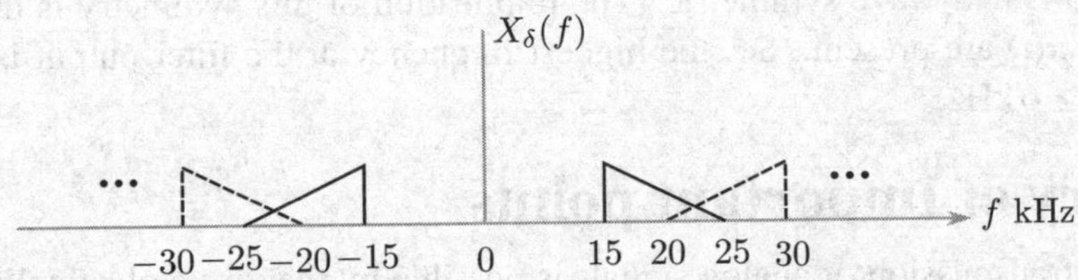

We cannot recover $x(t)$ from $x_\delta(t)$ by filtering, due to aliasing.

(iii) $f_s = 25\,\text{kHz}$

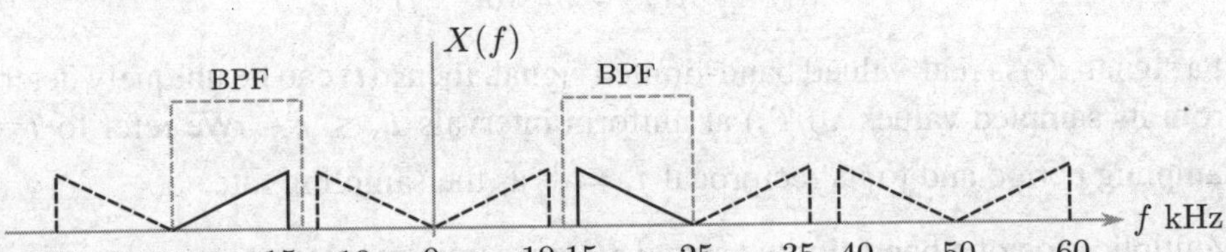

We can recover $x(t)$ from $x_\delta(t)$ by using an ideal BPF with upper cutoff frequency $f_{c2} = 25\,\text{kHz}$ and lower cutoff frequency f_{c1} such that $10 \le f_{c1} \le 15\,\text{kHz}$.

Example 7.12. A periodic square wave $x(t)$ with period $T_0 = 1$ ms, whose value alternates between +1 and −1 for ecah half period, is passed through an ideal low-pass filter with a cutoff frequency of 4 kHz. The filter output is to be sampled using a Dirac Comb. What is the smallest sampling rate we can choose? Consider both the symmetry of the signal as well as Nyquist criterion.

Solution: For analysis, refer the sketch shown in Fig. Ex. 7.12.

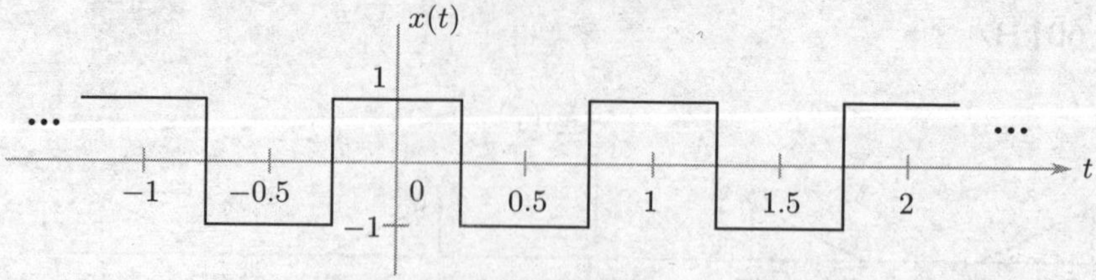

Fig. Ex. 7.12

We find that fundamental period, $T_0 = 1$ ms and hence fundamental frequency, $f_0 = 1/T_0 = 1$ kHz. Hence, the harmonics are 1 kHz apart.

For a filter with cutoff frequency of 4 kHz, upto and including 4th harmonics of $x(t)$ are passed and so, $f_s \geq 8$ kHz.

Since there are two half-cycles per period and each is an inverted replica of the other, the signal $x(t)$ is half-wave symmetric. The implication of this symmetry is that only odd harmonics of $x(t)$ are present. So, the highest frequency at the filter output is 3 kHz (not 4 kHz), so $f_s \geq 6$ kHz.

Summary of Important points

- Digital transmission of analog signals is possible by the virtue of sampling theorem, and that sampling operation is performed in accordance with the sampling theorem.
- A band-limited signal $x(t)$ is a signal for which $X(f)$ is identically zero above a certain frequency f_x; that is,

$$x(t) \overset{\textbf{FT}}{\longleftrightarrow} X(f) = 0 \quad \text{for} \quad |f| > f_x$$

- If a signal $x(t)$ is real-valued band-limited signal, then $x(t)$ can be uniquely determined from its sampled values $x(nT_s)$ at uniform intervals $T_s \leq \frac{1}{2f_x}$. We refer to T_s as the sampling period and to its reciprocal $f_s = \frac{1}{T_s}$ as the sampling rate.
- Multiplication of a band-limited signal $x(t)$ by a unit impulse train results in an infinte sequence of samples $\{x(nT_s)\}$, where n takes on all possible integer values. This ideal form of sampling is called instantaneous sampling.
- The minimum sampling rate, $f_s = 2f_x$ is called Nyquist rate, its reciprocal, $\frac{1}{2f_x}$ is called Nyquist interval.
- If $x(t)$ is sampled by using a train of rectangular pulses, we get samples known as natural samples represented by $s(t)$. Since top of each pulse in $s(t)$ retains the shape of its corresponding analog segment during the pulse interval, they are called natural samples.

Exercise Problems

EP. 7.1 A low-pass signal has a spectrum $X(f)$ given by

$$X(f) = \begin{cases} 1 - |f|/100, & |f| \le 100 \\ 0, & |f| > 100 \end{cases}$$

(a) Assume that $x(t)$ is ideally sampled at $f_s = 150\,\text{kHz}$. Sketch the spectrum of $x_\delta(t)$.

(b) Repeat part (a) with $f_s = 200\,\text{Hz}$ and $f_s = 400\,\text{Hz}$.

EP. 7.2 A band-pass signal $x(t)$ with a spectrum shown in Fig. EP. 7.2 is ideally sampled using a Dirac Comb. Sketch the spectrum of the sampled signal $x_\delta(t)$ when $f_s = 20, 30$ and 40 Hz. Indicate if and how the signal can be recovered.

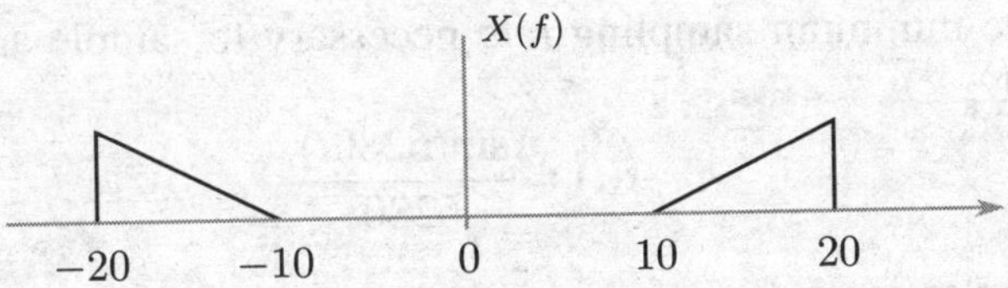

Fig. EP. 7.2

EP. 7.3 A band-pass signal with a spectrum shown in Fig. EP. 7.3 is ideally sampled using a Dirac Comb.

(a) Show that the signal can be reconstructed when $f_s = 2.5\,\text{Hz}$.

(b) Show that the signal can be reconstructed when $f_s = 5\,\text{Hz}$.

(c) Show that aliasing takes place when $f_s = 3.5\,\text{Hz}$.

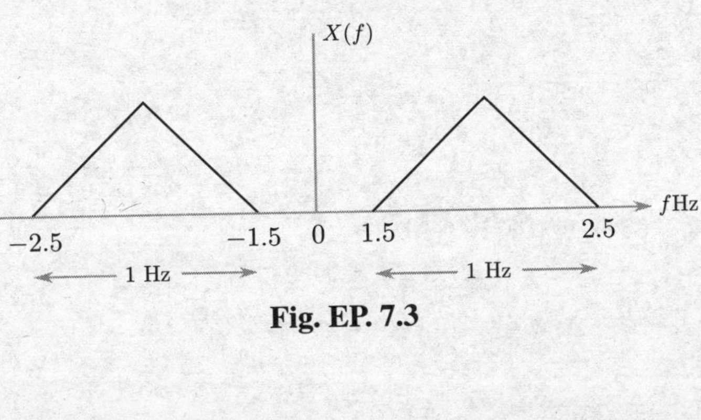

Fig. EP. 7.3

EP. 7.4 A signal $x(t) = \text{sinc}^2(5\pi t)$ is sampled at a rate of (i) 5 Hz; (ii) 10 Hz, (iii) 20 Hz. For each of three cases:

(a) Sketch the spectrum of the sampled signal.

(b) Explain whether you can recover signal $x(t)$ from the sampled signal.

(c) If the sampled signal is passed through an ideal low-pass filter of bandwidth 5 Hz, sketch the spectrum of the sampled signal.

EP. 7.5 A waveform $x(t) = 10\cos\left(100t + \frac{\pi}{3}\right) + 20\cos\left(200t + \frac{\pi}{6}\right)$ is to be ideally sampled for digital transmission.

(a) What is the maximum allowable time interval between sample values that will ensure perfect signal reconstruction ?

(b) If we want to reproduce 1 hour of this waveform, how many sample values need to be stored?

Ans: (a) 0.00157 seconds, (b) 2.29×10^6 samples

EP. 7.6 Determine the minimum sampling rate necessary to sample and perfectly reconstruct the signal

$$x(t) = \frac{\sin(6280t)}{6280t}$$

Ans: 2000 samples/second

EP. 7.7 Use the method of convolution to demonstrate the effect of undersampling the waveform $x(t) = \cos 2\pi f_0 t$ for a sampling rate of $f_s = \frac{3}{2} f_0$.

EP. 7.8 Prove that a signal cannot be simultaneously time-limited and band -limited.

EP. 7.9 Find a signal $x(t)$ that is band-limited to B_x Hz and whose samples are $x(0) = 1$, and $x(\pm T_s) = x(\pm 2T_s) = \pm x(3T_s) = \cdots = 0$, where the sampling interval T_s is the Nyquist interval for $x(t)$,that is, $T_s = \frac{1}{2} B_x$.

Ans: $\text{sinc}\,(2\pi B_x t)$

Chapter 8

The z-TRANSFORM

8.1 Introduction

In the study of discrete-time systems described by a linear difference equation with constant coefficients, the z-transform has a role similar to that of Laplace transform for continuous-time sysstems. The results of applying z-transform to a linear difference equation is that the linear difference equation gets transformed into an algebraic equation, thereby simplifying the analysis of discrete-time systems. In many cases, we do not know the input $x(n)$, but we can still obtain valuable information about the system's behaviour from the ratio of z-transform of output $y(n)$ and z-transform of input $x(n)$.

8.2 Definition of z-Transform

The z-Transform of a discrete-time signal or sequence $x(n)$ is defined by

$$\mathscr{Z}\{x(n)\} = X(z) = \sum_{n=-\infty}^{\infty} x(n)z^{-n} \tag{8.1}$$

The above Eq. (8.1) sometimes may also be represented as $x(n)\overset{\text{ZT}}{\longleftrightarrow}X(z)$. However, in this text we will adopt the notation $\mathscr{Z}\{x(n)\} = X(z)$, where $\mathscr{Z}\{.\}$ defines the operator for z-transforms. In the above equation, z is taken to be complex variable. This expression is sometimes referred to as the two-sided, or bilateral z-transform, since, the summation is over all the integers ($-\infty$ to $+\infty$). The set of z-values for which the summation converges is called the *region of convergence* (ROC) for the transform. In general, this region of convergence will be an annular region of the entire complex z-plane as given by

$$R_1 < |z| < R_2$$

The lower limit, R_1 may be zero and the upper limit, R_2 could possibly be infinity. If

$x(n)$ is a causal sequence, that is, $x(n) = 0$ for $n < 0$, then its z-transform is

$$X_1(z) = \sum_{n=0}^{\infty} x(n)z^{-n}$$

This expression is sometimes referred to as the one-sided or unilateral z-transform. Clearly, the bilateral and unilateral z-transforms are identical only if $x(n) = 0$ for $n < 0$. The unilateral z-transform is discussed later in the chapter. We will omit the word *bilateral* except where it is needed to avoid ambiguity.

8.3 z-Transform and ROC of Finite Duration Sequences

The range of values of the complex variable z for which the z-transform converges is called the *region of convergence*. Let us find the z-transform of the following finite duration sequences and the associated ROC.

8.3.1 Right-sided sequence

A right-sided sequence is one for which $x(n) = 0$ for all $n < n_0$, where n_0 is positive or negative, but finite. If $n_0 \geq 0$, the resulting sequence, $x(n)$ is said to be either a *causal* sequence or a positive-time sequence. For a causal finite sequence, the ROC is the entire z-plane except for $z = 0$.

Example 8.1. Consider the sequence, $x(n) = (\underset{\Uparrow}{1}, 2, 2, 1)$. Find $X(z)$.

Solution: Using Eq. (8.1) the z-transform of $x(n)$ is

$$\begin{aligned} X(z) &= \sum_{n=0}^{3} x(n)z^{-n} \\ &= x[0] + x[1]z^{-1} + x[2]z^{-2} + x[3]z^{-3} \\ &= 1 + 2z^{-1} + 2z^{-2} + z^{-3}; \quad \text{ROC} : |z| > 0 \end{aligned}$$

From the above expression for $X(z)$, we find that $X(z)$ is finite for all values of z, except at $z = 0$. Hence, the ROC is the entire z-plane except $z = 0$. This fact is explained mathematically by writing the ROC as $|z| > 0$.

8.3.2 Left-sided sequence

A left-sided sequence $x(n)$ is one for which $x(n) = 0$ for all $n > n_0$, where n_0 is positive or negative, but finite. If $n_0 \leq 0$, the resulting sequence $x(n)$ is called *anticausal*. For such type of finite sequence, the ROC is the entire z-plane excluding $z = \infty$.

Example 8.2. Consider the sequence, $x(n) = (1,\ 1,\ 2,\ \underset{\Uparrow}{2})$. Find $X(z)$.

Solution: Using Eq. (8.1) the z-transform of $x(n)$ is

$$
\begin{aligned}
X(z) &= \sum_{n=-3}^{0} x(n) z^{-n} \\
&= x(-3)z^3 + x(-2)z^2 + x(-1)z^1 + x(0) \\
&= z^3 + z^2 + 2z + 2; \quad \text{ROC} : |z| < \infty
\end{aligned}
$$

The above expression for $X(z)$ becomes infinity at $z = \infty$. Hence, the ROC is the entire z-plane except $z = \infty$. This is explained mathematically by writing the ROC as $|z| < \infty$.

8.3.3 Double-sided sequence

A signal that has finite duration in both the left and right sides is known as double-sided sequence. In this case, the ROC is the entire z-plane except at $z = 0$ and $z = \infty$.

Example 8.3. Consider the sequence, $x(n) = (2,\ 1,\ \underset{\Uparrow}{1},\ 2)$. Find $X(z)$.

Solution: Using Eq. (8.1) the z-transform of $x(n)$ is

$$
\begin{aligned}
X(z) &= \sum_{n=-2}^{1} x(n) z^{-n} \\
&= x(-2)z^2 + x(-1)z^1 + x(0) + x(1)z^{-1} \\
&= 2z^2 + z^1 + 1 + 2z^{-1}; \quad \text{ROC} : 0 < |z| < \infty
\end{aligned}
$$

The above expression for $X(z)$ becomes infinity at $z = 0$ and $z = \infty$. Hence, the ROC is the entire z-plane except at $z = 0$ and $z = \infty$. This fact is explained mathematically by writing the ROC as $0 < |z| < \infty$.

8.4 z-Transform and ROC of Infinite Duration Sequences

8.4.1 Positive-time exponential sequence

A positive-time exponential sequence is defined by $x(n) = a^n u(n)$.

Then, by Eq. (8.1) the z-transform of $x(n)$ is

$$X(z) = \sum_{n=-\infty}^{\infty} x(n)z^{-n}$$

$$\Rightarrow \qquad X(z) = \sum_{n=0}^{\infty} a^n z^{-n} \quad \text{(because } u(n) = 0 \text{ for } n < 0)$$

$$= \sum_{n=0}^{\infty} (az^{-1})^n$$

We know that,

$$\sum_{n=0}^{\infty} a^n = \frac{1}{1-a}; \quad |a| < 1$$

Hence,

$$X(z) = \frac{1}{1-az^{-1}} = \frac{z}{z-a}; \text{ ROC : } |z| > |a|$$

The above result converges if $|az^{-1}| < 1$, which is equivalent to $|z| > |a|$. This ROC is shown in Fig. 8.1. Values of z for which $X(z) = 0$ are called *zeros of* $X(z)$, while the values of z for which $X(z) = \infty$ are called *poles of* $X(z)$. Poles are usually indicated by an $\times$ as at $z = a$ and zero by a O as indicated at $z = 0$. Poles and zeros of $X(z)$ are shown in Fig. 8.1.

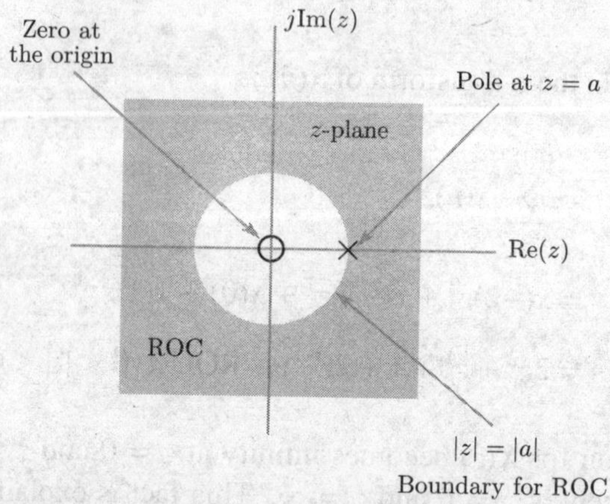

Fig. 8.1: *ROC for the z-transform of* $x(n) = a^n u(n)$.

8.4.2 Negative-time exponential sequence

A negative-time exponential sequence is defined by $x(n) = -b^n u(-n-1)$.

$$X(z) \triangleq \sum_{n=-\infty}^{+\infty} x(n)z^{-n}$$

$$\Rightarrow X(z) = \sum_{n=-\infty}^{\infty} -b^n u(-n-1) z^{-n}$$

We know that,

$$u(-n-1) = \begin{cases} 1, & n \le -1 \\ 0, & n > -1 \end{cases}$$

Therefore,

$$X(z) = -\sum_{n=-\infty}^{-1} b^n z^{-n}$$

$$= -\sum_{n=-\infty}^{-1} \left(\frac{b}{z}\right)^n$$

By substituting $n = -m$ in the above summation, we get

$$X(z) = -\sum_{m=\infty}^{1} \left(\frac{z}{b}\right)^m$$

$$= -\sum_{m=1}^{\infty} \left(\frac{z}{b}\right)^m = -\left[\sum_{m=1}^{\infty} \left(\frac{z}{b}\right)^m + 1 - 1\right]$$

$$= 1 - \sum_{m=0}^{\infty} \left(\frac{z}{b}\right)^m$$

We know that,

$$\sum_{n=0}^{\infty} a^n = \frac{1}{1-a}; \quad |a| < 1$$

Therefore,

$$X(z) = 1 - \frac{1}{1 - \dfrac{z}{b}} \quad \text{with} \quad \left|\frac{z}{b}\right| < 1$$

$$= \frac{z}{z-b}; \quad \text{ROC} : |z| < |b|$$

The ROC and pole-zero plot for this negative-time sequence are shown in Fig. 8.2. At this juncture, it is important to point out that the positive-time exponential sequence has a z-transform with ROC exterior to the circle, $|z| = |a|$, while the negative-time exponential sequence has a z-transform with ROC interior to the circle, $|z| = |b|$.

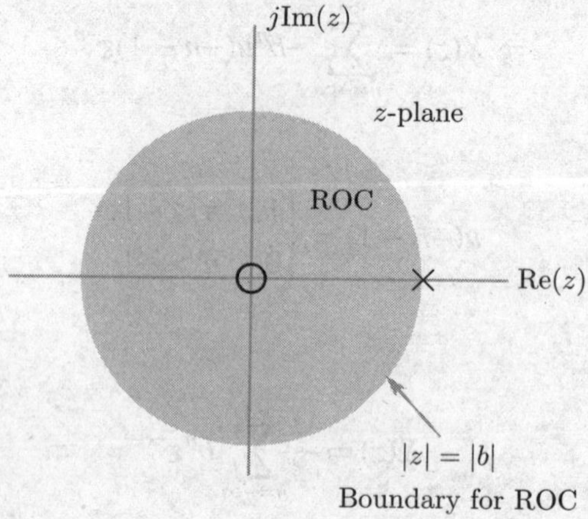

Fig. 8.2: *ROC for the z-transform of* $x(n) = -b^n u(-n-1)$.

8.4.3 Double-sided exponential sequence

A double-sided exponential sequence is the sum of positive and negative-time exponential sequences.

That is,

$$x(n) = a^n u(n) - b^n u(-n-1)$$

$$X(z) \triangleq \sum_{n=-\infty}^{\infty} x(n) z^{-n}$$

$$= \sum_{n=-\infty}^{\infty} [a^n u(n) - b^n u(-n-1)] z^{-n}$$

$$= \sum_{n=0}^{\infty} a^n z^{-n} - \sum_{n=-\infty}^{-1} b^n z^{-n}$$

The first power series converges, if $|az^{-1}| < 1$ or $|z| > |a|$, while the second power series converges, if $|b^{-1}z| < 1$ or $|z| < |b|$. Also, $X(z)$ is the sum of the transforms of positive and negative-time exponential sequences with ROC equal to the intersection of the two regions of convergence.

$$X(z) = \frac{z}{z-a} + \frac{z}{z-b}, \quad \text{ROC}: (|z| > |a|) \cap (|z| < |b|)$$

If $|b| < |a|$, the above intersection is the null set i.e., the transform does not converge, and if $|b| > |a|$, the transform converges in the annular region as shown in Fig. 8.3.

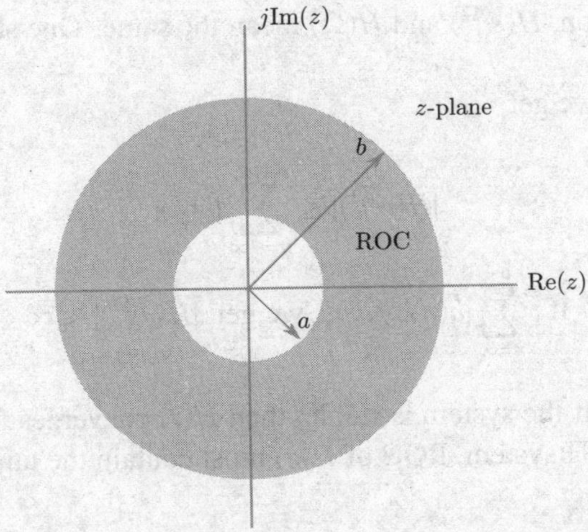

Fig. 8.3: *ROC of a two-sided exponential sequence for |b| > |a|.*

8.5 ROC and Stability

Let $h(n)$ be the impulse response of a causal or a non-causal system and $H(z)$ be the z-transform of $h(n)$. It should be noted that z-transform of $h(n)$ is defined as the system function.

Theorem 8.1. *If a discrete-time LTI system is BIBO stable, then the ROC of the system function $H(z)$ must contain the unit circle, that is, $|z| = 1$.*

Proof: A discrete-time LTI system is BIBO stable, if and only if its impulse response, $h(n)$ is absolutely summable. That is,

$$\sum_{n=-\infty}^{\infty} |h(n)| < \infty$$

We, know that,

$$H(z) = \sum_{n=-\infty}^{\infty} h(n)z^{-n}$$

Let,

$$z = e^{j\Omega} \quad \text{so that } |z| = |e^{j\Omega}| = 1$$

Then,

$$\left|H(e^{j\Omega})\right| = |H(\Omega)| = \left|\sum_{n=-\infty}^{\infty} h(n)e^{-j\Omega n}\right| \leq \sum_{n=-\infty}^{\infty} \left|h(n)e^{-j\Omega n}\right|$$

$$\Rightarrow |H(e^{j\Omega})| \leq \sum_{n=-\infty}^{\infty} |h(n)|\, |e^{-j\Omega n}|$$

In the above equation, $H(e^{j\Omega})$ and $H(\Omega)$ mean the same. One should not think that $e^{j\Omega}$ is replaced by Ω.

Since, $\left|e^{-j\Omega n}\right| = 1$, we get

$$|H(e^{j\Omega})| \leq \sum_{n=-\infty}^{\infty} |h(n)|$$

$$\text{If } \sum_{n=\infty}^{\infty} |h(n)| < \infty, \text{ we get } |H(e^{j\Omega})| < \infty$$

Hence, we see that if the system is stable, then $H(z)$ converges for $z = e^{j\Omega}$. That is, for a stable discrete-time LTI system, ROC of $H(z)$ must contain the unit circle, $|z| = 1$.

Example 8.4. Consider a discrete-time LTI system with impulse response $h(n)$ given by

$$h(n) = \alpha^n u(n)$$

Is this system BIBO stable?

Solution:

$$H(z) = \frac{z}{z - \alpha}; \quad \text{ROC : } |z| > |\alpha|$$

If $|\alpha| < 1$, the ROC of $H(z)$ contains the unit circle, $|z| = 1$ and accordingly the system will be stable. However, if $|\alpha| > 1$, ROC of $H(z)$ does not contain the unit circle, and consequently the system will not be stable.

8.6 Properties of ROC

The properties of ROC are very similar to the corresponding properties for the Laplace transform. The properties refer to a sequence $x(n)$ and its z-transform $X(z)$.

1. The region of convergence (ROC) of $X(z)$ consists of a ring in the z-plane centered about the origin (equivalent to a vertical strip in the s-plane).

2. The ROC does not contian any poles (same as in LT).

3. If $x(n)$ is of finite duration, then the ROC is the entire z-plane except possibly at $z = 0$ and/or $z = \infty$.

4. If $x(n)$ is a right-sided signal, the ROC lies outside a circle through the pole furthest from the origin. That is, all the poles of $X(z)$ are interior to the ROC.

5. If $x(n)$ is a left-sided signal, the ROC lies within a circle through the nearest pole to the origin. That is, all the poles of $X(z)$ are exterior to the ROC.

6. If $x(n)$ is bilateral, that is, the sum of left-sided and right-sided sequences, then the ROC is an annulus between two poles, if the left-sided and right-sided regions of convergence overlap.

7. $X(z)$ is analytic in the entire region of convergence. This means that $X(z)$ can be repeatedly differentiated, and can be interpreted as a Laurent series.

8. Discrete-time Fourier transform (DTFT) of a signal exists if the ROC of its z-transform includes the unit circle.

8.7 Properties of z-Transform

The regions of convergence of positive and negative-time exponential sequences were obtained by using the fundamental definition of z-transform. Infact, these regions of convergence were obtained by naturally imposing the restriction of convergence of an infinte series. In many cases, we will be able to obtain z-transform of sequences, including the ROC, without using the basic definition. The following properties give us the additional advantage in obtaining z-transform of various operations on the sequences whose transforms are known.

8.7.1 Linearity

If

$$\mathscr{Z}(\{x_1(n)\} = X_1(z) \quad \text{with ROC}: \; R_{x_1^-} < |z| < R_{x_1^+}$$

and,

$$\mathscr{Z}\{x_2(n)\} = X_2(z) \quad \text{with ROC}: \; R_{x_2^-} < |z| < R_{x_2^+}$$

then,

$$\mathscr{Z}\{a_1 x_1(n) + a_2 x_2(n)\} = a_1 X_1(z) + a_2 X_2(z)$$

with ROC being the intersection of the individual ROCs.

Proof:

$$\begin{aligned}
\mathscr{Z}\{a_1 x_1(n) + a_2 x_2(n)\} &\triangleq \sum_{n=-\infty}^{\infty} [a_1 x_1(n) + a_2 x_2(n)] z^{-n} \\
&= a_1 \sum_{n=-\infty}^{\infty} x_1(n) z^{-n} + a_2 \sum_{n=-\infty}^{\infty} x_2(n) z^{-n} \\
&= a_1 X_1(z) + a_2 X_2(z)
\end{aligned}$$

$$\text{ROC} : \; (R_{x_1^-} < |z| < R_{x_1^+}) \cap (R_{x_2^-} < |z| < R_{x_2^+})$$

8.7.2 Time shift or translation

If

$$\mathscr{Z}(\{x(n)\} = X(z) \quad \text{with ROC}: R_{x^-} < |z| < R_{x^+}$$

then,

$$\mathscr{Z}\{x(n-n_0)\} = z^{-n_0}X(z) \quad \text{with ROC} \; : \; R_{x^-} < |z| < R_{x^+}$$

Proof:

$$\mathscr{Z}\{x(n-n_0)\} \triangleq \sum_{n=-\infty}^{\infty} x(n-n_0)z^{-n}$$

$$\text{Put} \qquad n - n_0 = m$$

Then,

$$\begin{aligned}\mathscr{Z}\{x(n-n_0)\} &= \sum_{m=-\infty}^{\infty} x(m)z^{-(m+n_0)} \\ &= z^{-n_0}\sum_{m=-\infty}^{\infty} x(m)z^{-m} \\ &= z^{-n_0}X(z); \quad \text{ROC} \; : \; R_{x^-} < |z| < R_{x^+}\end{aligned}$$

8.7.3 Multiplication by an exponential

If

$$\mathscr{Z}\{x(n)\} = X(z) \quad \text{with ROC} \; : \; R_{x^-} < |z| < R_{x^+}$$

then,

$$\mathscr{Z}\{a^n x(n)\} = X(z)\Big|_{z \to \frac{z}{a}} \quad \text{with ROC} \; : \; |a|R_{x^-} < |z| < |a|R_{x^+}$$

Proof:

$$\begin{aligned}\mathscr{Z}\{a^n x(n)\} &\triangleq \sum_{n=-\infty}^{\infty} a^n x(n)z^{-n} \\ &= \sum_{n=-\infty}^{\infty} x(n)\left(\frac{z}{a}\right)^{-n} \\ &= X\left(\frac{z}{a}\right)\end{aligned}$$

$$\text{ROC} \; : \; R_{x^-} < \left|\frac{z}{a}\right| < R_{x^+} = |a|R_{x^-} < |z| < |a|R_{x^+}$$

8.7.4 Multiplication by a Ramp

If

$$\mathscr{Z}\{x(n)\} = X(z) \quad \text{with ROC} \; : R_{x^-} < |z| < R_{x^+}$$

then,

$$\mathscr{Z}\{nx(n)\} = -z\frac{dX(z)}{dz} \quad \text{with ROC} \; : \; R_{x^-} < |z| < R_{x^+}$$

Proof: We know that,

$$X(z) \triangleq \sum_{n=-\infty}^{\infty} x(n)z^{-n}$$

Differentiating both the sides with respect to z, we get

$$\frac{dX(z)}{dz} = -\sum_{n=-\infty}^{\infty} nx(n)z^{-n-1}$$

Multiplying both the sides by $-z$, we get

$$-z\frac{dX(z)}{dz} = \sum_{n=-\infty}^{\infty} [nx(n)]z^{-n}$$

Hence,

$$\mathscr{Z}\{nx(n)\} = -z\frac{dX(z)}{dz}$$

$$\text{ROC} \; : \; R_{x^-} < |z| < R_{x^+}$$

8.7.5 Convolution (Time-domain)

The response of a system, when the input is an impulse function, is defined as the impulse response or unit sample response of the system. The impulse response completely characterizes the behaviour of any LTI system.

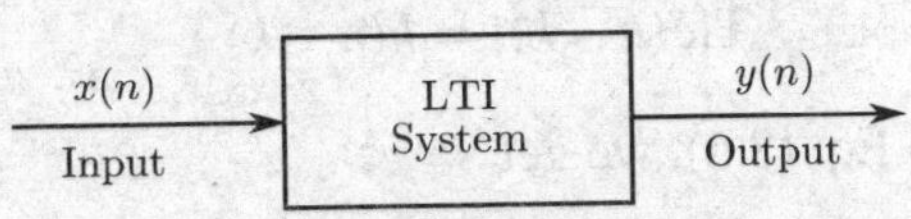

Fig. 8.4: *LTI System.*

Let us consider an LTI system represented by $\mathbf{T}\{\cdot\}$. The output of such a system is given by $y(n) = \mathbf{T}\{x(n)\}$. Let the input be an arbitrary sequence as shown in Fig. 8.5, that is, the input is made up delayed impulses having different weights and such an input may be represented mathematically as follows:

$$x(n) = \cdots + x(-2)\delta(n+2) + x(-1)\delta(n+1) + x(0)\delta(n) + x(1)\delta(n-1) + x(2)\delta(n-2) + \cdots \tag{8.3}$$

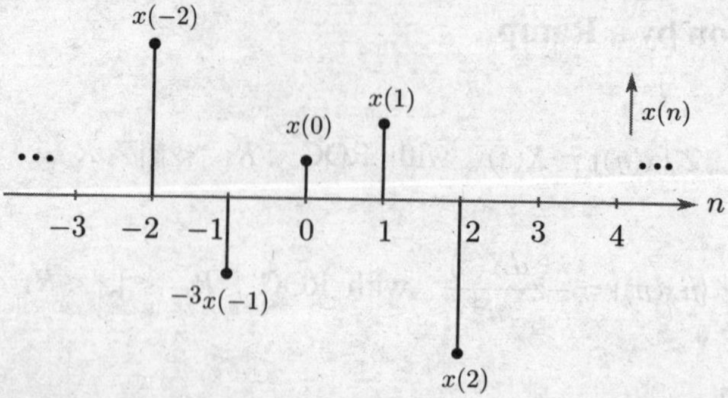

Fig. 8.5: *Graphical illustration of an arbitrary input sequence x(n).*

Representing the above Eq. (8.3) in a compact form, we get

$$x(n) = \sum_{k=-\infty}^{\infty} x(k)\delta(n-k) \tag{8.4}$$

The response of the system to an arbitrary input $x(n)$ is given by

$$y(n) = \mathbf{T}\{x(n)\} \tag{8.5}$$

Substituting Eq. (8.4) in Eq. (8.5), we get

$$\begin{aligned} y(n) &= \mathbf{T}\left\{\sum_{k=-\infty}^{\infty} x(k)\delta(n-k)\right\} \\ &= \sum_{k=-\infty}^{\infty} x(k)\mathbf{T}\{\delta(n-k)\} \end{aligned} \tag{8.6}$$

Also, $\mathbf{T}\{\delta(n)\} = h(n)$, where $h(n)$ is defined as the impulse response of the system. The system under investigation is time-invariant and hence, we can write,

$$\mathbf{T}\{\delta(n-k)\} = h(n-k) \tag{8.7}$$

Substituting Eq. (8.7) in Eq. (8.6), we get

$$y(n) = \sum_{k=-\infty}^{\infty} x(k)h(n-k) \tag{8.8}$$

Eq. (8.8) is defined as the convolution sum and is represented symbolically as

$$y(n) = x(n) * h(n) \tag{8.9}$$

The symbol '∗' stands for convolution operation. The above Eq. (8.9) means that the response of an LTI system is the convolution of the input, $x(n)$, with the impulse response, $h(n)$ of the system.

We will now prove that convolution in discrete time-domain corresponds to multiplication in z-domain.

Given,

$$x(n) * h(n) = \sum_{k=-\infty}^{\infty} x(k)h(n-k)$$

and

$$\mathscr{Z}\{x(n)\} = X(z) \quad \text{with ROC}: \; R_{x^-} < |z| < R_{x^+}$$

$$\mathscr{Z}\{h(n)\} = H(z) \quad \text{with ROC}: \; R_{h^-} < |z| < R_{h^+}$$

then,

$$\mathscr{Z}\{x(n) * h(n)\} = X(z)H(z) \quad \text{with ROC}: \; (R_{x^-} < |z| < R_{x^+}) \cap (R_{h^-} < |z| < R_{h^+})$$

Proof:

$$\mathscr{Z}\{x(n) * h(n)\} \triangleq [x(n) * h(n)]z^{-n}$$

$$= \sum_{n=-\infty}^{\infty} \sum_{k=-\infty}^{\infty} x(k)h(n-k)z^{-n}$$

Substitute $n - k = m$ and interchange the order of summations.

Then,

$$\mathscr{Z}\{x(n) * h(n)\} = \sum_{k=-\infty}^{\infty} x(k)z^{-k} \sum_{m=-\infty}^{\infty} h(m)z^{-m}$$

$$= X(z)H(z)$$

$$\text{ROC}: \; (R_{x^-} < |z| < R_{x^+}) \cap (R_{h^-} < |z| < R_{h^+})$$

8.7.6 Time-reversal

If

$$\mathscr{Z}\{x(n)\} = X(z) \qquad \text{with ROC}: \; z \in R_x$$

then,

$$\mathscr{Z}\{x(-n)\} = X\left(\frac{1}{z}\right) \quad \text{with ROC}: \; z \in \frac{1}{R_x}$$

Proof:

$$\mathscr{Z}\{x(-n)\} \triangleq \sum_{n=-\infty}^{\infty} x(-n)z^{-n}$$

Substituting $-n = l$ on the RHS, we get

$$\mathscr{Z}\{x(-n)\} = \sum_{l=+\infty}^{-\infty} x(l)z^{l}$$

$$= \sum_{l=-\infty}^{\infty} x(l)(z^{-1})^{-l}$$

$$= X\left(\frac{1}{z}\right); \quad \text{ROC} : z^{-1} \in R_x \quad \text{or} \quad z \in \frac{1}{R_x}$$

8.7.7 Convolution (z-domain)

If

$$\mathscr{Z}\{x(n)\} = X(z) \qquad \text{with ROC} : R_{x^-} < |z| < R_{x^+}$$

and

$$\mathscr{Z}\{y(n)\} = Y(z) \qquad \text{with ROC} : R_{y^-} < |z| < R_{y^+}$$

then,

$$\mathscr{Z}\{x(n)y(n)\} = \frac{1}{2\pi j}\oint_C X(v)Y\left(\frac{z}{v}\right)v^{-1}\,dv$$

where $\oint_C$ is a complex contour integral and C is a closed contour in the intersection of the ROCs of $X(v)$ and $Y\left(\frac{z}{v}\right)$. The region of convergence for the resulting transform is $R_{x^-}R_{y^-} < |z| < R_{x^+}R_{y^+}$.

Proof:

$$\mathscr{Z}\{x(n)y(n)\} \triangleq \sum_{n=-\infty}^{\infty} x(n)y(n)z^{-n} \tag{8.10}$$

Using complex inversion integral (discussed later), we have

$$x(n) = \frac{1}{2\pi j}\oint_C X(z)\frac{z^n}{z}dz \tag{8.11}$$

Substituting Eq. (8.11) with z replaced by v in Eq. (8.10), we get

$$\mathscr{Z}\{x(n)y(n)\} = \sum_{n=-\infty}^{\infty} \frac{1}{2\pi j}\oint_C X(v)\frac{v^n}{v}\,dv\; y(n)z^{-n}$$

$$= \frac{1}{2\pi j} \oint_C \frac{X(v)}{v}\, dv \sum_{n=-\infty}^{\infty} y(n) v^n z^{-n}$$

$$= \frac{1}{2\pi j} \oint_C \frac{X(v)}{v}\, dv \sum_{n=-\infty}^{\infty} y(n) \left(\frac{z}{v}\right)^{-n}$$

$$= \frac{1}{2\pi j} \oint_C X(v) Y\left(\frac{z}{v}\right) v^{-1}\, dv \tag{8.12}$$

If C is the unit circle, we can write Eq. (8.12) in a different form so that it looks more like a convolution. To make it happen, let $v = e^{j\theta}$ and $z = e^{j\Omega}$. Then, we have

$$dv = je^{j\theta}\, d\theta$$

Hence,

$$\mathscr{Z}\{x(n)y(n)\} = \frac{1}{2\pi j} \oint_C X\left(e^{j\theta}\right) Y\left(e^{j(\Omega-\theta)}\right) e^{-j\theta} je^{j\theta}\, d\theta$$

$$= \frac{1}{2\pi} \int_{-\pi}^{\pi} X\left(e^{j\theta}\right) Y\left(e^{j(\Omega-\theta)}\right) d\theta$$

Denoting $X\left(e^{j\theta}\right)$ by $X(\theta)$ and $Y\left(e^{j(\Omega-\theta)}\right)$ by $Y(\Omega-\theta)$, we get

$$\mathscr{Z}\{x(n)y(n)\} = \frac{1}{2\pi} \int_{-\pi}^{\pi} X(\theta)\hat{Y}(\Omega-\theta)\, d\theta$$

8.7.8 Initial value theorem

If $x(n)$ is a causal sequence with z-transform $X(z)$, then

$$x(0) = \lim_{z\to\infty} X(z)$$

Proof:

$$X(z) \triangleq \sum_{n=-\infty}^{\infty} x(n) z^{-n}$$

Since, $x(n)$ is a causal sequence, the lower limit in the above summation is taken as zero. Hence,

$$X(z) = \sum_{n=0}^{\infty} x(n) z^{-n} = x(0) + \sum_{n=1}^{\infty} x(n) z^{-n} \tag{8.13}$$

Taking the limit, $z \to \infty$ on both the sides of Eq. (8.13), we get

$$\lim_{z\to\infty} X(z) = x(0) + 0$$
$$= x(0)$$

Hence,

$$x(0) = \lim_{z\to\infty} X(z)$$

8.8 Transforms of some Useful Sequences

Sequence 1: Unit impulse function

(i) Let

$$x(n) = \delta(n)$$

Then,

$$X(z) \triangleq \sum_{n=-\infty}^{\infty} x(n)z^{-n}$$
$$= \sum_{n=-\infty}^{\infty} \delta(n)z^{-n}$$
$$= z^{-n}\Big|_{n=0}$$
$$= 1, \quad \text{ROC : All} \quad z$$

(ii) Let

$$x(n) = \delta(n - n_0), \quad n_0 > 0$$

Then,

$$X(z) = \sum_{n=-\infty}^{\infty} \delta(n - n_0)z^{-n}$$
$$= z^{-n}\Big|_{n=n_0} = z^{-n_0}; \quad \text{ROC : } |z| > 0$$

(iii) Let

$$x(n) = \delta(n + n_0), \quad n_0 > 0$$
$$X(z) = \sum_{n=-\infty}^{\infty} \delta(n + n_0)z^{-n}$$
$$= z^{-n}\Big|_{n=-n_0}$$
$$= z^{n_0}; \quad \text{ROC : } |z| < \infty$$

Sequence 2: Positive-time exponential sequence

$$x(n) = a^n u(n)$$

As derived in article 8.4.1, we have,

$$X(z) = \frac{z}{z-a}; \qquad \text{ROC} : |z| > |a|$$

Sequence 3: Negative-time exponential sequence

$$x(n) = -b^n u(-n-1)$$

As derived in article 8.4.2, we have,

$$X(z) = \frac{z}{z-b}; \quad \text{ROC} : |z| < |b|$$

Sequence 4: Unit step sequence

Let

$$x(n) = u(n)$$

We know that,

$$\mathscr{Z}\{a^n u(n)\} = \frac{z}{z-a}; \quad \text{ROC} : |z| > |a|$$

Assuming $a = 1$, we get

$$X(z) = \mathscr{Z}\{u(n)\} = \frac{z}{z-1}; \quad \text{ROC} : |z| > 1$$

Sequence 5: Negative-time unit step sequence

Let

$$x(n) = -u(-n-1)$$

We know that,

$$\mathscr{Z}\{-b^n u(-n-1)\} = \frac{z}{z-b}; \quad \text{ROC} : |z| < |b|$$

Now, assuming $b = 1$, we get

$$X(z) = \mathscr{Z}\{-u(-n-1)\} = \frac{z}{z-1}; \quad \text{ROC} : |z| < 1$$

Sequence 6: Sinusoidal sequence

(i) Let

$$x(n) = \sin \Omega_0 n \, u(n)$$

Then,

$$\begin{aligned}
X(z) &\triangleq \sum_{n=-\infty}^{\infty} \sin \Omega_0 n \, u(n) z^{-n} \\
&= \sum_{n=0}^{\infty} \sin \Omega_0 n \, z^{-n} \\
&= \sum_{n=0}^{\infty} \frac{1}{2j} \left[e^{j\Omega_0 n} - e^{-j\Omega_0 n} \right] z^{-n} \\
&= \frac{1}{2j} \left[\sum_{n=0}^{\infty} \left(e^{j\Omega_0} z^{-1} \right)^n - \sum_{n=0}^{\infty} \left(e^{-j\Omega_0} z^{-1} \right)^n \right] \\
&= \frac{1}{2j} \left[\underbrace{\frac{1}{1 - e^{j\Omega_0} z^{-1}}}_{\substack{\text{ROC}: |e^{j\Omega_0} z^{-1}| < 1 \\ \text{or } |z| > 1}} - \underbrace{\frac{1}{1 - e^{-j\Omega_0} z^{-1}}}_{\substack{\text{ROC}: |e^{-j\Omega_0} z^{-1}| < 1 \\ \text{or } |z| > 1}} \right] \\
&= \frac{1}{2j} \left[\frac{z}{z - e^{j\Omega_0}} - \frac{z}{z - e^{-j\Omega_0}} \right] \\
&= \frac{1}{2j} \left[\frac{z^2 - z e^{-j\Omega_0} - z^2 + z e^{j\Omega_0}}{z^2 - z e^{-j\Omega_0} - z e^{j\Omega_0} + 1} \right] \\
&= \frac{z \sin \Omega_0}{z^2 - 2z \cos \Omega_0 + 1}; \quad \text{ROC} : |z| > 1
\end{aligned}$$

(ii) Let

$$x(n) = \cos \Omega_0 n \, u(n)$$

Then,

$$\begin{aligned}
X(z) &\triangleq \sum_{n=-\infty}^{\infty} x(n) z^{-n} \\
&= \sum_{n=0}^{\infty} \cos \Omega_0 n \, z^{-n}
\end{aligned}$$

$$= \sum_{n=0}^{\infty} \frac{1}{2}\left[e^{j\Omega_0 n} + e^{-j\Omega_0 n}\right] z^{-n}$$

$$= \frac{1}{2}\left[\sum_{n=0}^{\infty}\left(e^{j\Omega_0} z^{-1}\right)^n + \sum_{n=0}^{\infty}\left(e^{-j\Omega_0} z^{-1}\right)^n\right]$$

$$= \frac{1}{2}\left[\underbrace{\frac{1}{1 - e^{j\Omega_0} z^{-1}}}_{\substack{\text{ROC}: |e^{j\Omega_0} z^{-1}|<1 \\ \text{or } |z|>1}} - \underbrace{\frac{1}{1 - e^{-j\Omega_0} z^{-1}}}_{\substack{\text{ROC}: |e^{-j\Omega_0} z^{-1}|<1 \\ \text{or } |z|>1}}\right]$$

$$= \frac{1}{2}\left[\frac{z}{z - e^{j\Omega_0}} + \frac{z}{z - e^{-j\Omega_0}}\right]$$

$$= \frac{1}{2}\left[\frac{z^2 - ze^{-j\Omega_0} + z^2 - ze^{j\Omega_0}}{z^2 - ze^{-j\Omega_0} - ze^{j\Omega_0} + 1}\right]$$

$$= \frac{z^2 - z\cos\Omega_0}{z^2 - 2z\cos\Omega_0 + 1}; \quad \text{ROC} : |z| > 1$$

Sequence 7: Unit ramp sequence

Let

$$x(n) = n\, u(n)$$

Then,

$$X(z) \triangleq \sum_{n=-\infty}^{\infty} n\, u(n)\, z^{-n}$$

$$= \sum_{n=0}^{\infty} n\, z^{-n}$$

$$= \sum_{n=0}^{\infty} n\left(z^{-1}\right)^n$$

we know that,

$$\sum_{n=0}^{\infty} n\, a^n = \frac{a}{(1-a)^2}; \ |a| < 1$$

Hence,

$$X(z) = \frac{z^{-1}}{(1 - z^{-1})^2}; \quad \text{ROC} : |z^{-1}| < 1 \quad \text{or } |z| > 1$$

$$= \frac{z}{(z - 1)^2}$$

Summarizing the results of various sequences obtained in article 8.7, Table 8.1 is generated.

8.1 The z-Transform of a few useful sequences

Time-domain sequence	**Sequence description**	**z-Transform**	**ROC**
Unit sample	$\delta(n)$	1	All z
or	$\delta(n - n_0),\ n_0 > 0$	z^{-n_0}	$\|z\| > 0$
Unit impulse	$\delta(n + n_0), n_0 > 0$	z^{n_0}	$\|z\| < \infty$
Unit step	$u(n)$	$\frac{z}{(z-1)}$	$\|z\| > 1$
	$-u(-n - 1)$	$\frac{z}{(z-1)}$	$\|z\| < 1$
Exponential	$a^n\ u(n)$	$\frac{z}{(z-a)}$	$\|z\| > \|a\|$
	$-b^n u(-n - 1)$	$\frac{z}{(z-b)}$	$\|z\| < \|b\|$
Sinusoidal	$\sin \Omega_0 n\ u(n)$	$\frac{z \sin \Omega_0}{z^2 - 2z \cos \Omega_0 + 1}$	$\|z\| > 1$
	$\cos \Omega_0 n\ u(n)$	$\frac{z^2 - z \cos \Omega_0}{z^2 - 2z \cos \Omega_0 + 1}$	$\|z\| > 1$
Unit ramp	$n\, u(n)$	$\frac{z}{(z-1)^2}$	$\|z\| > 1$

Example 8.5. Find the z-transform of

$$x(n) = \alpha^{-n} u(-n - 1)$$

Solution:

$$X(z) \triangleq \sum_{n=-\infty}^{\infty} x(n) z^{-n}$$

$$= \sum_{n=-\infty}^{\infty} \alpha^{-n} u(-n - 1) z^{-n}$$

$$= \sum_{n=-\infty}^{-1} (\alpha z)^{-n}$$

Substituting, $n = -m$ in the above summation, we get

$$X(z) = \sum_{m=\infty}^{1} (\alpha z)^m$$
$$= \sum_{m=1}^{\infty} (\alpha z)^m$$
$$= \sum_{m=0}^{\infty} (\alpha z)^m - 1$$

Recall: $\sum_{n=0}^{\infty} a^n = \frac{1}{1-a}$; $|a| < 1$

Hence,

$$X(z) = \frac{1}{1 - \alpha z} - 1;\ |\alpha z| < |\ \text{ or }\ |z| < \frac{1}{|\alpha|}$$

Thus,

$$X(z) = \frac{1}{1 - \alpha z} - 1 = \frac{-z}{z - \frac{1}{\alpha}};\ \text{ROC} : |z| < \frac{1}{|\alpha|}$$

Example 8.6. A finite sequence $x(n)$ is defined as

$$x(n) = (5,\ 2,\ \underset{\Uparrow}{-2},\ 1,\ 1,\ -3)$$

Find $X(z)$ and its ROC.

Solution:

$$X(z) \triangleq \sum_{n=-\infty}^{\infty} x(n) z^{-n}$$
$$= \sum_{n=-2}^{3} x(n) z^{-n}$$
$$= 5z^2 + 2z^1 - 2 + z^{-1} + z^{-2} - 3z^{-3}$$

For z not equal to zero or infinity, each term in $X(z)$ will be finite and consequently $X(z)$ will converge. Hence, the ROC is the entire z-plane except $z = 0$ and $z = \infty$. In other words, the ROC is given by $0 < |z| < \infty$.

Example 8.7. Consider the sequence

$$x(n) = \begin{cases} a^n, & 0 \le n \le N - 1, \quad a > 0 \\ 0, & \text{otherwise} \end{cases}$$

Find $X(z)$ and plot the poles and zeros of $X(z)$.

Solution:

$$X(z) = \sum_{n=0}^{N-1} a^n z^{-n} = \sum_{n=0}^{N-1} (a\,z^{-1})^n$$

We know that,

$$\sum_{n=0}^{N-1} b^n = \frac{1-b^N}{1-b}, \quad b \neq 1$$

Hence,

$$\begin{aligned} X(z) &= \frac{1-(az^{-1})^N}{1-az^{-1}} \\ &= \frac{1}{z^{N-1}} \frac{z^N - a^N}{z-a} \end{aligned} \tag{8.14}$$

From Eq. (8.14), we see that there are $(N-1)$ poles at $z = 0$ and a pole at $z = a$. Since, $x(n)$ is a finite sequence and is zero for $n < 0$, the ROC is $|z| > 0$. The N zeros of $X(z)$ are at

$$z_k = ae^{j\left(\frac{2\pi k}{N}\right)}, \quad k = 0, 1, \ldots, (N-1).$$

The zero for $k = 0$ cancels the pole at $z = a$. The remaining zeros are at

$$z_k = ae^{j\left(\frac{2\pi k}{N}\right)}, \quad k = 1, 2, \ldots, (N-1).$$

The pole-zero plot is shown in Fig. Ex. 8.7 for $N = 8$.

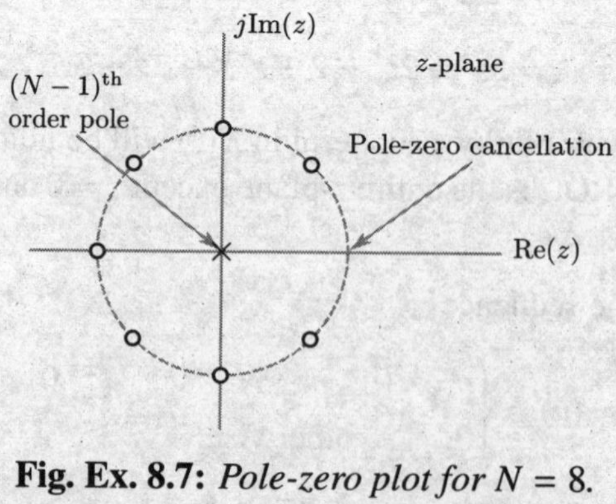

Fig. Ex. 8.7: *Pole-zero plot for* $N = 8$.

Example 8.8. Find the z-transform of

$$x(n) = a^n \cos(\Omega_0 n)\, u(n)$$

Solution: We know that,

$$\mathscr{Z}\{\cos(\Omega_0(n)\, u(n)\} = \frac{z^2 - z\cos\Omega_0}{z^2 - 2z\cos\Omega_0 + 1}; \quad \text{ROC} : |z| > 1$$

Recall: multiplication by exponential property:

$$\mathscr{Z}\{a^n g(n)\} = G\left(\frac{z}{a}\right)$$

Hence,

$$\mathscr{Z}\{a^n \cos\Omega_0 n\, u(n)\} = \frac{\left(\frac{z}{a}\right)^2 - \left(\frac{z}{a}\right)\cos\Omega_0}{\left(\frac{z}{a}\right)^2 - 2\left(\frac{z}{a}\right)\cos\Omega_0 + 1}$$

$$= \frac{1 - az^{-1}\cos\Omega_0}{1 - 2az^{-1}\cos\Omega_0 + a^2z^{-2}}$$

$$\text{ROC} : \left|\frac{z}{a}\right| > 1 \text{ or } |z| > |a|$$

Example 8.9. Find the z-transform of the following discrete-time signals, remembering to include the ROC for each one.

(a) $x(n) = 2^n u(n) + 3\left(\frac{1}{2}\right)^n u(n)$,

(b) $x(n) = 3\left(-\frac{1}{2}\right)^n u(n) - 2[3^n u(-n-1)]$,

(c) $x(n) = 3^{n+1}u(n) - 2\left(\frac{1}{2}\right)^n u(-n-1)$,

(d) $x(n) = 3e^{-2n}u(n) + 2[4^n u(-n-1)] + 5\delta(n)$,

(e) $x(n) = 2\delta(n-3) - 2\delta(n+3)$.

Solution:

(a) Given,

$$x(n) = 2^n u(n) + 3\left(\frac{1}{2}\right)^n u(n)$$

Then, $$X(z) = \mathscr{Z}\{2^n u(n)\} + 3\mathscr{Z}\left\{\left(\frac{1}{2}\right)^n u(n)\right\}$$

$$= \frac{z}{z-2} + 3\frac{z}{z-\frac{1}{2}} \tag{8.15}$$

The first term on the RHS has an ROC : $|z| > 2$ and the second term on the RHS has an ROC : $|z| > \frac{1}{2}$. Hence, the ROC of $X(z)$ is the intersection of the two ROCs for the individual transforms on the RHS of Eq. (8.15).

Therefore,

$$X(z) = \frac{z\left(z - \frac{1}{2}\right) + 3z(z-2)}{(z-2)\left(z - \frac{1}{2}\right)}; \quad \text{ROC} : (|z| > 2) \cap \left(|z| > \frac{1}{2}\right) = |z| > 2$$

(b) Given,

$$x(n) = 3\left(-\frac{1}{2}\right)^n u(n) - 2[3^n u(-n-1)]$$

Then,

$$X(z) = 3\mathscr{Z}\left\{\left(-\frac{1}{2}\right)^n u(n)\right\} + 2\mathscr{Z}\{-3^n u(-n-1)\}$$

$$= 3\frac{z}{z + \frac{1}{2}} + 2\frac{z}{z-3} \tag{8.16}$$

The ROC of the first term on the RHS of Eq. (8.16) is $|z| > \frac{1}{2}$ and that of the second term on the RHS of Eq. (8.16) is $|z| < 3$.

Therefore, the ROC of $X(z)$ is the intersection of the ROCs $|z| > \frac{1}{2}$ and $|z| < 3$.

Thus,

$$X(z) = \frac{3z(z-3) + 2z\left(z + \frac{1}{2}\right)}{\left(z + \frac{1}{2}\right)(z-3)}$$

$$\text{ROC} : \left(|z| > \frac{1}{2}\right) \cap (|z| < 3) = \frac{1}{2} < |z| < 3$$

(c) Given,

$$x(n) = 3[3^n u(n)] - 2\left(\frac{1}{2}\right)^n u(-n-1)$$

Then,

$$X(z) = 3\mathscr{Z}\{3^n u(n)\} + 2\mathscr{Z}\left\{-\left(\frac{1}{2}\right)^n u(-n-1)\right\} \tag{8.17}$$

The ROC of the first term on the RHS of Eq. (8.17) is $|z| > 3$ and that of the second term on the RHS of Eq. (8.17) is $|z| < \frac{1}{2}$. Hence, the ROC of $X(z)$ is the intersection of the two ROCs, that is,

$$\text{ROC} : (|z| > 3) \cap \left(|z| < \frac{1}{2}\right) = \emptyset = \text{nullset}.$$

Hence, $X(z)$ does not exist for the given sequence $x(n)$.

(d) Given,

$$x(n) = 3e^{-2n}u(n) + 2[4^n u(-n-1)] + 5\delta(n)$$

Then,

$$X(z) = 3\mathscr{Z}\{(e^{-2})^n u(n)\} - 2\mathscr{Z}\{-4^n u(-n-1)\} + 5\delta(n)$$

$$= \underbrace{3\left(\frac{z}{z-e^{-2}}\right)}_{\text{ROC: } |z|>e^{-2}} -2 \underbrace{\left(\frac{z}{z-4}\right)}_{\text{ROC: } |z|<4} + \underbrace{5}_{\text{ROC: All } z}$$

Hence,

$$X(z) = \frac{3z(z-4) - 2z(z-e^{-2}) + 5(z-e^{-2})(z-4)}{(z-e^{-2})(z-4)};$$

$$\text{ROC}: \left(|z| > e^{-2}\right) \cap (|z| < 4) \cap (\text{All } z) = e^{-2} < |z| < 4$$

(e) Given,

$$x(n) = 2\delta(n-3) - 2\delta(n+3)$$

Then,

$$X(z) = \underbrace{2\mathscr{Z}\{\delta(n-3)\}}_{\text{ROC: } |z|>0} - \underbrace{2\mathscr{Z}\{\delta(n+3)\}}_{\text{ROC: } |z|<\infty}$$

Hence,

$$X(z) = 2z^{-3} - 2z^3;$$

$$\text{ROC}: (|z| > 0) \cap (|z| < \infty) = 0 < |z| < \infty$$

Example 8.10. Find the z-transform of the following discrete-time signals and remember to include the ROC for each one. Use the definition of z-transform and its properties.

(a) $x(n) = u(n+2)$, (b) $x(n) = u(-n)$, (c) $x(n) = nu(n)$.

Solution:

(a) Given, $x(n) = u(n+2)$. Using time-shift property, we get,

$$X(z) = \mathscr{Z}\{u(n+2)\}$$

$$= z^2 \mathscr{Z}\{u(n)\}$$

$$= z^2 \underbrace{\left[\frac{z}{z-1}\right]}_{\text{ROC: } |z|>1}$$

Therefore,

$$X(z) = \frac{z^3}{z-1}; \quad \text{ROC}: \ |z| > 1$$

(b) Given,

$$x(n) = u(-n)$$

We know that,

$$\mathscr{Z}\{u(n)\} = \frac{z}{z-1}; \ \text{ROC} \ : \ |z| > 1$$

Now, using time-reversal property, we get

$$X(z) = \frac{z^{-1}}{z^{-1}-1}$$

$$= \frac{1}{1-z}; \quad \text{ROC} \ : \ |z^{-1}| > 1 \text{ or } |z| < 1$$

(c) Given,

$$x(n) = nu(n)$$

We know that,

$$\mathscr{Z}\{u(n)\} = \underbrace{\left[\frac{z}{z-1}\right]}_{\text{ROC}:\ |z|>1}$$

Also,

$$n \equiv -z\frac{d}{dz}$$

Hence,

$$X(z) = -z\frac{d}{dz}[\mathscr{Z}\{u(n)\}]$$

$$= -z\frac{d}{dz}\left[\frac{z}{z-1}\right]$$

$$= \frac{z}{(z-1)^2}; \ \text{ROC}: \ |z| > 1$$

Example 8.11. Find the z-transform of each of the following sequences:

(a) $x(n) = na^n u(n)$ (b) $x(n) = na^{n-1}u(n)$

Solution:

(a) Given,

$$x(n) = na^n u(n)$$

We know that,

$$\mathscr{Z}\{a^n u(n)\} = \frac{z}{z-a}; \quad \text{ROC}: |z| > |a|$$

Also,

$$n \equiv -z\frac{d}{dz}$$

Hence,

$$X(z) = -z\frac{d}{dz}\left[\frac{z}{z-a}\right]$$
$$= \frac{az}{(z-a)^2}; \quad \text{ROC}: |z| > |a|$$

(b) Given,

$$x(n) = na^{n-1}u(n)$$

Then,

$$\mathscr{Z}\{a^{n-1}u(n)\} = a^{-1}\mathscr{Z}\{a^n u(n)\}$$
$$= a^{-1}\frac{z}{z-a}; \quad \text{ROC}: |z| > |a|$$

Also,

$$n \equiv -z\frac{d}{dz}$$

Therefore,

$$X(z) = -z\frac{d}{dz}\left[a^{-1}\left(\frac{z}{z-a}\right)\right]$$
$$= \frac{-z}{a}\left[\frac{(z-a)\times 1 - z\times 1}{(z-a)^2}\right]$$
$$= \frac{-z}{a}\left[\frac{-a}{(z-a)^2}\right]$$
$$= \frac{z}{(z-a)^2}; \quad \text{ROC}: |z| > |a|$$

Example 8.12. Let $x(n) = \left(\frac{1}{2}\right)^{|n|}$

(a) Sketch $x(n)$.

(b) Find $X(z)$.

Solution:

(a) The sequence $x(n)$ is sketched in Fig. Ex. 8.12.

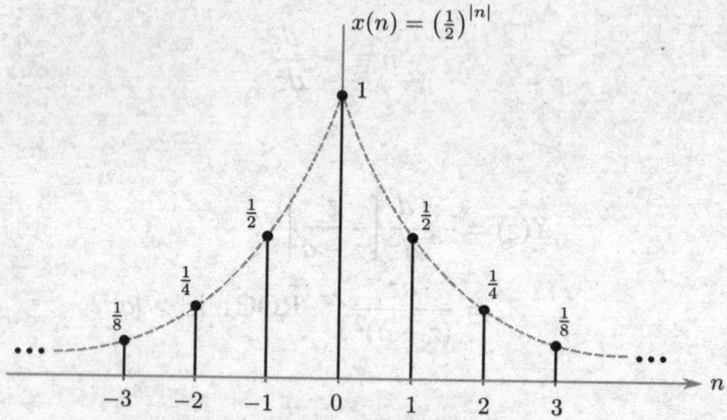

Fig. Ex. 8.12

(b) Since, $x(n)$ is a two-sided sequence, we can express it as

$$x(n) = \left(\frac{1}{2}\right)^n u(n) + \left(\frac{1}{2}\right)^{-n} u(-n-1)$$

We know that,

$$\mathscr{Z}\left\{\left(\frac{1}{2}\right)^n u(n)\right\} = \frac{z}{z-\frac{1}{2}}; \quad \text{ROC} : |z| > \frac{1}{2}$$

and $\quad \mathscr{Z}\{-2^n u(-n-1)\} = \dfrac{z}{z-2}; \quad \text{ROC}: |z| < 2$

Therefore,

$$X(z) = \frac{z}{z-\frac{1}{2}} - \frac{z}{z-2}$$

$$= \frac{z(z-2) - z\left(z-\frac{1}{2}\right)}{\left(z-\frac{1}{2}\right)(z-2)}$$

$$= \frac{-3z}{2\left(z-\frac{1}{2}\right)(z-2)}; \quad \text{ROC} : \left(|z| > \frac{1}{2}\right) \cap (|z| < 2) = \frac{1}{2} < |z| < 2$$

Example 8.13. Find the z-transform of $x(n) = n\alpha^n u(-n)$.

Solution: We know that,

$$\mathscr{Z}\{\alpha^n u(n)\} = \frac{z}{z-\alpha}; \quad \text{ROC}: |z| > |\alpha|$$

Therefore,

$$\mathscr{Z}\{\alpha^{-n} u(n)\} = \frac{z}{z-\frac{1}{\alpha}}; \quad \text{ROC}: |z| > \frac{1}{|\alpha|}$$

Now, using time-reversal property, we get

$$\mathscr{Z}\{\alpha^n u(-n)\} = \frac{z^{-1}}{z^{-1}-\frac{1}{\alpha}}; \quad \text{ROC}: |z^{-1}| > \frac{1}{|\alpha|} \quad \text{or} \quad |z| < |\alpha|$$

$$= \frac{1}{1-\alpha^{-1}z}$$

From multiplication by ramp property, we know that,

$$n \equiv -z\frac{d}{dz}$$

Therefore,

$$\begin{aligned} X(z) &= \mathscr{Z}\{n\alpha^n u(-n)\} \\ &= -z\frac{d}{dz}\left[\frac{1}{1-\alpha^{-1}z}\right] \\ &= \frac{\alpha^{-1}z}{(1-\alpha^{-1}z)^2}; \quad \text{ROC}: |z| < |\alpha| \end{aligned}$$

Example 8.14. Without explicitly solving for $X(z)$, find the ROC of the z-transform of each of the following sequences.

(a) $x(n) = \left[\left(\frac{1}{2}\right)^n + \left(\frac{3}{4}\right)^n\right] u(n-10)$

(b) $x(n) = \begin{cases} 1, & -10 \le n \le +10 \\ 0, & \text{otherwise} \end{cases}$

(c) $x(n) = 2^n u(-n)$.

Solution:

(a) Since, the first sequence is a right-sided infinite length sequence, the ROC is the exterior of a circle. The transform $X(z)$ has a pole at $z = \frac{1}{2}$ coming from the term $\left(\frac{1}{2}\right)^n$ and a pole at $z = \frac{3}{4}$ coming from the term $\left(\frac{3}{4}\right)^n$. Therefore, $X(z)$ has an ROC given by $\left(|z| > \frac{1}{2}\right) \cap \left(|z| > \frac{3}{4}\right) = |z| > \frac{3}{4}$.

(b) This is a finite-length sequence. The transform $X(z)$ has both positive and negative powers of z. The positive powers of z become unbounded as $z \to \infty$ and negative powers of z become unbounded as $z \to 0$. Hence, the ROC of $X(z)$ is $0 < |z| < \infty$.

(c) Since, this sequence is left-sided infinite length, the ROC is the interior of a circle. With a pole at $z = 2$, it follows that the ROC is $|z| < 2$.

Example 8.15. Use the z-transform to perform the convolution of the following two sequences.

$$h(n) = \begin{cases} \left(\frac{1}{2}\right)^n, & 0 \le n \le 2 \\ 0, & \text{elsewhere} \end{cases}$$

$$x(n) = \delta(n) + \delta(n-1) + 4\delta(n-2)$$

Solution: The convolution theorem for z-trasforms states that if $y(n) = h(n) * x(n)$, then the z-transform of $y(n)$ is $Y(z) = H(z)X(z)$.

In the present context,

$$\begin{aligned} H(z) &= \sum_{n=0}^{2} h(n)z^{-n} \\ &= \sum_{n=0}^{2} \left(\frac{1}{2}\right)^n z^{-n} \\ &= 1 + \frac{1}{2}z^{-1} + \frac{1}{4}z^{-2} \end{aligned}$$

and

$$\begin{aligned} X(z) &= \sum_{n=-\infty}^{n=+\infty} x(n)z^{-n} \\ &= \sum_{n=-\infty}^{\infty} [\delta(n) + \delta(n-1) + 4\delta(n-2)]z^{-n} \\ &= z^{-n}\Big|_{n=0} + z^{-n}\Big|_{n=1} + 4z^{-n}\Big|_{n=2} \\ &= 1 + z^{-1} + 4z^{-2} \end{aligned}$$

Therefore,

$$\begin{aligned}Y(z) &= H(z)X(z)\\ &= \left[1+\frac{1}{2}z^{-1}+\frac{1}{4}z^{-2}\right]\left[1+z^{-1}+4z^{-2}\right]\\ &= 1+\frac{3}{2}z^{-1}+\frac{19}{4}z^{-2}+\frac{9}{4}z^{-3}+z^{-4}\end{aligned}$$

Taking inverse transform, we get

$$y(n) = \delta(n) + \frac{3}{2}\delta(n-1) + \frac{19}{4}\delta(n-2) + \frac{9}{4}\delta(n-3) + \delta(n-4)$$

Example 8.16. Find the value of $x(0)$ for the sequence that has the z-transform

$$X(z) = \frac{1}{1-\frac{1}{2}z^{-1}}; \quad \text{ROC}: \ |z| > \frac{1}{2}$$

Solution:

$$X(z) = \frac{z}{z-\frac{1}{2}}$$

Since the ROC of $X(z)$ is $|z| > \frac{1}{2}$, it implies that $x(n)$ is a casual sequence. For a causal sequence $x(n)$, the value $x(0)$ is found by using the inital value theorem.

$$x(0) = \lim_{z\to\infty} X(z) = 1$$

Example 8.17. Let $x(n)$ be a left-sided sequence, that is, $x(n)$ is equal to zero for $n > 0$.

If

$$X(z) = \frac{3z^{-1}+2z^{-2}}{3-z^{-1}+z^{-2}},$$

find $x(0)$.

Solution: Since $x(n)$ is not a causal sequence, the inital value of $x(n)$, namely $x(0)$ cannot be found using intial value theorem. For a left-sided sequence that is zero for $n > 0$, the z-transform is

$$\begin{aligned}X(z) &= \sum_{n=-\infty}^{0} x(n)\, z^{-n}\\ &= x(0) + x(-1)z + x(-2)z^2 + \cdots\end{aligned}$$

Taking the limit $z \to 0$ on both sides, we get

$$\lim_{z\to 0} X(z) = x(0)$$

Therefore,

$$x(0) = \lim_{z\to 0} X(z)$$
$$= \lim_{z\to 0} \frac{3z^{-1} + 2z^{-2}}{3 - z^{-1} + z^{-2}}$$
$$= \lim_{z\to 0} \frac{3z + 2}{3z^2 - z + 1}$$
$$= 2$$

Example 8.18. Find the z-transform of the sequence given below:

$$x(n) = \frac{1}{n}(-2)^{-n}u(-n-1)$$

Solution: Finding $X(z)$ is difficult due to the factor n^{-1} in $x(n)$. Let us define a new sequence, $y(n)$, as follows:

$$y(n) = nx(n) = (-2)^{-n}u(-n-1)$$

$$\Rightarrow \quad Y(z) = \frac{-z}{z + \frac{1}{2}}; \quad \text{ROC}: |z| < \frac{1}{2} \tag{8.18}$$

Noting the relationship between $x(n)$ and $y(n)$, we can apply multiplication by ramp property to set up a differential equation for $X(z)$.

We know that,

$$y(n) = nx(n)$$

Taking z-transform on both the sides, we get

$$Y(z) = -z\frac{dX(z)}{dz} \tag{8.19}$$

Equating Eq. (8.18) and (8.19), we get

$$-z\frac{dX(z)}{dz} = \frac{-z}{z + \frac{1}{2}}$$

$$\Rightarrow \quad \frac{dX(z)}{dz} = \frac{1}{z + \frac{1}{2}}$$

The solution of this differential equation is,

$$X(z) = \log_e\left(z + \frac{1}{2}\right); \quad \text{ROC}: |z| < \frac{1}{2}$$

8.9 Inverse z-Transforms

There are three possible approaches for finding the inverse z-transform.

8.9.1 Partial fraction expansion method

Let $X(z)$ be a rational function of z, that is, a ratio of polynomials in z. Then, $x(n)$ can be found from $X(z)$ by using partial fraction expansion approach. It is noticed that z appears in the numerator of all functions, hence, partial fraction expansion is performed on $\frac{X(z)}{z}$. The expansion obtained is then multiplied by z to obtain the expansion of original function $X(z)$.

Table 8.2 given below, helps in remembering important z-transform pairs that are very useful in finding inverse z-transforms.

Table 8.2 Useful inverse z-transform pairs for partial fraction expansion method

k	$X_k(z)$	$x_k(n) = \mathscr{Z}^{-1}\{X_k(z)\}$ for ROC $\lvert z\rvert > \lvert a\rvert$
1	$\frac{z}{(z-a)}$	$a^n u(n)$
2	$\frac{z}{(z-a)^2}$	$na^{n-1}u(n)$
3	$\frac{z}{(z-a)^3}$	$\frac{n(n-1)a^{n-2}}{2!}u(n)$
4	$\frac{z}{(z-a)^4}$	$\frac{n(n-1)(n-2)a^{n-3}}{3!}u(n)$
$\vdots$	$\vdots$	$\vdots$
k	$\frac{z}{(z-a)^k}$	$\frac{n(n-1)\cdots(n-(k-2))a^{n-k+1}}{(k-1)!}u(n)$

Example 8.19. Find the inverse z-transform of the sequence

$$X(z) = \frac{z}{3z^2 - 4z + 1}$$

for the following ROCs:

(a) $|z| > 1$ (b) $|z| < \frac{1}{3}$ (c) $\frac{1}{3} < |z| < 1$

Solution: We have,

$$\frac{X(z)}{z} = \frac{1}{3z^2 - 4z + 1}$$

$$\Rightarrow \quad \frac{X(z)}{z} = \frac{1}{3\left(z - \frac{1}{3}\right)(z - 1)} \tag{8.20}$$

Let

$$\frac{X(z)}{z} = \frac{K_1}{z-\frac{1}{3}} + \frac{K_2}{z-1}$$

where K_1 and K_2 are the constants associated with partial fraction expansion.

$$K_1 = \left(z-\frac{1}{3}\right)\frac{1}{3\left(z-\frac{1}{3}\right)(z-1)}\Bigg|_{z=\frac{1}{3}} = -\frac{1}{2}$$

$$K_2 = (z-1)\frac{1}{3\left(z-\frac{1}{3}\right)(z-1)}\Bigg|_{z=1} = \frac{1}{2}$$

Substituting the values of K_1 and K_2 in Eq. (8.20), we get

$$\frac{X(z)}{z} = -\frac{1}{2}\frac{1}{z-\frac{1}{3}} + \frac{1}{2}\frac{1}{z-1}$$

$$\Rightarrow \quad X(z) = -\frac{1}{2}\frac{z}{z-\frac{1}{3}} + \frac{1}{2}\frac{z}{z-1} \tag{8.21}$$

(a) ROC : $|z| > 1$. This ROC is sketched in Fig. Ex. 8.19(a).

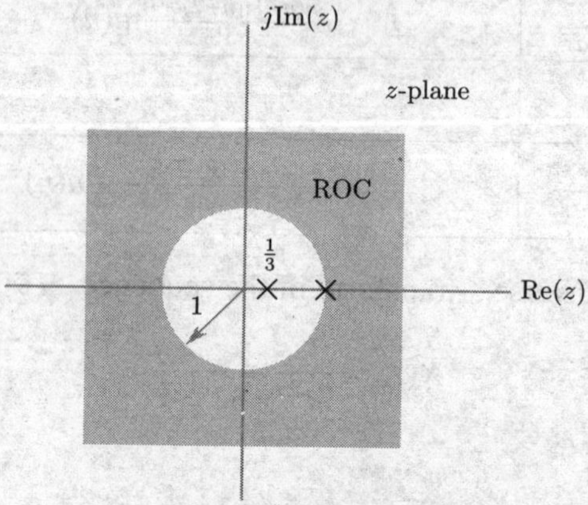

Fig. Ex. 8.19(a)

The poles at $z = \frac{1}{3}$ and $z = 1$ are interior to the ROC and hence will give positive-time sequence terms in $x(n)$.

Therefore,

$$x(n) = -\frac{1}{2}\left(\frac{1}{3}\right)^n u(n) + \frac{1}{2}u(n)$$

(b) ROC: $|z| < \frac{1}{3}$. This ROC is sketched in Fig. Ex. 8.19(b).

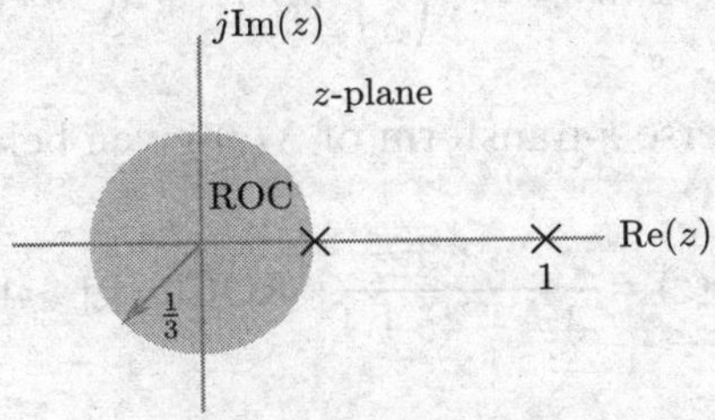

Fig. Ex. 8.19(b)

Both the poles are outside the ROC and hence will give negative-time sequence terms in $x(n)$.

Therefore,

$$x(n) = \frac{1}{2}\left(\frac{1}{3}\right)^n u(-n-1) - \frac{1}{2}u(-n-1)$$

(c) ROC: $\frac{1}{3} < |z| < 1$. The ROC is in the form of an annular ring and is as shown in Fig. Ex. 8.19(c).

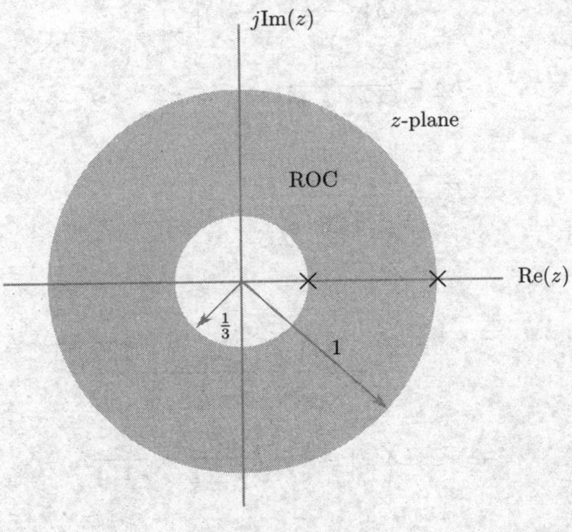

Fig. Ex. 8.19(c)

The pole at $z = \frac{1}{3}$ is interior to the ROC and hence provides the positive-time sequence part of $x(n)$. However, the pole at $z = 1$ is outside the ROC and hence provides the negative-time sequence part of $x(n)$.

Therefore,

$$x(n) = -\frac{1}{2}\left(\frac{1}{3}\right)^n u(n) - \frac{1}{2}u(-n-1)$$

Example 8.20. Find the inverse z-transform of $X(z)$ given below by using partial fraction expansion approach.

$$X(z) = \frac{z+1}{3z^2 - 4z + 1}; \text{ ROC : } |z| > 1$$

Solution:

$$\begin{aligned}\frac{X(z)}{z} &= \frac{(z+1)}{z(3z^2 - 4z + 1)} \\ &= \frac{(z+1)}{3z\left(z - \frac{1}{3}\right)(z-1)} \\ &= \frac{K_1}{z} + \frac{K_2}{z - \frac{1}{3}} + \frac{K_3}{z-1}\end{aligned} \tag{8.22}$$

where K_1, K_2 and K_3 are the constants associated with partial fraction expansion.

$$K_1 = \left.\frac{z+1}{3\left(z-\frac{1}{3}\right)(z-1)}\right|_{z=0} = 1$$

$$K_2 = \left.\frac{z+1}{3z(z-1)}\right|_{z=\frac{1}{3}} = -2$$

$$K_3 = \left.\frac{z+1}{3z\left(z-\frac{1}{3}\right)}\right|_{z=1} = 1$$

Therefore,

$$\frac{X(z)}{z} = \frac{1}{z} - \frac{2}{z-\frac{1}{3}} + \frac{1}{z-1}$$

$$\Rightarrow \quad X(z) = 1 - 2\frac{z}{z-\frac{1}{3}} + \frac{z}{z-1} \tag{8.23}$$

The poles at $z = \frac{1}{3}$ and $z = 1$ are interior to the ROC (shown in Fig. Ex. 8.20(a)) and hence provide the causal part of the sequence $x(n)$.

Therefore,

$$x(n) = \delta(n) - 2\left(\frac{1}{3}\right)^n u(n) + u(n)$$

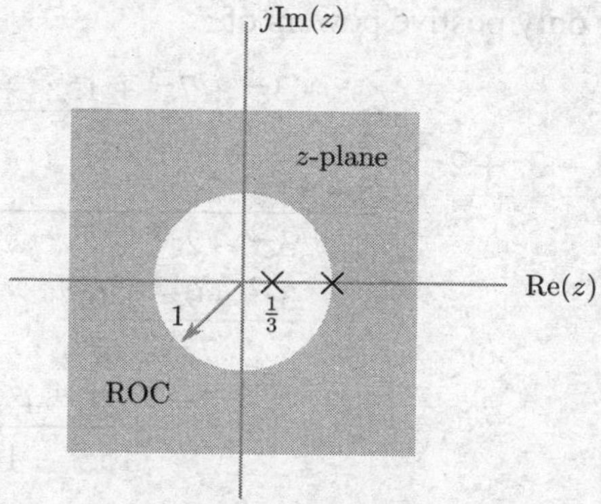

Fig. Ex. 8.20(a)

8.9.2 Inverse z-transform by long division method

The z-transform of an arbitrary sequence $x(n)$ is defined as follows:

$$X(z) \triangleq \sum_{n=-\infty}^{\infty} x(n)z^{-n}$$

$$= \cdots + x(-1)z^{1} + x(0)z^{0} + x(1)z^{-1} + \cdots \quad (8.24)$$

Hence, if $X(z)$ is expanded in a power series, the coefficients of exponents of z represent the inverse sequence values. Thus, the coefficients of z^{-m} is the mth term in the sequence. This approach will not provide a solution in the closed form, and the ROC will determine whether the series has positive or negative exponents. For causal sequences, the $X(z)$ will contain only negative exponents and for noncausal sequences the $X(z)$ will contain only positive exponents. For annular regions of convergence, a Laurent expansion would give both positive and negative-time terms. This method is useful when partial fraction expansion is difficult to perform and for obtaining a quick answer, we look at the first couple of values of the inverse transform of a given $X(z)$.

Example 8.21. Using long division method or power series expansion technique, find the inverse z-transform of the following $X(z)$:

(a) $X(z) = \dfrac{z}{2z^2 - 3z + 1}$; ROC : $|z| < \dfrac{1}{2}$

(b) $X(z) = \dfrac{z}{2z^2 - 3z + 1}$; ROC : $|z| > 1$

Solution:

(a) $X(z)$ has poles at $z = \dfrac{1}{2}$ and $z = 1$. These poles are exterior to the ROC : $|z| < \dfrac{1}{2}$. The implication is that $x(n)$ is a negative-time sequence. Thus, the long division must give a quotient having only postive powers of z.

$$\begin{array}{r|l} & z + 3z^2 + 7z^3 + 15z^4 + \cdots \\ \hline 1 - 3z + 2z^2 & z \\ & \underline{z - 3z^2 + 2z^3} \\ & \quad 3z^2 - 2z^3 \\ & \quad \underline{3z^2 - 9z^3 + 6z^4} \\ & \qquad 7z^3 - 6z^4 \\ & \qquad \underline{7z^3 - 21z^4 + 14z^5} \\ & \qquad\quad 15z^4 - 14z^5 \\ & \qquad\quad \vdots \end{array}$$

Thus,

$$\begin{aligned} X(z) &= \text{Quotient of long division} \\ &= \cdots + 15z^4 + 7z^3 + 3z^2 + z \end{aligned}$$

Hence,

$$x(n) = (\ldots,\ 15,\ 7,\ 3,\ 1,\ \underset{\Uparrow}{0})$$

(b) In this case, the poles at $z = \frac{1}{2}$ and $z = 1$ are interior to the ROC: $|z| > 1$. The implications is that $x(n)$ is a positive-time sequence term. Hence, the long division must give a quotient having only negative powers of z.

$$\begin{array}{r|l} & \frac{1}{2}z^{-1} + \frac{3}{4}z^{-2} + \frac{7}{8}z^{-3} + \cdots \\ \hline 2z^2 - 3z + 1 & z \\ & \underline{z - \frac{3}{2} + \frac{1}{2}z^{-1}} \\ & \quad \frac{3}{2} - \frac{1}{2}z^{-1} \\ & \quad \underline{\frac{3}{2} - \frac{9}{4}z^{-1} + \frac{3}{4}z^{-2}} \\ & \qquad \frac{7}{4}z^{-1} - \frac{3}{4}z^{-2} \\ & \qquad \vdots \end{array}$$

Thus,

$$\begin{aligned} X(z) &= \text{Quotient of long division} \\ &= \frac{1}{2}z^{-1} + \frac{3}{4}z^{-2} + \frac{7}{8}z^{-3} + \cdots \end{aligned}$$

Hence,

$$x(n) = \left(\underset{\Uparrow}{0},\ \frac{1}{2},\ \frac{3}{4},\ \frac{7}{8},\ \ldots\right)$$

Example 8.22. Rework the previous Example 8.21 using partial fraction expansion technique.

Solution: Given,

$$X(z) = \frac{z}{2(z-1)\left(z-\frac{1}{2}\right)}$$

$$\Rightarrow \quad \frac{X(z)}{z} = \frac{1}{2(z-1)\left(z-\frac{1}{2}\right)}$$

$$= \frac{K_1}{(z-1)} + \frac{K_2}{\left(z-\frac{1}{2}\right)}$$

where K_1 and K_2 are the constants associated with partial fraction expansion technique.

$$K_1 = \left.\frac{1}{2\left(z-\frac{1}{2}\right)}\right|_{z=1} = 1$$

$$K_2 = \left.\frac{1}{2(z-1)}\right|_{z=\frac{1}{2}} = -1$$

Hence,
$$\frac{X(z)}{z} = \frac{1}{z-1} - \frac{1}{z-\frac{1}{2}}$$

$$\Rightarrow \quad X(z) = \frac{z}{z-1} - \frac{z}{z-\frac{1}{2}} \tag{8.25}$$

(a) Since, the poles at $z = 1$ and $z = \frac{1}{2}$ are exterior to the ROC, $x(n)$ must be a noncausal sequence.

$$x(n) = -u(-n-1) + \left(\frac{1}{2}\right)^n u(-n-1)$$

$$= \left[\left(\frac{1}{2}\right)^n - 1\right] u(-n-1)$$

(b) Since the ROC is $|z| > 1$, both the poles are interior to the ROC. Hence, $x(n)$ must be a causal sequence.

$$x(n) = u(n) - \left(\frac{1}{2}\right)^n u(n)$$

$$= \left[1 - \left(\frac{1}{2}\right)^n\right] u(n)$$

8.9.3 Complex inversion integral

We know that z-transform of a sequence $x(n)$ is given by

$$X(z) \triangleq \sum_{n=-\infty}^{\infty} x(n) z^{-n}; \quad \text{ROC} : R \tag{8.26}$$

Multiplying $X(z)$ by $\frac{z^k}{2\pi jz}\,dz$ and integrating around a closed contour C which lies totally within the ROC: R, we get,

$$\begin{aligned}
\oint_C X(z)\frac{z^k}{2\pi jz}\,dz &= \frac{1}{2\pi j}\oint_C \frac{z^k}{z}X(z)\,dz \\
&= \frac{1}{2\pi j}\oint_C \sum_{n=-\infty}^{\infty} x(n)z^{-n}\frac{z^k}{z}\,dz \\
&= \frac{1}{2\pi j}\oint_C \sum_{n=-\infty}^{\infty} x(n)z^{-n+k-1}\,dz
\end{aligned} \tag{8.27}$$

The above contour integral can be evaluated by using Cauchy's integral theorem which states that if contour C encircles the origin in the anticlockwise direction, then

$$\frac{1}{2\pi j}\oint_C z^n\,dz = \begin{cases} 1, & \text{if} \quad n = -1 \\ 0, & \text{if} \quad n \neq -1 \end{cases} \tag{8.28}$$

Hence, only $n = k$ on the RHS of Eq. (8.27) gives a value other than zero.
Thus, when $n = k$, Eq. (8.27) becomes

$$\begin{aligned}
\oint_C X(z)\frac{z^k}{j2\pi z}\,dz &= \frac{1}{2\pi j}\oint_C x(k)z^{-1}\,dz \\
\Rightarrow \quad \frac{1}{2\pi j}\oint_C X(z)\frac{z^k}{z}\,dz &= x(k)\frac{1}{2\pi j}\oint_C z^{-1}\,dz \\
\Rightarrow \quad \frac{1}{2\pi j}\oint_C X(z)\frac{z^k}{z}\,dz &= x(k)
\end{aligned}$$

Replacing k by n, we get,

$$x(n) = \frac{1}{2\pi j}\oint_C X(z)\frac{z^n}{z}\,dz \tag{8.29}$$

The above Eq. (8.29) is known as the *Complex inversion integral* for evaluating inverse z-transforms. To evaulate the complex integral given by Eq. (8.29), it is necessary to use Cauchy's formula that is described below.

Cauchy's Formula

(i) Let C be a simple closed path and $f'(z)$ exists on and inside C.

Then,

$$\frac{1}{2\pi j}\oint_C \frac{f(z)}{(z-z_0)}\,dz = \begin{cases} f(z_0), & \text{for } z_0 \text{ inside } C \\ 0, & \text{for } z_0 \text{ outside } C \end{cases}$$

(ii) For poles of multiplicity k that are enclosed by the contour C, we have the following:

$$\frac{1}{2\pi j}\oint_C \frac{f(z)}{(z-z_0)^k}\,dz = \begin{cases} \frac{1}{(k-1)!}\frac{d^{k-1}}{dz^{k-1}}[f(z)]\Big|_{z=z_0}, & \text{for } z_0 \text{ inside } C \\ 0, & \text{for } z_0 \text{ outside } C \end{cases}$$

(iii) If there are n unique poles $z_1, z_2 \ldots z_n$ and $f(z)$ contains no poles inside C, it can be shown that

$$\frac{1}{2\pi j}\oint_C \frac{f(z)\,dz}{(z-z_1)(z-z_2)\ldots(z-z_n)} = \sum_{i=1}^{n} \lim_{z\to z_i} \phi_i(z)$$

where $$\phi_i(z) = \frac{f(z)}{(z-z_1)(z-z_2)\ldots(z-z_n)}$$

Example 8.23. Evaluate the contour integral given below for C defined by the circle, $|z| = 3$.

$$I = \frac{1}{2\pi j}\oint_C \frac{(z+2)}{(z-1)^2(z-2)(z-5)}\,dz$$

Solution:

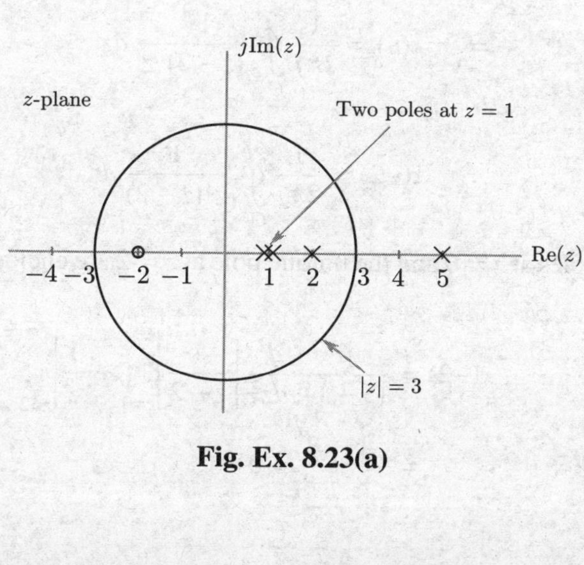

Fig. Ex. 8.23(a)

Refer Fig. Ex. 8.23(a). The poles at $z = 1$ and $z = 2$ are inside the contour C and hence, will contribute to the value of the integral. The pole at $z = 5$ is outside the contour and will not contribute to the integral.

The contribution to the integral I due to multiple poles at $z = 1$ is given by

$$I_1 = \frac{1}{(2-1)!}\frac{d^{2-1}}{dz^{2-1}}\left[\frac{z+2}{(z-2)(z-5)}\right]_{z=1}$$

$$= \frac{d}{dz}\left[\frac{z+2}{(z-2)(z-5)}\right]_{z=1} = \frac{19}{16}$$

The contribution to the integral I due to the unique pole $z = 2$ is given by

$$I_2 = \left.\frac{z+2}{(z-1)^2(z-5)}\right|_{z=2} = -\frac{4}{3}$$

Therefore,

$$I = I_1 + I_2$$

$$= \frac{19}{16} - \frac{4}{3} = -\frac{7}{48}$$

Example 8.24. Evaluate $x(-2)$, $x(-1)$, $x(-0)$ and $x(1)$ using complex inversion integral for

$$X(z) = \frac{1}{(z-2)} \quad \text{with ROC : } \quad |z| > 2$$

Solution: We know from Eq. (8.29) that,

$$x(n) = \frac{1}{2\pi j}\oint_C \frac{X(z)z^n}{z}\,dz$$

where C is within the ROC.

$$\Rightarrow \quad x(n) = \frac{1}{2\pi j}\oint_C \frac{1}{(z-2)}\frac{z^n}{z}\,dz$$

(i) When $n = -2$,

$$x(-2) = \frac{1}{2\pi j}\oint_C \frac{1}{z^3(z-2)}\,dz$$

The multiple poles at $z = 0$ and the unique pole at $z = 2$ are enclosed by the contour C.

Hence,

$$x(-2) = \frac{1}{(3-1)!}\frac{d^2}{dz^2}\left[\frac{1}{z-2}\right]_{z=0} + \left.\frac{1}{z^3}\right|_{z=2}$$

$$= -\frac{1}{8} + \frac{1}{8} = 0$$

(ii) When $n = -1$,

$$x(-1) = \frac{1}{2\pi j} \oint_C \frac{1}{z^2(z-2)}\, dz$$

$$= \frac{1}{(2-1)!} \frac{d}{dz}\left[\frac{1}{z-2}\right]_{z=0} + \left.\frac{1}{z^2}\right|_{z=2}$$

$$= -\frac{1}{4} + \frac{1}{4} = 0$$

(iii) When $n = 0$,

$$x(0) = \frac{1}{2\pi j} \oint_C \frac{1}{z(z-2)}\, dz$$

$$= \left.\frac{1}{(z-2)}\right|_{z=0} + \left.\frac{1}{z}\right|_{z=2}$$

$$= -\frac{1}{2} + \frac{1}{2} = 0$$

(iv) When $n = 1$,

$$x(1) = \frac{1}{2\pi j} \oint_C \frac{1}{z-2}\, dz$$

We know that,

$$\frac{1}{2\pi j} \oint_C \frac{f(z)}{z - z_0}\, dz = \begin{cases} f(z_0), & \text{for } z_0 \text{ inside } C \\ 0, & \text{for } z_0 \text{ outside } C \end{cases}$$

Here, $f(z) = 1$ and $z_0 = 2$.

Hence,

$$x(1) = f(z_0)$$

$$= 1.$$

Example 8.25. Using the complex inversion integral, find the inverse z-transform of $X(z)$ given by

$$X(z) = \frac{z^2}{\left(z - \frac{1}{2}\right)\left(z - \frac{1}{4}\right)}; \qquad \text{ROC}: \quad |z| > \frac{1}{2}$$

Solution: We know from Eq. (8.29) that

$$x(n) = \frac{1}{2\pi j} \oint_C X(z) \frac{z^n}{z}\, dz$$

Hence,

$$x(n) = \frac{1}{2\pi j} \oint_C \frac{z^2}{\left(z - \frac{1}{2}\right)\left(z - \frac{1}{4}\right)} \frac{z^n}{z}\, dz$$

$$= \frac{1}{2\pi j} \oint_C \frac{z^{n+1}}{\left(z - \frac{1}{2}\right)\left(z - \frac{1}{4}\right)}\, dz$$

Case 1: $n + 1 \geq 0$ or $n \geq -1$

$$x(n) = \left.\frac{z^{n+1}}{z - \frac{1}{4}}\right|_{z=\frac{1}{2}} + \left.\frac{z^{n+1}}{z - \frac{1}{2}}\right|_{z=\frac{1}{4}}$$

$$= 4\left(\frac{1}{2}\right)^{n+1} - 4\left(\frac{1}{4}\right)^{n+1}$$

Case 2: $n + 1 < 0$ or $n < -1$

Let $n = -2$.

Then,

$$x(-2) = \frac{1}{2\pi j} \oint_C \frac{z^{-1}}{\left(z - \frac{1}{2}\right)\left(z - \frac{1}{4}\right)}\, dz$$

$$= \frac{1}{2\pi j} \oint_C \frac{1}{z\left(z - \frac{1}{2}\right)\left(z - \frac{1}{4}\right)}\, dz$$

The poles at $z = 0$, $z = \frac{1}{2}$ and $z = \frac{1}{4}$ are enclosed by the contour C and hence will contribute to the complex integral.

$$x(-2) = \left.\frac{1}{\left(z - \frac{1}{2}\right)\left(z - \frac{1}{4}\right)}\right|_{z=0} + \left.\frac{1}{z\left(z - \frac{1}{4}\right)}\right|_{z=\frac{1}{2}} + \left.\frac{1}{z\left(z - \frac{1}{2}\right)}\right|_{z=\frac{1}{4}}$$

$$= 8 + 8 - 16 = 0$$

As a second example, let $n = -1$.

Then,

$$x(n) = \frac{1}{2\pi j}\oint_C \frac{1}{z^2\left(z-\frac{1}{2}\right)\left(z-\frac{1}{4}\right)}\,dz$$

$$= \frac{1}{(2-1)!}\frac{d}{dz}\left[\frac{1}{\left(z-\frac{1}{2}\right)\left(z-\frac{1}{4}\right)}\right]_{z=0} + \left.\frac{1}{z^2\left(z-\frac{1}{4}\right)}\right|_{z=\frac{1}{2}} + \left.\frac{1}{z^2\left(z-\frac{1}{2}\right)}\right|_{z=\frac{1}{4}}$$

$$= 0$$

If this process is continued for all negative values of n, we arrive at the result that $x(n)$ is zero for all negative n.

A general approach to prove that $x(n) = 0$ for $n < 0$ is to use a transformation, $z = \frac{1}{p}$. Please note that this transformation will change the ROC as $\left|\frac{1}{p}\right| > \frac{1}{2}$ or $|p| < 2$ and hence, the contour C will also change to a new contour C'.

$$x(n) = \frac{1}{2\pi j}\oint_{C'} \frac{\left(\frac{1}{p}\right)^{n+1}}{\left(\frac{1}{p}-\frac{1}{2}\right)\left(\frac{1}{p}-\frac{1}{4}\right)}\left(-\frac{1}{p^2}\right)dp$$

$$= -\frac{1}{2\pi j}\oint_{C'} \frac{p^{-(n+1)}}{\left(1-\frac{p}{2}\right)\left(1-\frac{p}{4}\right)}\,dp$$

For $n+1 < 0$, the numerator $p^{-(n+1)}$ will not contribute to poles. Also, the poles at $p = 2$ and $p = 4$ are outside the contour C'. Hence, $x(n) = 0$ for $n+1 < 0$ or $n < -1$.

Example 8.26. Find $x(n)$ for the following regions of convergence, using partial fraction expansion method.

$$X(z) = \frac{z(z^2-4z+5)}{(z-3)(z-1)(z-2)}$$

ROCs :

(a) $2 < |z| < 3$ (b) $|z| > 3$ (c) $|z| < 1$

Solution:

$$\frac{X(z)}{z} = \frac{z^2-4z+5}{(z-3)(z-1)(z-2)}$$

$$= \frac{K_1}{z-3} + \frac{K_2}{z-1} + \frac{K_3}{z-2}$$

where,

$$K_1 = \left.\frac{z^2 - 4z + 5}{(z-1)(z-2)}\right|_{z=3} = 1$$

$$K_2 = \left.\frac{z^2 - 4z + 5}{(z-3)(z-2)}\right|_{z=1} = 1$$

$$K_3 = \left.\frac{z^2 - 4z + 5}{(z-3)(z-1)}\right|_{z=2} = -1$$

Hence,

$$\frac{X(z)}{z} = \frac{1}{z-3} + \frac{1}{z-1} - \frac{1}{z-2}$$

$$\Rightarrow \quad X(z) = \frac{z}{z-3} + \frac{z}{z-1} - \frac{z}{z-2}$$

(a) ROC : $2 < |z| < 3$. This ROC is depicted in Fig. Ex. 8.26(a).

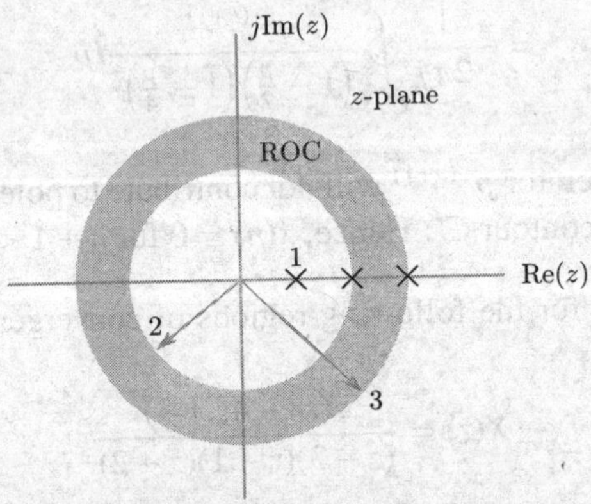

Fig. Ex. 8.26(a)

The poles at $z = 2$ and $z = 1$ are interior to the ROC and hence will provide the causal part of $x(n)$. However, the pole at $z = 3$ is exterior to the ROC and hence, will provide the negative-time sequence part of $x(n)$.

Therefore,

$$x(n) = -(3)^n u(-n-1) + u(n) - (2)^n u(n)$$

(b) ROC : $|z| > 3$. This ROC is shown in Fig. Ex. 8.26(b).

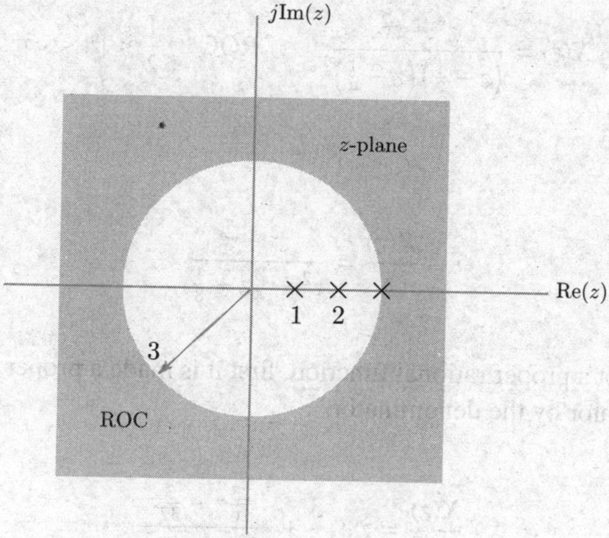

Fig. Ex. 8.26(b)

All the three poles are interior to the ROC and hence, $x(n)$ will contain only positive-time sequence terms.

Thus,

$$x(n) = (3)^n u(n) + u(n) - (2)^n u(n)$$

(c) ROC : $|z| < 1$

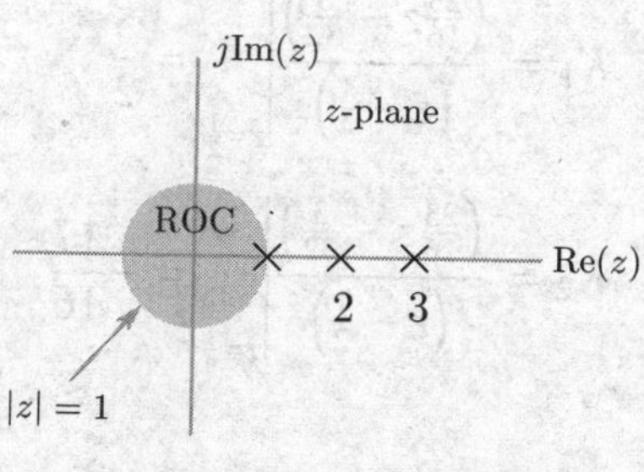

In this case, all the three poles are outside the ROC shown in Fig. Ex. 8.26(b) and hence, will contain only negative-time sequence terms.

Therefore,

$$x(n) = -(3)^n u(-n-1) - u(-n-1) + (2)^n u(-n-1)$$

Example 8.27. Find the inverse z-transform of

$$X(z) = \frac{z^4 + z^2}{\left(z - \frac{1}{2}\right)\left(z - \frac{1}{4}\right)}; \qquad \text{ROC} : \frac{1}{2} < |z| < \infty$$

Solution:

$$\frac{X(z)}{z} = \frac{z^3 + z}{z^2 - \frac{3}{4}z + \frac{1}{8}}$$

Since, $\frac{X(z)}{z}$ is not a proper rational function, first it is made a proper rational function by dividing the numerator by the denominator.

Consequently,

$$\frac{X(z)}{z} = z + \frac{3}{4} + \frac{\frac{23}{16}z - \frac{3}{32}}{z^2 - \frac{3}{4}z + \frac{1}{8}}$$

The proper rational function in z is then expanded using partial fraction technique.

$$\frac{\frac{23}{16}z - \frac{3}{32}}{z^2 - \frac{3}{4}z + \frac{1}{8}} = \frac{K_1}{z - \frac{1}{2}} + \frac{K_2}{z - \frac{1}{4}}$$

where,

$$K_1 = \left.\frac{\left(\frac{23}{16}z - \frac{3}{32}\right)}{\left(z - \frac{1}{4}\right)}\right|_{z=\frac{1}{2}} = \frac{5}{2}$$

$$K_2 = \left.\frac{\left(\frac{23}{16}z - \frac{3}{32}\right)}{\left(z - \frac{1}{2}\right)}\right|_{z=\frac{1}{4}} = -\frac{17}{16}$$

Therefore,

$$\frac{X(z)}{z} = z + \frac{3}{4} + \frac{5}{2}\frac{1}{z - \frac{1}{2}} - \frac{17}{16}\frac{1}{z - \frac{1}{4}}$$

$$\Rightarrow \quad X(z) = z^2 + \frac{3}{4}z + \frac{5}{2}\frac{z}{z - \frac{1}{2}} - \frac{17}{16}\frac{z}{z - \frac{1}{4}}$$

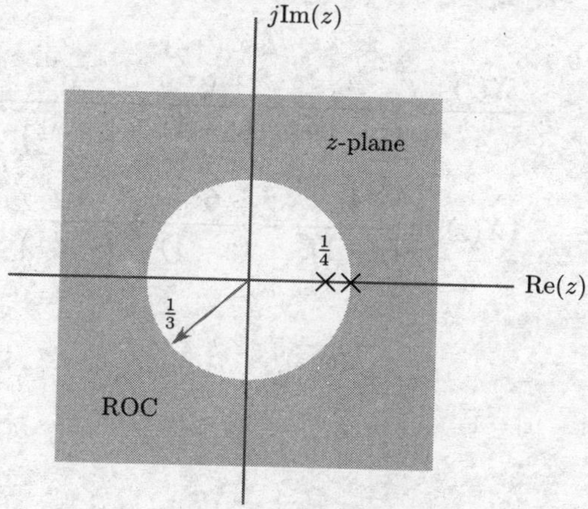

Fig. Ex. 8.27(a)

The poles at $z = \frac{1}{4}$ and $z = \frac{1}{2}$ are interior to ROC shown in Fig. Ex. 8.27(a) and hence, they will give positive-time sequence terms upon taking the inverse.

Hence,

$$x(n) = \delta(n+2) + \frac{3}{4}\delta(n+1) + \frac{5}{2}\left(\frac{1}{2}\right)^n u(n) - \frac{17}{6}\left(\frac{1}{4}\right)^n u(n)$$

Example 8.28. Find the inverse z-transform of

$$X(z) = \frac{z(z+1)}{(z-1)^2\left(z-\frac{1}{2}\right)}; \quad \text{ROC} : |z| > 1$$

Solution:

$$\frac{X(z)}{z} = \frac{(z+1)}{(z-1)^2\left(z-\frac{1}{2}\right)} = \frac{A}{(z-1)^2} + \frac{B}{z-1} + \frac{C}{z-\frac{1}{2}}$$

where

$$A = \left.\frac{z+1}{\left(z-\frac{1}{2}\right)}\right|_{z=1} = 4$$

$$B = \frac{1}{(2-1)!}\left(\frac{d}{dz}\left[\frac{z+1}{z-\frac{1}{2}}\right]\right)_{z=1}$$

$$= -6$$

$$C = \left.\frac{z+1}{(z-1)^2}\right|_{z=\frac{1}{2}}$$

$$= 6$$

Therefore,

$$\frac{X(z)}{z} = \frac{4}{(z-1)^2} - \frac{6}{(z-1)} + \frac{6}{\left(z-\frac{1}{2}\right)}$$

$$\Rightarrow \quad X(z) = \frac{4z}{(z-1)^2} - \frac{6z}{(z-1)} + \frac{6z}{\left(z-\frac{1}{2}\right)} \tag{8.30}$$

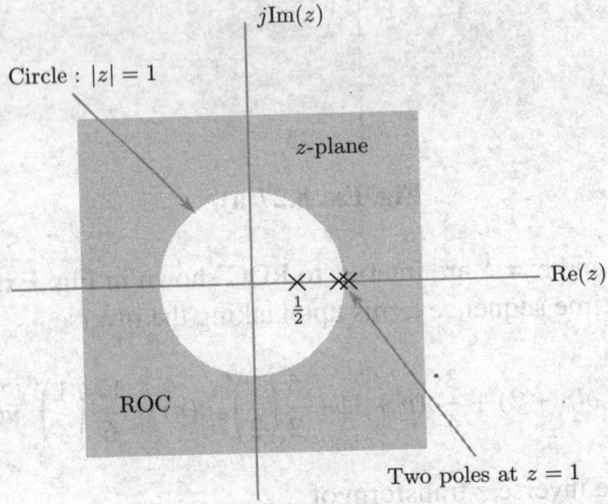

Fig. Ex. 8.28(a)

The multiple poles at $z = 1$ and the single pole at $z = \frac{1}{2}$ are interior to the ROC shown in Fig. Ex. 8.28(a). Hence, $x(n)$ will be a causal sequence. Therefore, inverse z-transform of Eq. (8.30) yields

$$x(n) = 4nu(n) - 6u(n) + 6\left(\frac{1}{2}\right)^n u(n)$$

Example 8.29. Find the inverse z-transform of

$$X(z) = \frac{z^2+z}{\left(z-\frac{1}{2}\right)^3\left(z-\frac{1}{4}\right)}; \quad \text{ROC}: |z| > \frac{1}{2}$$

Solution:

$$\frac{X(z)}{z} = \frac{z+1}{\left(z-\frac{1}{2}\right)^3\left(z-\frac{1}{4}\right)}$$

$$= \frac{K_1}{z-\frac{1}{2}} + \frac{K_2}{\left(z-\frac{1}{2}\right)^2} + \frac{K_3}{\left(z-\frac{1}{2}\right)^3} + \frac{K_4}{z-\frac{1}{4}}$$

where,

$$K_1 = \frac{1}{2!}\frac{d^2}{dz^2}\left[\frac{(z+1)}{\left(z-\frac{1}{4}\right)}\right]\Bigg|_{z=\frac{1}{2}} = 80$$

$$K_2 = \frac{d}{dz}\left[\frac{(z+1)}{\left(z-\frac{1}{4}\right)}\right]\Bigg|_{z=\frac{1}{2}} = -20$$

$$K_3 = \frac{(z+1)}{\left(z-\frac{1}{4}\right)}\Bigg|_{z=\frac{1}{2}} = 6$$

$$K_4 = \frac{(z+1)}{\left(z-\frac{1}{2}\right)^3}\Bigg|_{z=\frac{1}{4}} = -80$$

Therefore,

$$\frac{X(z)}{z} = \frac{80}{\left(z-\frac{1}{2}\right)} - \frac{20}{\left(z-\frac{1}{2}\right)^2} + \frac{6}{\left(z-\frac{1}{2}\right)^3} - \frac{80}{z-\frac{1}{4}}$$

$$\Rightarrow \quad X(z) = 80\frac{z}{\left(z-\frac{1}{2}\right)} - 20\frac{z}{\left(z-\frac{1}{2}\right)^2} + 6\frac{z}{\left(z-\frac{1}{2}\right)^3} - 80\frac{z}{\left(z-\frac{1}{4}\right)}$$

The multiple poles at $z = \frac{1}{2}$ and the single pole at $z = \frac{1}{4}$ are interior to the ROC. Hence, $x(n)$ is a causal sequence.

Thus,

$$\begin{aligned} x(n) &= 80\left(\frac{1}{2}\right)^n u(n) - 20n\left(\frac{1}{2}\right)^{n-1} u(n) \\ &\quad + 6\left[\frac{n(n-1)}{2!}\right]\left(\frac{1}{2}\right)^{n-1} u(n-1) - 80\left(\frac{1}{4}\right)^n u(n) \\ &= \left[80\left(\frac{1}{2}\right)^n - 20n\left(\frac{1}{2}\right)^{n-1} + 6\left[\frac{n(n-1)}{2}\right]\left(\frac{1}{2}\right)^{n-1} - 80\left(\frac{1}{4}\right)^n\right] u(n) \end{aligned}$$

Example 8.30. Find the inverse z-transform of $X(z)$ given below:

$$X(z) = \frac{z^{-1}}{-2z^{-2} - z^{-1} + 1}; \quad \text{ROC}: 1 < |z| < 2$$

Solution: Given,

$$X(z) = \frac{z^{-1}}{-2z^{-2} - z^{-1} + 1}$$

$$\Rightarrow \quad X(Z) = \frac{z}{-2 - z + z^2}$$

$$= \frac{z}{(z-2)(z+1)}$$

$$\Rightarrow \quad \frac{X(z)}{z} = \frac{1}{(z-2)(z+1)} = \frac{K_1}{(z-2)} + \frac{K_2}{z+1}$$

where,

$$K_1 = \left.\frac{1}{z+1}\right|_{z=2} = \frac{1}{3}$$

$$K_2 = \left.\frac{1}{z-2}\right|_{z=-1} = -\frac{1}{3}$$

Thus,

$$\frac{X(z)}{z} = \frac{1}{3}\frac{1}{(z-2)} - \frac{1}{3}\frac{1}{z+1}$$

$$\Rightarrow \quad X(z) = \frac{1}{3}\frac{z}{z-2} - \frac{1}{3}\frac{z}{z-(-1)}$$

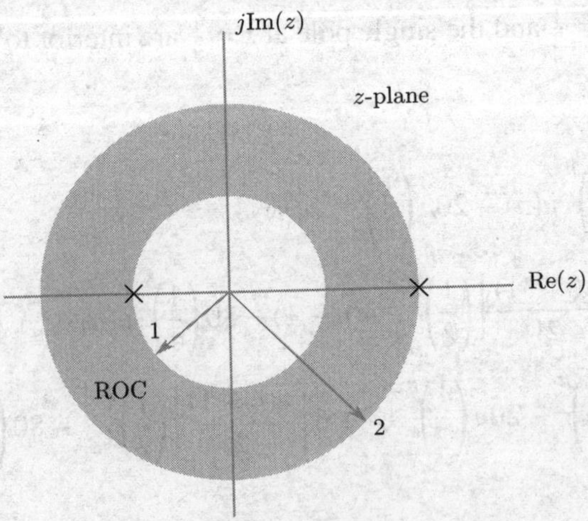

Fig. Ex. 8.30(a)

The pole at $z = -1$ is interior to the ROC shown in Fig. Ex. 8.30(a) and hence, yields the causal part of $x(n)$. While the pole at $z = 2$ is outside the ROC and hence yields the noncausal part of the sequence $x(n)$.

Therefore,

$$x(n) = -\frac{1}{3}(2)^n u(-n-1) - \frac{1}{3}(-1)^n u(n)$$

Example 8.31. Find the inverse z-transform of the following $X(z)$.

(a) $X(z) = \log_e\left(\frac{1}{1-az^{-1}}\right)$; ROC : $|z| > |a|$

(b) $X(z) = \log_e\left(\frac{1}{1-a^{-1}z}\right)$; ROC : $|z| < |a|$

Solution:

(a) The power series expansion for $\log(1-r)$ is given by

$$\log(1-r) = -\sum_{n=1}^{\infty} \frac{1}{n} r^n, \quad |r| < 1 \tag{8.31}$$

Now,
$$X(z) = \log_e\left(\frac{1}{1-az^{-1}}\right)$$
$$= -\log_e\left(1-az^{-1}\right), \quad |z| > |a| \quad \text{or} \quad \left|az^{-1}\right| < 1$$

Since, $|az^{-1}| < 1$, Eq. (8.31) has the power series expansion

$$X(z) = -\sum_{n=1}^{\infty} -\frac{1}{n}\left(az^{-1}\right)^n$$
$$= +\sum_{n=1}^{\infty} \frac{1}{n} a^n z^{-n}$$

from which we can identify $x(n)$ as

$$x(n) = \begin{cases} \frac{1}{n}a^n, & n \geq 1 \\ 0, & n \leq 0 \end{cases}$$

$$\text{or} \quad x(n) = \frac{1}{n} a^n u(n-1)$$

Alternate method:

$$X(z) = -\log_e\left(1-az^{-1}\right); \ \text{ROC}: \ |z| > |a|$$
$$= -\log_e\left(\frac{z-a}{z}\right)$$
$$\Rightarrow \quad \frac{dX(z)}{dz} = -\frac{1}{\frac{z-a}{z}}\left[\frac{z \times 1 - (z-a) \times 1}{z^2}\right]$$

$$\Rightarrow \qquad \frac{dX(z)}{dz} = \frac{-z}{z-a}\left(\frac{a}{z^2}\right)$$

$$\Rightarrow \qquad -z\frac{dX(z)}{dz} = \frac{a}{z-a}$$

$$\Rightarrow \qquad -z\frac{dX(z)}{dz} = az^{-1}\left[\frac{z}{z-a}\right]$$

Taking inverse z-transform, we get

$$nx(n) = a[a^n u(n)]_{n\to n-1}$$

Therefore,

$$x(n) = \frac{a^n}{n}u(n-1)$$

(b)

$$X(z) = \log_e\left[\frac{1}{1-a^{-1}z}\right] = -\log_e[1-a^{-1}z];\ \text{ROC}:\ |z| < |a|$$

Since, the ROC is $|z| < |a|$, that is, $|a^{-1}z| < 1$, use of Eq. (8.31) indicates $X(z)$ has the power series expansion

$$X(z) = \sum_{n=1}^{\infty}\frac{1}{n}\left(a^{-1}z\right)^n$$

Putting $n = -m$, we get

$$X(z) = \sum_{m=-1}^{-\infty} -\frac{1}{m}\left(a^{-1}z\right)^{-m}$$

$$= \sum_{m=-1}^{-\infty} -\frac{1}{m}a^m z^{-m}$$

from which we can identify $x(n)$ as

$$x(n) = \begin{cases} 0, & n \geq 0 \\ -\left(\frac{1}{n}\right)a^n, & n \leq -1 \end{cases}$$

$$\text{or} \qquad x(n) = -\frac{1}{n}a^n u(-n-1)$$

Alternate method:

$$X(z) = -\log_e\left(1 - a^{-1}z\right), \quad \text{ROC} : |z| < |a|$$

$$\Rightarrow \quad \frac{dX(z)}{dz} = -\left[\frac{1}{1 - a^{-1}z} \times -a^{-1}\right]$$

$$\Rightarrow \quad \frac{dX(z)}{dz} = -\left[\frac{a}{a - z} \times -a^{-1}\right]$$

$$\Rightarrow \quad \frac{dX(z)}{dz} = -\frac{1}{z - a}$$

$$\Rightarrow \quad -z\frac{dX(z)}{dz} = \left[\frac{z}{z - a}\right]; \quad \text{ROC}: |z| < |a|$$

Taking inverse z-transforms on both the sides, we get

$$nx(n) = -a^n u(-n - 1)$$

Therefore,

$$x(n) = -\frac{a^n}{n}u(-n - 1)$$

Example 8.32. Find the inverse z-transform of

$$X(z) = \frac{(2 + 3z^{-2} + 2z^{-3})}{z^2 + 4z + 3}, \quad \text{ROC} : |z| > 3$$

Solution: Let

$$X(z) = [2z^{-1} + 3z^{-3} + 2z^{-4}]X_A(z) \tag{8.32}$$

where,

$$X_A(z) = \frac{z}{z^2 + 4z + 3}$$

$$= \frac{z}{(z + 1)(z + 3)}$$

Let

$$\frac{X_A(z)}{z} = \frac{1}{(z + 1)(z + 3)} = \frac{K_1}{z + 1} + \frac{K_2}{z + 3}$$

where,

$$K_1 = \left.\frac{1}{z+3}\right|_{z=-1} = \frac{1}{2}$$

$$K_2 = \left.\frac{1}{z+1}\right|_{z=-3} = -\frac{1}{2}$$

Then,

$$X_A(z) = \frac{1}{2}\frac{z}{z-(-1)} - \frac{1}{2}\frac{z}{z-(-3)}$$

Since, the ROC of $X_A(z)$ is $|z| > 3$, $x_A(n)$ is positive-time sequence. Therefore,

$$x_A(n) = \frac{1}{2}(-1)^n u(n) - \frac{1}{2}(-3)^n u(n) \tag{8.33}$$

Taking inverse z-transforms on both the sides of Eq. (8.32), we get

$$x(n) = 2x_A(n-1) + 3x_A(n-3) + 2x_A(n-4)$$

$$\Rightarrow \quad x(n) = 2\left[\frac{1}{2}(-1)^{n-1}u(n-1) - \frac{1}{2}(-3)^{n-1}u(n-1)\right]$$
$$+ 3\left[\frac{1}{2}(-1)^{n-3}u(n-3) - \frac{1}{2}(-3)^{n-3}u(n-3)\right]$$
$$+ 2\left[\frac{1}{2}(-1)^{n-4}u(n-4) - \frac{1}{2}(-3)^{n-4}u(n-4)\right]$$

Example 8.33. Find the inverse z-transform of $X(z)$ given below:

$$X(z) = e^{\frac{1}{z}}$$

Take $x(n)$ to be a causal sequence.

Solution: Given,

$$X(z) = e^{\frac{1}{z}}$$

$$\Rightarrow \quad \log_e X(z) = \frac{1}{z}$$

Differentiating both sides with respect to z, we get

$$\frac{1}{X(z)}\frac{dX(z)}{dz} = -\frac{1}{z^2}$$

$$\Rightarrow \quad \frac{dX(z)}{dz} = -z^{-2}X(z)$$

Multiplying both the sides by $(-z)$, we get

$$-z\frac{dX(z)}{dz} = z^{-1}X(z)$$

Taking inverse z-transform, gives

$$nx(n) = x(n-1) \tag{8.34}$$

Eq. (8.34) is a recursive equation and to solve this equation we need an initial condition. Because $x(n)$ is a causal sequence, we may use initial value theorem to find $x(0)$. Specifically,

$$\begin{aligned} x(0) &= \lim_{z\to\infty} X(z) \\ &= \lim_{z\to\infty} e^{\frac{1}{z}} \\ &= e^0 = 1 \end{aligned}$$

Using Eq. (8.34), we get the following results:

(i) When $n = 1$,

$$x(1) = x(0) = 1$$

(ii) When $n = 2$,

$$x(2) = \frac{1}{2}x(1) = \frac{1}{2}$$

Generalizing, we get

$$x(n) = \frac{1}{n!} \quad \text{for } n > 0$$

Therefore,

$$x(n) = \delta(n) + \frac{1}{n!}u(n-1)$$

Example 8.34. Find the inverse z-transform of $X(z) = \sin z$.

Solution: Expansion of $X(z)$ in a Taylor's series about $z = 0$ gives the following result.

$$\begin{aligned} X(z) &= X(z)|_{z=0} + z\frac{dX(z)}{dz}\bigg|_{z=0} + \frac{z^2}{2!}\frac{d^2X(z)}{dz^2} + \cdots \\ &= z - \frac{z^3}{3!} + \frac{z^5}{5!} - \cdots \end{aligned}$$

Because,

$$X(z) = \sum_{n=-\infty}^{\infty} x(n)z^{-n}$$

the coefficients of z^{-n} in Taylor's series expansion can be associated with $x(n)$.

Thus,

$$x(n) = \left(\ldots, \frac{1}{5!}, 0, \frac{-1}{3!}, 0, 1, \underset{\Uparrow}{0}\right)$$

Example 8.35. Find the inverse z-transform of $X(z)$ given below:

$$X(z) = \frac{z^2 + ze^{-3T} - 2ze^{-2T}}{z^2 - [e^{-2T} + e^{-3T}]z + e^{-5T}}; \quad \text{ROC}: |z| > e^{-2T}$$

Solution: We have,

$$\frac{X(z)}{z} = \frac{z + e^{-3T} - 2e^{-2T}}{(z - e^{-3T})(z - e^{-2T})} = \frac{K_1}{z - e^{-3T}} + \frac{K_2}{z - e^{-2T}}$$

where,

$$K_1 = \left.\frac{z + e^{-3T} - 2e^{-2T}}{z - e^{-2T}}\right|_{z=e^{-3T}} = 2$$

$$K_2 = \left.\frac{z + e^{-3T} - 2e^{-2T}}{z - e^{-3T}}\right|_{z=e^{-2T}} = -1$$

Thus,

$$X(z) = 2\frac{z}{z - e^{-3T}} - \frac{z}{z - e^{-2T}}$$

The poles at $z = e^{-3T}$ and $z = e^{-2T}$ are interior to the ROC shown in Fig. Ex. 8.35(a). Hence, $x(n)$ will be a causal sequence.

$$x(n) = 2\left(e^{-3T}\right)^n u(n) - \left(e^{-2T}\right)^n u(n) = 2e^{-3nT}u(n) - e^{-2nT}u(n)$$

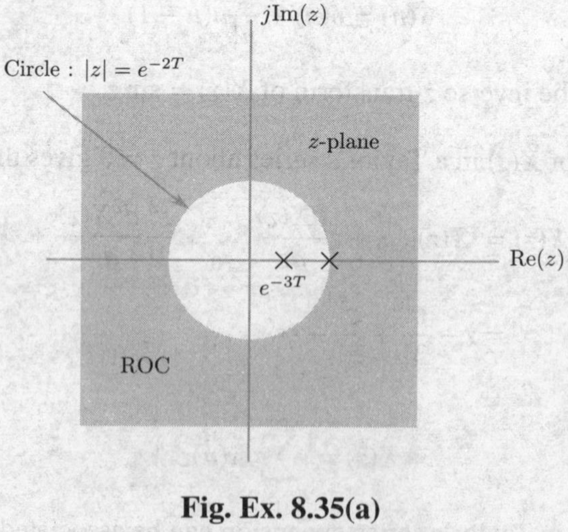

Fig. Ex. 8.35(a)

8.10 Unilateral or Single-sided z-Transform

8.10.1 Definition

The unilateral z-transform $X_1(z)$ of a sequence $x(n)$ is defined as

$$\mathscr{Z}_u\{x(n)\} = X_1(z) = \sum_{n=0}^{\infty} x(n)z^{-n}$$

and differs from the double-sided z-transform because, the lower limit of the summation is zero. Thus, the unilateral z-transform of $x(n)$ can be thought of as the bilateral transform of $x(n)\ u(n)$. Since, $x(n)\ u(n)$ is a causal sequence, the ROC of $X_1(z)$ is always the exterior to the circle passing through the furthest pole from the origin in the z-plane.

8.10.2 Useful properties

Most of the properties of bilateral and unilateral z-transforms are identical, however, the following time-shifting property is different for both the transforms.

$$\text{If} \qquad \mathscr{Z}_u\{x(n)\} = X_1(z),$$

then for $m \geq 0$, $\mathscr{Z}_u\{x(n-m)\} = z^{-m}X_1(z) + x(-1)z^{-m+1} + x(-2)z^{-m+2} + \cdots + x(-m)$

Proof: The unilateral z-transform of $x(n)$ is

$$\mathscr{Z}_u\{x(n-m)\} \triangleq \sum_{n=0}^{\infty} x(n-m)z^{-n}$$

Putting $n - m = k$, we get

$$\begin{aligned}
\mathscr{Z}_u\{x(n-m)\} &= \sum_{k=-m}^{\infty} x(k)z^{-(m+k)} \\
&= z^{-m} \sum_{k=-m}^{\infty} x(k)z^{-k} \\
&= z^{-m} \left[\sum_{k=-m}^{-1} x(k)z^{-k} + \sum_{k=0}^{\infty} x(k)z^{-k} \right] \\
&= z^{-m} \left[x(-1)z + x(-2)z^2 + \cdots + x(-m)z^m + X_1(z) \right]
\end{aligned}$$

where,

$$X_1(z) = \mathscr{Z}_u\{x(k)\}, \quad k \geq 0$$
$$= \sum_{k=0}^{\infty} x(k)z^{-k}$$

Hence,

$$\mathscr{Z}_u\{x(n-m)\} = z^{-m}X_1(z) + x(-1)z^{-m+1} + x(-2)z^{-m+2} + \cdots + x(-m) \tag{8.35}$$

Please remember that the double-sided z-transform of sequence $x(n)$ has the following time-shifting property:

$$\mathscr{Z}\{x(n-m)\} = z^{-m}X(z)$$

Corollary:

1. $\mathscr{Z}_u\{x(n-1)\} = x(-1) + z^{-1}X_1(z)$,
2. $\mathscr{Z}_u\{x(n-2)\} = x(-2) + x(-1)z^{-1} + X_1(z)z^{-2}$,
3. $\mathscr{Z}_u\{x(n-3)\} = x(-3) + x(-2)z^{-1} + x(-1)z^{-2} + X_1(z)z^{-3}$ and so on.

8.10.3 Utility of unilateral z-transform

The unilateral z-transform is useful for calculating the response of a causal system to a causal input when the system is described by a linear constant-coefficient difference equation with nonzero initial conditions.

8.11 Relations between System Representations

The system function $H(z)$ of an LTI system is defined as the ratio of z-transform of the output $y(n)$ to z-transform of the input $x(n)$.

That is,

$$H(z) = \frac{Y(z)}{X(z)}$$

Also, $H(z)$ is defined as the ratio of unilateral z-transform of the output, $y(n)$ to the unilateral z-transform of the input, $x(n)$ under the assumption that all initial conditions are zero.

That is,

$$H(z) = \frac{Y_1(z)}{X_1(z)}$$

The inverse z-transform of $H(z)$ is defined as the unit impulse or unit sample response of the system and is denoted by $h(n)$.

From $H(z)$, it is possible to obtain the difference equation realization of the system. $H(z)$ is first written in terms of negative powers of z and set equal to $\frac{Y(z)}{X(z)}$. Then, cross-multiplying and finding the inverse transform gives the difference equation. The frequency response of a stable LTI system is obtained by letting $z = e^{j\Omega}$ in $H(z)$.

That is,

$$H\left(e^{j\Omega}\right) = H(\Omega) = H(z)\Big|_{z=e^{j\Omega}}$$

8.12 Linear constant coefficient Difference equations

Difference equation

An equation that shows the relationships between consecutive values of a sequence and the differences among them. They are often rearranged as a recursive formula so that a system's output can be computed from the input signal and past outputs.

A subclass of linear time-invariant systems are those for which the input $x(n)$ and output $y(n)$ satisfy an Nth order linear constant coefficient equation, given by

$$\sum_{k=0}^{N} a_k y(n-k) = \sum_{r=0}^{M} b_r x(n-r), \quad a_0 \neq 0$$

If the sytem is causal, then we can arrange the above equation to provide a computational realization of the system in the following way:

$$y(n) = -\sum_{k=1}^{N} \frac{a_k}{a_0} y(n-k) + \sum_{r=0}^{M} \frac{b_r}{a_0} x(n-r)$$

The above equation is solved iteratively. The iterative procedure allows the calculation of present output at time n in terms of the present input at time n, the past inputs at times $n-1, \ldots, n-M$, and the past outputs at times $n-1, \ldots, n-N$. In general, the analytical expression is difficult to recognize, and therefore we go for solution by using z-transforms.

Example 8.36. Find the unilateral z-transform of the following sequences:

(a) $x(n) = 2^n u(n)$, (b) $x(n) = \left(\frac{1}{2}\right)^{n+1} u(n+1)$.

Solution:

(a) Since, $x(n)$ is a causal sequence, that is, $x(n) = 0$ for $n < 0$, it implies that $X(z) = X_1(z)$.

$$X_1(z) \triangleq \sum_{n=0}^{\infty} x(n) z^{-n}$$

$$= \sum_{n=0}^{\infty} 2^n z^{-n}$$

$$= \sum_{n=0}^{\infty} \left(2z^{-1}\right)^n$$

$$= \frac{1}{1-2z^{-1}}; \quad \text{ROC}: \ |2z^{-1}| < 1 \text{ or } |z| > 2$$

$$= \frac{z}{z-2}$$

(b) In this case, $x(n)$ is not a causal sequence. Hence, $X_1(z) \neq X(z)$.

$$X_1(z) \triangleq \sum_{n=0}^{\infty} x(n) z^{-n}$$

$$= \sum_{n=0}^{\infty} \left(\frac{1}{2}\right)^{n+1} u(n+1)\, z^{-n}$$

$$= \frac{1}{2} \sum_{n=0}^{\infty} \left(\frac{1}{2} z^{-1}\right)^n$$

$$= \frac{1}{2}\left[\frac{1}{1-\frac{1}{2}z^{-1}}\right]; \quad \text{ROC}: \ \left|\frac{1}{2}z^{-1}\right| < 1 \text{ or } |z| > \frac{1}{2}$$

$$= \frac{1}{2}\left[\frac{z}{z-\frac{1}{2}}\right]$$

It may be noted that when finding $X_1(z)$, the sample of $x(n)$ at $n = -1$ is omitted. This is the reason why $X_1(z)$ is different from $X(z)$ for a noncausal sequence.

Example 8.37. Consider the system shown in Fig. Ex. 8.37. Find the system function $H(z)$ and the impulse response $h(n)$.

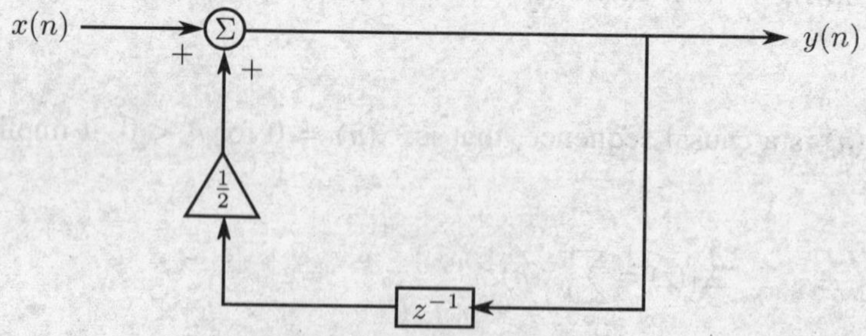

Fig. Ex. 8.37

Solution:

$$y(n) = x(n) + \frac{1}{2}y(n-1)$$

Taking z-transform on the both the sides keeping in mind the time-shift property, we get

$$Y(z) = X(z) + \frac{1}{2}z^{-1}Y(z)$$

Therefore,

$$H(z) = \frac{Y(z)}{X(z)} = \frac{1}{1-\frac{1}{2}z^{-1}} = \frac{z}{z-\frac{1}{2}}$$

Since, the system is causal, the ROC is $|z| > \frac{1}{2}$.

Taking inverse z-transform of $H(z)$, we get, $h(n) = \left(\frac{1}{2}\right)^n u(n)$.

Example 8.38. Solve the following difference equation for $y(n)$ making use of the specified initial condition and the input $x(n)$.

$$y(n) = \frac{1}{2}y(n-1) + x(n), \quad n \geq 0$$

$$y(-1) = \frac{1}{4}$$

and $$x(n) = \left(\frac{1}{2}\right)^n u(n)$$

Solution: Taking unilateral z-transform on both the sides of the given difference equation, we have

$$Y_1(z) = \frac{1}{2}\left[y(-1) + Y_1(z)z^{-1}\right] + X_1(z)$$

$$= \frac{1}{2}\left[\frac{1}{4} + z^{-1}Y_1(z)\right] + \frac{z}{z-\frac{1}{2}}$$

$$\Rightarrow \quad Y_1(z)\left[1 - \frac{1}{2}z^{-1}\right] = \frac{1}{8} + \frac{z}{z-\frac{1}{2}}$$

$$\Rightarrow \quad Y_1(z) = \frac{1}{8}\frac{1}{\left(1-\frac{1}{2}z^{-1}\right)} + \left(\frac{1}{1-\frac{1}{2}z^{-1}}\right)\left(\frac{z}{z-\frac{1}{2}}\right)$$

$$\Rightarrow \quad Y_1(z) = \frac{1}{8}\frac{z}{\left(z-\frac{1}{2}\right)} + z\left[\frac{z}{\left(z-\frac{1}{2}\right)^2}\right]$$

Taking inverse z-transform of $Y_1(z)$, we get

$$y(n) = \frac{1}{8}\left(\frac{1}{2}\right)^n u(n) + n\left(\frac{1}{2}\right)^{n-1} u(n)\Big|_{n\to n+1}$$

$$= \frac{1}{8}\left(\frac{1}{2}\right)^n u(n) + (n+1)\left(\frac{1}{2}\right)^n u(n+1)$$

Since, $n \geq 0$, we replace $u(n+1)$ by $u(n)$.
Hence,

$$y(n) = \frac{1}{8}\left(\frac{1}{2}\right)^n u(n) + (n+1)\left(\frac{1}{2}\right)^n u(n)$$

$$= \frac{1}{8}\left(\frac{1}{2}\right)^n + (n+1)\left(\frac{1}{2}\right)^n, \quad n \geq 0$$

Example 8.39. The output of a discrete-time LTI system is found to be $y(n) = 2\left(\frac{1}{3}\right)^n u(n)$ when the input $x(n)$ is $u(n)$.

(a) Find the impulse response, $h(n)$ of the sytem.

(b) Find the output when the input $x(n)$ is $\left(\frac{1}{2}\right)^n u(n)$.

Solution:

(a) Given,

$$y(n) = 2\left(\frac{1}{3}\right)^n u(n)$$

$$\Rightarrow \quad Y(z) = 2\frac{z}{z - \frac{1}{3}}; \quad \text{ROC} : |z| > \frac{1}{3}$$

and $\quad x(n) = u(n)$

$$\Rightarrow \quad X(z) = \frac{z}{z-1}; \quad \text{ROC} : |z| > 1$$

Hence, the system function $H(z)$ is

$$H(z) = \frac{Y(z)}{X(z)} = \frac{2(z-1)}{z - \frac{1}{3}}; \quad \text{ROC} : |z| > \frac{1}{3}$$

Let,

$$\frac{H(z)}{z} = \frac{2(z-1)}{z\left(z - \frac{1}{3}\right)} = \frac{K_1}{z} + \frac{K_2}{z - \frac{1}{3}}$$

where,

$$K_1 = \left.\frac{2(z-1)}{\left(z - \frac{1}{3}\right)}\right|_{z=0} = 6$$

$$K_2 = \left.\frac{2(z-1)}{z}\right|_{z=\frac{1}{3}} = -4$$

Hence,

$$\frac{H(z)}{z} = \frac{6}{z} - \frac{4}{z - \frac{1}{3}}$$

$$\Rightarrow \qquad H(z) = 6 - \frac{4z}{\left(z - \frac{1}{3}\right)}; \quad \text{ROC}: |z| > \frac{1}{3}$$

Taking inverse z-transform of $H(z)$, we get

$$y(n) = 6\delta(n) - 4\left(\frac{1}{3}\right)^n u(n)$$

(b) Given,

$$x(n) = \left(\frac{1}{2}\right)^n u(n)$$

Then,

$$X(z) = \frac{z}{z - \frac{1}{2}}; \quad \text{ROC}: |z| > \frac{1}{2}$$

We know that,

$$H(z) = \frac{Y(z)}{X(z)}$$

$$\Rightarrow \qquad Y(z) = H(z)X(z)$$

$$= \frac{2z(z-1)}{\left(z - \frac{1}{2}\right)\left(z - \frac{1}{3}\right)}; \quad \text{ROC}: |z| > \frac{1}{2}$$

Therefore,

$$\frac{Y(z)}{z} = \frac{2(z-1)}{\left(z - \frac{1}{2}\right)\left(z - \frac{1}{3}\right)}$$

$$= \frac{K_1}{\left(z - \frac{1}{2}\right)} + \frac{K_2}{\left(z - \frac{1}{3}\right)}$$

where,

$$K_1 = \left.\frac{2(z-1)}{\left(z - \frac{1}{3}\right)}\right|_{z=\frac{1}{2}} = -6$$

$$K_2 = \left.\frac{2(z-1)}{\left(z - \frac{1}{2}\right)}\right|_{z=\frac{1}{3}} = 8$$

Hence,

$$Y(z) = -6\frac{z}{\left(z - \frac{1}{2}\right)} + 8\frac{z}{\left(z - \frac{1}{3}\right)}$$

Taking inverse z-transform of $Y(z)$, we obtain

$$y(n) = -6\left(\frac{1}{2}\right)^n u(n) + \left(\frac{1}{3}\right)^n u(n)$$

It may be noted that for the causal sequences $X(z) = X_1(z)$ and $Y(z) = Y_1(z)$.

Example 8.40. Given that $H(z) = \dfrac{z+1}{z^2 - 2z + 3}$ represents a causal system. Find a difference equation realization and the frequency response of the system.

Solution: Because the system is causal, the numerator and the denominator of $H(z)$ must contain negative powers of z and this can be accomplished by dividing the numerator and denominator by the highest power of z in the denominator. Hence,

$$H(z) = \frac{Y(z)}{X(z)} = \frac{z^{-1} + z^{-2}}{1 - 2z^{-1} + 3z^{-2}}$$

Cross-multiplying, we get

$$Y(z) - 2z^{-1}Y(z) + 3z^{-2}Y(z) = z^{-1}X(z) + z^{-2}X(z)$$

Taking inverse z-transform yields the difference equation

$$y(n) = 2y(n-1) - 3y(n-2) + x(n-1) + x(n-2)$$

The frequency response $H\left(e^{j\Omega}\right)$ is obtained by letting $z = e^{j\Omega}$ in $H(z)$.

$$\begin{aligned} H\left(e^{j\Omega}\right) &= \left.\frac{z+1}{z^2 - 2z + 3}\right|_{z=e^{j\Omega}} \\ &= \frac{e^{j\Omega} + 1}{e^{j2\Omega} - 2e^{j\Omega} + 3} \\ &= \frac{(1 + \cos\Omega) + j\sin\Omega}{(\cos 2\Omega - 2\cos\Omega + 3) + j(\sin 2\Omega - 2\sin\Omega)} \end{aligned}$$

Quite often, it is advantageous to remember the various relationships between difference equation, system function, impulse response, and frequency response for stable causal systems represented by linear, constant coefficient difference equations. This is often facilitated by a flow diagram shown in Fig. Ex. 8.40.

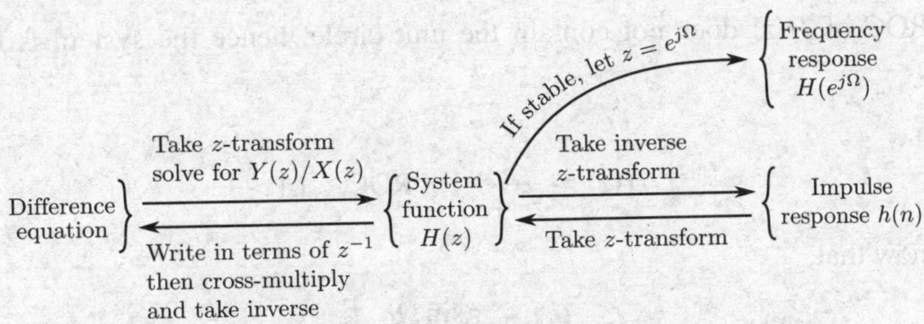

Fig. Ex. 8.40: *Relationship between system representations.*

Example 8.41. Given

$$H(z) = \frac{z}{z^2 + 1}; \qquad \text{ROC}: |z| > 1$$

(a) Is this a causal system? why or why not?

(b) Is the system BIBO stable? why or why not?

(c) Find the unit sample response of the system.

(d) Find the difference equation realization of the system connecting the input $x(n)$ and output $y(n)$.

(e) Is the system linear or nonlinear?

(f) Is the system shift-invariant? Why or why not?

Solution:

(a) The system function $H(z)$ has poles at $z = \pm j$ and they are interior to the ROC : $|z| > 1$. Hence, $h(n)$ is a causal sequence and the system in question is causal.

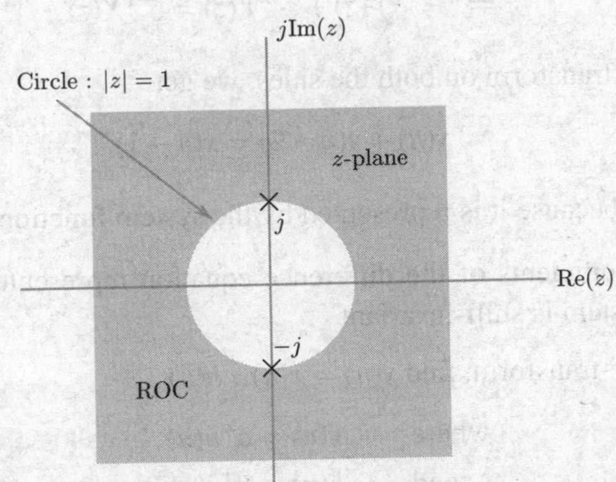

(b) The ROC of $H(z)$ does not contain the unit circle, hence the system is not BIBO stable.

(c) Given,

$$H(z) = \frac{z}{z^2 + 1}; \quad \text{ROC} : |z| > 1$$

We know that,

$$\mathscr{Z}\{\sin \Omega_0 n u(n)\} = \frac{z \sin \Omega_0}{z^2 - 2z \cos \Omega_0 + 1}; \quad \text{ROC} : |z| > 1$$

Comparing $z^2 + 1$ with $z^2 - 2z \cos \Omega_0 + 1$, we get

$$-2z \cos \Omega_0 = 0$$

$$\Rightarrow \quad \Omega_0 = \frac{\pi}{2}$$

Hence, taking inverse z-transform of $H(z)$, we get

$$h(n) = \sin\left(\frac{\pi}{2} n\right) u(n)$$

(d) Let

$$H(z) = \frac{Y(z)}{X(z)} = \frac{z}{z^2 + 1}$$

For a causal system, the numerator and the denominator of $H(z)$ must contain negative powers of z.

Accordingly,

$$\frac{Y(z)}{X(z)} = \frac{z^{-1}}{1 + z^{-2}}$$

$$\Rightarrow \quad Y(z) + z^{-2} Y(z) = z^{-1} X(z)$$

Taking inverse z-transform on both the sides, we get,

$$y(n) + y(n - 2) = x(n - 1) \tag{8.36}$$

(e) System is linear because it is represented by the system function $H(z)$.

(f) Since, all the coefficients of the difference equation represented by Eq. (8.36) are constants, the system is shift-invariant.

Example 8.42. Using z-transform, find $y(n) = x(n) * h(n)$,

$$\text{where} \quad x(n) = \alpha^n u(n)$$

$$\text{and} \quad h(n) = \beta^n u(n)$$

Solution: We have,

$$X(z) = \frac{z}{z-\alpha}; \quad \text{ROC}: |z| > |\alpha|$$

and $$H(z) = \frac{z}{z-\beta}; \quad \text{ROC}: |z| > |\beta|$$

Given,

$$y(n) = x(n) * h(n)$$

$$\Rightarrow \quad Y(z) = X(z)H(z)$$

$$= \frac{z^2}{(z-\alpha)(z-\beta)}; \quad \text{ROC}: |z| > \max(\alpha,\beta)$$

Let

$$\frac{Y(z)}{z} = \frac{z}{(z-\alpha)(z-\beta)}$$

$$= \frac{K_1}{(z-\alpha)} + \frac{K_2}{(z-\beta)}$$

where

$$K_1 = \left.\frac{z}{z-\beta}\right|_{z=\alpha} = \frac{\alpha}{(\alpha-\beta)}$$

$$K_2 = \left.\frac{z}{z-\alpha}\right|_{z=\beta} = \frac{-\beta}{(\alpha-\beta)}$$

Thus,

$$Y(z) = \frac{\alpha}{(\alpha-\beta)}\frac{z}{(z-\alpha)} - \frac{\beta}{(\alpha-\beta)}\frac{z}{(z-\beta)}$$

Therefore,

$$y(n) = \frac{\alpha}{(\alpha-\beta)}\alpha^n u(n) - \frac{\beta}{(\alpha-\beta)}\beta^n u(n)$$

$$= \left(\frac{\alpha^{n+1}-\beta^{n+1}}{\alpha-\beta}\right)u(n), \quad \alpha \neq \beta$$

when $\alpha = \beta$,

$$Y(z) = \frac{z^2}{(z-\alpha)^2}$$

Let

$$\frac{Y(z)}{z} = \frac{z}{(z-\alpha)^2}$$
$$= \frac{A_1}{z-\alpha} + \frac{A_2}{(z-\alpha)^2}$$

where,

$$A_2 = z\Big|_{z=\alpha} = \alpha$$
$$A_1 = \frac{d[z]}{dz}\Big|_{z=\alpha} = 1$$

Therefore,

$$Y(z) = \frac{z}{z-\alpha} + \frac{\alpha z}{(z-\alpha)^2}; \quad \text{ROC} : |z| > \alpha$$
$$\Rightarrow \quad y(n) = \alpha^n u(n) + \alpha n \alpha^{n-1} u(n)$$
$$= [\alpha^n + n\alpha^n]u(n)$$

Example 8.43. Using z-transform, find

$$y(n) = x(n) * h(n)$$

where,

$$x(n) = a^n u(n)$$
$$\text{and} \quad h(n) = a^{-n} u(-n), \quad 0 < a < 1.$$

Solution: Given,

$$x(n) = a^n u(n)$$
$$\Rightarrow \quad X(z) = \frac{z}{z-a}; \quad \text{ROC} : |z| > |a|$$

Since $h(n) = x(-n)$, applying the time-reversal property of z-transform, we get

$$H(z) = \frac{z^{-1}}{z^{-1}-a}; \quad \text{ROC}: |z^{-1}| > |a|$$
$$\Rightarrow \quad H(z) = \frac{1}{1-az}; \quad \text{ROC}: |z| < \frac{1}{|a|}$$

We want to find $y(n) = x(n) * h(n)$. Taking z-transform on both the sides, we get

$$Y(z) = X(z)H(z); \quad \text{ROC}: (|z| > |a|) \cap \left(|z| < \frac{1}{|a|}\right)$$
$$= \frac{-1}{a} \frac{z}{(z-a)\left(z-\frac{1}{a}\right)}; \quad \text{ROC} : a < |z| < \frac{1}{a}$$

Let

$$\frac{Y(z)}{z} = \frac{-1}{a}\frac{1}{(z-a)\left(z-\frac{1}{a}\right)} = \frac{K_1}{(z-a)} + \frac{K_2}{\left(z-\frac{1}{a}\right)}$$

where,

$$K_1 = -\frac{1}{a}\frac{1}{\left(z-\frac{1}{a}\right)}\Bigg|_{z=a} = +\frac{1}{(1-a^2)}$$

$$K_2 = -\frac{1}{a}\frac{1}{z-a}\Bigg|_{z=\frac{1}{a}} = -\frac{1}{(1-a^2)}$$

Thus,

$$Y(z) = \frac{1}{1-a^2}\frac{z}{(z-a)} - \frac{1}{1-a^2}\frac{z}{z-\frac{1}{a}}; \quad \text{ROC} : a < |z| < \frac{1}{a}$$

The pole at $z = \frac{1}{a}$ is exterior to the ROC and hence will give the negative-time sequence part of $y(n)$. However, the pole at $z = a$ is interior to the ROC and hence provides the causal part of the sequence $y(n)$.

Therefore,

$$\begin{aligned} y(n) &= \frac{1}{1-a^2}a^n u(n) + \frac{1}{(1-a^2)}\left(\frac{1}{a}\right)^n u(-n-1) \\ &= \frac{1}{1-a^2}a^{|n|} \end{aligned}$$

Example 8.44. Find the frequency response of the system having the system function,

$$H(z) = \frac{1-2z+3z^2}{3-2z+z^2}$$

Show that the magnitude of frequency response is constant for all values of Ω.

Solution: The frequency response of the system is obtained by putting $z = e^{j\Omega}$ in $H(z)$.

$$H(e^{j\Omega}) = \frac{1-2e^{j\Omega}+3e^{j2\Omega}}{3-2e^{j\Omega}+e^{j2\Omega}}$$

$$\Rightarrow \quad H(e^{j\Omega}) = \frac{e^{j\Omega}(e^{-j\Omega}-2+3e^{j\Omega})}{e^{j\Omega}(3e^{-j\Omega}-2+e^{j\Omega})}$$

$$= \frac{(\cos\Omega - 2 + 3\cos\Omega) + j(3\sin\Omega - \sin\Omega)}{(\cos\Omega - 2 + 3\cos\Omega) - j(3\sin\Omega - \sin\Omega)}$$

Therefore,

$$|H(e^{j\Omega})| = \frac{\sqrt{(\cos\Omega - 2 + 3\cos\Omega)^2 + (3\sin\Omega - \sin\Omega)^2}}{\sqrt{(\cos\Omega - 2 + 3\cos\Omega)^2 + (3\sin\Omega - \sin\Omega)^2}}$$

$$= 1$$

Example 8.45. Solve the following difference equation:

$$y(n) + y(n-2) = \delta(n), \ n \geq 0$$

$$y(-2) = 0, \ y(-1) = 1$$

Solution: Taking unilateral z-transform on both the sides of the given difference equation, we get,

$$Y_1(z) + \left[y(-2) + y(-1)z^{-1} + Y_1(z)z^{-2}\right] = 1$$

$$\Rightarrow \quad Y_1(z) + \left[0 + 1z^{-1} + z^{-2}Y_1(z)\right] = 1$$

$$\Rightarrow \quad Y_1(z) = \frac{1 - z^{-1}}{1 + z^{-2}}$$

$$= \frac{z^2}{z^2+1} - \frac{z}{z^2+1}$$

$$\text{ROC} : |z| > |\pm j| \quad \text{or } |z| > 1$$

We know that,

$$\mathscr{Z}_u\{\cos(\Omega_0 n)\, u(n)\} = \frac{z^2 - z\cos\Omega_0}{z^2 - 2z\cos\Omega_0 + 1}; \quad \text{ROC} : |z| > 1$$

$$\text{and} \quad \mathscr{Z}_u\{\sin(\Omega_0 n)\, u(n)\} = \frac{z\sin\Omega_0}{z^2 - 2z\cos\Omega_0 + 1}; \quad \text{ROC} : |z| > 1$$

Comparing $z^2 + 1$ with $z^2 - 2z\cos\Omega_0 + 1$, we get

$$\cos\Omega_0 = 0 \quad \text{or} \quad \Omega_0 = \frac{\pi}{2}$$

Rewrite $Y_1(z)$ as

$$Y_1(z) = \frac{z^2 - z\cos\frac{\pi}{2}}{z^2 - 2z\cos\frac{\pi}{2} + 1} - \frac{z\sin\frac{\pi}{2}}{z^2 - 2z\cos\frac{\pi}{2} + 1}; \quad \text{ROC} : |z| > 1$$

Taking inverse z-transform of $Y_1(z)$, we get

$$y(n) = \cos\left(\frac{\pi}{2}n\right)u(n) - \sin\left(\frac{\pi}{2}n\right)u(n)$$

Example 8.46. Solve the following difference equation for $y(n)$ using z-transform and the specified initial conditions.

$$y(n) = y(n-1) - y(n-2) + 2, \quad n \geq 0$$

$$y(-2) = 1, \; y(-1) = 2$$

Solution: Taking unilateral z-transform on both the sides of the given difference equation, we get

$$Y_1(z) = [y(-1) + z^{-1}Y_1(z)] - [y(-2) + y(-1)z^{-1} + z^{-2}Y_1(z)] + 2\frac{z}{z-1}$$

$$\Rightarrow \quad Y_1(z) = [2 + z^{-1}Y_1(z)] - [1 + 2z^{-1} + z^{-2}Y_1(z)] + 2\frac{z}{z-1}$$

Therefore,

$$Y_1(z) = \frac{z(3z^2 - 3z + 2)}{(z-1)(z^2 - z + 1)}$$

Let

$$\frac{Y_1(z)}{z} = \frac{3z^2 - 3z + 2}{(z-1)(z^2 - z + 1)} = \frac{K_1}{z-1} + \frac{K_2 z + K_3}{z^2 - z + 1}$$

By performing partial fraction expansion, we get, $K_1 = 2$, $K_2 = 1$ and $K_3 = 0$.
Hence,

$$Y_1(z) = 2\frac{z}{z-1} + \frac{z^2}{z^2 - z + 1}; \quad \text{ROC} : |z| > 1$$

Note that for causal sequences, the bilateral and unilateral z-transforms give identical results.

Hence,

$$\mathscr{Z}_u\{\cos\Omega_0\, u(n)\} = \frac{z^2 - z\cos\Omega_0}{z^2 - 2z\cos\Omega_0 + 1}; \quad \text{ROC} : |z| > 1$$

$$\mathscr{Z}_u\{\sin\Omega_0 n\, u(n)\} = \frac{z\sin\Omega_0}{z^2 - 2z\cos\Omega_0 + 1}; \quad \text{ROC} : |z| > 1$$

Comparing $z^2 - z + 1$ with $z^2 - 2z\cos\Omega_0 + 1$, we get

$$2\cos\Omega_0 = 1$$

$$\Rightarrow \quad \Omega_0 = \frac{\pi}{3}$$

Rewriting $Y_1(z)$, we get

$$Y_1(z) = 2\frac{z}{(z-1)} + \frac{\left(z^2 - \frac{z}{2} + \frac{z}{2}\right)}{z^2 - \frac{2z}{2} + 1}$$

$$= 2\frac{z}{(z-1)} + \frac{z^2 - z\cos\frac{\pi}{3}}{\left(z^2 - 2z\cos\frac{\pi}{3} + 1\right)} + \frac{1}{2}\frac{2}{\sqrt{3}}\frac{z\frac{\sqrt{3}}{2}}{\left(z^2 - 2z\cos\frac{\pi}{3} + 1\right)}$$

$$= 2\frac{z}{(z-1)} + \frac{z^2 - z\cos\frac{\pi}{3}}{\left(z^2 - 2z\cos\frac{\pi}{3} + 1\right)} + \frac{1}{\sqrt{3}}\frac{z\sin\frac{\pi}{3}}{\left(z^2 - 2z\cos\frac{\pi}{3} + 1\right)}; \quad \text{ROC} : |z| > 1$$

Taking inverse z-transform of $Y_1(z)$, we get,

$$y(n) = 2u(n) + \cos\left(\frac{\pi}{3}n\right)u(n) + \frac{1}{\sqrt{3}}\sin\left(\frac{\pi}{3}n\right)u(n)$$

Example 8.47. A causal discrete-time LTI system is described by

$$y(n) - \frac{3}{4}y(n-1) + \frac{1}{8}y(n-2) = x(n)$$

where $x(n)$ and $y(n)$ are the input and the output of the system respectively.

(a) Determine the system function $H(z)$.

(b) Find the impulse response $h(n)$.

(c) Find the step response of the system.

(d) Find the BIBO stability of the system.

(e) Find the frequency response of the system.

Solution:

(a) Taking z-transform on both the sides of the given difference equation, and using time-shift property, we get

$$Y(z) - \frac{3}{4}z^{-1}Y(z) + \frac{1}{8}z^{-2}Y(z) = X(z)$$

$$\Rightarrow \qquad H(z) = \frac{Y(z)}{X(z)} = \frac{1}{1 - \frac{3}{4}z^{-1} + \frac{1}{8}z^{-2}}$$

$$\Rightarrow \qquad H(z) = \frac{z^2}{\left(z - \frac{1}{2}\right)\left(z - \frac{1}{4}\right)}; \quad \text{ROC} : |z| > \frac{1}{2}$$

(b) Taking inverse z-transform of $H(z)$, we get $h(n)$. Now,

$$\frac{H(z)}{z} = \frac{z}{\left(z-\frac{1}{2}\right)\left(z-\frac{1}{4}\right)} = \frac{K_1}{\left(z-\frac{1}{2}\right)} + \frac{K_2}{\left(z-\frac{1}{4}\right)}$$

where,

$$K_1 = \left.\frac{z}{\left(z-\frac{1}{4}\right)}\right|_{z=\frac{1}{2}} = 2$$

$$K_2 = \left.\frac{z}{\left(z-\frac{1}{2}\right)}\right|_{z=\frac{1}{4}} = -1$$

Thus,

$$H(z) = 2\frac{z}{\left(z-\frac{1}{2}\right)} - \frac{z}{\left(z-\frac{1}{4}\right)}; \quad \text{ROC : } |z| > \frac{1}{2}$$

Taking inverse z-transform of $H(z)$, we get

$$h(n) = 2\left(\frac{1}{2}\right)^n u(n) - \left(\frac{1}{4}\right)^n u(n)$$

(c) Given,

$$x(n) = u(n)$$

$$\Rightarrow \quad X(z) = \frac{z}{z-1} \quad \text{ROC : } |z| > 1$$

We know that,

$$Y(z) = H(z)X(z)$$

$$\Rightarrow \quad Y(z) = \frac{z^3}{\left(z-\frac{1}{2}\right)\left(z-\frac{1}{4}\right)(z-1)}$$

$$\text{ROC : } \left(|z| > \frac{1}{2}\right) \cap (|z| > 1) = |z| > 1$$

Now,

$$\frac{Y(z)}{z} = \frac{z^2}{\left(z-\frac{1}{2}\right)\left(z-\frac{1}{4}\right)(z-1)} = \frac{K_1}{\left(z-\frac{1}{2}\right)} + \frac{K_2}{\left(z-\frac{1}{4}\right)} + \frac{K_3}{(z-1)}$$

where,

$$K_1 = \left.\frac{z^2}{\left(z-\frac{1}{4}\right)(z-1)}\right|_{z=\frac{1}{2}} = -2$$

$$K_2 = \left.\frac{z^2}{\left(z-\frac{1}{2}\right)(z-1)}\right|_{z=\frac{1}{4}} = \frac{1}{3}$$

$$K_3 = \left.\frac{z^2}{\left(z-\frac{1}{2}\right)\left(z-\frac{1}{4}\right)}\right|_{z=1} = \frac{8}{3}$$

Thus,

$$Y(z) = -2\frac{z}{\left(z-\frac{1}{2}\right)} + \frac{1}{3}\frac{z}{\left(z-\frac{1}{4}\right)} + \frac{8}{3}\frac{z}{(z-1)}; \qquad \text{ROC} : |z| > 1$$

Therefore,

$$y(n) = -2\left(\frac{1}{2}\right)^n u(n) + \frac{1}{3}\left(\frac{1}{4}\right)^n u(n) + \frac{8}{3}u(n)$$

(d) Since, the ROC of $H(z)$ contains the unit circle, the system is BIBO stable.

(e) Since system is stable, the frequency response is obtained by letting $z = e^{j\Omega}$ in $H(z)$. That is,

$$H\left(e^{j\Omega}\right) = H(\Omega) = \frac{e^{j2\Omega}}{\left(e^{j\Omega}-\frac{1}{2}\right)\left(e^{j\Omega}-\frac{1}{4}\right)}$$

Example 8.48. Find $x(n)$ for $X(z)$ given below:

$$X(z) = \log_n\left(1+\alpha z^{-1}\right)$$

Solution: Noting that

$$\log_n(1+x) = \sum_{n=1}^{\infty}\left(\frac{(-1)^{n+1}x^n}{n}\right)$$

Then

$$X(z) = \sum_{n=1}^{\infty}\left(\frac{(-1)^{n+1}\alpha^n z^{-n}}{n}\right)$$

Comparing the above equation with

$$X(z) = \sum_{n=-\infty}^{\infty} x(n)z^{-n}, \qquad \text{we get}$$

$$x(n) = \begin{cases} \frac{(-1)^{n+1}\alpha^n}{n} & \text{if } n \geq 1 \\ 0 & \text{if } n \leq 0 \end{cases}$$

Example 8.49. Determine the impulse response corresponding to the following transfer function if

(i) System is stable, (ii) System is causal.

$$H(z) = \frac{3z^{-1}}{(1 - 2z^{-1})^2}$$

Solution: Rewrite $H(z)$ as

$$H(z) = \frac{3z}{(z - 2)^2}$$

(i) If the system is stable, then the ROC of $H(z)$ must contain the unit circle, $|z| = 1$. Hence, the ROC of $H(z)$ is $|z| < 2$.

$H(z)$ has two poles at $z = 2$ and these poles are interior to the ROC and they give negative-time sequence terms.

It may be noted that if $x(n) = x_+(n)u(n)$ is a positive-time sequence, whose z-transform is $X(z)$. Then for the same $X(z)$, a noncausal sequence is given by $-x_+(n)u(-n-1)$. As an example, let us take a positive-time sequence given by

$$x(n) = \underbrace{na^{n-1}}_{x_+(n)}u(n) \overset{\text{ZT}}{\longleftrightarrow} X(z) = \frac{z}{(z-a)^2}; \quad |z| > |a|$$

For the same z-transform $X(z)$, the corresponding negative-time sequence is

$$\underbrace{-na^{n-1}}_{-x_+(n)}u(-n-1) \overset{\text{ZT}}{\longleftrightarrow} \frac{z}{(z-a)^2}; \quad |z| < |a|$$

Hence, $h(n) = -3n2^{n-1}u(-n-1)$

(ii) For system to be causal both the poles of $H(z)$ must be interior to the ROC. Hence, the ROC of $H(z)$ is $z > 2$.

Thus,

$$h(n) = 3n2^{n-1}\, u(n)$$

8.13 Realization of system Function

A digital filter transfer function can be realized in a variety of ways. There are two types of realizations: 1. Recursive and 2. Non-recursive.

1. For recursive realization, current output $y(n)$ is a function of past outputs, past and present inputs.

2. For non-recursive realization, current output $y(n)$ is a function of only past and present inputs.

We restrict our discussion only for recursive realization of $H(z)$. That is, for transfer functions having both poles and zeros.

8.13.1 Direct form I realization

Let us consider an LTI system characterized by the following system function with $M \leq N$.

$$H(z) = \frac{Y(z)}{X(z)} = \frac{\sum_{k=0}^{M} b_k z^{-k}}{1 + \sum_{k=1}^{N} a_k z^{-k}} \tag{8.37}$$

Cross-multiplying, we get

$$Y(z) + \sum_{k=1}^{N} a_k z^{-k} Y(z) = \sum_{k=0}^{M} b_k z^{-k} X(z) \tag{8.38}$$

Taking inverse z-transform on both the sides of Eq. (8.38), we get

$$y(n) = -\sum_{k=1}^{N} a_k y(n-k) + \sum_{k=0}^{M} b_k x(n-k) \tag{8.39}$$

A realization of the system function using Eq. (8.39) as shown in Fig. 8.6 is called direct form-I realization. The present output $y(n)$ is seen to be a weighted sum of the input $x(n)$ at the present time n, past inputs $x(n-k)$ for $k = 1, 2, \ldots, M$, and past outputs $y(n-k)$ for $k = 1, 2, \ldots N$. Hence, this form of realization is recursive in nature. The delay blocks represent a form of storage and delay.

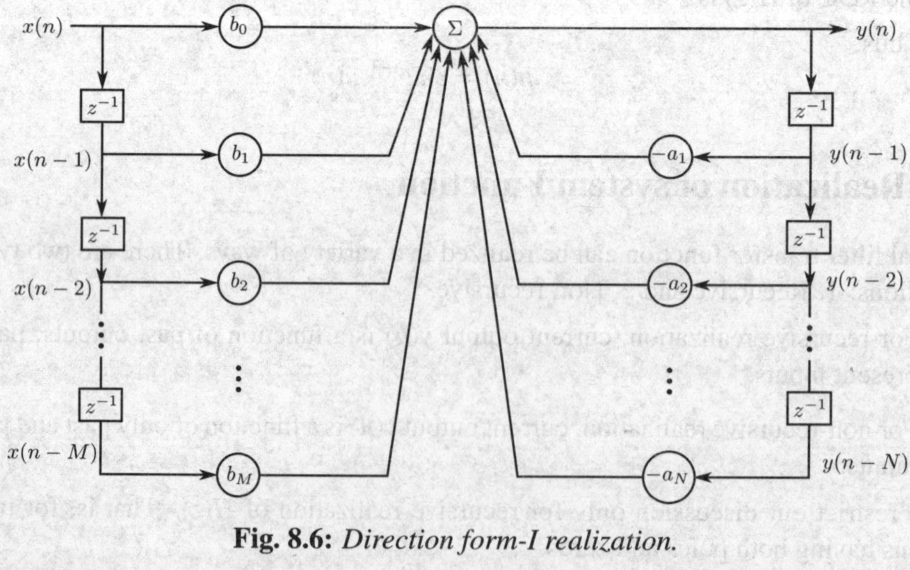

Fig. 8.6: *Direction form-I realization.*

8.13.2 Direct form II realization

Another realization of $H(z)$ can be obtained by breaking $H(z)$ into a product of two transfer functions $H_1(z)$ and $H_2(z)$, where $H_1(z)$, contains only the denominator or poles and $H_2(z)$ contains only the numerator or zeros. That is,

$$H(z) = H_1(z)H_2(z) \tag{8.40}$$

where,

$$H_1(z) = \frac{1}{\left(1 + \sum_{k=1}^{N} a_k z^{-k}\right)} \tag{8.41}$$

and $$H_2(z) = \sum_{k=0}^{M} b_k z^{-k} \tag{8.42}$$

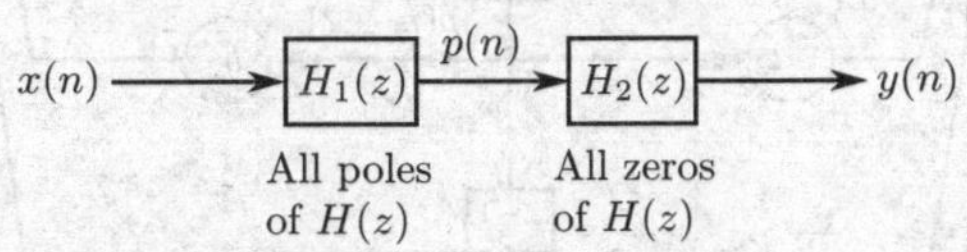

Fig. 8.7: *Decomposition for direct form-II realization*

From Fig. 8.7, we can write

$$P(z) = X(z)H_1(z)$$

$$\Rightarrow \quad P(z) = X(z)\left(\frac{1}{1 + \sum_{k=1}^{N} a_k z^{-k}}\right)$$

$$\Rightarrow \quad P(z) + \sum_{k=1}^{N} a_k z^{-k} P(z) = X(z)$$

Taking inverse z-transform on both the sides, we get

$$p(n) = -\sum_{k=1}^{N} a_k p(n-k) + x(n) \tag{8.43}$$

Again from Fig. 8.7, we can write

$$Y(z) = P(z)H_2(z)$$

$$\Rightarrow \quad Y(z) = P(z)\left(\sum_{k=0}^{M} b_k z^{-k}\right)$$

Taking inverse z-transform on both sides, we get

$$y(n) = \sum_{k=0}^{M} b_k p(n-k) \qquad (8.44)$$

Using Eqs. (8.43) and (8.44), we can set up the block diagram of direct form-II realization. The same is shown in Fig. 8.8.

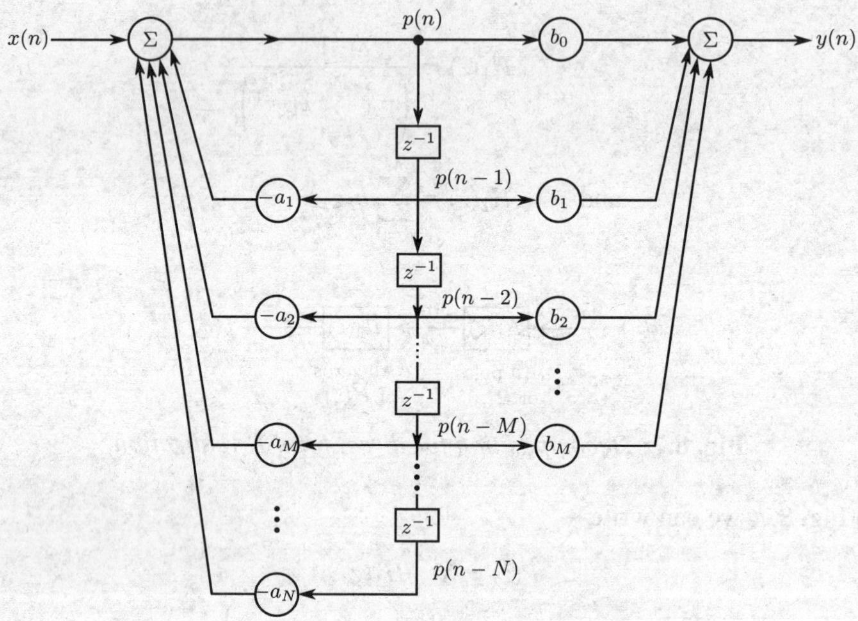

Fig. 8.8: *Direct form-II Canonical realization.*

In this form of realization, the number of delay blocks is seen to be N, the order of the difference equation. This is one of the best forms of realization as it contains the minimum number of delay blocks and hence is known as canonical realization.

Example 8.50. Find the direct form-I and direct form-II realizations of a discrete-time system represented by the transfer function $H(z)$ given below:

$$H(z) = \frac{8z^3 - 4z^2 + 11z - 2}{\left(z - \frac{1}{4}\right)\left(z^2 - z + \frac{1}{2}\right)}$$

Solution: Since, the system in the question is causal, the numerator and the denominator of $H(z)$ must contain negative powers of z.

That is,

$$H(z) = \frac{8 - 4z^{-1} + 11z^{-2} - 2z^{-3}}{1 - \frac{5}{4}z^{-1} + \frac{3}{4}z^{-2} - \frac{1}{8}z^{-3}}$$

Direct form-I realization:

Let

$$H(z) = \frac{Y(z)}{X(z)} = \frac{8 - 4z^{-1} + 11z^{-2} - 2z^{-3}}{1 - \frac{5}{4}z^{-1} + \frac{3}{4}z^{-2} - \frac{1}{8}z^{-3}}$$

$$\Rightarrow \quad Y(z) - \frac{5}{4}z^{-1}Y(z) + \frac{3}{4}z^{-2}Y(z) - \frac{1}{8}z^{-3}Y(z)$$

$$= 8X(z) - 4z^{-1}X(z) + 11z^{-2}X(z) - 2z^{-3}X(z)$$

Taking inverse z-transform on both the sides and rearranging, we get

$$y(n) = 8x(n) - 4x(n-1) + 11x(n-2) - 2x(n-3) + \frac{5}{4}y(n-1) - \frac{3}{4}y(n-2) + \frac{1}{8}y(n-3)$$

Using the above equation, the block diagram of Direct form-I realization is drawn and is as shown in Fig. Ex. 8.50(a).

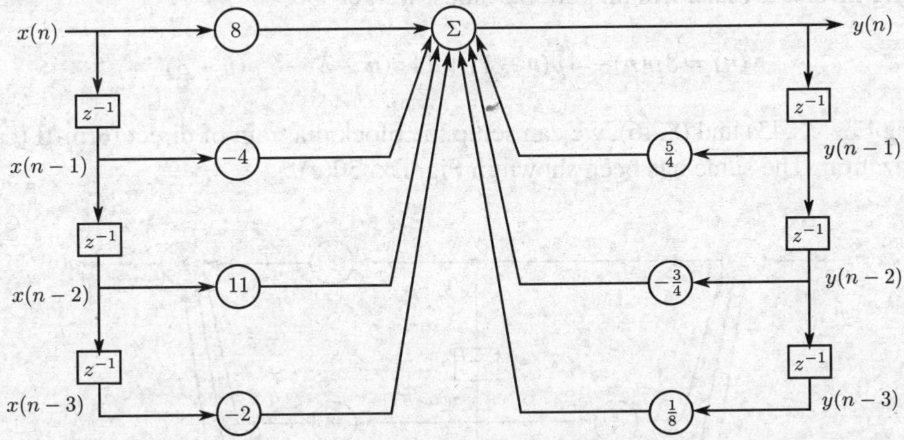

Fig. Ex. 8.50(a)

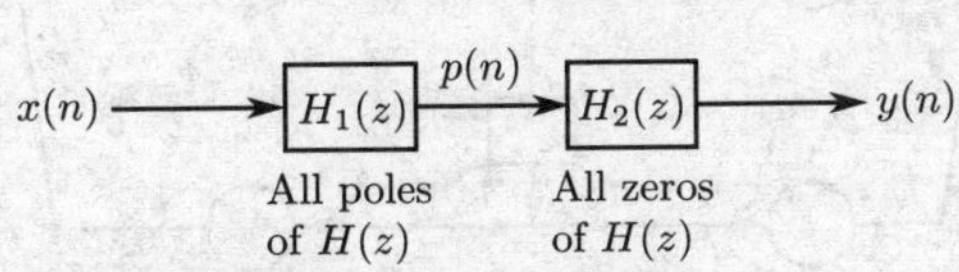

Fig. Ex. 8.50(b)

Referring Fig. Ex. 8.50(b), we can write

$$P(z) = X(z)\, H_1(z)$$

$$\Rightarrow \qquad P(z) = X(z)\left(\frac{1}{1 - \frac{5}{4}z^{-1} + \frac{3}{4}z^{-2} - \frac{1}{8}z^{-3}}\right)$$

$$\Rightarrow \qquad P(z) - \frac{5}{4}z^{-1}P(z) + \frac{3}{4}z^{-2}P(z) - \frac{1}{8}z^{-3}P(z) = X(z)$$

Taking inverse z-transform on both the sides, we get

$$p(n) = \frac{5}{4}p(n-1) - \frac{3}{4}p(n-2) + \frac{1}{8}p(n-3) + x(n) \tag{8.45}$$

Also, referring Fig. Ex. 50(b), we can write

$$Y(z) = P(z)H_2(z)$$

$$\Rightarrow \qquad Y(z) = P(z)\left(8 - 4z^{-1} + 11z^{-2} - 2z^{-3}\right)$$

$$\Rightarrow \qquad Y(z) = 8P(z) - 4z^{-1}P(z) + 11z^{-2}P(z) - 2z^{-3}P(z)$$

Taking inverse z-transform on both the sides, we get

$$y(n) = 8p(n) - 4p(n-1) + 11p(n-2) - 2p(n-3) \tag{8.46}$$

Using Eqs. (8.45) and (8.46), we can setup the block diagram of direct form-II (canonical) realization. The same has been shown in Fig. Ex. 50(c).

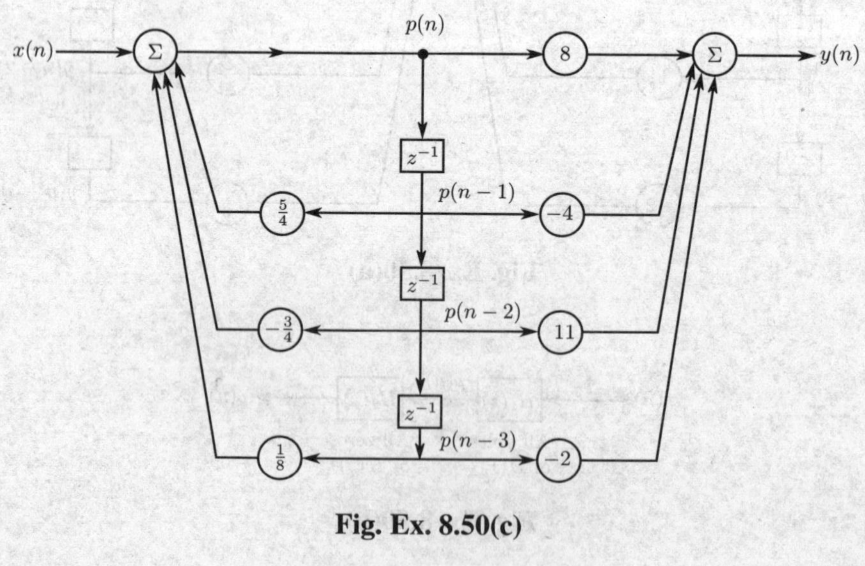

Fig. Ex. 8.50(c)

8.14 Relation between z-Transform and Discrete-Time Fourier Transform

When z-transform of a sequence is only calculated for values of z on the unit circle $z = e^{j\Omega}$, the Fourier transform of this sequence is immediately obtained.

We know that,

$$\mathscr{Z}\{x(n)\} \triangleq \sum_{n=-\infty}^{\infty} x(n)z^{-n} \tag{8.47}$$

Also we define descrete-time Fourier transform of $x(n)$ as

$$\textbf{DTFT}\{x(n)\} = \sum_{n=-\infty}^{\infty} x(n)e^{-j\Omega n} \tag{8.48}$$

From Eqs. (8.47) and (8.48), we find that

$$\textbf{DTFT}\{x(n)\} = \mathscr{Z}\{x(n)\}\Big|_{z=e^{j\Omega}} \tag{8.49}$$

The above relationship can be generalized if we evaluate z-transform on a circle of radius r described by $z = re^{j\Omega}$, $0 < r < \infty$.

$$X(z)\Big|_{z=re^{j\Omega}} = X\left(re^{j\Omega}\right) = \sum_{n=-\infty}^{\infty} x(n)\left(re^{-j\Omega}\right)^{n}$$

$$= \sum_{n=-\infty}^{\infty} x(n)r^{-n}e^{-j\Omega n} \tag{8.50}$$

This yields therefore,

$$\textbf{DTFT}\left\{x(n)r^{-n}\right\} = \mathscr{Z}\{x(n)\}|_{z=re^{j\Omega}} \tag{8.51}$$

The relationship represented by Eqs. (8.49) and (8.51) have meaning if and only if, the Fourier transforms of $x(n)$ and $x(n)r^{-n}$ exist and the unit circle of z-plane or circle with radius r, belong to the ROC of the z-transform. If the unit circle does not belong to the ROC of the z-transform, that does not necessarily mean that Fourier transform does not exist, instead that the Fourier transform is not analytic. There are actually sequences that have a Fourier transform but not a z-transform, for example, $x(n) = 1$.

Example 8.51. What is the z-transform of the series which has the Fourier transform $X\left(e^{j\Omega}\right) = 1 + \cos\Omega$? Also find $x(n)$.

Solution: $X\left(e^{j\Omega}\right)$ can be differentiated with respect to Ω any number of times, and is therefore analytic. When we replace $e^{j\Omega}$ by z, we get $X(z)$.

With,

$$\begin{aligned} X\left(e^{j\Omega}\right) &= 1 + \cos\Omega \\ &= 1 + \frac{1}{2}e^{j\Omega} + \frac{1}{2}e^{-j\Omega} \end{aligned} \tag{8.52}$$

we obtain

$$X(z) = 1 + \frac{1}{2}z + \frac{1}{2}z^{-1}$$

The ROC is the whole z-plane excepting $z = 0$ and $z = \infty$, and we must therefore be dealing with a series of finite length.

Comparing Eq. (8.52) with

$$\begin{aligned} \mathbf{DTFT}\{x(n)\} &= X\left(e^{j\Omega}\right) \\ &= \sum_{n=-\infty}^{\infty} x(n)e^{-j\Omega n} \end{aligned}$$

we get,

$$x(n) = \begin{cases} \frac{1}{2}, & n = -1 \\ 1, & n = 0 \\ \frac{1}{2}, & n = 1 \\ 0, & \text{otherwise} \end{cases}$$

8.15 Relation between z-transform and Laplace transform

We can obtain a discrete-time signal from a continuous-time signal, by multiplying it by a train of impulses. Let us denote the continuous-time signal by $f(t)$ and the train of impulses by $s_\delta(t)$. The time-domain expression for the train of impulses depicted in Fig. 8.9(b) is

$$s_\delta(t) = \sum_{n=-\infty}^{\infty} \delta(t - nT_s) \tag{8.53}$$

Multiplication of $f(t)$ by $s_\delta(t)$ produce a signal $x(t)$ defined as

$$x(t) = f(t)s_\delta(t) \tag{8.54}$$

$$= f(t)\sum_{n=-\infty}^{\infty}\delta(t-nT_s)$$

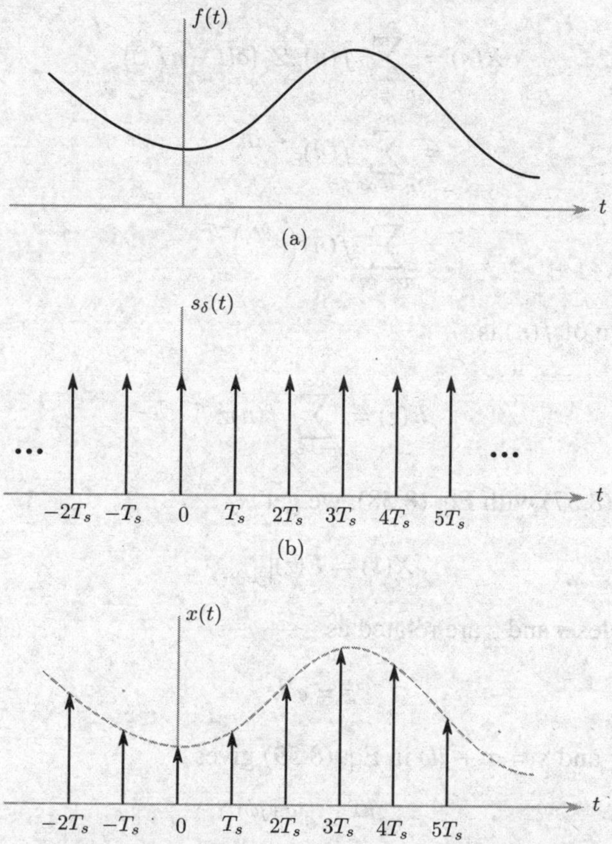

Fig. 8.9: *Formation of discrete-time signals*

Applying product property of an impulse function, we get

$$x(t) = \sum_{n=-\infty}^{\infty} f(nT_s)\delta(t-nT_s) \tag{8.55}$$

For the sake of simplicity, let us denote $f(nT_s)$ by $f(n)$. Then Eq. (8.55) becomes

$$x(t) = \sum_{n=-\infty}^{\infty} f(n)\delta(t-nT_s) \tag{8.56}$$

Recall the standard Laplace transform pairs:

1. $\delta(t) \overset{\text{LT}}{\longleftrightarrow} 1$

2. $\delta(t-t_0) \overset{\text{LT}}{\longleftrightarrow} e^{-st_0}$

Hence, taking Laplace transform of both sides of Eq. (8.56) gives

$$\begin{aligned} X(s) &= \sum_{n=-\infty}^{\infty} f(n)\, \mathscr{L}\{\delta(t-nT_s)\} \\ &= \sum_{n=-\infty}^{\infty} f(n) e^{-snT_s} \\ &= \sum_{n=-\infty}^{\infty} f(n) \left(e^{sT_s}\right)^{-n} \end{aligned} \tag{8.57}$$

Also, z-transform of $f(n)$ is

$$F(z) = \sum_{n=-\infty}^{\infty} f(n) z^{-n} \tag{8.58}$$

Comparing Eq. (8.57) with Eq. (8.58), we get

$$X(s) = F(z)\Big|_{z=e^{sT_s}}$$

Thus, the variables s and z are related as

$$z = e^{sT_s} \tag{8.59}$$

Letting $z = re^{j\Omega}$ and $s = \sigma + j\omega$ in Eq. (8.59) gives

$$re^{j\Omega} = e^{(\sigma+j\omega)T_s}$$

Thus,

$$r = e^{\sigma T_s} \tag{8.60}$$

$$\text{and} \qquad \Omega = \omega T_s \tag{8.61}$$

It may be noted that $z = re^{j\Omega}$ represents the equation of a circle whose radius is r in the complex z-plane.

Let us examine the behaviour of z when $\sigma < 0$, $\sigma = 0$ and $\sigma > 0$.

- When $\sigma < 0$, Eq. (8.60) implies that $r < 1$. Hence, left half of the s-plane ($\sigma < 0$) maps inside the unit circle of the z-plane, and for different negative values of σ, we get concentric circles of radius less than unity.

- When $\sigma > 0$, Eq. (8.60) gives $r > 1$. Thus, the right-half of the s-plane maps outside the unit circle of z-plane, and for different positive values of σ we get concentric circles with radius greater than unity.
- When $\sigma = 0$, Eq. (8.60) gives $r = 1$. This means image of $j\omega$ axis of the s-plane is the circle of unit radius in the z-plane.

  The mapping of z-plane to s-plane is a multivalued transformation since $s = \frac{1}{T}\log_e z$ and

$$\log_e z = \log_e z + j2n\pi$$

Example 8.52. State and prove Final value theorem related to z-transforms.

Statement: If $\mathscr{Z}_u\{x(n)\} = X_1(z)$ and the poles of $(z-1)X_1(z)$ are all inside the unit circle, then the final value of $x(n)$ (as $n \to \infty$) is given by

$$\lim_{n\to\infty} x(n) = \lim_{z\to 1}[(z-1)X_1(z)]$$

Proof: Form the sequence $x(n+1) - x(n)$ and take unilateral z-transform on it.

$$\mathscr{Z}_u\{x(n+1) - x(n)\} \triangleq \sum_{n=0}^{\infty}[x(n+1) - x(n)]z^{-n}$$

$$\Rightarrow \mathscr{Z}_u\{x(n+1) - x(n)\} = \lim_{n\to\infty}\left[\sum_{k=0}^{n}[x(k+1) - x(k)]z^{-k}\right]$$

$$= \lim_{n\to\infty}\left[\sum_{k=0}^{n} x(k+1)z^{-k} - \sum_{k=0}^{n} x(k)z^{-k}\right]$$

$$= \lim_{n\to\infty}\left[-x(0) + x(1)\left(1 - z^{-1}\right) + x(2)\left(z^{-1} - z^{2}\right) + \cdots + x(n+1)z^{-n}\right]$$

Now, taking the limit $z \to 1$ on both the sides, we get

$$\lim_{z\to 1}\mathscr{Z}_u\{x(n+1) - x(n)\} = \lim_{n\to\infty}[-x(0) + x(n+1)]$$

$$= -x(0) + \lim_{n\to\infty} x(n) \qquad (8.62)$$

Also,

$$\mathscr{Z}_u\{x(n+1) - x(n)\} = \mathscr{Z}_u\{x(n+1)\} - \mathscr{Z}_u\{x(n)\}$$

$$= zX_1(z) - zx(0) - X_1(z)$$

$$= (z-1)X_1(z) - zx(0)$$

Taking the limit $z \to 1$ on both the sides of the above equation, we get

$$\lim_{z\to 1} \mathscr{Z}_u\{x(n+1) - x(n)\} = \lim_{z\to 1}(z-1)X_1(z) - x(0) \tag{8.63}$$

Equating Eqs. (8.62) and (8.63) we get

$$-x(0) + \lim_{n\to\infty} x(n) = -x(0) + \lim_{z\to 1}(z-1)X_1(z)$$

Thus,

$$\lim_{n\to\infty} x(n) = \lim_{z\to 1}(z-1)X_1(z)$$

Hence the proof.

Summary of Important points

- The z-transform is the discrete-time counterpart of the Laplace transform.
- Some signals which do not have a DTFT do have a z-transform.
- The bilateral z-transform of the discrete-time sequence $x(n)$ is defined as

$$X(z) = \sum_{n=-\infty}^{\infty} x(n)z^{-n}$$

- The unilateral z-transform of a signal $x(n)$ is defined as

$$X_1(z) = \sum_{n=0}^{\infty} x(n)z^{-n}$$

- Every z-transform has an associated region of convergence (ROC) in the z-plane. The ROC of the z-transform consists of those values of z, for which the sum converges.
- For causal sequences, the ROC in the z-plane lies exterior to the circle, but all the poles of $X(z)$ are interior to this circle. For noncausal signals, the ROC is interior to the circle such that all the poles of $X(z)$ are exterior to this circle. If $x(n)$ consists of both a causal and an anticausal part, then ROC is an annular ring, such that the poles outside this region correspond to the noncausal part of $x(n)$, and the poles interior to the annulus correspond to the causal part.
- The inverse z-transform can be found by complex inversion integral, long division, or partial fraction expansion. Usage of complex inversion integral is rare, and long division does not provide a closed-form of result. Hence, partial-fraction expansion is usually preferred.

- The properties of z-transform are similar to those of the Laplace transform. Among the application of z-transform, are the solution of difference equations and the evaluation of the convolution of two discrete-time sequences.

- The unilateral z-transform can be used to solve difference equations with inital conditions.

- If $y(n)$ represents the convolution of two discrete sequences $x(n)$ and $h(n)$, then $Y(z) = H(z)X(z)$.

- The transfer function $H(z)$ of a system with input $x(n)$, impulse response $h(n)$, and output $y(n)$ is

$$H(z) = \mathscr{Z}\{h(n)\} = \frac{Y(z)}{X(z)}$$

Exercise Problems

EP. 8.1 Using the definition of z-transform, find the z-transform of two discrete-time sequences $x(n)$ and $y(n)$ given below:

(a) $x(n) = \begin{cases} \left(\frac{1}{2}\right)^n, & n \geq 0 \\ 0, & n < 0 \end{cases}$

(b) $y(n) = \begin{cases} -\left(\frac{1}{2}\right)^n, & n < 0 \\ 0, & n \geq 0 \end{cases}$

Ans:

(a) $X(z) = \dfrac{z}{z - \frac{1}{2}}$; ROC: $|z| > \dfrac{1}{2}$

(b) $Y(z) = \dfrac{z}{z - \frac{1}{2}}$; ROC: $|z| < \dfrac{1}{2}$

EP. 8.2 Find the z-transform of the sequence, $x(n) = \left(\dfrac{1}{2}\right)^n u(n) + \left(\dfrac{1}{3}\right)^n u(n)$.

Ans: $X(z) = \dfrac{2z^2 - \frac{5}{6}z}{\left(z - \frac{1}{2}\right)\left(z - \frac{1}{3}\right)}$; ROC : $|z| > \dfrac{1}{2}$

EP. 8.3 Find the z-transform of the sequence $x(n)$ defined below:

$$x(n) = \begin{cases} 3^n, & n < 0 \\ \left(\frac{1}{3}\right)^n, & n = 0, 2, 4, \ldots \\ \left(\frac{1}{2}\right)^n, & n = 1, 3, 5, \ldots \end{cases}$$

Ans: $X(z) = -\left(\dfrac{z}{z-3}\right) + \dfrac{z^2}{z^2 - \frac{1}{9}} + \dfrac{2z}{z^2 - \frac{1}{4}}$; ROC: $\dfrac{1}{2} < |z| < 3$

EP. 8.4 Find the z-transform of the sequence $x(n)$ given below:

$$x(n) = \left(\frac{1}{2}\right)^{|n|}$$

Ans: $X(z) = \dfrac{z}{z - \frac{1}{2}} - \dfrac{z}{z-2}$; ROC: $\dfrac{1}{2} < |z| < 2$

EP. 8.5 Sketch the ROC (if exists) in the z-plane of the bilateral z-transform of $x(n)$ and $y(n)$.

(a) $x(n) = u(n) + u(-n)$

(b) $y(n) = u(n) - u(n-10)$

Ans:

(a) Does not exist

(b) All z

EP. 8.6 Using the time-shift property, find the z-transforms of the following signals:

(a) $x_1(n) = u(n-5)$

(b) $x_2(n) = u(n+2)$

(c) $x_3(n) = \left(\dfrac{2}{3}\right)^n u(n+2)$

Ans:

(a) $\dfrac{z^{-4}}{z-1}$; ROC : $|z| > 1$

(b) $\dfrac{z^3}{z-1}$; ROC : $|z| > 1$

(c) $\dfrac{z^3}{z - \frac{2}{3}}$; ROC : $|z| > \dfrac{2}{3}$

EP. 8.7 Using the z-domian differentiation property, find the z-transform of

$$x(n) = n(n+1)\,u(n)$$

Ans: $X(z) = \dfrac{2z^2}{(z-1)^3}$; ROC : $|z| > 1$

EP. 8.8 Find the initial value $x(0)$ and the final value $x(\infty)$ of discrete-time causal sequence $x(n)$, whose z-transform is given below:

$$X(z) = \frac{z^3 - \frac{3}{4}z^2 + 2z - \frac{5}{4}}{(z-1)\left(z - \frac{1}{3}\right)\left(z^2 - \frac{1}{2}z + 1\right)}$$

Verify the result by finding $x(n)$ and then taking the limits $n \to 0$ and $n \to \infty$.

Ans: $x(0) = 1, \quad x(\infty) = 1$

$$x(n) = u(n) + \left(\frac{1}{3}\right)^n u(n) - \left(\frac{1}{2}\right)^n \cos\left(\frac{\pi}{3}n\right)$$

EP. 8.9 Decide whether the following z-transforms belong to a finite series or an infinite series. Give the ROC for each under the assumption that it is a right-sided series being dealt with.

(a) $H_1(z) = z^3 - 1$

(b) $H_2(z) = \dfrac{1}{z^3} - 1$

(c) $H_3(z) = \dfrac{1}{z^2}$

(d) $H_4(z) = 1 + \dfrac{z+1}{z-0.5}$

(e) $H_5(z) = \dfrac{2z^2 - 4z + 2}{z(z-1)}$

Ans:

(a) No poles except at infinity
$\Rightarrow$ Finite, ROC : $|z| < \infty$

(b) & (c) No poles at the origin
$\Rightarrow$ Finite, ROC : $|z| > 0$

(d) Infinite, ROC : $|z| > 0.5$

(e) Finite, ROC : $|z| > 0$

EP. 8.10 Using the relation

$$a^n u(n) \overset{\textbf{ZT}}{\longleftrightarrow} \frac{z}{z-a}; \quad \text{ROC} : |z| > |a|$$

find the z-transform of the following $x(n)$:

(a) $x(n) = na^{n-1}u(n)$

(b) $x(n) = n(n-1)a^{n-2}u(n)$

(c) $x(n) = n(n-1)\ldots(n-k+1)a^{n-k}u(n)$

Ans:

(a) $\dfrac{z}{(z-a)^2}$; ROC : $|z| > |a|$

(b) $\dfrac{2z}{(z-a)^3}$; ROC : $|z| > |a|$

(c) $\dfrac{k!\,z}{(z-a)^{k+1}}$; ROC : $|z| > |a|$

EP. 8.11 Given,

$$X(z) = \frac{(z-10)(z+10)}{(z-1)(z-2)(z-3)}$$

(a) State all the possible regions of convergence.

(b) For which ROC is $X(z)$, the z-transform of a causal sequence?

Ans:

(a) $0 < |z| < 1,\quad 1 < |z| < 2,\quad 2 < |z| < 3,\quad |z| > 3$

(b) $|z| > 3$

EP. 8.12 Find the z-transform of each of the following sequences:

(a) $x(n) = \left(\dfrac{1}{2}\right)^n u(-n)$

(b) $x(n) = \left(\dfrac{1}{2}\right)^n u(n) + (3)^n u(-n-1)$

(c) $x(n) = \left(\dfrac{1}{3}\right)^n \cos(\Omega_0 n)u(n)$

Ans:

(a) $X(z) = \dfrac{1}{1-2z}$; ROC: $|z| < \dfrac{1}{2}$

(b) $X(z) = \dfrac{4z^3}{z - \dfrac{1}{2}}$; ROC: $\frac{1}{2} < |z| < 3$

(c) $X(z) = \dfrac{z^2 - \dfrac{1}{3}z\cos\Omega_0}{z^2 - \dfrac{2}{3}\cos\Omega_0 z + \dfrac{1}{9}}$; ROC : $|z| > \dfrac{1}{3}$

EP. 8.13 Let $x(n)$ be a finite-length sequence that is nonzero only for $0 \le n \le N-1$, and consider the one-sided periodic sequence, $y(n)$, that is formed by periodically extending $x(n)$ as follows:

$$y(n) = \sum_{k=0}^{\infty} x(n - kN)$$

Express z-transform of $y(n)$ in terms of $X(z)$ and find the ROC of $Y(z)$.

Ans: $Y(z) = \dfrac{X(z)}{1 - z^{-N}}$; ROC : $|z| > 1$

EP. 8.14 Let

$$X(z) = \frac{z(z-2)}{(z-1)(z-0.5)}$$

Find the initial value $x(0)$ and the final value $x(\infty)$ of the sequence $x(n)$.

Ans: $x(0) = 1$, $x(\infty) = -2$

EP. 8.15 Find the z-transforms and ROC for the following:

(a) $x(n) = \cos\left(\dfrac{n\pi}{4} - \dfrac{\pi}{4}\right)u(n)$

(b) $y(n) = n(0.5)^n \cos(0.25n\pi)u(n)$

Ans:

(a) $X(z) = \dfrac{0.707z^2}{z^2 - z\sqrt{2} + 1}$; ROC : $|z| > 1$

(b) $Y(z) = \dfrac{4\sqrt{2}z^2 - 8z^2 + \sqrt{2}z}{(4z^2 - 2\sqrt{2}z + 1)^2}$; ROC : $|z| > 0.5$

EP. 8.16 Find the z-transform $X(z)$ and its ROC for the following:

(a) $x(n) = (0.5)^{|n|}$

(b) $x(n) = (0.5)^{|n|} + (-0.5)^{|n|}$

Ans:

(a) $X(z) = \dfrac{z}{z-0.5} + \dfrac{2}{2-z} - 1;$ ROC : $0.5 < |z| < 2$

(b) $X(z) = \dfrac{2z^2}{z^2-0.25} + \dfrac{4}{4-z^2} - 2;$ ROC : $0.5 < |z| < 2$

EP. 8.17 The z-transform of $x(n)$ is $X(z) = \dfrac{4z}{(z+0.5)^2}$, $|z| > 0.5$. Find the z-transform of the following, using properties and specify the ROC.

(a) $y_1(n) = x(n) * x(n)$

(b) $y_2(n) = x(n) - x(n-1)$

(c) $y_3(n) = n^2 x(n)$

Ans:

(a) $x(n) * x(n) \overset{\text{ZT}}{\longleftrightarrow} X^2(z) = \dfrac{16z^2}{(z+0.5)^4};$ ROC : $|z| > 0.5$

(b) $y_2(n) \overset{\text{ZT}}{\longleftrightarrow} Y_2(z) = \dfrac{4(z-1)}{(z+0.5)^2};$ ROC : $|z| > 0.5$

(c) $y_3(n) \overset{\text{ZT}}{\longleftrightarrow} Y_3(z) = \dfrac{4z^3 - 8z^2 + z}{(z+0.5)^4};$ ROC : $|z| > 0.5$

EP. 8.18 Use the z-transform to perform the convolution of the following two sequences:

$$h(n) = \begin{cases} \left(\frac{1}{2}\right)^n, & 0 \le n \le 2 \\ 0, & \text{elsewhere} \end{cases}$$

$$x(n) = \delta(n) + \delta(n-1) + 4\delta(n-2)$$

Ans:

$$\begin{aligned} y(n) &= x(n) * h(n) \\ &= \delta(n) + \frac{3}{2}\delta(n-1) + \frac{19}{4}\delta(n-2) + \frac{9}{4}\delta(n-3) + \delta(n-4) \end{aligned}$$

EP. 8.19 Use the z-transform to perform the convolution of the following two sequences:

$$x(n) = \delta(n) - \delta(n-2)$$
$$h(n) = 2\delta(n) - 2\delta(n-1) + 3\delta(n-2) + \delta(n-3)$$

Ans:

$$\begin{aligned} y(n) &= x(n) * h(n) \\ &= 2\delta(n) - 2\delta(n-1) - \delta(n-2) + 5\delta(n-3) - 6\delta(n-4) - 2\delta(n-5) \end{aligned}$$

EP. 8.20 Use derivative property to find the z-transform of $x(n)$.

$$x(n) = \frac{1}{n}(-2)^{-n}u(-n-1)$$

Ans: $X(z) = \log_e\left(z + \frac{1}{2}\right)$; ROC : $|z| < \frac{1}{2}$

EP. 8.21 Determine the z-transform for each of the following sequences. Sketch the pole-zero plot and indicate whether or not Fourier transform of the sequence exists.

(a) $x(n) = \delta(n-1)$

(b) $x(n) = -\left(\frac{1}{2}\right)^n u(-n-1)$

Ans:

(a) $X(z) = z^{-1}$; ROC : All z except $z = 0$, Fourier transform exists.

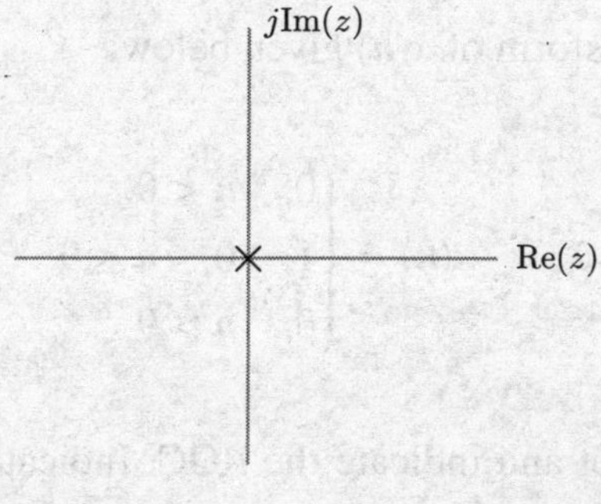

(b) $X(z) = \dfrac{z}{z - \frac{1}{2}}$; ROC : $|z| < \frac{1}{2}$, Fourier transform does not exist.

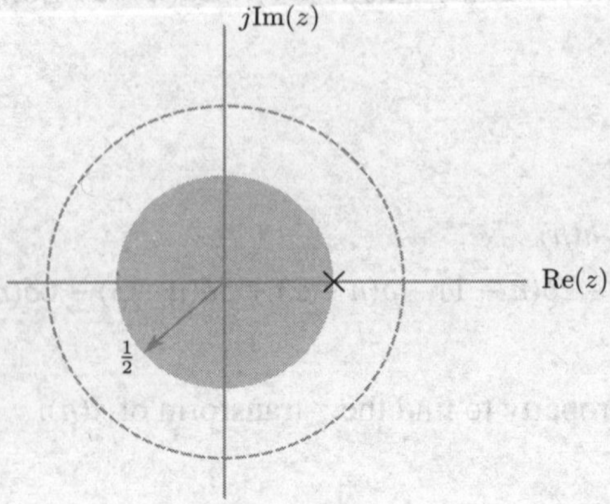

EP. 8.22 Determine the inverse z-transforms of the following functions:

(a) $X_1(z) = \dfrac{z-1}{z(z+0.5)}$; ROC : $|z| > 0.5$

(b) $X_2(z) = \dfrac{z-1}{(z+0.5)^2}$; ROC : $|z| > 0.5$

Ans:

(a) $x_1(n) = 6\delta(n) - 2\delta(n-1) - 6(-0.5)^n u(n)$

(b) $x_2(n) = -4\delta(n) + (4 - 6n)(-0.5)^n u(n)$

EP. 8.23 Determine the z-transform of $x(n)$ given below:

$$x(n) = \begin{cases} 0, & n < 0 \\ 1, & 0 \leq n \leq 9 \\ 0, & n > 9 \end{cases}$$

Sketch the pole-zero plot and indicate the ROC. Indicate whether the Fourier transform of the sequence exists.

Ans: $X(z) = \dfrac{z^{10}-1}{z^9(z-1)}$; ROC : All z expect $z = 0$, Fourier Transform exists.

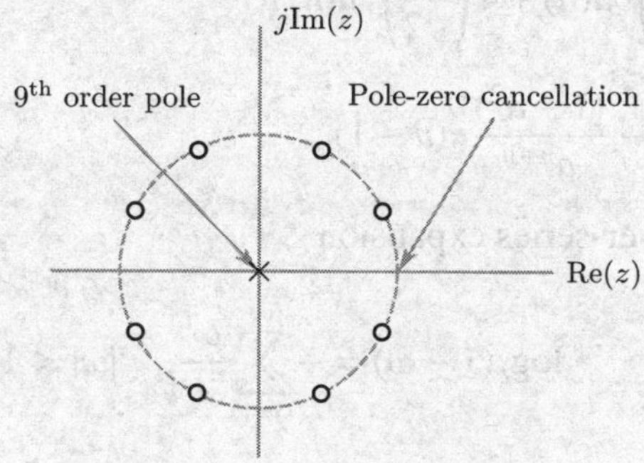

EP. 8.24 Using the method indicated, determine the sequence associated with each of the z-transform given below:

(a) Partial fractions:

$$X(z) = \frac{1-2z^{-1}}{1-\frac{5}{2}z^{-1}+z^{-2}}$$

and $x(n)$ is absolutely summable.

(b) Long division:

$$X(z) = \frac{1-\frac{1}{2}z^{-1}}{1+\frac{1}{2}z^{-1}}$$

and $x(n)$ is right-sided.

Ans:

(a) $x(n) = \left(\dfrac{1}{2}\right)^n u(n)$

(b) $x(n) = \delta(n) - \left(\dfrac{1}{2}\right)^{n-1} u(n-1)$

EP. 8.25 Listed here are two z-transforms. For each one, determine the inverse z-transform based on the partial fraction expansion and Taylor's series method based on the use of long division.

(a) $X(z) = \dfrac{1-\dfrac{1}{2}z^{-1}}{1+\dfrac{3}{4}z^{-1}+\dfrac{1}{8}z^{-2}}$; ROC : $|z| > \dfrac{1}{2}$

(b) $X(z) = \dfrac{1-az^{-1}}{z^{-1}-a}$; ROC : $|z| > \left|\dfrac{1}{a}\right|$

Ans:

(a) $x(n) = -3\left(-\frac{1}{4}\right)^n u(n) + 4\left(-\frac{1}{2}\right)^n u(n)$

(b) $x(n) = -\frac{1}{a}\delta(n) - \frac{(1-a^2)}{a^{n+1}}u(n-1)$

EP. 8.26 By using the Power-series expansion

$$\log_e(1-\omega) = -\sum_{i=1}^{\infty}\frac{\omega^i}{i}, \quad |\omega| < 1$$

determine the inverse of each of the following two z-transforms:

(a) $X(z) = \log_e(1-2z), \quad |z| < \frac{1}{2}$

(b) $X(z) = \log_e\left(1 - \frac{1}{2}z^{-1}\right), \quad |z| > \frac{1}{2}$

Ans:

(a) $x(n) = \frac{2^{-n}}{n}u(-n-1)$

(b) $x(n) = \frac{-2^n}{n}u(n-1)$

EP. 8.27 Figure below shows the pole-zero pattern of a discrete-time system. Demonstrate algebraically that $|H(e^{j\Omega})|$ is constant. $ROC: |z| > a$.

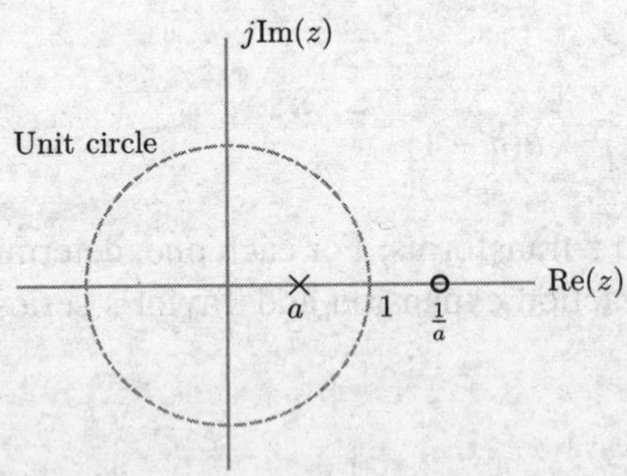

EP. 8.28 A causal LTI system is described by the difference equation

$$y(n) = y(n-1) + y(n-2) + x(n-1)$$

(a) Find the system function $H(z) = \dfrac{Y(z)}{X(z)}$ for this system.

(b) Find the unit impulse response of this system.

(c) Find the system stability.

Ans:

(a) $H(z) = \dfrac{z}{z^2 - z - 1}$; ROC: $|z| > \dfrac{(1 + \sqrt{5})}{2}$

(b) $h(n) = -\dfrac{1}{\sqrt{5}}\left(\dfrac{1+\sqrt{5}}{2}\right)^n u(n) + \dfrac{1}{\sqrt{5}}\left(\dfrac{1-\sqrt{5}}{2}\right)^n u(n)$

(c) Unstable system (refer figure shown below).

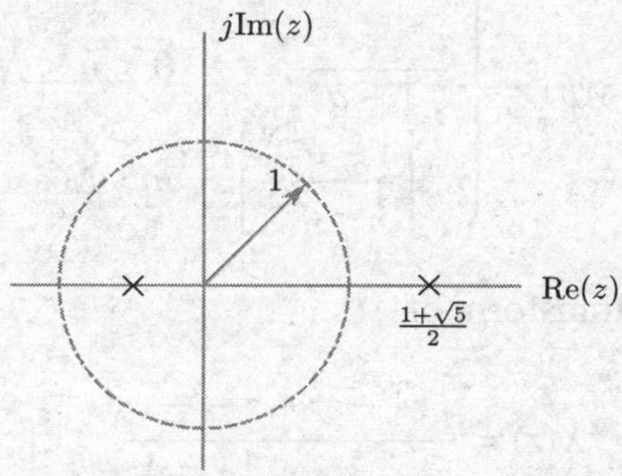

EP. 8.29 Consider an LTI system with input $x(n)$ and output $y(n)$ for which

$$y(n-1) - \frac{5}{2}y(n) + y(n+1) = x(n)$$

The system may or may not be stable and/or causal.

By considering pole-zero pattern associated with the difference equation, determine the three possible choices for the unit sample response of the system.

Ans:

(a) $h_1(n) = -\dfrac{2}{3}\left(\dfrac{1}{2}\right)^n u(n) + \dfrac{2}{3}(2)^n u(n)$

(b) $h_2(n) = -\dfrac{2}{3}\left(\dfrac{1}{2}\right)^n u(n) + \left(-\dfrac{2}{3}\right)2^n u(-n-1)$

(c) $h_3(n) = +\dfrac{2}{3}\left(\dfrac{1}{2}\right)^n u(-n-1) + \left(-\dfrac{2}{3}\right)2^n u(-n-1)$

EP. 8.30 Find the discrete convolution sum:

$$y(n) = h(n) * x(n)$$

where

$$h(n) = \begin{cases} \alpha^n, & n \geq 0 \\ 0, & n < 0 \end{cases}$$

$$x(n) = \begin{cases} 1, & 0 \leq n \leq N-1 \\ 0, & \text{elsewhere} \end{cases}$$

Ans:

$$y(n) = \begin{cases} 0, & n < 0 \\ \dfrac{a^n - a^{-1}}{1-a}, & 0 \leq n \leq N-1 \\ a^n \left[\dfrac{1-a^{-N}}{1-a^{-1}}\right], & n > N-1 \end{cases}$$

EP. 8.31 Find the inverse z-transform of

$$X(z) = \frac{z^3 - z^2 + z - \frac{1}{16}}{z^3 - \frac{5}{4}z^2 + \frac{1}{2}z - \frac{1}{16}}, \quad |z| > \frac{1}{2}$$

Ans: $x(n) = \delta(n) - 9\left(\frac{1}{2}\right)^n u(n) + 5n\left(\frac{1}{2}\right)^n u(n) + 9\left(\frac{1}{4}\right)^n u(n)$

EP. 8.32 Solve the difference equation

$$y(n) - \frac{3}{4}y(n-1) + \frac{1}{8}y(n-2) = 2\sin\left(\frac{n\pi}{2}\right)$$

with the initial conditions $y(-1) = 2$ and $y(-2) = 4$.

Ans: $x(n) = \frac{13}{5}\left(\frac{1}{2}\right)^n - \frac{8}{17}\left(\frac{1}{4}\right)^n + \frac{112}{85}\sin\frac{n\pi}{2} - \frac{96}{85}\cos\frac{n\pi}{2}, n \geq 0$

EP. 8.33 The step response of a linear, time-invariant, causal system be

$$y(n) = \frac{6}{5}u(n) - \frac{1}{3}\left(\frac{1}{2}\right)^n u(n) + \frac{2}{15}\left(-\frac{1}{4}\right)^n u(n)$$

(a) Find the transfer function of the system.

(b) Find the impulse response of the system.

(c) Find a difference equation realization for the system.

Ans:

(a) $H(z) = \dfrac{z^2 - \frac{1}{4}z}{\left(z - \frac{1}{2}\right)\left(z - \frac{1}{4}\right)}$; $ROC: \quad |z| > \dfrac{1}{2}$

(b) $h(n) = \dfrac{2}{3}\left(-\dfrac{1}{4}\right)^n u(n) + \dfrac{1}{3}\left(\dfrac{1}{2}\right)^n u(n)$

(c) $y(n) - \frac{1}{4}y(n-1) - \frac{1}{8}y(n-2) = x(n) - \frac{1}{4}x(n-1)$

EP. 8.34 Consider the system described by the difference equation

$$y(n) - 2y(n-1) + 2y(n-2) = x(n) + \frac{1}{2}x(n-1)$$

(a) Find the system function $H(z)$.

(b) Find the stability of the system.

(c) Determine the unit sample response of the system.

Ans:

(a) $H(z) = \dfrac{z^2 + \frac{1}{2}z}{z^2 - 2z + 2}$; ROC : $|z| > \sqrt{2}$

(b) Unstable system.

(c) $h(n) = \left(\sqrt{2}\right)^n \cos\left(\dfrac{\pi}{4}n\right)u(n) + \dfrac{3}{2}\left(\sqrt{2}\right)^n \sin\left(\dfrac{\pi}{4}n\right)u(n).$

EP. 8.35 If $X_1(z)$ denotes the unilateral z-transform of $x(n)$, determine interms of $X_1(z)$, the unilateral z-transform of:

(a) $x(n+1)$

(b) $x(n-3)$

Ans:

(a) $\mathscr{Z}_u\{x(n+1)\} = z[X_1(z) - x(0)]$

(b) $\mathscr{Z}_u\{x(n-3)\} = z^{-3}X_1(z) + z^{-2}x(-1) + z^{-1}x(-2) + x(-3)$

EP. 8.36 Find the unilateral z-transform of the following sequences:

(a) $x(n) = \delta(n)$

(b) $x(n) = \delta(n-1)$

(c) $x(n) = \delta(n+1)$

(d) $x(n) = \left(\frac{1}{2}\right)^n u(n)$

(e) $x(n) = -\left(\frac{1}{2}\right)^n u(-n-1)$

Ans: (a) $X_1(z) = 1$ (b) $X_1(z) = z^{-1}$ (c) 0 (d) $X_1(z) = \frac{z}{z-\frac{1}{2}}$ (e) $X_1(z) = 0$.

EP. 8.37 Find the inverse z-transform of the following system functions:

(a) $H_1(z) = \dfrac{z}{(z-1)^2(z-2)}$

(b) $H_2(z) = \dfrac{z^2-3z}{(z-2)(z^2-2z+2)}$

Ans:

(a) $h_1(n) = (2)^n u(n) - nu(n) - u(n)$

(b) $h_2(n) = -0.5(2)^n u(n) + 2(0.7906)\left(\sqrt{2}\right)^n \cos\left(\frac{n\pi}{4} - 71.56°\right)$

EP. 8.38 Find the inverse z-transform of the following system function:

$$H(z) = \frac{z^2+z}{z^2-2z+4}$$

Ans: $h(n) = (2)^n \cos\left(\frac{n\pi}{3}\right) u(n) + 1.1547(2)^n \sin\left(\frac{n\pi}{3}\right) u(n)$

EP. 8.39 A linear time-invariant system is represented by its system function:

$$H(z) = \frac{z^2}{z^2 - \frac{1}{6}z - \frac{1}{6}}$$

Find the system response, when the input $x(n) = 4u(n)$. The initial conditions are $y(-1) = 0$ and $y(-2) = 12$.

Ans: $y(n) = -1.2\left(\frac{1}{2}\right)^n u(n) + 0.8\left(-\frac{1}{3}\right)^n u(n)$

EP. 8.40 The transfer function of a system is

$$H(z) = \frac{2z(z-1)}{z^2+4z+4}$$

Find the system response $y(n)$ for the input $x(n) = nu(n)$.

Ans: $y(n) = \frac{1}{9}[2 - 2(-2)^n + 12n(-2)^{n-1}]u(n)$

EP. 8.41 Find the response $y(n)$ using z-transform.

$$y(n) - 0.4y(n-1) = (0.4)^n u(n) + 2(0.5)^{n-1}u(n-1)$$
$$y(-1) = 2.5$$

Ans: $y(n) = [-3n(0.4)^n - 26(0.4)^n + 30(0.5)^n]u(n)$

EP. 8.42 How can you identify whether a system is causal and/or stable system from the following information?

(a) Its impulse response $h(n)$.

(b) Its transfer function $H(z)$ and its ROC.

(c) Its system difference equation.

(d) Its pole-zero plot.

EP. 8.43 In evaluating the inverse z-transform by complex inversion integral, a contour integral must be computed. Evaluate the following typical contour integrals for contour C specified:

(a) $\oint_C \frac{1}{z^2}\,dz; \quad C: |z| = 1$

(b) $\oint_C \frac{z^{-2}}{z-1}\,dz; \quad C: |z| = 2$

(c) $\frac{1}{2\pi j}\oint_C \frac{1}{z^5}\,dz; \quad C: |z| = \frac{1}{5}$

(d) $\oint_C \frac{1}{(z-1)^2(z+2)}\,dz; \quad C: |z| = 3$

(e) $\oint_C \frac{z-2}{z^2(z+2)}\,dz; \quad C: |z| = 1$

Ans: (a) 0 (b) 0 (c) 0 (d) 0 (e) $2\pi j$

EP. 8.44 Evaluate the inverse z-transform of the following $X(z)$ for the ROC indicated by using the complex inversion integral.

(a) $X(z) = \dfrac{z^2}{\left(z - \frac{1}{2}\right)\left(z - \frac{1}{4}\right)}$ for ROC : $|z| > \dfrac{1}{2}$ and

(b) $X(z) = \dfrac{z^2}{z - \frac{1}{2}}$ for ROC : $|z| > \dfrac{1}{2}$

Ans: (a) $x(n) = \left[2\left(\frac{1}{2}\right)^n - \left(\frac{1}{4}\right)^n\right] u(n)$ (b) $x(n) = \left(\frac{1}{2}\right)^{n+1} u(n+1)$

Chapter 9

Random processes and their temporal characteristics

Signals are broadly classified into deterministic and nondeterministic signals. Deterministic signals which are functions of time are predictable. By knowing their past and present values, the future values can be predicted. In the case of nondeterministic signals, by knowing their past and present values, the future values cannot be entirely predicted. Nondeterministic signals are also known as random signals. The signals or waveforms we encounter in engineering are generally random. Examples are, waveform or signals which carry information of speech, text, image, video, digital data, signals from sensors, command or control signals etc. These signals are all random, which are examples of information bearing signals and hence called desired signals. The desired signal is generally corrupted by undesired random signals, called interference signals or noise. Even though the desired signals and noise are unpredictable and random they can be described in terms of their average properties. Characterisation of random signals and noise are needed for evaluating the performance of communication systems. This chapter deals with statistical characterisation of random signals.

9.1 Concept of random process or stochastic process

Random process or Stochastic process can be regarded as extension of concept of random variable. Random variable X is defined as a function of possible outcomes s of an experiment. It is functional mapping from the set S of all outcomes s, called sample space to numbers. Random process is a function of both outcome s and time t. Outcome s_i is assigned a time function $x(t, s_i)$

Let S be the sample space and $s_1, s_2, \ldots s_k$ represent possible outcomes. Let the corresponding time functions assigned to these outcomes be $x(t, s_1)$, $x(t, s_2)$ ..., $x(t, s_k)$. The first variable t corresponds to observation time and the second variable s corresponds to outcome. The family of all such functions is denoted by $X(t, s)$ and is called a random process or stochastic process. Thus random process $X(t, s)$ denotes a family of time waveforms, called family or ensemble of sample functions. $x(t, s_i)$ denotes a particular waveform called sample function or ensemble member associated with outcome s_i. Figure 9.1 illustrates the correspondence between sample points, s_i and waveforms $x(t, s_i)$, i = a, 2,..., k.

Consider the following cases.

Case (i) t and s are both variables then $X(t, s)$ represents family or ensemble of time functions $x(t, s)$

Case(ii) Suppose t is variable and s is fixed and is equal to s_i say. Then $X(t, s_i)$ represents a single time function $x(t. s_i)$ and is called sample function associated with outcome s_i.

Case (iii) Suppose t is fixed and is equal to t_1 and s is variable. Then $X(t_1, s)$ represents a set of numbers, $x(t_1, s_1)$, $x(t_1, s_2)$, ..., $x(t_1, s_k)$ depending on the outcome s. Hence $X(t_1, s)$ is random variable, generally denoted by $X(t_1)$.

Case(iv) Suppose both t and s are fixed say $t = t_1$, and $s = s_k$. Then, $X(t_1, s_k)$ represents a number $x(t_1, s_k)$ It is the value of the sample function $x(t_1, s_k)$ at $t = 1$ and hence $x(t, s_k)$ is a number.

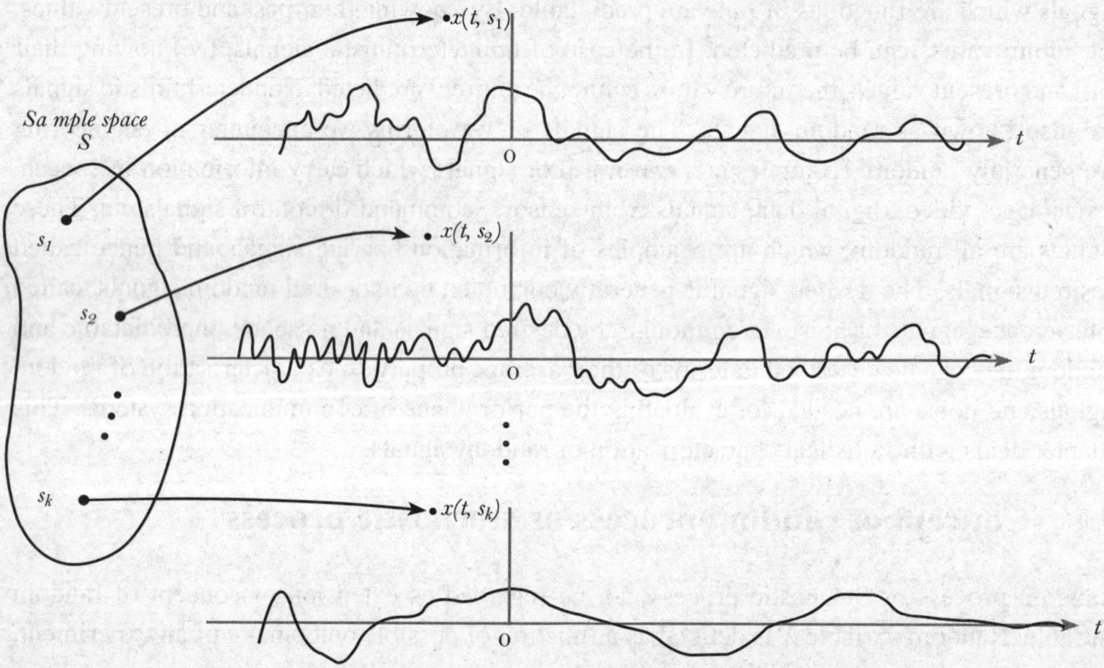

Figure 9.1: *A continuous random process where the correspondence between sample points s_i- and waveforms $x(t,s_1)$ are illustrated.*

Usually the second variable s is omitted and the random process is denoted by just $X(t)$.

The distinction between random variable and random process may be summarised as follows:

(i) a random variable X is a rule or a mapping which assigns a number to each outcome s.

(ii) a random process is a family of time functions $X(t, s)$. To every outcome s_i according to some rule a time function $X(t, s_i)$ is assigned.

9.2 Classification of processes

Random processes are classified according to the characteristics of time t and the type of random variable X at time t. It is assumed that both t and X takes values in the range $-\infty$ to $+\infty$. Four types of random processes, based on nature of values of time t and random variable X, are considered.

(i) Continuous random process

If X is a continuous random variable and t can have any continuous values then the random process $X(t)$ is called continuous random process. The random process depicted in Fig. 9.1 is an example of continuous random process. Noise voltage across a resistor is an example of sample function of continuous random process. The family of noise voltages across many resistors of same value at a given temperature is an example of continuous random process.

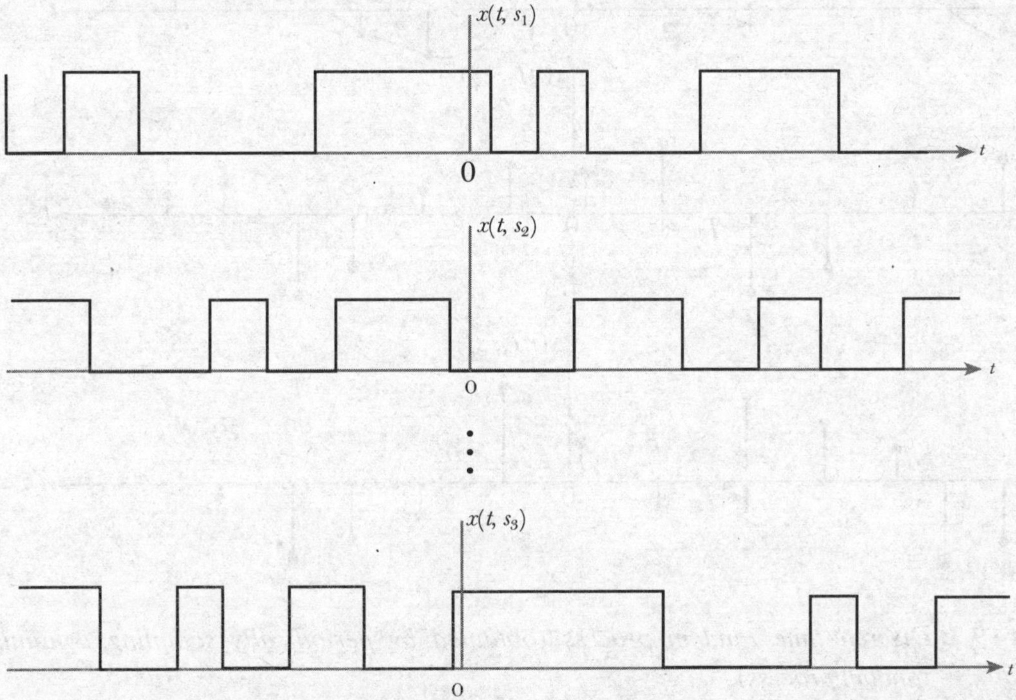

Figure 9.2: *Discrete random process X(t).*

(ii) Discrete random process

If X takes only discrete values and t is continuous, then the random process is called a discrete random process. The output from a digital system has only two values. If a sample function of random process of the type shown in Fig. 9.1 is passed through a comparator, the output is random but has only two values. An example of discrete random process is shown in Fig. 9.2.

(iii) Continuous random sequence or Discrete-time random process

If X is a continuous random variable but time t takes only discrete values, then $X(t)$ is called continuous random sequence. A continuous random sequence is generated when the sample functions of a continuous random process of the type given in Fig. 9.1 are periodically sampled. A continuous random sequence is defined at only discrete-time hence it is also called as discrete-time random process. Its sample function is called discrete-time random signal or discrete-time sequence. The interval between samples is denoted by T_s and $X(nTs)$, $n = 0, \pm 1, \pm 2, \ldots$, are a set of random variables. Figure 9.3 depicts example of discrete-time random process.

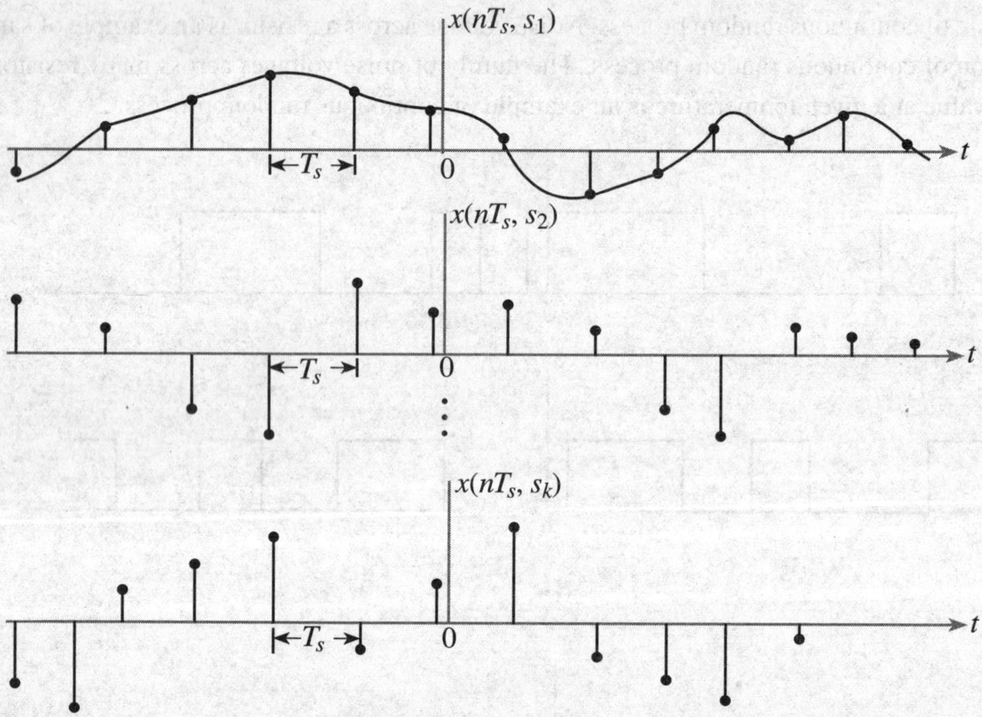

Figure 9.3: *Discrete-time random process (obtained by periodically sampling continuous random process).*

(iv) Discrete random sequence

If both random variable and time take discrete values then the random process is called a discrete random sequence. Figure 9.4 depicts a discrete random sequence. A discrete random sequence is obtained by periodically sampling a discrete random process.

Quantised samples of a discrete-time random process also results in discrete random sequence. Such sequences result whenever the sample values are represented by finite and fixed number of bits.

9.3 Deterministic and nondeterministic processes

A random process can also be classified based on the type of its sample functions. Suppose $x(t, s_k)$ is a sample function of random process $X(t)$. If its future values cannot be predicted exactly from its observed past values, the process is called nondeterministic. On the other hand, suppose the individual sample functions $x(t, s_i) = 1, 2, ...,$ are different. But once a sample function is chosen, say $x(t, s_k)$, if its future values can be predicted exactly by knowing its past values, then the random process $X(t)$ is called deterministic random process. Consider the random process

$$X(t) = A_c \sin(\omega_c t + \theta_c)$$

where A_c, ω_c and θ_c are random variables. Suppose sample function $x(t, s_k)$ corresponding to outcome s_k has particular values $A_c = 2$, $\omega_c = 200\pi$ and $\theta_c = \pi/4$. That is

$$x(t, s_k) = 2\sin(200\pi t + \pi/4)$$

Then if $x(t, s_k)$ is known prior to any time instant $t = t_0$, its values can be predicted for all future times $> t_0$.

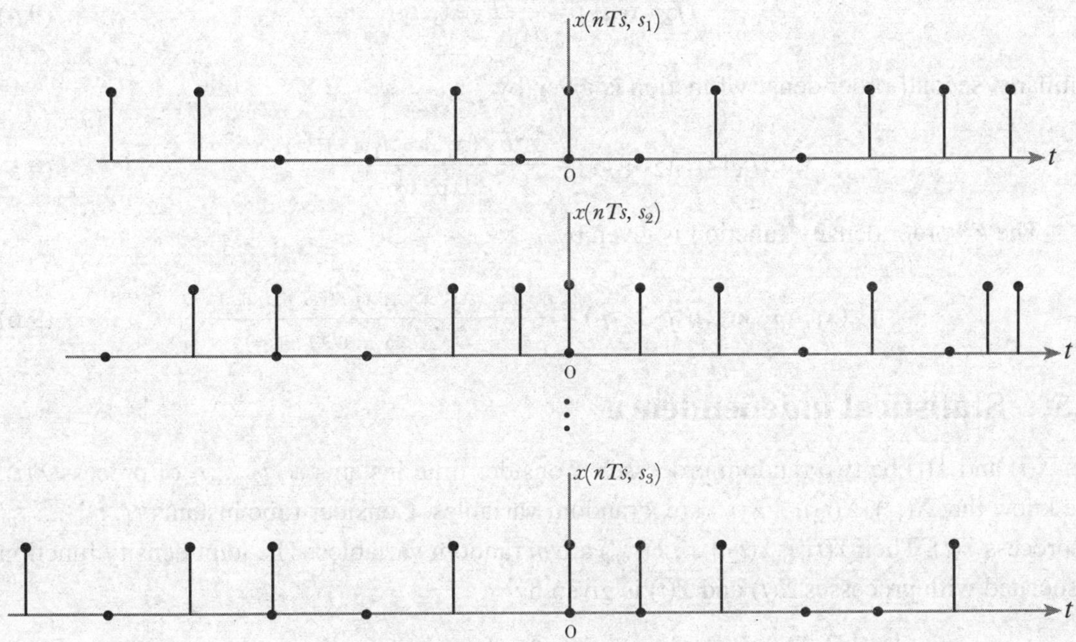

Figure 9.4: *Discrete random sequence.*

9.4 Distribution and density functions

Consider a random process $X(t)$. For $t = t_1$, $X(t_1)$ is a random variable. The probability distribution function associated with random variable $X(t_1)$is defined as

$$F_X(x_1; t_1) = P(X(t_1) \leq x_1) \tag{9.1}$$

where x_1 is any real number.

In general the distribution function is a function of time which is indicated by the second variable t_1. Suppose we consider two time instants t_1 and t_2. Then $X(t_1)$ and $X(t_2)$ are random variables. The second-order joint distribution function is defined as

$$F_X(x_1, x_2; t_1, t_2 = P(X(t_1) \leq x_1, X(t_2) \leq x_2) \tag{9.2}$$

In a similar manner we can define k^{th} order joint distribution function. Consider k time instants $t_1, t_2, ..., t_k$. We have k random variables $X(t_1), X(t_2), ..., X(t_k)$. The kth order joint distribution function is given by

$$Fx(x_1, x_2, ..., x_k ; t_1, t_2 ..., t_k) = P[X(t_1) \leq x_1, X(t_2) \leq x_2 ..., X(t_k) \leq x_k] \tag{9.3}$$

The joint probability density functions can be determined from the joint distribution functions. First-order density function is given by

$$f_X(x_1;t_1) = \frac{d}{dx_1} F_X(x_1;t_1) \tag{9.4}$$

Similarly, second-order density function is given by

$$f_X(x_1, x_2;t_1,t_2) = \frac{\partial^2 F_X(x_1, x_2;t_1,t_2)}{\partial x_1 \partial x_2} \tag{9.5}$$

The kth order density function is given by

$$f_X(x_1, x_2 ... x_k;t_1,t_2 ... t_k) = \frac{\partial^k F_X(x_1, x_2 ... x_k;t_1,t_2 ... t_k)}{\partial x_1 \partial x_2 ... \partial x_k} \tag{9.6}$$

9.5 Statistical independence

Let $X(t)$ and $Y(t)$ be two random processes. Consider time instants $t_1, t_2 ..., t_k$ of process $X(t)$. We know that $X(t_1), X(t_2)..., X(t_k)$ are k random variables. Consider time instants $t_1', t_2', ..., t_m'$ of process $Y(t)$. Then $Y(t_1'), Y(t_2') ..., Y(t_m')$ are m random variables. The joint density function associated with processes $X(t)$ and $Y(t)$ is given by

$$f_{XY}(x_1,x_2 ... x_k, y_1, y_2 ... y_m;t_1,t_2 ... t_k,t_1't_2' ...t_m') \tag{9.7}$$

If it turns out that

$$\begin{aligned} &f_{XY}(x_1,x_2 ... x_k, y_1, y_2 ... y_m;t_1,t_2 ... t_k,t'_1,t'_2 ...t'_m) \\ &= f_X(x_1,x_2 ... x_k,t_1,t_2 ... t_k)\, f_Y(y_1, y_2 ... y_m,t'_1t'_2 ...t'_m) \end{aligned} \tag{9.8}$$

for all choices of $k, m, t_1, t_2, ..., t_k, t_1', t_2' ..., t_m'$ then the two processes are said to be strictly independent.

9.6 Stationarity

Let $X(t)$ be a random process defined for t > – ∞. Consider instants of time $t_1, t_2..., t_k$ at which $X(t_1), X(t_2)..., X(t_k)$ are random variables. Let the kth order joint probability density of this set of k random variables be denoted by

$$f_{X_1 X_2 ..., X_k}(x_1, x_2, ..., x_k; t_1, t_2, ..., t_k) \tag{9.9}$$

where X_i denotes random variable $X(t_i)$, $i = 1, 2, ..., k$. Suppose the observation time is shifted by a fixed amount τ. Then the new set of random variables are $X(t_i + \tau)$. The kth order joint probability density function of this new set of random variables is denoted by

$$f_{X'_1 X'_2 ..., X'_k}(x'_1, x'_2, ..., x'_k\,;\, t_1 + \tau, t_2 + \tau\, ..., t_k + \tau) \tag{9.10}$$

where X_i' denotes random variable $X_i(t_1 + \tau)$, $i = 1,2, ..., k$.

The random process $X(t)$ is said to be stationary in the strict-sense if

$$f_{X'_1 X'_2 ..., X'_k}(x_1, x_2, ..., x_k\,;\, t_1\, t_2\, ..., t_k)$$
$$= f_{X'_1 X'_2 ..., X'_k}(x'_1, x'_2, ..., x'_k\,;\, t_1 + \tau, t_2 + \tau, t_3 + \tau, ..., t_k + \tau) \tag{9.11}$$

for all k and τ and for all observation times $t_1, t_2, ..., t_k$. Thus, a random process $X(t)$ is said to be stationary in the strict-sense, if the joint probability density function of all orders are independent of shift in time. That is, the joint probability density function is invariant to a time shift. In other words, a random process $X(t)$ is stationary in the strict-sense if its statistics does not change with shift in time. Determining stationarity of higher order in the strict-sense may be too complex.

We next consider characterisation of first-order and second-order stationary processes.

9.7 First-order stationary process

Consider the first-order density function $f_X(x_1; t_1)$ of random process $X(t)$.

$$\text{If } f_X(x_1{:}t_1) = f_X(x_1{:}t_1{+}\tau) \tag{9.12}$$

for any t_1, and any real number τ, then $X(t)$ is said to be first-order stationary process. For a first-order stationary process the mean value of the process is independent of time.

Consider $E[X(t_1)]$. By definition

$$E[X(t_1)] = \int_{-\infty}^{\infty} x_1 f_X(x_1;t_1)\,dx_1 \tag{9.13}$$

Now consider $E[X(t_1 + \tau)]$. By definition

$$E[X(t_1 + \tau)] = \int_{-\infty}^{\infty} x_1 f_X(x_1;t_1 + \tau)\,dx_1 \tag{9.14}$$

For a first-order stationary process $f_X(x_1; t_1) = f_X(x_1; t_1 + \tau)$ for all τ and t_1

Hence from Eq. s (9.13) and (9.14) we see that

$$E[X(t_1)] = E[X(t_1 + \tau)] \tag{9.15}$$

for all t_1 and τ.

This implies that mean value of first-order stationary process is independent of time. Hence, it is a constant.

9.8 Second-order stationary process

A process $X(t)$ is said to have second-order stationarity if its second-order probability density function satisfies

$$f_X(x_1, x_2; t_1, t_2) = f_X(x_1, x_2; t_1 + \lambda, t_2 + \lambda) \tag{9.16}$$

for all t_1, t_2 and λ.

Since the process is stationary, the joint density function is invariant to shift in time λ. Hence, it is a function of only time difference. The time difference in the left hand side of density function in Eq. (9.16) is $(t_2 - t_1)$. The time difference in the right hand side of Eq. (9.16) is,

$$(t_2 + \lambda) - (t_1 + \lambda) = (t_2 - t_1)$$

Hence the second-order joint density function depends only on time difference $(t_2 - t_1)$, for all (t_1, t_2) and not on absolute time.

A second-order stationary process is also first-order stationary. Autocorrelation function of $X(t)$ is defined as

$$R_{XX}(t_1, t_2) = E[X(t_1)X(t_2)]$$

$$= \int_{-\infty}^{\infty}\int_{-\infty}^{\infty} f_{X_1X_2}(x_1, x_2; t_1, t_2)\,dx_1 dx_2 \tag{9.17}$$

If $X(t)$ is of second-order stationarity, the joint density function is a function of only time difference $(t_2 - t_1)$. This means right hand side of Eq. (9.17) is a function of only $(t_2 - t_1)$. Hence, the left hand side of Eq. (9.17), $R_{XX}(t_1, t_2)$ is also a function of $(t_2 - t_1)$.

9.9 Wide-sense stationary (WSS) process

In many of the practical problems, the knowledge of only mean value and autocorrelation of the random process are required. The analysis becomes simple if these quantities are independent of absolute time.

A random process $X(t)$ for which mean value $E[X(t)] = \overline{X}$ is a constant independent of time and autocorrelation function $E[X(t_1)X(t_2)] = R_{XX}(t_2 - t_1)$ is dependent on time difference only, is called **Wide-sense Stationary** (WSS). A random process which is stationary of order two is also wide-sense stationary. However the converse need not necessarily be true.

Random processes $X(t)$ are said to be jointly WSS. if the following hold good:

(i) mean value X,

$$E[X(t)] = \overline{X} \tag{9.18}$$

(ii) autocorrelation of X,

$$E[X(t)X(t+\tau)] = R_{XX}(\tau) \tag{9.19}$$

(iii) mean value of Y,

$$E[Y(t)] = \overline{Y} \tag{9.20}$$

(iv) autocorrelation of Y,

$$E[Y(t)Y(t+\tau)] = R_{YY}(\tau) \tag{9.21}$$

(v) and cross-correlations of X and Y

$R_{XY}(t_1, t_2) = E[X(t_1)Y(t_2)]$ is a function of only time difference $(t_2 - t_1) = \tau$ and not on absolute time. This means that

$$\begin{aligned} R_{XY}(t_1, t_2 + \tau) &= E[X(t)X(t+\tau)] \\ &= R_{XY}(\tau) \end{aligned} \tag{9.22}$$

9.10 Ergodicity

Let $X(t)$ be a stationary random process. Then at any time t_1 or t_2, $X(t_1)$ or $X(t_2)$ are random variables. The expected values $E[X(t_1)]$ or $E[X(t_2)]$ are called ensemble or statistical averages. Since $X(t)$ is a stationary random process, $E[X(t_1)]$ or $E[X(t_2)]$ are equal and constants independent of time.

Let us denote a single sample function $X(t_1, s_i)$ by $x(t)$. Then time average of $x(t)$ is given by

$$\langle x(t) \rangle = \lim_{T\to\infty} \frac{1}{2T} \int_{-T}^{T} x(t)\,dt \tag{9.23}$$

For some stationary processes it may turn out that the time average given by (9.23), denoted by $\langle x(t) \rangle$, and ensemble (statistical) average denoted by $E[X(t_1)]$ are same, that is

$$\langle x(t) \rangle = E[X(t_1)] = \overline{X} \tag{9.24}$$

Suppose this equality holds good for higher order averages also like second moment, second central moment etc. That is, second moment:

$$\text{Ensemble average } E[X^2(t)] = \langle x^2(t) \rangle \tag{9.25}$$

where $\langle x^2(t)\rangle$ is time average given by

$$\langle x^2(t)\rangle = \lim_{T\to\infty} \frac{1}{2T}\int_{-T}^{T} x^2(t)\,dt \qquad ...,(9.26)$$

Second central moment:

Ensemble average $E[(X(t)-\overline{X})^2] = \langle (x(t)-\langle x(t)\rangle)^2\rangle$ (9.27)

where $\langle x(t) - \langle x(t)\rangle)^2\rangle$ is time average given by

$$\langle (x(t)-\langle x(t)\rangle)^2\rangle = \lim_{T\to\infty} \frac{1}{2T}\int_{-T}^{T} (x(t)-\langle x(t)\rangle)^2\,dt \qquad (9.28)$$

Then the process is called an ergodic process. Thus a stationary process is called an ergodic process if all higher order ensemble or statistical averages are equal to corresponding time averages.

9.11 Joint ergodic processes

Let $X(t)$ and $Y(t)$ be two stationary random processes which are individually ergodic. Let their time cross-correlation function be

$$\mathbb{R}_{XY}(\tau) = \lim_{T\to\infty} \frac{1}{2T}\int_{-T}^{T} x(t)y(t+\tau)\,dt \qquad (9.29)$$

and statistical correlation function be

$$R_{XY}(\tau) = E[X(t)X(t+\tau)] \qquad (9.30)$$

If

$$\mathbb{R}_{XY}(\tau) = R_{XY}(\tau) \qquad (9.31)$$

then the two ergodic processes $X(t)$ and $Y(t)$ are called jointly ergodic.

Thus two stationary random processes $X(t)$ and $Y(t)$ are called jointly ergodic, if they are individually ergodic and also have a time cross-correlation function that is equal to the ensemble or statistical cross-correlation function.

9.12 Mean-ergodic process

Let $X(t)$ be a random process. Let its mean value (ensemble or statistical average) $\overline{X}$ be a constant. Let $\langle x(t)\rangle$ be time average of any sample function $x(t)$. If the statistical average $\overline{X}$ of the process equal to the time average $\langle x(t)\rangle$ of any sample function $x(t)$ with probability 1 for all sample functions, the process $X(t)$ is called mean-ergodic or ergodic in the mean. That is, for a mean-ergodic process, the statistical average $\overline{X}$ = time average $\langle x(t)\rangle$ with probability 1 for all sample functions of $X(t, s)$.

9.13 Correlation-ergodic process

Let $X(t)$ be a stationary continuous process. Let $R_{XX}(\tau)$ be its autocorrelation function (statistical average), where

$$R_{XX}(\tau) = E[X(t)X(t+\tau)] \tag{9.32}$$

which is also called statistical autocorrelation.

The time autocorrelation denoted by $\mathbb{R}_{XX}(\tau)$ is given by

$$\mathbb{R}_{XX}(\tau) = \lim_{T\to\infty} \frac{1}{2T}\int_{-T}^{T} x(t)y(t+\tau)\,dt \tag{9.33}$$

$X(t)$ is called autocorrelation-ergodic or ergodic in the autocorrelation if and only if for all τ, statistical autocorrelation $R_{XX}(\tau)$ is equal to the time autocorrelation $\mathbb{R}_{XX}(\tau)$.

Consider two processes $X(t)$ and $Y(t)$. Their statistical cross-correlation function is given by

$$R_{XY}(\tau) = E[X(t)Y(t+\tau)] \tag{9.34}$$

and time cross-correlation function is given by

$$\mathbb{R}_{XY}(\tau) = \langle x(t)y(t+\tau)\rangle = \lim_{T\to\infty} \frac{1}{2T}\int_{-T}^{T} x(t)y(t+\tau)\,dt \tag{9.35}$$

$X(t)$ and $Y(t)$ are called cross-correlation-ergodic or ergodic in the cross-correlation function if

$$R_{XY}(\tau) = \mathbb{R}_{XY}(\tau) \tag{9.36}$$

Let us next discuss the properties of autocorrelation and cross-correlation functions of random processes.

9.14 Autocorrelation function

Let $X(t)$ be a random process. Consider two time instants t_1 and t_2. Then $X(t_1)$ and $X(t_2)$ are two random variables. Autocorrelation function of random process $X(t)$ is defined as the correlation, $E[X(t_1)X(t_2)]$ of two random variables $X(t_1)$ and $X(t_2)$ and is denoted by $R_{XX}(t_1, t_2)$.

We have

$$R_{XX}(t_1, t_2) = E[X(t_1)X(t_2)] \tag{9.37}$$

Let $t_2 = t_1 + \tau$, where τ is a real number. Then the general form of autocorrelation function is

$$R_{XX}(t, t+\tau) = E[X(t)X(t+\tau)] \tag{9.38}$$

If $X(t)$ is second-order stationary or at least wide-sense stationary, then it is seen that autocorrelation function must be a function of only time difference $\tau = (t_2 - t_1)$. Hence, for a wide-sense stationary or second-order stationary processes,

$$R_{XX}(\tau) = E[X(t)X(t+\tau)] \tag{9.39}$$

9.15 Properties of autocorrelation function of wide-sense stationary or second-order stationary processes

Property 1 :

$$|R_{XX}(\tau)| \leq R_{XX}(0) \tag{9.40}$$

That is autocorrelation value has its peak value at $\tau = 0$ and $R_{XX}(\tau)$ is bounded by its peak value.

Proof: Consider $E[(X(t) - X(t + \tau))^2]$; $\tau \neq 0$. For $\tau \neq 0$ the quantity $(X(t) - X(t + \tau))$ may be zero or positive or negative. However its square is always zero or positive. Hence its ensemble average is always greater than or equal to zero. That is,

$$E[(X(t) - X(t + \tau))^2] \geq 0 \tag{9.41}$$

Expanding the squared term inside expectation, we have

$$E[X^2(t) + X^2(t + \tau) - 2X(t)X(t) + \tau] \geq 0 \tag{9.42}$$

Since expectation is a linear operation, the expected value of sum is equal to sum of expected values. Hence, Eq. (9.42) is equal to

$$E[X^2(t)] + \mathrm{E}[X^2(t + \tau)] - 2E[X(t)X(t + \tau)] \geq 0 \tag{9.43}$$

From the definition of autocorrelation function, the first two terms in Eq. (9.43) is equal to $R_{XX}(0)$ and the third term is $-2R_{XX}(\tau)$. Hence, Eq. (9.43) becomes

$$2R_{XX}(0) - 2R_{XX}(\tau) \geq 0$$

Rearranging, we get,

$$R_{XX}(0) \geq R_{XX}(\tau) \tag{9.44}$$

Similarly by considering

$$E[(X(t) + X(t + \tau))^2] \geq 0 \tag{9.45}$$

we can show that,

$$R_{XX}(0) \geq -R_{XX}(\tau) \tag{9.46}$$

Combining the results (9.44) and (9.46) we have,

$$R_{XX}(0) \geq | R_{XX}(\tau) | \tag{9.47}$$

Property 2:

$$R_{XX}(0) = E[X^2(t)] \tag{9.48}$$

Proof: This follows from the definition of autocorrelation function as given in (9.39) with $\tau = 0$,

$$R_{XX}(0) = E[X(t)X(t)] = E[X^2(t)]$$

Hence, the peak value of autocorrelation function is equal to the mean squared value of the process which is also called the power in the process.

Property 3:

$$R_{XX}(-\tau) = R_{XX}(\tau) \tag{9.49}$$

An autocorrelation function has an even symmetry about $\tau = 0$.

Proof:

$$R_{XX}(\tau) = E[X(t)X(t+\tau)]$$

$$R_{XX}(-\tau) = E[X(t)X(t-\tau)] \tag{9.50}$$

Put $t - \tau = t'$. Then, $t = (t' + \tau)$. Substituting these values in (9.50), we have,

$$R_{XX}(-\tau) = E[X(t'+\tau)\,X(t')]$$

$$= R_{XX}(\tau) \tag{9.51}$$

Property 4: If $E[X(t)] = \overline{X} \neq 0$ and $X(t)$ is ergodic with no periodic components, then

$$\lim_{|\tau|\to\infty} R_{XX}(\tau) = (\overline{X})^2 \tag{9.52}$$

Proof: We have

$$R_{XX}(\tau) = E[X(t)X(t+\tau)]$$

$$\lim_{|\tau|\to\infty} R_{XX}(\tau) = \lim_{|\tau|\to\infty} E[X(t)X(t+\tau)] \tag{9.53}$$

When $|\tau|$ is large, it is justified to assume $X(t)$ and $X(t+\tau)$ are independent if no periodic components are present.

Hence, in Eq. (9.53), the expectation of product can be written as product of expectations. This allows Eq. (9.53) to be written as

$$\lim_{|\tau|\to\infty} R_{XX}(\tau) = \lim_{|\tau|\to\infty} E[X(t)]E[X(t+\tau)] \tag{9.54}$$

Since the process is stationary,

$$E[X(t)] = E[X(t+\tau)]$$

Hence,

$$\lim_{|\tau|\to\infty} R_{XX}(\tau) = (E[X(t)])^2 = (\overline{X})^2 \tag{9.55}$$

Property 5: If $X(t)$ is periodic with period T, then $R_{XX}(\tau)$ is also periodic with the same period.

Proof: $X(t)$ is periodic with period T. This implies that $X(t + kT) = X(t)$ for all T and all integer k.

Consider, $R_{XX}(\tau) = E[X(t)X(t+\tau)]$.

Replacing τ by $\tau + kT$ gives

$$R_{XX}(\tau + kT) = E[X(t)X(t + \tau + kT)] \tag{9.56}$$

Since $X(t)$ is periodic,

$$X(t + \tau + kT) = X(t + \tau)$$

Making use of the above equation in Eq. (9.56) gives

$$\begin{aligned} R_{XX}(\tau + kT) &= E[X(t)X(t + \tau)] \\ &= R_{XX}(\tau) \end{aligned} \tag{9.57}$$

Thus, it is seen that if $X(t)$ has a periodic component with period T, then $R_{XX}(\tau)$ will also have a periodic component with the same period T.

Property 6: If $X(t)$ is ergodic zero mean process and has no periodic component, then

$$\lim_{|\tau|\to\infty} R_{XX}(\tau) = 0$$

Proof: As discussed in the proof of property 4, when $|\tau|\to\infty$, $X(t)$ and $X(t + \tau)$ become independent and from Eq. (9.52), we have,

$$\lim_{|\tau|\to\infty} R_{XX}(\tau) = (\overline{X})^2$$

The above equation is true, when $X(t)$ has no periodic component. If the process is zero mean, $\overline{X} = 0$. Then

$$\lim_{|\tau|\to\infty} R_{XX}(\tau) = 0$$

Property 7: $R_{XX}(\tau)$ cannot have an arbitrary shape.

Proof: The proof is based on Wiener-Kinchine theorem which says that Fourier Transform of $R_{XX}(\tau)$ gives the power spectral density of process $X(t)$. The power spectral density $S_X(f)$ of process $X(t)$ is nonnegative. Hence, only those $R_{XX}(\tau)$ for which the Fourier transform $S_X(f)$ is nonnegative, can be a valid autocorrelation function. It may also be noted that, many processes may have same autocorrelation function, even though a given process has unique autocorrelation function.

9.16 Cross-correlation function

Cross-correlation function of two random processes $X(t)$ and $Y(t)$ is defined as

$$R_{XY}(t_1,t_2) = E[X(t_1)Y(t_2)] \tag{9.58}$$

Letting $t_1 = t$ and $t_2 - t = \tau$, we have,

$$R_{XX}(t,t+\tau) = E[X(t)Y(t+\tau)] \tag{9.59}$$

If the two processes are statistically independent, the cross-correlation function becomes,

$$R_{XY}(t,t+\tau) = E[X(t)]\, E[Y(t+\tau)] \tag{9.60}$$

If $X(t)$ and $Y(t)$ are statistically independent and jointly wide-sense stationary, cross-correlation function is given by,

$$R_{XY}(\tau) = \overline{X}\,\overline{Y} \tag{9.61}$$

which is a constant.

If

$$R_{XX}(t, t+\tau) = E[X(t)Y(t+\tau)] = 0 \tag{9.62}$$

then $X(t)$ and $Y(t)$ are called orthogonal processes.

9.17 Properties of cross-correlation function of second-order stationary or wide-sense stationary processes

Property 1: $R_{XY}(-\tau) = R_{YX}(\tau)$

Proof: By definition,

$$R_{XY}(\tau) = E[X(t)\ Y(t+\tau)] \tag{9 63}$$

Consider,

$$R_{XY}(-\tau) = E[X(t)Y(t-\tau)] \tag{9.64}$$

Put $t - \tau = u$. Then, $t = u + \tau$. Substituting these in Eq. (9.64) we have,

$$\begin{aligned} R_{XY}(-\tau) &= E[X(u+\tau)Y(u)] \\ &= E[Y(u)X(u+\tau)] \end{aligned} \tag{9.65}$$

Comparing with Eq. (9.63), we have right hand side of Eq. (9.65) equal to $R_{YX}(\tau)$.

Hence

$$R_{XY}(-\tau) = R_{YX}(\tau) \tag{9.66}$$

Property 2:

$$|R_{XY}(\tau)| \le \frac{1}{2}\left[R_{XX}(0) + R_{YY}(0)\right]$$

Proof: Consider $[X(t) - Y(t+\tau)]^2$; which is always greater than or equal to zero. Hence

$$E[X(t) - Y(t+\tau))^2] \ge 0 \tag{9.67}$$

Expanding the squared term in the expectation, we get

$$E[X^2(t) + Y^2(t+\tau) - 2X(t)Y(t+\tau)] \ge 0$$

Since expectation operation is linear operation, we get,

$$E[X^2(t)] + E[Y^2(t+\tau)] \ge 2E[X(t)Y(t+\tau)]$$

$$\Rightarrow \qquad \frac{1}{2}\,[R_{XX}(0) + R_{YY}(0)] \ge R_{XY}(\tau) \tag{9.68}$$

Similarly considering

$$(X(t) + Y(t + \tau))^2 \geq 0$$

and taking expectation we can show that

$$\frac{1}{2}[R_{XX}(0) + R_{YY}(0)] \geq -R_{XY}(\tau) \tag{9.69}$$

Combining Eq. s (9.68) and (9.69) we have

$$|R_{XY}(\tau)| \leq \frac{1}{2}[R_{XX}(0) + R_{YY}(0)] \tag{9.70}$$

Property 3:

$$|R_{XY}(\tau)| \leq \sqrt{R_{XX}(0) \times R_{YY}(0)}$$

Proof: Consider the inequality,

$$E[(Y(t+\tau) + aX(t))^2] \geq 0 \tag{9.71}$$

where a is a real number. Expanding the terms inside the expectation, we have

$$E[Y^2(t+\tau)] + a^2 E[X^2(t)] + 2aE[X(t)Y(t+\tau)] \geq 0 \tag{9.72}$$

We have,

$$E[Y^2(t+\tau)] = R_{YY}(0)$$
$$E[X^2(t)] = R_{XX}(0)$$
$$E[X(t)Y(t+\tau)] = R_{XY}(\tau)$$

Substituting these values in the inequality (9.72), we get

$$R_{YY}(0) + a^2 R_{XX}(0) + 2aR_{XY}(\tau) \geq 0 \tag{9.73}$$

The inequality is true for any value of a. We can have a tighter bound by choosing the value of a that minimises the left hand side of the inequality (9.73). Let the left hand side be denoted by F, the value of a which minimises F can be obtained by differentiating F with respect to a and setting it equal to zero.

$$\frac{dF}{da} = 2aR_{XX}(0) + 2R_{XY}(\tau) = 0$$

Solving for a which minimises F, we have

$$a = \frac{-R_{XY}(\tau)}{R_{XX}(0)} \tag{9.74}$$

Substituting this value of a in the inequality (9.73), we have,

$$R_{YY}(0)+\left[\frac{R_{XY}(\tau)}{R_{XX}(0)}\right]^2 R_{XX}(0)-\frac{2R_{XY}(\tau)}{R_{XX}(0)}R_{XY}(\tau)\geq 0$$

$$R_{YY}(0)+\frac{[R_{XY}(\tau)]^2}{R_{XX}(0)}-2\frac{[R_{XY}(\tau)]^2}{R_{XX}(0)}\geq 0$$

Rearranging we have,

$$R_{YY}(0)\geq \frac{[R_{XY}(\tau)]^2}{R_{XX}(0)}$$

$$[R_{XY}(\tau)]^2 \leq R_{XX}(0)R_{YY}(0)$$

That is,

$$|R_{XY}(\tau)| \leq \sqrt{R_{XX}(0)R_{YY}(0)} \quad (9.75)$$

9.18 Covariance function of random processes

The concept of the covariance of two random variables may be extended to random processes. The autocovariance function of random process $X(t)$ is defined as

$$C_{XX}(t, t+\tau) = E[\{X(t) - E[X(t)]\}\{X(t+\tau) - E[X(t+\tau)]\}] \quad (9.76)$$

Expanding the terms, we get

$$C_{XX}(t, t+\tau) = E[X(t)\, X(t+\tau) - X(t)\, E[X(t+\tau)]$$
$$-E[X(t)]\, X(t+\tau) + E[X(t))\, E[X(t+\tau)]] \quad (9.77)$$

which simplifies to

$$C_{XX}(t, t+\tau) = R_{XX}(t, t+\tau) - E[X(\tau)]E[X(t+\tau)] \quad (9.78)$$

The cross covariance function for two processes $X(t)$ and $Y(t)$ is defined by

$$C_{XY}(t, t+\tau) = E[(X(t) - E[X(t)])(Y(t+\tau) - E[Y(t+\tau)])] \quad (9.79)$$

Simplifying gives,

$$C_{XY}(t, t+\tau) = R_{XY}(t, t+\tau) - E[X(t)]\, E[Y(t+\tau)] \quad (9.80)$$

For processes which are second-order stationary or wide-sense stationary, the expression for autocovariance and cross covariance becomes

$$C_{XX}(\tau) = R_{XX}(\tau) - (\overline{X})^2 \quad (9.81)$$

and

$$C_{XY}(\tau) = R_{XY}(\tau) - \overline{X}\,\overline{Y} \quad (9.82)$$

The variance of random process $X(t)$ is given by $C_{XX}(0)$. For a wide-sense stationary process, σ_X^2 is obtained by letting $\tau = 0$ in Eq. (9.81).

$$\sigma_X^2 = C_{XX}(0) = R_{XX}(0) - (\overline{X})^2 \quad (9.83)$$

For two random processes $X(t)$ and $Y(t)$, if cross covariance function

$$C_{XY}(t, t+\tau) = 0 \tag{9.84}$$

they are called uncorrelated. Using Eq. (9.84) in Eq. (9.80), we get

$$R_{XY}(t, t+\tau) = E[X(t)]E[Y(t+\tau)] \tag{9.85}$$

Eq. (9.85) means the processes $X(t)$ and $Y(t)$ are statistically independent and are uncorrelated. The converse is not necessarily true. However in case of joint gaussian processes, uncorrelatedness also implies statistical independency.

9.19 Gaussian random process

Gaussian Random Process is one of the important processes from engineering point of view.

Consider a continuous random process $X(t)$. Consider N time instants $t_1, t_2, ..., t_N$. Then $X(t_1) = X_1, X(t_2) = X_2, ..., X(t_N) = X_N$ are N random variables. If for any $N = 1,2, ...,$ and for any time instants $t_1, t_2, ..., t_N$, these random variables are jointly gaussian, then the process $X(t)$ is called gaussian random process. The joint probability density function of the N random variables $X_1, X_2, ..., X_N$, is then given by

$$f_X(x_1, x_2, ..., x_N; t_1, t_2 ..., t_N) = \frac{\exp\{-(1/2)[x-\overline{X}]^T [C_X]^{-1} [x-\overline{X}]\}}{\sqrt{(2\pi)^N |[C_X]|}} \tag{9.86}$$

where $[x-\overline{X}] = \begin{bmatrix} x_1 - \overline{X}_1 \\ x_2 - \overline{X}_2 \\ \vdots \\ x_N - \overline{X}_N \end{bmatrix}$ is a $N \times 1$ matrix

and $[C_X] = \begin{bmatrix} C_{11} & C_{12} & \cdots & C_{1N} \\ C_{21} & C_{22} & \cdots & C_{2N} \\ \vdots & \vdots & \vdots & \vdots \\ C_{N1} & C_{N2} & \cdots & C_{NN} \end{bmatrix}$ is a $N \times N$ matrix called covarience matrix

The mean values of $X(t_i)$ are denoted by $\overline{X}_i$ and is given by

$$\overline{X}_i = E[X_i] = E[X(t_i)] \tag{9.87}$$

The elements of the covariance of matrix $[C_X]$ are given by

$$\begin{aligned} C_{ik} = C_{X_i X_k} &= E[(X_i - \overline{X}_i)(X_k - \overline{X}_k)] \\ &= E[(X(t_i) - E[X(t_i)])(X(t_k) - E[X(t_k)])] \\ &= C_{XX}(t_i, t_k) \end{aligned} \tag{9.88}$$

$C_{XX}(t_i, t_k)$ is the autocovariance of random variables $X(t_i)$ and $X(t_k)$. A gaussian random process can be completely specified, if the mean value $\overline{X}_i$ and the autocovariance functions are known. Autocovariance function $C_{XX}(t_i, t_k)$can also be expressed in terms of autocorrelation and mean values. Consider $C_{XX}(t_i, t_k)$,

$$\begin{aligned} C_{XX}(t_i,t_k) &= E[\{X(t_i)-E[X(t_i)]\}\{X(t_k)-E[X(t_k)]\}] \\ &= E[\{X(t_i)-\overline{X}_i\}\{X(t_k)-\overline{X}_k\}] \\ &= E[X(t_i)X(t_k)-\overline{X}_k\, E[X(t_i)-\overline{X}_i E[X(t_k)]+\overline{X}_i\overline{X}_k \\ &= R_{XX}(t_i,t_k)-\overline{X}_k\,\overline{X}_i-\overline{X}_i\,\overline{X}_k+\overline{X}_i\,\overline{X}_k \\ &= R_{XX}(t_i,t_k)-\overline{X}_i\,\overline{X}_k \end{aligned} \tag{9.89}$$

If the gaussian process is wide-sense stationary, then the mean value $\overline{X}_i = E[X(t_i)] = \overline{X}$, a constant and does not depend on time. On the other hand, the autocovariance and autocorrelation functions will depend only on time differences and not absolute time.

$$C_{XX}(t_i, t_k) = C_{XX}(t_k - t_i) \tag{9.90}$$

$$R_{XX}(t_i, t_k) = R_{XX}(t_k - t_i) \tag{9.91}$$

Gaussian density function of any order involves only mean values and covariance functions. Since mean values are constant and covariance functions are dependent only on time differences but not absolute time, a wide-sense gaussian random process is also a strictly stationary process.

9.20 Poisson random process

Poisson Random Process is one of the important discrete random processes. Consider an example of occurrence of some event at random instants of time. Poisson process $X(t)$ describes the number of times the event has occurred as a function of time in the interval $(0, t)$. The examples of Poisson processes are

- telephone calls arriving at a certain switch in a public telephone network,
- packets arriving at a node in a packet switched communication network.
- customers entering into a supermarket or bank,
- failure of component in an electronic system,
- the occurrence of lightning strike in a region.

In the above examples, individual events occur at random instants of time. The process gives the count of such occurrences with time. Hence, Poisson process is also called sometimes as Poisson counting process.

Let $X(t)$ represent the number of occurrences of an event with time. Since the event occurs at random instants of time, the number of occurrences as a function of time is a random value which depends on time. Hence, $X(t)$ is a random process. $X(t)$ has integer valued non decreasing sample functions. Even though $X(t)$ can be defined for $t < 0$, for convenience, we assume $X(t) = 0$ at $t = 0$ and for $t < 0$. For $t > 0$, $X(t)$ is the number of occurrences in the interval $(0, t)$. This assumption is justified, since in many situations, one is interested in the processes defined only for $t > 0$.

Definition of Poisson process is based on the condition that only one event occurs in any small interval of time. Two events may occur at instants of times which are very close. But they do not occur simultaneously at a given instant. Further, the times of occurrence of events are statistically independent. That is, if we consider non overlapping time intervals, the number of occurrences of events in one time interval should be independent of the number of occurrences of events in the other time interval. The number of occurrences of an event in any finite interval of time is described by Poisson distribution function. The average rate of occurrence of event with respect to time is denoted by λ.

9.21 Probability density function

Consider a time interval $(0, t)$. The average number of occurrences of events in a time interval $(0, t)$ is given by λt.

The probability of exactly k occurrences over a time interval $(0, t)$ is given by

$$P[X(t)=k]=\frac{(\lambda t)^k e^{-\lambda t}}{k!}; \quad t \geq 0, \quad k=0,1,2,... \tag{9.92}$$

The probability density of the number of occurrences is given by

$$f_X(x)=\sum_{k=0}^{\infty}\frac{(\lambda t)^k e^{-\lambda t}}{k!}\delta(x-k) \tag{9.93}$$

The mean value of the process is given by

$$E[X(t)]=\int_{-\infty}^{\infty} x f_X(x)dx=\int_{-\infty}^{\infty} x\sum_{k=0}^{\infty}\frac{(\lambda t)^k e^{-\lambda t}}{k!}\delta(x-k)\,dx$$

$$=\sum_{k=0}^{\infty}\frac{k(\lambda t)^k e^{-\lambda t}}{k!}=\lambda t \qquad \left(\because \int_{-\infty}^{\infty} x\delta(x-k)dx = x\,|_{x=k}=k\right) \tag{9.94}$$

λt is the average number of occurrences or arrivals in the interval $[0, t)$. Parameter λ is called as mentioned earlier, the average rate of arrival or arrival rate of the Poisson process. It is to be noted that Poisson process is an example of nonstationary process; since the mean value of the process depends on time. Second moment of $X(t)$, is given by

$$
\begin{aligned}
E[X^2(t)] &= \int_{-\infty}^{\infty} x^2 f_X(x)\,dx \\
&= \sum_{k=0}^{\infty} \frac{k^2 (\lambda t)^k e^{-\lambda t}}{k!} \qquad \left(\because \int_{-\infty}^{\infty} x^2 \delta(x-k)\,dx = x^2 \mid_{x=k} = k^2 \right) \\
&= \lambda t[1+\lambda t] \qquad (9.95)
\end{aligned}
$$

Variance of $X(t)$ is given by

$$
\begin{aligned}
\sigma_X^2 &= E[X^2(t)] - \{E[X(t)]\}^2 \\
&= \lambda t + \lambda^2 t^2 - \lambda^2 t^2 \\
&= \lambda t \qquad (9.96)
\end{aligned}
$$

Hence, mean and variance of a Poisson random variable are same at λt.

9.22 Joint probability density

Consider two time instants t_1 and t_2 such that $0 < t_1 < t_2$. The probability of occurrence of k_1 events in the time interval $(0, t_1)$ is

$$
P[X(t_1) = k_1] = \frac{(\lambda t_1)^{k_1} e^{-\lambda t_1}}{k_1!}, \; k_1 = 0, 1, 2, \ldots \qquad (9.97)
$$

The joint probability density is obtained by considering the conditional probability. The conditional probability of k_2 occurrences in the interval $(0, t_2)$ given that k_1 events occurred in the interval $(0, t_1)$ is the probability that $(k_2 - k_1)$ events occurred in the interval t_1, t_2). This is given by

$$
P[X(t_2) = k_2 \mid X(t_1) = k_1] = \frac{[\lambda(t_2 - t_1)]^{k_2 - k_1} e^{-\lambda(t_2 - t_1)}}{(k_2 - k_1)!}, \; k_2 \geq k_1 \qquad (9.98)
$$

The joint probability of k_2 occurrences at time t_2 and k_1 occurrences at time t_1 is given by the product of (9.97) and (9.98).

$$
\begin{aligned}
P(k_1, k_2) &= P[X(t_2) = k_2 \mid X(t_1) = k_1] P[X(t_1) = k_1] \\
&= \frac{(\lambda t_1)^{k_1} [\lambda(t_2 - t_1)]^{k_2 - k_1} e^{-\lambda t_2}}{k_1!(k_2 - k_1)!}, \; k_2 \geq k_1 \qquad (9.99)
\end{aligned}
$$

For the process random variables $X(t_1)$ and $X(t_2)$, the joint probability density function, then becomes,

$$
f_X(x_1, x_2) = \sum_{k_1=0}^{\infty} \sum_{k_2=0}^{\infty} P(k_1, k_2) \delta(x_1 - k_1) \delta(x_2 - k_2) \qquad (9.100)
$$

9.23 Autocorrelation function

The autocorrelation function is given by

$$R_{XX}(t_1,t_2) = E[X(t_1)X(t_2)]$$

Let us write $X(t_2)$ as,

$$X(t_2) = X(t_1) + X(t_2) - X(t_1)$$

Note that by adding and subtracting $X(t_1)$ from $X(t_2)$, the value is unchanged. Substituting this $X(t_2)$ in the expression for autocorrelation, we have,

$$\begin{aligned} R_{XX}(t_1,t_2) &= E[X(t_1)\{X(t_1) + X(t_2) - X(t_1)\}] \\ &= E[X^2(t_1)] + E[X(t_1)\{X(t_2) - X(t_1)\}] \end{aligned}$$

Assuming $t_2 > t_1$ $X(t_1)$ represents the number of occurrences of events in the interval (0, t_1). While $X(t_2) - X(t_1)$ is the number of occurrences of events in the interval (t_1, t_2). These two intervals 0 to t_1 and t_1 to t_2 being non overlapping, the number of occurrences of events in the two intervals are independent. Hence,

$$\begin{aligned} R_{XX}(t_1,t_2) &= E[X^2(t_1)] + E[X(t_1)]E[(X(t_2) - X(t_1))] \\ &= E[X^2(t_1)] + E[X(t_1)E[X(t_2)] - \{E[X(t_1)]\}^2 \end{aligned}$$

But $E[X^2(t_1)] - \{E[X(t_1)]\}^2$ is the variance of $X(t_1)$. Making use of this fact in the above expression, we get

$$R_{XX}(t_1,t_2) = \sigma^2_{X(t_1)} + E[X(t_1)]E[X(t_2)] \tag{9.101}$$

We have,

$$\begin{aligned} \sigma^2_{X(t_1)} &= \lambda t_1 \\ E[X(t_1)] &= \lambda t_1 \\ E[X(t_2)] &= \lambda t_2 \end{aligned}$$

Hence

$$\begin{aligned} R_{XX}(t_1,t_2) &= \lambda t_1 + \lambda t_1\, \lambda t_2 \\ &= \lambda t_1[1 + \lambda t_2] \text{ for } t_2 > t_1 \end{aligned} \tag{9.102}$$

When $t_1 > t_2$, we can show similarly, that

$$R_{XX}(t_1,t_2) = \lambda t_2[1 + \lambda t_1] \tag{9.103}$$

Autocovariance function is given by

$$C_{XX}(t_1,t_2) = \text{Var}[X(t_1)] = \sigma^2_{X(t_1)} = \lambda t_1 \text{ for } t_2 > t_1 \tag{9.104}$$

and

$$C_{XX}(t_1,t_2) = \text{Var}[X(t_2)] = \sigma^2_{X(t_2)} = \lambda t_2 \text{ for } t_1 > t_2 \quad (9.105)$$

In general for the Poisson process we have

$$C_{XX}(t_1,t_2) = \lambda \min(t_1,t_2) \quad (9.106)$$

Example 9.1: Let $X(t)$ be a random process such that $X(t)$ has an identical distribution for any t and the random variables $X(t_i)$, $i = 1, 2, ...$ are statistically independent for any $t_1, t_2 ..., t_k$, where $t_i \neq t_j$, $i, j = 1, 2, ..., k$. Show that random process $X(t)$ is stationary in the strict-sense.

***Solution*:** At any set of time instants $t_1, t_2, ..., t_k$, $X(t_1), X(t_2)..., X(t_k)$ are random variables and has identical first-order probability density functions. Hence the first-order probability density function is invariant to time shift. The kth order joint probability density function is given by

$$f_{X_1,X_2,...,X_k}(x_1,x_2,...,x_k;t_1,t_2,...,t_k) \quad (9.107)$$

Since the k random variables are statistically independent, the kth order joint probability density function given by (9.107) can be expressed as product of k first-order (marginal) probability density functions as

$$\prod_{i=1}^{k} f_{X_i}(x_i,t_i) \quad (9.108)$$

The first-order probability density function is invariant to a time shift. Hence the kth order probability density function is invariant to a time shift. Hence $X(t)$ has k^{th} order stationarity. The result is true for any $n > k$. Hence random process $X(t)$ is stationary in the strict-sense.

Example 9.2: $X(t) = A \sin \omega_0 t$ is a random process where ω_0 is constant and A is a uniform random variable over the interval (0,1). Determine mean value of $X(t)$. Is $X(t)$ a stationary process?

***Solution*:** By definition,

$$E[X(t)] = E[A \sin \omega_0 t] = E[A] \sin \omega_0 t$$

$$E[A] = \int_a^1 a.1.da = \frac{a^2}{2}\Bigg|_{a=0}^{a=1} = \frac{1}{2}$$

Hence, $$E[X(t) = \frac{1}{2} \sin \omega_0 t$$

$E[X(t)]$ is a function of time. Hence $X(t)$ is nonstationary process.

Example 9.3: Consider the random process defined by

$$X(t) = A\cos(\omega_0 t + \theta)$$

Assume A and ω_0 are constants. θ is a uniformly distributed random variable in the interval

$(0, 2\pi)$. Show that $X(t)$ is a wide-sense stationary process.

Solution: For $X(t)$ to be wide-sense stationary process, its mean value $E[X(t)]$ should be a constant independent of time and autocorrelation function $R_{XX}(t, t+\tau)$ should be a function of time difference. Mean value of $X(t)$ is

$$E[X(t)] = \int_0^{2\pi} X(t) f_\Theta(\theta)\, d\theta$$

Since θ is uniformly distributed its probability density function $f_\Theta(\theta)$ is given by

$$f_\Theta(\theta) = \frac{1}{2\pi};\ 0 < \theta < 2\pi$$
$$= 0 \text{ otherwise}$$

Substituting the values for $X(t)$ and $f_\Theta(\theta)$ in the expression for $E[X(t)]$,

$$E[X(t)] = \int_0^{2\pi} A\cos(\omega_0 t + \theta)\frac{1}{2\pi}\, d\theta$$
$$= \frac{A}{2\pi}\sin(\omega_0 t + \theta)\Big|_{\theta=0}^{\theta=2\pi} = 0$$

Consider two instants of time $t_1 = t$ and $t_2 = t + \tau$. The autocorrelation function is

$$R_{XX}(t, t+\tau) = E[A\cos(\omega_0 t + \theta) A\cos(\omega_0 t + \omega_0\tau + \theta)]$$

Using the trigonometric identity,

$$\cos\alpha\, \cos\beta = \frac{1}{2}[\cos(\alpha+\beta) + \cos(\alpha-\beta)]$$

and identifying or $\alpha = \omega_0(t+\tau) + \theta$ and $\beta = (\omega_0 t + \theta)$, and noting that A and ω_0 are constants,

$$R_{XX}(t, t+\tau) = \frac{A^2}{2}E[\cos(2\omega_0 t + \omega_0\tau + \theta) + \cos\omega_0\tau]$$
$$= \frac{A^2}{2}\cos\omega_0\tau + \frac{A^2}{2}E[\cos(2\omega_0 t + \omega_0\tau + \theta)]$$

Consider the second term.

$$\frac{A^2}{2}E[\cos(2\omega_0 t + \omega_0\tau + \theta)]$$
$$= \frac{A^2}{2}\int_0^{2\pi} \cos(2\omega_0 t + \omega_0\tau + \theta)\frac{1}{2\pi}\, d\theta$$
$$= \frac{A^2}{2}\frac{1}{2\pi}\sin(2\omega_0 t + \omega_0\tau + \theta)\Big|_{\theta=0}^{\theta=2\pi} = 0$$

Hence autocorrelation function is,

$$R_{XX}(t,\tau)=\frac{A^2}{2}\cos\omega_0\tau$$

which depends only on time difference τ. Thus, the mean value of process $X(t)$ is zero, which is a constant and autocorrelation function $R_{XX}(t, t+\tau)$ depends only on time difference τ. Hence, $X(t)$ is wide-sense stationary.

Example 9.4: Consider a random process $X(t) = \cos 2\pi Ft$, where F is a random variable uniformly distributed over some interval $(0, f_0)$. Determine mean value of $X(t)$. Is $X(t)$ stationary?

Solution:

$E[X(t)]=\int_0^{f_0}\frac{\cos 2\pi ft}{f_0}df$, where $\frac{1}{f_0}$ is the probability density function of random variable F.

$$E[X(t)]=\frac{1}{f_0}\frac{\sin 2\pi ft}{2\pi t}\bigg|_{f=0}^{f=f_0}$$

$$=\frac{\sin(2\pi f_0 t)}{2\pi f_0 t}=\sin c(2f_0 t)$$

Since $E[X(t)]$ is not constant, $X(t)$ is a nonstationary process.

Example 9.5: Consider the random process

$$X(t)=A\cos(\omega t+\theta)$$

where ω_0 and θ are constants and A is a random variable uniformly distributed in the interval $(-\alpha, \alpha)$. Determine (i) the mean of $X(t)$ (ii) autocorrelation function $R_{XX}(t_1, t_2)$ of the process $X(t)$ and (iii) whether $X(t)$ is a wide-sense stationary process.

Solution: Random variable A is uniformly distributed in the interval $(-\alpha, \alpha)$. Hence probability density function of random variable A is given by

$$f_A(a)=\begin{cases}\frac{1}{2\alpha}; & -\alpha<a<\alpha\\ 0\ ; & \text{otherwise}\end{cases}$$

Now $$E[X(t)]=E[A\cos(\omega t+\theta)]$$

ω_0 and θ are constants. Hence $\cos(\omega_0+\theta)$ is a constant, and $E[X(t)]$ can be written as

$$E[X(t)]=\cos(\omega t+\theta)E[A]$$

$$=[\cos(\omega t+\theta)]\int_{-\alpha}^{\alpha}af_A(a)da$$

$$= [\cos(\omega t + \theta)] \int_{-\alpha}^{\alpha} a \frac{1}{2\alpha} da$$
$$= [\cos(\omega t + \theta)] \times 0 = 0$$

(ii) Autocorrelation function of $X(t)$:

$$R_{XX}(t_1, t_2) = E[(A\cos(\omega t_1 + \theta))(A\cos(\omega t_2 + \theta))]$$
$$= \cos(\omega t_1 + \theta)\cos(\omega t_2 + \theta)E[A^2]$$
$$= \cos(\omega t_1 + \theta)\cos(\omega t_2 + \theta)\int_{-\alpha}^{\alpha} a^2 \frac{1}{2\alpha} da$$
$$= \frac{\alpha^2}{4}[\cos(\omega(t_1 + t_2) + 2\theta) + \cos\omega(t_2 - t_1)]$$

Let $t_1 = t$ and $t_2 = t + \tau$, then

$$R_{XX}(t, t + \tau) = \frac{\alpha^2}{4}\cos(\omega(2t + \tau) + 2\theta) + \cos\omega\tau$$

The right hand side is not a function of $\tau = (t_2 - t_1)$ alone. Hence, $X(t)$ is not a wide-sense stationary process.

Example 9.6: $X(t) = A\sin(\omega_0 t)$ is a random process, ω_0 is constant and A is a uniformly distributed random variable in the interval (0,1). Is $X(t)$ stationary?

Determine the autocorrelation function of $X(t)$.

***Solution*:**

$$R_{XX}(t_1, t_2) = E[X(t_1)X(t_2)] = E[A^2\sin(\omega_0 t_1)\sin(\omega_0 t_2)]$$
$$= E[A^2]\sin(\omega_0 t_1)\sin(\omega_0 t_2)$$

Since $$E[A^2] = \int_0^1 a^2 da = \left.\frac{a^3}{3}\right|_0^1 = \frac{1}{3}$$

we get $$R_{XX}(t_1, t_2) = \frac{1}{3}\sin(\omega_0 t_1)\sin(\omega_0 t_2)$$

$R_{XX}(t_1, t_2)$ is a function of time. Hence $X(t)$ is not a stationary process.

Example 9.7: Check whether the processes having following autocorrelation functions are wide-sense stationary processes.

(*i*) $R_{XX}(t, t + \tau) = \cos 8t\, e^{-(t+\tau)}$

(*ii*) $R_{XX}(t, t+\tau) = \dfrac{\sin 2\tau}{1+\tau^2}$

(*iii*) $R_{XX}(t, t+\tau) = -10^{-|\tau|}$

(*iv*) $R_{XX}(t, t+\tau) = \dfrac{5\tau}{1+2\tau^2+3\tau^4}$

(*v*) $R_{XX}(t, t+\tau) = 5e^{-|\tau|}$

***Solution*:**

(i) Since $R_{XX}(t, t+\tau)$ is a function of t, not that of WSS process.

(ii) Autocorrelation function does not have peak value at $\tau = 0$. $R_{XX}(t, t+\tau)$ cannot be a valid autocorrelation function.

(iii) It is not a valid autocorrelation function, since at $\tau = 0$ it does not have peak value.

(iv) It is not a valid autocorrelation function since it is not having even symmetry about $\tau = 0$.

(v) $R_{XX}(t, t+\tau)$ is a valid autocorrelation function and depends only on time difference τ.

Example 9.8: The autocorrelation function $R_{XX}(\tau)$ for a stationary ergodic process $X(t)$ with no periodic component is given by

$$R_{XX}(\tau) = 36 + \frac{5}{1+7\tau^2}$$

Find the magnitude of the mean value and variance of the process $X(t)$.

***Solution*:** From the properties of autocorrelation function, we know that if $X(t)$ is ergodic with no periodic components then $\lim_{\tau\to\infty} R_{XX}(\tau) = (\overline{X})^2$.

Hence, $(\overline{X})^2 = \lim_{\tau\to\infty} R_{XX}(\tau) = 36$ and $(\overline{X}) = 6$.

$$E[X^2(t)] = R_{XX}(0) = 36 + 5 = 4$$

The variance is given by

$$\sigma_X^2 = E[X^2(t)] - (E[X(t)])^2$$
$$= 41 - 36 = 5$$

Example 9.9: $X(t)$ is a wide-sense stationary random process with autocorrelation function

$$R_{XX}(\tau) = e^{-a|\tau|};\ a > 0 \text{ is a constant}$$

and

$$Y(t) = X(t)\ \cos(\omega_0 t + \theta)$$

where ω_0 is a constant and θ is a random variable uniformly distributed in the range $(-\pi, \pi)$. θ is statistically independent of $X(t)$. Determine the autocorrelation function $R_{YY}(t, t+\tau)$ of $Y(t)$.

Solution:

$$R_{YY}(t, t+\tau) = E[Y(t)Y(t+\tau)]$$
$$= E[X(t)X(t+\tau)\cos(\omega_0 t+\theta)\cos(\omega_0 t+\omega_0\tau+\theta)]$$

Since θ and $X(t)$ are statistically independent, $R_{YY}(t, t+\tau)$ can be written as the product of expected values involving $X(t)$ and θ separately.

Using the relation

$$\cos A\cos B = \frac{\cos(B-A)+\cos(B+A)}{2}$$

$$R_{YY}(t, t+\tau) = E[X(t)X(t+\tau)]\frac{E[\cos\omega_0\tau+\cos(2\omega_0 t+\omega_0\tau+2\theta)]}{2}$$

$$= \frac{1}{2}R_{XX}(\tau)[\cos\omega_0\tau + E[\cos(2\omega_0 t+\omega_0\tau)+2\theta)]]$$

But

$$E[\cos(2\omega_0 t+\omega_0\tau+2\theta)] = \int_{-\pi}^{\pi}\frac{1}{2\pi}\cos(2\omega_0 t+\omega_0\tau+2\theta)\,d\theta = 0$$

Hence we have

$$R_{YY}(t, t+\tau) = \frac{1}{2}R_{XX}(\tau)\cos\omega_0\tau$$

which is a function of τ alone. Hence

$$R_{YY}(t, t+\tau) = R_{YY}(\tau) = \frac{1}{2}R_{XX}(\tau)\cos\omega_0\tau = \frac{1}{2}e^{-a|\tau|}\cos\omega_0\tau$$

Example 9.10: $X(t)$ is a random process given by $X(t) = A\cos\omega_0 t + B\sin\omega_0 t$, where A and B are random variables and ω_0 is a constant. Show that if A and B are uncorrelated, zero mean random variables having same variances σ^2 but different density functions, $X(t)$ is wide-sense stationary.

Solution:

$$X(t) = A\cos\omega_0 t + B\sin\omega_0 t$$
$$E[X(t)] = E[A\cos\omega_0 t + B\sin\omega_0 t]$$
$$= E[A]\cos\omega_0 t + E[B]\sin\omega_0 t$$

Given $E[A] = E[B] = 0$. Hence,

$$E[X(t)] = 0$$

Autocorrelation function of $X(t)$ is next computed.

$$
\begin{aligned}
R_{XX}(t, t+\tau) &= E[X(t)X(t+\tau) \\
&= E\{[A\cos\omega_0 t + B\sin\omega_0 t][A\cos\omega_0(t+\tau) + B\sin\omega_0(t+\tau)]) \\
&= E\{A^2 \cos\omega_0 t\ \cos(\omega_0 t + \omega_0\tau) + AB\cos\omega_0 t\ \sin(\omega_0 t + \omega_0\tau) \\
&\quad + AB\sin\omega_0 t\ \cos(\omega_0 t + \omega_0\tau) + B^2 \sin\omega_0 t \sin(\omega_0 t + \omega_0\tau)\} \\
&= \frac{E[A^2]}{2}[\cos(2\omega_0 t + \omega_0\tau) + \cos\omega_0\tau] \\
&\quad + \frac{E[AB]}{2}[\sin(2\omega_0 t + \omega_0\tau) + \sin\omega_0\tau] \\
&\quad + \frac{E[AB]}{2}[\sin(2\omega_0 t + \omega_0\tau) - \sin\omega_0\tau] \\
&\quad + \frac{E[B^2]}{2}[\cos\omega_0\tau - \cos(2\omega_0 t + \omega_0\tau)]
\end{aligned}
$$

Since A and B are zero mean random variables, with identical variances,

$$E[A^2] = E[B^2] = \sigma^2$$

Since A and B are uncorrelated

$$
\begin{aligned}
E(AB) &= E[A]E[B] \\
&= 0
\end{aligned}
$$

Substituting these values in the expression for $R_{XX}(t, t+\tau)$ we have

$$
\begin{aligned}
R_{XX}(t, t+\tau) &= \frac{\sigma^2}{2}[\cos(2\omega_0 t + \omega_0\tau) + \cos\omega_0\tau] \\
&\quad + \frac{\sigma^2}{2}[\cos\omega_0\tau - \cos(2\omega_0 t + \omega_0\tau)] \\
&= \sigma^2 \cos\omega_0\tau
\end{aligned}
$$

We have $E[X(t)\ ] = 0$ and $R_{XX}(t, t+\tau) = \sigma^2 \cos\ \omega_0\tau$, a function of τ alone and independent of t. Hence, $X(t)$ is a wide-sense stationary process.

Example 9.11: A random process is described by $X(t) = r\cos(\omega t + \phi)$. Random variables r and ϕ are statistically independent. ϕ is uniform in the interval $(-\pi, \pi)$. Determine the autocorrelation function $R_{XX}(t_1, t_2)$ of the process $X(t)$.

Solution:

$$
\begin{aligned}
R_{XX}(t_1, t_2) &= E[X(t_1)X(t_2)] \\
&= E[r\cos(\omega t_1 + \phi) r\cos(t_2 + \phi)] \\
&= \frac{1}{2}E[r^2]\ E[\cos\omega(t_1 - t_2) + \cos(\omega t_1 + \omega t_2 + 2\phi)]
\end{aligned}
$$

We find that,

$$E[\cos(\omega t_1+\omega t_2+2\phi)]=\frac{1}{2\pi}\int_{-\pi}^{\pi}\cos(\omega t_1+\omega t_2+2\phi)\,d\phi$$
$$=0$$

Hence,

$$R_{XX}(t_1,t_2)=\frac{1}{2}E[r^2]\cos\omega(t_1-t_2)$$

Example 9.12: $X(t)$ is a random process with mean value $E[X(t)\ ] = 3$ and autocorrelation function given by

$$R_{XX}(t_1,t_2)=9+4e^{-0.2(t_2-t_1)}$$

Determine the (i) mean value, (ii) variance and (iii) covariance of random variables $Y = X(t_1)$ and $Z = X(t_2)$, when $t_1 = 5$ and $t_2 = 8$.

***Solution*:**

(i) $E[X(t)\ ] = 3$ is independent of time. Hence, $E[Y] = E[X(t_1)] = E[X(5)] = 3$ and $E[Z] = E[X(t_2)] = E[X(8)] = 3$. Both Y and Z have same mean value.

(ii) $E(Y^2)=E[X(t_1)X(t_1)]=R_{XX}(t_1,\ t_1)=9+4=13\cdot$

Similarly $\quad E[Z^2]=E[X(t_2)X(t_2)]=R_{XX}(t_2,t_2)=9+4=13.$

(iii) $E(YZ)=E[X(t_1)X(t_2)]=R_{XX}(t_1,t_2)=9+4e^{-0.2(t_2-t_1)}$, $t_1 = 5$ and $t_2 = 8$. Substituting these values, we get $E(YZ)=R_{XX}(5,8)=9+4e^{-0.6}=11.195$.

Variance of Y is
$$\sigma_Y^2=E[Y^2]-\{E[Y]\}^2$$
$$=13-9=4$$

Variance of Z is
$$\sigma_Z^2=E[Z^2]-\{E[Z]\}^2$$
$$=13-9=4$$

Both Y and Z have same variance 4. Covariance of Y and Z is given by

$$C_{YZ}=E[(Y-\overline{Y})(Z-\overline{Z})]=E[YZ-Y\overline{Z}-Z\overline{Y}-\overline{Y}\ \overline{Z}]$$

$$\Rightarrow \qquad C_{YZ}=E(YZ)-\overline{Y}E[Z]-\overline{Z}E[Y]-\overline{Y}\ \overline{Z}$$
$$=E[X(t_1)X(t_2)]-\overline{Y}\ \overline{Z}-\overline{Z}\ \overline{Y}-\overline{Y}\ \overline{Z}$$
$$=R_X(t_1,t_2)-\overline{Y}\ \overline{Z}=C_{XX}(t_1,\ t_2)$$

Thus $C_{YZ} = C_{XX}(t_1, t_2) = R_{XX}(t_1, t_2) - \overline{X(t_1)}\ \overline{X(t_2)}$. Substituting the values for $t_1 = 5$ and $t_2 = 8$, we get

$$C_{YZ} = C_{XX}(5,8) = 11.195 - (3)(3) = 2.195$$

Example 9.13: Two random processes $X(t)$ and $Y(t)$ are given by

$$X(t) = A\cos\omega_0 t + B\sin\omega_0 t$$
$$Y(t) = B\cos\omega_0 t - A\sin\omega_0 t$$

where A and B are random variables and ω_0 is a constant. Show that if A and B are uncorrelated, zero mean random variables with the same variance then,

(i) $X(t)$ and $Y(t)$ are individually wide-sense stationary processes.

(ii) $X(t)$ and $Y(t)$ are jointly wide-sense stationary.

***Solution*:** If the mean value of any random process is constant and autocorrelation value is a function of only time difference, then the process is wide-sense stationary. It is seen in Example 9.10, that $E[X(t)] = 0$, constant and $R_{XX}(t, t + \tau) = \sigma^2\cos\omega_0\tau$, where σ^2 is equal variance of A and B. Hence $X(t)$ is a wide-sense stationary process.

Process $Y(t)$ is considered next.

$$E[Y(t)] = E[B\cos\omega_0 t - A\sin\omega_0 t]$$
$$= E[B]\cos\omega_0 t - E[A]\sin\omega_0 t$$

It is given, $E[B] = E[A] = 0$

Hence, $E[Y(t)] = 0$, a constant. Let us next find $R_{YY}(t, t + \tau)$

$$R_{YY}(t,\ t+\tau) = E[Y(t)Y(t+\tau)]$$
$$= E[B\cos\omega_0 t - A\sin\omega_0 t)(B\cos\omega_0(t+\tau) - A\sin\omega_0(t+\tau))]$$
$$= E[B^2\cos\omega_0 t\cos\omega_0(t+\tau) - BA\cos\omega_0 t\sin\omega_0(t+\tau)$$
$$- AB\sin\omega_0 t\cos\omega_0(t+\tau) + A^2\sin\omega_0 t\sin\omega_0(t+\tau)]$$
$$= E[B^2]\cos\omega_0 t\cos\omega_0(t+\tau) + E[A^2]\sin\omega_0 t\sin\omega_0(t+\tau)$$

The other two terms are zero because A and B are uncorrelated and are of zero mean, implying

$$E[AB] = E[BA] = E[A]E[B] = 0\,.$$

Recall the identities:

$$\cos\alpha\cos\beta = \frac{1}{2}[\cos(\alpha+\beta) + \cos(\alpha-\beta)]\text{and}$$
$$\sin\alpha\sin\beta = \frac{1}{2}[\cos(\alpha-\beta) - \cos(\alpha+\beta)]$$

Identifying, $\alpha = \omega_0 (t+\tau)$ and $\beta = \omega_0 t$, we have

$$\cos\omega_0(t+\tau)\cos\omega_0 t = \frac{1}{2}[\cos(2\omega_0 t+\omega_0\tau)+\cos\omega_0\tau]$$

and

$$\sin\omega_0(t+\tau)\sin\omega_0 t = \frac{1}{2}[\cos\omega_0\tau-\cos(2\omega_0 t+\omega_0\tau)]$$

Substituting these expressions and noting that $E[A^2]=E[B^2]=\sigma^2$ where σ^2 is variance of *A* and *B*, we get

$$R_{YY}(t,t+\tau) = \frac{\sigma^2}{2}[\cos(2\omega_0 t+\omega_0\tau)+\cos\omega_0\tau]$$
$$+\frac{\sigma^2}{2}[\cos\omega_0\tau-\cos(2\omega_0 t+\omega_0\tau)] = \sigma^2\cos\omega_0\tau$$

Since $E[Y(t)] = 0$, a constant and $R_{YY}(t, t+\tau) = \sigma^2 \cos \omega_0\tau$ is a function of τ, the time difference, $Y(t)$ is a wide-sense stationary process.

We next consider cross-correlation function $R_{XY}(t, t+\tau)$.

$$\begin{aligned} R_{XY}(t,t+\tau) &= E[X(t)Y(t+\tau)] \\ &= E[(A\cos\omega_0 t+B\sin\omega_0 t)(B\cos\omega_0(t+\tau)-A\sin\omega_0(t+\tau))] \\ &= E[AB\cos\omega_0 t\cos\omega_0(t+\tau)-A^2\cos\omega_0 t\sin\omega_0(t+\tau) \\ &\quad +B^2\sin\omega_0 t\cos\omega_0(t+\tau)-AB\sin\omega_0 t\sin\omega_0(t+\tau)] \end{aligned}$$

Since, it is given that *A* and *B* are uncorrelated and are of zero mean, we have $E[AB]=E[BA]=E[A]\,E[B]=0$. Therefore,

$$R_{XY}(t,t+\tau) = -E[A^2]\cos\omega_0 t\sin\omega_0(t+\tau)+E[B^2]\sin\omega_0 t\cos\omega_0(t+\tau)$$

It is given that variance of *A* and *B* are equal. Hence,

$$E[A^2]=E[B^2]=\sigma^2$$

Using trigonometric identities as given earlier, we have,

$$R_{XY}(t,t+\tau) = -\sigma^2\sin(\omega_0\tau)$$

Since $R_{XY}(t, t+\tau)$ depends only on time difference τ, $X(t)$ and $Y(t)$ are jointly wide-sense stationary.

Example 9.14:

$$Y(t) = X(t)\cos(\omega_0 t + \theta)$$

where $X(t)$ is a random process and θ is uniformly distributed over the interval $(0, 2\pi)$. Determine under what conditions is $Y(t)$ wide-sense stationary. Assume θ and $X(t)$ are statistically independent and ω_0 is constant.

Solution: $E[Y(t)] = E[X(t)\cos(\omega_0 t + \theta)]$. Since $X(t)$ and θ are independent, the mean of products is equal to products of mean. Hence, $E[Y(t)] = E[X(t)]E[\cos(\omega_0 t + \theta)]$.

But

$$E[\cos(\omega_0 t + \theta) = \int_0^{2\pi} \frac{\cos(\omega_0 t + \theta)}{2\pi} d\theta \quad \left(\because \text{PDF} = \frac{1}{2\pi}\right)$$

$$= \frac{1}{2\pi}\sin(\omega_0 t + \theta)\Big|_{\theta=0}^{\theta=2\pi} = 0$$

$$E[Y(t)] = 0, \text{ a constant}$$

The autocorrelation of $Y(t)$ is

$$R_{YY}(t, t+\tau) = E(Y(t)Y(t+\tau)$$

$$= E\{[X(t)\cos(\omega_0 t + \theta)][X(t+\tau)\cos\omega_0(t+\tau) + \theta)]\}$$

Since $X(t)$ and θ are uncorrelated, we can write,

$$R_{YY}(t, t+\tau) = E[X(t)X(t+\tau)\}E\{\cos(\omega_0 t + \theta)\cos\omega_0(t+\tau) + \theta)\}$$

$$= R_{XX}(t, t+\tau)E\left\{\frac{\cos\omega_0\tau + \cos(2\omega_0 t + \omega_0\tau + 2\theta)}{2}\right\}$$

$$= \frac{1}{2}R_{XX}(t, t+\tau)\cos\omega_0\tau + E\left[\frac{\cos(2\omega_0 t + \omega_0\tau + 2\theta)}{2}\right]$$

The second term is zero. The reason is given below.

$$E[\cos(2\omega_0 t + \omega_0\tau + 2\theta)] = \int_0^{2\pi} \frac{1}{2\pi}\cos(2\omega_0 t + \omega_0\tau + 2\theta)\, d\theta$$

$$= \frac{1}{2\pi}\frac{\sin(2\omega_0 t + \omega_\tau + 2\theta)}{2}\Big|_{\theta=0}^{2\pi} = 0$$

Hence

$$R_{YY}(t, t+\tau) = \frac{1}{2}R_{XX}(t, t+\tau)\cos\omega_0\tau$$

The process $Y(t)$ will be wide-sense stationary, if $X(t)$ will be wide-sense stationary.

Example 9.15: Let $X(t) = At + B$, where A and B are statistically independent random variables, both uniformly distributed over the interval (–1, 1). Determine whether $X(t)$ is wide-sense stationary.

Solution: A and B are uniformly distributed over the interval (–1, 1).

Hence,

$$E[A] = \int_{-1}^{1} a f_A(a) da$$

$$= \int_{-1}^{1} (a)\left(\frac{1}{2}\right) da$$

$$= \frac{a^2}{4}\Big|_{a=-1}^{1} = 0$$

Similarly, $E[B] = 0$.

Since A and B are statistically independent, $E[AB] = E[A]E[B]$ and hence equal to zero.

Autocorrelation function $R_{XX}(t, t+\tau)$ of $X(t)$ is,

$$R_{XX}(t, t+\tau) = E[X(t)X(t+\tau)]$$

$$= E[(At+B)(A(t+\tau)+B)]$$

$$= E[A^2 t(t+\tau) + ABt + AB(t+\tau) + B^2]$$

Expectation of sum is equal to sum of expectation. Taking term by term expectation,

$$R_{XX}(t, t+\tau) = E[A^2]t(t+\tau) + E[AB]t$$

$$+ E[AB](t+\tau) + E[B^2]$$

We have

$$E[A^2] = \int_{-1}^{1} a^2 \frac{1}{2} da = \frac{a^2}{(3)(2)}\Big|_{a=-1}^{1} = \frac{1}{3}$$

Similarly

$$E[B^2] = \frac{1}{3}$$

It is seen that $E[AB] = 0$. Substituting these values,

$$R_{XX}(t, t+\tau) = \frac{1}{3}(t)(t+\tau) + \frac{1}{3};$$

We find that $R_{XX}(t, t+\tau)$ is a function of t. Hence, $X(t)$ is not wide-sense stationary.

Exercise Problems

EP. 9.1: For a random process $X(t)$ it is verified that

$$f_X(x_1, x_2, x_3;\ t_1, t_2, t_3) = f_X(x_1, x_2, x_3;\ t_1 + \Delta, t_2 + \Delta, t_3 + \Delta)$$

for all t_1, t_2, t_3 and Δ.

(i) Is $X(t)$ a first-order stationary process?

(ii) Is $X(t)$ a second-order stationary process?

(iii) Is $X(t)$ a third order stationary process?

(iv) Is $X(t)$ a wide-sense stationary process?

(v) Is $X(t)$ a strictly stationary process?

(vi) Is $X(t)$ an ergodic process?

Justify your answer.

Ans: (i) yes (ii) yes (iii) yes (iv) it is also wide-sense stationary.

EP. 9.2: A random process is defined by $X(t) = At$ where A is a continuous random variable uniformly distributed in the range (0, 1). Classify the random process $X(t)$.

Ans: E[At] m E[A]t hence non stationary process.

EP. 9.3: A random process is defined by $X(t) = At + B$ where A and B are statistically independent random variables both uniformly distributed in the interval (–1, 1). Find (a) the mean (b) the autocorrelation (c) autocovariance functions of $X(t)$. Is $X(t)$ wide-sense stationary?

Ans: Refer Example 9.15.

EP. 9.4: A random process is defined as $X(t) = A\cos(\pi t)$ where A is a gaussian random variable with zero mean and variance σ_A^2.

(a) Find the density functions of random variables $X(0)$ and $X(1)$

(b) Find $E[X(t)]$ and $R_{XX}(t, t+\tau)$

(c) Is $X(t)$ wide-sense stationary?

EP. 9.5: A random process $Y(t)$ is defined in terms of a process $X(t)$ which is wide-sense stationary as $Y(t) = X(t) - X(t+\tau)$. Show that (a) $E[Y(t)] = 0$ and (b) $\sigma_A^2 = 2[R_{XX}(0) - R_{XX}(\tau)]$.

EP. 9.6: $X(t)$ and $Y(t)$ are two random processes and $W(t) = X(t) + Y(t)$ Find expression for $R_{WW}(t, t+\tau)$ if

(a) $X(t)$ and $Y(t)$ are correlated

(b) they are uncorrelated

(c) they are uncorrelated and with zero mean.

EP. 9.7: Random process $X(t)$ and $Y(t)$ are defined as

$X(t) = A \cos(\omega_0 t + \theta)$

$Y(t) = B \cos(\omega_0 t + \theta)$

where A, B and ω_0 are constants while θ is a random variable uniform on $(0, 2\pi)$.

Determine

(i) $E[X(t)]$ and $E[Y(t)]$

(ii) $R_{XX}(\tau)$ and $R_{YY}(\tau)$

Show that $X(t)$ and $Y(t)$ are individually wide-sense stationary.

Find cross correction function $R_{XY}(t, t + \tau)$. Show that $X(t)$ and $Y(t)$ are jointly wide-sense stationary.

Ans: Refer Example 9.3.

EP. 9.8: $X(t)$ is a random process such that for all n and all $(t_1, t_2, ..., t_n)$ vector $(X(t_1), X(t_2) ...,$ $X(t_n)$ is an n dimensional gaussian random variable with mean zero and covariance matrix is $n \times n$ identity matrix. Show that $X(t)$ is strictly stationary.

EP. 9.9: State and prove properties of autocorrelation function of random process $X(t)$.

EP. 9.10: Random process $X(t)$ is defined by $X(t) = A$, where A is a random variable uniformly distributed on $(-1, 1)$. Is the process stationary? Is the process ergodic?

EP. 9.11: Let $X(t)$ be an ergodic process. Show that Fourier transform of autocorrelation function $R_{XX}(\tau)$ is power spectra density of $X(t)$.

EP. 9.12: Random process $X(t) = A + Bt$ where A and B are independent random variables each uniformly distributed on $(-1,1)$. Find $E[X(t)]$ and $R_{XX}(t_1, t_2)$.

Ans: Refer Example 9.15.

EP. 9.13: Show that for $k \geq 2$, any k^{th} order stationary process is wide-sense stationary.

EP. 9.14: Show that for jointly stationary process $X(t)$ and $Y(t)$,

$$R_{XY}(\tau) = R_{YX}(-\tau)$$

EP. 9.15: Let $Y(t) = X(t) - X(t + T)$ where $X(t)$ is a stationary non periodic process and T is a constant. Determine $R_{XX}(\tau)$. Find the mean and variance of $Y(t)$ in terms of $R_{XX}(\tau)$.

EP. 9.16: Let $X(t)$ and $Y(t)$ be jointly stationary so that

$$R_{XY}(t_1, t_2) = R_{XY}(t_1 - t_2)$$

Show that $R_{XY}(\tau) = R_{YX}(-\tau)$

EP. 9.17: *A* random process $X(t)$ is defined as $X(t) = A\cos(\omega_0 + \theta)$ where ω and θ are constants and A is a random variable uniformly distributed over (–1, 1). Determine whether $X(t)$ is wide-sense stationary.

Ans: Refer Example 9.5.

EP. 9.18: If $X(t)$ and $Y(t)$ are wide-sense stationary random process show that

$$|R_{XY}(\tau)| \leq \sqrt{R_{XX}(0)R_{YY}(0)}$$

EP. 9.19: Let $r(t) = A_0 \cos \omega_0 t + n(t)$, where A_0 and ω_0 are constants. Assume $n(t)$ is a wide-sense stationary random noise process with a zero mean value and an autocorrelation of $R_n(\tau)$.

(a) Find $E[r(t)]$ and $R_{rr}(t_1, t_2)$

(b) Find whether $r(t)$ is wide-sense stationary.

EP. 9.20: Random process $Y(t)$ is given by $Y(t) = X(t)\cos(\omega t + \theta)$; ω_0 is a constant. Where ω_0 is a stationary random process with $E[X(t)] = 0$ and $E[X^2(t)] = \sigma\frac{2}{x}$.

(a) If $\theta = 0$ find $E[Y(t)]$ and $E[Y^2(t)]$. Is $Z(t)$ stationary?

(b) If θ is a random variable independent of $X(t)$ and uniformly distributed in the interval $(-\pi, \pi)$ show that

$$E[Y(t) = 0 \text{ and } E[Y^2(t)] = \frac{\sigma_x^2}{2}.$$

Chapter 10

Spectral characteristic of random process

In the previous chapters, we have studied the time domain characteristics of random process like autocorrelation, cross-correlation and covariance function without considering the spectral properties of the process. As we all know that both time domain and frequency domain analysis is needed for analyzing linear systems for deterministic waveforms, similarly we can describe the random process in frequency domain also.

The spectral description of a deterministic waveform is obtained by applying Fourier transform and hence Fourier transform plays an important role in spectral characterization of random waveform also. But, the direct transformation approach is not attractive for random waveforms, since the transform may not exist for all these waveforms. Hence, the spectral analysis of random processes needs more attention than the deterministic signals/processes. For this reason, this chapter focus more on the important concepts that apply to characterizing random process in frequency domain.

10.1 Power spectral density

The Fourier transform $X(\omega)$ of a deterministic signal $x(t)$ is given by

$$X(\omega) = \int_{-\infty}^{\infty} x(t)e^{-j\omega t}\,dt \tag{10.1}$$

The function $X(\omega)$ gives the information regarding the spectral characteristics of the signal $x(t)$ and is simply called as spectrum of $x(t)$. Both the amplitude and phase of frequencies present in $x(t)$ are described by $X(\omega)$. Hence if $X(\omega)$ is known, we can recover $x(t)$ by taking inverse Fourier transform given by

$$x(t) = \frac{1}{2\pi}\int_{-\infty}^{\infty} X(\omega)e^{j\omega t}\,d\omega \tag{10.2}$$

But, the Fourier transform may not exist for all signals. As we know that for any function $x(t)$ to be Fourier transformable, it is sufficient that $x(t)$ satisfies Dirichlets conditions presented below for convenience:

(i) The function of $x(t)$ is a single valued bounded function with a finite number of maxima and minima and a finite number of discontinuities in any finite time interval.

(ii) The function $x(t)$ is absolutely integrable or summable. Mathematically

$$\int_{-\infty}^{\infty} |x(t)|\, dt < \infty \tag{10.3}$$

These conditions are only sufficient but not necessary for $X(\omega)$ to exist. Many signals of practical interest do not satisfy these conditions but do have transforms.

Examples are: (i) Unit impulse function $\delta(t)$ has transform $\delta(\omega)$.

(ii) Unit step function $u(t)$ defined by

$$x(t) = u(t) = \begin{cases} 1, & t > 0 \\ 0, & t < 0 \end{cases}$$

has the transform $X(\omega) = \pi\delta(\omega) + (1/j\omega)$.

If we try to apply Eq. (10.1) to a random process, we may encounter problems. The main problem is the fact that $X(\omega)$ may not exist, because the Dirichlets conditions may not be satisfied by most sample functions of the process. Hence we conclude that we cannot describe the spectrum of a random process by simply applying. Eq. (10.1) on the sample functions of the process.

For a sample function $x(t)$ of a random process $X(t)$ to be Fourier transformable, let us define $x_T(t)$ as that portion of $x(t)$ that exists between $-T$ and T, i.e.

$$x_T(t) = \begin{cases} x(t), & -T < t < T \\ 0, & \text{elsewhere} \end{cases} \tag{10.4}$$

Now, as long as T is finite, we assume that $x_T(t)$ has bounded variation and hence will satisfy the condition.

$$\int_{-\infty}^{\infty} |x_T(t)|\, dt = \int_{-T}^{T} |x_T(t)|\, dt < \infty \tag{10.5}$$

and hence will have Fourier transform (since Dirichlets condition is satisfied) which is denoted by $X_T(\omega)$ and is given by

$$X_T(\omega) = \int_{-T}^{T} x_T(t)\, e^{-j\omega t} dt = \int_{-T}^{T} x(t)\, e^{-j\omega t} dt \tag{10.6}$$

since in the bounded interval $(-T, T)$, $x_T(t) = x(t)$.

Now, the energy contained in $x(t)$ in the interval $(-T, T)$ is equal to energy in $x_T(t)$ and is given by

$$E(T) = \int_{-T}^{T} x^2(t)dt = \int_{-T}^{T} x_T^2(t)dt \tag{10.7}$$

Using Parseval's theorem for energy, Eq. (10.7) can be written as

$$E(T)=\int_{-T}^{T}x^2(t)dt=\frac{1}{2\pi}\int_{-\infty}^{\infty}\left|X_T(\omega)\right|^2 d\omega \tag{10.8}$$

By dividing expressions in (10.8) by $2T$, we obtain average power $P(T)$ in $x(t)$ over the interval $(-T, T)$ given by

$$P(T)=\frac{E(T)}{2T}=\frac{1}{2T}\int_{-T}^{T}x^2(t)dt=\frac{1}{2\pi}\int_{-\infty}^{\infty}\frac{\left|X_T(\omega)\right|^2}{2T}d\omega \tag{10.9}$$

At this point we observe that the expression $\left|X_T(\omega)\right|^2/2T$ is the power density spectrum/power spectral density (PSD), since power evaluation of $P(T)$ has resulted in finding the area under this expression.

However, Eq. (10.9) represents the power of the sample function $x(t)$ in the interval $(-T, T)$. If we need to calculate the total power of the ensemble member $x(t)$ we have to extend this time interval T to ∞. So, the power of one sample function $x(t)$ of random process $X(t)$ is given by

$$P(T)=\lim_{T\to\infty}\frac{1}{2T}\int_{-T}^{T}x^2(t)dt=\lim_{T\to\infty}\frac{1}{2\pi}\int_{-\infty}^{\infty}\frac{\left|X_T(\omega)\right|^2}{2T}d\omega \tag{10.10}$$

$P(T)$ is a random variable with respect to a random process. So, by taking the expected value of Eq. (10.10), we can obtain the average power P_{XX} for the random process $X(t)$, given by

$$P_{XX}=E\left[P(T)\right]=\lim_{T\to\infty}\frac{1}{2T}\int_{-T}^{T}E\left[X^2(t)\right]dt=\frac{1}{2\pi}\int_{-\infty}^{\infty}\lim_{T\to\infty}\frac{E\left[\left|X_T(\omega)\right|^2\right]}{2T}d\omega \tag{10.11}$$

We can conclude two important points form Eq. (10.11).

(1) The average power P_{XX} in a random processes $X(t)$ is equal to time average of its second moment:

$$P_{XX}=\lim_{T\to\infty}\frac{1}{2T}\int_{-T}^{T}E\left[X^2(t)\right]dt=A\left\{E\left[X^2(t)\right]\right\} \tag{10.12}$$

where $A\{f(t)\}=\lim_{T\to\infty}\frac{1}{2T}\int_{-T}^{T}f(t)dt$ is the time average of the function $f(t)$. If the random process is at least wide-sense stationary, then $E\left[X^2(t)\right]=\overline{X}^2$, a constant and hence $P_{XX}=\overline{X}^2$, a constant.

(2) The expression $\lim_{T\to\infty}\frac{E\left[\left|X_T(\omega)\right|^2\right]}{2T}$ represents power density spectrum/power spectral density of the random process $X(t)$, since the power P_{XX} of $X(t)$ is obtained by taking the area under the above expression.

Hence, if we define

$$S_{XX}(\omega) = \lim_{T\to\infty} \frac{E\left[\left|X_T(\omega)\right|^2\right]}{2T} \tag{10.13}$$

Then the formula for power is

$$P_{XX}(\omega) = \frac{1}{2T}\int_{-\infty}^{\infty} S_{XX}(\omega)d\omega \tag{10.14}$$

where $S_{XX}(\omega)$ is the PSD of random process $X(t)$.

Example 10.1: Consider a random process $X(t) = \cos(\omega_0 t + \theta)$ where ω_0 is a real constant and θ is a uniform random variable in the interval $(0, \pi/2)$.

(a) Show that $X(t)$ is not a WSS process.

(b) Also find the average power in the process.

Solution:

(a)

$$E[X(t)] = \int_0^{\pi/2} \frac{2}{\pi}\cos(\omega_0 t + \theta)d\theta$$

$$= \frac{2}{\pi}\left[\sin(\omega_0 t + \theta)\right]_0^{\pi/2}$$

$$= \frac{2}{\pi}\left[\cos(\omega_0 t) - \sin(\omega_0 t)\right]$$

For a process $X(t)$ to be at least WSS, the mean $E[X(t)]$ should be constant and auto-correlation function $R_{XX}(t_1, t_2)$ should depend on $\tau = t_2 - t_1$ and not on t_1, t_2.

In the above problem $E[X(t)]$ is not constant but depends on time t. So, we need not check whether auto-correlation function depends on 't' or not, and say that the random process $X(t)$ is not WSS process.

(b) This problem can be solved using either Eq. (10.12) or Eqs. (10.13) and (10.14). First we shall find the average power P_{XX} in $X(t)$ using Eq. (10.12).

Method I: Mean squared value of $X(t)$ is

$$E\left[X^2(t)\right] = E\left[\cos^2(\omega_0 t + \theta)\right]$$

$$= E\left[\frac{1}{2}\right] + E\left[\frac{\cos 2(\omega_0 t + \theta)}{2}\right]$$

$$= \frac{1}{2} + \frac{1}{2}\frac{2}{\pi}\int_0^{\pi/2} \cos 2(\omega_0 t + \theta)d\theta$$

$$= \frac{1}{2} + \frac{1}{\pi}\left[\frac{\sin 2\left(\omega_0 t + \theta\right)}{2}\right]_0^{\pi/2}$$

$$= \frac{1}{2} + \frac{1}{\pi}\left[\frac{-\sin\left(2\omega_0 t\right)}{2} - \frac{\sin 2\omega_0 t}{2}\right]$$

$$= \frac{1}{2} - \frac{\sin 2\omega_0 t}{\pi}$$

Therefore, the average power P_{XX} in the process $X(t)$ is

$$P_{XX} = \lim_{T\to\infty} \frac{1}{2T}\int_{-T}^{T} E\left[X^2\left(t\right)\right]dt$$

$$= \lim_{T\to\infty} \frac{1}{2T}\int_{-T}^{T}\left[\frac{1}{2} - \frac{\sin 2\omega_0 t}{\pi}\right]dt$$

$$= \lim_{T\to\infty} \frac{1}{2T}\left[\frac{1}{2}\int_{-T}^{T} dt - \frac{1}{\pi}\int_{-T}^{T} \sin 2\omega_0 t dt\right]$$

$$= \lim_{T\to\infty} \frac{1}{2T}\left\{\left[\frac{1}{2}(2T)\right] - \frac{1}{\pi}\left[\frac{-\cos 2\omega_0 t}{2\omega_0}\right]_{-T}^{T}\right\}$$

$$= \lim_{T\to\infty} \frac{1}{2T}\left[T - \frac{1}{\pi}\left[-1+1\right]\right] = \frac{1}{2}$$

Method II: We can also calculate P_{XX} of $X(t)$ using Eqs.(10.13) and (10.14). First let us find $X_T(\omega)$.

$$X_T(\omega) = \int_{-T}^{T} \cos\left(\omega_0 t + \theta\right)\exp\left(-j\omega t\right)dt$$

$$= \frac{1}{2}\exp\left(j\theta\right)\int_{-T}^{T} \exp\left[j\left(\omega_0 - \omega\right)t\right]dt$$

$$+ \frac{1}{2}\exp\left(-j\theta\right)\int_{-T}^{T} \exp\left[-j\left(\omega_0 + \omega\right)t\right]dt$$

Using $\cos\omega_0 t = \dfrac{e^{j\omega_0 t} + e^{-j\omega_0 t}}{2}$, we get

$$X_T(\omega) = \frac{\exp\left(j\theta\right)}{2}\left[\frac{\exp\left[j\left(\omega_0 - \omega\right)t\right]}{j\left(\omega_0 - \omega\right)}\right]_{-T}^{T} + \frac{\exp\left(-j\theta\right)}{2}\left[\frac{\exp\left[-j\left(\omega_0 + \omega\right)t\right]}{-j\left(\omega_0 + \omega\right)}\right]_{-T}^{T}$$

$$= \exp(j\theta)\left[\frac{\exp\left[j\left(\omega_0-\omega\right)T\right]-\exp\left[j\left(\omega_0-\omega\right)T\right]}{2j\left(\omega_0-\omega\right)}\right]$$

$$+\exp(-j\theta)\left[\frac{\exp\left[-j\left(\omega_0+\omega\right)T\right]-\exp\left[j\left(\omega_0+\omega\right)T\right]}{-2j\left(\omega_0-\omega\right)}\right]$$

$$= T\exp(j\theta)\frac{\sin(\omega-\omega_0)T}{(\omega-\omega_0)T}+T\exp\left(-j\theta\right)\frac{\sin(\omega+\omega_0)T}{(\omega+\omega_0)T}$$

Next we find $\left|X_T\left(\omega\right)\right|^2 = X_T\left(\omega\right)X_T^*\left(\omega\right)$:

$$\left|X_T\left(\omega\right)^2\right| = \left[T\exp(j\theta)\frac{\sin(\omega-\omega_0)T}{(\omega-\omega_0)}+T\exp(-j\theta)\frac{\sin(\omega+\omega_0)T}{(\omega+\omega_0)}\right]$$

$$\times\left[T\exp(-j\theta)\frac{\sin(\omega-\omega_0)T}{(\omega-\omega_0)T}+T\exp(j\theta)\frac{\sin(\omega+\omega_0)T}{(\omega+\omega_0)T}\right]$$

$$= T^2\left\{\frac{\sin^2(\omega-\omega_0)T}{(\omega-\omega_0)^2T^2}+\frac{\sin^2(\omega-\omega_0T)}{(\omega-\omega_0)^2T^2}\right.$$

$$\left.+\frac{\sin(\omega+\omega)T\sin(\omega-\omega_0)T}{(\omega-\omega_0)(\omega+\omega_0)T^2}\left[\exp\left(2j\theta\right)+\exp\left(-2j\theta\right)\right]\right.$$

Now taking expectation on both sides:

$$E\left[\left|X_T(\omega)\right|^2\right] = \frac{2T^2}{\pi}\int_0^{\pi/2}\frac{\left[\sin^2(\omega-\omega_0)T\right]}{(\omega-\omega_0)^2T^2}d\theta+\frac{2T^2}{\pi}\int_0^{\pi/2}\frac{\left[\sin^2(\omega-\omega_0)T\right]}{(\omega-\omega_0)^2T^2}d\theta$$

$$+\frac{2T^2}{\pi}\int_0^{\pi/2}\frac{\left[\sin(\omega+\omega_0)T\sin(\omega-\omega_0)T\right]}{(\omega-\omega_0)(\omega+\omega_0)T^2}\left(e^{2j\theta}+e^{-2j\theta}\right)d\theta$$

$$= \frac{2T^2}{\pi}\frac{\sin^2(\omega-\omega_0)T}{(\omega-\omega_0)^2T^2}\left[\theta\right]_0^{\pi/2}+\frac{2T^2}{\pi}\frac{\sin^2(\omega-\omega_0)T}{(\omega+\omega_0)^2T^2}\left[\theta\right]_0^{\pi/2}$$

$$+\frac{2T^2}{\pi}\frac{\sin(\omega+\omega_0)T\sin(\omega-\omega_0)T}{(\omega^2-\omega_0^2)T^2}\left[\frac{e^{2j\theta}-e^{-2j\theta}}{2j}\right]_0^{\pi/2}$$

$$= \left\{\frac{\sin^2(\omega-\omega_0)T}{(\omega-\omega_0)^2}+\frac{\sin^2(\omega+\omega_0)T}{(\omega+\omega_0)^2}+0\right\}$$

$$\left[\because \frac{e^{2j\theta}-e^{-2j\theta}}{2j} = \sin 2\theta \text{ and } \left[\sin 2\theta\right]_0^{\pi/2} = 0\right]$$

Therefore,

$$E\left[\left|X_T(\omega)\right|^2\right]=\left\{\frac{\sin^2(\omega-\omega_0)T]}{(\omega-\omega_0)^2}+\frac{\sin^2(\omega+\omega_0)T]}{(\omega+\omega_0)^2}\right\}$$

Since $S_{XX}(\omega)=\lim\limits_{T\to\infty}\dfrac{E\left[\left|X_T(\omega)\right|^2\right]}{2T}$, we first calculate

$$\frac{E\left[\left|X_T(\omega)\right|^2\right]}{2T}=\frac{1}{2}\left\{\frac{T\sin^2(\omega-\omega_0)T}{\left[(\omega-\omega_0)T\right]^2}+\frac{T\sin^2(\omega+\omega_0)T]}{\left[(\omega-\omega_0)T\right]^2}\right\}$$

It is known that

$$\lim_{T\to\infty}\left[\frac{\sin\alpha T}{\alpha T}\right]^2=\frac{\pi\delta(\alpha)}{T}$$

Hence,

$$S_{XX}(\omega)=\lim_{T\to\infty}\frac{E\left[\left|X_T(\omega)\right|\right]^2}{2T} \qquad \text{(using Eq.(10.13))}$$

$$=\frac{1}{2}\left\{T\frac{\pi\delta\left(\omega-\omega_0\right)}{T}+T\frac{\pi\delta\left(\omega-\omega_0\right)}{T}\right\}$$

$$=\frac{\pi}{2}\left[\delta\left(\omega+\omega_0\right)+\delta\left(\omega-\omega_0\right)\right]$$

Finally the average power is obtained using Eq. (10.14):

$$P_{XX}=\frac{1}{2\pi}\int_{-\infty}^{\infty}\left[\frac{\pi}{2}\delta\left(\omega+\omega_0\right)+\delta\left(\omega-\omega_0\right)\right]d\omega=\frac{1}{4}[1+1]=\frac{1}{2}$$

Thus, the P_{XX} found here agrees with P_{XX} found using Method I.

10.2 Properties of power density spectrum

The power density spectrum $S_{XX}(\omega)$ of a random process $X(t)$ possesses very important properties.

Property 1: The power spectral density and time average of auto-correlation function (ACF) form a Fourier transform pair. Mathematically:

$$\frac{1}{2\pi}\int_{-\infty}^{\infty}S_{XX}\left(\omega\right)e^{j\omega\tau}d\omega=A\left[R_{XX}\left(t,t+\tau\right)\right] \qquad (10.15a)$$

and

$$S_{XX}(\omega) = \int_{-\infty}^{\infty} A\left[R_{XX}(t, t+\tau)\right]e^{-j\omega\tau}d\tau \tag{10.15b}$$

Proof: Starting with

$$X_T(\omega) = \int_{-T}^{T} x(t)e^{-j\omega t}dt$$

and

$$S_{XX}(\omega) = \lim_{T\to\infty} \frac{E\left[\left|X_T(\omega)\right|^2\right]}{2T}$$

Substituting for $X_T(\omega)$:

$$S_{XX}(\omega) = \lim_{T\to\infty} E\left[\frac{1}{2T}\int_{-T}^{T} x(t_1)e^{j\omega t_1}dt_1 \int_{-T}^{T} x(t_2)e^{-j\omega t_2}dt_2\right]$$

where t_1, t_2 are dummy variables and $x(t_1)$ and $x(t_2)$ are sample functions of the random process $X(t)$ observed at time t_1 and t_2. We can use $X(t_1)$ and $X(t_2)$ so that operations can be performed on the processes instead of sample functions.

Thus

$$S_{XX}(\omega) = \lim_{T\to\infty} E\left[\frac{1}{2T}\int_{-T}^{T} X(t_1)e^{j\omega t_1}dt_1 \int_{-T}^{T} X(t_2)e^{-j\omega t_2}dt_2\right]$$

$$= \lim_{T\to\infty} \frac{1}{2T}\left[\int_{-T}^{T}\int_{-T}^{T} E\left[X(t_1)X(t_2)\right]e^{-j\omega(t_2-t_1)}dt_2dt_1\right] \tag{10.16}$$

Using

$$R_{XX}(t_1,t_2) = E[X(t_1)X(t_2)]$$

$$S_{XX}(\omega) = \lim_{T\to\infty} \frac{1}{2T}\left[\int_{-T}^{T}\int_{-T}^{T} R_{XX}(t_1,t_2)e^{-j\omega(t_2-t_1)}dt_1dt_2\right] \tag{10.17}$$

Next we find inverse Fourier transform of Eq. (10.17). The transform will exist since power spectrum is real, non-negative and is absolutely integrable from Eqs. (10.13) and (10.14). So, the inverse transform is

$$\frac{1}{2\pi}\int_{-\infty}^{\infty} S_{XX}(\omega)e^{j\omega\tau}d\omega$$

$$= \frac{1}{2\pi}\int_{-\infty}^{\infty} \lim_{T\to\infty} \frac{1}{2T}\int_{-T}^{T}\int_{-T}^{T} R_{XX}(t_1, t)\, e^{-j\omega(t-t_1)}dt dt_1 e^{j\omega\tau}d\omega \quad (\text{Letting } t_2 = t)$$

$$= \lim_{T\to\infty} \frac{1}{2T}\int_{-T}^{T}\int_{-T}^{T} R_{XX}(t_1,t)\frac{1}{2\pi}\int_{-\infty}^{\infty} e^{-j\omega(\tau-t+t_1)}d\omega dt_1 dt \tag{10.18}$$

The inner integral of Eq. (10.18) is the impulse

$$2\pi\delta\left(\tau - t + t_1\right) = 2\pi\delta\left(t - t_1 - \tau\right)$$

Hence

$$\frac{1}{2\pi}\int_{-\infty}^{\infty} S_{XX}(\omega)e^{j\omega\tau}d\omega = \lim_{t\to\infty}\frac{1}{2T}\int_{-T}^{T}\int_{-T}^{T} R_{XX}(t, t_1)\delta(t - t_1 - \tau)dt_1 dt$$

$$= \lim_{t\to\infty}\frac{1}{2T}\int_{-T}^{T} R_{XX}(t, t+\tau)dt - T < (t+\tau) < T \quad (10.19)$$

The quantity on the RHS of Eq. (10.19) is the time average of the autocorreclation function of the random process. That is

$$A\left\{R_{XX}\left(t, t+\tau\right)\right\} = \lim_{T\to\infty}\frac{1}{2T}\int_{-T}^{T} R_{XX}(t, t+\tau)dt \quad (10.20)$$

From Eqs. (10.19) and (10.20), we can write

$$\frac{1}{2\pi}\int_{-\infty}^{\infty} S_{XX}(\omega)e^{j\omega\tau}d\omega = A\left\{R_{XX}(t, t+\tau)\right\} \quad (10.21)$$

which means the inverse Fourier transform of power spectral density $S_{XX}(\omega)$ of the random process $X(t)$ is time average of the process autocorrelation function. Hence direct transform also holds good. That is:

$$S_{XX}(\omega) = \int_{-\infty}^{\infty} A\left\{R_{XX}(t, t+\tau)\right\}e^{j\omega\tau}\ d\tau \quad (10.22)$$

which shows that $S_{XX}(\omega)$ and $A\{R_{XX}(t, t+\tau)\}$ form a Fourier transform pair.

$$A\{R_{XX}(t, t+\tau)\} \leftrightarrow S_{XX}(\omega) \quad (10.23)$$

Hence the property is proved.

For the special case, where $X(t)$ is at least wide-sense stationary (WSS) $A\{R_{XX}(t, t+\tau)\} = R_{XX}(\tau)$ and hence we get

$$S_{XX}(\omega) = \int_{-\infty}^{\infty} R_{XX}(\tau)e^{j\omega\tau}d\tau \quad (10.24)$$

and

$$R_{XX}(\tau) = \frac{1}{2\pi}\int_{-\infty}^{\infty} S_{XX}(\omega)e^{j\omega\tau}d\omega \quad (10.25)$$

or

$$R_{XX}(\tau) \leftrightarrow S_{XX}(\omega) \quad (10.26)$$

From Eq. (10.26) it is clear that knowledge of power spectrum of a process allows complete recovery of the auto-correlation function when $X(t)$ is at least WSS process. Only for non-stationary process, the time average of ACF is recoverable from Eq. (10.23).

Eq. s (10.15a) and (10.15b) are usually called **Wiener-Khintchine relations** after the great American mathematician Norbert Wiener (1894 - 1964) and Russian mathematician A.I. Khintchine (1894 -1959). They have given a powerful link between time domain property of the random process (correlation functions) and its frequency domain property (power spectrum).

Property 2: The PSD of the random process $X(t)$ is always real non-negative, i.e.,

$$S_{XX}(\omega) \geq 0 \text{ and real} \tag{10.27}$$

Proof: From the definition of $S_{XX}(\omega)$:

$$S_{XX}(\omega) = \lim_{T\to\infty} \frac{E\left[\left|X_T(\omega)\right|^2\right]}{2T} \tag{10.13}$$

Here $\left|X_T(\omega)\right|^2$ is always non-negative and real since it gives the square of the magnitude of the random process $X(t)$ and expectation of non-negative function is non-negative. Hence, from Eq. (10.13) we can say that $S_{XX}(\omega)$ is always non-negative.

Property 3: The PSD of a real valued random process is an even function of frequency i.e.

$$S_{XX}(-\omega) = S_{XX}(\omega) \text{ if } X(t) \text{ is real} \tag{10.28}$$

Proof: Let $X_T(t)$ be any Fourier transformable signal and let $X_T(\omega)$ be its Fourier transform:

$$X_T(\omega) = \int_{-\infty}^{\infty} x_T(t)e^{-j\omega t}\,dt$$

$$X_T(-\omega) = \int_{-\infty}^{\infty} x_T(t)e^{j\omega t}\,dt$$

$$= X_T^*(\omega) \text{ for real } x_T(t)$$

$$X_T(-\omega) = X_T^*(\omega) \tag{10.29}$$

$$\Rightarrow X_T(\omega) = X_T^*(-\omega) \text{ for real } x_T(t) \tag{10.30}$$

We know that

$$S_{XX}(\omega) = \lim_{T\to\infty} \frac{E[X_T^*(\omega)X_T(\omega)]}{2T}$$

$$S_{XX}(-\omega) = \lim_{T\to\infty} \frac{E[X_T^*(-\omega)X_T(-\omega)]}{2T}$$

$$= \lim_{T\to\infty} \frac{E[X_T(\omega)X_T^*(\omega)]}{2T}$$

$$= S_{XX}(\omega) \text{ [from Eq. (10.30) and (10.29)]}$$

$$S_{XX}(-\omega) = S_{XX}(\omega) \text{ for real } X(t)$$

Property 4: The area under the graph of PSD of a random process which is at least WSS gives the mean square value of the process, i.e.

$$E[X^2(t)] = \int_{-\infty}^{\infty} S_{XX}(\omega)d\omega \tag{10.31}$$

Proof: From Wiener-Khinchin relation, we have

$$R_{XX}(\tau) = \frac{1}{2\pi}\int_{-\infty}^{\infty} S_{XX}(\omega)e^{j\omega\tau}d\omega$$

On substituting $\tau = 0$ we get

$$R_{XX}(0) = \frac{1}{2\pi}\int_{-\infty}^{\infty} S_{XX}(\omega)d\omega \tag{10.32}$$

We know that for at least WSS random process

$$R_{XX}(\tau) = E[X(t)X(t+\tau)] \tag{10.33}$$

On substituting $\tau = 0$ in Eq. (10.33), we get

$$R_{XX}(0) = E[X^2(t)] \tag{10.34}$$

From Eqs. (10.32) and (10.34), we get

$$E\left[X^2(t)\right] = \int_{-\infty}^{\infty} S_{XX}(\omega)d\omega$$

Hence proved.

Property 5: The zero frequency value of PSD of a WSS random process equals the total area under the graph of ACF i.e.

$$S_{XX}(0) = \int_{-\infty}^{\infty} R_{XX}(\tau)d\tau \tag{10.35}$$

Proof: From Wiener-Khintchine relation (10.24)

$$S_{XX}(\omega) = \int_{-\infty}^{\infty} R_{XX}(\tau)e^{-j\omega\tau}d\tau$$

If $\omega = 0$, we get

$$S_{XX}(\omega) = \int_{-\infty}^{\infty} R_{XX}(\tau)\,d\tau$$

which represents the total area under the curve of ACF.

Property 6: The PSD of the derivative of the random process $X(t)$ (i.e. $\dfrac{dX(t)}{dt}$) is ω^2 times the PSD of the random process $X(t)$, i.e

$$S_{XX}(\omega) = \omega^2 S_{XX}(\omega) \tag{10.36}$$

Proof: Let

$$X_T(\omega) = \mathscr{F}\left[\frac{dX_T(t)}{dt}\right] = j\omega X_T(\omega)$$

$$S_{XX}(\omega) = \lim_{T\to\infty} \frac{E\left[\left|X_T(\omega)\right|^2\right]}{2T}$$

$$= \lim_{T\to\infty} \frac{E\left[\left|j\omega X_T(\omega)\right|^2\right]}{2T}$$

$$= \omega^2 \lim_{T\to\infty} \frac{E\left[\left|X_T(\omega)\right|^2\right]}{2T}$$

$$= \omega^2 S_{XX}(\omega)$$

$$S_{XX}(\omega) = \omega^2 S_{XX}(\omega)$$

Hence proved.

10.3 Bandwidth of Power Density Spectrum

The power spectral density $S_{XX}(\omega)$ is similar to probability density function (PDF) in characteristics like non-negativeness and being real, but in terms of area under the curve of the function, the area of PSD is not unity, where as it is unity for the PDF function. Indeed, by dividing PSD by its area, i.e. normalized PSD $S_{XX_N}(\omega)$ is analogous to PDF which is given by:

$$S_{XX_N}(\omega) = \frac{S_{XX}(\omega)}{\int_{-\infty}^{\infty} S_{XX}(\omega)d\omega} \tag{10.37}$$

In density function the measure of spread is given by standard deviation. In the same way, the analogous quantity for normalized PSD is called **rms bandwidth** which is denoted by W_{rms}(rad/sec).

Since $S_{XX}(\omega)$ and hence $S_{XX_N}(\omega)$ are even functions for real process their mean is zero and hence the standard deviation for normalized power spectrum i.e. rms bandwidth is given by

$$W_{\text{rms}} = \left[\int_{-\infty}^{\infty} \omega^2 S_{XX_N}(\omega)d\omega\right]^{1/2}. \tag{10.38}$$

$$W_{\text{rms}}^2 = \int_{-\infty}^{\infty} \omega^2 \frac{S_{XX}(\omega)}{\int_{-\infty}^{\infty} S_{XX}(\omega)d\omega} d\omega$$

$$= \frac{\int_{-\infty}^{\infty} \omega^2 S_{XX}(\omega) d\omega}{\int_{-\infty}^{\infty} S_{XX}(\omega) d\omega} \tag{10.39}$$

10.4 Cross power spectral density

The power spectral density of a random process gives the frequency domain information of a single random process. Similarly the cross power spectral density provides information regarding frequency interrelationship between two random processes.

Consider a real random process *Z*(*t*) which is given by the sum of two real random processes *X*(*t*) and *Y*(*t*):

$$Z(t) = X(t) + Y(t) \tag{10.40}$$

The auto-correlation function of *Z*(*t*) is given by

$$\begin{aligned} R_{ZZ}(t, t+\tau) &= E[Z(t)Z(t+\tau)] \\ &= E\{[X(t)+Y(t)][X(t+\tau)+Y(t+\tau)]\} \\ &= E\{[X(t)+X(t+\tau)+X(t)Y(t+\tau) \\ &\quad +Y(t)X(t+\tau)+Y(t)Y(t+\tau)]\} \\ &= R_{XX}(t, t+\tau)+R_{XY}(t, t+\tau) \\ &\quad +R_{YX}(t, t+\tau)+R_{YY}(t, t+\tau) \end{aligned} \tag{10.41}$$

Taking time average on both sides of Eq. (10.41) and applying Fourier transform on both sides, we get:

$$\begin{aligned} \mathscr{F}\{A[R_{ZZ}(t, t+\tau)]\} &= \mathscr{F}\{A[R_{XX}(t, t+\tau)]\} + \mathscr{F}\{A[R_{XY}(t, t+\tau)]\} \\ &\quad + \mathscr{F}\{A[R_{YX}(t, t+\tau)]\} + \mathscr{F}\{A[R_{YY}(t, t+\tau)]\} \end{aligned} \tag{10.42}$$

Using Eq. (10.23), Eq. (10.42) can be written as

$$S_{ZZ}(\omega) = S_{XX}(\omega) + \mathscr{F}\{\mathrm{A}[R_{XY}(t, t+\tau)]\} + \mathscr{F}\{\mathrm{A}[R_{YX}(t, t+\tau)]\} + S_{YY}(\omega) \tag{10.43}$$

The terms $S_{ZZ}(\omega)$, $S_{XX}(\omega)$, $S_{YY}(\omega)$ represents PSD of random processes $Z(t)$, $X(t)$ and $Y(t)$, respectively. The remaining two terms on the RHS of Eq. (10.41) are new quantities and they are the topics of the next section.

For two random processes *X*(*t*) and *Y*(*t*) to be Fourier transformable, let us define $x_T(t)$ and $y_T(t)$ as truncated ensemble members, i.e.

$$x_T(t) = \begin{cases} x(t), & -T < t < T \\ 0, & \text{elsewhere} \end{cases}$$

$$y_T(t) = \begin{cases} y(t) & -T < t < T \\ 0, & \text{elsewhere} \end{cases} \tag{10.44}$$

Now, as long as T is finite, we assume that $x_T(t)$ and $y_T(t)$ has bounded variations so that Dirichlets conditions (10.3) are satisfied. Hence they possess Fourier transforms, which are denoted by $X_T(\omega)$ and $Y_T(\omega)$, respectively. Mathematically

$$x_T(t) \leftrightarrow X_T(\omega)$$
$$y_T(t) \leftrightarrow Y_T(\omega) \tag{10.45}$$

We now define the cross power $P_{XY}(T)$ in the two processes within the interval $(-T,T)$ as

$$\begin{aligned} P_{XY}(T) &= \frac{1}{2T}\int_{-T}^{T} x_T(t)y_T(t)\,dt \\ &= \frac{1}{2T}\int_{-T}^{T} x(t)y(t)\,dt \end{aligned} \tag{10.46}$$

since $x(t) = x_T(t)$ and $y(t) = y_T(t)$ in the limit $(-T,T)$. Applying Parseval's theorem on Eq. (10.46), we get

$$\begin{aligned} P_{XY}(T) &= \frac{1}{2T}\int_{-T}^{T} x(t)y(t)dt \\ &= \frac{1}{2\pi}\int_{-\infty}^{\infty} \frac{X_T^*(\omega)Y_T(\omega)d\omega}{2T} \end{aligned} \tag{10.47}$$

This power is the cross power of ensemble members of random processes $X(t)$ and $Y\{t)$. Hence, to find the average cross power of random processes themselves which is denoted by $\overline{P}_{XY}(T)$, we need to take the average values of the ensemble member's cross power. Hence taking expected value of Eq. (10.47) results in

$$\begin{aligned} \overline{P}_{XY}(T) &= \frac{1}{2T}\int_{-T}^{T} x(t)y(t)dt \\ &= \frac{1}{2\pi}\int_{-\infty}^{\infty} \frac{X_T^*(\omega)Y_T(\omega)d\omega}{2T} \end{aligned} \tag{10.48}$$

Finally we find the total average cross power P_{XY} of the random processes $X(t)$ and $Y(t)$ over the entire interval by letting $T \to \infty$:

$$\begin{aligned} P_{XY} &= \lim_{T\to\infty} \frac{1}{2T}\int_{-T}^{T} R_{XY}(t, t+\tau)dt \\ &= \frac{1}{2\pi}\int_{-\infty}^{\infty} \lim_{T\to\infty} \frac{E[X_T^*(\omega)Y_T(\omega)]}{2T} d\omega \end{aligned} \tag{10.49}$$

It is clear from Eq. (10.49) that the cross power is given by the area under the integrand which involves ω, and hence the integrand is defined as cross power density spectrum and denoted by

$$S_{XY}(\omega) = \lim_{T\to\infty} \frac{E[X_T^*(\omega)Y_T(\omega)]}{2T} \tag{10.50}$$

So, using Eqs. (10.49) and (10.50), we call the formula for cross power as

$$P_{XY} = \frac{1}{2\pi}\int_{-\infty}^{\infty} S_{XY}(\omega)d\omega \tag{10.51}$$

This cross power or the corresponding cross power spectral density can also be calculated between $Y(t)$ and $X(t)$. Then we define P_{YX} and $S_{YX}(\omega)$ respectively as

$$P_{YX} = \frac{1}{2\pi}\int_{-\infty}^{\infty} S_{YX}(\omega)d\omega \tag{10.52}$$

where
$$S_{YX}(\omega) = \lim_{T\to\infty} \frac{E[Y_T^*(\omega)X_T(\omega)]}{2T} \tag{10.53}$$

From Eqs. (10.50), (10.51), (10.52) and (10.53), we can say that

$$P_{YX} = P_{XY}^* \tag{10.54}$$

10.5 Properties of cross power spectral density

The following are some of the important properties of cross power spectral density of real random processes $X(t)$ and $Y(t)$:

Property 1:

$$S_{XY}(\omega) = S_{YX}(-\omega) = S_{YX}^*(\omega) \tag{10.55}$$

Proof: We have

$$S_{XY}(\omega) = \lim_{T\to\infty} \frac{E[X_T^*(\omega)Y_T(\omega)]}{2T} \text{ and } S_{YX}(\omega) = \lim_{T\to\infty} \frac{E[X_T(\omega)Y_T^*(\omega)]}{2T} \tag{10.56}$$

Also for real random processes $X(t)$ and $Y(t)$:

$$X(\omega) = \int_{-\infty}^{\infty} X(t)e^{-j\omega t}dt$$

$$\Rightarrow X^*(\omega) = \int_{-\infty}^{\infty} X^*(t)e^{+j\omega t}dt$$

$$= \int_{-\infty}^{\infty} X(t)e^{j\omega t}dt \text{ [Since } X(t) \text{ is real]}$$

$$= \int_{-\infty}^{\infty} X(t)e^{-j(-\omega)t}dt$$

Taking complex conjugates on both sides, we have

$$X^*(\omega) = X(-\omega) \Rightarrow X(\omega) = X^*(-\omega) \tag{10.57a}$$

Similarly,

$$Y^*(\omega) = Y(-\omega) \Rightarrow Y(\omega) = Y^*(-\omega) \tag{10.57b}$$

Now from Eq. (10.56)

$$\begin{aligned} S_{YX}(-\omega) &= \lim_{T\to\infty} \frac{E[X_T(-\omega)Y_T^*(-\omega)]}{2T} \\ &= \lim_{T\to\infty} \frac{E[X_T^*(\omega)Y_T(\omega)]}{2T} && \{\text{from (10.57)}\} \\ &= S_{XY}(\omega) && \{\text{from (10.56)}\} \end{aligned}$$

Again from Eq. (10.56)

$$\begin{aligned} S_{YX}^*(\omega) &= \lim_{T\to\infty} \frac{E\left[(Y_T^*(\omega))^* X_T^*(\omega)\right]}{2T} \\ &= \lim_{T\to\infty} \frac{E\left[Y_T(\omega)X_T^*(\omega)\right]}{2T} \\ &= S_{XY}(\omega) \end{aligned}$$

Hence,

$$S_{XY}(\omega) = S_{YX}(-\omega) = S_{YX}^*(\omega)$$

Property 2: Real part of cross-correlation function is an even function of ω, i.e.,

$$\text{Re}\left[S_{XY}(\omega)\right] = \text{Re}\left[S_{XY}(-\omega)\right] \tag{10.58}$$

Similarly,

$$\text{Re}\left[S_{YX}(\omega)\right] = \text{Re}\left[S_{YX}(-\omega)\right]$$

where $\text{Re}\left[\,.\,\right]$ represents the real part.

Proof: We have

$$\begin{aligned} S_{XY}(\omega) &= \int_{-\infty}^{\infty} R_{XY}(\tau)e^{-j\omega\tau}\,d\tau \\ &= \int_{-\infty}^{\infty} R_{XY}(\tau)\left(\cos\omega\tau - j\sin\omega\tau\right)d\tau \end{aligned}$$

$$\text{Re}\left[S_{XY}(\omega)\right] = \int_{-\infty}^{\infty} R_{XY}(\tau)\cos\omega\tau\,d\tau \tag{10.59}$$

Now consider

$$\text{Re}\left[S_{XY}(-\omega)\right] = \int_{-\infty}^{\infty} R_{XY}(\tau)\cos\left[(-\omega)\tau\right]d\tau$$

$$= \int_{-\infty}^{\infty} R_{XY}(\tau)\cos\omega\tau\, d\tau \tag{10.60}$$

From Eqs. (10.59) and (10.60) it is proved that

$$\text{Re}\left[S_{XY}(\omega)\right] = \text{Re}\left[S_{XY}(-\omega)\right]$$

Similarly starting with $S_{XY}(\omega)$ we can prove

$$\text{Re}\left[S_{YX}(\omega)\right] = \text{Re}\left[S_{YX}(-\omega)\right]$$

Property 3: Imaginary part of $S_{XY}(\omega)$ is an odd function of ω, i.e.

$$\text{Im}\left[S_{XY}(-\omega)\right] = -\text{Im}\left[S_{XY}(\omega)\right] \text{ or}$$

$$\text{Im}\left[S_{YX}(-\omega)\right] = -\text{Im}\left[S_{YX}(\omega)\right] \tag{10.61}$$

where Im[-] represents imaginary part.

Proof: We have

$$S_{XY}(\omega) = \int_{-\infty}^{\infty} R_{XY}(\tau)e^{-j\omega\tau}\, d\tau$$

$$= \int_{-\infty}^{\infty} R_{XY}(\tau)\left[\cos\omega\tau - j\sin\omega\tau\right]d\tau$$

$$\text{Im}\left[S_{XY}(\omega)\right] = -\int_{-\infty}^{\infty} R_{XY}(\tau)\sin\omega\tau\, d\tau \tag{10.62}$$

Now

$$\text{Im}\left[S_{XY}(-\omega)\right] = -\int_{-\infty}^{\infty} R_{XY}(\tau)\sin\left[(-\omega)\tau\right]d\tau$$

$$= \int_{-\infty}^{\infty} R_{XY}(\tau)\sin\omega\tau d\tau \tag{10.63}$$

From Eqs. (10.62) and (10.63), we have

$$\text{Im}\left[S_{XY}(-\omega)\right] = -\text{Im}\left[S_{XY}(\omega)\right]$$

Similarly, starting from $S_{YX}(\omega)$, we can prove that

$$\text{Im}\left[S_{YX}(-\omega)\right] = -\text{Im}\left[S_{YX}(\omega)\right]$$

Property 4: If *X*(*t*) and *Y*(*t*) are orthogonal then

$$S_{XY}(\omega) = 0 \text{ and } S_{YX}(\omega) = 0, \tag{10.64}$$

Proof: We have

$$S_{XY}(\omega) = \int_{-\infty}^{\infty} R_{XY}(\tau)e^{-j\omega\tau}\, d\tau$$

If *X*(*t*) and *Y*(*t*) are orthogonal, then $R_{XY}(\tau) = 0$. Therefore $S_{XY}(\omega) = 0$. Similarly we can prove $S_{YX}(\omega) = 0$, if *X*(*t*) and *Y*(*t*) are orthogonal.

Property 5: If $X(t)$ and $Y(t)$ are uncorrelated wide-sense stationary (WSS) random processes, then

$$S_{XY}(\omega) = S_{YX}(\omega) = 2\pi E[X]E[Y]\delta(\omega) \tag{10.65}$$

Proof: We have

$$S_{XY}(\omega) = \int_{-\infty}^{\infty} R_{XY}(\tau)e^{-j\omega\tau}d\tau$$

$$= \int_{-\infty}^{\infty} E[X(t)Y(t+\tau)]e^{-j\omega\tau}d\tau \tag{10.66}$$

If $X(t)$ and $Y(t)$ are uncorrelated, then

$$E[X(t)Y(t)] = E[X(t)]\, E[Y(t)]$$

Therefore, Eq. (10.66) can be written as

$$S_{XY}(\omega) = \int_{-\infty}^{\infty} E[X(t)]\, E[Y(t+\tau)]\, e^{-j\omega\tau}d\tau \tag{10.67}$$

For WSS random process

$$E[Y(t+\tau)] = E[Y] \text{ and } E[X(t)] = E[X]$$

Therefore Eq. (10.67) can be written as

$$S_{XY}(\omega) = \int_{-\infty}^{\infty} E[X]\, E[Y]e^{-j\omega\tau}d\tau$$

$$= E[X]\, E[Y]\int_{-\infty}^{\infty} e^{-j\omega\tau}d\tau$$

$$= 2\pi E[X]\, E[Y]\delta(\omega) \qquad \left(\because \int_{-\infty}^{\infty} e^{-j\omega\tau}d\tau = 2\pi\,\delta(\omega)\right)$$

Similarly starting with $S_{XY}(\omega)$, we can prove that

$$S_{YX}(\omega) = 2\pi E[X]\, E[Y]\delta(\omega)$$

Hence proved.

Property 6: The cross power spectral density and time average of cross-correlation function form a Fourier transform pair:

$$A[R_{XY}(t, t+\tau) \leftrightarrow S_{XY}(\omega) \text{ or } A[R_{YX}(t, t+\tau) \leftrightarrow S_{YX}(\omega) \tag{10.68}$$

For the case of jointly WSS random processes, Eq. (10.68) reduces to

$$R_{XY}(\tau) \leftrightarrow S_{XY}(\omega) \text{ or } R_{YX}(\tau) \leftrightarrow S_{YX}(\omega) \tag{10.69}$$

Proof: In the following discussion, we will show that

$$\frac{1}{2\pi}\int_{-\infty}^{\infty} S_{XY}(\omega)e^{j\omega\tau}d\omega = \lim_{T\to\infty}\frac{1}{2T}\int_{-T}^{T} R_{XY}(t,t+\tau)dt$$

The Fourier transform of the truncated signal $X_T(t)$ and $Y_T(t)$ of $X(t)$ and $Y(t)$ in the interval $(-T, T)$ is given by

$$X_T(\omega) = \int_{-T}^{T} X(t)e^{-j\omega t}dt \tag{10.70}$$

$$Y_T(\omega) = \int_{-T}^{T} Y(t_1)e^{-j\omega t_1}dt_1 \tag{10.71}$$

$$X_T^*(\omega)Y_T(\omega) = \int_{-T}^{T} X(t)e^{+j\omega t}dt \int_{-T}^{T} Y(t_1)e^{-j\omega t}dt_1 \tag{10.72}$$

The cross power spectral density from Eq.(10.50),

$$S_{XY}(\omega) = \lim_{T\to\infty} \frac{E[X_T^*(\omega)Y_T(\omega)]}{2T}$$

Substituting Eq. (10.72) in the above Eq. , we get

$$S_{XY}(\omega) = \lim_{T\to\infty} \frac{E\left[\int_{-T}^{T} X(t)e^{j\omega t}dt \int_{-T}^{T} Y(t_1)e^{-j\omega t_1}dt_1\right]}{2T}$$

$$= \lim_{T\to\infty} \frac{1}{2T}\int_{-T}^{T}\int_{-T}^{T} R_{XY}(t,t_1)e^{-j\omega(t_1-t)}dtdt_1 \tag{10.73}$$

Next we use the same procedure as in Section 10.2. First we take inverse Fourier transform on both sides of Eq. (10.73). So, we get

$$\frac{1}{2\pi}\int_{-\infty}^{\infty} S_{XY}(\omega)e^{j\omega\tau}d\omega = \frac{1}{2\pi}\int_{-\infty}^{\infty} \lim_{T\to\infty}\frac{1}{2T}\int_{-T}^{T}\int_{-T}^{T} R_{XY}(t,t_1)e^{-j\omega(t_1-t)}dt\,dt_1 e^{j\omega\tau}d\omega$$

$$= \lim_{T\to\infty}\frac{1}{2T}\int_{-T}^{T}\int_{-T}^{T} R_{XY}(t,t_1)\frac{1}{2\pi}\int_{-\infty}^{\infty} e^{j\omega(\tau-t_1+t)}d\omega\,dt_1 dt$$

$$= \lim_{T\to\infty}\frac{1}{2T}\int_{-T}^{T}\int_{-T}^{T} R_{XY}(t,t_1)\delta(t_1-\tau-t)dt_1\,dt$$

$$= \lim_{T\to\infty}\frac{1}{2T}\int_{-T}^{T} R_{XY}(t,t+\tau)dt \tag{10.74}$$

Therefore, $$\frac{1}{2\pi}\int_{-\infty}^{\infty} S_{XY}(\omega)e^{j\omega\tau}d\omega = \lim_{T\to\infty}\frac{1}{2T}\int_{-T}^{T} R_{XY}(t,t+\tau)dt \tag{10.75}$$

Therefore, Eq. (10.75) indicates that cross power spectral density and time average cross-correlation function form a Fourier transform pair.

For at least wide-sense stationary processes $X(t)$ and $Y(t)$ the time average of cross-correlation is just $R_{XY}(\tau)$ instead of $R_{XY}(t, t_1)$, where $\tau = t - t_1$. Hence for this case Eq. (10.75) can be written as

$$\frac{1}{2\pi}\int_{-\infty}^{\infty} S_{XY}(\omega)e^{j\omega\tau}d\omega = R_{XY}(\tau) \tag{10.76}$$

Similarly, starting with $S_{YX}(\omega)$ we can prove that

$$\frac{1}{2\pi}\int_{-\infty}^{\infty} S_{YX}(\omega)e^{j\omega\tau}d\omega = \lim_{T\to\infty}\frac{1}{2T}\int_{-T}^{T} R_{YX}(t,t+\tau)dt \tag{10.77}$$

and for at least WSS processes

$$\frac{1}{2\pi}\int_{-\infty}^{\infty} S_{YX}(\omega)e^{-j\omega\tau}d\omega = R_{YX}(\tau) \tag{10.78}$$

Example 10.1: If the autocorrelation function of a WSS process $X(t)$ is $R_{XX}(\tau)=K\,e^{-K|\tau|}$, show that its power spectral density is $S_{XX}(\omega)=\dfrac{2}{1+(\omega/K)^2}$.

***Solution*:**

$$S_{XX}(\omega)=\int_{-\infty}^{\infty} Ke^{-K|\tau|}e^{-j\omega\tau}d\tau$$

$$=\int_{-\infty}^{0} Ke^{+K\tau}e^{-j\omega\tau}d\tau+\int_{0}^{\infty} Ke^{-K\tau}e^{-j\omega\tau}d\tau$$

$$=K\left[\frac{e^{(K-j\omega)\tau}}{(K-j\omega)}\right]_{-\infty}^{0}+\left[\frac{e^{-(K+j\omega)\tau}}{-(K+j\omega)}\right]_{0}^{\infty}$$

$$=\frac{K}{K-j\omega}+\frac{K}{K+j\omega}$$

$$=\frac{2K^2}{K^2+\omega^2}=\frac{2}{1+\left(\dfrac{\omega}{K}\right)^2}$$

Example 10.2: If a random process

$$X(t)=A\cos\left(\omega_0 t+\theta\right)$$

where A and ω_0 are constants and θ is a uniformly distributed random variable in the interval $(0,\pi)$. Find

(a) whether $X(t)$ is WSS process

(b) power in $X(t)$ by time averaging of its second moment.

(c) the power spectral density of $X(t)$.

***Solution*:**

(a) For a process to be WSS we have to prove that

(i) The ensemble mean of the process is constant

(ii) The ACF is independent of absolute value of time 't', but ACF can depend on time difference 'τ'

Hence
$$E[X(t)] = \int_{-\infty}^{\infty} x(t) f_\theta(\theta) d\theta$$
$$= \int_0^{\pi} A\cos(\omega_0 t + \theta)\frac{1}{\pi} d\theta$$
$$= \frac{A}{\pi}\int_0^{\pi} \cos(\omega_0 t + \theta) d\theta$$
$$= \frac{A}{\pi}[\sin(\omega_0 t + \pi) - \sin\omega_0 t)$$
$$= -\frac{2A}{\pi}\sin(\omega_0 t)$$

The mean is time dependent and hence $X(t)$ is not WSS process.

(b) Mean square value of $X(t)$ is
$$\overline{X}^2(t) = E[X^2(t)] = E[A^2\cos(\omega_0 t + \theta)]$$
$$= \frac{A^2}{2}E[1 + \cos(2\omega_0 t + 2\theta)]$$
$$= \frac{A^2}{2} + \frac{A^2}{2}\int_0^{\pi}\cos(2\omega_0 t + 2\theta)\frac{1}{\pi}d\theta$$
$$= \frac{A^2}{2}$$

Hence power of $X(t)$ is
$$P_{XX} = \lim_{T\to\infty}\frac{1}{2T}\int_{-T}^{T}\overline{X}^2(t)dt$$
$$= \lim_{T\to\infty}\frac{1}{2T}\int_{-T}^{T}\frac{A^2}{2}dt = \frac{A^2}{2}$$

(c) The PSD of $X(t)$ is given by
$$S_{XX}(\omega) = \lim_{T\to\infty}\frac{E\left[\left|X_T(\omega)\right|^2\right]}{2T}$$

We have
$$X(t) = A\cos(\omega_0 t + \theta)$$

and
$$X_T(t) = \begin{cases} A\cos(\omega_0 t + \theta), & -T < t < T \\ 0, & \text{elsewhere} \end{cases}$$

Therefore
$$X_T(\omega) = A\int_{-\infty}^{\infty} X_T(t)e^{-j\omega\tau}\,dt$$
$$= A\int_{-T}^{T} \cos(\omega_0 t + \theta)e^{-j\omega t}\,dt$$
$$= A\int_{-T}^{T} \frac{e^{j(\omega_0 t+\theta)} + e^{-j(\omega_0 t+\theta)}}{2}\, e^{-j\omega t}\,dt$$
$$= ATe^{j\theta}\mathrm{Sa}\left[(\omega-\omega_0)T\right] + ATe^{-j\theta}\mathrm{Sa}\left[(\omega-\omega_0)T\right]$$
$$\therefore \left|X_T(\omega)\right|^2 = X_T(\omega)\cdot X_T^*(\omega)$$
$$= A^2T^2\left\{\mathrm{Sa}^2\left[(\omega-\omega_0)T\right] + \mathrm{Sa}^2\left[(\omega+\omega_0)T\right] + \left(e^{-j2\theta}+e^{j2\theta}\right)\mathrm{Sa}\left[(\omega-\omega_0)T\right]\mathrm{Sa}\left[(\omega+\omega_0)T\right]\right\}$$
$$= A^2T^2\left\{\mathrm{Sa}^2\left[(\omega-\omega_0)T\right] + \mathrm{Sa}^2\left[(\omega+\omega_0)T\right] + 2\mathrm{Sa}\left[(\omega-\omega_0)T\right]\mathrm{Sa}\left[(\omega+\omega_0)T\right]\cos 2\theta\right\}$$
$$\therefore E\left[\left|X_T(\omega)\right|^2\right] = A^2T^2\left\{\mathrm{Sa}^2\left[(\omega-\omega_0)T\right] + \mathrm{Sa}^2\left[(\omega+\omega_0)T\right] + 0\right\}$$

Hence the PSD of $X(t)$ is

$$S_{XX}(\omega) = \lim_{T\to\infty} \frac{E\left[\left|X_T(\omega)\right|^2\right]}{2T}$$
$$= \lim_{T\to\infty}\left\{\frac{A^2\pi}{2}\left\{\frac{T}{\pi}\mathrm{Sa}^2\left[(\omega-\omega_0)T\right] + \frac{T}{\pi}\mathrm{Sa}^2\left[(\omega-\omega_0)T\right]\right\}\right\}$$
$$= \frac{A^2\pi}{2}\left[\delta(\omega-\omega_0) + \delta(\omega+\omega_0)\right] \quad \left[\because \lim_{T\to\infty}\frac{T}{\pi}\mathrm{Sa}^2(\omega\mp\omega_0) = \delta(\omega\mp\omega_0)\right]$$

Example 10.3: Determine which of the following functions can be valid power density spectra. Explain the reason.

(a) $\dfrac{\omega^2}{\omega^6 + 3\omega^2 + 3}$

(b) $\exp\left[-(\omega-1)^2\right]$

(c) $\dfrac{\omega^2}{\omega^4+1} - \delta(\omega)$

(d) $\dfrac{\omega^4}{1+\omega^2+j\omega^6}$

Solution: For a function to be a valid PSD it should satisfy the following properties.

(1) It should be every where non negative.

(2) It should be an even function of 'ω'.

(3) It should be always real.

(a) This function satisfies all the above properties. Hence it is valid PSD.

(b) This is not even function of 'ω'. Hence not a valid PSD.

(c) This is not everywhere non-negative. Hence not a valid PSD.

Eg.: At $\omega = 0$, its value is

$$0 - 1 = -1$$

(d) It is not a real function. Hence not a valid PSD.

Example 10.4: Given that $X(t) = \sum_{i=1}^{N} a_i X_i(t)$ where $\{a_i\}$ is a set of real constants and the processes $X_i(t)$ are stationary and orthogonal show that

$$S_{XX}(\omega) = \sum_{i=1}^{N} a_i^2 S_{X_i X_i}(\omega)$$

Solution: We have

$$R_{XX}(T) = E\left[X(t)X(t+\tau)\right]$$

$$= E\left[\sum_{i=1}^{N} a_i X_i(t) \sum_{i=1}^{N} a_j X_j(t+\tau)\right]$$

$$= E\left[\sum_{i=1}^{N}\sum_{j=1}^{N} a_i\, a_j\, X_i(t) X_j(t+\tau)\right]$$

$$= \sum_{i=1}^{N}\sum_{j=1}^{N} a_i a_j E\left[X_i(t)X_j(t+T)\right]$$

since the processes $X_i(t), i = 1, \ldots\ldots N$ are orthogonal

$$E\left[X_i(t)X_j(t+\tau)\right] = 0, \quad \forall\ i \neq j$$

and

$$E\left[X_i(t)X_j(t+\tau)\right] = R_{X_iX_i}(\tau) \quad \text{for } i = j$$

Hence

$$R_{XX}(t, t+\tau) = \sum_{i=1}^{N} a_i^2 R_{X_iX_i}(\tau)$$
$$= R_{XX}(\tau)$$

Taking Fourier transform of above Eq. we get

$$S_{XX}(\omega) = \sum_{i=1}^{N} a_i^2 S_{X_iX_i}(\omega)$$

Hence proved.

Example 10.5: Find the rms bandwidth of the power spectrum

$$S_{XX}(\omega) = \begin{cases} A\cos\left(\dfrac{\pi\omega}{2W}\right), & |\omega| \le W \\ 0, & |\omega| > W \end{cases}$$

where $\omega > 0$ and $p > 0$ are constants.

Solution:

$$\int_{-\infty}^{\infty} S_{XX}(\omega)d\omega = \int_{-W}^{W} A\cos\left(\frac{\pi\omega}{2W}\right)d\omega$$

$$= A\left[\frac{\sin\left(\frac{\pi\omega}{2W}\right)}{\frac{\pi}{2W}}\right]_{-W}^{W} - \frac{2WA}{\pi}[1+1] = \frac{4WA}{\pi}$$

$$\int_{-\infty}^{\infty} \omega^2 S_{XX}(\omega)d\omega = \int_{-W}^{W} \omega^2 A\cos\left(\frac{\pi\omega}{2W}\right)d\omega$$

Let $y = \dfrac{\pi\omega}{2W}$ then, $dy = \cos\left(\dfrac{\pi}{2W}\right)d\omega$

Therefore, $$\int_{-\infty}^{\infty} \omega^2 S_{XX}(\omega)d\omega = A\left(\frac{2W}{\pi}\right)^3 \int_{-\pi/2}^{\pi/2} y^2 \cos y\, dy$$

$$= A\left(\frac{2W}{\pi}\right)^3 \left[\frac{\pi^2 - 8}{2}\right]$$

So, RMS bandwidth of the power spectrum is

$$W_{rms}^2 = \frac{\int_{-\infty}^{\infty} \omega^2 S_{XX}(\omega) d\omega}{\int_{-\infty}^{\infty} S_{XX}(\omega) d\omega}$$

$$= W^2\left(1-\frac{8}{\pi^2}\right)$$

Example 10.6: The autocorrelation function of a random process $X(t)$ is

$$R_{XX}(\tau) = 3 + 2\exp(-4\tau^2)$$

(a) Find the power spectrum of $X(t)$ (b) Find the average power in $X(t)$.

(c) What fraction of power lies in the frequency band $-\frac{1}{\sqrt{2}} \le \omega \le \frac{1}{\sqrt{2}}$

Solution:

(a)

$$S_{XX}(\omega) = \int_{-\infty}^{\infty} R_{XX}(\tau) e^{-j\omega\tau} d\tau$$

$$= \int_{-\infty}^{\infty} \left[3 + 2e^{-4\tau^2}\right] e^{-j\omega\tau} d\tau$$

$$= 6\pi\delta(\omega) + \sqrt{\pi} e^{-\omega^2/16}$$

We have used Fourier transform of constant $K = 2\pi K\delta(\omega)$ and

$$\mathscr{F}\left[e^{-t^2/2\sigma^2}\right] = \sigma\sqrt{2\pi} e^{-\sigma^2\omega^2/2}, \qquad \sigma > 0$$

(b) Power $= R_{XX}(0) = 5\text{W}$

(c) Power in the limit $\left(\frac{-1}{\sqrt{2}} \le \omega \le \frac{1}{\sqrt{2}}\right)$ is

$$P = \frac{1}{2\pi} \int_{-1/\sqrt{2}}^{1/\sqrt{2}} \left[6\pi\delta(\omega) + \sqrt{\pi} e^{-\omega^2/16}\right] d\omega$$

$$= 3 + 2\int_{-1/\sqrt{2}}^{1/\sqrt{2}} \frac{e^{-\omega^2/2(8)}}{\sqrt{2\pi 8}} d\omega$$

$$= 3 + 2\left\{F\left(\frac{1/\sqrt{2}}{2\sqrt{2}}\right) - F\left(-\frac{1/\sqrt{2}}{2\sqrt{2}}\right)\right\} \quad \text{(from gaussian distribution tables)}$$

$$= 3 + 2\left\{F\left(\frac{1}{4}\right) - 1 + F\left(\frac{1}{4}\right)\right\}$$

$$= 3 + 2\left[2(0.5987) - 1\right] = 3.3948$$

The fraction of power that lies in the frequency band $-\frac{1}{\sqrt{2}} \le \omega \le \frac{1}{\sqrt{2}}$ is

$$\frac{3.3948}{5} = 0.6790 = 67.9\%$$

SP 10.7: If $S_{XX}(\omega)$ is a valid power spectral density of the random process $X(t)$. Check whether the functions $\frac{\partial S_{XX}(\omega)}{d\omega}$ and $\frac{\partial^2 S_{XX}(\omega)}{d\omega^2}$ can be valid power spectra.

Solution: The PSD $S_{XX}(\omega)$ need not be a monotonically non-decreasing function of 'ω'. Hence the derivative of $S_{XX}(\omega)$ i.e., $\partial S_{XX}(\omega)/d\omega$ need not be always positive.

Similarly, the derivative $\frac{\partial S_{XX}(\omega)}{d\omega}$ need not be a monotonically non-decreasing function of

'ω'. Hence its derivative i.e., $\frac{\partial^2 S_{XX}(\omega)}{d\omega}$ need not be always positive.

Hence, $\frac{\partial S_{XX}(\omega)}{d\omega}$ and $\frac{\partial^2 S_{XX}(\omega)}{d\omega}$ need not be valid PSDs.

SP 10.8: Assume a random process has power spectrum

$$S_{XX}(\omega) = \begin{cases} 4 - \left(\frac{\omega^2}{9}\right), & |\omega| \le 6 \\ 0, & \text{elsewhere} \end{cases}$$

Find

(a) Average power

(b) RMS bandwidth

Solution:

(a) $$P_{XX} = \frac{1}{2\pi}\int_{-\infty}^{\infty} S_{XX}(\omega)d\omega = \frac{1}{2\pi}\int_{-6}^{6}\left[4 - \frac{\omega^2}{9}\right]d\omega = \frac{16}{\pi}$$

(b) $$\int_{-\infty}^{\infty} \omega^2 S_{XX}(\omega)d\omega = \int_{-6}^{6}\left[4\omega^2 - \frac{\omega^4}{9}\right]d\omega$$

$$= \left[\frac{4\omega^3}{3} - \frac{\omega^5}{45}\right]_{-6}^{6}$$

$$= \frac{4\times 6^3}{3} - \frac{6^5}{45} + \frac{4\times 6^3}{3} - \frac{6^5}{45}$$

$$= \frac{8\times 6^3}{3} - \frac{2\times 6^5}{45} = \frac{32\times 36}{5}$$

$$\therefore W_{rms}^2 = \frac{\int_{-\infty}^{\infty} \omega^2 S_{XX}(\omega)d\omega}{\int_{-\infty}^{\infty} S_{XX}(\omega)d\omega}$$

$$= \frac{\frac{32\times 36}{5}}{\frac{16}{\pi}\times 2\pi} = \frac{32\times 36}{5\times 32}$$

$$= \frac{36}{5}$$

Example 10.9: Assume a random process $X(t)$ has power spectrum

$$S_{XX}(\omega) = \frac{6|\omega|}{\left(\omega^2 + W^2\right)^2}$$

where W > 0 is a constant. At what positive value of ω denoted by $\omega_{\max}$ does $S_{XX}(\omega)$ reach a maximum value.

Solution: If we make $\frac{dS_{XX}(\omega)}{d\omega} = 0$, then the corresponding value of 'ω' denoted as $\omega_{\max}$. Hence,

$$\frac{dS_{XX}(\omega)}{d\omega} = \frac{d}{d\omega}\left[\frac{6|\omega|}{\left(W^2 + \omega^2\right)^2}\right] = 0$$

$$\Rightarrow \frac{\left(W^2 + \omega^2\right)^2 6 - 12\omega\left(W^2 + \omega^2\right)2\omega}{\left(W^2 + \omega^2\right)^4} = 0$$

$$\Rightarrow \frac{\left(W^2+\omega^2\right)\left[6\left(W^2+\omega^2\right)-24\omega^2\right]}{\left(W^2+\omega^2\right)^4}=0$$

$$\Rightarrow 6W^2+6\omega^2-24\omega^2=0$$

$$\Rightarrow 6W^2-18\omega^2=0$$

$$\Rightarrow \omega=\omega_{\max}=\frac{W}{\sqrt{3}}$$

Example 10.10: Find the autocorrelation function corresponding to the power density spectrum.

$$S_{XX}(\omega)=\frac{157+12\omega^2}{\left(16+\omega^2\right)\left(9+\omega^2\right)}$$

Solution: Using partial fractions

$$S_{XX}(\omega)=\frac{A}{16+\omega^2}+\frac{B}{9+\omega^2}=\frac{(9A+16B)+(A+B)\omega^2}{\left(16+\omega^2\right)\left(9+\omega^2\right)}$$

Comparing coefficients of expression for $S_{XX}(\omega)$ on both sides we have

$$9A+16B=157$$

$$A+B=12$$

Solving for *A* and *B,* we get

$$A=5,\quad B=7$$

Therefore $$S_{XX}(\omega)=\frac{5}{\left(16+\omega^2\right)}+\frac{7}{\left(9+\omega^2\right)}$$

We know that inverse Fourier transform of $\left[\frac{1}{\alpha^2+\omega^2}\right]$ is $\frac{1}{2\alpha}e^{-\alpha|t|}$

Taking inverse fourier transform of $S_{XX}(\omega)$, we get

$$R_{XX}(\tau)=\frac{5}{8}e^{-4|\tau|}+\frac{7}{6}e^{-3|\tau|}$$

Example 10.11: Joint wide-sense stationary random process *X*(*t*) and *Y*(*t*) define a process *W*(*t*) as

$$W(t)=X(t)\cos\omega_0 t+Y(t)\sin\omega_0 t$$

where ω_0 is a real positive constant.

(a) Develop some conditions on the mean values and correlation functions of $X(t)$ and $Y(t)$ such that $W(t)$ is a wide-sense stationary.

(b) With the conditions of part (a) applied to $W(t)$, find its power spectrum in terms of power spectrum of $X(t)$ and $Y(t)$.

(c) If $X(t)$ and $Y(t)$ are also uncorrelated, what is the power spectrum of $W(t)$?

Solution:

(a) For $W(t)$ to be WSS its mean should be constant and ACF should not depend on time.

$$E\left[W(t)\right] = E\left[X(t)\cos\omega_0 t + Y(t)\sin\omega_0 t\right]$$
$$= \overline{X}\cos\omega_0 t + \overline{Y}\sin\omega_0 t$$

For this to be constant $\overline{X}$ and $\overline{Y}$ should be zero.

$$R_{WW}(t, t+\tau) = E\Big[\left\{X(t)\cos\omega_0 t + Y(t)\sin\omega_0 t\right\}$$
$$\times\left\{X(t+\tau)\cos\left(\omega_0 t + \omega_0\tau\right) + Y(t+\tau)\sin\left(\omega_0 t + \omega_0\tau\right)\right\}\Big]$$

$$= R_{XX}(\tau)\cos\omega_0 t\cos\left(\omega_0 t + \omega_0\tau\right)$$
$$+ R_{YY}(\tau)\sin\omega_0 t\sin\left(\omega_0 t + \omega_0\tau\right)$$
$$+ R_{YX}(\tau)\sin\left(\omega_0 t\right)\cos\left(\omega_0 t + \omega_0\tau\right)$$
$$+ R_{XY}(\tau)\cos\omega_0 t\sin\left(\omega_0 t + \omega_0\tau\right)$$

$$= \left[R_{XX}(\tau) + R_{YY}(\tau)\right]\frac{1}{2}\cos\omega_0\tau$$
$$+ \left[R_{XX}(\tau) - R_{YX}(\tau)\right]\frac{1}{2}\sin\left(\omega_0\tau\right)$$
$$+ \left[R_{XX}(\tau) - R_{YY}(\tau)\right]\frac{1}{2}\cos\left(2\omega_0 t + \omega_0\tau\right)$$
$$+ \left[R_{XX}(\tau) + R_{YX}(\tau)\right]\frac{1}{2}\sin\left(2\omega_0 t + \omega_0\tau\right)$$

This function is independent of t only if

$$R_{XX}(\tau) = R_{YY}(\tau) \quad \text{and} \quad R_{YX}(\tau) = -R_{XY}(\tau)$$

The function now becomes

$$R_{WW}(t, t+\tau) = \frac{1}{2}\left[R_{XX}(\tau) + R_{YY}(\tau)\right]\cos\omega_0\tau$$
$$+ \frac{1}{2}\left[R_{XY}(\tau) - R_{YX}(\tau)\right]\sin\omega_0\tau$$

which is independent of 't'

Hence for $W(t)$ to be WSS the conditions to be satisfied are

(1) $E\left[X(t)\right]=0,\ E\left[Y(t)\right]=0$ (2) $R_{XX}(\tau)=R_{YY}(\tau)$ (3) $R_{XY}(\tau)=-R_{YX}(\tau)$

(b)

$$R_{WW}(\tau)=\frac{1}{2}\left[R_{XX}(\tau)\cos\omega_0\tau+R_{YY}(\tau)\cos\omega_0\tau \right.$$
$$\left.+R_{XY}(\tau)\sin\omega_0\tau-R_{XY}(\tau)\sin\omega_0\tau\right]$$

Taking Fourier transform we get

$$S_{WW}(\omega)=\frac{1}{4}\{S_{XX}(\omega-\omega_0)+S_{XX}(\omega+\omega_0)$$
$$+S_{YY}(\omega-\omega_0)+S_{YY}(\omega+\omega_0)$$
$$+j[S_{XY}(\omega+\omega_0)-S_{XY}(\omega-\omega_0)$$
$$-S_{YX}(\omega+\omega_0)+S_{YX}(\omega-\omega_0)]\}$$

(c) If $X(t)$ and $Y(t)$ are uncorrelated. Then

$$R_{XY}(\tau)=\overline{X}\,\overline{Y}=0$$

Hence, if $X(t)$ and $Y(t)$ are uncorrelated,

$$S_{XY}(\omega)=0 \text{ and } S_{YX}(\omega)=0 \text{ for all } \omega$$

Thus,

$$S_{WW}(\omega)=\frac{1}{4}[S_{XX}(\omega-\omega_0)+S_{XX}(\omega+\omega_0)$$
$$+S_{YY}(\omega-\omega_0)+S_{YY}(\omega+\omega_0)]$$

Example 10.12: Given the random process

$$X(t)=A\cos(\omega_0 t+\theta)$$

$$Y(t)=W(t)\cos(\omega_0 t+\theta)$$

where A and ω_0 are real positive constants θ is a random variable independent of the random process $W(t)$, whose mean is $\overline{W}$ a constant, then prove that

$$S_{XY}(\omega)=\frac{A\overline{W}\pi}{2}\left[\delta(\omega-\omega_0)+\delta(\omega+\omega_0)\right]$$

irrespective of the pdf of θ.

Solution: We have

$$X_T(\omega)=\int_{-T}^{T}X(t)e^{-j\omega t}dt=\int_{-T}^{T}\frac{A}{2}\left[e^{j\omega_0 t+j\theta}+e^{-j\omega_0 t-j\theta}\right]e^{j\omega t}dt$$
$$=ATe^{j\theta}\mathrm{Sa}\left[(\omega-\omega_0)T\right]+ATe^{-j\theta}\mathrm{Sa}\left[(\omega+\omega_0)T\right]$$

Similarly

$$Y_T(\omega) = \int_{-T}^{T} y(t)e^{-j\omega t}dt = \int_{-T}^{T} \frac{W(t)}{2}\left[e^{j\omega_0 t + j\theta} + e^{-j\omega_0 t - j\theta}\right]e^{j\omega t}dt$$

$$= \frac{e^{j\theta}}{2}\int_{-T}^{T} W(t)e^{-j(\omega-\omega_0)t}dt + \frac{e^{-j\theta}}{2}\int_{-T}^{T} W(t)e^{-j(\omega+\omega_0)t}dt$$

Hence

$$E\left[X_T^*(\omega)Y_T(\omega)\right] = E\left[\frac{AT}{2}\text{Sa}\left[(\omega-\omega_0)T\right]\int_{-T}^{T} W(t)e^{-j(\omega-\omega_0)t}dt\right.$$

$$+\frac{AT}{2}e^{-j2\theta}\text{Sa}\left[(\omega-\omega_0)T\right]\int_{-T}^{T} W(t)e^{-j(\omega+\omega_0)t}dt$$

$$+\frac{AT}{2}e^{j2\theta}\text{Sa}\left[(\omega+\omega_0)T\right]\int_{-T}^{T} W(t)e^{-j(\omega-\omega_0)t}dt$$

$$\left.+\frac{AT}{2}\text{Sa}\left[(\omega+\omega_0)T\right]\int_{-T}^{T} W(t)e^{-j(\omega+\omega_0)t}dt\right]$$

$$= \frac{AT\overline{W}}{2}\left[2T\ \text{Sa}^2\left[(\omega-\omega_0)\right]T\right]$$

$$+2T\ \text{Sa}^2\left[(\omega+\omega_0)T\right] + 2T\ E\left[e^{-j2\theta}\right]\text{Sa}\left[(\omega-\omega_0)T\right]$$

$$\times\text{Sa}\left[(\omega+\omega_0)T\right] + 2T\ E\left[e^{j2\theta}\right]\text{Sa}\left[(\omega-\omega_0)T\right]$$

$$\times\text{Sa}\left[(\omega+\omega_0)T\right]$$

$$\lim_{T\to\infty}\frac{E\left[X_T^*(\omega)Y_T(\omega)\right]}{2T} = \frac{A\overline{W}}{2}\left\{\lim_{T\to\infty} T\ \text{Sa}^2\left[(\omega-\omega_0)T\right]\right.$$

$$+\lim_{T\to\infty} T\,\text{Sa}^2\left[(\omega+\omega_0)T\right] + E\left[e^{j2\theta}+e^{-j2\theta}\right]$$

$$\left.\lim_{T\to\infty} T\ \text{Sa}\left[(\omega-\omega_0)T\right]\lim_{T\to\infty}\text{Sa}\left[(\omega+\omega_0)T\right]\right\}$$

We know that

$$\lim_{T\to\infty}\text{Sa}\left[\omega\right] = \begin{cases} 0, & \omega \neq 0 \\ 1, & \omega = 0 \end{cases}$$

$$\lim_{T\to\infty}\frac{T}{\pi}\text{Sa}^2\left[\alpha T\right] = \delta(\alpha)$$

$$\lim_{T\to\infty}\frac{T}{\pi}\mathrm{Sa}\left[\alpha T\right]=\delta\left(\alpha\right)$$

Therefore, $S_{XY}(\omega)=\lim_{T\to\infty}\dfrac{E\left[X_T^*(\omega)Y_T(\omega)\right]}{2T}=\dfrac{\pi A\overline{W}}{2}\left[\delta(\omega-\omega_0)+\delta(\omega+\omega_0)\right]$

Here we need not to calculate $E\left[e^{j2\theta}+e^{-j2\theta}\right]$. Hence the result is independent of density of θ.

Example 10.13: If we decompose gross power spectrums into real and imaginary parts like

$$S_{XY}(\omega)=\mathrm{Re}\left[S_{XY}(\omega)\right]+j\,\mathrm{Im}\left[S_{XY}(\omega)\right]$$

$$S_{YX}(\omega)=\mathrm{Re}\left[S_{YX}(\omega)\right]+j\mathrm{Im}\left[S_{YX}(\omega)\right]$$

Then prove that

$$\mathrm{Re}\left[S_{XY}(\omega)\right]=\mathrm{Re}\left[S_{YX}(-\omega)\right]=\mathrm{Re}\left[S_{YX}(\omega)\right]$$

and

$$\mathrm{Im}\left[S_{XY}(\omega)\right]=\mathrm{Im}\left[S_{YX}(-\omega)\right]=-\mathrm{Im}\left[S_{YX}(\omega)\right]$$

Solution: We have

$$\mathrm{Re}\left[S_{XY}(\omega)\right]=\frac{1}{2}\left[S_{XY}(\omega)+S_{XY}^*(\omega)\right]$$

$$=\frac{1}{2}\lim_{T\to\infty}\frac{1}{2T}E\left[X_T^*(\omega)Y_T(\omega)+X_T(\omega)Y_T^*(\omega)\right] \tag{10.79}$$

$$\mathrm{Re}\left[S_{YX}(\omega)\right]=\frac{1}{2}\left[S_{YX}(\omega)+S_{YX}^*(\omega)\right]$$

$$=\frac{1}{2}\lim_{T\to\infty}\frac{1}{2T}E\left[Y_T^*(\omega)X_T(\omega)+Y_T(\omega)X_T^*(\omega)\right] \tag{10.80}$$

From (10.79) and (10.80) we have

$$\mathrm{Re}\left[S_{XY}(\omega)\right]=\mathrm{Re}\left[S_{YX}(\omega)\right]$$

$$\mathrm{Re}[S_{YX}(-\omega]=\frac{1}{2}\lim_{T\to\infty}\frac{1}{2T}E[Y_T^*(-\omega)X_T(-\omega)+Y_T(-\omega)X^*{}_T(-\omega)]$$

$$=\frac{1}{2}\lim_{T\to\infty}\frac{1}{2T}E\left[Y_T(\omega)X_T^*(\omega)+Y_T^*(\omega)X_T(\omega)\right]$$

$$=\mathrm{Re}\left[S_{YX}(\omega)\right]$$

[Since $Y_T^*(\omega) = Y_T(-\omega), Y(t)$ and hence $Y(\omega)$ are real].

Therefore, $$\text{Re}\left[S_{XY}(\omega)\right] = \text{Re}\left[S_{YX}(\omega)\right] = \text{Re}\left[S_{YX}(-\omega)\right]$$

$$\text{Im}\left[S_{XY}(\omega)\right] = \frac{1}{2j}\left[S_{XY}(\omega) - S_{XY}^*(\omega)\right]$$

$$= \frac{1}{2j}\lim_{T\to\infty}\frac{1}{2T}E\left[X_T^*(\omega)Y_T(\omega) - X_T(\omega)Y_T^*(\omega)\right] \quad (10.81)$$

$$\text{Im}\left[S_{XY}(\omega)\right] = \frac{1}{2j}\left[S_{YX}(\omega) - S_{YX}^*(\omega)\right]$$

$$= \frac{1}{2j}\lim_{T\to\infty}\frac{1}{2T}E\left[Y_T^*(\omega)X_T(\omega) - Y_T(\omega)X_T^*(\omega)\right] \quad (10.82)$$

From Eqs. (10.81) and (10.82) $\text{Im}\left[S_{XY}(\omega)\right] = -\text{Im}\left[S_{YX}(\omega)\right]$

Example 10.14: If $X(t)$ and $Y(t)$ are real random processes, determine which of the following functions are valid and give reasons.

(a) $R_{XY}(\tau) = \exp\left(-|\tau|\right)$

(b) $\left|R_{XY}(\tau)\right| \le j\sqrt{R_{XX}(0)R_{YY}(0)}$

(c) $R_{XX}(\tau) = 2\sin(3\tau)$

(d) $S_{XX}(\omega) = \dfrac{6}{6+7\omega^3}$

(e) $S_{XX}(\omega) = \dfrac{4e^{-3|\tau|}}{1+\omega^2}$

(f) $S_{XX}(\omega) = 3 + j\omega^2$

(g) $S_{XY}(\omega) = 18\delta(\omega)$

Solution: For a function to be valid correlation function it should satisfy the following conditions:

(1) It should be real

(2) It should be an even function of ω

(3) $\lim\limits_{T\to 0} R_{XX}(T) = 0$

For a function to be a PSD, the conditions to be followed are given in Example 10.3.

(a) Yes, it is ACF
(b) No, since ACF cannot be imaginary
(c) No, since ACF cannot be odd
(d) No, since PSD cannot be negative
(e) No, since PSD cannot depend on 'τ'
(f) No, since imaginary part of PSD must be odd
(g) Yes, it is valid PSD.

Example 10.15: A WSS process *X*(*t*) is applied at an ideal differentiator having the response $Y(t) = \dfrac{dX(t)}{dt}$. The cross-correlation of the input, output processes is known to be

$$R_{XY}(\tau) = \frac{\partial R_{XX}(\tau)}{\partial \tau}$$

Find $S_{XY}(\omega)$ and $S_{YX}(\omega)$ interms of power spectrum $S_{XX}(\omega)$ of $X(t)$.

Solution: We have

$$R_{XY}(\tau) = \frac{\partial R_{XX}(\tau)}{\partial \tau}$$

Taking Fourier transform

$$S_{XY}(\omega) = j\omega S_{XX}(\omega)$$

We know that

$$S_{YX}(\omega) = S_{XY}(-\omega)$$

Therefore,

$$S_{YX}(\omega) = -j\omega S_{XX}(-\omega)$$
$$= -j\omega S_{XX}(\omega)$$

[because $S_{XX}(\omega)$ has even symmetry.]

Example 10.16: The cross-correlation of jointly wide-sense stationary processes $X(t)$ and $Y(t)$ is assumed to be

$$R_{XY}(\tau) = Bu(\tau)\exp\left(-W\tau\right)$$

where $B > 0$ and $W > 0$ are constants.

(a) Find $R_{YX}(\tau)$ (b) Find $S_{XY}(\omega)$ and $S_{YX}(\omega)$

Solution:

(a)

$$R_{YX}(\tau) = R_{XY}(-\tau) = Bu(-\tau)\exp(W\tau)$$

(b) Denoting Fourier transform operator by $\mathscr{F}$, we have

$$S_{XY}(\omega) = \mathscr{F}\left[R_{XY}(\tau)\right] = \int_{-\infty}^{\infty} Bu(\tau)\exp(-W\tau)e^{-j\omega\tau}d\tau$$

$$= B\int_{0}^{\infty} \exp(-W\tau)e^{-j\omega\tau}d\tau = B\int_{0}^{\infty} e^{-(W+j\omega)\tau}d\tau$$

$$= \frac{B}{(W+j\omega)}$$

Therefore,

$$S_{YX}(\omega) = S_{XY}(-\omega) = \frac{B}{W-j\omega}$$

Example 10.17: Find the cross-correlation function corresponding to the cross-power spectrum

$$S_{XY}(\omega) = \frac{6}{\left(9+\omega^2\right)\left(3+j\omega\right)^2}$$

Solution: Let

$$S_{XY}(\omega) = S_1(\omega)\times S_2(\omega)$$

where

$$S_1(\omega) = \frac{6}{9+\omega^2}, \quad S_2(\omega) = \frac{1}{\left(3+j\omega\right)^2}$$

If

$$S_1(\omega) \leftrightarrow s_1(\tau) \text{ and } S_2(\omega) \leftrightarrow s_2(\tau)$$

then

$$s_1(\tau) = e^{-3|\tau|} \text{ and } s_2(\tau) = u(\tau)\tau e^{-3\tau}$$

Therefore,

$$R_{XY}(\tau) = \mathscr{F}^{-1}\left[S_{XY}(\omega)\right]$$

$$= \mathscr{F}^{-1}\left[S_1(\omega)S_2(\omega)\right]$$

$$= s_1(\tau) * s_2(\tau)$$

$$= e^{-3|\tau|} * u(\tau)\tau e^{-3\tau}$$

For $\tau < 0$

$$R_{XY}(\tau) = \int_{-\infty}^{\tau} e^{3\tau_1}(\tau-\tau_1)e^{-3\tau+3\tau_1}d\tau_1$$

$$= \frac{1}{36}e^{3\tau}, \quad \tau < 0$$

For $\tau \geq 0$

$$R_{YX}(\tau) = \int_{-\infty}^{0} e^{3\tau_1}(\tau - \tau_1)e^{-3\tau+3\tau_1}d\tau_1$$

$$+ \int_{0}^{T} e^{-3\tau_1}e^{-3\tau+3\tau_1}(\tau - \tau_1)d\tau_1$$

$$= \left(\frac{9\tau - 1}{36}\right)e^{-3\tau}, \quad \tau \geq 0$$

Therefore,

$$R_{XY}(\tau) = \frac{e^{-3|\tau|}}{36}\left[(9\tau - 1)u(\tau) + u(-\tau)\right]$$

Exercise Problems

EP. 10.1 If a random process is defined by

$$x(t) = u(t)A_0 \cos(\omega_0 t + \theta)$$

where $u(t)$ is a step function, A_0, ω_0 are constants and θ is a random variable uniformly distributed in the interval $(0, \pi)$ then

(a) Check whether $x(t)$ is wide-sense stationary

(b) Find power in $x(t)$ and also power spectral density of $X(t)$.

EP. 10.2 Work out the above problem (EP. 10.1) by making θ a random variable uniformly distributed in the interval $\left(0, \frac{\pi}{2}\right)$.

EP. 10.3 Work out problem (10.1) if the random process is

$$x(t) = A\,x_0^2 \cos^2\left(\omega_0 t + \theta\right)$$

EP. 10.4 If A and B are two random variables in the random process

$$x(t) = A\cos\omega_0 t + B\sin\omega_0 t$$

and ω_0 is a real constant.

(a) Show that $x(t)$ is a wide-sense stationary if A and B are uncorrelated with zero means and equal variances.

(b) Find the autocorrelation function of $x(t)$

(c) Find the power spectral density

EP. 10.5 Check whether the following functions arc valid power spectral densities by giving appropriate reasons

(a) $\dfrac{\cos 2\omega}{1+\omega^2}$

(b) $\dfrac{1}{\left(1+\omega^2\right)^2}$

(c) $\dfrac{|\omega|}{1+2\omega+\omega^2}$

(d) $\dfrac{1}{\sqrt{1-3\omega^2}}$

EP. 10.6 A random process $x(t) = A\cos\left(Wt+\theta\right)$

where A_0 is a real constant, W is a random variable with power spectral density function $S_W(\omega)$ and θ is a uniformly distributed random variable in the interval $(0,2\pi)$ and is independent of W. Show that the power spectrum of $x(t)$ is

$$S_{XX}(\omega) = \frac{\pi A_0^2}{2}\left[S_W(\omega) + S_W(-\omega)\right]$$

EP. 10.7 If $x(t)$ is a stationary random processes, find the power spectrum of

$$Y(t) = A + B\,X(t)$$

interms of power spectrum of $X(t)$. Assume A and B are real constants.

EP. 10.8 Find the RMS bandwidth of the power spectrums

(a) $S_{XX}(\omega) = \begin{cases} A, & |\omega| < W \\ 0, & |\omega| > W \end{cases}$

(b) $S_{XX}(\omega) = \begin{cases} A\left(1-|\omega/W|\right) & |\omega| \le W \\ 0, & |\omega| > W \end{cases}$

(c) $S_{XX}(\omega) = \begin{cases} \dfrac{A}{1+(\omega/W)^2}, & |\omega| \le KW \\ 0, & |\omega| > KW \end{cases}$

where A, W, K are real positive constants.

EP. 10.9 Find the

(a) Average power

(b) RMS bandwidth

(c) Autocorrelation function of the following random processes:

(1) $S_{XX}(\omega) = \begin{cases} 4-(\omega^2/9), & |\omega| \le 6 \\ 0, & \text{elsewhere} \end{cases}$

(2) $S_{XX}(\omega) = \dfrac{\omega^2}{\left[1+(\omega/W)^2\right]^N}$

where $N \geq 3$ is an integer,

(3) $S_{XX}(\omega) = \dfrac{1}{\left[1+(\omega/W)^2\right]^N}$

where $N \geq 2$ is an integer.

EP. 10.10 If a random process $x(t)$ has autocorrelation function.

$$R_{XX}(\tau) = Ae^{-\alpha|\tau|}\cos(\omega_0\tau)$$

where A, α are real positive constants and ω_0 is real constant, find the power spectrum of $X(t)$.

EP. 10.11 Find the autocorrelation function of the random processes, whose power spectral densities are

(a) $S_{XX}(\omega) = \dfrac{6\omega^2}{(1+\omega^2)^3}$

(b) $S_{XX}(\omega) = \begin{cases} P\cos\left(\dfrac{\pi\omega}{2W}\right), & |\omega| \le W \\ 0, & |\omega| > W \end{cases}$

(c) $S_{XX}(\omega) = \delta(\omega) + \delta(\omega - \omega_0) + \delta(\omega + \omega_0)$

(d) $S_{XX}(\omega) = \dfrac{8}{(9+\omega^2)^2}$

EP. 10.12 Find the power spectrum corresponding to the autocorrelation function

$$R_{XX}(\tau) = \left[\cos(\alpha\tau) + \sin\left(\alpha|\tau|\right)\right] e^{-\omega|\tau|}$$

where $\alpha > 0$ is a constant

EP. 10.13 A random process

$$Z(t) = Ax(t) + By(t)$$

where $x(t)$ and $y(t)$ are jointly wide-sense stationary processes and A, B are real constants.

(a) Find the power spectrum $S_{ZZ}(\omega)$ of $Z(t)$

(b) Find $S_{ZZ}(\omega)$ if $x(t)$ and $y(t)$ are uncorrelated.

(c) Find the cross powers spectrums $S_{XZ}(\omega)$ and $S_{YZ}(\omega)$.

EP. 10.14 For the function

$$R_{XY}(\tau) = Bu(\tau)\tau\exp(-W\tau)$$

where B and W are positive constants.

(a) Find $R_{YX}(\tau)$

(b) Find $S_{XY}(\omega)$ and $S_{YX}(\omega)$

EP. 10.15 Find the cross-correlation function corresponding to cross power spectrum

$$S_{XY}(\omega) = \frac{6}{\left(9+\omega^2\right)\left(3+j\omega\right)^2}$$

Appendix–A

Symmetry of the function $x(t)$

Four types of symmetry can be readily recognized and then classified to simplify the task of calculating the Fourier coefficients. They are

1. Even-function symmetry.
2. Odd-function symmetry.
3. Half-wave symmetry.
4. Quarter-wave symmetry.

A function is *even* when $x(t) = x(-t)$, and a function is *odd* when $x(t) = -x(-t)$. The function shown in Fig. A.1 is an even function. For even functions all $b_n = 0$ and

$$a_n = \frac{4}{T}\int_0^{T/2} f(t)\cos n\omega_0 t\, dt$$

For odd functions all $a_n = 0$ and

$$b_n = \frac{4}{T}\int_0^{T/2} x(t)\sin n\omega_0 t\, dt$$

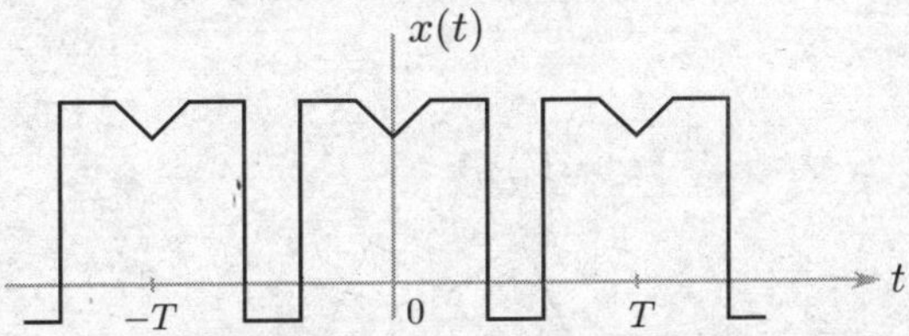

Fig. A.1 An even periodic function, $x(t) = x(-t)$

An example of an odd function is $\sin\omega_0 t$. Another odd function is shown in Fig. A.2. *Half-wave symmetry* for a function $x(t)$ is obtained when

$$x(t) = -x\left(t - \frac{T}{2}\right) \tag{A.1}$$

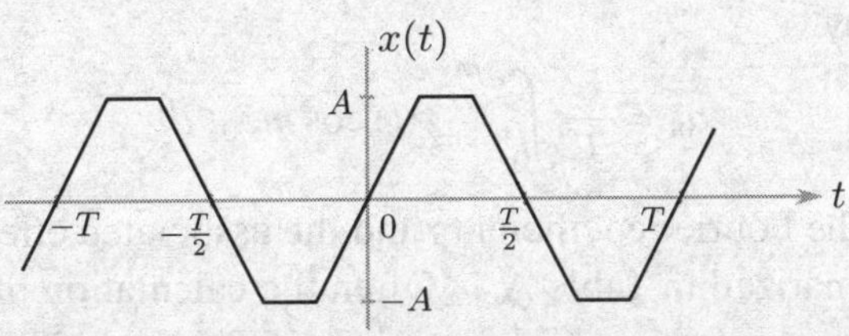

Fig. A.2 An odd periodic function, $x(t) = x(-t)$

These half-wave symmetric waveforms have the property that the second half of each period looks like the first half turned upside down. The function shown in Fig. A.3 has half-wave symmetry. If a function has half-wave symmetry, then both a_n and b_n are zero for even values of n. We see that $a_0 = 0$ for half-wave symmetry since the average value of the function over one period is zero.

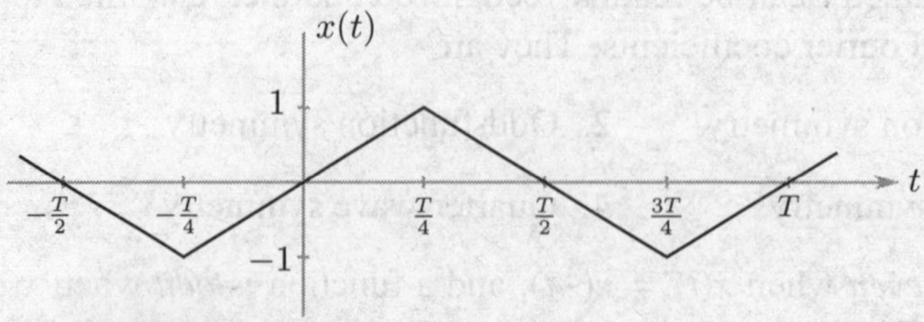

Fig. A.3: *An example of an odd function with half-wave symmetry*

Quarter-wave symmetry is a term used to describe a function that has half-wave symmetry and, in addition, has symmetry about the midpoint of the positive and negative half-cycles. An example of an odd function with quarter-wave symmetry is shown in Fig. A.4. If a function is odd and quarter-wave, then $a_0 = 0$, $a_n = 0$ for all n, $b_n = 0$ for even n. For odd n, b_n is given by

$$b_n = \frac{8}{T}\int_0^{T/4} x(t)\sin n\omega_0 t\,dt$$

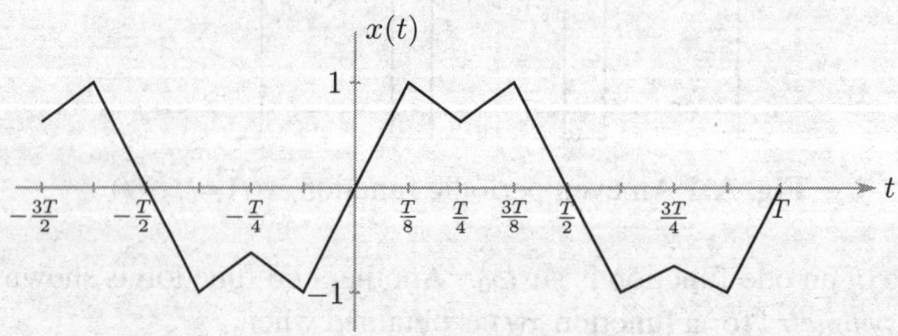

Fig. A.4: *An odd function with quarter-wave symmetry*

If a function is even and quarter-wave, then $a_0 = 0$, $b_n = 0$ for all n, $a_n = 0$ for even n. For odd n, a_n is given by

$$a_n = \frac{8}{T}\int_0^{T/4} x(t)\cos n\omega_0 t\,dt$$

The calculation of the Fourier coefficients and the associated effects of symmetry of the waveform $x(t)$ are summarized in Table A.1. Often the calculation of the Fourier series can be simplified by intelligent selection of the origin ($t = 0$), since the analyst usually has the choice to arbitrarily select this point.

Table A.1: Fourier Series and Symmetry

Symmetry	Fourier Coefficients
1. Odd function $x(t) = -x(-t)$	$a_n = 0$ for all n; $b_n = \frac{4}{T}\int_0^{T/2} x(t)\sin n\omega_0 t\,dt$
2. Even function $x(t) = x(-t)$	$b_n = 0$ for all n; $a_n = \frac{4}{T}\int_0^{T/2} x(t)\cos n\omega_0 t\,dt$
3. Half-wave symmetry $x(t) = -x\left(t - \frac{T}{2}\right)$	$a_0 = 0$; $a_n = 0$ for even n; $b_n = 0$ for even n; $a_n = \frac{4}{T}\int_0^{T/2} x(t)\cos n\omega_0 t\,dt$ for odd n; $b_n = \frac{4}{T}\int_0^{T/2} x(t)\sin n\omega_0 t\,dt$ for odd n
4. Quarter-wave symmetry Half-wave symmetric and symmetric about the midpoints of the positive and negative half-cycles	A. Odd function: $a_0 = 0, a_n = 0$ for all n; $b_n = 0$ for even n; $b_n = \frac{8}{T}\int_0^{T/4} x(t)\sin n\omega_0 t\,dt$ for odd n. B. Even function: $a_0 = 0, b_n = 0$ for all n; $a_n = 0$ for even n; $a_n = \frac{8}{T}\int_0^{T/4} x(t)\cos n\omega_0 t\,dt$ for odd n

Example 8.1. Find the Fourier series representation for the waveform shown in Fig. Ex. A.1.

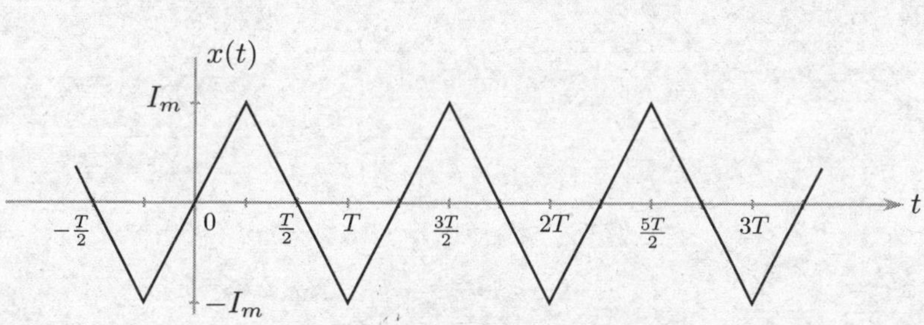

Fig. Ex. A.1

Solution: We begin by looking for degrees of symmetry in the waveform. We find that the function is odd and, in additon, has half-wave and quarter-wave symmetry. Because the function is odd, all the a coefficients are zero; that is, $a_0 = 0$ and $a_n = 0$ for all n. Because the function has half-wave symmetry, $b_n = 0$ for even values of n. Because the function has quarter-wave symmetry, the expression for b_n for odd values of n is

$$b_n = \frac{8}{T}\int_0^{T/4} x(t)\sin n\omega_0 t\, dt$$

In the interval $0 \le t \le T/4$, the expression for $x(t)$ is

$$x(t) = \frac{4I_m}{T}t$$

Thus,

$$\begin{aligned} b_n &= \frac{8}{T}\int_0^{T/4} \frac{4I_m}{T} t \sin n\omega_0 t\, dt \\ &= \frac{32I_m}{T^2}\left(\frac{\sin n\omega_0 t}{n^2\omega_0^2} - \frac{t\cos n\omega_0 t}{n\omega_0}\bigg|_0^{T/4}\right) \\ &= \frac{8I_m}{\pi^2 n^2}\sin\frac{n\pi}{2}; \quad n \text{ is odd.} \end{aligned}$$

The Fourier series representation of $x(t)$ is

$$\begin{aligned} x(t) &= \frac{8I_m}{\pi^2}\sum_{n=1,3,5,\ldots}^{\infty} \frac{1}{n^2}\sin\frac{n\pi}{2}\sin\omega_0 t \\ &= \frac{8I_m}{\pi^2}\left(\sin\omega_0 t - \frac{1}{9}\sin 3\omega_0 t + \frac{1}{25}\sin 5\omega_0 t - \frac{1}{49}\sin 7\omega_0 t + \cdots\right) \end{aligned}$$

Appendix B

Response of linear system to random input signal

We assume that the system is linear time invariant (LTI) system and the random input signal is a sample function of wide-sense stationary (WSS) process.

An LTI system is completely characterised by its impulse response $h(t)$. Hence, its output can be found using convolution integral.

Let $x(t)$ be a sample function of random process $X(t)$ assumed to be WSS. and also $x(t)$ be a random signal made the input to the LTI system having the impulse response $h(t)$. Then, the output signal $y(t)$ is given by the convolution of $x(t)$ and $h(t)$. That is,

$$y(t) = \int_{-\infty}^{\infty} x(\tau)h(t-\tau)d\tau \tag{B.1a}$$

or

$$y(t) = \int_{-\infty}^{\infty} h(\tau)x(t-\tau)d\tau \tag{B.1b}$$

It may be noted that $x(t)$ and $y(t)$ can be regarded as the sample functions of random process $X(t)$ and $Y(t)$, respectively. Thus, random process $Y(t)$ is given by

$$Y(t) = \int_{-\infty}^{\infty} X(\tau)\, h(t-\tau)d\tau \tag{B.2a}$$

or

$$Y(t) = \int_{-\infty}^{\infty} h(\tau)\, X(t-\tau)d\tau \tag{B.2b}$$

where $X(t)$ is a WSS random process

B.1 Mean value of system response $Y(t)$

Mean value of random process $Y(t)$ is given by expected value of $Y(t)$ denoted by $E[Y(t)]$. Substituting for $Y(t)$ from Eq.(B.2), we get

$$E[Y(t)] = E\left[\int_{-\infty}^{\infty} h(\tau)\, X(t-\tau)\, d\tau\right]$$
$$= \int_{-\infty}^{\infty} h(\tau)\, E\big[X(t-\tau)\big] d\tau \tag{B.3}$$

Since $X(t)$ is assumed to be a WSS process, $E[X(t)]$ is independent of time and is a constant. Hence, $E[X(t-\tau)] = E[X]$. Substituting this in Eq. (B.3), we have

$$E[Y(t)] = E[X]\int_{-\infty}^{\infty} h(\tau) d\tau = E[Y] \tag{B.4}$$

Mean value of $Y(t)$ is also independent of time and is a constant. $E[Y]$ is equal to the product of the mean value of $X(t)$ and the area under impulse response of the LTI system as given in Eq.(B.4).

B.2 Mean square value of system response *Y*(*t*)

Mean squared value of $Y(t)$ is given by $E[Y^2(t)]$. Using Eq.(B.2), we have

$$E[Y^2(t)] = E\left[\int_{-\infty}^{\infty} h(\tau_1) X(t-\tau_1) d\tau_1 \int_{-\infty}^{\infty} h(\tau_2) X(t-\tau_2) d\tau_2\right]$$

Rearranging right hand side,

$$E[Y^2(t)] = \int_{-\infty}^{\infty}\int_{-\infty}^{\infty} E\big[X(t-\tau_1)X(t-\tau_2)\big] h(\tau_1)\, h(\tau_2) d\tau_1\, d\tau_2 \tag{B.5}$$

Since $X(t)$ is assumed to be WSS process, $E\big[X(t-\tau_1)X(t-\tau_2)\big]$ depends only on time difference and is nothing but $R_{XX}(\tau_1 - \tau_2)$.

Thus, $$E\big[Y^2(t)\big] = \int_{-\infty}^{\infty}\int_{-\infty}^{\infty} R_{XX}(\tau_1-\tau_2)\, h(\tau_1) h(\tau_2) d\tau_1\, d\tau_2 \tag{B.6}$$

B.3 Autocorrelation function of system response *Y*(*t*)

Auto correlation function of $Y(t)$ is defined as $E\big[Y(t)Y(t+\tau)\big]$ and is denoted by $R_{YY}(t, t+\tau)$.

$$R_{YY}(t, t+\tau) = E\big[Y(t), Y(t+\tau)\big]$$

Substituting for $Y(t)$ from Eq.(B.2), we have

$$R_{YY}(t, t+\tau) = E\left[\int_{-\infty}^{\infty} h(\tau_1) X(t-\tau_1) d\tau_1 \int_{-\infty}^{\infty} h(\tau_2) X(t+\tau-\tau_2) d\tau_2\right]$$
$$= \int_{-\infty}^{\infty}\int_{-\infty}^{\infty} E\big[X(t-\tau_1) X(t+\tau-\tau_2\big]. h(\tau_1) h(\tau_2) d\tau_1 d\tau_2 \tag{B.7}$$

Since $X(t)$ is assumed to be WSS process, the first term (expectation) inside the integral depends only on time difference and is given by $R_{XX}(\tau + \tau_1 - \tau_2)$. Substituting this in Eq.(B.7), we have,

$$R_{YY}(t, t+\tau) = \int_{-\infty}^{\infty}\int_{-\infty}^{\infty} R_{XX}(t+\tau_1-\tau_2)h(\tau_1)h(\tau_2)d\tau_1 d\tau_2 \tag{B.8}$$

From Eq. (B.8), it is seen that auto correlation function of $Y(t)$ also depends only on time difference. Hence $R_{YY}(t, t+\tau) = R_{YY}(\tau)$ and $Y(t)$ is also WSS process.

Eq.(B.8) can be regarded as a two fold convolution integral and hence, we may write

$$R_{YY}(\tau) = R_{XX}(\tau) * h(-\tau) * h(\tau) \tag{B.9}$$

Thus, auto correlation of output process $Y(t)$ is a two fold convolution of input process $X(t)$ and system impulse response $h(t)$.

B.4 Cross-correlation function of input process *X*(*t*) and output process *Y*(*t*)

Cross-correlation function of processes $X(t)$ and $Y(t)$ is denoted by $R_{XY}(t, t+\tau)$ and is given by

$$R_{XY}(t, t+\tau) = E[X(t)\ Y(t+\tau)]$$

Substituting for $Y(t+\tau)$ from Eq.(B.2) we get,

$$R_{XY}(t, t+\tau) = E\left[X(t)\int_{-\infty}^{\infty} h(\tau_1)X(t+\tau-\tau_1)d\tau_1\right]$$

$$= \int_{-\infty}^{\infty} E\left[X(t).X(t+\tau-\tau_1)\right]h(\tau_1)d\tau_1 \tag{B.10}$$

Since $X(t)$ is assumed to be WSS process, the first term inside the integral is a function of time difference and written as $R_{XX}(\tau - \tau_1)$. Hence, the cross-correlation function is also a function of time difference and may be written as $R_{XY}(\tau)$. Thus Eq. (B.10) can be written as

$$R_{XY}(\tau) = \int_{-\infty}^{\infty} R_{XX}(\tau-\tau_1)h(\tau_1)d\tau_1$$

or

$$R_{XY}(\tau) = R_{XX}(\tau) * h(\tau) \tag{B.11}$$

We can similarly show that

$$R_{YX}(\tau) = \int_{-\infty}^{\infty} R_{XX}(\tau-\tau_1)h(-\tau_1)d\tau_1$$

or

$$R_{YX}(\tau) = R_{XX}(\tau) * h(-\tau) \tag{B.12}$$

It is seen that if $X(t)$ is WSS proces. $E[X(t)$ is independent of time and is a constant and the autocorrelation function $R_{XX}(\tau)$ of $X(t)$ depends only on the time difference τ. In this case with $X(t)$ as input process, the output process $Y(t)$ is such that $E[Y(t)]$ is a constant and autocorrelation function $R_{YY}(\tau)$ of process $Y(t)$ also depends only on time difference τ. Further, cross-correlation function $R_{XY}(\tau)$ and $R_{YX}(\tau)$ also depend on time difference. This implies that if $X(t)$ is WSS process, $Y(t)$ is also WSS process and $X(t)$ and $Y(t)$ are jointly WSS processes.

Using the Eqs. (B.11), and (B.12), we can establish relation between autocorrelation and cross-correlation functions.

We have from Eq.(B.9),

$$R_{YY}(\tau) = R_{XX}(\tau) * h(-\tau) * h(\tau) \tag{B.13}$$

Using Eq.(B.11) in Eq.(B.13), we have,

$$R_{YY}(\tau) = R_{XY}(\tau) * h(-\tau) \tag{B.14}$$

Using Eq.(B.12) in Eq.(B.13), we have

$$R_{YY}(\tau) = R_{YX}(\tau) * h(\tau) \tag{B.15}$$

Eqs. (B.14) and (B.15) relate the auto correlation of $Y(t)$ with cross-correlation functions $R_{XY}(\tau)$ or $R_{YX}(\tau)$.

B.5 Power spectral density of system response $Y(t)$

Let the input to LTI system be WSS process $X(t)$ having the auto correlation function $R_{XX}(\tau)$. Let the corresponding output process be $Y(t)$, having the autocorrelation function $R_{YY}(\tau)$.

Power spectral density (PSD) of process $X(t)$ is the Fourier Transform of $R_{XX}(\tau)$. Similarly, PSD of process $Y(t)$ is Fourier Transform of $R_{YY}(\tau)$.

PSD of process $X(t)$ is

$$S_{XX}(\omega) = \int_{-\infty}^{\infty} R_{XX}(\tau) e^{-j\omega\tau}\, d\tau \tag{B.16}$$

PSD of output process $Y(t)$ is

$$S_{YY}(\omega) = \int_{-\infty}^{\infty} R_{YY}(\tau) e^{-j\omega\tau}\, d\tau \tag{B.17}$$

Substituting for $R_{YY}(\tau)$ from Eq.(B.8)

$$S_{YY}(\omega) = \int_{-\infty}^{\infty} \left[\int_{-\infty}^{\infty} \int_{-\infty}^{\infty} R_{XX}(t + \tau_1 - \tau_2)\, h(\tau_1) h(\tau_2) d\tau_1\, d\tau_2 \right] e^{-j\omega\tau}\, d\tau \tag{B.18}$$

$$= \int_{-\infty}^{\infty} h(\tau_1) \int_{-\infty}^{\infty} h(\tau_2) \int_{-\infty}^{\infty} R_{XX}(t + \tau_1 - \tau_2)\, e^{-j\omega\tau}\, d\tau\, d\tau_1\, d\tau_2 \tag{B.19}$$

Put $\alpha = \tau + \tau_1 - \tau_2$ in Eq.(B.19). Then $d\alpha = d\tau$

and $$S_{YY}(\omega) = \int_{-\infty}^{\infty} h(\tau_1) \int_{-\infty}^{\infty} h(\tau_2) \int_{-\infty}^{\infty} R_{XX}(\alpha) e^{-j\omega(\alpha - \tau_1 + \tau_2)}\, d\tau_1\, d\tau_2\, d\tau$$

$$= \int_{-\infty}^{\infty} h(\tau_1) e^{j\omega\tau_1} d\tau_1 \int_{-\infty}^{\infty} h(\tau_2) e^{-j\omega\tau_2} d\tau_2 \int_{-\infty}^{\infty} R_{XX}(\alpha)\, e^{-j\omega\alpha}\, d\alpha$$

The three integrals in Eq.(B.20) are $H^*(\omega)$, $H(\omega)$ and $S_{XX}(\omega)$, respectively, where $H(\omega)$ is Fourier Transform of $h(t)$.

Hence, $$S_{YY}(\omega) = H^*(\omega) H(\omega) S_{XX}(\omega) = |H(\omega)|^2\, S_{XX}(\omega) \tag{B.21}$$

Average input signal power in $X(t)$

$$P_X = \frac{1}{2\pi} \int_{-\infty}^{\infty} S_{XX}(\omega)\, d\omega \tag{B.22}$$

Similarly average power in output Signal $Y(t)$ is

$$P_{YY} = \frac{1}{2\pi} \int_{-\infty}^{\infty} S_{YY}(\omega)\, d\omega$$

$$P_{YY} = \frac{1}{2\pi} \int_{-\infty}^{\infty} |H(\omega)|^2\, S_{XX}(\omega)\, d\omega \quad \text{(B.23}$$

B.6 Cross power spectral density of input and output

The cross-correlation functions $R_{XY}(\tau)$ or $R_{YX}(\tau)$ of input $X(t)$ and $Y(t)$ are given in Eqs. (B.11) and (B.12), which are rewritten here for reference.

$$R_{XY}(\tau) = R_{XX}(\tau) * h(\tau) \tag{B.24}$$

and $$R_{YX}(\tau) = R_{XX}(\tau) * h(-\tau) \tag{B.25}$$

Taking Fourier Transform on both sides of Eq.(B.24) and noting (i) Fourier Transform of auto correlation function is PSD and (ii) Convolution in τ domain is multiplication in frequency domain, we have cross power spectral density

$$S_{XY}(\omega) = S_{XX}(\omega) H(\omega) \tag{B.26}$$

where $H(\omega)$ is Fourier Transform of $h(t)$ and $S_{XX}(\omega)$ is PSD of $X(t)$.

Similarly taking Fourier Transform on both sides of Eq.(B.25), we have cross power spectral density

$$S_{YX}(\omega) = S_{XX}(\omega) H(-\omega) \tag{B.27}$$

where $H(-\omega)$ is Fourier Transform of $h(-\tau)$.

Example B.1:

A WSS process $X(t)$ has auto correlation function $R_{XX}(\tau) = a^2 + b^2 e^{-|\tau|}$, where a and b are positive constants. $Y(t)$ is the output for input $X(t)$ to LTI system with impulse response

$$h(t) = te^{-\alpha t} \quad t > 0$$
$$= 0 \quad t < 0$$

where α is real positive constant. Find mean values of $X(t)$ and $Y(t)$.

Solution:

We know that $$R_{XX}(\tau) = E[X(t)X(t+\tau)]$$

Consider $$\lim_{|\tau|\to\infty} R_{XX}(\tau) = \lim_{|\tau|\to\infty} E[X(t)X(t+\tau)]$$

when $\tau \to \infty$, $X(t)$ and $X(t+\tau)$ become independent and

$$\lim_{|\tau|\to\infty} E\left[X(t)X(t+\tau)\right] = E\left[X(t)\right]E\left[X(t+\tau)\right]$$

Since the process $X(t)$ is WSS, $E[X(t)]$ is constant and hence is independent of time t.

Hence, $$E[X(t)] = E[X(t+\tau)] = \overline{X}$$

and $$\lim_{|\tau|\to\infty} R_{XX}(\tau) = \left[\overline{X}\right]^2 = \left[E[X(t)]\right]^2$$

Thus, mean value of $X(t)$ squared $= \left[E[X(t)]\right]^2 = \lim_{|\tau|\to\infty} R_{XX}(\tau) = a^2$

Therefore, mean value of $X(t) = a$.

Mean value of $Y(t)$ is obtained using Eq. (B.4.)

$$E\left[Y(t)\right] = E[X]\int_{-\infty}^{\infty} h(\tau)\,d\tau$$

$$= a\int_{0}^{\infty} t\,e^{-\alpha t}\,dt = \frac{a}{\alpha^2}$$

(using table of integrals to get $\frac{1}{\alpha^2}$)

Example B. 2:

Impulse response of a LTI system is

$$h(t) = e^{-\alpha t} \quad t > 0$$
$$= 0 \quad t < 0$$

The input to the system is white noise process, $X(t)$ with power spectral density N_o Watts/Hz. Determine the PSD of output process $Y(t)$.

Solution:

Recall:

$$S_{YY}(\omega) = S_{XX}(\omega)\,\left|H(\omega)\right|^2$$

We find $H(\omega)$ as follows:

$$H(\omega) = \int_{-\infty}^{\infty} h(t)e^{-j\omega t}\,dt$$

$$= \int_{0}^{\infty} e^{-\alpha t}e^{-\omega t}\,dt = \frac{1}{\alpha + j\omega}$$

$$\Rightarrow \left|H(\omega)\right|^2 = \frac{1}{\left(\omega^2 + \alpha^2\right)}$$

Hence,

$$S_{YY}(\omega) = \frac{N_o}{\omega^2 + \alpha^2}\ \text{Watts / Hz}$$

Index

T